Introduction to Probability

Second Edition

CHAPMAN & HALL/CRC
Texts in Statistical Science Series

Recently Published Titles

Extending the Linear Model with R
Generalized Linear, Mixed Effects and Nonparametric Regression Models, Second Edition
J.J. Faraway

Modeling and Analysis of Stochastic Systems, Third Edition
V.G. Kulkarni

Pragmatics of Uncertainty
J.B. Kadane

Stochastic Processes
From Applications to Theory
P.D. Moral and S. Penev

Modern Data Science with R
B.S. Baumer, D.T. Kaplan, and N.J. Horton

Generalized Additive Models
An Introduction with R, Second Edition
S. Wood

Design of Experiments
An Introduction Based on Linear Models
Max Morris

Introduction to Statistical Methods for Financial Models
T.A. Severini

Statistical Regression and Classification
From Linear Models to Machine Learning
Norman Matloff

Introduction to Functional Data Analysis
Piotr Kokoszka and Matthew Reimherr

Stochastic Processes
An Introduction, Third Edition
P.W. Jones and P. Smith

Theory of Stochastic Objects
Probability, Stochastic Processes and Inference
Athanasios Christou Micheas

Linear Models and the Relevant Distributions and Matrix Algebra
David A. Harville

An Introduction to Generalized Linear Models, Fourth Edition
Annette J. Dobson and Adrian G. Barnett

Graphics for Statistics and Data Analysis with R
Kevin J. Keen

Statistics in Engineering, Second Edition
With Examples in MATLAB and R
Andrew Metcalfe, David A. Green, Tony Greenfield, Mahayaudin Mansor, Andrew Smith, and Jonathan Tuke

Introduction to Probability, Second Edition
Joseph K. Blitzstein and Jessica Hwang

For more information about this series, please visit: https://www.crcpress.com/go/textsseries

Introduction to Probability

Second Edition

Joseph K. Blitzstein
Harvard University
Cambridge, Massachusetts

Jessica Hwang
Stanford University
Stanford, California

CRC Press is an imprint of the
Taylor & Francis Group, an **informa** business
A CHAPMAN & HALL BOOK

CRC Press
Taylor & Francis Group
6000 Broken Sound Parkway NW, Suite 300
Boca Raton, FL 33487-2742

Printed on acid-free paper
Version Date: 20190116

International Standard Book Number-13: 978-1-1383-6991-7 (Hardback)

Visit the Taylor & Francis Web site at
http://www.taylorandfrancis.com

and the CRC Press Web site at
http://www.crcpress.com

To our mothers, Steffi and Min

Contents

Preface

This book provides a modern introduction to probability and develops a foundation for understanding statistics, randomness, and uncertainty. A variety of applications and examples are explored, from basic coin-tossing and the study of coincidences to Google PageRank and Markov chain Monte Carlo. As probability is often considered to be a counterintuitive subject, many intuitive explanations, diagrams, and practice problems are given. Each chapter ends with a section showing how to explore the ideas of that chapter in R, a free software environment for statistical calculations and simulations.

Lecture videos from Stat 110 at Harvard, the course which gave rise to this book, are freely available at `http://stat110.net`. Additional supplementary materials such as R code, animations, and solutions to the exercises marked with Ⓢ, are also available at this site.

Calculus is a prerequisite for this book; there is no statistics prerequisite. The main mathematical challenge lies not in performing technical calculus derivations, but in translating between abstract concepts and concrete examples. Some major themes and features are listed below.

1. *Stories.* Throughout this book, definitions, theorems, and proofs are presented through stories: real-world interpretations that preserve mathematical precision and generality. We explore probability distributions using the generative stories that make them widely used in statistical modeling. When possible, we refrain from tedious derivations and instead aim to give interpretations and intuitions for why key results are true. Our experience is that this approach promotes long-term retention of the material by providing insight instead of demanding rote memorization.

2. *Pictures.* Since pictures are thousand-word stories, we supplement definitions with illustrations so that key concepts are associated with memorable diagrams. In many fields, the difference between a novice and an expert has been described as follows: the novice struggles to memorize a large number of seemingly disconnected facts and formulas, whereas the expert sees a unified structure in which a few principles and ideas connect these facts coherently. To help students see the structure of probability, we emphasize the connections between ideas (both verbally and visually), and at the end of most chapters we present recurring, ever-expanding maps of concepts and distributions.

3. *Dual teaching of concepts and strategies.* Our intent is that in reading this book, students will learn not only the concepts of probability, but also a set of problem-solving strategies that are widely applicable outside of probability. In the worked examples, we explain each step of the solution but also comment on how we knew to take the approach we did. Often we present multiple solutions to the same problem.

   We explicitly identify and name important strategies such as symmetry and pattern recognition, and we proactively dispel common misunderstandings, which are marked with the ☣ (biohazard) symbol.

4. *Practice problems.* The book contains about 600 exercises of varying difficulty. The exercises are intended to reinforce understanding of the material and strengthen problem-solving skills instead of requiring repetitive calculations. Some are *strategic practice problems*, grouped by theme to facilitate practice of a particular topic, while others are *mixed practice*, in which several earlier topics may need to be synthesized. About 250 exercises have detailed online solutions for practice and self-study.

5. *Simulation, Monte Carlo, and R.* Many probability problems are too difficult to solve exactly, and in any case it is important to be able to check one's answer. We introduce techniques for exploring probability via simulation, and show that often a few lines of R code suffice to create a simulation for a seemingly complicated problem.

6. *Focus on real-world relevance and statistical thinking.* Examples and exercises in this book have a clear real-world motivation, with a particular focus on building a strong foundation for further study of statistical inference and modeling. We preview important statistical ideas such as sampling, simulation, Bayesian inference, and Markov chain Monte Carlo; other application areas include genetics, medicine, computer science, and information theory. Our choice of examples and exercises is intended to highlight the power, applicability, and beauty of probabilistic thinking.

The second edition benefits from hundreds of comments, questions, and reviews from students who took courses using the book, faculty who taught with the book, and readers using the book for self-study. We have added many new examples, exercises, and explanations based on our experience teaching with the book and the feedback we have received.

New supplementary materials have also been added at `http://stat110.net`, including animations and interactive visualizations that were created in connection with the edX online version of Stat 110. These are intended to help make probability feel more intuitive, visual, and tangible.

Acknowledgments

We thank our colleagues, the Stat 110 teaching assistants, and several thousand Stat 110 students for their comments and ideas related to the course and the book. In particular, we thank Alvin Siu, Angela Fan, Anji Tang, Anqi Zhao, Arman Sabbaghi, Carolyn Stein, David Jones, David Rosengarten, David Watson, Dennis Sun, Hyungsuk Tak, Johannes Ruf, Kari Lock, Keli Liu, Kelly Bodwin, Kevin Bartz, Lazhi Wang, Martin Lysy, Michele Zemplenyi, Miles Ott, Peng Ding, Rob Phillips, Sam Fisher, Sebastian Chiu, Sofia Hou, Sushmit Roy, Theresa Gebert, Valeria Espinosa, Viktoriia Liublinska, Viviana Garcia, William Chen, and Xander Marcus.

We also thank Ella Maru Studio for helping to create the cover image for the second edition. The image illustrates the interplay between two-dimensional and one-dimensional probability distributions.

We especially thank Bo Jiang, Raj Bhuptani, Shira Mitchell, Winston Lin, and the anonymous reviewers for their detailed comments, and Andrew Gelman, Carl Morris, Persi Diaconis, Stephen Blyth, Susan Holmes, and Xiao-Li Meng for countless insightful discussions about probability.

John Kimmel at Chapman & Hall/CRC Press provided wonderful editorial expertise throughout the writing of this book. We greatly appreciate his support.

Finally, we would like to express our deepest gratitude to our families for their love and encouragement.

Joe Blitzstein and Jessica Hwang
Cambridge, MA and Stanford, CA
January 2019

1

Probability and counting

Luck. Coincidence. Randomness. Uncertainty. Risk. Doubt. Fortune. Chance. You've probably heard these words countless times, but chances are that they were used in a vague, casual way. Unfortunately, despite its ubiquity in science and everyday life, probability can be deeply counterintuitive. If we rely on intuitions of doubtful validity, we run a serious risk of making inaccurate predictions or overconfident decisions. The goal of this book is to introduce probability as a logical framework for quantifying uncertainty and randomness in a principled way. We'll also aim to strengthen intuition, both when our initial guesses coincide with logical reasoning and when we're not so lucky.

1.1 Why study probability?

Mathematics is the logic of certainty; probability is the logic of uncertainty. Probability is extremely useful in a wide variety of fields, since it provides tools for understanding and explaining variation, separating signal from noise, and modeling complex phenomena. To give just a small sample from a continually growing list of applications:

1. *Statistics*: Probability is the foundation and language for statistics, enabling many powerful methods for using data to learn about the world.
2. *Physics*: Einstein famously said "God does not play dice with the universe", but current understanding of quantum physics heavily involves probability at the most fundamental level of nature. Statistical mechanics is another major branch of physics that is built on probability.
3. *Biology*: Genetics is deeply intertwined with probability, both in the inheritance of genes and in modeling random mutations.
4. *Computer science*: Randomized algorithms make random choices while they are run, and in many important applications they are simpler and more efficient than any currently known deterministic alternatives. Probability also plays an essential role in studying the performance of algorithms, and in machine learning and artificial intelligence.

5. *Meteorology*: Weather forecasts are (or should be) computed and expressed in terms of probability.

6. *Gambling*: Many of the earliest investigations of probability were aimed at answering questions about gambling and games of chance.

7. *Finance*: At the risk of redundancy with the previous example, it should be pointed out that probability is central in quantitative finance. Modeling stock prices over time and determining "fair" prices for financial instruments are based heavily on probability.

8. *Political science*: In recent years, political science has become more and more quantitative and statistical, with applications such as analyzing surveys of public opinion, assessing gerrymandering, and predicting elections.

9. *Medicine*: The development of randomized clinical trials, in which patients are randomly assigned to receive treatment or placebo, has transformed medical research in recent years. As the biostatistician David Harrington remarked, "Some have conjectured that it could be the most significant advance in scientific medicine in the twentieth century.... In one of the delightful ironies of modern science, the randomized trial 'adjusts' for both observed and unobserved heterogeneity in a controlled experiment by introducing chance variation into the study design." [16]

10. *Life*: Life is uncertain, and probability is the logic of uncertainty. While it isn't practical to carry out a formal probability calculation for every decision made in life, thinking hard about probability can help us avert some common fallacies, shed light on coincidences, and make better predictions.

Probability provides procedures for principled problem-solving, but it can also produce pitfalls and paradoxes. For example, we'll see in this chapter that even Gottfried Wilhelm von Leibniz and Sir Isaac Newton, the two people who independently discovered calculus in the 17th century, were not immune to basic errors in probability. Throughout this book, we will use the following strategies to help avoid potential pitfalls.

1. *Simulation*: A beautiful aspect of probability is that it is often possible to study problems via *simulation*. Rather than endlessly debating an answer with someone who disagrees with you, you can run a simulation and see empirically who is right. Each chapter in this book ends with a section that gives examples of how to do calculations and simulations in R, a free statistical computing environment.

2. *Biohazards*: Studying common mistakes is important for gaining a stronger understanding of what is and is not valid reasoning in probability. In this book, common mistakes are called *biohazards* and are denoted by ☣ (since making such mistakes can be hazardous to one's health!).

3. *Sanity checks*: After solving a problem one way, we will often try to solve the same problem in a different way or to examine whether our answer makes sense in simple and extreme cases.

1.2 Sample spaces and Pebble World

The mathematical framework for probability is built around *sets*. Imagine that an experiment is performed, resulting in one out of a set of possible outcomes. Before the experiment is performed, it is unknown which outcome will be the result; after, the result "crystallizes" into the actual outcome.

Definition 1.2.1 (Sample space and event). The *sample space* S of an experiment is the set of all possible outcomes of the experiment. An *event* A is a subset of the sample space S, and we say that A *occurred* if the actual outcome is in A.

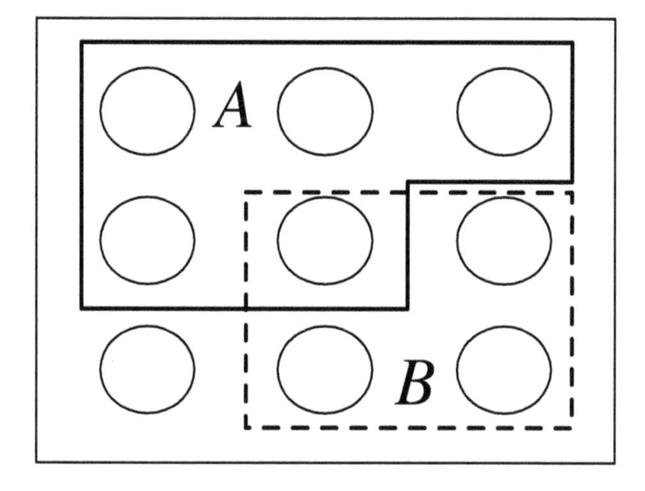

FIGURE 1.1
A sample space as Pebble World, with two events A and B spotlighted.

The sample space of an experiment can be finite, countably infinite, or uncountably infinite (see Section A.1.5 of the math appendix for an explanation of countable and uncountable sets). When the sample space is finite, we can visualize it as *Pebble World*, as shown in Figure 1.1. Each pebble represents an outcome, and an event is a set of pebbles.

Performing the experiment amounts to randomly selecting one pebble. If all the pebbles are of the same mass, all the pebbles are equally likely to be chosen. This special case is the topic of the next two sections. In Section 1.6, we give a general definition of probability that allows the pebbles to differ in mass.

Set theory is very useful in probability, since it provides a rich language for expressing and working with events; Section A.1 of the math appendix provides a review of set theory. Set operations, especially unions, intersections, and complements, make it easy to build new events in terms of already-defined events. These concepts also let us express an event in more than one way; often, one expression for an event is much easier to work with than another expression for the same event.

For example, let S be the sample space of an experiment and let $A, B \subseteq S$ be events. Then the union $A \cup B$ is the event that occurs if and only if *at least one* of A, B occurs, the intersection $A \cap B$ is the event that occurs if and only if *both* A and B occur, and the complement A^c is the event that occurs if and only if A does *not* occur. We also have *De Morgan's laws*:

$$(A \cup B)^c = A^c \cap B^c \text{ and } (A \cap B)^c = A^c \cup B^c,$$

since saying that it is *not* the case that at least one of A and B occur is the same as saying that A does not occur and B does not occur, and saying that it is *not* the case that both occur is the same as saying that at least one does not occur. Analogous results hold for unions and intersections of more than two events.

In the example shown in Figure 1.1, A is a set of 5 pebbles, B is a set of 4 pebbles, $A \cup B$ consists of the 8 pebbles in A or B (including the pebble that is in both), $A \cap B$ consists of the pebble that is in both A and B, and A^c consists of the 4 pebbles that are not in A.

The notion of sample space is very general and abstract, so it is important to have some concrete examples in mind.

Example 1.2.2 (Coin flips). A coin is flipped 10 times. Writing Heads as H and Tails as T, a possible outcome (pebble) is $HHHTHHTTHT$, and the sample space is the set of all possible strings of length 10 of H's and T's. We can (and will) encode H as 1 and T as 0, so that an outcome is a sequence $(s_1, \dots, s_{10})$ with $s_j \in \{0, 1\}$, and the sample space is the set of all such sequences. Now let's look at some events:

1. Let A_1 be the event that the first flip is Heads. As a set,

$$A_1 = \{(1, s_2, \dots, s_{10}) : s_j \in \{0, 1\} \text{ for } 2 \leq j \leq 10\}.$$

This is a subset of the sample space, so it is indeed an event; saying that A_1 occurs is the same thing as saying that the first flip is Heads. Similarly, let A_j be the event that the jth flip is Heads for $j = 2, 3, \dots, 10$.

2. Let B be the event that at least one flip was Heads. As a set,

$$B = \bigcup_{j=1}^{10} A_j.$$

3. Let C be the event that all the flips were Heads. As a set,

$$C = \bigcap_{j=1}^{10} A_j.$$

4. Let D be the event that there were at least two consecutive Heads. As a set,

$$D = \bigcup_{j=1}^{9} (A_j \cap A_{j+1}). \qquad \square$$

Example 1.2.3 (Pick a card, any card). Pick a card from a standard deck of 52 cards. The sample space S is the set of all 52 cards (so there are 52 pebbles, one for each card). Consider the following four events:

- A: card is an ace.
- B: card has a black suit.
- D: card is a diamond.
- H: card is a heart.

As a set, H consists of 13 cards:

$$\{\text{Ace of Hearts, Two of Hearts, } \dots, \text{ King of Hearts}\}.$$

We can create various other events in terms of the events A, B, D, and H. Unions, intersections, and complements are especially useful for this. For example:

- $A \cap H$ is the event that the card is the Ace of Hearts.
- $A \cap B$ is the event {Ace of Spades, Ace of Clubs}.
- $A \cup D \cup H$ is the event that the card is red or an ace.
- $(A \cup B)^c = A^c \cap B^c$ is the event that the card is a red non-ace.

Also, note that $(D \cup H)^c = D^c \cap H^c = B$, so B can be expressed in terms of D and H. On the other hand, the event that the card is a spade can't be written in terms of A, B, D, H since none of them are fine-grained enough to be able to distinguish between spades and clubs.

There are *many* other events that could be defined using this sample space. In fact, the counting methods introduced later in this chapter show that there are $2^{52} \approx 4.5 \times 10^{15}$ events in this problem, even though there are only 52 pebbles.

What if the card drawn were a joker? That would indicate that we had the wrong sample space; we are assuming that the outcome of the experiment is guaranteed to be an element of S. $\square$

As the preceding examples demonstrate, events can be described in English or in set notation. Sometimes the English description is easier to interpret while the set notation is easier to manipulate. Let S be a sample space and s_{actual} be the actual outcome of the experiment (the pebble that ends up getting chosen when the experiment is performed). A mini-dictionary for converting between English and sets is given on the next page.

English	Sets
Events and occurrences	
sample space	S
s is a possible outcome	$s \in S$
A is an event	$A \subseteq S$
A occurred	$s_{\text{actual}} \in A$
something must happen	$s_{\text{actual}} \in S$
New events from old events	
A or B (inclusive)	$A \cup B$
A and B	$A \cap B$
not A	A^c
A or B, but not both	$(A \cap B^c) \cup (A^c \cap B)$
at least one of $A_1, \dots, A_n$	$A_1 \cup \dots \cup A_n$
all of $A_1, \dots, A_n$	$A_1 \cap \dots \cap A_n$
Relationships between events	
A implies B	$A \subseteq B$
A and B are mutually exclusive	$A \cap B = \emptyset$
$A_1, \dots, A_n$ are a partition of S	$A_1 \cup \dots \cup A_n = S, A_i \cap A_j = \emptyset$ for $i \neq j$

1.3 Naive definition of probability

Historically, the earliest definition of the probability of an event was to count the number of ways the event could happen and divide by the total number of possible outcomes for the experiment. We call this the *naive definition* since it is restrictive and relies on strong assumptions; nevertheless, it is important to understand, and useful when not misused.

Definition 1.3.1 (Naive definition of probability). Let A be an event for an experiment with a finite sample space S. The *naive probability* of A is

$$P_{\text{naive}}(A) = \frac{|A|}{|S|} = \frac{\text{number of outcomes favorable to } A}{\text{total number of outcomes in } S}.$$

(We use $|A|$ to denote the size of A; see Section A.1.5 of the math appendix.)

In terms of Pebble World, the naive definition just says that the probability of A is the fraction of pebbles that are in A. For example, in Figure 1.1 it says

$$P_{\text{naive}}(A) = \frac{5}{9},\ P_{\text{naive}}(B) = \frac{4}{9},\ P_{\text{naive}}(A \cup B) = \frac{8}{9},\ P_{\text{naive}}(A \cap B) = \frac{1}{9}.$$

For the complements of the events just considered,

$$P_{\text{naive}}(A^c) = \frac{4}{9},\ P_{\text{naive}}(B^c) = \frac{5}{9},\ P_{\text{naive}}((A \cup B)^c) = \frac{1}{9},\ P_{\text{naive}}((A \cap B)^c) = \frac{8}{9}.$$

In general,

$$P_{\text{naive}}(A^c) = \frac{|A^c|}{|S|} = \frac{|S| - |A|}{|S|} = 1 - \frac{|A|}{|S|} = 1 - P_{\text{naive}}(A).$$

In Section 1.6, we will see that this result about complements *always* holds for probability, even when we go beyond the naive definition. A good strategy when trying to find the probability of an event is to start by thinking about whether it will be easier to find the probability of the event or the probability of its complement. De Morgan's laws are especially useful in this context, since it may be easier to work with an intersection than a union, or vice versa.

The naive definition is very restrictive in that it requires S to be finite, with equal mass for each pebble. It has often been misapplied by people who assume equally likely outcomes without justification and make arguments to the effect of "either it will happen or it won't, and we don't know which, so it's 50-50". In addition to sometimes giving absurd probabilities, this type of reasoning isn't even internally consistent. For example, it would say that the probability of life on Mars is 1/2 ("either there is or there isn't life there"), but it would also say that the probability of *intelligent* life on Mars is 1/2, and it is clear intuitively—and by the properties of probability developed in Section 1.6—that the latter should have strictly lower probability than the former. But there are several important types of problems where the naive definition *is* applicable:

- when there is *symmetry* in the problem that makes outcomes equally likely. It is common to assume that a coin has a 50% chance of landing Heads when tossed, due to the physical symmetry of the coin.[1] For a standard, well-shuffled deck of cards, it is reasonable to assume that all orders are equally likely. There aren't certain overeager cards that especially like to be near the top of the deck; any particular location in the deck is equally likely to house any of the 52 cards.

- when the outcomes are equally likely *by design*. For example, consider conducting a survey of n people in a population of N people. A common goal is to obtain a *simple random sample*, which means that the n people are chosen randomly with all subsets of size n being equally likely. If successful, this ensures that the naive definition is applicable, but in practice this may be hard to accomplish because of various complications, such as not having a complete, accurate list of contact information for everyone in the population.

[1] See Diaconis, Holmes, and Montgomery [7] for a physical argument that the chance of a tossed coin coming up the way it started is about 0.51 (close to but slightly more than 1/2), and Gelman and Nolan [11] for an explanation of why the probability of Heads is close to 1/2 even for a coin that is manufactured to have different weights on the two sides (for standard coin-tossing; allowing the coin to spin is a different matter).

- when the naive definition serves as a useful *null model*. In this setting, we *assume* that the naive definition applies just to see what predictions it would yield, and then we can compare observed data with predicted values to assess whether the hypothesis of equally likely outcomes is tenable.

1.4 How to count

Calculating the naive probability of an event A involves counting the number of pebbles in A and the number of pebbles in the sample space S. Often the sets we need to count are extremely large. This section introduces some fundamental methods for counting; further methods can be found in books on *combinatorics*, the branch of mathematics that studies counting.

1.4.1 Multiplication rule

In some problems, we can directly count the number of possibilities using a basic but versatile principle called the *multiplication rule*. We'll see that the multiplication rule leads naturally to counting rules for *sampling with replacement* and *sampling without replacement*, two scenarios that often arise in probability and statistics.

Theorem 1.4.1 (Multiplication rule). Consider a compound experiment consisting of two sub-experiments, Experiment A and Experiment B. Suppose that Experiment A has a possible outcomes, and for each of those outcomes Experiment B has b possible outcomes. Then the compound experiment has ab possible outcomes.

To see why the multiplication rule is true, imagine a tree diagram as in Figure 1.2. Let the tree branch a ways according to the possibilities for Experiment A, and for each of those branches create b further branches for Experiment B. Overall, there are $\underbrace{b + b + \cdots + b}_{a} = ab$ possibilities.

☣ **1.4.2.** It is often easier to think about the experiments as being in chronological order, but there is no requirement in the multiplication rule that Experiment A has to be performed before Experiment B.

Example 1.4.3 (Runners). Suppose that 10 people are running a race. Assume that ties are not possible and that all 10 will complete the race, so there will be well-defined first place, second place, and third place winners. How many possibilities are there for the first, second, and third place winners?

Solution: There are 10 possibilities for who gets first place, then once that is fixed there are 9 possibilities for who gets second place, and once these are both fixed

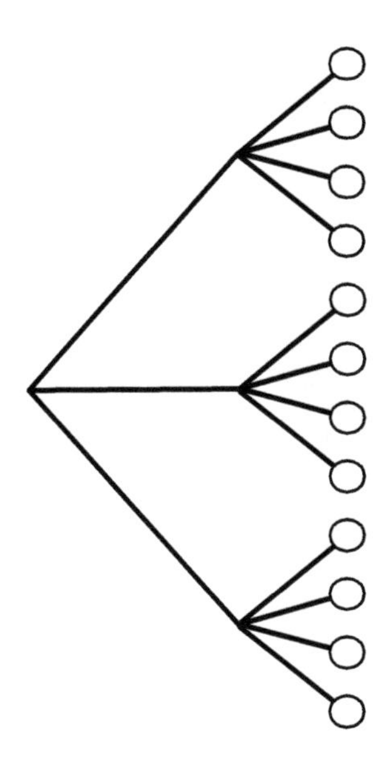

FIGURE 1.2
Tree diagram illustrating the multiplication rule. If Experiment A has 3 possible outcomes, for each of which Experiment B has 4 possible outcomes, then overall there are $3 \cdot 4 = 12$ possible outcomes.

there are 8 possibilities for third place. So by the multiplication rule, there are $10 \cdot 9 \cdot 8 = 720$ possibilities.

We did not have to consider the first place winner first. We could just as well have said that there are 10 possibilities for who got third place, then once that is fixed there are 9 possibilities for second place, and once those are both fixed there are 8 possibilities for first place. Or imagine that there are 3 platforms, which the first, second, and third place runners will stand on after the race. The platforms are gold, silver, and bronze, allocated to the first, second, and third place runners, respectively. Again there are $10 \cdot 9 \cdot 8 = 720$ possibilities for how the platforms will be occupied after the race, and there is no reason that the platforms must be considered in the order (gold, silver, bronze). □

Example 1.4.4 (Chessboard)**.** How many squares are there in an 8×8 chessboard, as in Figure 1.3? Even the name "8×8 chessboard" makes this easy: there are $8 \cdot 8 = 64$ squares on the board. The grid structure makes this clear, but we can also think of this as an example of the multiplication rule: to specify a square, we can specify which row and which column it is in. There are 8 choices of row, for each of which there are 8 choices of column.

Furthermore, we can see without doing any calculations that half the squares are white and half are black. Imagine rotating the chessboard 90 degrees clockwise. Then all the positions that had a white square now contain a black square, and vice versa, so the number of white squares must equal the number of black squares. We

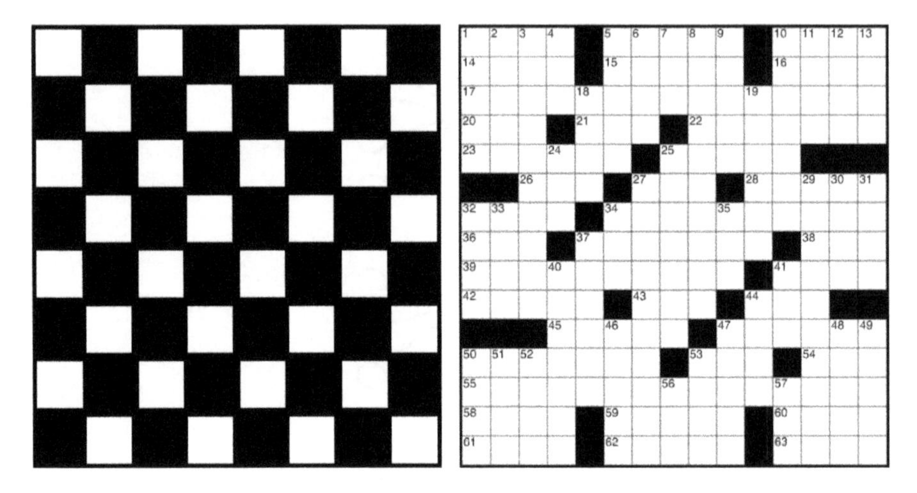

FIGURE 1.3
An 8×8 chessboard (left) and a crossword puzzle grid (right). The chessboard has $8 \cdot 8 = 64$ squares, whereas counting the number of white squares in the crossword puzzle grid requires more work.

can also count the number of white squares using the multiplication rule: in each of the 8 rows there are 4 white squares, giving a total of $8 \cdot 4 = 32$ white squares.

In contrast, it would require more effort to count the number of white squares in the crossword puzzle grid shown in Figure 1.3. The multiplication rule does not apply, since different rows sometimes have different numbers of white squares. □

Example 1.4.5 (Ice cream cones). Suppose you are buying an ice cream cone. You can choose whether to have a cake cone or a waffle cone, and whether to have chocolate, vanilla, or strawberry as your flavor. This decision process can be visualized with a tree diagram, as in Figure 1.4.

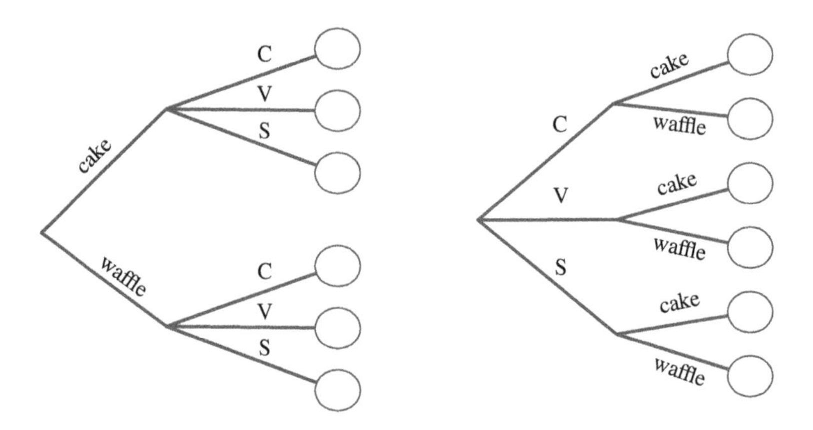

FIGURE 1.4
Tree diagram for choosing an ice cream cone. Regardless of whether the type of cone or the flavor is chosen first, there are $2 \cdot 3 = 3 \cdot 2 = 6$ possibilities.

By the multiplication rule, there are $2 \cdot 3 = 6$ possibilities. This is a very simple example, but is worth thinking through in detail as a foundation for thinking about and visualizing more complicated examples. Soon we will encounter examples where drawing the tree in a legible size would take up more space than exists in the known universe, yet where conceptually we can still think in terms of the ice cream example. Some things to note:

1. It doesn't matter whether you choose the type of cone first ("I'd like a waffle cone with chocolate ice cream") or the flavor first ("I'd like chocolate ice cream on a waffle cone"). Either way, there are $2 \cdot 3 = 3 \cdot 2 = 6$ possibilities.

2. It doesn't matter whether the same flavors are available on a cake cone as on a waffle cone. What matters is that there are exactly 3 flavor choices for each cone choice. If for some strange reason it were forbidden to have chocolate ice cream on a waffle cone, with no substitute flavor available (aside from vanilla and strawberry), there would be $3 + 2 = 5$ possibilities and the multiplication rule wouldn't apply. In larger examples, such complications could make counting the number of possibilities vastly more difficult.

Now suppose you buy *two* ice cream cones on a certain day, one in the afternoon and the other in the evening. Write, for example, (cakeC, waffleV) to mean a cake cone with chocolate in the afternoon, followed by a waffle cone with vanilla in the evening. By the multiplication rule, there are $6^2 = 36$ possibilities in your delicious compound experiment.

But what if you're only interested in what kinds of ice cream cones you had that day, not the order in which you had them, so you don't want to distinguish, for example, between (cakeC, waffleV) and (waffleV, cakeC)? Are there now $36/2 = 18$ possibilities? No, since possibilities like (cakeC, cakeC) were already only listed once each. There are $6 \cdot 5 = 30$ ordered possibilities (x, y) with $x \neq y$, which turn into 15 possibilities if we treat (x, y) as equivalent to (y, x), plus 6 possibilities of the form (x, x), giving a total of 21 possibilities. Note that if the 36 original ordered pairs (x, y) are equally likely, then the 21 possibilities here are *not* equally likely. □

Example 1.4.6 (Subsets). A set with n elements has 2^n subsets, including the empty set $\emptyset$ and the set itself. This follows from the multiplication rule since for each element, we can choose whether to include it or exclude it. For example, the set $\{1, 2, 3\}$ has the 8 subsets $\emptyset, \{1\}, \{2\}, \{3\}, \{1, 2\}, \{1, 3\}, \{2, 3\}, \{1, 2, 3\}$. This result explains why in Example 1.2.3 there are 2^{52} events that can be defined. □

We can use the multiplication rule to arrive at formulas for sampling with and without replacement. Many experiments in probability and statistics can be interpreted in one of these two contexts, so it is appealing that both formulas follow directly from the same basic counting principle.

Theorem 1.4.7 (Sampling with replacement). Consider n objects and making k choices from them, one at a time *with replacement* (i.e., choosing a certain object does not preclude it from being chosen again). Then there are n^k possible outcomes

(where order matters, in the sense that, e.g., choosing object 3 and then object 7 is counted as a different outcome than choosing object 7 and then object 3.)

For example, imagine a jar with n balls, labeled from 1 to n. We sample balls one at a time with replacement, meaning that each time a ball is chosen, it is returned to the jar. Each sampled ball is a sub-experiment with n possible outcomes, and there are k sub-experiments. Thus, by the multiplication rule there are n^k ways to obtain a sample of size k.

Theorem 1.4.8 (Sampling without replacement). Consider n objects and making k choices from them, one at a time *without replacement* (i.e., choosing a certain object precludes it from being chosen again). Then there are $n(n-1)\cdots(n-k+1)$ possible outcomes for $1 \leq k \leq n$, and 0 possibilities for $k > n$ (where order matters). By convention, $n(n-1)\cdots(n-k+1) = n$ for $k = 1$.

This result also follows directly from the multiplication rule: each sampled ball is again a sub-experiment, and the number of possible outcomes decreases by 1 each time. Note that for sampling k out of n objects without replacement, we need $k \leq n$, whereas in sampling with replacement the objects are inexhaustible.

Example 1.4.9 (Permutations and factorials). A *permutation* of $1, 2, \ldots, n$ is an arrangement of them in some order, e.g., $3, 5, 1, 2, 4$ is a permutation of $1, 2, 3, 4, 5$. By Theorem 1.4.8 with $k = n$, there are $n!$ permutations of $1, 2, \ldots, n$. For example, there are $n!$ ways in which n people can line up for ice cream. (Recall that $n!$ is $n(n-1)(n-2)\cdots 1$ for any positive integer n, and $0! = 1$.) □

Theorems 1.4.7 and 1.4.8 are theorems about *counting*, but when the naive definition applies, we can use them to calculate *probabilities*. This brings us to our next example, a famous problem in probability called the *birthday problem*. The solution incorporates both sampling with replacement and sampling without replacement.

Example 1.4.10 (Birthday problem). There are k people in a room. Assume each person's birthday is equally likely to be any of the 365 days of the year (we exclude February 29), and that people's birthdays are independent (we will define *independence* formally later, but intuitively it means that knowing some people's birthdays gives us no information about other people's birthdays; this would not hold if, e.g., we knew that two of the people were twins). What is the probability that at least one pair of people in the group have the same birthday?

Solution:

There are 365^k ways to assign birthdays to the people in the room, since we can imagine the 365 days of the year being sampled k times, with replacement. By assumption, all of these possibilities are equally likely, so the naive definition of probability applies.

Used directly, the naive definition says we just need to count the number of ways to assign birthdays to k people such that there are two people who share a birthday.

But this counting problem is hard, since it could be Emma and Steve who share a birthday, or Steve and Naomi, or all three of them, or the three of them could share a birthday while two others in the group share a different birthday, or various other possibilities.

Instead, let's count the complement: the number of ways to assign birthdays to k people such that no two people share a birthday. This amounts to sampling the 365 days of the year *without* replacement, so the number of possibilities is $365 \cdot 364 \cdot 363 \cdots (365 - k + 1)$ for $k \leq 365$. Therefore the probability of no birthday matches in a group of k people is

$$P(\text{no birthday match}) = \frac{365 \cdot 364 \cdots (365 - k + 1)}{365^k},$$

and the probability of at least one birthday match is

$$P(\text{at least 1 birthday match}) = 1 - \frac{365 \cdot 364 \cdots (365 - k + 1)}{365^k}.$$

Figure 1.5 plots the probability of at least one birthday match as a function of k. The first value of k for which the probability of a match exceeds 0.5 is $k = 23$. Thus, in a group of 23 people, there is a better than 50% chance that there is at least one birthday match. At $k = 57$, the probability of a match already exceeds 99%.

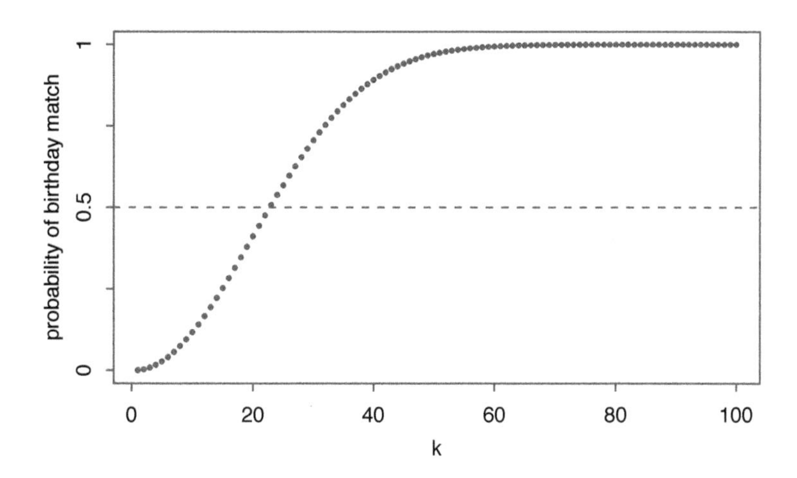

FIGURE 1.5
Probability that in a room of k people, at least two were born on the same day. This probability first exceeds 0.5 when $k = 23$.

Of course, for $k = 366$ we are *guaranteed* to have a match, but it's surprising that even with a much smaller number of people it's overwhelmingly likely that there is a birthday match. For a quick intuition into why it should not be so surprising, note that with 23 people there are $\binom{23}{2} = 253$ *pairs* of people, any of which could be a birthday match.

Problems 26 and 27 show that the birthday problem is much more than a fun party game, and much more than a way to build intuition about coincidences; there are also important applications in statistics and computer science. Problem 62 explores the more general setting in which the probability is not necessarily $1/365$ for each day. It turns out that in the non-equal probability case, having at least one match becomes even *more* likely. □

☣ **1.4.11** (Labeling objects). Drawing a sample from a population is a very fundamental concept in statistics. It is important to think of the objects or people in the population as *named* or *labeled.* For example, if there are n balls in a jar, we can imagine that they have labels from 1 to n, even if the balls look the same to the human eye. In the birthday problem, we can give each person an ID (identification) number, rather than thinking of the people as indistinguishable particles or a faceless mob.

A related example is an instructive blunder made by Leibniz in a seemingly simple problem (see Gorroochurn [14] for discussion of this and a variety of other probability problems from a historical perspective).

Example 1.4.12 (Leibniz's mistake). If we roll two fair dice, which is more likely: a sum of 11 or a sum of 12?

Solution:

Label the dice A and B, and consider each die to be a sub-experiment. By the multiplication rule, there are 36 possible outcomes for ordered pairs of the form (value of A, value of B), and they are equally likely by symmetry. Of these, $(5, 6)$ and $(6, 5)$ are favorable to a sum of 11, while only $(6, 6)$ is favorable to a sum of 12. Therefore a sum of 11 is twice as likely as a sum of 12; the probability is $1/18$ for the former, and $1/36$ for the latter.

However, Leibniz wrongly argued that a sum of 11 and a sum of 12 are equally likely, claiming that each of these sums can be attained in only one way. Here Leibniz was making the mistake of treating the two dice as indistinguishable objects, viewing $(5, 6)$ and $(6, 5)$ as the same outcome.

What are the antidotes to Leibniz's mistake? First, as explained in ☣ 1.4.11, we should *label* the objects in question instead of treating them as indistinguishable. If Leibniz had labeled his dice A and B, or green and orange, or left and right, he would not have made this mistake. Second, before we use counting for probability, we should ask ourselves whether the naive definition applies (see ☣ 1.4.23 for another example showing that caution is needed before applying the naive definition). □

1.4.2 Adjusting for overcounting

In many counting problems, it is not easy to directly count each possibility once and only once. If, however, we are able to count each possibility exactly c times for some c, then we can adjust by dividing by c. For example, if we have exactly double-counted each possibility, we can divide by 2 to get the correct count. We call this *adjusting for overcounting.*

Example 1.4.13 (Committees and teams). Consider a group of four people.

(a) How many ways are there to choose a two-person committee?

(b) How many ways are there to break the people into two teams of two?

Solution:

(a) One way to count the possibilities is by listing them out: labeling the people as 1, 2, 3, 4, the possibilities are $\boxed{12}$, $\boxed{13}$, $\boxed{14}$, $\boxed{23}$, $\boxed{24}$, $\boxed{34}$.

Another approach is to use the multiplication rule with an adjustment for overcounting. By the multiplication rule, there are 4 ways to choose the first person on the committee and 3 ways to choose the second person on the committee, but this counts each possibility twice, since picking 1 and 2 to be on the committee is the same as picking 2 and 1 to be on the committee. Since we have overcounted by a factor of 2, the number of possibilities is $(4 \cdot 3)/2 = 6$.

(b) Here are 3 ways to see that there are 3 ways to form the teams. Labeling the people as $1, 2, 3, 4$, we can directly list out the possibilities: $\boxed{12\,|\,34}$, $\boxed{13\,|\,24}$, and $\boxed{14\,|\,23}$. Listing out all possibilities would quickly become tedious or infeasible with more people though. Another approach is to note that it suffices to specify person 1's teammate (and then the other team is determined). A third way is to use (a) to see that there are 6 ways to choose one team. This overcounts by a factor of 2, since picking 1 and 2 to be a team is equivalent to picking 3 and 4 to be a team. So again the answer is $6/2 = 3$. □

A *binomial coefficient* counts the number of subsets of a certain size for a set, such as the number of ways to choose a committee of size k from a set of n people. Sets and subsets are by definition *unordered*, e.g., $\{3, 1, 4\} = \{4, 1, 3\}$, so we are counting the number of ways to choose k objects out of n, without replacement and without distinguishing between the different orders in which they could be chosen.

Definition 1.4.14 (Binomial coefficient). For any nonnegative integers k and n, the *binomial coefficient* $\binom{n}{k}$, read as "n choose k", is the number of subsets of size k for a set of size n.

For example, $\binom{4}{2} = 6$, as shown in Example 1.4.13. The binomial coefficient $\binom{n}{k}$ is sometimes called a *combination*, but we do not use that terminology here since "combination" is such a useful general-purpose word. Algebraically, binomial coefficients can be computed as follows.

Theorem 1.4.15 (Binomial coefficient formula). For $k \leq n$, we have

$$\binom{n}{k} = \frac{n(n-1)\cdots(n-k+1)}{k!} = \frac{n!}{(n-k)!k!}.$$

For $k > n$, we have $\binom{n}{k} = 0$.

Proof. Let A be a set with $|A| = n$. Any subset of A has size at most n, so $\binom{n}{k} = 0$ for $k > n$. Now let $k \leq n$. By Theorem 1.4.8, there are $n(n-1)\cdots(n-k+1)$ ways to make an *ordered* choice of k elements without replacement. This overcounts each subset of interest by a factor of $k!$ (since we don't care how these elements are ordered), so we can get the correct count by dividing by $k!$. ■

☣ **1.4.16.** The binomial coefficient $\binom{n}{k}$ is often defined in terms of factorials, but keep in mind that $\binom{n}{k}$ is 0 if $k > n$, even though the factorial of a negative number is undefined. Also, the middle expression in Theorem 1.4.15 is often better for computation than the expression with factorials, since factorials grow *extremely* fast. For example,

$$\binom{100}{2} = \frac{100 \cdot 99}{2} = 4950$$

can even be done by hand, whereas computing $\binom{100}{2} = 100!/(98!\cdot 2!)$ by first calculating 100! and 98! would be wasteful and possibly dangerous because of the extremely large numbers involved ($100! \approx 9.33 \times 10^{157}$).

Example 1.4.17 (Club officers)**.** In a club with n people, there are $n(n-1)(n-2)$ ways to choose a president, vice president, and treasurer, and there are

$$\binom{n}{3} = \frac{n(n-1)(n-2)}{3!}$$

ways to choose 3 officers without predetermined titles. □

Example 1.4.18 (Permutations of a word)**.** How many ways are there to permute the letters in the word LALALAAA? To determine a permutation, we just need to choose where the 5 A's go (or, equivalently, just decide where the 3 L's go). So there are

$$\binom{8}{5} = \binom{8}{3} = \frac{8 \cdot 7 \cdot 6}{3!} = 56 \text{ permutations.}$$

How many ways are there to permute the letters in the word STATISTICS? Here are two approaches. We could choose where to put the S's, then where to put the T's (from the remaining positions), then where to put the I's, then where to put the A (and then the C is determined). Alternatively, we can start with 10! and then adjust for overcounting, dividing by 3!3!2! to account for the fact that the S's can be permuted among themselves in any way, and likewise for the T's and I's. This gives

$$\binom{10}{3}\binom{7}{3}\binom{4}{2}\binom{2}{1} = \frac{10!}{3!3!2!} = 50400 \text{ possibilities.}$$ □

Example 1.4.19 (Binomial theorem)**.** The *binomial theorem* states that

$$(x+y)^n = \sum_{k=0}^{n} \binom{n}{k} x^k y^{n-k},$$

for any nonnegative integer n. To prove the binomial theorem, expand out the product

$$\underbrace{(x+y)(x+y)\dots(x+y)}_{n \text{ factors}}.$$

Just as $(a+b)(c+d) = ac+ad+bc+bd$ is the sum of terms where we pick the a or the b from the first factor (but not both) and the c or the d from the second factor (but not both), the terms of $(x+y)^n$ are obtained by picking either the x or the y (but not both) from each factor. There are $\binom{n}{k}$ ways to choose exactly k of the x's, and each such choice yields the term $x^k y^{n-k}$. The binomial theorem follows. □

We can use binomial coefficients to calculate probabilities in many problems for which the naive definition applies.

Example 1.4.20 (Full house in poker). A 5-card hand is dealt from a standard, well-shuffled 52-card deck. The hand is called a *full house* in poker if it consists of three cards of some rank and two cards of another rank, e.g., three 7's and two 10's (in any order). What is the probability of a full house?

Solution:

All of the $\binom{52}{5}$ possible hands are equally likely by symmetry, so the naive definition is applicable. To find the number of full house hands, use the multiplication rule (and imagine the tree). There are 13 choices for what rank we have three of; for concreteness, assume we have three 7's and focus on that branch of the tree. There are $\binom{4}{3}$ ways to choose which 7's we have. Then there are 12 choices for what rank we have two of, say 10's for concreteness, and $\binom{4}{2}$ ways to choose two 10's. Thus,

$$P(\text{full house}) = \frac{13\binom{4}{3}12\binom{4}{2}}{\binom{52}{5}} = \frac{3744}{2598960} \approx 0.00144.$$

The decimal approximation is more useful when playing poker, but the answer in terms of binomial coefficients is exact and *self-annotating* (seeing $\binom{52}{5}$ is a much bigger hint about its origin than seeing 2598960). □

Example 1.4.21 (Newton-Pepys problem). Isaac Newton was consulted about the following problem by Samuel Pepys, who wanted the information for gambling purposes. Which of the following events has the highest probability?

A: At least one 6 appears when 6 fair dice are rolled.

B: At least two 6's appear when 12 fair dice are rolled.

C: At least three 6's appear when 18 fair dice are rolled.

Solution:

The three experiments have 6^6, 6^{12}, and 6^{18} possible outcomes, respectively, and by symmetry the naive definition applies in all three experiments.

A: Instead of counting the number of ways to obtain at least one 6, it is easier to count the number of ways to get no 6's. Getting no 6's is equivalent to sampling the numbers 1 through 5 with replacement 6 times, so 5^6 outcomes are favorable to A^c (and $6^6 - 5^6$ are favorable to A). Thus

$$P(A) = 1 - \frac{5^6}{6^6} \approx 0.67.$$

B: Again we count the outcomes in B^c first. There are 5^{12} ways to get no 6's in 12 die rolls. There are $\binom{12}{1}5^{11}$ ways to get exactly one 6: we first choose which die lands 6, then sample the numbers 1 through 5 with replacement for the other 11 dice. Adding these, we get the number of ways to fail to obtain at least two 6's. Then

$$P(B) = 1 - \frac{5^{12} + \binom{12}{1}5^{11}}{6^{12}} \approx 0.62.$$

C: We count the outcomes in C^c, i.e., the number of ways to get zero, one, or two 6's in 18 die rolls. There are 5^{18} ways to get no 6's, $\binom{18}{1}5^{17}$ ways to get exactly one 6, and $\binom{18}{2}5^{16}$ ways to get exactly two 6's (choose which two dice will land 6, then decide how the other 16 dice will land).

$$P(C) = 1 - \frac{5^{18} + \binom{18}{1}5^{17} + \binom{18}{2}5^{16}}{6^{18}} \approx 0.60.$$

Therefore A has the highest probability.

Newton arrived at the correct answer using similar calculations. Newton also provided Pepys with an intuitive argument for why A was the most likely of the three; however, his intuition was invalid. As explained in Stigler [24], using loaded dice could result in a different ordering of A, B, C, but Newton's intuitive argument did not depend on the dice being fair. □

In this book, we care about counting not for its own sake, but because it sometimes helps us to find probabilities. Here is an example of a neat but treacherous counting problem; the solution is elegant, but it is rare that the result can be used with the naive definition of probability.

Example 1.4.22 (Bose-Einstein). How many ways are there to choose k times from a set of n objects with replacement, if order doesn't matter (we only care about how many times each object was chosen, not the order in which they were chosen)?

Solution:

When order does matter, the answer is n^k by the multiplication rule, but this problem is much harder. We will solve it by solving an *isomorphic* problem (the same problem in a different guise).

Let us find the number of ways to put k indistinguishable particles into n distinguishable boxes. That is, swapping the particles in any way is not considered a separate possibility: all that matters are the *counts* for how many particles are in each box. This scenario is known as a *Bose-Einstein* problem, since the physicists Satyendra Nath Bose and Albert Einstein studied related problems about indistinguishable particles in the 1920s, using their ideas to successfully predict the existence of a strange state of matter known as a Bose-Einstein condensate.

Any configuration can be encoded as a sequence of $|$'s and $\bullet$'s in a natural way, as illustrated in Figure 1.6.

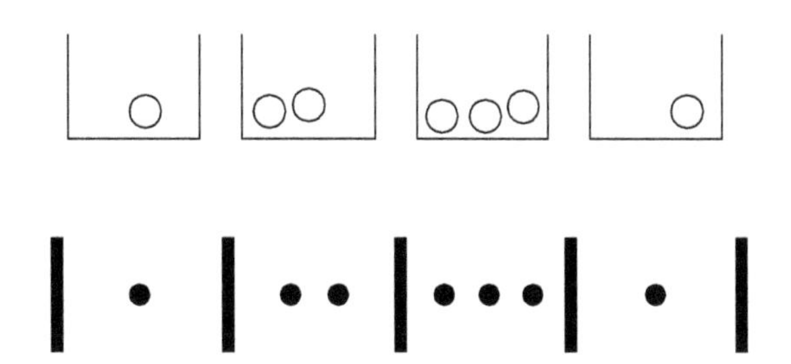

FIGURE 1.6
Bose-Einstein encoding: putting $k = 7$ indistinguishable particles into $n = 4$ distinguishable boxes can be expressed as a sequence of $|$'s and $\bullet$'s, where $|$ denotes a wall and $\bullet$ denotes a particle.

To be valid, a sequence must start and end with a $|$, and have exactly $n - 1$ $|$'s and exactly k $\bullet$'s in between the starting and ending $|$'s; conversely, any such sequence is a valid encoding for some configuration of particles in boxes. Imagine that we have written down the starting and ending $|$'s, which represent the outer walls, and in between there are $n + k - 1$ fill-in-the-blank slots in between the outer walls. We need only choose where to put the k $\bullet$'s (since then where the $n + k - 1$ interior $|$'s go is completely determined). So the number of possibilities is $\binom{n+k-1}{k}$. This counting method is sometimes called the *stars and bars* argument, where here we have dots in place of stars.

To relate this result back to the original question, we can let each box correspond to one of the n objects and use the particles as "check marks" to tally how many times each object is selected. For example, if a certain box contains exactly 3 particles, that means the object corresponding to that box was chosen exactly 3 times. The particles being indistinguishable corresponds to the fact that we don't care about the order in which the objects are chosen. Thus, the answer to the original question is also $\binom{n+k-1}{k}$.

Another isomorphic problem is to count the number of solutions $(x_1, \ldots, x_n)$ to the equation $x_1 + x_2 + \cdots + x_n = k$, where the x_i are nonnegative integers. This is equivalent since we can think of x_i as the number of particles in the ith box.

☣ **1.4.23.** The Bose-Einstein result should *not* be used in the naive definition of probability except in very special circumstances. For example, consider a survey where a sample of size k is collected by choosing people from a population of size n

one at a time, with replacement and with equal probabilities. Then the n^k *ordered* samples are equally likely, making the naive definition applicable, but the $\binom{n+k-1}{k}$ unordered samples (where all that matters is how many times each person was sampled) are *not* equally likely.

As another example, with $n = 365$ days in a year and k people, how many possible *unordered* birthday lists are there? For example, for $k = 3$, we want to count lists like (May 1, March 31, April 11), where all permutations are considered equivalent. We can't do a simple adjustment for overcounting such as $n^k/3!$ since, e.g., there are 6 permutations of (May 1, March 31, April 11) but only 3 permutations of (March 31, March 31, April 11). By Bose-Einstein, the number of lists is $\binom{n+k-1}{k}$. But the ordered birthday lists are equally likely, not the unordered lists, so the Bose-Einstein value should not be used in calculating birthday probabilities. □

1.5 Story proofs

A *story proof* is a proof by interpretation. For counting problems, this often means counting the same thing in two different ways, rather than doing tedious algebra. A story proof often avoids messy calculations and goes further than an algebraic proof toward *explaining* why the result is true. The word "story" has several meanings, some more mathematical than others, but a story proof (in the sense in which we're using the term) is a fully valid mathematical proof. Here are some examples of story proofs, which also serve as further examples of counting.

Example 1.5.1 (Choosing the complement). For any nonnegative integers n and k with $k \leq n$, we have

$$\binom{n}{k} = \binom{n}{n-k}.$$

This is easy to check algebraically (by writing the binomial coefficients in terms of factorials), but a story proof makes the result easier to understand intuitively.

Story proof: Consider choosing a committee of size k in a group of n people. We know that there are $\binom{n}{k}$ possibilities. But another way to choose the committee is to specify which $n - k$ people are *not* on the committee; specifying who is on the committee determines who is *not* on the committee, and vice versa. So the two sides are equal, as they are two ways of counting the same thing. □

Example 1.5.2 (The team captain). For any positive integers n and k with $k \leq n$,

$$n\binom{n-1}{k-1} = k\binom{n}{k}.$$

This is again easy to check algebraically (using the fact that $m! = m(m-1)!$ for any positive integer m), but a story proof is more insightful.

Story proof: Consider a group of n people, from which a team of k will be chosen, one of whom will be the team captain. To specify a possibility, we could first choose the team captain and then choose the remaining $k-1$ team members; this gives the left-hand side. Equivalently, we could first choose the k team members and then choose one of them to be captain; this gives the right-hand side. □

Example 1.5.3 (Vandermonde's identity). A famous relationship between binomial coefficients, called *Vandermonde's identity*,[2] says that

$$\binom{m+n}{k} = \sum_{j=0}^{k} \binom{m}{j}\binom{n}{k-j}.$$

This identity will come up several times in this book. Trying to prove it with a brute force expansion of all the binomial coefficients would be a nightmare. But a story proves the result elegantly and makes it clear *why* the identity holds.

Story proof: Consider a student organization consisting of m juniors and n seniors, from which a committee of size k will be chosen. There are $\binom{m+n}{k}$ possibilities. If there are j juniors in the committee, then there must be $k-j$ seniors in the committee. The right-hand side of the identity sums up the cases for j. □

Example 1.5.4 (Partnerships). Let's use a story proof to show that

$$\frac{(2n)!}{2^n \cdot n!} = (2n-1)(2n-3)\cdots 3 \cdot 1.$$

Story proof: We will show that both sides count the number of ways to break $2n$ people into n partnerships. Take $2n$ people, and give them ID numbers from 1 to $2n$. We can form partnerships by lining up the people in some order and then saying the first two are a pair, the next two are a pair, etc. This overcounts by a factor of $n! \cdot 2^n$ since the order of pairs doesn't matter, nor does the order within each pair. Alternatively, count the number of possibilities by noting that there are $2n-1$ choices for the partner of person 1, then $2n-3$ choices for person 2 (or person 3, if person 2 was already paired to person 1), and so on. □

1.6 Non-naive definition of probability

We have now seen several methods for counting outcomes in a sample space, allowing us to calculate probabilities if the naive definition applies. But the naive definition can only take us so far, since it requires equally likely outcomes and can't handle

[2]Vandermonde's identity is named after the 18th century French mathematician Alexandre-Théophile Vandermonde, but it was discovered much earlier, and stated in 1303 by the Chinese mathematician Zhu Shijie.

an infinite sample space. To generalize the notion of probability, we'll write down a short wish list of how we want probability to behave (in math, the items on the wish list are called *axioms*), and then define a probability function to be something that satisfies the properties we want!

Here is the general definition of probability that we'll use for the rest of this book. It requires just two axioms, but from these axioms it is possible to prove a vast array of results about probability.

Definition 1.6.1 (General definition of probability). A *probability space* consists of a sample space S and a *probability function* P which takes an event $A \subseteq S$ as input and returns $P(A)$, a real number between 0 and 1, as output. The function P must satisfy the following axioms:

1. $P(\emptyset) = 0$, $P(S) = 1$.
2. If $A_1, A_2, \dots$ are disjoint events, then

$$P\left(\bigcup_{j=1}^{\infty} A_j\right) = \sum_{j=1}^{\infty} P(A_j).$$

(Saying that these events are *disjoint* means that they are *mutually exclusive*: $A_i \cap A_j = \emptyset$ for $i \neq j$.)

In Pebble World, the definition says that probability behaves like mass: the mass of an empty pile of pebbles is 0, the total mass of all the pebbles is 1, and if we have non-overlapping piles of pebbles, we can get their combined mass by adding the masses of the individual piles. Unlike in the naive case, we can now have pebbles of differing masses, and we can also have a countably infinite number of pebbles as long as their total mass is 1.

We can even have uncountable sample spaces, such as having S be an area in the plane. In this case, instead of pebbles, we can visualize mud spread out over a region, where the total mass of the mud is 1.

Any function P (mapping events to numbers in the interval $[0, 1]$) that satisfies the two axioms is considered a valid probability function. However, the axioms don't tell us how probability should be *interpreted*; different schools of thought exist.

The *frequentist* view of probability is that it represents a long-run frequency over a large number of repetitions of an experiment: if we say a coin has probability $1/2$ of Heads, that means the coin would land Heads 50% of the time if we tossed it over and over and over.

The *Bayesian* view of probability is that it represents a degree of belief about the event in question, so we can assign probabilities to hypotheses like "candidate A will win the election" or "the defendant is guilty" even if it isn't possible to repeat the same election or the same crime over and over again.

The Bayesian and frequentist perspectives are complementary, and both will be helpful for developing intuition in later chapters. Regardless of how we choose to interpret probability, we can use the two axioms to derive other properties of probability, and these results will hold for *any* valid probability function.

Theorem 1.6.2 (Properties of probability). Probability has the following properties, for any events A and B.

1. $P(A^c) = 1 - P(A)$.
2. If $A \subseteq B$, then $P(A) \leq P(B)$.
3. $P(A \cup B) = P(A) + P(B) - P(A \cap B)$.

Proof.

1. Since A and A^c are disjoint and their union is S, the second axiom gives

$$P(S) = P(A \cup A^c) = P(A) + P(A^c),$$

But $P(S) = 1$ by the first axiom. So $P(A) + P(A^c) = 1$.

2. If $A \subseteq B$, then we can write B as the union of A and $B \cap A^c$, where $B \cap A^c$ is the part of B not also in A. This is illustrated in the Venn diagram below.

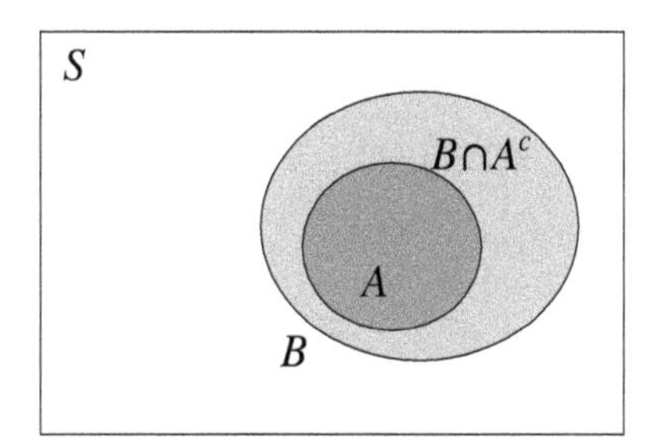

Since A and $B \cap A^c$ are disjoint, we can apply the second axiom:

$$P(B) = P(A \cup (B \cap A^c)) = P(A) + P(B \cap A^c).$$

Probability is nonnegative, so $P(B \cap A^c) \geq 0$, proving that $P(B) \geq P(A)$.

3. The intuition for this result can be seen using a Venn diagram like the one below.

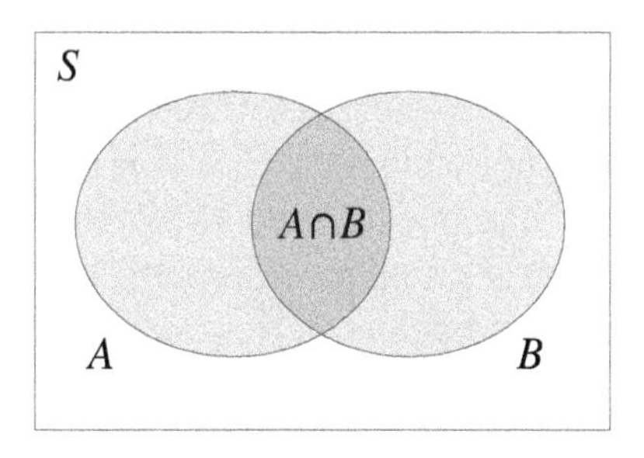

The shaded region represents $A \cup B$, but the probability of this region is not the sum $P(A) + P(B)$, because that would count the football-shaped region $A \cap B$ twice. To correct for this, we subtract $P(A \cap B)$. This is a useful intuition, but not a proof.

For a proof using the axioms of probability, we can write $A \cup B$ as the union of the disjoint events A and $B \cap A^c$. Then by the second axiom,

$$P(A \cup B) = P(A \cup (B \cap A^c)) = P(A) + P(B \cap A^c).$$

So it suffices to show that $P(B \cap A^c) = P(B) - P(A \cap B)$. Since $A \cap B$ and $B \cap A^c$ are disjoint and their union is B, another application of the second axiom gives us

$$P(A \cap B) + P(B \cap A^c) = P(B).$$

So $P(B \cap A^c) = P(B) - P(A \cap B)$, as desired. ■

The third property is a special case of *inclusion-exclusion*, a formula for finding the probability of a union of events when the events are not necessarily disjoint. We showed above that for two events A and B,

$$P(A \cup B) = P(A) + P(B) - P(A \cap B).$$

For three events, inclusion-exclusion says

$$\begin{aligned} P(A \cup B \cup C) = {} & P(A) + P(B) + P(C) \\ & - P(A \cap B) - P(A \cap C) - P(B \cap C) \\ & + P(A \cap B \cap C). \end{aligned}$$

For intuition, consider a triple Venn diagram like the one below.

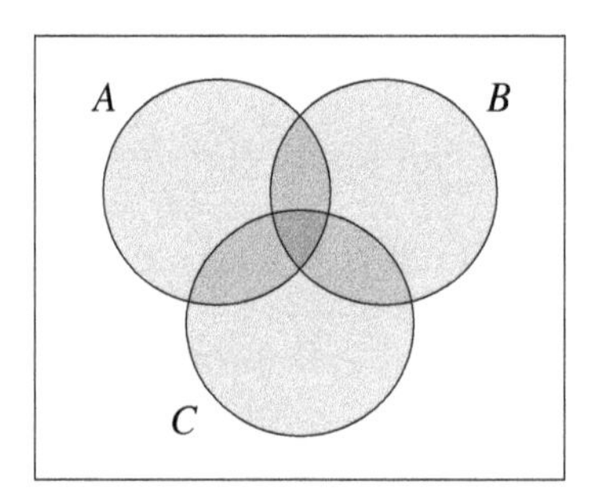

To get the total area of the shaded region $A \cup B \cup C$, we start by adding the areas of the three circles, $P(A) + P(B) + P(C)$. The three football-shaped regions have each been counted twice, so we then subtract $P(A \cap B) + P(A \cap C) + P(B \cap C)$. Finally, the region in the center has been added three times and subtracted three times, so in order to count it exactly once, we must add it back again. This ensures that each region of the diagram has been counted once and exactly once.

Now we can write inclusion-exclusion for n events.

Theorem 1.6.3 (Inclusion-exclusion). For any events $A_1, \dots, A_n$,

$$P\left(\bigcup_{i=1}^{n} A_i\right) = \sum_i P(A_i) - \sum_{i<j} P(A_i \cap A_j) + \sum_{i<j<k} P(A_i \cap A_j \cap A_k) - \dots$$
$$+ (-1)^{n+1} P(A_1 \cap \dots \cap A_n).$$

This formula can be proven by induction using just the axioms, but instead we'll present a shorter proof in Chapter 4 after introducing some additional tools. The rationale behind the alternating addition and subtraction in the general formula is analogous to the special cases we've already considered.

The next example, *de Montmort's matching problem*, is a famous application of inclusion-exclusion. Pierre Rémond de Montmort was a French mathematician who studied probability in the context of gambling and wrote a treatise [19] devoted to the analysis of various card games. He posed the following problem in 1708, based on a card game called Treize.

Example 1.6.4 (de Montmort's matching problem). Consider a well-shuffled deck of n cards, labeled 1 through n. You flip over the cards one by one, saying the numbers 1 through n as you do so. You win the game if, at some point, the number you say aloud is the same as the number on the card being flipped over (for example, if the 7th card in the deck has the label 7). What is the probability of winning?

Solution:

Let A_i be the event that the ith card in the deck has the number i written on it. We are interested in the probability of the union $A_1 \cup \dots \cup A_n$: as long as at least one of the cards has a number matching its position in the deck, you will win the game. (An ordering for which you lose is called a *derangement*, though hopefully no one has ever become deranged due to losing at this game.)

To find the probability of the union, we'll use inclusion-exclusion. First,

$$P(A_i) = \frac{1}{n}$$

for all i. One way to see this is with the naive definition of probability, using the full sample space: there are $n!$ possible orderings of the deck, all equally likely, and $(n-1)!$ of these are favorable to A_i (fix the card numbered i to be in the ith position in the deck, and then the remaining $n-1$ cards can be in any order). Another way to see this is by symmetry: the card numbered i is equally likely to be in any of the n positions in the deck, so it has probability $1/n$ of being in the ith spot. Second,

$$P(A_i \cap A_j) = \frac{(n-2)!}{n!} = \frac{1}{n(n-1)},$$

since we require the cards numbered i and j to be in the ith and jth spots in the deck and allow the remaining $n-2$ cards to be in any order, so $(n-2)!$ out of $n!$ possibilities are favorable to $A_i \cap A_j$. Similarly,

$$P(A_i \cap A_j \cap A_k) = \frac{1}{n(n-1)(n-2)},$$

and the pattern continues for intersections of 4 events, etc.

In the inclusion-exclusion formula, there are n terms involving one event, $\binom{n}{2}$ terms involving two events, $\binom{n}{3}$ terms involving three events, and so forth. By the symmetry of the problem, all n terms of the form $P(A_i)$ are equal, all $\binom{n}{2}$ terms of the form $P(A_i \cap A_j)$ are equal, and the whole expression simplifies considerably:

$$\begin{aligned} P\left(\bigcup_{i=1}^{n} A_i\right) &= \frac{n}{n} - \frac{\binom{n}{2}}{n(n-1)} + \frac{\binom{n}{3}}{n(n-1)(n-2)} - \cdots + (-1)^{n+1} \cdot \frac{1}{n!} \\ &= 1 - \frac{1}{2!} + \frac{1}{3!} - \cdots + (-1)^{n+1} \cdot \frac{1}{n!}. \end{aligned}$$

Comparing this to the Taylor series for $1/e$ (see Section A.8 of the math appendix),

$$e^{-1} = 1 - \frac{1}{1!} + \frac{1}{2!} - \frac{1}{3!} + \ldots,$$

we see that for large n, the probability of winning the game is extremely close to $1-1/e$, or about 0.63. Interestingly, as n grows, the probability of winning approaches $1-1/e$ instead of going to 0 or 1. With a lot of cards in the deck, the number of possible locations for matching cards increases while the probability of any particular match decreases, and these two forces offset each other and balance to give a probability of about $1-1/e$. □

Inclusion-exclusion is a very general formula for the probability of a union of events, but it helps us the most when there is symmetry among the events A_j; otherwise the sum can be extremely tedious. In general, when symmetry is lacking, we should try to use other tools before turning to inclusion-exclusion as a last resort.

1.7 Recap

Probability allows us to quantify uncertainty and randomness in a principled way. Probabilities arise when we perform an experiment: the set of all possible outcomes of the experiment is called the sample space, and a subset of the sample space is called an event. It is useful to be able to go back and forth between describing events in English and writing them down mathematically as sets (often using unions, intersections, and complements).

Pebble World can help us visualize sample spaces and events when the sample space is finite. In Pebble World, each outcome is a pebble, and an event is a set of pebbles. If all the pebbles have the same mass (i.e., are equally likely), we can apply the naive definition of probability, which lets us calculate probabilities by counting.

To this end, we discussed several tools for counting. When counting the number of possibilities, we often use the multiplication rule. For example, there are $n!$ permutations of the numbers $1, 2, \ldots, n$ and there are 2^n subsets of a set with n elements. If we can't directly use the multiplication rule, we can sometimes count each possibility exactly c times for some c, and then divide by c to get the actual number of possibilities. For example, this strategy is useful for finding an expression for binomial coefficients in terms of factorials.

An important pitfall to avoid is misapplying the naive definition of probability, implicitly or explicitly assuming equally likely outcomes without justification. One technique to help avoid this is to *give objects labels*, for precision and so that we are not tempted to treat them as indistinguishable.

Moving beyond the naive definition, we define probability to be a function that takes an event and assigns to it a real number between 0 and 1. We require a valid probability function to satisfy two axioms:

1. $P(\emptyset) = 0$, $P(S) = 1$.
2. If $A_1, A_2, \ldots$ are disjoint events, then

$$P\left(\bigcup_{j=1}^{\infty} A_j\right) = \sum_{j=1}^{\infty} P(A_j).$$

Many useful properties can be derived from these axioms. For example,

$$P(A^c) = 1 - P(A)$$

for any event A, and we have the inclusion-exclusion formula

$$P\left(\bigcup_{i=1}^{n} A_i\right) = \sum_i P(A_i) - \sum_{i<j} P(A_i \cap A_j) + \sum_{i<j<k} P(A_i \cap A_j \cap A_k) - \ldots + (-1)^{n+1} P(A_1 \cap \cdots \cap A_n)$$

for any events $A_1, \ldots, A_n$. For $n = 2$, this is the much nicer-looking result

$$P(A_1 \cup A_2) = P(A_1) + P(A_2) - P(A_1 \cap A_2).$$

Figure 1.7 illustrates how a probability function maps events to numbers between 0 and 1. We'll add many new concepts to this diagram as we continue our journey through the field of probability.

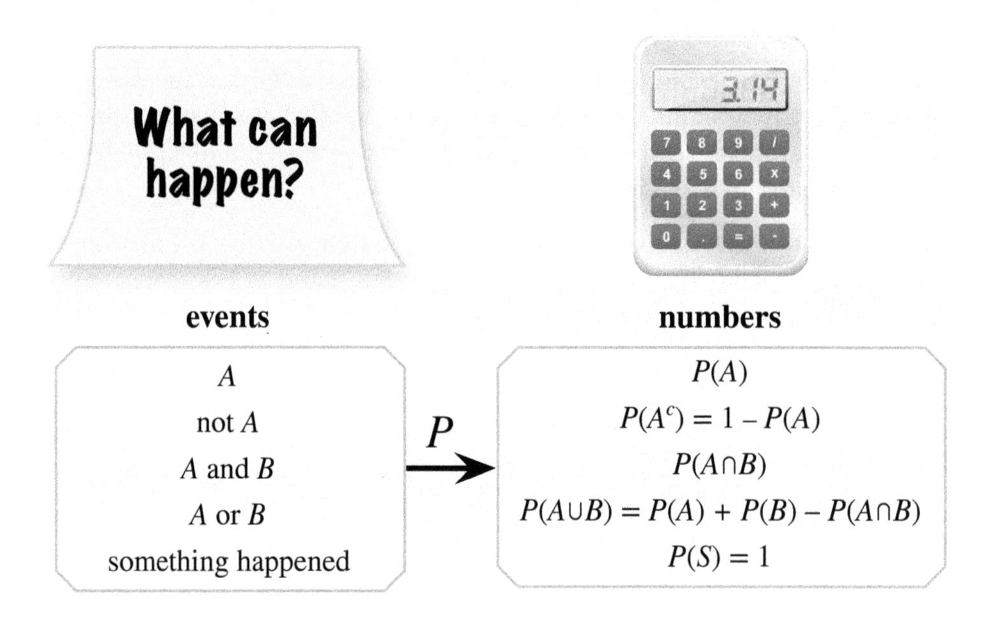

FIGURE 1.7
It is important to distinguish between *events* and *probabilities*. The former are sets, while the latter are numbers. Before the experiment is done, we generally don't know whether or not a particular event will occur (happen). So we assign it a probability of happening, using a probability function P. We can use set operations to define new events in terms of old events, and the properties of probabilities to relate the probabilities of the new events to those of the old events.

1.8 R

R is a very powerful, popular environment for statistical computing and graphics, freely available for Mac OS X, Windows, and UNIX systems. Knowing how to use R is an extremely useful skill. R and various supporting information can be obtained at `https://www.r-project.org`. RStudio is an excellent alternative interface for R, freely available at `https://www.rstudio.com`.

In the R section at the end of each chapter, we provide R code to let you try out some of the examples from the chapter, especially via simulation. These sections are not intended to be a full introduction to R; many R tutorials are available for free online, and many books on R are available. But the R sections show how to implement various simulations, computations, and visualizations that naturally accompany the material from each chapter. The R code at the end of each chapter is also available at `http://stat110.net`.

Vectors

R is built around *vectors*, and getting familiar with "vectorized thinking" is very important for using R effectively. To create a vector, we can use the `c` command (which stands for *combine* or *concatenate*). For example,

```
v <- c(3,1,4,1,5,9)
```

defines `v` to be the vector $(3, 1, 4, 1, 5, 9)$. (The left arrow `<-` is typed as `<` followed by `-`. The symbol `=` can be used instead, but the arrow is more suggestive of the fact that the variable on the left is being set equal to the value on the right.) Similarly, `n <- 110` sets n equal to 110; R views n as a vector of length 1.

```
sum(v)
```

adds up the entries of `v`, `max(v)` gives the largest value, `min(v)` gives the smallest value, and `length(v)` gives the length.

A shortcut for getting the vector $(1, 2, \ldots, n)$ is to type `1:n`; more generally, if m and n are integers then `m:n` gives the sequence of integers from m to n (in increasing order if $m \leq n$ and in decreasing order otherwise).

To access the ith entry of a vector `v`, use `v[i]`. We can also get subvectors easily:

```
v[c(1,3,5)]
```

gives the vector consisting of the 1st, 3rd, and 5th entries of `v`. It's also possible to get a subvector by specifying what to exclude, using a minus sign:

```
v[-(2:4)]
```

gives the vector obtained by removing the 2nd through 4th entries of $\mathbf{v}$ (the parentheses are needed since `-2:4` would be $(-2, -1, \ldots, 4)$).

Many operations in R are interpreted *componentwise*. For example, in math the cube of a vector doesn't have a standard definition, but in R typing `v^3` simply cubes each entry individually. Similarly,

```
1/(1:100)^2
```

is a very compact way to get the vector $(1, \frac{1}{2^2}, \frac{1}{3^2}, \ldots, \frac{1}{100^2})$.

In math, $\mathbf{v} + \mathbf{w}$ is undefined if $\mathbf{v}$ and $\mathbf{w}$ are vectors of different lengths, but in R the shorter vector gets "recycled"! For example, `v+3` adds 3 to each entry of $\mathbf{v}$.

Factorials and binomial coefficients

We can compute $n!$ using `factorial(n)` and $\binom{n}{k}$ using `choose(n,k)`. As we have seen, factorials grow extremely quickly. What is the largest n for which R returns a number for `factorial(n)`? Beyond that point, R will return `Inf` (infinity), with a warning message. But it may still be possible to use `lfactorial(n)` for larger values of n, which computes $\log(n!)$. Similarly, `lchoose(n,k)` computes $\log \binom{n}{k}$.

Sampling and simulation

The `sample` command is a useful way of drawing random samples in R. (Technically, they are *pseudo-random* since there is an underlying deterministic algorithm, but they "look like" random samples for almost all practical purposes.) For example,

```
n <- 10; k <- 5
sample(n,k)
```

generates an ordered random sample of 5 of the numbers from 1 to 10, without replacement, and with equal probabilities given to each number. To sample with replacement instead, just add in `replace = TRUE`:

```
n <- 10; k <- 5
sample(n,k,replace=TRUE)
```

To generate a random permutation of $1, 2, \ldots, n$ we can use `sample(n,n)`, which because of R's default settings can be abbreviated to `sample(n)`.

We can also use `sample` to draw from a non-numeric vector. For example, `letters` is built into R as the vector consisting of the 26 lowercase letters of the English alphabet, and `sample(letters,7)` will generate a random 7-letter "word" by sampling from the alphabet, without replacement.

The `sample` command also allows us to specify general probabilities for sampling each number. For example,

```
sample(4, 3, replace=TRUE, prob=c(0.1,0.2,0.3,0.4))
```

samples three numbers between 1 and 4, with replacement, and with probabilities given by $(0.1, 0.2, 0.3, 0.4)$. If the sampling is without replacement, then at each stage the probability of any not-yet-chosen number is *proportional* to its original probability.

Generating *many* random samples allows us to perform a *simulation* for a probability problem. The `replicate` command, which is explained below, is a convenient way to do this.

Matching problem simulation

Let's show by simulation that the probability of a matching card in Example 1.6.4 is approximately $1 - 1/e$ when the deck is sufficiently large. Using R, we can perform the experiment a bunch of times and see how many times we encounter at least one matching card:

```
n <- 100
r <- replicate(10^4,sum(sample(n)==(1:n)))
sum(r>=1)/10^4
```

In the first line, we choose how many cards are in the deck (here, 100 cards). In the second line, let's work from the inside out:

- `sample(n)==(1:n)` is a vector of length `n`, the ith element of which equals 1 if the ith card matches its position in the deck and 0 otherwise. That's because for two numbers a and b, the expression `a==b` is TRUE if $a = b$ and FALSE otherwise, and TRUE is encoded as 1 and FALSE is encoded as 0.
- `sum` adds up the elements of the vector, giving us the number of matching cards in this run of the experiment.
- `replicate` does this 10^4 times. We store the results in `r`, a vector of length 10^4 containing the numbers of matched cards from each run of the experiment.

In the last line, we add up the number of times where there was at least one matching card, and we divide by the number of simulations.

To explain what the code is doing within the code rather than in separate documentation, we can add *comments* using the `#` symbol to mark the start of a comment. Comments are ignored by R but can make the code much easier to understand for the reader (who could be you—even if you will be the only one using your code, it is often hard to remember what everything means and how the code is supposed to work when looking at it a month after writing it). Short comments can be on the same line as the corresponding code; longer comments should be on separate lines. For example, a commented version of the above simulation is:

```
n <- 100                                   # number of cards
r <- replicate(10^4,sum(sample(n)==(1:n))) # shuffle; count matches
sum(r>=1)/10^4                             # proportion with a match
```

What did you get when you ran the code? We got 0.63, which is quite close to $1 - 1/e$.

Birthday problem calculation and simulation

The following code uses `prod` (which gives the product of a vector) to calculate the probability of at least one birthday match in a group of 23 people:

```
k <- 23
1-prod((365-k+1):365)/365^k
```

Better yet, R has built-in functions, `pbirthday` and `qbirthday`, for the birthday problem! `pbirthday(k)` returns the probability of at least one match if the room has `k` people. `qbirthday(p)` returns the number of people needed in order to have probability `p` of at least one match. For example, `pbirthday(23)` is 0.507 and `qbirthday(0.5)` is 23.

We can also find the probability of having at least one *triple birthday match*, i.e., three people with the same birthday; all we have to do is add `coincident=3` to say we're looking for triple matches. For example, `pbirthday(23,coincident=3)` returns 0.014, so 23 people give us only a 1.4% chance of a triple birthday match. `qbirthday(0.5,coincident=3)` returns 88, so we'd need 88 people to have at least a 50% chance of at least one triple birthday match.

To simulate the birthday problem, we can use

```
b <- sample(1:365,23,replace=TRUE)
tabulate(b)
```

to generate random birthdays for 23 people and then tabulate the counts of how many people were born on each day (the command `table(b)` creates a prettier table, but is slower). We can run 10^4 repetitions as follows:

```
r <- replicate(10^4, max(tabulate(sample(1:365,23,replace=TRUE))))
sum(r>=2)/10^4
```

If the probabilities of various days are not all equal, the calculation becomes much more difficult, but the simulation can easily be extended since `sample` allows us to specify the probability of each day (by default `sample` assigns equal probabilities, so in the above the probability is $1/365$ for each day).

1.9 Exercises

Exercises marked with Ⓢ have detailed solutions at `http://stat110.net`.

Counting

1. How many ways are there to permute the letters in the word MISSISSIPPI?

2. (a) How many 7-digit phone numbers are possible, assuming that the first digit can't be a 0 or a 1?

   (b) Re-solve (a), except now assume also that the phone number is not allowed to start with 911 (since this is reserved for emergency use, and it would not be desirable for the system to wait to see whether more digits were going to be dialed after someone has dialed 911).

3. Fred is planning to go out to dinner each night of a certain week, Monday through Friday, with each dinner being at one of his ten favorite restaurants.

   (a) How many possibilities are there for Fred's schedule of dinners for that Monday through Friday, if Fred is not willing to eat at the same restaurant more than once?

   (b) How many possibilities are there for Fred's schedule of dinners for that Monday through Friday, if Fred is willing to eat at the same restaurant more than once, but is not willing to eat at the same place twice in a row (or more)?

4. A *round-robin tournament* is being held with n tennis players; this means that every player will play against every other player exactly once.

   (a) How many possible outcomes are there for the tournament (the outcome lists out who won and who lost for each game)?

   (b) How many games are played in total?

5. A *knock-out tournament* is being held with 2^n tennis players. This means that for each round, the winners move on to the next round and the losers are eliminated, until only one person remains. For example, if initially there are $2^4 = 16$ players, then there are 8 games in the first round, then the 8 winners move on to round 2, then the 4 winners move on to round 3, then the 2 winners move on to round 4, the winner of which is declared the winner of the tournament. (There are various systems for determining who plays whom within a round, but these do not matter for this problem.)

   (a) How many rounds are there?

   (b) Count how many games in total are played, by adding up the numbers of games played in each round.

   (c) Count how many games in total are played, this time by directly thinking about it without doing almost any calculation.

   Hint: How many players need to be eliminated?

6. There are 20 people at a chess club on a certain day. They each find opponents and start playing. How many possibilities are there for how they are matched up, assuming that in each game it *does* matter who has the white pieces (in a chess game, one player has the white pieces and the other player has the black pieces)?

7. Two chess players, A and B, are going to play 7 games. Each game has three possible outcomes: a win for A (which is a loss for B), a draw (tie), and a loss for A (which is a win for B). A win is worth 1 point, a draw is worth 0.5 points, and a loss is worth 0 points.

   (a) How many possible outcomes for the individual games are there, such that overall player A ends up with 3 wins, 2 draws, and 2 losses?

   (b) How many possible outcomes for the individual games are there, such that A ends up with 4 points and B ends up with 3 points?

   (c) Now assume that they are playing a best-of-7 match, where the match will end when either player has 4 points or when 7 games have been played, whichever is first. For example, if after 6 games the score is 4 to 2 in favor of A, then A wins the match and they don't play a 7th game. How many possible outcomes for the individual games are there, such that the match lasts for 7 games and A wins by a score of 4 to 3?

8. Ⓢ (a) How many ways are there to split a dozen people into 3 teams, where one team has 2 people, and the other two teams have 5 people each?

   (b) How many ways are there to split a dozen people into 3 teams, where each team has 4 people?

9. Ⓢ (a) How many paths are there from the point $(0, 0)$ to the point $(110, 111)$ in the plane such that each step either consists of going one unit up or one unit to the right?

   (b) How many paths are there from $(0, 0)$ to $(210, 211)$, where each step consists of going one unit up or one unit to the right, and the path has to go through $(110, 111)$?

10. To fulfill the requirements for a certain degree, a student can choose to take any 7 out of a list of 20 courses, with the constraint that at least 1 of the 7 courses must be a statistics course. Suppose that 5 of the 20 courses are statistics courses.

    (a) How many choices are there for which 7 courses to take?

    (b) Explain intuitively why the answer to (a) is *not* $\binom{5}{1} \cdot \binom{19}{6}$.

11. Let A and B be sets with $|A| = n, |B| = m$.

    (a) How many functions are there from A to B (i.e., functions with domain A, assigning an element of B to each element of A)?

    (b) How many one-to-one functions are there from A to B? (See Section A.2.1 of the math appendix for information about one-to-one functions.)

12. Four players, named A, B, C, and D, are playing a card game. A standard, well-shuffled deck of cards is dealt to the players (so each player receives a 13-card hand).

    (a) How many possibilities are there for the hand that player A will get? (Within a hand, the order in which cards were received doesn't matter.)

    (b) How many possibilities are there overall for what hands everyone will get, assuming that it matters which player gets which hand, but not the order of cards within a hand?

    (c) Explain intuitively why the answer to Part (b) is not the fourth power of the answer to Part (a).

13. A certain casino uses 10 standard decks of cards mixed together into one big deck, which we will call a *superdeck*. Thus, the superdeck has $52 \cdot 10 = 520$ cards, with 10 copies of each card. How many different 10-card hands can be dealt from the superdeck? The order of the cards does not matter, nor does it matter which of the original 10 decks the cards came from. Express your answer as a binomial coefficient.

    Hint: Bose-Einstein.

14. You are ordering two pizzas. A pizza can be small, medium, large, or extra large, with any combination of 8 possible toppings (getting no toppings is allowed, as is getting all 8). How many possibilities are there for your two pizzas?

Story proofs

15. Ⓢ Give a story proof that $\sum_{k=0}^{n} \binom{n}{k} = 2^n$.

16. Ⓢ Show that for all positive integers n and k with $n \geq k$,

$$\binom{n}{k} + \binom{n}{k-1} = \binom{n+1}{k},$$

doing this in two ways: (a) algebraically and (b) with a story, giving an interpretation for why both sides count the same thing.

Hint for the story proof: Imagine an organization consisting of $n+1$ people, with one of them pre-designated as the president of the organization.

17. Give a story proof that

$$\sum_{k=0}^{n} \binom{n}{k}^2 = \binom{2n}{n},$$

for all positive integers n.

18. Give a story proof that

$$\sum_{k=1}^{n} k\binom{n}{k}^2 = n\binom{2n-1}{n-1},$$

for all positive integers n.

Hint: Consider choosing a committee of size n from two groups of size n each, where only one of the two groups has people eligible to become the chair of the committee.

19. Give a story proof that

$$\sum_{k=2}^{n} \binom{k}{2}\binom{n-k+2}{2} = \binom{n+3}{5},$$

for all integers $n \geq 2$.

Hint: Consider the middle number in a subset of $\{1, 2, \dots, n+3\}$ of size 5.

20. Ⓢ (a) Show using a story proof that

$$\binom{k}{k} + \binom{k+1}{k} + \binom{k+2}{k} + \cdots + \binom{n}{k} = \binom{n+1}{k+1},$$

where n and k are positive integers with $n \geq k$. This is called the *hockey stick identity.*

Hint: Imagine arranging a group of people by age, and then think about the oldest person in a chosen subgroup.

(b) Suppose that a large pack of Haribo gummi bears can have anywhere between 30 and 50 gummi bears. There are 5 delicious flavors: pineapple (clear), raspberry (red), orange (orange), strawberry (green, mysteriously), and lemon (yellow). There are 0 non-delicious flavors. How many possibilities are there for the composition of such a pack of gummi bears? You can leave your answer in terms of a couple binomial coefficients, but not a sum of lots of binomial coefficients.

21. Define $\left\{ {n \atop k} \right\}$ as the number of ways to partition $\{1, 2, \dots, n\}$ into k nonempty subsets, or the number of ways to have n students split up into k groups such that each group has at least one student. For example, $\left\{ {4 \atop 2} \right\} = 7$ because we have the following possibilities.

- {1}, {2, 3, 4}
- {2}, {1, 3, 4}
- {3}, {1, 2, 4}
- {4}, {1, 2, 3}

- {1, 2}, {3, 4}
- {1, 3}, {2, 4}
- {1, 4}, {2, 3}

Prove the following identities:

(a)

$$\left\{ {n+1 \atop k} \right\} = \left\{ {n \atop k-1} \right\} + k \left\{ {n \atop k} \right\}.$$

Hint: I'm either in a group by myself or I'm not.

(b)

$$\sum_{j=k}^{n} \binom{n}{j} \left\{ {j \atop k} \right\} = \left\{ {n+1 \atop k+1} \right\}.$$

Hint: First decide how many people are not going to be in my group.

22. The Dutch mathematician R.J. Stroeker remarked:

Every beginning student of number theory surely must have marveled at the miraculous fact that for each natural number n the sum of the first n positive consecutive cubes is a perfect square. [26]

Furthermore, it is the square of the sum of the first n positive integers! That is,

$$1^3 + 2^3 + \cdots + n^3 = (1 + 2 + \cdots + n)^2.$$

Usually this identity is proven by induction, but that does not give much insight into why the result is true, nor does it help much if we wanted to compute the left-hand side but didn't already know this result. In this problem, you will give a story proof of the identity.

(a) Give a story proof of the identity

$$1 + 2 + \cdots + n = \binom{n+1}{2}.$$

Hint: Consider a round-robin tournament (see Exercise 4).

(b) Give a story proof of the identity

$$1^3 + 2^3 + \cdots + n^3 = 6\binom{n+1}{4} + 6\binom{n+1}{3} + \binom{n+1}{2}.$$

It is then just basic algebra (not required for this problem) to check that the square of the right-hand side in (a) is the right-hand side in (b).

Hint: Imagine choosing a number between 1 and n and then choosing 3 numbers between 0 and n smaller than the original number, with replacement. Then consider cases based on how many distinct numbers were chosen.

Naive definition of probability

23. Three people get into an empty elevator at the first floor of a building that has 10 floors. Each presses the button for their desired floor (unless one of the others has already pressed that button). Assume that they are equally likely to want to go to floors 2 through 10 (independently of each other). What is the probability that the buttons for 3 consecutive floors are pressed?

24. Ⓢ A certain family has 6 children, consisting of 3 boys and 3 girls. Assuming that all birth orders are equally likely, what is the probability that the 3 eldest children are the 3 girls?

25. Ⓢ A city with 6 districts has 6 robberies in a particular week. Assume the robberies are located randomly, with all possibilities for which robbery occurred where equally likely. What is the probability that some district had more than 1 robbery?

26. A survey is being conducted in a city with 1 million residents. It would be far too expensive to survey all of the residents, so a random sample of size 1000 is chosen (in practice, there are many challenges with sampling, such as obtaining a complete list of everyone in the city, and dealing with people who refuse to participate). The survey is conducted by choosing people one at a time, *with* replacement and with equal probabilities.

    (a) Explain how sampling with vs. without replacement here relates to the birthday problem.

    (b) Find the probability that at least one person will get chosen more than once.

27. A *hash table* is a commonly used data structure in computer science, allowing for fast information retrieval. For example, suppose we want to store some people's phone numbers. Assume that no two of the people have the same name. For each name x, a *hash function* h is used, letting $h(x)$ be the location that will be used to store x's phone number. After such a table has been computed, to look up x's phone number one just recomputes $h(x)$ and then looks up what is stored in that location.

    The hash function h is deterministic, since we don't want to get different results every time we compute $h(x)$. But h is often chosen to be *pseudorandom*. For this problem, assume that true randomness is used. Let there be k people, with each person's phone number stored in a random location (with equal probabilities for each location, independently of where the other people's numbers are stored), represented by an integer between 1 and n. Find the probability that at least one location has more than one phone number stored there.

28. Ⓢ A college has 10 time slots for its courses, and blithely assigns courses to completely random time slots, independently. The college offers exactly 3 statistics courses. What is the probability that 2 or more of the statistics courses are in the same time slot?

29. Ⓢ For each part, decide whether the blank should be filled in with $=$, $<$, or $>$, and give a clear explanation.

    (a) (probability that the total after rolling 4 fair dice is 21) ____ (probability that the total after rolling 4 fair dice is 22)

    (b) (probability that a random 2-letter word is a palindrome[3]) ____ (probability that a random 3-letter word is a palindrome)

30. With definitions as in the previous problem, find the probability that a random n-letter word is a palindrome for $n = 7$ and for $n = 8$.

31. Ⓢ Elk dwell in a certain forest. There are N elk, of which a simple random sample of size n are captured and tagged ("simple random sample" means that all $\binom{N}{n}$ sets of n elk are equally likely). The captured elk are returned to the population, and then a new sample is drawn, this time with size m. This is an important method that is widely used in ecology, known as *capture-recapture*. What is the probability that exactly k of the m

[3]A *palindrome* is an expression such as "A man, a plan, a canal: Panama" that reads the same backwards as forwards (ignoring spaces, capitalization, and punctuation). Assume for this problem that all words of the specified length are equally likely, that there are no spaces or punctuation, and that the alphabet consists of the lowercase letters a, b, ..., z. A word is any string of letters from the alphabet; it does not need to be a word that has a meaning in the English language.

elk in the new sample were previously tagged? (Assume that an elk that was captured before doesn't become more or less likely to be captured again.)

32. Four cards are face down on a table. You are told that two are red and two are black, and you need to guess which two are red and which two are black. You do this by pointing to the two cards you're guessing are red (and then implicitly you're guessing that the other two are black). Assume that all configurations are equally likely, and that you do not have psychic powers. Find the probability that exactly j of your guesses are correct, for $j = 0, 1, 2, 3, 4$.

33. Ⓢ A jar contains r red balls and g green balls, where r and g are fixed positive integers. A ball is drawn from the jar randomly (with all possibilities equally likely), and then a second ball is drawn randomly.

    (a) Explain intuitively why the probability of the second ball being green is the same as the probability of the first ball being green.

    (b) Define notation for the sample space of the problem, and use this to compute the probabilities from (a) and show that they are the same.

    (c) Suppose that there are 16 balls in total, and that the probability that the two balls are the same color is the same as the probability that they are different colors. What are r and g (list all possibilities)?

34. Ⓢ A random 5-card poker hand is dealt from a standard deck of cards. Find the probability of each of the following possibilities (in terms of binomial coefficients).

    (a) A flush (all 5 cards being of the same suit; do not count a royal flush, which is a flush with an ace, king, queen, jack, and 10).

    (b) Two pair (e.g., two 3's, two 7's, and an ace).

35. A random 13-card hand is dealt from a standard deck of cards. What is the probability that the hand contains at least 3 cards of every suit?

36. A group of 30 dice are thrown. What is the probability that 5 of each of the values $1, 2, 3, 4, 5, 6$ appear?

37. A deck of cards is shuffled well. The cards are dealt one by one, until the first time an ace appears.

    (a) Find the probability that no kings, queens, or jacks appear before the first ace.

    (b) Find the probability that exactly one king, exactly one queen, and exactly one jack appear (in any order) before the first ace.

38. Tyrion, Cersei, and ten other people are sitting at a round table, with their seating arrangement having been randomly assigned. What is the probability that Tyrion and Cersei are sitting next to each other? Find this in two ways:

    (a) using a sample space of size $12!$, where an outcome is fully detailed about the seating;

    (b) using a much smaller sample space, which focuses on Tyrion and Cersei.

39. An organization with $2n$ people consists of n married couples. A committee of size k is selected, with all possibilities equally likely. Find the probability that there are exactly j married couples within the committee.

40. There are n balls in a jar, labeled with the numbers $1, 2, \ldots, n$. A total of k balls are drawn, one by one *with replacement*, to obtain a sequence of numbers.

    (a) What is the probability that the sequence obtained is strictly increasing?

    (b) What is the probability that the sequence obtained is increasing (but not necessarily strictly increasing, i.e., there can be repetitions)?

41. Each of n balls is independently placed into one of n boxes, with all boxes equally likely. What is the probability that exactly one box is empty?

42. Ⓢ A *norepeatword* is a sequence of at least one (and possibly all) of the usual 26 letters a,b,c,...,z, with repetitions not allowed. For example, "course" is a norepeatword, but "statistics" is not. Order matters, e.g., "course" is not the same as "source".

    A norepeatword is chosen randomly, with all norepeatwords equally likely. Show that the probability that it uses all 26 letters is very close to $1/e$.

Axioms of probability

43. Show that for any events A and B,

$$P(A) + P(B) - 1 \leq P(A \cap B) \leq P(A \cup B) \leq P(A) + P(B).$$

For each of these three inequalities, give a simple criterion for when the inequality is actually an equality (e.g., give a simple condition such that $P(A \cap B) = P(A \cup B)$ if and only if the condition holds).

44. Let A and B be events. The *difference* $B - A$ is defined to be the set of all elements of B that are not in A. Show that if $A \subseteq B$, then

$$P(B - A) = P(B) - P(A),$$

directly using the axioms of probability.

45. Let A and B be events. The *symmetric difference* $A \triangle B$ is defined to be the set of all elements that are in A or B but not both. In logic and engineering, this event is also called the *XOR* (*exclusive or*) of A and B. Show that

$$P(A \triangle B) = P(A) + P(B) - 2P(A \cap B),$$

directly using the axioms of probability.

46. Let $A_1, A_2, \ldots, A_n$ be events. Let B_k be the event exactly k of the A_i occur, and C_k be the event that at least k of the A_i occur, for $0 \leq k \leq n$. Find a simple expression for $P(B_k)$ in terms of $P(C_k)$ and $P(C_{k+1})$.

47. Events A and B are *independent* if $P(A \cap B) = P(A)P(B)$ (independence is explored in detail in the next chapter).

(a) Give an example of independent events A and B in a finite sample space S (with neither equal to $\emptyset$ or S), and illustrate it with a Pebble World diagram.

(b) Consider the experiment of picking a random point in the rectangle

$$R = \{(x, y) : 0 < x < 1, 0 < y < 1\},$$

where the probability of the point being in any particular region contained within R is the area of that region. Let A_1 and B_1 be rectangles contained within R, with areas not equal to 0 or 1. Let A be the event that the random point is in A_1, and B be the event that the random point is in B_1. Give a geometric description of when it is true that A and B are independent. Also, give an example where they are independent and another example where they are not independent.

(c) Show that if A and B are independent, then

$$P(A \cup B) = P(A) + P(B) - P(A)P(B) = 1 - P(A^c)P(B^c).$$

48. Ⓢ Arby has a belief system assigning a number $P_{\text{Arby}}(A)$ between 0 and 1 to every event A (for some sample space). This represents Arby's degree of belief about how likely A is to occur. For any event A, Arby is willing to pay a price of $1000 \cdot P_{\text{Arby}}(A)$ dollars to buy a certificate such as the one shown below:

Certificate
The owner of this certificate can redeem it for \$1000 if A occurs. No value if A does not occur, except as required by federal, state, or local law. No expiration date.

Likewise, Arby is willing to sell such a certificate at the same price. Indeed, Arby is willing to buy or sell any number of certificates at this price, as Arby considers it the "fair" price.

Arby stubbornly refuses to accept the axioms of probability. In particular, suppose that there are two *disjoint* events A and B with

$$P_{\text{Arby}}(A \cup B) \neq P_{\text{Arby}}(A) + P_{\text{Arby}}(B).$$

Show how to make Arby go bankrupt, by giving a list of transactions Arby is willing to make that will *guarantee* that Arby will lose money (you can assume it will be known whether A occurred and whether B occurred the day after any certificates are bought/sold).

Inclusion-exclusion

49. A fair die is rolled n times. What is the probability that at least 1 of the 6 values never appears?

50. Ⓢ A card player is dealt a 13-card hand from a well-shuffled, standard deck of cards. What is the probability that the hand is void in at least one suit ("void in a suit" means having no cards of that suit)?

51. Ⓢ For a group of 7 people, find the probability that all 4 seasons (winter, spring, summer, fall) occur at least once each among their birthdays, assuming that all seasons are equally likely.

52. A certain class has 20 students, and meets on Mondays and Wednesdays in a classroom with exactly 20 seats. In a certain week, everyone in the class attends both days. On both days, the students choose their seats completely randomly (with one student per seat). Find the probability that no one sits in the same seat on both days of that week.

53. Fred needs to choose a password for a certain website. Assume that he will choose an 8-character password, and that the legal characters are the lowercase letters a, b, c, . . . , z, the uppercase letters A, B, C, . . . , Z, and the numbers 0, 1, . . . , 9.

(a) How many possibilities are there if he is required to have at least one lowercase letter in his password?

(b) How many possibilities are there if he is required to have at least one lowercase letter and at least one uppercase letter in his password?

(c) How many possibilities are there if he is required to have at least one lowercase letter, at least one uppercase letter, and at least one number in his password?

54. Ⓢ Alice attends a small college in which each class meets only once a week. She is deciding between 30 non-overlapping classes. There are 6 classes to choose from for each day of the week, Monday through Friday. Trusting in the benevolence of randomness, Alice decides to register for 7 randomly selected classes out of the 30, with all choices equally likely. What is the probability that she will have classes every day, Monday through Friday? (This problem can be done either directly using the naive definition of probability, or using inclusion-exclusion.)

55. A club consists of 10 seniors, 12 juniors, and 15 sophomores. An organizing committee of size 5 is chosen randomly (with all subsets of size 5 equally likely).

 (a) Find the probability that there are exactly 3 sophomores in the committee.

 (b) Find the probability that the committee has at least one representative from each of the senior, junior, and sophomore classes.

Mixed practice

56. For each part, decide whether the blank should be filled in with $=$, $<$, or $>$, and give a clear explanation. In (a) and (b), order doesn't matter.

 (a) (number of ways to choose 5 people out of 10) ____ (number of ways to choose 6 people out of 10)

 (b) (number of ways to break 10 people into 2 teams of 5) ____ (number of ways to break 10 people into a team of 6 and a team of 4)

 (c) (probability that all 3 people in a group of 3 were born on January 1) ____ (probability that in a group of 3 people, 1 was born on each of January 1, 2, and 3)

 Martin and Gale play an exciting game of "toss the coin", where they toss a fair coin until the pattern HH occurs (two consecutive Heads) or the pattern TH occurs (Tails followed immediately by Heads). Martin wins the game if and only if HH occurs before TH occurs.

 (d) (probability that Martin wins) ____ 1/2

57. Take a deep breath before attempting this problem. In the book *Innumeracy* [20], John Allen Paulos writes:

 > Now for better news of a kind of immortal persistence. First, take a deep breath. Assume Shakespeare's account is accurate and Julius Caesar gasped ["Et tu, Brute!"] before breathing his last. What are the chances you just inhaled a molecule which Caesar exhaled in his dying breath?

 Assume that one breath of air contains 10^{22} molecules, and that there are 10^{44} molecules in the atmosphere. (These are slightly simpler numbers than the estimates that Paulos gives; for the purposes of this problem, assume that these are exact. Of course, in reality there are many complications such as different types of molecules in the atmosphere, chemical reactions, variation in lung capacities, etc.)

 Suppose that the molecules in the atmosphere now are the same as those in the atmosphere when Caesar was alive, and that in the 2000 years or so since Caesar, these molecules have been scattered completely randomly through the atmosphere. Also assume that Caesar's last breath was sampled *without* replacement but that your breathing is sampled *with* replacement (without replacement makes more sense but with replacement is easier to work with, and is a good approximation since the number of molecules in the atmosphere is so much larger than the number of molecules in one breath).

Find the probability that at least one molecule in the breath you just took was shared with Caesar's last breath, and give a simple approximation in terms of e.

Hint: As discussed in the math appendix, $(1+\frac{x}{n})^n \approx e^x$ for n large.

58. A widget inspector inspects 12 widgets and finds that exactly 3 are defective. Unfortunately, the widgets then get all mixed up and the inspector has to find the 3 defective widgets again by testing widgets one by one.

(a) Find the probability that the inspector will now have to test at least 9 widgets.

(b) Find the probability that the inspector will now have to test at least 10 widgets.

59. There are 15 chocolate bars and 10 children. In how many ways can the chocolate bars be distributed to the children, in each of the following scenarios?

(a) The chocolate bars are fungible (interchangeable).

(b) The chocolate bars are fungible, and each child must receive at least one.

Hint: First give each child a chocolate bar, and then decide what to do with the rest.

(c) The chocolate bars are not fungible (it matters which particular bar goes where).

(d) The chocolate bars are not fungible, and each child must receive at least one.

Hint: The strategy suggested in (b) does not apply. Instead, consider *randomly* giving the chocolate bars to the children, and apply inclusion-exclusion.

60. Given $n \geq 2$ numbers $(a_1, a_2, \ldots, a_n)$ with no repetitions, a *bootstrap sample* is a sequence $(x_1, x_2, \ldots, x_n)$ formed from the a_j's by sampling with replacement with equal probabilities. Bootstrap samples arise in a widely used statistical method known as the *bootstrap*. For example, if $n = 2$ and $(a_1, a_2) = (3, 1)$, then the possible bootstrap samples are $(3,3), (3,1), (1,3)$, and $(1,1)$.

(a) How many possible bootstrap samples are there for $(a_1, \ldots, a_n)$?

(b) How many possible bootstrap samples are there for $(a_1, \ldots, a_n)$, if order does not matter (in the sense that it only matters how many times each a_j was chosen, not the order in which they were chosen)?

(c) One random bootstrap sample is chosen (by sampling from $a_1, \ldots, a_n$ with replacement, as described above). Show that not all unordered bootstrap samples (in the sense of (b)) are equally likely. Find an unordered bootstrap sample $\mathbf{b}_1$ that is as likely as possible, and an unordered bootstrap sample $\mathbf{b}_2$ that is as unlikely as possible. Let p_1 be the probability of getting $\mathbf{b}_1$ and p_2 be the probability of getting $\mathbf{b}_2$ (so p_i is the probability of getting the *specific* unordered bootstrap sample $\mathbf{b}_i$). What is p_1/p_2? What is the ratio of the probability of getting an unordered bootstrap sample whose probability is p_1 to the probability of getting an unordered sample whose probability is p_2?

61. Ⓢ There are 100 passengers lined up to board an airplane with 100 seats (with each seat assigned to one of the passengers). The first passenger in line crazily decides to sit in a randomly chosen seat (with all seats equally likely). Each subsequent passenger takes their assigned seat if available, and otherwise sits in a random available seat. What is the probability that the last passenger in line gets to sit in their assigned seat? (This is a common interview problem, and a beautiful example of the power of symmetry.)

Hint: Call the seat assigned to the jth passenger in line "seat j" (regardless of whether the airline calls it seat 23A or whatever). What are the possibilities for which seats are available to the last passenger in line, and what is the probability of each of these possibilities?

62. In the birthday problem, we assumed that all 365 days of the year are equally likely (and excluded February 29). In reality, some days are slightly more likely as birthdays than others. For example, scientists have long struggled to understand why more babies are born 9 months after a holiday. Let $\mathbf{p} = (p_1, p_2, \dots, p_{365})$ be the vector of birthday probabilities, with p_j the probability of being born on the jth day of the year (February 29 is still excluded, with no offense intended to Leap Dayers).

    The kth *elementary symmetric polynomial* in the variables $x_1, \dots, x_n$ is defined by

$$e_k(x_1, \dots, x_n) = \sum_{1 \le j_1 < j_2 < \dots < j_k \le n} x_{j_1} \dots x_{j_k}.$$

    This just says to add up all of the $\binom{n}{k}$ terms we can get by choosing and multiplying k of the variables. For example, $e_1(x_1, x_2, x_3) = x_1 + x_2 + x_3$, $e_2(x_1, x_2, x_3) = x_1x_2 + x_1x_3 + x_2x_3$, and $e_3(x_1, x_2, x_3) = x_1x_2x_3$.

    Now let $k \ge 2$ be the number of people.

    (a) Find a simple expression for the probability that there is at least one birthday match, in terms of $\mathbf{p}$ and an elementary symmetric polynomial.

    (b) Explain intuitively why it makes sense that P(at least one birthday match) is minimized when $p_j = \frac{1}{365}$ for all j, by considering simple and extreme cases.

    (c) The famous *arithmetic mean-geometric mean inequality* says that for $x, y \ge 0$,

$$\frac{x+y}{2} \ge \sqrt{xy}.$$

    This inequality follows from adding $4xy$ to both sides of $x^2 - 2xy + y^2 = (x-y)^2 \ge 0$.

    Define $\mathbf{r} = (r_1, \dots, r_{365})$ by $r_1 = r_2 = (p_1 + p_2)/2$, $r_j = p_j$ for $3 \le j \le 365$. Using the arithmetic mean-geometric mean bound and the fact, which you should verify, that

$$e_k(x_1, \dots, x_n) = x_1x_2e_{k-2}(x_3, \dots, x_n) + (x_1 + x_2)e_{k-1}(x_3, \dots, x_n) + e_k(x_3, \dots, x_n),$$

    show that

$$P(\text{at least one birthday match}|\mathbf{p}) \ge P(\text{at least one birthday match}|\mathbf{r}),$$

    with strict inequality if $\mathbf{p} \ne \mathbf{r}$, where the "given $\mathbf{r}$" notation means that the birthday probabilities are given by $\mathbf{r}$. Using this, show that the value of $\mathbf{p}$ that minimizes the probability of at least one birthday match is given by $p_j = \frac{1}{365}$ for all j.

2

Conditional probability

We have introduced probability as a language for expressing our degrees of belief or uncertainties about events. Whenever we observe new evidence (i.e., obtain *data*), we acquire information that may affect our uncertainties. A new observation that is consistent with an existing belief could make us more sure of that belief, while a surprising observation could throw that belief into question. *Conditional probability* is the concept that addresses this fundamental question: how should we update our beliefs in light of the evidence we observe?

2.1 The importance of thinking conditionally

Conditional probability is essential for scientific, medical, and legal reasoning, since it shows how to incorporate evidence into our understanding of the world in a logical, coherent manner. In fact, a useful perspective is that *all probabilities are conditional*; whether or not it's written explicitly, there is always background knowledge (or assumptions) built into every probability.

Suppose, for example, that one morning we are interested in the event R that it will rain that day. Let $P(R)$ be our assessment of the probability of rain before looking outside. If we then look outside and see ominous clouds in the sky, then presumably our probability of rain should increase; we denote this new probability by $P(R|C)$ (read as "probability of R given C"), where C is the event of there being ominous clouds. When we go from $P(R)$ to $P(R|C)$, we say that we are "conditioning on C". As the day progresses, we may obtain more and more information about the weather conditions, and we can continually update our probabilities. If we observe that events $B_1, \ldots, B_n$ occurred, then we write our new conditional probability of rain given this evidence as $P(R|B_1, \ldots, B_n)$. If eventually we observe that it does start raining, our conditional probability becomes 1.

Furthermore, we will see that conditioning is a very powerful problem-solving strategy, often making it possible to solve a complicated problem by decomposing it into manageable pieces with case-by-case reasoning. Just as in computer science a common strategy is to break a large problem up into bite-size pieces (or even byte-size pieces), in probability a common strategy is to reduce a complicated probability problem to a bunch of simpler conditional probability problems. In particular, we will discuss a strategy known as *first-step analysis*, which often allows us to obtain recursive solutions to problems where the experiment has multiple stages.

Due to the central importance of conditioning, both as the means by which we update beliefs to reflect evidence and as a problem-solving strategy, we say that

Conditioning is the soul of statistics.

2.2 Definition and intuition

Definition 2.2.1 (Conditional probability). If A and B are events with $P(B) > 0$, then the *conditional probability* of A given B, denoted by $P(A|B)$, is defined as

$$P(A|B) = \frac{P(A \cap B)}{P(B)}.$$

Here A is the event whose uncertainty we want to update, and B is the evidence we observe (or want to treat as given). We call $P(A)$ the *prior* probability of A and $P(A|B)$ the *posterior* probability of A ("prior" means before updating based on the evidence, and "posterior" means after updating based on the evidence).

It is important to interpret the event appearing after the vertical conditioning bar as the evidence that we have observed or that is being conditioned on: $P(A|B)$ is the probability of A given the evidence B, *not* the probability of some entity called $A|B$. As discussed in ☣ 2.4.1, there is no such event as $A|B$.

For any event A, $P(A|A) = P(A \cap A)/P(A) = 1$. Upon observing that A has occurred, our updated probability for A is 1. If this weren't the case, we would demand a new definition of conditional probability!

Example 2.2.2 (Two cards). A standard deck of cards is shuffled well. Two cards are drawn randomly, one at a time without replacement. Let A be the event that the first card is a heart, and B be the event that the second card is red. Find $P(A|B)$ and $P(B|A)$.

Solution:

By the naive definition of probability and the multiplication rule,

$$P(A \cap B) = \frac{13 \cdot 25}{52 \cdot 51} = \frac{25}{204},$$

since a favorable outcome is determined by choosing any of the 13 hearts and then any of the remaining 25 red cards. Also, $P(A) = 1/4$ since the 4 suits are equally likely, and

$$P(B) = \frac{26 \cdot 51}{52 \cdot 51} = \frac{1}{2}$$

since there are 26 favorable possibilities for the *second* card, and for each of those, the first card can be any other card (recall from Chapter 1 that chronological order is not needed in the multiplication rule).

A neater way to see that $P(B) = 1/2$ is by *symmetry*: from a vantage point before having done the experiment, the second card is equally likely to be any card in the deck.

We now have all the pieces needed to apply the definition of conditional probability:

$$P(A|B) = \frac{P(A \cap B)}{P(B)} = \frac{25/204}{1/2} = \frac{25}{102},$$
$$P(B|A) = \frac{P(B \cap A)}{P(A)} = \frac{25/204}{1/4} = \frac{25}{51}.$$

This is a simple example, but already there are several things worth noting.

1. It's extremely important to be careful about which events to put on which side of the conditioning bar. In particular, $P(A|B) \neq P(B|A)$. The next section explores how $P(A|B)$ and $P(B|A)$ are related in general. Confusing these two quantities is called the *prosecutor's fallacy* and is discussed in Section 2.8. If instead we had defined B to be the event that the second card is a heart, then the two conditional probabilities would have been equal.

2. Both $P(A|B)$ and $P(B|A)$ make sense (intuitively and mathematically); the chronological order in which cards were chosen does not dictate which conditional probabilities we can look at. When we calculate conditional probabilities, we are considering what *information* observing one event provides about another event, not whether one event *causes* another. For further intuition, imagine that someone spreads out the cards and draws one card with their left hand and another card with their right hand, at the same time. Defining A and B based on the left hand's card and the right hand's card rather than the first card and second card would not change the structure of the problem in any important way.

3. We can also see that $P(B|A) = 25/51$ by a direct interpretation of what conditional probability means: if the first card drawn is a heart, then the remaining cards consist of 25 red cards and 26 black cards (all of which are equally likely to be drawn next), so the conditional probability of getting a red card is $25/(25 + 26) = 25/51$. It is harder to find $P(A|B)$ in this way: if we learn that the second card is red, we might think "that's nice to know, but what we really want to know is whether it's a heart!" The conditional probability results from later sections in this chapter give us methods for getting around this issue. □

To shed more light on what conditional probability means, here are two intuitive interpretations.

Intuition 2.2.3 (Pebble World). Consider a finite sample space, with the outcomes visualized as pebbles with total mass 1. Since A is an event, it is a set of pebbles, and likewise for B. Figure 2.1(a) shows an example.

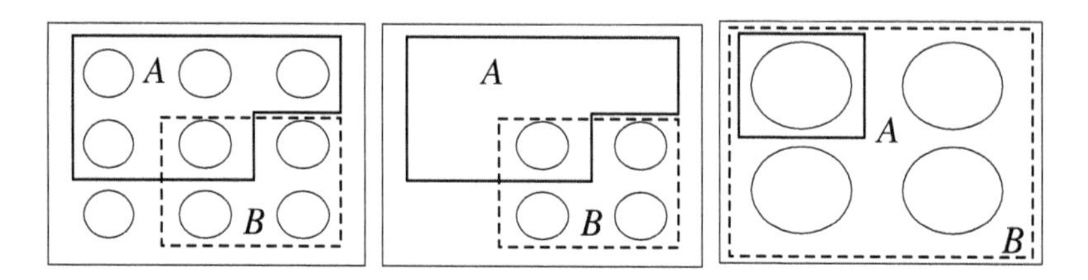

FIGURE 2.1
Pebble World intuition for $P(A|B)$. From left to right: (a) Events A and B are subsets of the sample space. (b) Because we know B occurred, get rid of the outcomes in B^c. (c) In the restricted sample space, renormalize so the total mass is still 1.

Now suppose that we learn that B occurred. In Figure 2.1(b), upon obtaining this information, we get rid of all the pebbles in B^c because they are incompatible with the knowledge that B has occurred. Then $P(A \cap B)$ is the total mass of the pebbles remaining in A. Finally, in Figure 2.1(c), we *renormalize*, that is, divide all the masses by a constant so that the new total mass of the remaining pebbles is 1. This is achieved by dividing by $P(B)$, the total mass of the pebbles in B. The updated mass of the outcomes corresponding to event A is the conditional probability $P(A|B) = P(A \cap B)/P(B)$.

In this way, our probabilities have been updated in accordance with the observed evidence. Outcomes that contradict the evidence are discarded, and their mass is redistributed among the remaining outcomes, preserving the relative masses of the remaining outcomes. For example, if pebble 2 weighs twice as much as pebble 1 initially, and both are contained in B, then after conditioning on B it is still true that pebble 2 weighs twice as much as pebble 1. But if pebble 2 is not contained in B, then after conditioning on B its mass is updated to 0. □

Intuition 2.2.4 (Frequentist interpretation). Recall that the frequentist interpretation of probability is based on relative frequency over a large number of repeated trials. Imagine repeating our experiment many times, generating a long list of observed outcomes. The conditional probability of A given B can then be thought of in a natural way: it is the fraction of times that A occurs, restricting attention to the trials where B occurs.

In Figure 2.2, our experiment has outcomes which can be written as a string of 0's and 1's; B is the event that the first digit is 1 and A is the event that the second digit is 1. Conditioning on B, we circle all the repetitions where B occurred, and then we look at the fraction of circled repetitions in which event A also occurred.

In symbols, let n_A, n_B, n_{AB} be the number of occurrences of $A, B, A \cap B$ respectively in a large number n of repetitions of the experiment. The frequentist interpretation is that

$$P(A) \approx \frac{n_A}{n}, \ P(B) \approx \frac{n_B}{n}, \ P(A \cap B) \approx \frac{n_{AB}}{n}.$$

Then $P(A|B)$ is interpreted as n_{AB}/n_B, which equals $(n_{AB}/n)/(n_B/n)$. This interpretation again translates to $P(A|B) = P(A \cap B)/P(B)$. $\square$

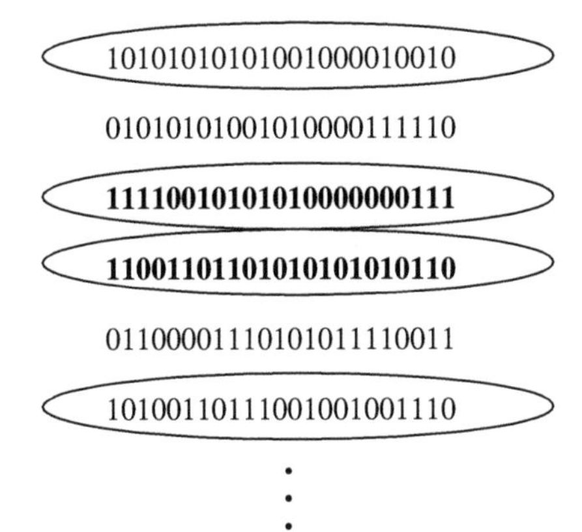

FIGURE 2.2
Frequentist intuition for $P(A|B)$. The repetitions where B occurred are circled; among these, the repetitions where A occurred are highlighted in bold. $P(A|B)$ is the long-run relative frequency of the repetitions where A occurs, within the subset of repetitions where B occurs.

For practice with applying the definition of conditional probability, let's do some more examples. The next three examples all start with the same basic scenario of a family with two children, but subtleties arise depending on the exact assumptions and the exact information we condition on.

Example 2.2.5 (Two children). Martin Gardner posed the following puzzle in the 1950s, in his column in Scientific American.

> *Mr. Jones has two children. The older child is a girl. What is the probability that both children are girls?*
>
> *Mr. Smith has two children. At least one of them is a boy. What is the probability that both children are boys?*

At first glance this problem seems like it should be a simple application of conditional probability, but for decades there have been controversies about whether or why the two parts of the problem should have different answers, and the extent to which the problem is ambiguous. Gardner gave the answers 1/2 and 1/3 to the two parts, respectively, which may seem paradoxical: why should it matter whether we learn the older child's gender, as opposed to just learning one child's gender?

It is important to clarify the assumptions of the problem. Several implicit assumptions are being made to obtain the answers that Gardner gave.

- It assumes that gender is binary, so that each child can be definitively categorized as a boy or a girl. In fact, many people don't neatly fit into either of the categories "male" or "female", and identify themselves as having a non-binary gender.
- It assumes that $P(\text{boy}) = P(\text{girl})$, both for the elder child and for the younger child. In fact, in most countries slightly more boys are born than girls. For example, in the United States it is commonly estimated that 105 boys are born for every 100 girls who are born.
- It assumes that the genders of the two children are independent, i.e., knowing the elder child's gender gives no information about the younger child's gender, and vice versa. This would be unrealistic if, e.g., the children were identical twins.

Under these (admittedly problematic) simplifying assumptions, we can solve the problem as follows.

Solution:

With the assumptions listed above, the definition of conditional probability gives

$$P(\text{both girls}|\text{elder is a girl}) = \frac{P(\text{both girls, elder is a girl})}{P(\text{elder is a girl})} = \frac{1/4}{1/2} = 1/2,$$

$$P(\text{both girls}|\text{at least one girl}) = \frac{P(\text{both girls, at least one girl})}{P(\text{at least one girl})} = \frac{1/4}{3/4} = 1/3.$$

(We solved the second part of the problem in terms of girls rather than boys to make it a bit easier to compare the two parts of the problem.) It may seem counterintuitive that the two results are different, since there is no reason for us to care whether the elder child is a girl as opposed to the younger child. Indeed, by symmetry,

$$P(\text{both girls}|\text{younger is a girl}) = P(\text{both girls}|\text{elder is a girl}) = 1/2.$$

However, there is no such symmetry between the conditional probabilities $P(\text{both girls}|\text{elder is a girl})$ and $P(\text{both girls}|\text{at least one girl})$. Saying that the elder child is a girl designates a *specific* child, and then the other child (the younger child) has a 50% chance of being a girl. "At least one" does *not* refer to a specific child. Conditioning on a specific child being a girl knocks away 2 of the 4 "pebbles" in the sample space $\{GG, GB, BG, BB\}$, where, for example, GB means the elder child is a girl and the younger child is a boy. In contrast, conditioning on at least one child being a girl knocks away only BB. □

Example 2.2.6 (Random child is a girl). A family has two children. You randomly run into one of the two, and learn that she is a girl. With assumptions as in the previous example, what is the conditional probability that both are girls? Also assume that you are equally likely to run into either child, and that which one you run into has nothing to do with gender.

Solution:

Intuitively, the answer should be 1/2: imagine that the child we encountered is in front of us and the other is at home. Both being girls just says that the child who is at home is a girl, which seems to have nothing to do with the fact that the child in front of us is a girl. But let us check this more carefully, using the definition of conditional probability. This is also good practice with writing events in set notation.

Let G_1, G_2, and G_3 be the events that the elder, younger, and random child is a girl, respectively. By assumption, $P(G_1) = P(G_2) = P(G_3) = 1/2$. By the naive definition, or by independence as explained in Section 2.5, $P(G_1 \cap G_2) = 1/4$. Thus,

$$P(G_1 \cap G_2 | G_3) = P(G_1 \cap G_2 \cap G_3)/P(G_3) = (1/4)/(1/2) = 1/2,$$

since $G_1 \cap G_2 \cap G_3 = G_1 \cap G_2$ (if both children are girls, it guarantees that the random child is a girl).

Keep in mind though that in order to arrive at 1/2, an assumption was needed about how the random child was selected. In statistical language, we say that we collected a *random sample*; here the sample consists of one of the two children. One of the most important principles in statistics is that it is essential to think carefully about how the sample was collected, not just stare at the raw data without understanding where they came from. To take a simple extreme case, suppose that a repressive law forbids a boy from leaving the house if he has a sister. Then "the random child is a girl" is equivalent to "at least one of the children is a girl", so the problem reduces to the first part of Example 2.2.5. □

Example 2.2.7 (A girl born in winter). A family has two children. Find the probability that both children are girls, given that at least one of the two is a girl who was born in winter. In addition to the assumptions from Example 2.2.5, assume that the four seasons are equally likely and that gender is independent of season. (This means that knowing the gender gives no information about the probabilities of the seasons, and vice versa; see Section 2.5 for much more about independence.)

Solution:

By definition of conditional probability,

$$P(\text{both girls}|\text{at least one winter girl}) = \frac{P(\text{both girls, at least one winter girl})}{P(\text{at least one winter girl})}.$$

Since the probability that a specific child is a winter-born girl is 1/8, the denominator equals

$$P(\text{at least one winter girl}) = 1 - (7/8)^2.$$

To compute the numerator, use the fact that "both girls, at least one winter girl" is the same event as "both girls, at least one winter child"; then use the assumption that gender and season are independent:

$$\begin{aligned}P(\text{both girls, at least one winter girl}) &= P(\text{both girls, at least one winter child})\\ &= (1/4)P(\text{at least one winter child})\\ &= (1/4)(1 - P(\text{both are non-winter}))\\ &= (1/4)(1 - (3/4)^2).\end{aligned}$$

Thus,

$$P(\text{both girls}|\text{at least one winter girl}) = \frac{(1/4)(1 - (3/4)^2)}{1 - (7/8)^2} = \frac{7/64}{15/64} = 7/15.$$

At first this result seems absurd! In Example 2.2.5, the result was that the conditional probability of both children being girls, given that at least one is a girl, is 1/3; why should it be any different when we learn that at least one is a winter-born girl? The point is that information about the birth season brings "at least one is a girl" closer to "a specific one is a girl". Conditioning on more and more specific information brings the probability closer and closer to 1/2.

For example, conditioning on "at least one is a girl who was born on a March 31 at 8:20 pm" comes very close to specifying a child, and learning information about a specific child does not give us information about the other child. The seemingly irrelevant information such as season of birth interpolates between the two parts of Example 2.2.5. Exercise 29 generalizes this example to an arbitrary characteristic that is independent of gender. □

2.3 Bayes' rule and the law of total probability

The definition of conditional probability is simple—just a ratio of two probabilities—but it has far-reaching consequences. The first consequence is obtained easily by moving the denominator in the definition to the other side of the equation.

Theorem 2.3.1 (Probability of the intersection of two events). For any events A and B with positive probabilities,

$$P(A \cap B) = P(B)P(A|B) = P(A)P(B|A).$$

This follows from taking the definition of $P(A|B)$ and multiplying both sides by $P(B)$, and then taking the definition of $P(B|A)$ and multiplying both sides by $P(A)$. At first sight this theorem may not seem very useful: it *is* the definition of conditional probability, just written slightly differently, and anyway it seems circular to use $P(A|B)$ to help find $P(A \cap B)$ when $P(A|B)$ was defined in terms of $P(A \cap B)$. But we will see that the theorem is in fact very useful, since it often turns out to be possible to find conditional probabilities without going back to the definition, and in such cases Theorem 2.3.1 can help us more easily find $P(A \cap B)$.

Applying Theorem 2.3.1 repeatedly, we can generalize to the intersection of n events.

Theorem 2.3.2 (Probability of the intersection of n events)**.** For any events $A_1, \ldots, A_n$ with $P(A_1, A_2, \ldots, A_{n-1}) > 0$,

$$P(A_1, A_2, \ldots, A_n) = P(A_1)P(A_2|A_1)P(A_3|A_1, A_2) \cdots P(A_n|A_1, \ldots, A_{n-1}),$$

The commas denote intersections, e.g., $P(A_3|A_1, A_2)$ is $P(A_3|A_1 \cap A_2)$.

In fact, this is $n!$ theorems in one, since we can permute $A_1, \ldots, A_n$ however we want without affecting the left-hand side. Often the right-hand side will be much easier to compute for some orderings than for others. For example,

$$P(A_1, A_2, A_3) = P(A_1)P(A_2|A_1)P(A_3|A_1, A_2) = P(A_2)P(A_3|A_2)P(A_1|A_2, A_3),$$

and there are 4 other expansions of this form too. It often takes practice and thought to be able to know which ordering to use.

We are now ready to introduce the two main theorems of this chapter—Bayes' rule and the law of total probability—which will allow us to compute conditional probabilities in a wide range of problems. Bayes' rule is an extremely famous, extremely useful result that relates $P(A|B)$ to $P(B|A)$.

Theorem 2.3.3 (Bayes' rule)**.**

$$P(A|B) = \frac{P(B|A)P(A)}{P(B)}.$$

This follows immediately from Theorem 2.3.1, which in turn followed immediately from the definition of conditional probability. Yet Bayes' rule has important implications and applications in probability and statistics, since it is so often necessary to find conditional probabilities, and often $P(B|A)$ is much easier to find directly than $P(A|B)$ (or vice versa).

Another way to write Bayes' rule is in terms of *odds* rather than probability.

Definition 2.3.4 (Odds)**.** The *odds* of an event A are

$$\text{odds}(A) = P(A)/P(A^c).$$

For example, if $P(A) = 2/3$, we say the odds in favor of A are 2 to 1. (This is sometimes written as $2 : 1$, and is sometimes stated as 1 to 2 odds against A; care is needed since some sources do not explicitly state whether they are referring to odds in favor or odds against an event.) Of course we can also convert from odds back to probability:

$$P(A) = \text{odds}(A)/(1 + \text{odds}(A)).$$

By taking the Bayes' rule expression for $P(A|B)$ and dividing it by the Bayes' rule expression for $P(A^c|B)$, we arrive at the odds form of Bayes' rule.

Theorem 2.3.5 (Odds form of Bayes' rule). For any events A and B with positive probabilities, the odds of A after conditioning on B are

$$\frac{P(A|B)}{P(A^c|B)} = \frac{P(B|A)}{P(B|A^c)}\frac{P(A)}{P(A^c)}.$$

In words, this says that the *posterior odds* $P(A|B)/P(A^c|B)$ are equal to the *prior odds* $P(A)/P(A^c)$ times the factor $P(B|A)/P(B|A^c)$, which is known in statistics as the *likelihood ratio*. Sometimes it is convenient to work with this form of Bayes' rule to get the posterior odds, and then if desired we can convert from odds back to probability.

The *law of total probability* (LOTP) relates conditional probability to unconditional probability. It is essential for fulfilling the promise that conditional probability can be used to decompose complicated probability problems into simpler pieces, and it is often used in tandem with Bayes' rule.

Theorem 2.3.6 (Law of total probability). Let $A_1, \dots, A_n$ be a partition of the sample space S (i.e., the A_i are disjoint events and their union is S), with $P(A_i) > 0$ for all i. Then

$$P(B) = \sum_{i=1}^{n} P(B|A_i)P(A_i).$$

Proof. Since the A_i form a partition of S, we can decompose B as

$$B = (B \cap A_1) \cup (B \cap A_2) \cup \dots \cup (B \cap A_n).$$

This is illustrated in Figure 2.3, where we have chopped B into the smaller pieces $B \cap A_1$ through $B \cap A_n$. By the second axiom of probability, because these pieces are disjoint, we can add their probabilities to get $P(B)$:

$$P(B) = P(B \cap A_1) + P(B \cap A_2) + \dots + P(B \cap A_n).$$

Now we can apply Theorem 2.3.1 to each of the $P(B \cap A_i)$:

$$P(B) = P(B|A_1)P(A_1) + \dots + P(B|A_n)P(A_n).$$ ■

The law of total probability tells us that to get the unconditional probability of B, we can divide the sample space into disjoint slices A_i, find the conditional probability of B within each of the slices, then take a weighted sum of the conditional probabilities, where the weights are the probabilities $P(A_i)$. The choice of how to divide up the sample space is crucial: a well-chosen partition will reduce a complicated problem into simpler pieces, whereas a poorly chosen partition will only exacerbate our problems, requiring us to calculate n difficult probabilities instead of just one!

The next few examples show how we can use Bayes' rule together with the law of total probability to update our beliefs based on observed evidence.

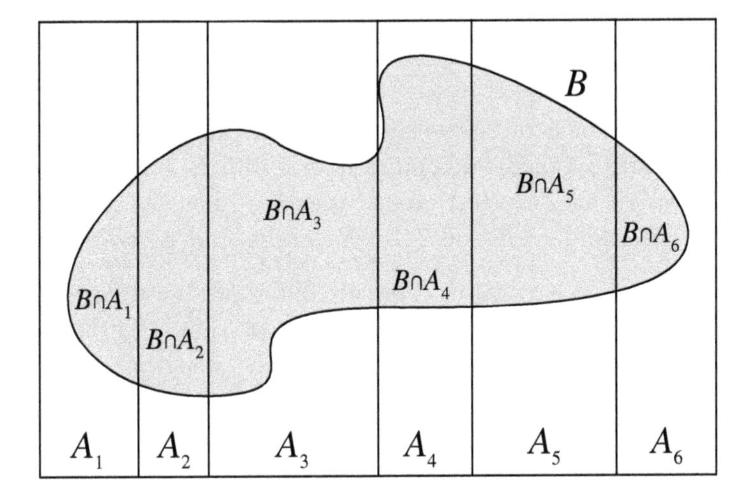

FIGURE 2.3
The A_i partition the sample space; $P(B)$ is equal to $\sum_i P(B \cap A_i)$.

Example 2.3.7 (Random coin). You have one fair coin, and one biased coin which lands Heads with probability 3/4. You pick one of the coins at random and flip it three times. It lands Heads all three times. Given this information, what is the probability that the coin you picked is the fair one?

Solution:

Let A be the event that the chosen coin lands Heads three times and let F be the event that we picked the fair coin. We are interested in $P(F|A)$, but it is easier to find $P(A|F)$ and $P(A|F^c)$ since it helps to know which coin we have; this suggests using Bayes' rule and the law of total probability. Doing so, we have

$$\begin{aligned} P(F|A) &= \frac{P(A|F)P(F)}{P(A)} \\ &= \frac{P(A|F)P(F)}{P(A|F)P(F) + P(A|F^c)P(F^c)} \\ &= \frac{(1/2)^3 \cdot 1/2}{(1/2)^3 \cdot 1/2 + (3/4)^3 \cdot 1/2} \\ &\approx 0.23. \end{aligned}$$

Before flipping the coin, we thought we were equally likely to have picked the fair coin as the biased coin: $P(F) = P(F^c) = 1/2$. Upon observing three Heads, however, it becomes more likely that we've chosen the biased coin than the fair coin, so $P(F|A)$ is only about 0.23. □

☣ **2.3.8** (Prior vs. posterior). It would *not* be correct in the calculation in the above example to say after the first step, "$P(A) = 1$ because we know A happened." It is true that $P(A|A) = 1$, but $P(A)$ is the *prior* probability of A and $P(F)$ is the *prior* probability of F—both are the probabilities before we observe any data in the experiment. These must not be confused with *posterior* probabilities conditional on the evidence A.

Example 2.3.9 (Testing for a rare disease). A patient named Fred is tested for a disease called conditionitis, a medical condition that afflicts 1% of the population. The test result is positive, i.e., the test claims that Fred has the disease. Let D be the event that Fred has the disease and T be the event that he tests positive.

Suppose that the test is "95% accurate"; there are different measures of the accuracy of a test, but in this problem it is assumed to mean that $P(T|D) = 0.95$ and $P(T^c|D^c) = 0.95$. The quantity $P(T|D)$ is known as the *sensitivity* or *true positive rate* of the test, and $P(T^c|D^c)$ is known as the *specificity* or *true negative rate.*

Find the conditional probability that Fred has conditionitis, given the evidence provided by the test result.

Solution:

Applying Bayes' rule and the law of total probability, we have

$$\begin{aligned} P(D|T) &= \frac{P(T|D)P(D)}{P(T)} \\ &= \frac{P(T|D)P(D)}{P(T|D)P(D) + P(T|D^c)P(D^c)} \\ &= \frac{0.95 \cdot 0.01}{0.95 \cdot 0.01 + 0.05 \cdot 0.99} \\ &\approx 0.16. \end{aligned}$$

So there is only a 16% chance that Fred has conditionitis, given that he tested positive, even though the test seems to be quite reliable!

Most people find it surprising to learn that the conditional probability of having the disease given a positive test result is only 16%, even though the test is 95% accurate (see Gigerenzer and Hoffrage [13]). The key to understanding this surprisingly low posterior probability of having the disease is to realize that there are two factors at play: the evidence from the test, and our *prior information* about the prevalence of the disease.

Although the test provides evidence in favor of disease, conditionitis is also a rare condition! The conditional probability $P(D|T)$ reflects a balance between these two factors, appropriately weighing the rarity of the disease against the rarity of a mistaken test result.

For further intuition, consider a population of 10000 people as illustrated in Figure 2.4, where 100 have conditionitis and 9900 don't; this corresponds to a 1% disease rate. If we tested everybody in the population, we'd expect that out of the 100 diseased individuals, 95 would test positive and 5 would test negative. Out of the 9900 healthy individuals, we'd expect $(0.95)(9900) \approx 9405$ to test negative and 495 to test positive.

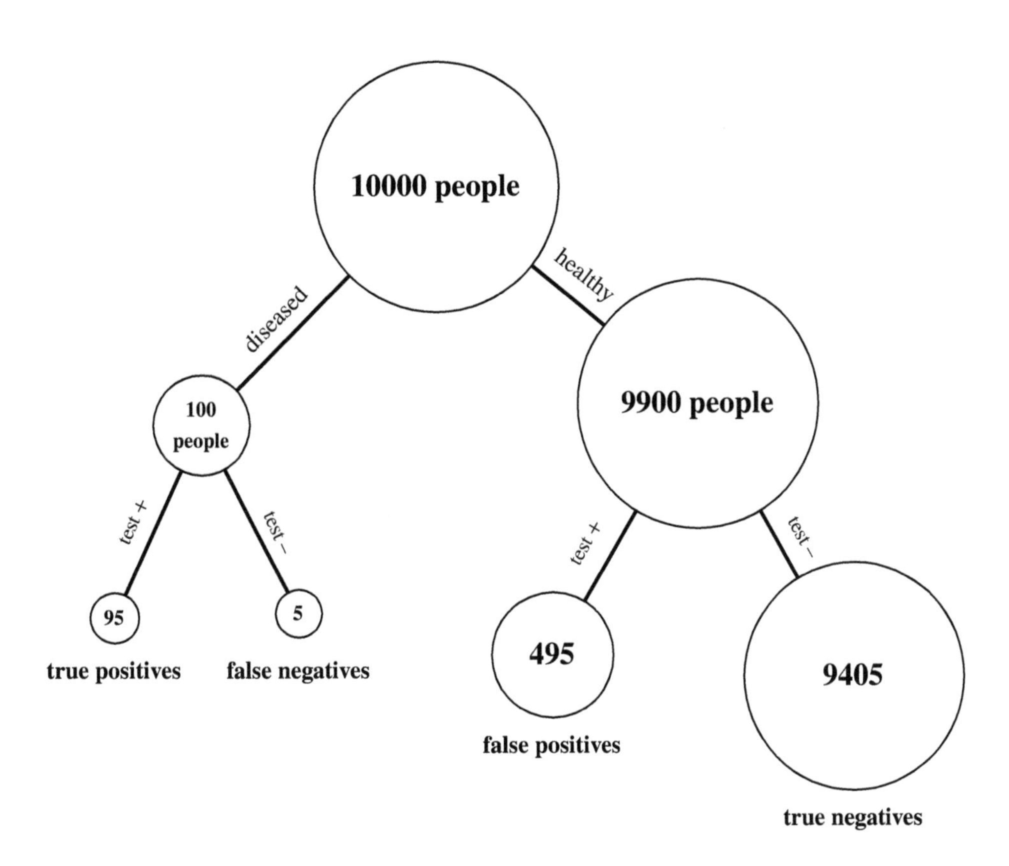

FIGURE 2.4
Testing for a rare disease in a population of 10000 people, where the prevalence of the disease is 1% and the true positive and true negative rates are both equal to 95%. Bubbles are not to scale.

Now let's focus in on those individuals who test positive; that is, let's condition on a positive test result. The 95 true positives (i.e., the individuals who test positive and have the disease) are far outnumbered by the 495 false positives (i.e., the individuals who test positive despite not having the disease). So most people who test positive for the disease don't actually have the disease! □

Example 2.3.10 (Six-fingered man). A crime has been committed in a certain country. The perpetrator is one (and only one) of the n men who live in the country. Initially, these n men are all deemed equally likely to be the perpetrator. An eyewitness then reports that the crime was committed by a man with six fingers on his right hand.

Let p_0 be the probability that an innocent man has six fingers on his right hand, and p_1 be the probability that the perpetrator has six fingers on his right hand, with $p_0 < p_1$. (We may have $p_1 < 1$, since eyewitnesses are not 100% reliable.) Let $a = p_0/p_1$ and $b = (1 - p_1)/(1 - p_0)$.

Rugen lives in the country. He is found to have six fingers on his right hand.

(a) Given this information, what is the probability that Rugen is the perpetrator?

(b) Now suppose that all n men who live in the country have their hands checked, and Rugen is the *only* one with six fingers on his right hand. Given this information, what is the probability that Rugen is the perpetrator?

Solution:

(a) Let R be the event that Rugen is guilty and M be the event that he has six fingers on his right hand. By Bayes' rule and LOTP,

$$P(R|M) = \frac{P(M|R)P(R)}{P(M|R)P(R) + P(M|R^c)P(R^c)} = \frac{p_1 \cdot \frac{1}{n}}{p_1 \cdot \frac{1}{n} + p_0\left(1 - \frac{1}{n}\right)} = \frac{1}{1 + a(n-1)}.$$

(b) Let N be the event that none of the men in the country other than Rugen have six fingers on their right hands. With notation as above,

$$\begin{aligned} P(R|M,N) &= \frac{P(M,N|R)P(R)}{P(M,N|R)P(R) + P(M,N|R^c)P(R^c)} \\ &= \frac{p_1(1-p_0)^{n-1} \cdot \frac{1}{n}}{p_1(1-p_0)^{n-1} \cdot \frac{1}{n} + p_0(1-p_1)(1-p_0)^{n-2}\left(1 - \frac{1}{n}\right)} \\ &= \frac{1}{1 + ab(n-1)}. \end{aligned}$$

□

2.4 Conditional probabilities are probabilities

When we condition on an event E, we update our beliefs to be consistent with this knowledge, effectively putting ourselves in a universe where we know that E occurred. Within our new universe, however, the laws of probability operate just as before. Conditional probability satisfies all the properties of probability! Therefore, any of the results we have derived about probability are still valid if we replace all unconditional probabilities with probabilities conditional on E. In particular:

- Conditional probabilities are between 0 and 1.
- $P(S|E) = 1$, $P(\emptyset|E) = 0$.
- If $A_1, A_2, \dots$ are disjoint, then $P(\cup_{j=1}^{\infty} A_j|E) = \sum_{j=1}^{\infty} P(A_j|E)$.
- $P(A^c|E) = 1 - P(A|E)$.
- Inclusion-exclusion: $P(A \cup B|E) = P(A|E) + P(B|E) - P(A \cap B|E)$.

☣ **2.4.1.** When we write $P(A|E)$, it does *not* mean that $A|E$ is an event and we're taking its probability; $A|E$ is not an event. Rather, $P(\cdot|E)$ is a probability function which assigns probabilities in accordance with the knowledge that E has occurred, and $P(\cdot)$ is a different probability function which assigns probabilities without regard for whether E has occurred or not. When we take an event A and plug it into the $P(\cdot)$ function, we'll get a number, $P(A)$; when we plug it into the $P(\cdot|E)$ function, we'll get another number, $P(A|E)$, which incorporates the information (if any) provided by knowing that E occurred.

To prove mathematically that conditional probabilities are probabilities, fix an event E with $P(E) > 0$, and for any event A, define $\tilde{P}(A) = P(A|E)$. This notation helps emphasize the fact that we are fixing E and treating $P(\cdot|E)$ as our new probability function. We just need to check the two axioms of probability. First,

$$\tilde{P}(\emptyset) = P(\emptyset|E) = \frac{P(\emptyset \cap E)}{P(E)} = 0, \tilde{P}(S) = P(S|E) = \frac{P(S \cap E)}{P(E)} = 1.$$

Second, if $A_1, A_2, \dots$ are disjoint events, then

$$\tilde{P}(A_1 \cup A_2 \cup \cdots) = \frac{P((A_1 \cap E) \cup (A_2 \cap E) \cup \cdots)}{P(E)} = \frac{\sum_{j=1}^{\infty} P(A_j \cap E)}{P(E)} = \sum_{j=1}^{\infty} \tilde{P}(A_j).$$

So $\tilde{P}$ satisfies the axioms of probability.

Conversely, *all* probabilities can be thought of as conditional probabilities: whenever we make a probability statement, there is always some background information that we are conditioning on, even if we don't state it explicitly. Consider the rain example from the beginning of this chapter. It would be natural to base the initial probability of rain today, $P(R)$, on the fraction of days in the past on which it rained. But which days in the past should we look at? If it's November 1, should we only count past rainy days in autumn, thus conditioning on the season? What about conditioning on the exact month, or the exact day? We could ask the same about location: should we look at days when it rained in our exact location, or is it enough for it to have rained somewhere nearby?

In order to determine the seemingly unconditional probability $P(R)$, we actually have to make decisions about what background information to condition on! These choices require careful thought and different people may come up with different *prior* probabilities $P(R)$ (though everyone can agree on how to update based on new evidence).

Since all probabilities are conditional on background information, we can imagine that there is always a vertical conditioning bar, with background knowledge K to the right of the vertical bar. Then the unconditional probability $P(A)$ is just shorthand for $P(A|K)$; the background knowledge is absorbed into the letter P instead of being written explicitly.

To summarize our discussion in a nutshell:

Conditional probabilities are probabilities, and all probabilities are conditional.

We now state conditional forms of Bayes' rule and the law of total probability. These are obtained by taking the ordinary forms of Bayes' rule and LOTP and adding E to the right of the vertical bar everywhere.

Theorem 2.4.2 (Bayes' rule with extra conditioning). Provided that $P(A \cap E) > 0$ and $P(B \cap E) > 0$, we have

$$P(A|B, E) = \frac{P(B|A, E)P(A|E)}{P(B|E)}.$$

Theorem 2.4.3 (LOTP with extra conditioning). Let $A_1, \dots, A_n$ be a partition of S. Provided that $P(A_i \cap E) > 0$ for all i, we have

$$P(B|E) = \sum_{i=1}^{n} P(B|A_i, E)P(A_i|E).$$

The extra conditioning forms of Bayes' rule and LOTP can be proved similarly to how we verified that $\tilde{P}$ satisfies the axioms of probability, but they also follow directly from the "metatheorem" that *conditional probabilities are probabilities.*

Example 2.4.4 (Random coin, continued). Continuing with the scenario from Example 2.3.7, suppose that we have now seen our chosen coin land Heads three times. If we toss the coin a fourth time, what is the probability that it will land Heads once more?

Solution:

As before, let A be the event that the chosen coin lands Heads three times, and define a new event H for the chosen coin landing Heads on the fourth toss. We are interested in $P(H|A)$. It would be very helpful to know whether we have the fair coin. LOTP with extra conditioning gives us $P(H|A)$ as a weighted average of $P(H|F, A)$ and $P(H|F^c, A)$, and within these two conditional probabilities we *do* know whether we have the fair coin:

$$\begin{aligned} P(H|A) &= P(H|F, A)P(F|A) + P(H|F^c, A)P(F^c|A) \\ &\approx \frac{1}{2} \cdot 0.23 + \frac{3}{4} \cdot (1 - 0.23) \\ &\approx 0.69. \end{aligned}$$

The posterior probabilities $P(F|A)$ and $P(F^c|A)$ are from our answer to Example 2.3.7.

An equivalent way to solve this problem is to define a new probability function $\tilde{P}$ such that for any event B, $\tilde{P}(B) = P(B|A)$. This new function assigns probabilities that are updated with the knowledge that A occurred. Then by the ordinary law of total probability,

$$\tilde{P}(H) = \tilde{P}(H|F)\tilde{P}(F) + \tilde{P}(H|F^c)\tilde{P}(F^c),$$

which is exactly the same as our use of LOTP with extra conditioning. This once again illustrates the principle that conditional probabilities are probabilities. □

Example 2.4.5 (Unanimous agreement). The article "Why too much evidence can be a bad thing" by Lisa Zyga [30] says:

> *Under ancient Jewish law, if a suspect on trial was unanimously found guilty by all judges, then the suspect was acquitted. This reasoning sounds counterintuitive, but the legislators of the time had noticed that unanimous agreement often indicates the presence of systemic error in the judicial process.*

There are n judges deciding a case. The suspect has prior probability p of being guilty. Each judge votes whether to convict or acquit the suspect. With probability s, a systemic error occurs (e.g., the defense is incompetent). If a systemic error occurs, then the judges unanimously vote to convict (i.e., all n judges vote to convict).

Whether a systemic error occurs is independent of whether the suspect is guilty. Given that a systemic error doesn't occur and that the suspect is guilty, each judge has probability c of voting to convict, independently. Given that a systemic error doesn't occur and that the suspect is not guilty, each judge has probability w of voting to convict, independently. Suppose that

$$0 < p < 1,\ 0 < s < 1,\ \text{and}\ 0 < w < \frac{1}{2} < c < 1.$$

(a) For this part only, suppose that exactly k out of n judges vote to convict, where $k < n$. Given this information, find the probability that the suspect is guilty.

(b) Now suppose that all n judges vote to convict. Given this information, find the probability that the suspect is guilty.

(c) Is the answer to (b), viewed as a function of n, an increasing function? Give a short, intuitive explanation in words.

Solution:

(a) Since $k < n$, a systemic error didn't occur. We will implicitly condition on this in this part. Let G be the event that the suspect is guilty and X be the number of judges who vote to convict. Using Bayes' rule, LOTP, and the Binomial PMF,

$$P(G|X=k) = \frac{P(X=k|G)P(G)}{P(X=k)} = \frac{pc^k(1-c)^{n-k}}{pc^k(1-c)^{n-k} + (1-p)w^k(1-w)^{n-k}}.$$

(b) Let U be the event $X = n$ and B be the event that a systemic error occurs. Then

$$P(G|U) = \frac{P(U|G)P(G)}{P(U)} = \frac{pP(U|G)}{pP(U|G) + (1-p)P(U|G^c)}.$$

By LOTP with extra conditioning,

$$P(U|G) = P(U|G,B)P(B|G) + P(U|G,B^c)P(B^c|G) = s + (1-s)c^n,$$
$$P(U|G^c) = P(U|G^c,B)P(B|G^c) + P(U|G^c,B^c)P(B^c|G^c) = s + (1-s)w^n.$$

Thus,

$$P(G|U) = \frac{p(s+(1-s)c^n)}{p(s+(1-s)c^n) + (1-p)(s+(1-s)w^n)}.$$

(c) No, since a large value of n yields a high chance of systemic error, and if a systemic error occurs then the judges' votes are uninformative about whether the suspect is guilty. The answer to (b) reverts to the prior probability p as $n \to \infty$. □

We often want to condition on more than one piece of information, and we now have several ways of doing that. For example, here are some approaches for finding $P(A|B,C)$:

1. We can think of B, C as the single event $B \cap C$ and use the definition of conditional probability to get

$$P(A|B,C) = \frac{P(A,B,C)}{P(B,C)}.$$

This is a natural approach if it's easiest to think about B and C in tandem. We can then try to evaluate the numerator and denominator. For example, we can use LOTP in both the numerator and the denominator, or we can write the numerator as $P(B,C|A)P(A)$ (which would give us a version of Bayes' rule) and use LOTP to help with the denominator.

2. We can use Bayes' rule with extra conditioning on C to get

$$P(A|B,C) = \frac{P(B|A,C)P(A|C)}{P(B|C)}.$$

This is a natural approach if we want to think of everything in our problem as being conditioned on C.

3. We can use Bayes' rule with extra conditioning on B to get

$$P(A|B,C) = \frac{P(C|A,B)P(A|B)}{P(C|B)}.$$

This is the same as the previous approach, except with the roles of B and C swapped. We mention it separately just to emphasize that it's a bad idea to plug into a formula without thinking about which event should play which role.

It is both challenging and powerful that there are a variety of ways to approach this kind of conditioning problem.

2.5 Independence of events

We have now seen several examples where conditioning on one event changes our beliefs about the probability of another event. The situation where events provide no information about each other is called *independence.*

Definition 2.5.1 (Independence of two events). Events A and B are *independent* if

$$P(A \cap B) = P(A)P(B).$$

If $P(A) > 0$ and $P(B) > 0$, then this is equivalent to

$$P(A|B) = P(A),$$

and also equivalent to $P(B|A) = P(B)$.

In words, two events are independent if we can obtain the probability of their intersection by multiplying their individual probabilities. Alternatively, A and B are independent if learning that B occurred gives us no information that would change our probabilities for A occurring (and vice versa).

Note that independence is a *symmetric relation*: if A is independent of B, then B is independent of A.

☣ **2.5.2.** Independence is completely different from disjointness. If A and B are disjoint, then $P(A \cap B) = 0$, so disjoint events can be independent only if $P(A) = 0$ or $P(B) = 0$. Knowing that A occurs tells us that B definitely did not occur, so A clearly conveys information about B, meaning the two events are not independent (except if A or B already has zero probability).

Intuitively, it makes sense that if A provides no information about whether or not B occurred, then it also provides no information about whether or not B^c occurred. We now prove a handy result along those lines.

Proposition 2.5.3. If A and B are independent, then A and B^c are independent, A^c and B are independent, and A^c and B^c are independent.

Proof. Let A and B be independent. We will first show that A and B^c are independent. If $P(A) = 0$, then A is independent of *every* event, including B^c. So assume $P(A) \neq 0$. Then

$$P(B^c|A) = 1 - P(B|A) = 1 - P(B) = P(B^c),$$

so A and B^c are independent. Swapping the roles of A and B, we have that A^c and B are independent. Using the fact that A, B independent implies A, B^c independent, with A^c playing the role of A, we also have that A^c and B^c are independent. ■

We also often need to talk about independence of three or more events.

Definition 2.5.4 (Independence of three events). Events A, B, and C are said to be *independent* if all of the following equations hold:

$$\begin{aligned}
P(A \cap B) &= P(A)P(B), \\
P(A \cap C) &= P(A)P(C), \\
P(B \cap C) &= P(B)P(C), \\
P(A \cap B \cap C) &= P(A)P(B)P(C).
\end{aligned}$$

If the first three conditions hold, we say that A, B, and C are *pairwise independent*. Pairwise independence does *not* imply independence: it is possible that just learning about A or just learning about B is of no use in predicting whether C occurred, but learning that *both* A and B occurred could still be highly relevant for C. Here is a simple example of this distinction.

Example 2.5.5 (Pairwise independence doesn't imply independence). Consider two fair, independent coin tosses, and let A be the event that the first is Heads, B the event that the second is Heads, and C the event that both tosses have the same result. Then A, B, and C are pairwise independent but not independent, since $P(A \cap B \cap C) = 1/4$ while $P(A)P(B)P(C) = 1/8$. The point is that just knowing about A or just knowing about B tells us nothing about C, but knowing what happened with *both* A and B gives us information about C (in fact, in this case it gives us perfect information about C). □

On the other hand, $P(A \cap B \cap C) = P(A)P(B)P(C)$ does not imply pairwise independence; this can be seen quickly by looking at the extreme case $P(A) = 0$, when the equation becomes $0 = 0$, which tells us nothing about B and C.

We can define independence of any number of events similarly. Intuitively, the idea is that knowing what happened with any particular subset of the events gives us no information about what happened with the events not in that subset.

Definition 2.5.6 (Independence of many events). For n events $A_1, A_2, \dots, A_n$ to be *independent*, we require any pair to satisfy $P(A_i \cap A_j) = P(A_i)P(A_j)$ (for $i \neq j$), any triplet to satisfy $P(A_i \cap A_j \cap A_k) = P(A_i)P(A_j)P(A_k)$ (for i, j, k distinct), and similarly for all quadruplets, quintuplets, and so on. This can quickly become unwieldy, but later we will discuss other ways to think about independence. For infinitely many events, we say that they are independent if every finite subset of the events is independent.

Conditional independence is defined analogously to independence.

Definition 2.5.7 (Conditional independence). Events A and B are said to be *conditionally independent* given E if $P(A \cap B|E) = P(A|E)P(B|E)$.

☣ **2.5.8.** It is easy to make terrible blunders stemming from confusing independence and conditional independence. Two events can be conditionally independent given E, but not independent given E^c. Two events can be conditionally independent given E, but not independent. Two events can be independent, but not conditionally independent given E.

In particular, $P(A, B) = P(A)P(B)$ does *not* imply $P(A, B|E) = P(A|E)P(B|E)$; we can't just insert "given E" everywhere, as we did in going from LOTP to LOTP with extra conditioning. This is because LOTP *always* holds (it is a consequence of the axioms of probability), whereas $P(A, B)$ may or may not equal $P(A)P(B)$, depending on what A and B are.

The next few examples illustrate these distinctions. Great care is needed in working with conditional probabilities and conditional independence!

Example 2.5.9 (Conditional independence given E vs. given E^c). Suppose there are two types of classes: good classes and bad classes. In good classes, if you work hard, you are very likely to get an A. In bad classes, the professor randomly assigns grades to students regardless of their effort. Let G be the event that a class is good, W be the event that you work hard, and A be the event that you receive an A. Then W and A are conditionally independent given G^c, but they are not conditionally independent given G. □

Example 2.5.10 (Conditional independence doesn't imply independence). Returning once more to the scenario from Example 2.3.7, suppose we have chosen either a fair coin or a biased coin with probability 3/4 of Heads, but we do not know which one we have chosen. We flip the coin a number of times. Conditional on choosing the fair coin, the coin tosses are independent, with each toss having probability 1/2 of Heads. Similarly, conditional on choosing the biased coin, the tosses are independent, each with probability 3/4 of Heads.

However, the coin tosses are not unconditionally independent, because if we don't know which coin we've chosen, then observing the sequence of tosses gives us information about whether we have the fair coin or the biased coin in our hand. This in turn helps us to predict the outcomes of future tosses from the same coin.

To state this formally, let F be the event that we've chosen the fair coin, and let A_1 and A_2 be the events that the first and second coin tosses land Heads. Conditional on F, A_1 and A_2 are independent, but A_1 and A_2 are not unconditionally independent because A_1 provides information about A_2. □

Example 2.5.11 (Independence doesn't imply conditional independence)**.** My friends Alice and Bob are the only two people who ever call me on the phone. Each day, they decide independently whether to call me that day. Let A be the event that Alice calls me next Friday and B be the event that Bob calls me next Friday. Assume A and B are unconditionally independent with $P(A) > 0$ and $P(B) > 0$.

However, given that I receive exactly one call next Friday, A and B are no longer independent: the call is from Alice if and only if it is not from Bob. In other words, letting C be the event that I receive exactly one call next Friday, $P(B|C) > 0$ while $P(B|A, C) = 0$, so A and B are not conditionally independent given C. □

Example 2.5.12. (Why is the baby crying?) A certain baby cries if and only if she is hungry, tired, or both. Let C be the event that the baby is crying, H be the event that she is hungry, and T be the event that she is tired. Let $P(C) = c, P(H) = h$, and $P(T) = t$, where none of c, h, t are equal to 0 or 1. Let H and T be independent.

(a) Find c, in terms of h and t.

(b) Find $P(H|C), P(T|C)$, and $P(H, T|C)$.

(c) Are H and T conditionally independent given C? Explain in two ways: algebraically using the quantities from (b), and with an intuitive explanation in words.

Solution:

(a) Since H and T are independent, we have

$$P(C) = P(H \cup T) = P(H) + P(T) - P(H \cap T) = h + t - ht.$$

(b) By Bayes' rule,

$$\begin{aligned} P(H|C) &= \frac{P(C|H)P(H)}{P(C)} = \frac{h}{c}, \\ P(T|C) &= \frac{P(C|T)P(T)}{P(C)} = \frac{t}{c}, \\ P(H,T|C) &= \frac{P(C|H,T)P(H,T)}{P(C)} = \frac{ht}{c}. \end{aligned}$$

(c) No, H and T are not conditionally independent given C, since

$$P(H, T|C) = \frac{ht}{c} < \frac{ht}{c^2} = P(H|C)P(T|C).$$

We can also see intuitively why they are not conditionally independent given C: if the baby is crying but not hungry, she must be tired. □

2.6 Coherency of Bayes' rule

An important property of Bayes' rule is that it is *coherent*: if we receive multiple pieces of information and wish to update our probabilities to incorporate all the information, it does not matter whether we update sequentially, taking each piece of evidence into account one at a time, or simultaneously, using all the evidence at once. Suppose, for example, that we're conducting a weeklong experiment that yields data at the end of each day. We could use Bayes' rule every day to update our probabilities based on the data from that day. Or we could go on vacation for the week, come back on Friday afternoon, and update using the entire week's worth of data. Either method will give the same result.

Let's look at a concrete application of this principle.

Example 2.6.1 (Testing for a rare disease, continued). Fred, who tested positive for conditionitis in Example 2.3.9, decides to get tested a second time. The new test is independent of the original test (given his disease status) and has the same sensitivity and specificity. Unfortunately for Fred, he tests positive a second time. Find the probability that Fred has the disease, given the evidence, in two ways: in one step, conditioning on both test results simultaneously, and in two steps, first updating the probabilities based on the first test result, and then updating again based on the second test result.

Solution:

Let D be the event that he has the disease, T_1 that the first test result is positive, and T_2 that the second test result is positive. In Example 2.3.9, we used Bayes' rule and the law of total probability to find $P(D|T_1)$. Another quick solution uses the odds form of Bayes' rule:

$$\frac{P(D|T_1)}{P(D^c|T_1)} = \frac{P(D)}{P(D^c)}\frac{P(T_1|D)}{P(T_1|D^c)} = \frac{1}{99}\cdot\frac{0.95}{0.05} \approx 0.19.$$

Since $P(D|T_1)/(1 - P(D|T_1)) = 0.19$, we have $P(D|T_1) = 0.19/(1 + 0.19) \approx 0.16$, in agreement with our answer from before. The odds form of Bayes' rule is faster in this case because we don't need to compute the unconditional probability $P(T_1)$ in the denominator of the ordinary Bayes' rule. Now, again using the odds form of Bayes' rule, let's find out what happens if Fred tests positive a second time.

One-step method: Updating based on both test results at once, we have

$$\begin{aligned}\frac{P(D|T_1 \cap T_2)}{P(D^c|T_1 \cap T_2)} &= \frac{P(D)}{P(D^c)}\frac{P(T_1 \cap T_2|D)}{P(T_1 \cap T_2|D^c)}\\ &= \frac{1}{99}\cdot\frac{0.95^2}{0.05^2} = \frac{361}{99} \approx 3.646,\end{aligned}$$

which corresponds to a probability of 0.78.

Two-step method: After the first test, the posterior odds of Fred having the disease are

$$\frac{P(D|T_1)}{P(D^c|T_1)} = \frac{1}{99} \cdot \frac{0.95}{0.05} \approx 0.19,$$

from the above. These posterior odds become the new prior odds, and then updating based on the second test gives

$$\begin{aligned}\frac{P(D|T_1 \cap T_2)}{P(D^c|T_1 \cap T_2)} &= \frac{P(D|T_1)}{P(D^c|T_1)} \frac{P(T_2|D, T_1)}{P(T_2|D^c, T_1)} \\ &= \left(\frac{1}{99} \cdot \frac{0.95}{0.05}\right) \frac{0.95}{0.05} = \frac{361}{99} \approx 3.646,\end{aligned}$$

which is the same result as above.

Note that with a second positive test result, the probability that Fred has the disease jumps from 0.16 to 0.78, making us much more confident that Fred is actually afflicted with conditionitis. The moral of the story is that getting a second opinion is a good idea! □

2.7 Conditioning as a problem-solving tool

Conditioning is a powerful tool for solving problems because it lets us engage in *wishful thinking*: when we encounter a problem that would be made easier if only we knew whether E happened or not, we can condition on E and then on E^c, consider these possibilities separately, then combine them using LOTP.

2.7.1 Strategy: condition on what you wish you knew

Example 2.7.1 (Monty Hall). On the game show Let's Make a Deal, hosted by Monty Hall, a contestant chooses one of three closed doors, two of which have a goat behind them and one of which has a car. Monty, who knows where the car is, then opens one of the two remaining doors. The door he opens always has a goat behind it (he never reveals the car!). If he has a choice, then he picks a door at random with equal probabilities. Monty then offers the contestant the option of switching to the other unopened door. If the contestant's goal is to get the car, should she switch doors?

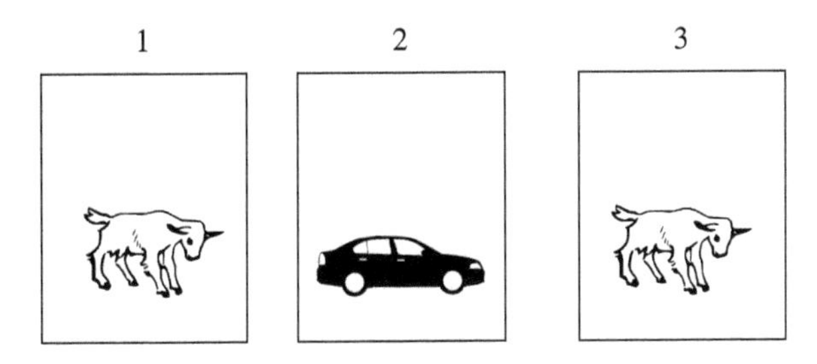

Solution:

Let's label the doors 1 through 3. Without loss of generality, we can assume the contestant picked door 1 (if she didn't pick door 1, we could simply relabel the doors, or rewrite this solution with the door numbers permuted). Monty opens a door, revealing a goat. As the contestant decides whether or not to switch to the remaining unopened door, what does she really wish she knew? Naturally, her decision would be a lot easier if she knew where the car was! This suggests that we should condition on the location of the car. Let C_i be the event that the car is behind door i, for $i = 1, 2, 3$. By the law of total probability,

$$P(\text{get car}) = P(\text{get car}|C_1) \cdot \frac{1}{3} + P(\text{get car}|C_2) \cdot \frac{1}{3} + P(\text{get car}|C_3) \cdot \frac{1}{3}.$$

Suppose the contestant employs the switching strategy. If the car is behind door 1, then switching will fail, so $P(\text{get car}|C_1) = 0$. If the car is behind door 2 or 3, then because Monty always reveals a goat, the remaining unopened door must contain the car, so switching will succeed. Thus,

$$P(\text{get car}) = 0 \cdot \frac{1}{3} + 1 \cdot \frac{1}{3} + 1 \cdot \frac{1}{3} = \frac{2}{3},$$

so the switching strategy succeeds 2/3 of the time. The contestant should switch to the other door.

Figure 2.5 is a tree diagram of the argument we have just outlined: using the switching strategy, the contestant will win as long as the car is behind doors 2 or 3, which has probability 2/3. We can also give an intuitive frequentist argument in favor of switching. Imagine playing this game 1000 times. Typically, about 333 times your initial guess for the car's location will be correct, in which case switching will fail. The other 667 or so times, you will win by switching.

There's a subtlety though, which is that when the contestant chooses whether to switch, she also knows which door Monty opened. We showed that the *unconditional* probability of success is 2/3 (when following the switching strategy), but let's also show that the *conditional* probability of success for switching, given the information that Monty provides, is also 2/3.

Let M_j be the event that Monty opens door j, for $j = 2, 3$. Then

$$P(\text{get car}) = P(\text{get car}|M_2)P(M_2) + P(\text{get car}|M_3)P(M_3),$$

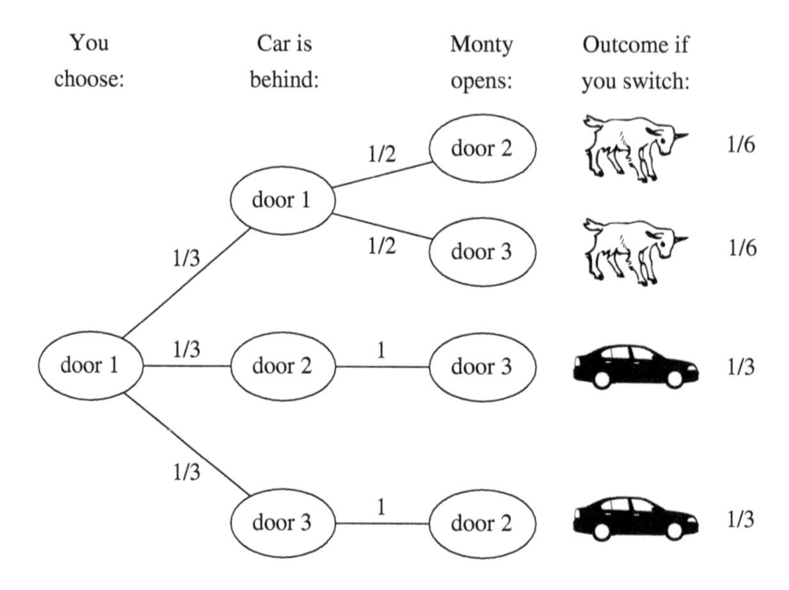

FIGURE 2.5
Tree diagram of Monty Hall problem. Switching gets the car 2/3 of the time.

where by symmetry $P(M_2) = P(M_3) = 1/2$ and $P(\text{get car}|M_2) = P(\text{get car}|M_3)$. The symmetry here is that there is nothing in the statement of the problem that distinguishes between door 2 and door 3; in contrast, Problem 40 considers a scenario where Monty enjoys opening door 2 more than he enjoys opening door 3.

Let $x = P(\text{get car}|M_2) = P(\text{get car}|M_3)$. Plugging in what we know,

$$\frac{2}{3} = P(\text{get car}) = \frac{x}{2} + \frac{x}{2} = x,$$

as claimed.

Bayes' rule also works nicely for finding the conditional probability of success using the switching strategy, given the evidence. Suppose that Monty opens door 2. Using the notation and results above,

$$P(C_1|M_2) = \frac{P(M_2|C_1)P(C_1)}{P(M_2)} = \frac{(1/2)(1/3)}{1/2} = \frac{1}{3}.$$

So given that Monty opens door 2, there is a 1/3 chance that the contestant's original choice of door has the car, which means that there is a 2/3 chance that the switching strategy will succeed.

Many people, upon seeing this problem for the first time, argue that there is no advantage to switching: "There are two doors remaining, and one of them has the car, so the chances are 50-50." After the last chapter, we recognize that this argument misapplies the naive definition of probability. Yet the naive definition, even when inappropriate, has a powerful hold on people's intuitions. When Marilyn vos Savant presented a correct solution to the Monty Hall problem in her column for *Parade* magazine in 1990, she received thousands upon thousands of letters from readers (even mathematicians) insisting that she was wrong.

To build correct intuition, let's consider an extreme case. Suppose that there are a million doors, 999,999 of which contain goats and 1 of which has a car. After the contestant's initial pick, Monty opens 999,998 doors with goats behind them and offers the choice to switch. In this extreme case, it becomes clear that the probabilities are *not* 50-50 for the two unopened doors; very few people would stubbornly stick with their original choice. The same is true for the three-door case.

Just as we had to make assumptions about how we came across the random girl in Example 2.2.6, here the 2/3 success rate of the switching strategy depends on the assumptions we make about how Monty decides which door to open. In the exercises, we consider several variants and generalizations of the Monty Hall problem, some of which change the desirability of the switching strategy. □

2.7.2 Strategy: condition on the first step

In problems with a recursive structure, it can often be useful to condition on the first step of the experiment. The next two examples apply this strategy, which we call *first-step analysis.*

Example 2.7.2 (Branching process). A single amoeba, Bobo, lives in a pond. After one minute Bobo will either die, split into two amoebas, or stay the same, with equal probability, and in subsequent minutes all living amoebas will behave the same way, independently. What is the probability that the amoeba population will eventually die out?

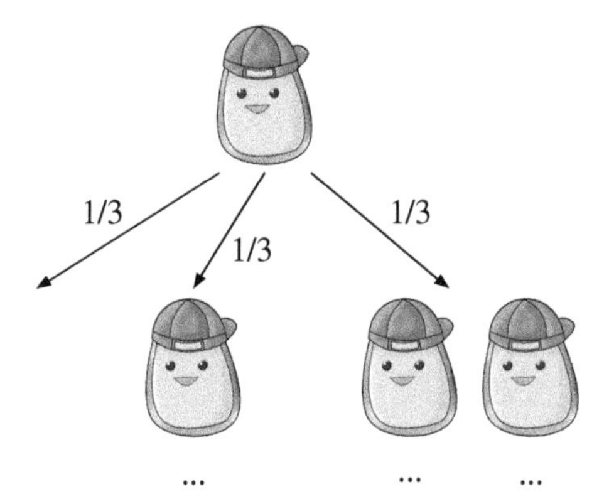

Solution:

Let D be the event that the population eventually dies out; we want to find $P(D)$. We proceed by conditioning on the outcome at the first step: let B_i be the event that Bobo turns into i amoebas after the first minute, for $i = 0, 1, 2$. We know $P(D|B_0) = 1$ and $P(D|B_1) = P(D)$ (if Bobo stays the same, we're back to where we started). If Bobo splits into two, then we just have two independent versions of our original problem! We need both of the offspring to eventually die out, so $P(D|B_2) = P(D)^2$. Now we have exhausted all the possible cases and can combine them with the law of total probability:

$$
\begin{aligned}
P(D) &= P(D|B_0) \cdot \frac{1}{3} + P(D|B_1) \cdot \frac{1}{3} + P(D|B_2) \cdot \frac{1}{3} \\
&= 1 \cdot \frac{1}{3} + P(D) \cdot \frac{1}{3} + P(D)^2 \cdot \frac{1}{3}.
\end{aligned}
$$

Solving for $P(D)$ gives $P(D) = 1$: the amoeba population will die out with probability 1.

The strategy of first-step analysis works here because the problem is self-similar in nature: when Bobo continues as a single amoeba or splits into two, we end up with another version or another two versions of our original problem. Conditioning on the first step allows us to express $P(D)$ in terms of itself. □

Example 2.7.3 (Gambler's ruin). Two gamblers, A and B, make a sequence of \$1 bets. In each bet, gambler A has probability p of winning, and gambler B has probability $q = 1 - p$ of winning. Gambler A starts with i dollars and gambler B starts with $N - i$ dollars; the total wealth between the two remains constant since every time A loses a dollar, the dollar goes to B, and vice versa.

We can visualize this game as a *random walk* on the integers between 0 and N, where p is the probability of going to the right in a given step: imagine a person who starts at position i and, at each time step, moves one step to the right with probability p and one step to the left with probability $q = 1 - p$. The game ends when either A or B is ruined, i.e., when the random walk reaches 0 or N. What is the probability that A wins the game (walking away with all the money)?

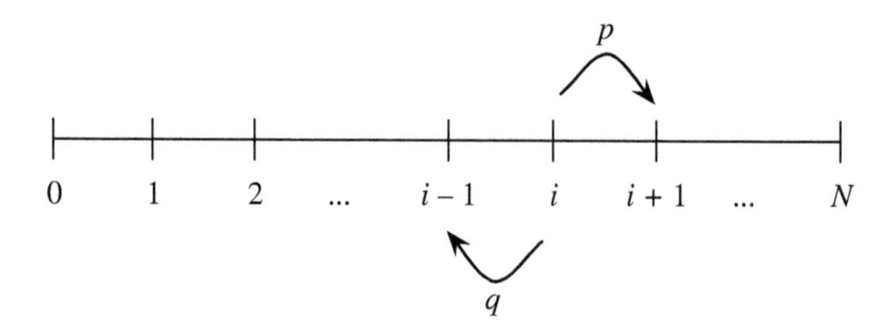

Solution:

We recognize that this game, like Bobo's reproductive process, has a recursive structure: after the first step, it's exactly the same game, except that A's wealth is now either $i + 1$ or $i - 1$. Let p_i be the probability that A wins the game, given that A starts with i dollars. We will use first-step analysis to solve for the p_i. Let W be the event that A wins the game. By LOTP, conditioning on the outcome of the first round, we have

$$\begin{aligned} p_i &= P(W|\text{A starts at } i, \text{ wins round 1}) \cdot p + P(W|\text{A starts at } i, \text{ loses round 1}) \cdot q \\ &= P(W|\text{A starts at } i+1) \cdot p + P(W|\text{A starts at } i-1) \cdot q \\ &= p_{i+1} \cdot p + p_{i-1} \cdot q. \end{aligned}$$

This must be true for all i from 1 to $N-1$, and we also have the boundary conditions $p_0 = 0$ and $p_N = 1$. Now we can solve this equation, called a *difference equation*, to obtain the p_i. Section A.4 of the math appendix discusses how to solve difference equations, so we will omit some of the steps here.

The characteristic equation of the difference equation is $px^2 - x + q = 0$, which has roots 1 and q/p. If $p \neq 1/2$, these roots are distinct, and the general solution is

$$p_i = a \cdot 1^i + b \cdot \left(\frac{q}{p}\right)^i.$$

Using the boundary conditions $p_0 = 0$ and $p_N = 1$, we get

$$a = -b = \frac{1}{1 - \left(\frac{q}{p}\right)^N},$$

and we simply plug these back in to get the specific solution. If $p = 1/2$, the roots of the characteristic polynomial are not distinct, so the general solution is

$$p_i = a \cdot 1^i + b \cdot i \cdot 1^i.$$

The boundary conditions give $a = 0$ and $b = 1/N$.

In summary, the probability of A winning with a starting wealth of i is

$$p_i = \begin{cases} \frac{1-\left(\frac{q}{p}\right)^i}{1-\left(\frac{q}{p}\right)^N} & \text{if } p \neq 1/2, \\ \frac{i}{N} & \text{if } p = 1/2. \end{cases}$$

The $p = 1/2$ case is consistent with the $p \neq 1/2$ case, in the sense that

$$\lim_{p \to \frac{1}{2}} \frac{1 - (\frac{q}{p})^i}{1 - (\frac{q}{p})^N} = \frac{i}{N}.$$

To see this, let $x = q/p$ and let x approach 1. By L'Hôpital's rule,

$$\lim_{x \to 1} \frac{1 - x^i}{1 - x^N} = \lim_{x \to 1} \frac{ix^{i-1}}{Nx^{N-1}} = \frac{i}{N}.$$

The answer for the $p = 1/2$ case has a simple interpretation: A's probability of winning equals the proportion of the wealth that A starts out with. So if $p = 1/2$

and A starts out with much less money than B, then A's chance of winning the game is low. Having $p < 1/2$ may also make A's chance of winning low, even if p is only a little bit less than 1/2 and the players start out with the same amount of money. For example, if $p = 0.49$ and each player starts out with \$100, then A has only about a 1.8% chance of winning the game.

We have focused on the probability of A winning the game, but what about B? Rather than starting from scratch, we can use *symmetry*: aside from notation, there is nothing in the description of the game to distinguish A from B. By symmetry, the probability of B winning from a starting wealth of $N - i$ is obtained by switching the roles of q and p, and of i and $N - i$. This gives

$$P(\text{B wins}|\text{B starts at } N - i) = \begin{cases} \frac{1-\left(\frac{p}{q}\right)^{N-i}}{1-\left(\frac{p}{q}\right)^{N}} & \text{if } p \neq 1/2, \\ \frac{N-i}{N} & \text{if } p = 1/2. \end{cases}$$

It can then be verified that for all i and all p, $P(\text{A wins}) + P(\text{B wins}) = 1$, so the game is guaranteed to end: the probability is 0 that it will oscillate forever. □

2.8 Pitfalls and paradoxes

The next two examples are fallacies of conditional thinking that have arisen in the legal context. The prosecutor's fallacy is the confusion of $P(A|B)$ with $P(B|A)$; the defense attorney's fallacy is the failure to condition on *all* the evidence.

☣ **2.8.1** (Prosecutor's fallacy). In 1998, Sally Clark was tried for murder after two of her sons died shortly after birth. During the trial, an expert witness for the prosecution testified that the probability of a newborn dying of sudden infant death syndrome (SIDS) was 1/8500, so the probability of two deaths due to SIDS in one family was $(1/8500)^2$, or about one in 73 million. Therefore, he continued, the probability of Clark's innocence was one in 73 million.

There are at least two major problems with this line of reasoning. First, the expert witness found the probability of the intersection of "first son dies of SIDS" and "second son dies of SIDS" by multiplying the individual event probabilities; as we know, this is only valid if deaths due to SIDS are *independent* within a family. This independence would not hold if genetic or other family-specific risk factors cause all newborns within certain families to be at increased risk of SIDS.

Second, the so-called expert has confused two different conditional probabilities: $P(\text{innocence}|\text{evidence})$ is different from $P(\text{evidence}|\text{innocence})$. The witness claims that the probability of observing two newborn deaths if the defendant were innocent is extremely low; that is, $P(\text{evidence}|\text{innocence})$ is small. What we are interested in, however, is $P(\text{innocence}|\text{evidence})$, the probability that the defendant is innocent given all the evidence. By Bayes' rule,

$$P(\text{innocence}|\text{evidence}) = \frac{P(\text{evidence}|\text{innocence})P(\text{innocence})}{P(\text{evidence})},$$

so to calculate the conditional probability of innocence given the evidence, we must take into account $P(\text{innocence})$, the prior probability of innocence. This probability is extremely high: although double deaths due to SIDS are rare, so are double infanticides! Expanding the denominator as

$$P(\text{evidence}|\text{innocence})P(\text{innocence}) + P(\text{evidence}|\text{guilt})P(\text{guilt}),$$

note that if $P(\text{guilt})$ is small enough so that the second term is negligible compared to the first term, then the denominator of $P(\text{innocence}|\text{evidence})$ is approximately equal to the numerator, making $P(\text{innocence}|\text{evidence})$ close to 1.

The posterior probability of innocence given the evidence depends strongly on both $P(\text{evidence}|\text{innocence})$, which is very low, and $P(\text{innocence})$, which is very high. The expert's probability of $(1/8500)^2$, questionable in and of itself, is only part of the equation.

Sadly, Clark was convicted of murder and sent to prison, partly based on the expert's wrongheaded testimony, and spent over three years in jail before her conviction was overturned. The outcry over the misuse of conditional probability in the Sally Clark case led to the review of hundreds of other cases where similar faulty logic was used by the prosecution.

☣ **2.8.2** (Defense attorney's fallacy)**.** A woman has been murdered, and her husband is put on trial for this crime. Evidence comes to light that the defendant had a history of abusing his wife. The defense attorney argues that the evidence of abuse should be excluded on grounds of irrelevance, since only 1 in 10,000 men with wives they abuse subsequently murder their wives. Should the judge grant the defense attorney's motion to bar this evidence from trial?

Suppose that the defense attorney's 1-in-10,000 figure is correct, and further assume the following for a relevant population of husbands and wives: 1 in 10 husbands abuse their wives, 1 in 5 murdered wives were murdered by their husbands, and 50% of husbands who murder their wives previously abused them. Also, assume that if the husband of a murdered wife is *not* guilty of the murder, then the probability that he abused his wife reverts to the unconditional probability of abuse.

How to define the "relevant population" and how to estimate such probabilities are difficult issues. For example, should we look at citywide, statewide, national, or international statistics? How should we account for unreported abuse and unsolved murders? What if murder rates are changing over time? For this problem, assume that a reasonable choice of the relevant population has been agreed on, and that the stated probabilities are known to be correct.

Let A be the event that the husband commits abuse against his wife, and let G be the event that the husband is guilty. The defense's argument is that $P(G|A) = 1/10{,}000$, so guilt is still extremely unlikely conditional on a previous history of abuse.

However, the defense attorney fails to condition on a crucial fact: in this case, *we know that the wife was murdered.* Therefore, the relevant probability is not $P(G|A)$, but $P(G|A, M)$, where M is the event that the wife was murdered.

Bayes' rule with extra conditioning gives

$$\begin{aligned} P(G|A, M) &= \frac{P(A|G, M)P(G|M)}{P(A|G, M)P(G|M) + P(A|G^c, M)P(G^c|M)} \\ &= \frac{0.5 \cdot 0.2}{0.5 \cdot 0.2 + 0.1 \cdot 0.8} \\ &= \frac{5}{9}. \end{aligned}$$

So the posterior probability of guilt, $P(G|A, M)$, is over 5,000 times as large as the quantity $P(G|A)$ that the defense attorney focused on. Conditioning on the evidence of abuse increases the probability of guilt from $P(G|M) = 0.2$ to $P(G|A, M) \approx 0.56$, so the defendant's history of abuse gives very important information, contrary to the defense attorney's argument.

In the above calculation of $P(G|A, M)$, we did not use the defense attorney's $P(G|A)$ number anywhere; it is irrelevant to our calculation because it does not account for the fact that the wife was murdered. We must condition on *all* the evidence.

We end this chapter with a paradox about conditional probability and aggregation of data.

Example 2.8.3 (Simpson's paradox)**.** Two doctors, Dr. Hibbert and Dr. Nick, each perform two types of surgeries: heart surgery and Band-Aid removal. Each surgery can be either a success or a failure. The two doctors' respective records are given in the following tables, and shown graphically in Figure 2.6, where white dots represent successful surgeries and black dots represent failed surgeries.

	Heart	**Band-Aid**
Success	70	10
Failure	20	0

Dr. Hibbert

	Heart	**Band-Aid**
Success	2	81
Failure	8	9

Dr. Nick

Dr. Hibbert had a higher success rate than Dr. Nick in heart surgeries: 70 out of 90 versus 2 out of 10. Dr. Hibbert also had a higher success rate in Band-Aid removal: 10 out of 10 versus 81 out of 90. But if we *aggregate* across the two types of surgeries to compare overall surgery success rates, Dr. Hibbert was successful in 80 out of 100 surgeries while Dr. Nick was successful in 83 out of 100 surgeries: Dr. Nick's overall success rate is higher!

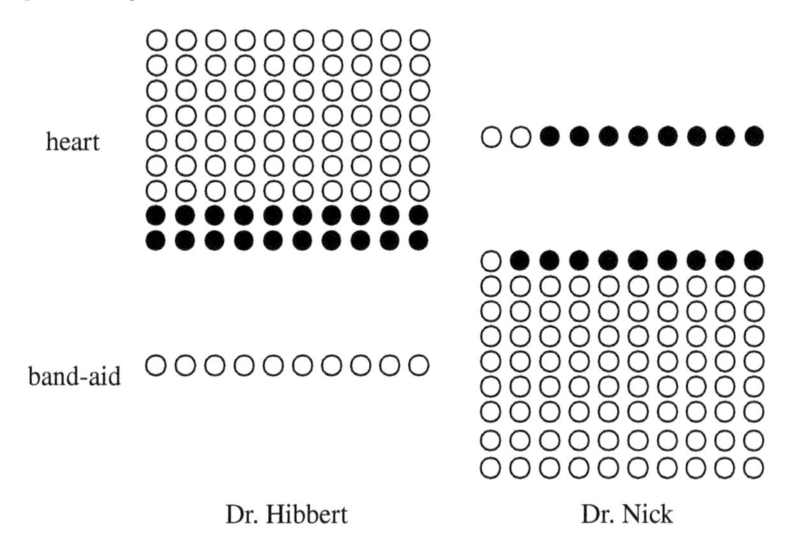

FIGURE 2.6
An example of Simpson's paradox. White dots represent successful surgeries and black dots represent failed surgeries. Dr. Hibbert is better in both types of surgery but has a lower overall success rate, because he is performing the harder type of surgery much more often than Dr. Nick is.

What's happening is that Dr. Hibbert, presumably due to his reputation as the superior doctor, is performing a greater number of heart surgeries, which are inherently riskier than Band-Aid removals. His overall success rate is lower not because of lesser skill on any particular type of surgery, but because a larger fraction of his surgeries are risky.

Let's use event notation to make this precise. For events A, B, and C, we say that we have a *Simpson's paradox* if

$$P(A|B,C) < P(A|B^c,C)$$
$$P(A|B,C^c) < P(A|B^c,C^c),$$

but

$$P(A|B) > P(A|B^c).$$

In this case, let A be the event of a successful surgery, B be the event that Dr. Nick is the surgeon, and C be the event that the surgery is a heart surgery. The conditions for Simpson's paradox are fulfilled because the probability of a successful surgery is lower under Dr. Nick than under Dr. Hibbert whether we condition on heart surgery or on Band-Aid removal, but the overall probability of success is higher for Dr. Nick.

The law of total probability tells us mathematically why this can happen:

$$P(A|B) = P(A|C,B)P(C|B) + P(A|C^c,B)P(C^c|B)$$
$$P(A|B^c) = P(A|C,B^c)P(C|B^c) + P(A|C^c,B^c)P(C^c|B^c).$$

The above equations express $P(A|B)$ as a weighted average of $P(A|C, B)$ and $P(A|C^c, B)$, and $P(A|B^c)$ as a weighted average of $P(A|C, B^c)$ and $P(A|C^c, B^c)$. If the corresponding weights were the same in both of these weighted averages, then Simpson's paradox could not occur. But the weights here are *different*:

$$P(C|B) < P(C|B^c) \text{ and } P(C^c|B) > P(C^c|B^c),$$

since Dr. Nick is much less likely than Dr. Hibbert to be performing a heart surgery.

Although we have

$$P(A|C, B) < P(A|C, B^c)$$

and

$$P(A|C^c, B) < P(A|C^c, B^c),$$

the fact that the weights are so different results in the inequality flipping when we do not condition on whether or not C occurred:

$$P(A|B) > P(A|B^c).$$

Numerically, the two weighted averages are

$$P(A|B) = 0.83 = (2/10) \cdot 0.1 + (81/90) \cdot 0.9$$
$$P(A|B^c) = 0.80 = (70/90) \cdot 0.9 + (10/10) \cdot 0.1.$$

The first equation (corresponding to Dr. Nick) puts much more weight on the second term (corresponding to the easier surgery) than does the second equation.

Aggregation across different types of surgeries presents a misleading picture of the doctors' abilities because we lose the information about which doctor tends to perform which type of surgery. When we think *confounding variables* like surgery type could be at play, we should examine the disaggregated data to see what is really going on.

Simpson's paradox arises in many real-world contexts. In the following examples, you should try to identify the events A, B, and C that create the paradox.

- *Gender discrimination in college admissions*: In the 1970s, men were significantly more likely than women to be admitted for graduate study at the University of California, Berkeley, leading to charges of gender discrimination. Yet within most individual departments, women were admitted at a higher rate than men. It was found that women tended to apply to the departments with more competitive admissions, while men tended to apply to less competitive departments.

- *Baseball batting averages*: It is possible for player 1 to have a higher batting average than player 2 in the first half of a baseball season *and* a higher batting average than player 2 in the second half of the season, yet have a lower overall batting average for the entire season. It depends on how many at-bats the players have in each half of the season. (An *at-bat* is when it's a player's turn to try to hit the ball; the player's *batting average* is the number of hits the player gets divided by the player's number of at-bats.)

- *Health effects of smoking*: Cochran [4] found that within any age group, cigarette smokers had higher mortality rates than cigar smokers, but because cigarette smokers were on average younger than cigar smokers, overall mortality rates were lower for cigarette smokers. □

2.9 Recap

The conditional probability of A given B is

$$P(A|B) = \frac{P(A \cap B)}{P(B)}.$$

Conditional probability has exactly the same properties as probability, but $P(\cdot|B)$ updates our uncertainty about events to reflect the observed evidence B. Events whose probabilities are unchanged after observing the evidence B are said to be independent of B. Two events can also be conditionally independent given a third event E. Conditional independence does not imply unconditional independence, nor does unconditional independence imply conditional independence.

Two important results about conditional probability are Bayes' rule, which relates $P(A|B)$ to $P(B|A)$, and the law of total probability, which allows us to get unconditional probabilities by partitioning the sample space and calculating conditional probabilities within each slice of the partition.

Bayes' rule says that

$$P(A|B) = \frac{P(B|A)P(A)}{P(B)},$$

while LOTP says that

$$P(B) = \sum_{i=1}^{n} P(B|A_i)P(A_i),$$

for any partition $A_1, \dots, A_n$ of the sample space. Bayes' rule and LOTP are often used in tandem.

Conditioning is extremely helpful for problem-solving because it allows us to break a problem into smaller pieces, consider all possible cases separately, and then combine them. When using this strategy, we should try to condition on the information that, if known, would make the problem simpler, hence the saying *condition on what you wish you knew*. When a problem involves multiple stages, it can be helpful to *condition on the first step* to obtain a recursive relationship.

Common mistakes in thinking conditionally include:

- confusion of the prior probability $P(A)$ with the posterior probability $P(A|B)$;

- the prosecutor's fallacy, confusing $P(A|B)$ with $P(B|A)$;
- the defense attorney's fallacy, failing to condition on all the evidence;
- unawareness of Simpson's paradox and the importance of thinking carefully about whether to aggregate data.

Figure 2.7 illustrates how probabilities can be updated as new evidence comes in sequentially. Imagine that there is some event A that we are interested in. On Monday morning, for example, our prior probability for A is $P(A)$. If we observe on Monday afternoon that B occurred, then we can use Bayes' rule (or the definition of conditional probability) to compute the posterior probability $P(A|B)$.

We use this posterior probability for A as the new prior on Tuesday morning, and then we continue to collect evidence. Suppose that on Tuesday we observe that C occurred. Then we can compute the new posterior probability $P(A|B, C)$ in various ways (in this context, probably the most natural way is to use Bayes' rule with extra conditioning on B). This in turn becomes the new prior if we are going to continue to collect evidence.

2.10 R

Simulating the frequentist interpretation

Recall that the frequentist interpretation of conditional probability based on a large number n of repetitions of an experiment is $P(A|B) \approx n_{AB}/n_B$, where n_{AB} is the number of times that $A \cap B$ occurs and n_B is the number of times that B occurs. Let's try this out by simulation, and verify the results of Example 2.2.5. So let's simulate n families, each with two children.

```
n <- 10^5
child1 <- sample(2,n,replace=TRUE)
child2 <- sample(2,n,replace=TRUE)
```

Here `child1` is a vector of length n, where each element is a 1 or a 2. Letting 1 stand for "girl" and 2 stand for "boy", this vector represents the gender of the elder child in each of the n families. Similarly, `child2` represents the gender of the younger child in each family.

Alternatively, we could have used

```
sample(c("girl","boy"),n,replace=TRUE)
```

but it is more convenient working with numerical values.

Let A be the event that both children are girls and B the event that the elder is a girl.

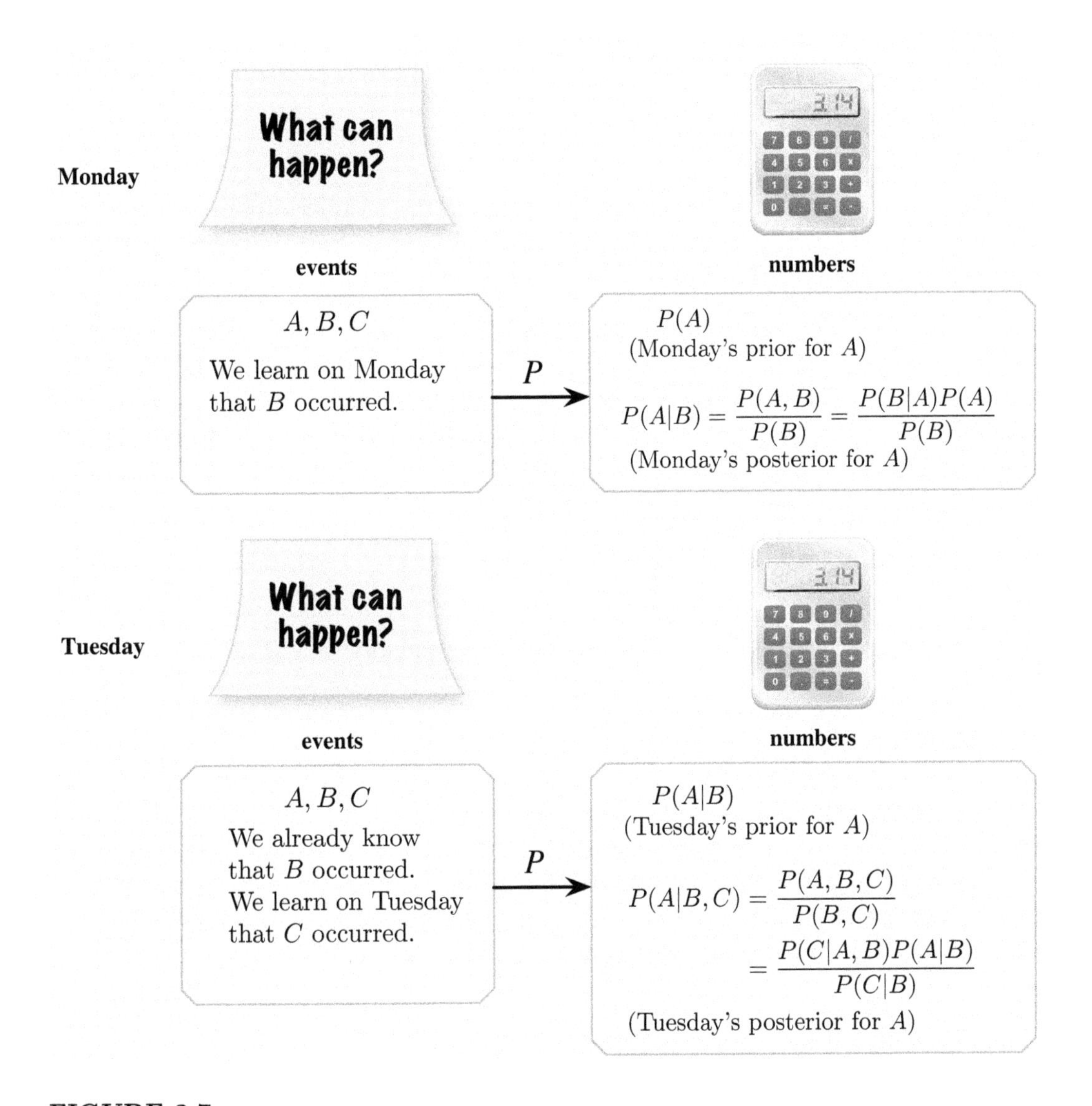

FIGURE 2.7
Conditional probability tells us how to update probabilities as new evidence comes in. Shown are the probabilities for an event A initially, after obtaining one piece of evidence B, and after obtaining a second piece of evidence C. The posterior for A after observing the first piece of evidence becomes the new prior before observing the second piece of evidence. After both B and C are observed, a new posterior for A can be found in various ways. This then becomes the new prior if more evidence will be collected.

Following the frequentist interpretation, we count the number of repetitions where B occurred and name it `n.b`, and we also count the number of repetitions where $A \cap B$ occurred and name it `n.ab`. Finally, we divide `n.ab` by `n.b` to approximate $P(A|B)$.

```
n.b <- sum(child1==1)
n.ab <- sum(child1==1 & child2==1)
n.ab/n.b
```

The ampersand `&` is an elementwise AND, so `n.ab` is the number of families where both the first child and the second child are girls. When we ran this code, we got 0.50, which agrees with $P(\text{both girls}|\text{elder is a girl}) = 1/2$.

Now let A be the event that both children are girls and B the event that at least one of the children is a girl. Then $A \cap B$ is the same, but `n.b` needs to count the number of families where at least one child is a girl. This is accomplished with the elementwise OR operator `|` (this is *not* a conditioning bar; it is an inclusive OR, returning TRUE if at least one element is TRUE).

```
n.b <- sum(child1==1 | child2==1)
n.ab <- sum(child1==1 & child2==1)
n.ab/n.b
```

We got 0.33, which agrees with $P(\text{both girls}|\text{at least one girl}) = 1/3$.

Monty Hall simulation

Many long, bitter debates about the Monty Hall problem could have been averted by *trying it out* with a simulation. To study how well the never-switch strategy performs, let's generate 10^5 runs of the Monty Hall game. To simplify notation, assume the contestant always chooses door 1. Then we can generate a vector specifying which door has the car for each repetition:

```
n <- 10^5
cardoor <- sample(3,n,replace=TRUE)
```

At this point we could generate the vector specifying which doors Monty opens, but that's unnecessary since the never-switch strategy succeeds if and only if door 1 has the car! So the fraction of times when the never-switch strategy succeeds is `sum(cardoor==1)/n`, which was 0.334 in our simulation. This is very close to the true value, 1/3.

What if we want to *play* the Monty Hall game interactively? We can do this by programming a *function*. Entering the following code in R defines a function called `monty`, which can then be invoked by entering the command `monty()` any time you feel like playing the game!

```
monty <- function() {
    doors <- 1:3

    # randomly pick where the car is
    cardoor <- sample(doors,1)

    # prompt player
    print("Monty Hall says 'Pick a door, any door!'")

    # receive the player's choice of door (should be 1,2, or 3)
    chosen <- scan(what = integer(), nlines = 1, quiet = TRUE)

    # pick Monty's door (can't be the player's door or the car door)
    if (chosen != cardoor) montydoor <- doors[-c(chosen, cardoor)]
    else montydoor <- sample(doors[-chosen],1)

    # find out whether the player wants to switch doors
    print(paste("Monty opens door ", montydoor, "!", sep=""))
    print("Would you like to switch (y/n)?")
    reply <- scan(what = character(), nlines = 1, quiet = TRUE)

    # interpret what player wrote as "yes" if it starts with "y"
    if (substr(reply,1,1) == "y"){
        chosen <- doors[-c(chosen,montydoor)]
    }

    # announce the result of the game!
    if (chosen == cardoor) print("You won!")
    else print("You lost!")
}
```

The `print` command prints its argument to the screen. We combine this with the `paste` command since if we used `print("Monty opens door montydoor")` then R would literally print "Monty opens door montydoor". The `scan` command interactively requests input from the user; we use `what = integer()` when we want the user to enter an integer and `what = character()` when we want the user to enter text. Using `substr(reply,1,1)` extracts the first character of `reply`, in case the user replies with "yes" or "yep" or "yeah!" rather than with "y".

2.11 Exercises

Exercises marked with Ⓢ have detailed solutions at `http://stat110.net`.

Conditioning on evidence

1. Ⓢ A spam filter is designed by looking at commonly occurring phrases in spam. Suppose that 80% of email is spam. In 10% of the spam emails, the phrase "free money" is used, whereas this phrase is only used in 1% of non-spam emails. A new email has just arrived, which does mention "free money". What is the probability that it is spam?

2. Ⓢ A woman is pregnant with twin boys. Twins may be either identical or fraternal. Suppose that 1/3 of twins born are identical, that identical twins have a 50% chance of being both boys and a 50% chance of being both girls, and that for fraternal twins each twin independently has a 50% chance of being a boy and a 50% chance of being a girl. Given the above information, what is the probability that the woman's twins are identical?

3. According to the CDC (Centers for Disease Control and Prevention), men who smoke are 23 times more likely to develop lung cancer than men who don't smoke. Also according to the CDC, 21.6% of men in the U.S. smoke. What is the probability that a man in the U.S. is a smoker, given that he develops lung cancer?

4. Fred is answering a multiple-choice problem on an exam, and has to choose one of n options (exactly one of which is correct). Let K be the event that he knows the answer, and R be the event that he gets the problem right (either through knowledge or through luck). Suppose that if he knows the right answer he will definitely get the problem right, but if he does not know then he will guess completely randomly. Let $P(K) = p$.

   (a) Find $P(K|R)$ (in terms of p and n).

   (b) Show that $P(K|R) \geq p$, and explain why this makes sense intuitively. When (if ever) does $P(K|R)$ equal p?

5. Three cards are dealt from a standard, well-shuffled deck. The first two cards are flipped over, revealing the Ace of Spades as the first card and the 8 of Clubs as the second card. Given this information, find the probability that the third card is an ace in two ways: using the definition of conditional probability, and by symmetry.

6. A hat contains 100 coins, where 99 are fair but one is double-headed (always landing Heads). A coin is chosen uniformly at random. The chosen coin is flipped 7 times, and it lands Heads all 7 times. Given this information, what is the probability that the chosen coin is double-headed? (Of course, another approach here would be to *look at both sides of the coin*—but this is a metaphorical coin.)

7. A hat contains 100 coins, where *at least* 99 are fair, but there may be one that is double-headed (always landing Heads); if there is no such coin, then all 100 are fair. Let D be the event that there is such a coin, and suppose that $P(D) = 1/2$. A coin is chosen uniformly at random. The chosen coin is flipped 7 times, and it lands Heads all 7 times.

   (a) Given this information, what is the probability that one of the coins is double-headed?

   (b) Given this information, what is the probability that the chosen coin is double-headed?

8. The screens used for a certain type of cell phone are manufactured by 3 companies, A, B, and C. The proportions of screens supplied by A, B, and C are 0.5, 0.3, and 0.2, respectively, and their screens are defective with probabilities 0.01, 0.02, and 0.03, respectively. Given that the screen on such a phone is defective, what is the probability that Company A manufactured it?

9. (a) Show that if events A_1 and A_2 have the same *prior* probability $P(A_1) = P(A_2)$, A_1 implies B, and A_2 implies B, then A_1 and A_2 have the same *posterior* probability $P(A_1|B) = P(A_2|B)$ if it is observed that B occurred.

   (b) Explain why (a) makes sense intuitively, and give a concrete example.

10. Fred is working on a major project. In planning the project, two milestones are set up, with dates by which they should be accomplished. This serves as a way to track Fred's progress. Let A_1 be the event that Fred completes the first milestone on time, A_2 be the event that he completes the second milestone on time, and A_3 be the event that he completes the project on time.

    Suppose that $P(A_{j+1}|A_j) = 0.8$ but $P(A_{j+1}|A_j^c) = 0.3$ for $j = 1, 2$, since if Fred falls behind on his schedule it will be hard for him to get caught up. Also, assume that the second milestone supersedes the first, in the sense that once we know whether he is on time in completing the second milestone, it no longer matters what happened with the first milestone. We can express this by saying that A_1 and A_3 are conditionally independent given A_2 and they're also conditionally independent given A_2^c.

    (a) Find the probability that Fred will finish the project on time, given that he completes the first milestone on time. Also find the probability that Fred will finish the project on time, given that he is late for the first milestone.

    (b) Suppose that $P(A_1) = 0.75$. Find the probability that Fred will finish the project on time.

11. An *exit poll* in an election is a survey taken of voters just after they have voted. One major use of exit polls has been so that news organizations can try to figure out as soon as possible who won the election, before the votes are officially counted. This has been notoriously inaccurate in various elections, sometimes because of *selection bias*: the sample of people who are invited to and agree to participate in the survey may not be similar enough to the overall population of voters.

    Consider an election with two candidates, Candidate A and Candidate B. Every voter is invited to participate in an exit poll, where they are asked whom they voted for; some accept and some refuse. For a randomly selected voter, let A be the event that they voted for A, and W be the event that they are willing to participate in the exit poll. Suppose that $P(W|A) = 0.7$ but $P(W|A^c) = 0.3$. In the exit poll, 60% of the respondents say they voted for A (assume that they are all honest), suggesting a comfortable victory for A. Find $P(A)$, the true proportion of people who voted for A.

12. Alice is trying to communicate with Bob, by sending a message (encoded in binary) across a channel.

    (a) Suppose for this part that she sends only one bit (a 0 or 1), with equal probabilities. If she sends a 0, there is a 5% chance of an error occurring, resulting in Bob receiving a 1; if she sends a 1, there is a 10% chance of an error occurring, resulting in Bob receiving a 0. Given that Bob receives a 1, what is the probability that Alice actually sent a 1?

    (b) To reduce the chance of miscommunication, Alice and Bob decide to use a *repetition code*. Again Alice wants to convey a 0 or a 1, but this time she repeats it two more times, so that she sends 000 to convey 0 and 111 to convey 1. Bob will decode the message by going with what the majority of the bits were. Assume that the error probabilities are as in (a), with error events for different bits independent of each other. Given that Bob receives 110, what is the probability that Alice intended to convey a 1?

13. Company A has just developed a diagnostic test for a certain disease. The disease afflicts 1% of the population. As defined in Example 2.3.9, the *sensitivity* of the test is the probability of someone testing positive, given that they have the disease, and the *specificity* of the test is the probability that of someone testing negative, given that they don't have the disease. Assume that, as in Example 2.3.9, the sensitivity and specificity are both 0.95.

    Company B, which is a rival of Company A, offers a competing test for the disease. Company B claims that their test is faster and less expensive to perform than Company A's test, is less painful (Company A's test requires an incision), and yet has a higher overall success rate, where overall success rate is defined as the probability that a random person gets diagnosed correctly.

(a) It turns out that Company B's test can be described and performed very simply: no matter who the patient is, diagnose that they do not have the disease. Check whether Company B's claim about overall success rates is true.

(b) Explain why Company A's test may still be useful.

(c) Company A wants to develop a new test such that the overall success rate is higher than that of Company B's test. If the sensitivity and specificity are equal, how high does the sensitivity have to be to achieve their goal? If (amazingly) they can get the sensitivity equal to 1, how high does the specificity have to be to achieve their goal? If (amazingly) they can get the specificity equal to 1, how high does the sensitivity have to be to achieve their goal?

14. Consider the following scenario, from Tversky and Kahneman [27]:

    Let A be the event that before the end of next year, Peter will have installed a burglar alarm system in his home. Let B denote the event that Peter's home will be burglarized before the end of next year.

    (a) Intuitively, which do you think is bigger, $P(A|B)$ or $P(A|B^c)$? Explain your intuition.

    (b) Intuitively, which do you think is bigger, $P(B|A)$ or $P(B|A^c)$? Explain your intuition.

    (c) Show that for *any* events A and B (with probabilities not equal to 0 or 1), the inequality $P(A|B) > P(A|B^c)$ is equivalent to $P(B|A) > P(B|A^c)$.

    (d) Tversky and Kahneman report that 131 out of 162 people whom they posed (a) and (b) to said that $P(A|B) > P(A|B^c)$ and $P(B|A) < P(B|A^c)$. What is a plausible explanation for why this was such a popular opinion despite (c) showing that it is impossible for these inequalities both to hold?

15. Let A and B be events with $0 < P(A \cap B) < P(A) < P(B) < P(A \cup B) < 1$. You are hoping that *both* A and B occurred. Which of the following pieces of information would you be happiest to observe: that A occurred, that B occurred, or that $A \cup B$ occurred?

16. Show that $P(A|B) \leq P(A)$ implies $P(A|B^c) \geq P(A)$, and give an intuitive explanation of why this makes sense.

17. In deterministic logic, the statement "A implies B" is equivalent to its *contrapositive*, "not B implies not A". In this problem we will consider analogous statements in probability, the logic of uncertainty. Let A and B be events with probabilities not equal to 0 or 1.

    (a) Show that if $P(B|A) = 1$, then $P(A^c|B^c) = 1$.

    Hint: Apply Bayes' rule and LOTP.

    (b) Show however that the result in (a) does not hold in general if $=$ is replaced by $\approx$. In particular, find an example where $P(B|A)$ is very close to 1 but $P(A^c|B^c)$ is very close to 0.

    Hint: What happens if A and B are independent?

18. Show that if $P(A) = 1$, then $P(A|B) = 1$ for any B with $P(B) > 0$. Intuitively, this says that if someone dogmatically believes something with absolute certainty, then no amount of evidence will change their mind. The principle of avoiding assigning probabilities of 0 or 1 to any event (except for mathematical certainties) was named *Cromwell's rule* by the statistician Dennis Lindley, due to Cromwell saying to the Church of Scotland, "Think it possible you may be mistaken."

    Hint: Write $P(B) = P(B \cap A) + P(B \cap A^c)$, and then show that $P(B \cap A^c) = 0$.

19. Explain the following Sherlock Holmes saying in terms of conditional probability, carefully distinguishing between prior and posterior probabilities: "It is an old maxim of mine that when you have excluded the impossible, whatever remains, however improbable, must be the truth."

20. The Jack of Spades (with cider), Jack of Hearts (with tarts), Queen of Spades (with a wink), and Queen of Hearts (without tarts) are taken from a deck of cards. These four cards are shuffled, and then two are dealt. Note: Literary references to cider, tarts, and winks do not need to be considered when solving this problem.

    (a) Find the probability that both of these two cards are queens, given that the first card dealt is a queen.

    (b) Find the probability that both are queens, given that at least one is a queen.

    (c) Find the probability that both are queens, given that one is the Queen of Hearts.

21. A fair coin is flipped 3 times. The toss results are recorded on separate slips of paper (writing "H" if Heads and "T" if Tails), and the 3 slips of paper are thrown into a hat.

    (a) Find the probability that all 3 tosses landed Heads, given that at least 2 were Heads.

    (b) Two of the slips of paper are randomly drawn from the hat, and both show the letter H. Given this information, what is the probability that all 3 tosses landed Heads?

22. Ⓢ A bag contains one marble which is either green or blue, with equal probabilities. A green marble is put in the bag (so there are 2 marbles now), and then a random marble is taken out. The marble taken out is green. What is the probability that the remaining marble is also green?

23. Ⓢ Let G be the event that a certain individual is guilty of a certain robbery. In gathering evidence, it is learned that an event E_1 occurred, and a little later it is also learned that another event E_2 also occurred. Is it possible that individually, these pieces of evidence increase the chance of guilt (so $P(G|E_1) > P(G)$ and $P(G|E_2) > P(G)$), but together they decrease the chance of guilt (so $P(G|E_1, E_2) < P(G)$)?

24. Is it possible to have events A_1, A_2, B, C with $P(A_1|B) > P(A_1|C)$ and $P(A_2|B) > P(A_2|C)$, yet $P(A_1 \cup A_2|B) < P(A_1 \cup A_2|C)$? If so, find an example (with a "story" interpreting the events, as well as giving specific numbers); otherwise, show that it is impossible for this phenomenon to happen.

25. Ⓢ A crime is committed by one of two suspects, A and B. Initially, there is equal evidence against both of them. In further investigation at the crime scene, it is found that the guilty party had a blood type found in 10% of the population. Suspect A does match this blood type, whereas the blood type of Suspect B is unknown.

    (a) Given this new information, what is the probability that A is the guilty party?

    (b) Given this new information, what is the probability that B's blood type matches that found at the crime scene?

26. Ⓢ To battle against spam, Bob installs two anti-spam programs. An email arrives, which is either legitimate (event L) or spam (event L^c), and which program j marks as legitimate (event M_j) or marks as spam (event M_j^c) for $j \in \{1, 2\}$. Assume that 10% of Bob's email is legitimate and that the two programs are each "90% accurate" in the sense that $P(M_j|L) = P(M_j^c|L^c) = 9/10$. Also assume that given whether an email is spam, the two programs' outputs are conditionally independent.

    (a) Find the probability that the email is legitimate, given that the 1st program marks it as legitimate (simplify).

    (b) Find the probability that the email is legitimate, given that both programs mark it as legitimate (simplify).

(c) Bob runs the 1st program and M_1 occurs. He updates his probabilities and then runs the 2nd program. Let $\tilde{P}(A) = P(A|M_1)$ be the updated probability function after running the 1st program. Explain briefly in words whether or not $\tilde{P}(L|M_2) = P(L|M_1 \cap M_2)$: is conditioning on $M_1 \cap M_2$ in one step equivalent to first conditioning on M_1, then updating probabilities, and then conditioning on M_2?

27. Suppose that there are 5 blood types in the population, named type 1 through type 5, with probabilities $p_1, p_2, \dots, p_5$. A crime was committed by two individuals. A suspect, who has blood type 1, has prior probability p of being guilty. At the crime scene, blood evidence is collected, which shows that one of the criminals has type 1 and the other has type 2.

    Find the posterior probability that the suspect is guilty, given the evidence. Does the evidence make it more likely or less likely that the suspect is guilty, or does this depend on the values of the parameters $p, p_1, \dots, p_5$? If it depends on these values, give a simple criterion for when the evidence makes it more likely that the suspect is guilty.

28. Fred has just tested positive for a certain disease.

    (a) Given this information, find the posterior odds that he has the disease, in terms of the prior odds, the sensitivity of the test, and the specificity of the test.

    (b) Not surprisingly, Fred is much more interested in $P(\text{have disease}|\text{test positive})$, known as the *positive predictive value*, than in the sensitivity $P(\text{test positive}|\text{have disease})$. A handy rule of thumb in biostatistics and epidemiology is as follows:

    For a rare disease and a reasonably good test, specificity matters much more than sensitivity in determining the positive predictive value.

    Explain intuitively why this rule of thumb works. For this part you can make up some specific numbers and interpret probabilities in a frequentist way as proportions in a large population, e.g., assume the disease afflicts 1% of a population of 10000 people and then consider various possibilities for the sensitivity and specificity.

29. A family has two children. Let C be a characteristic that a child can have, and assume that each child has characteristic C with probability p, independently of each other and of gender. For example, C could be the characteristic "born in winter" as in Example 2.2.7. Under the assumptions of Example 2.2.5, show that the probability that both children are girls given that at least one is a girl with characteristic C is $\frac{2-p}{4-p}$. Note that this is 1/3 if $p = 1$ (agreeing with the first part of Example 2.2.5) and approaches 1/2 from below as $p \to 0$ (agreeing with Example 2.2.7).

Independence and conditional independence

30. Ⓢ A family has 3 children, creatively named A, B, and C.

    (a) Discuss intuitively (but clearly) whether the event "A is older than B" is independent of the event "A is older than C".

    (b) Find the probability that A is older than B, given that A is older than C.

31. Ⓢ Is it possible that an event is independent of itself? If so, when is this the case?

32. Ⓢ Consider four nonstandard dice (the *Efron dice*), whose sides are labeled as follows (the 6 sides on each die are equally likely).

    A: 4, 4, 4, 4, 0, 0

    B: 3, 3, 3, 3, 3, 3

    C: 6, 6, 2, 2, 2, 2

D: $5, 5, 5, 1, 1, 1$

These four dice are each rolled once. Let A be the result for die A, B be the result for die B, etc.

(a) Find $P(A > B), P(B > C), P(C > D)$, and $P(D > A)$.

(b) Is the event $A > B$ independent of the event $B > C$? Is the event $B > C$ independent of the event $C > D$? Explain.

33. Alice, Bob, and 100 other people live in a small town. Let C be the set consisting of the 100 other people, let A be the set of people in C who are friends with Alice, and let B be the set of people in C who are friends with Bob. Suppose that for each person in C, Alice is friends with that person with probability $1/2$, and likewise for Bob, with all of these friendship statuses independent.

(a) Let $D \subseteq C$. Find $P(A = D)$.

(b) Find $P(A \subseteq B)$.

(c) Find $P(A \cup B = C)$.

34. Suppose that there are two types of drivers: good drivers and bad drivers. Let G be the event that a certain man is a good driver, A be the event that he gets into a car accident next year, and B be the event that he gets into a car accident the following year. Let $P(G) = g$ and $P(A|G) = P(B|G) = p_1, P(A|G^c) = P(B|G^c) = p_2$, with $p_1 < p_2$. Suppose that given the information of whether or not the man is a good driver, A and B are independent (for simplicity and to avoid being morbid, assume that the accidents being considered are minor and wouldn't make the man unable to drive).

(a) Explain intuitively whether or not A and B are independent.

(b) Find $P(G|A^c)$.

(c) Find $P(B|A^c)$.

35. Ⓢ You are going to play 2 games of chess with an opponent whom you have never played against before (for the sake of this problem). Your opponent is equally likely to be a beginner, intermediate, or a master. Depending on which, your chances of winning an individual game are 90%, 50%, or 30%, respectively.

(a) What is your probability of winning the first game?

(b) Congratulations: you won the first game! Given this information, what is the probability that you will also win the second game (assume that, given the skill level of your opponent, the outcomes of the games are independent)?

(c) Explain the distinction between assuming that the outcomes of the games are independent and assuming that they are conditionally independent given the opponent's skill level. Which of these assumptions seems more reasonable, and why?

36. (a) Suppose that in the population of college applicants, being good at baseball is independent of having a good math score on a certain standardized test (with respect to some measure of "good"). A certain college has a simple admissions procedure: admit an applicant if and only if the applicant is good at baseball or has a good math score on the test.

Give an intuitive explanation of why it makes sense that among students that the college admits, having a good math score is *negatively associated* with being good at baseball, i.e., conditioning on having a good math score decreases the chance of being good at baseball.

(b) Show that if A and B are independent and $C = A \cup B$, then A and B are conditionally dependent given C (as long as $P(A \cap B) > 0$ and $P(A \cup B) < 1$), with

$$P(A|B, C) < P(A|C).$$

This phenomenon is known as *Berkson's paradox*, especially in the context of admissions to a school, hospital, etc.

37. Two different diseases cause a certain weird symptom; anyone who has either or both of these diseases will experience the symptom. Let D_1 be the event of having the first disease, D_2 be the event of having the second disease, and W be the event of having the weird symptom. Suppose that D_1 and D_2 are independent with $P(D_j) = p_j$, and that a person with neither of these diseases will have the weird symptom with probability w_0. Let $q_j = 1 - p_j$, and assume that $0 < p_j < 1$.

(a) Find $P(W)$.

(b) Find $P(D_1|W), P(D_2|W)$, and $P(D_1, D_2|W)$.

(c) Determine algebraically whether or not D_1 and D_2 are conditionally independent given W.

(d) Suppose for this part only that $w_0 = 0$. Give a clear, convincing intuitive explanation in words of whether D_1 and D_2 are conditionally independent given W.

38. We want to design a spam filter for email. As described in Exercise 1, a major strategy is to find phrases that are much more likely to appear in a spam email than in a non-spam email. In that exercise, we only consider one such phrase: "free money". More realistically, suppose that we have created a list of 100 words or phrases that are much more likely to be used in spam than in non-spam.

Let W_j be the event that an email contains the jth word or phrase on the list. Let

$$p = P(\text{spam}), p_j = P(W_j|\text{spam}), r_j = P(W_j|\text{not spam}),$$

where "spam" is shorthand for the event that the email is spam.

Assume that $W_1, \dots, W_{100}$ are conditionally independent given that the email is spam, and conditionally independent given that it is not spam. A method for classifying emails (or other objects) based on this kind of assumption is called a *naive Bayes classifier*. (Here "naive" refers to the fact that the conditional independence is a strong assumption, not to Bayes being naive. The assumption may or may not be realistic, but naive Bayes classifiers sometimes work well in practice even if the assumption is not realistic.)

Under this assumption we know, for example, that

$$P(W_1, W_2, W_3^c, W_4^c, \dots, W_{100}^c|\text{spam}) = p_1 p_2 (1 - p_3)(1 - p_4) \dots (1 - p_{100}).$$

Without the naive Bayes assumption, there would be vastly more statistical and computational difficulties since we would need to consider $2^{100} \approx 1.3 \times 10^{30}$ events of the form $A_1 \cap A_2 \cdots \cap A_{100}$ with each A_j equal to either W_j or W_j^c. A new email has just arrived, and it includes the 23rd, 64th, and 65th words or phrases on the list (but not the other 97). So we want to compute

$$P(\text{spam}|W_1^c, \dots, W_{22}^c, W_{23}, W_{24}^c, \dots, W_{63}^c, W_{64}, W_{65}, W_{66}^c, \dots, W_{100}^c).$$

Note that we need to condition on *all* the evidence, not just the fact that $W_{23} \cap W_{64} \cap W_{65}$ occurred. Find the conditional probability that the new email is spam (in terms of p and the p_j and r_j).

Monty Hall

39. Ⓢ (a) Consider the following 7-door version of the Monty Hall problem. There are 7 doors, behind one of which there is a car (which you want), and behind the rest of which there are goats (which you don't want). Initially, all possibilities are equally likely for where the car is. You choose a door. Monty Hall then opens 3 goat doors, and offers you the option of switching to any of the remaining 3 doors.

    Assume that Monty Hall knows which door has the car, will always open 3 goat doors and offer the option of switching, and that Monty chooses with equal probabilities from all his choices of which goat doors to open. Should you switch? What is your probability of success if you switch to one of the remaining 3 doors?

    (b) Generalize the above to a Monty Hall problem where there are $n \geq 3$ doors, of which Monty opens m goat doors, with $1 \leq m \leq n-2$.

40. Ⓢ Consider the Monty Hall problem, except that Monty enjoys opening door 2 more than he enjoys opening door 3, and if he has a choice between opening these two doors, he opens door 2 with probability p, where $\frac{1}{2} \leq p \leq 1$.

    To recap: there are three doors, behind one of which there is a car (which you want), and behind the other two of which there are goats (which you don't want). Initially, all possibilities are equally likely for where the car is. You choose a door, which for concreteness we assume is door 1. Monty Hall then opens a door to reveal a goat, and offers you the option of switching. Assume that Monty Hall knows which door has the car, will always open a goat door and offer the option of switching, and as above assume that if Monty Hall has a choice between opening door 2 and door 3, he chooses door 2 with probability p (with $\frac{1}{2} \leq p \leq 1$).

    (a) Find the unconditional probability that the strategy of always switching succeeds (unconditional in the sense that we do not condition on which of doors 2 or 3 Monty opens).

    (b) Find the probability that the strategy of always switching succeeds, given that Monty opens door 2.

    (c) Find the probability that the strategy of always switching succeeds, given that Monty opens door 3.

41. The ratings of Monty Hall's show have dropped slightly, and a panicking executive producer complains to Monty that the part of the show where he opens a door lacks suspense: Monty always opens a door with a goat. Monty replies that the reason is so that the game is never spoiled by him revealing the car, but he agrees to update the game as follows.

    Before each show, Monty secretly flips a coin with probability p of Heads. If the coin lands Heads, Monty resolves to open a door with a goat (with equal probabilities if there is a choice). Otherwise, Monty resolves to open a random door, with equal probabilities. Of course, Monty will not open the door that the contestant initially chooses. The contestant knows p but does not know the outcome of the coin flip. When the show starts, the contestant chooses a door. Monty (who knows where the car is) then opens a door. If the car is revealed, the game is over; if a goat is revealed, the contestant is offered the option of switching. Now suppose it turns out that the contestant chooses door 1 and then Monty opens door 2, revealing a goat. What is the contestant's probability of success if they switch to door 3?

42. Consider the following variation of the Monty Hall problem, where in some situations Monty may *not* open a door and give the contestant the choice of whether to switch doors. Specifically, there are 3 doors, with 2 containing goats and 1 containing a car. The car is equally likely to be anywhere, and Monty knows where the car is. Let $0 \leq p \leq 1$.

The contestant chooses a door. If this initial choice has the car, Monty *will* open another door, revealing a goat (choosing with equal probabilities among his two choices of door), and then offer the contestant the choice of whether to switch to the other unopened door.

If the contestant's initial choice has a goat, then with probability p Monty *will* open another door, revealing a goat, and then offer the contestant the choice of whether to switch to the other unopened door; but with probability $1 - p$, Monty will *not* open a door, and the contestant must stick with their initial choice.

The contestant decides in advance to use the following strategy: initially choose door 1. Then, if Monty opens a door and offers the choice of whether to switch, do switch.

(a) Find the unconditional probability that the contestant will get the car. Also, check what your answer reduces to in the extreme cases $p = 0$ and $p = 1$, and briefly explain why your answer makes sense in these two cases.

(b) Monty now opens door 2, revealing a goat. So the contestant switches to door 3. Given this information, find the conditional probability that the contestant will get the car.

43. You are the contestant on the Monty Hall show. Monty is trying out a new version of his game, with rules as follows. You get to choose one of three doors. One door has a car behind it, another has a computer, and the other door has a goat (with all permutations equally likely). Monty, who knows which prize is behind each door, will open a door (but not the one you chose) and then let you choose whether to switch from your current choice to the other unopened door.

Assume that you prefer the car to the computer, the computer to the goat, and (by transitivity) the car to the goat.

(a) Suppose for this part only that Monty always opens the door that reveals your less preferred prize out of the two alternatives, e.g., if he is faced with the choice between revealing the goat or the computer, he will reveal the goat. Monty opens a door, revealing a goat (this is again for this part only). Given this information, should you switch? If you do switch, what is your probability of success in getting the car?

(b) Now suppose that Monty reveals your less preferred prize with probability p, and your more preferred prize with probability $q = 1 - p$. Monty opens a door, revealing a computer. Given this information, should you switch (your answer can depend on p)? If you do switch, what is your probability of success in getting the car (in terms of p)?

44. Monty Hall has introduced a new twist in his game, by generalizing the assumption that the initial probabilities for where the car is are $(\frac{1}{3}, \frac{1}{3}, \frac{1}{3})$. Specifically, there are three doors, behind one of which there is a car (which the contestant wants), and behind the other two of which there are goats (which the contestant doesn't want). Initially, door i has probability p_i of having the car, where p_1, p_2, p_3 are known constants such that $0 < p_1 \leq p_2 \leq p_3 < 1$ and $p_1 + p_2 + p_3 = 1$. The contestant chooses a door. Then Monty opens a door (other than the one the contestant chose) and offers the contestant the option of switching to the other unopened door.

(a) Assume for this part that Monty knows in advance which door has the car. He always opens a door to reveal a goat, and if he has a choice of which door to open he chooses with equal probabilities. Suppose for this part that the contestant initially chooses door 3, and then Monty opens door 2, revealing a goat. Given the above information, find the conditional probability that door 3 has the car. Should the contestant switch doors? (If whether to switch depends on the p_i's, give a fully simplified criterion in terms of the p_i's.)

(b) Now assume instead that Monty does *not* know in advance where the car is. He randomly chooses which door to open (other than the one the contestant chose), with equal probabilities. (The game is spoiled if he reveals the car.) Suppose again that the contestant initially chooses door 3, and then Monty opens door 2, revealing a goat. Given the above information, find the conditional probability that door 3 has the car. Should the contestant switch doors? (If whether to switch depends on the p_i's, give a fully simplified criterion in terms of the p_i's.)

(c) Repeat (a), except with the contestant initially choosing door 1 rather than door 3.

(d) Repeat (b), except with the contestant initially choosing door 1 rather than door 3.

45. Monty Hall is trying out a new version of his game. In this version, instead of there always being 1 car and 2 goats, the prizes behind the doors are generated *independently*, with each door having probability p of having a car and $q = 1 - p$ of having a goat. In detail: There are three doors, behind each of which there is one prize: either a car or a goat. For each door, there is probability p that there is a car behind it and $q = 1 - p$ that there is a goat, independent of the other doors.

The contestant chooses a door. Monty, who knows the contents of each door, then opens one of the two remaining doors. In choosing which door to open, Monty will always reveal a goat if possible. If both of the remaining doors have the same kind of prize, Monty chooses randomly (with equal probabilities). After opening a door, Monty offers the contestant the option of switching to the other unopened door.

The contestant decides in advance to use the following strategy: first choose door 1. Then, after Monty opens a door, switch to the other unopened door.

(a) Find the unconditional probability that the contestant will get a car.

(b) Monty now opens door 2, revealing a goat. Given this information, find the conditional probability that the contestant will get a car.

46. Monty Hall is trying out a new version of his game, with rules as follows. The contestant gets to choose one of *four* doors. One door has a car behind it, another has an apple, another has a book, and another has a goat. All 24 permutations for which door has which prize are equally likely. In order from least preferred to most preferred, the contestant's preferences are: goat, apple, book, car.

Monty, who knows which prize is behind each door, will open a door (other than the contestant's initial choice) and then let the contestant choose whether to switch to another unopened door. Monty will reveal the least preferred prize (among the 3 doors other than the contestant's initial choice) with probability p, the intermediately preferred prize with probability $1 - p$, and the most preferred prize never.

The contestant decides in advance to use the following strategy: Initially choose door 1. After Monty opens a door, switch to one of the other two unopened doors, randomly choosing between them (with probability $1/2$ each).

(a) Find the unconditional probability that the contestant will get the car.

Hint: Condition on where the car is.

(b) Find the unconditional probability that Monty will reveal the apple.

Hint: Condition on what is behind door 1.

(c) Monty now opens a door, revealing the apple. Given this information, find the conditional probability that the contestant will get the car.

47. You are the contestant on Monty Hall's game show. Hoping to double the excitement of the game, Monty will offer you *two* opportunities to switch to another door. Specifically, the new rules are as follows. There are *four* doors. Behind one door there is a car (which you want); behind the other three doors there are goats (which you don't want). Initially, all possibilities are equally likely for where the car is. Monty knows where the car is, and when he has a choice of which door to open, he chooses with equal probabilities.

You choose a door, which for concreteness we assume is door 1. Monty opens a door (other than door 1), revealing a goat, and then offers you the option to switch to another door. Monty then opens *another* door (other than your currently selected door), revealing another goat. So now there are two open doors (with goats) and two unopened doors. Again Monty offers you the option to switch.

You decide in advance to use one of the following four strategies: stay-stay, stay-switch, switch-stay, switch-switch, where, for example, "stay-switch" means that the first time Monty offers you the choice of switching, you stay with your current selection, but then the second time Monty offers you the choice, you do switch doors. In each part below the goal is to find or compare *unconditional* probabilities, i.e., from a vantage point of before the game has started.

(a) Find the probability of winning the car if you follow the stay-stay strategy.

(b) Find the probability of winning the car if you follow the stay-switch strategy.

(c) Find the probability of winning the car if you follow the switch-stay strategy.

(d) Find the probability of winning the car if you follow the switch-switch strategy.

(e) Which of these four strategies is the best?

First-step analysis and gambler's ruin

48. Ⓢ A fair die is rolled repeatedly, and a running total is kept (which is, at each time, the total of all the rolls up until that time). Let p_n be the probability that the running total is ever *exactly* n (assume the die will always be rolled enough times so that the running total will eventually exceed n, but it may or may not ever equal n).

(a) Write down a recursive equation for p_n (relating p_n to earlier terms p_k in a simple way). Your equation should be true for all positive integers n, so give a definition of p_0 and p_k for $k < 0$ so that the recursive equation is true for small values of n.

(b) Find p_7.

(c) Give an intuitive explanation for the fact that $p_n \to 1/3.5 = 2/7$ as $n \to \infty$.

49. A sequence of $n \geq 1$ independent trials is performed, where each trial ends in "success" or "failure" (but not both). Let p_i be the probability of success in the ith trial, $q_i = 1 - p_i$, and $b_i = q_i - 1/2$, for $i = 1, 2, \dots, n$. Let A_n be the event that the number of successful trials is even.

(a) Show that for $n = 2$, $P(A_2) = 1/2 + 2b_1b_2$.

(b) Show by induction that

$$P(A_n) = 1/2 + 2^{n-1}b_1b_2 \dots b_n.$$

(This result is very useful in cryptography. Also, note that it implies that if n coins are flipped, then the probability of an even number of Heads is $1/2$ if and only if at least one of the coins is fair.) Hint: Group some trials into a supertrial.

(c) Check directly that the result of (b) is true in the following simple cases: $p_i = 1/2$ for some i; $p_i = 0$ for all i; $p_i = 1$ for all i.

50. Ⓢ Calvin and Hobbes play a match consisting of a series of games, where Calvin has probability p of winning each game (independently). They play with a "win by two" rule: the first player to win two games more than his opponent wins the match. Find the probability that Calvin wins the match (in terms of p), in two different ways:

(a) by conditioning, using the law of total probability.

(b) by interpreting the problem as a gambler's ruin problem.

51. Ⓢ A gambler repeatedly plays a game where in each round, he wins a dollar with probability 1/3 and loses a dollar with probability 2/3. His strategy is "quit when he is ahead by $2". Suppose that he starts with a million dollars. Show that the probability that he'll ever be ahead by $2 is less than 1/4.

52. As in the gambler's ruin problem, two gamblers, A and B, make a series of bets, until one of the gamblers goes bankrupt. Let A start out with i dollars and B start out with $N-i$ dollars, and let p be the probability of A winning a bet, with $0 < p < \frac{1}{2}$. Each bet is for $\frac{1}{k}$ dollars, with k a positive integer, e.g., $k = 1$ is the original gambler's ruin problem and $k = 20$ means they're betting nickels. Find the probability that A wins the game, and determine what happens to this as $k \to \infty$.

53. There are 100 equally spaced points around a circle. At 99 of the points, there are sheep, and at 1 point, there is a wolf. At each time step, the wolf randomly moves either clockwise or counterclockwise by 1 point. If there is a sheep at that point, he eats it. The sheep don't move. What is the probability that the sheep who is initially opposite the wolf is the last one remaining?

54. An immortal drunk man wanders around randomly on the integers. He starts at the origin, and at each step he moves 1 unit to the right or 1 unit to the left, with probabilities p and $q = 1-p$ respectively, independently of all his previous steps. Let S_n be his position after n steps.

    (a) Let p_k be the probability that the drunk ever reaches the value k, for all $k \geq 0$. Write down a difference equation for p_k (you do not need to solve it for this part).

    (b) Find p_k, fully simplified; be sure to consider all 3 cases: $p < 1/2, p = 1/2$, and $p > 1/2$. Feel free to assume that if $A_1, A_2, \ldots$ are events with $A_j \subseteq A_{j+1}$ for all j, then $P(A_n) \to P(\cup_{j=1}^{\infty} A_j)$ as $n \to \infty$ (because it is true; this is known as *continuity of probability*).

Simpson's paradox

55. Ⓢ (a) Is it possible to have events A, B, C such that $P(A|C) < P(B|C)$ and $P(A|C^c) < P(B|C^c)$, yet $P(A) > P(B)$? That is, A is less likely than B given that C is true, and also less likely than B given that C is false, yet A is more likely than B if we're given no information about C. Show this is impossible (with a short proof) or find a counterexample (with a story interpreting A, B, C).

    (b) If the scenario in (a) is possible, is it a special case of Simpson's paradox, equivalent to Simpson's paradox, or neither? If it is impossible, explain intuitively why it is impossible even though Simpson's paradox is possible.

56. Ⓢ Consider the following conversation from an episode of *The Simpsons*:

    > Lisa: *Dad, I think he's an ivory dealer! His boots are ivory, his hat is ivory, and I'm pretty sure that check is ivory.*
    >
    > Homer: *Lisa, a guy who has lots of ivory is* less *likely to hurt Stampy than a guy whose ivory supplies are low.*

    Here Homer and Lisa are debating the question of whether or not the man (named Blackheart) is likely to hurt Stampy the Elephant if they sell Stampy to him. They clearly disagree about how to use their observations about Blackheart to learn about the probability (conditional on the evidence) that Blackheart will hurt Stampy.

    (a) Define clear notation for the various events of interest here.

    (b) Express Lisa's and Homer's arguments (Lisa's is partly implicit) as conditional probability statements in terms of your notation from (a).

(c) Assume it is true that someone who has a lot of a commodity will have less desire to acquire more of the commodity. Explain what is wrong with Homer's reasoning that the evidence about Blackheart makes it less likely that he will harm Stampy.

57. (a) There are two crimson jars (labeled C_1 and C_2) and two mauve jars (labeled M_1 and M_2). Each jar contains a mixture of green gummi bears and red gummi bears. Show by example that it is possible that C_1 has a much higher percentage of green gummi bears than M_1, and C_2 has a much higher percentage of green gummi bears than M_2, yet if the contents of C_1 and C_2 are merged into a new jar and likewise for M_1 and M_2, then the combination of C_1 and C_2 has a lower percentage of green gummi bears than the combination of M_1 and M_2.

    (b) Explain how (a) relates to Simpson's paradox, both intuitively and by explicitly defining events A, B, C as in the statement of Simpson's paradox.

58. As explained in this chapter, Simpson's paradox says that it is possible to have events A, B, C such that $P(A|B, C) < P(A|B^c, C)$ and $P(A|B, C^c) < P(A|B^c, C^c)$, yet $P(A|B) > P(A|B^c)$.

    (a) Can Simpson's paradox occur if A and B are independent? If so, give a concrete example (with both numbers and an interpretation); if not, prove that it is impossible.

    (b) Can Simpson's paradox occur if A and C are independent? If so, give a concrete example (with both numbers and an interpretation); if not, prove that it is impossible.

    (c) Can Simpson's paradox occur if B and C are independent? If so, give a concrete example (with both numbers and an interpretation); if not, prove that it is impossible.

59. Ⓢ The book *Red State, Blue State, Rich State, Poor State* by Andrew Gelman [12] discusses the following election phenomenon: within any U.S. state, a wealthy voter is more likely to vote for a Republican than a poor voter, yet the wealthier states tend to favor Democratic candidates!

    (a) Assume for simplicity that there are only 2 states (called Red and Blue), each of which has 100 people, and that each person is either rich or poor, and either a Democrat or a Republican. Make up numbers consistent with the above, showing how this phenomenon is possible, by giving a 2×2 table for each state (listing how many people in each state are rich Democrats, etc.). So within each state, a rich voter is more likely to vote for a Republican than a poor voter, but the percentage of Democrats is higher in the state with the higher percentage of rich people than in the state with the lower percentage of rich people.

    (b) In the setup of (a) (not necessarily with the numbers you made up there), let D be the event that a randomly chosen person is a Democrat (with all 200 people equally likely), and B be the event that the person lives in the Blue State. Suppose that 10 people move from the Blue State to the Red State. Write P_{old} and P_{new} for probabilities before and after they move. Assume that people do not change parties, so we have $P_{\text{new}}(D) = P_{\text{old}}(D)$. Is it possible that *both* $P_{\text{new}}(D|B) > P_{\text{old}}(D|B)$ and $P_{\text{new}}(D|B^c) > P_{\text{old}}(D|B^c)$ are true? If so, explain how it is possible and why it does not contradict the law of total probability $P(D) = P(D|B)P(B) + P(D|B^c)P(B^c)$; if not, show that it is impossible.

Mixed practice

60. A patient is being given a blood test for the disease conditionitis. Let p be the prior probability that the patient has conditionitis. The blood sample is sent to one of two labs for analysis, lab A or lab B. The choice of which lab to use is made randomly, independent of the patient's disease status, with probability $1/2$ for each lab.

For lab A, the probability of someone testing *positive* given that they *do* have the disease is a_1, and the probability of someone testing *negative* given that they do *not* have the disease is a_2. The corresponding probabilities for lab B are b_1 and b_2.

(a) Find the probability that the patient has the disease, given that they tested positive.

(b) Find the probability that the patient's blood sample was analyzed by lab A, given that the patient tested positive.

61. Fred decides to take a series of n tests, to diagnose whether he has a certain disease (any individual test is not perfectly reliable, so he hopes to reduce his uncertainty by taking multiple tests). Let D be the event that he has the disease, $p = P(D)$ be the prior probability that he has the disease, and $q = 1 - p$. Let T_j be the event that he tests positive on the jth test.

(a) Assume for this part that the test results are conditionally independent given Fred's disease status. Let $a = P(T_j|D)$ and $b = P(T_j|D^c)$, where a and b don't depend on j. Find the posterior probability that Fred has the disease, given that he tests positive on all n of the n tests.

(b) Suppose that Fred tests positive on all n tests. However, some people have a certain gene that makes them *always* test positive. Let G be the event that Fred has the gene. Assume that $P(G) = 1/2$ and that D and G are independent. If Fred does *not* have the gene, then the test results are conditionally independent given his disease status. Let $a_0 = P(T_j|D, G^c)$ and $b_0 = P(T_j|D^c, G^c)$, where a_0 and b_0 don't depend on j. Find the posterior probability that Fred has the disease, given that he tests positive on all n of the tests.

62. A certain hereditary disease can be passed from a mother to her children. Given that the mother has the disease, her children independently will have it with probability 1/2. Given that she doesn't have the disease, her children won't have it either. A certain mother, who has probability 1/3 of having the disease, has two children.

(a) Find the probability that neither child has the disease.

(b) Is whether the elder child has the disease independent of whether the younger child has the disease? Explain.

(c) The elder child is found not to have the disease. A week later, the younger child is also found not to have the disease. Given this information, find the probability that the mother has the disease.

63. Three fair coins are tossed at the same time. Explain what is wrong with the following argument: "there is a 50% chance that the three coins all landed the same way, since obviously it is possible to find two coins that match, and then the other coin has a 50% chance of matching those two".

64. An urn contains red, green, and blue balls. Let r, g, b be the proportions of red, green, blue balls, respectively ($r + g + b = 1$).

(a) Balls are drawn randomly *with replacement.* Find the probability that the first time a green ball is drawn is before the first time a blue ball is drawn.

Hint: Explain how this relates to finding the probability that a draw is green, given that it is either green or blue.

(b) Balls are drawn randomly *without replacement.* Find the probability that the first time a green ball is drawn is before the first time a blue ball is drawn. Is the answer the same or different than the answer in (a)?

Hint: Imagine the balls all lined up, in the order in which they will be drawn. Note that where the red balls are standing in this line is irrelevant.

(c) Generalize the result from (a) to the following setting. Independent trials are performed, and the outcome of each trial is classified as being exactly one of type 1, type 2, ..., or type n, with probabilities $p_1, p_2, \ldots, p_n$, respectively. Find the probability that the first trial to result in type i comes before the first trial to result in type j, for $i \neq j$.

65. Marilyn vos Savant was asked the following question for her column in *Parade*:

> You're at a party with 199 other guests when robbers break in and announce that they are going to rob one of you. They put 199 blank pieces of paper in a hat, plus one marked "you lose." Each guest must draw, and the person who draws "you lose" will get robbed. The robbers offer you the option of drawing first, last, or at any time in between. When would you take your turn?

The draws are made *without replacement*, and for (a) are uniformly random.

(a) Determine whether it is optimal to draw first, last, or somewhere in between (or whether it does not matter), to maximize the probability of not being robbed. Give a clear, concise, and compelling explanation.

(b) More generally, suppose that there is one "you lose" piece of paper, with "weight" v, and there are n blank pieces of paper, each with "weight" w. At each stage, draws are made with probability proportional to weight, i.e., the probability of drawing a particular piece of paper is its weight divided by the sum of the weights of all the remaining pieces of paper. Determine whether it is better to draw first or second (or whether it does not matter); here $v > 0, w > 0$, and $n \geq 1$ are known constants.

66. A fair die is rolled repeatedly, until the running total is at least 100 (at which point the rolling stops). Find the most likely value of the final running total (i.e., the value of the running total at the first time when it is at least 100).
Hint: Consider the possibilities for what the running total is just before the last roll.

67. Homer has a box of donuts, which currently contains exactly c chocolate, g glazed, and j jelly donuts. Homer eats donuts one after another, each time choosing uniformly at random from the remaining donuts.

(a) Find the probability that the last donut remaining in the box is a chocolate donut.

(b) Find the probability of the following event: glazed is the first type of donut that Homer runs out of, and then jelly is the second type of donut that he runs out of.

Hint: Consider the last donut remaining, and the last donut that is either glazed or jelly.

68. Let D be the event that a person develops a certain disease, and C be the event that the person was exposed to a certain substance (e.g., D may correspond to lung cancer and C may correspond to smoking cigarettes). We are interested in whether exposure to the substance is related to developing the disease (and if so, how they are related).
The *odds ratio* is a very widely used measure in epidemiology of the association between disease and exposure, defined as

$$\text{OR} = \frac{\text{odds}(D|C)}{\text{odds}(D|C^c)},$$

where conditional odds are defined analogously to unconditional odds: $\text{odds}(A|B) = \frac{P(A|B)}{P(A^c|B)}$. The *relative risk* of the disease for someone exposed to the substance, another widely used measure, is

$$\text{RR} = \frac{P(D|C)}{P(D|C^c)}.$$

The relative risk is especially easy to interpret, e.g., $\text{RR} = 2$ says that someone exposed to the substance is twice as likely to develop the disease as someone who isn't exposed (though this does not necessarily mean that the substance *causes* the increased chance of getting the disease, nor is there necessarily a causal interpretation for the odds ratio).

(a) Show that if the disease is rare, both for exposed people and for unexposed people, then the relative risk is approximately equal to the odds ratio.

(b) Let p_{ij} for $i = 0, 1$ and $j = 0, 1$ be the probabilities in the following 2×2 table.

	D	D^c
C	p_{11}	p_{10}
C^c	p_{01}	p_{00}

For example, $p_{10} = P(C, D^c)$. Show that the odds ratio can be expressed as a *cross-product ratio*, in the sense that

$$\text{OR} = \frac{p_{11}p_{00}}{p_{10}p_{01}}.$$

(c) Show that the odds ratio has the neat symmetry property that the roles of C and D can be swapped without changing the value:

$$\text{OR} = \frac{\text{odds}(C|D)}{\text{odds}(C|D^c)}.$$

This property is one of the main reasons why the odds ratio is so widely used, since it turns out that it allows the odds ratio to be estimated in a wide variety of problems where relative risk would be hard to estimate well.

69. A researcher wants to estimate the percentage of people in some population who have used illegal drugs, by conducting a survey. Concerned that a lot of people would lie when asked a sensitive question like "Have you ever used illegal drugs?", the researcher uses a method known as *randomized response*. A hat is filled with slips of paper, each of which says either "I have used illegal drugs" or "I have not used illegal drugs". Let p be the proportion of slips of paper that say "I have used illegal drugs" (p is chosen by the researcher in advance).
Each participant chooses a random slip of paper from the hat and answers (truthfully) "yes" or "no" to whether the statement on that slip is true. The slip is then returned to the hat. The researcher does not know which type of slip the participant had. Let y be the probability that a participant will say "yes", and d be the probability that a participant has used illegal drugs.

(a) Find y, in terms of d and p.

(b) What would be the worst possible choice of p that the researcher could make in designing the survey? Explain.

(c) Now consider the following alternative system. Suppose that proportion p of the slips of paper say "I have used illegal drugs", but that now the remaining $1 - p$ say "I was born in winter" rather than "I have not used illegal drugs". Assume that $1/4$ of people are born in winter, and that a person's season of birth is independent of whether they have used illegal drugs. Find d, in terms of y and p.

70. At the beginning of the play *Rosencrantz and Guildenstern Are Dead* by Tom Stoppard [25], Guildenstern is spinning coins and Rosencrantz is betting on the outcome for each. The coins have been landing Heads over and over again, prompting the following remark:

> *Guildenstern*: A weaker man might be moved to re-examine his faith, if in nothing else at least in the law of probability.

The coin spins have resulted in Heads 92 times in a row.

(a) Fred and his friend are watching the play. Upon seeing the events described above, they have the following conversation:

Fred: That outcome would be incredibly unlikely with fair coins. They must be using trick coins (maybe with double-headed coins), or the experiment must have been rigged somehow (maybe with magnets).

Fred's friend: It's true that the string HH...H of length 92 is very unlikely; the chance is $1/2^{92} \approx 2 \times 10^{-28}$ with fair coins. But *any* other specific string of H's and T's with length 92 has *exactly* the same probability! The reason the outcome seems extremely unlikely is that the number of possible outcomes grows exponentially as the number of spins grows, so *any* outcome would seem extremely unlikely. You could just as well have made the same argument even without looking at the results of their experiment, which means you really don't have evidence against the coins being fair.

Discuss these comments, to help Fred and his friend resolve their debate.

(b) Suppose there are only two possibilities: either the coins are all fair (and spun fairly), or double-headed coins are being used (in which case the probability of Heads is 1). Let p be the prior probability that the coins are fair. Find the posterior probability that the coins are fair, given that they landed Heads in 92 out of 92 trials.

(c) Continuing from (b), for which values of p is the posterior probability that the coins are fair greater than 0.5? For which values of p is it less than 0.05?

71. There are n types of toys, which you are collecting one by one. Each time you buy a toy, it is randomly determined which type it has, with equal probabilities. Let p_{ij} be the probability that just after you have bought your ith toy, you have exactly j toy types in your collection, for $i \geq 1$ and $0 \leq j \leq n$. (This problem is in the setting of the *coupon collector* problem, a famous problem which we study in Example 4.3.12.)

(a) Find a recursive equation expressing p_{ij} in terms of $p_{i-1,j}$ and $p_{i-1,j-1}$, for $i \geq 2$ and $1 \leq j \leq n$.

(b) Describe how the recursion from (a) can be used to calculate p_{ij}.

72. *A/B testing* is a form of randomized experiment that is used by many companies to learn about how customers will react to different treatments. For example, a company may want to see how users will respond to a new feature on their website (compared with how users respond to the current version of the website) or compare two different advertisements.

As the name suggests, two different treatments, Treatment A and Treatment B, are being studied. Users arrive one by one, and upon arrival are randomly assigned to one of the two treatments. The trial for each user is classified as "success" (e.g., the user made a purchase) or "failure". The probability that the nth user receives Treatment A is allowed to depend on the outcomes for the previous users. This set-up is known as a *two-armed bandit*.

Many algorithms for how to randomize the treatment assignments have been studied. Here is an especially simple (but fickle) algorithm, called a *stay-with-a-winner* procedure:

(i) Randomly assign the first user to Treatment A or Treatment B, with equal probabilities.

(ii) If the trial for the nth user is a success, stay with the same treatment for the $(n+1)$st user; otherwise, switch to the other treatment for the $(n+1)$st user.

Let a be the probability of success for Treatment A, and b be the probability of success for Treatment B. Assume that $a \neq b$, but that a and b are unknown (which is why the test is needed). Let p_n be the probability of success on the nth trial and a_n be the probability that Treatment A is assigned on the nth trial (using the above algorithm).

(a) Show that

$$p_n = (a-b)a_n + b,$$
$$a_{n+1} = (a+b-1)a_n + 1 - b.$$

(b) Use the results from (a) to show that p_{n+1} satisfies the following recursive equation:

$$p_{n+1} = (a + b - 1)p_n + a + b - 2ab.$$

(c) Use the result from (b) to find the long-run probability of success for this algorithm, $\lim_{n\to\infty} p_n$, assuming that this limit exists.

73. In humans (and many other organisms), genes come in pairs. A certain gene comes in two types (*alleles*): type a and type A. The *genotype* of a person for that gene is the types of the two genes in the pair: AA, Aa, or aa (aA is equivalent to Aa). Assume that the Hardy-Weinberg law applies here, which means that the frequencies of AA, Aa, aa in the population are $p^2, 2p(1-p), (1-p)^2$ respectively, for some p with $0 < p < 1$.

When a woman and a man have a child, the child's gene pair has one gene contributed by each parent. Suppose that the mother is equally likely to contribute either of the two genes in her gene pair, and likewise for the father, independently. Also suppose that the genotypes of the parents are independent of each other (with probabilities given by the Hardy-Weinberg law).

(a) Find the probabilities of each possible genotype (AA, Aa, aa) for a child of two random parents. Explain what this says about stability of the Hardy-Weinberg law from one generation to the next.

Hint: Condition on the genotypes of the parents.

(b) A person of type AA or aa is called *homozygous* (for the gene under consideration), and a person of type Aa is called *heterozygous* (for that gene). Find the probability that a child is homozygous, given that both parents are homozygous. Also, find the probability that a child is heterozygous, given that both parents are heterozygous.

(c) Suppose that having genotype aa results in a distinctive physical characteristic, so it is easy to tell by looking at someone whether or not they have that genotype. A mother and father, neither of whom are of type aa, have a child. The child is also not of type aa. Given this information, find the probability that the child is heterozygous.

Hint: Use the definition of conditional probability. Then expand both the numerator and the denominator using LOTP, conditioning on the genotypes of the parents.

74. A standard deck of cards will be shuffled and then the cards will be turned over one at a time until the first ace is revealed. Let B be the event that the *next* card in the deck will also be an ace.

(a) Intuitively, how do you think $P(B)$ compares in size with $1/13$ (the overall proportion of aces in a deck of cards)? Explain your intuition. (Give an intuitive discussion rather than a mathematical calculation; the goal here is to describe your intuition explicitly.)

(b) Let C_j be the event that the first ace is at position j in the deck. Find $P(B|C_j)$ in terms of j, fully simplified.

(c) Using the law of total probability, find an expression for $P(B)$ as a sum. (The sum can be left unsimplified, but it should be something that could easily be computed in software such as R that can calculate sums.)

(d) Find a fully simplified expression for $P(B)$ using a symmetry argument.

Hint: If you were deciding whether to bet on the next card after the first ace being an ace or to bet on the last card in the deck being an ace, would you have a preference?

3

Random variables and their distributions

In this chapter, we introduce *random variables*, an incredibly useful concept that simplifies notation and expands our ability to quantify uncertainty and summarize the results of experiments. Random variables are essential throughout the rest of this book, and throughout statistics, so it is crucial to think through what they mean, both intuitively and mathematically.

3.1 Random variables

To see why our current notation can quickly become unwieldy, consider again the gambler's ruin problem from Chapter 2. In this problem, we may be very interested in how much wealth each gambler has at any particular time. So we could make up notation like letting A_{jk} be the event that gambler A has exactly j dollars after k rounds, and similarly defining an event B_{jk} for gambler B, for all j and k.

This is already too complicated. Furthermore, we may also be interested in other quantities, such as the difference in their wealths (gambler A's minus gambler B's) after k rounds, or the duration of the game (the number of rounds until one player is bankrupt). Expressing the event "the duration of the game is r rounds" in terms of the A_{jk} and B_{jk} would involve a long, awkward string of unions and intersections. And then what if we want to express gambler A's wealth as the equivalent amount in euros rather than dollars? We can multiply a *number* in dollars by a currency exchange rate, but we can't multiply an *event* by an exchange rate.

Instead of having convoluted notation that obscures how the quantities of interest are related, wouldn't it be nice if we could say something like the following?

> Let X_k be the wealth of gambler A after k rounds. Then $Y_k = N - X_k$ is the wealth of gambler B after k rounds (where N is the fixed total wealth); $X_k - Y_k = 2X_k - N$ is the difference in wealths after k rounds; $c_k X_k$ is the wealth of gambler A in euros after k rounds, where c_k is the euros per dollar exchange rate after k rounds; and the duration is $R = \min\{n : X_n = 0 \text{ or } Y_n = 0\}$.

The notion of a random variable will allow us to do exactly this! It needs to be introduced carefully though, to make it both conceptually and technically correct. Sometimes a definition of "random variable" is given that is a barely paraphrased version of "a random variable is a variable that takes on random values", but such a feeble attempt at a definition fails to say where the randomness come from. Nor does it help us to derive properties of random variables: we're familiar with working with algebraic equations like $x^2 + y^2 = 1$, but what are the valid mathematical operations if x and y are *random* variables? To make the notion of random variable precise, we define it as a *function* mapping the sample space to the real line. (See the math appendix for review of some concepts about functions.)

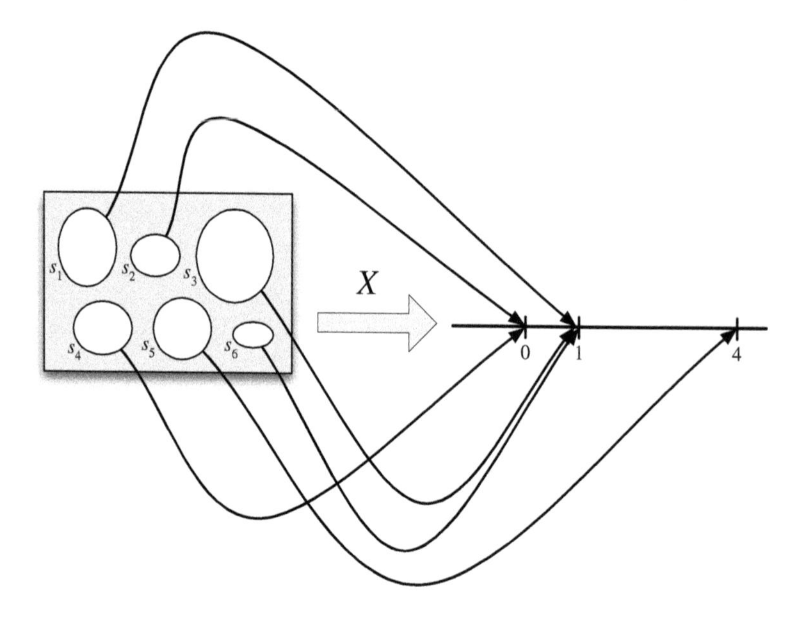

FIGURE 3.1
A random variable maps the sample space into the real line. The r.v. X depicted here is defined on a sample space with 6 elements, and has possible values 0, 1, and 4. The randomness comes from choosing a random pebble according to the probability function P for the sample space.

Definition 3.1.1 (Random variable). Given an experiment with sample space S, a *random variable* (r.v.) is a function from the sample space S to the real numbers $\mathbb{R}$. It is common, but not required, to denote random variables by capital letters.

Thus, a random variable X assigns a numerical value $X(s)$ to each possible outcome s of the experiment. The randomness comes from the fact that we have a random experiment (with probabilities described by the probability function P); the mapping itself is deterministic, as illustrated in Figure 3.1. The same r.v. is shown in a simpler way in the left panel of Figure 3.2, in which we inscribe the values inside the pebbles.

This definition is abstract but fundamental; one of the most important skills to develop when studying probability and statistics is the ability to go back and forth between abstract ideas and concrete examples. Relatedly, it is important to work on recognizing the essential pattern or structure of a problem and how it connects to problems you have studied previously. We will often discuss stories that involve tossing coins or drawing balls from urns because they are simple, convenient scenarios to work with, but many other problems are *isomorphic*: they have the same essential structure, but in a different guise.

To start, let's consider a coin-tossing example. The structure of the problem is that we have a sequence of trials where there are two possible outcomes for each trial. Here we think of the possible outcomes as H (Heads) and T (Tails), but we could just as well think of them as "success" and "failure" or as 1 and 0, for example.

Example 3.1.2 (Coin tosses). Consider an experiment where we toss a fair coin twice. The sample space consists of four possible outcomes: $S = \{HH, HT, TH, TT\}$. Here are some random variables on this space (for practice, you can think up some of your own). Each r.v. is a numerical summary of some aspect of the experiment.

- Let X be the number of Heads. This is a random variable with possible values 0, 1, and 2. Viewed as a function, X assigns the value 2 to the outcome HH, 1 to the outcomes HT and TH, and 0 to the outcome TT. That is,

$$X(HH) = 2, X(HT) = X(TH) = 1, X(TT) = 0.$$

- Let Y be the number of Tails. In terms of X, we have $Y = 2 - X$. In other words, Y and $2 - X$ are the same r.v.: $Y(s) = 2 - X(s)$ for all s.
- Let I be 1 if the first toss lands Heads and 0 otherwise. Then I assigns the value 1 to the outcomes HH and HT and 0 to the outcomes TH and TT. This r.v. is an example of what is called an *indicator random variable* since it indicates whether the first toss lands Heads, using 1 to mean "yes" and 0 to mean "no".

We can also encode the sample space as $\{(1,1), (1,0), (0,1), (0,0)\}$, where 1 is the code for Heads and 0 is the code for Tails. Then we can give explicit formulas for X, Y, I:

$$X(s_1, s_2) = s_1 + s_2, \ Y(s_1, s_2) = 2 - s_1 - s_2, \ I(s_1, s_2) = s_1,$$

where for simplicity we write $X(s_1, s_2)$ to mean $X((s_1, s_2))$, etc.

For most r.v.s we will consider, it is tedious or infeasible to write down an explicit formula in this way. Fortunately, it is usually unnecessary to do so, since (as we saw in this example) there are other ways to define an r.v., and (as we will see throughout the rest of this book) there are many ways to study the properties of an r.v. other than by doing computations with an explicit formula for what it maps each outcome s to. □

As in the previous chapters, for a sample space with a finite number of outcomes we can visualize the outcomes as pebbles, with the mass of a pebble corresponding to its probability, such that the total mass of the pebbles is 1. A random variable simply labels each pebble with a number. Figure 3.2 shows two random variables defined on the same sample space: the pebbles or outcomes are the same, but the real numbers assigned to the outcomes are different.

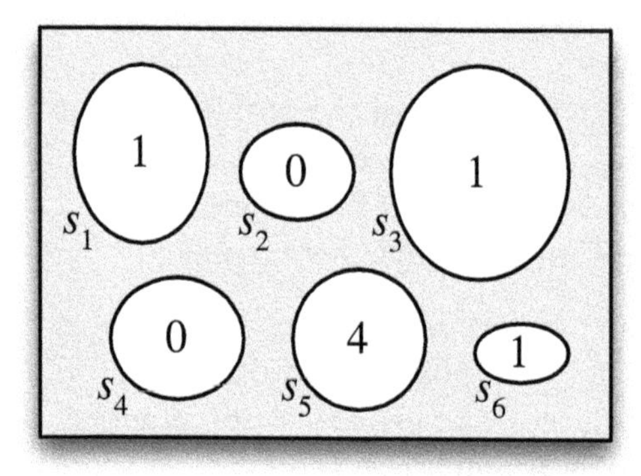

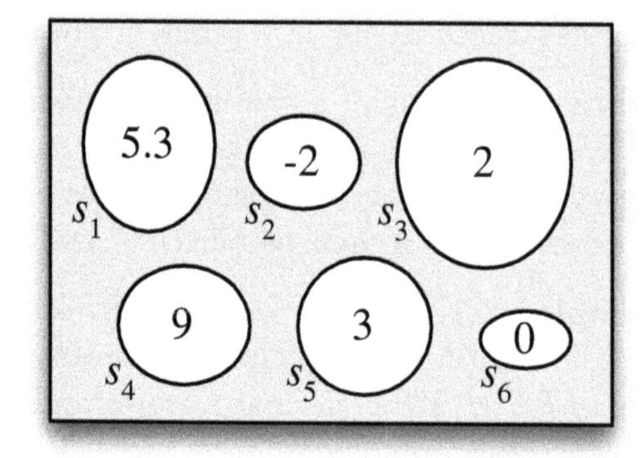

FIGURE 3.2
Two random variables defined on the same sample space.

As we've mentioned earlier, the source of the randomness in a random variable is the experiment itself, in which a sample outcome $s \in S$ is chosen according to a probability function P. Before we perform the experiment, the outcome s has not yet been realized, so we don't know the value of X, though we could calculate the probability that X will take on a given value or range of values. After we perform the experiment and the outcome s has been realized, the random variable crystallizes into the numerical value $X(s)$.

Random variables provide *numerical* summaries of the experiment in question. This is very handy because the sample space of an experiment is often incredibly complicated or high-dimensional, and the outcomes $s \in S$ may be non-numeric. For example, the experiment may be to collect a random sample of people in a certain city and ask them various questions, which may have numeric (e.g., age or height) or non-numeric (e.g., political party or favorite movie) answers. The fact that r.v.s take on numerical values is a very convenient simplification compared to having to work with the full complexity of S at all times.

3.2 Distributions and probability mass functions

There are two main types of random variables used in practice: *discrete* r.v.s and *continuous* r.v.s. In this chapter and the next, our focus is on discrete r.v.s. Continuous r.v.s are introduced in Chapter 5.

Definition 3.2.1 (Discrete random variable). A random variable X is said to be *discrete* if there is a finite list of values $a_1, a_2, \dots, a_n$ or an infinite list of values $a_1, a_2, \dots$ such that $P(X = a_j \text{ for some } j) = 1$. If X is a discrete r.v., then the finite or countably infinite set of values x such that $P(X = x) > 0$ is called the *support* of X.

Most commonly in applications, the support of a discrete r.v. is a set of integers. In contrast, a *continuous* r.v. can take on any real value in an interval (possibly even the entire real line); such r.v.s are defined more precisely in Chapter 5. It is also possible to have an r.v. that is a hybrid of discrete and continuous, such as by flipping a coin and then generating a discrete r.v. if the coin lands Heads and generating a continuous r.v. if the coin lands Tails. But the starting point for understanding such r.v.s is to understand discrete and continuous r.v.s.

Given a random variable, we would like to be able to describe its behavior using the language of probability. For example, we might want to answer questions about the probability that the r.v. will fall into a given range: if L is the lifetime earnings of a randomly chosen U.S. college graduate, what is the probability that L exceeds a million dollars? If M is the number of major earthquakes in California in the next five years, what is the probability that M equals 0?

The *distribution* of a random variable provides the answers to these questions; it specifies the probabilities of all events associated with the r.v., such as the probability of it equaling 3 and the probability of it being at least 110. We will see that there are several equivalent ways to express the distribution of an r.v. For a discrete r.v., the most natural way to do so is with a *probability mass function*, which we now define.

Definition 3.2.2 (Probability mass function). The *probability mass function* (PMF) of a discrete r.v. X is the function p_X given by $p_X(x) = P(X = x)$. Note that this is positive if x is in the support of X, and 0 otherwise.

☣ **3.2.3.** In writing $P(X = x)$, we are using $X = x$ to denote an *event*, consisting of all outcomes s to which X assigns the number x. This event is also written as $\{X = x\}$; formally, $\{X = x\}$ is defined as $\{s \in S : X(s) = x\}$, but writing $\{X = x\}$ is shorter and more intuitive. Going back to Example 3.1.2, if X is the number of Heads in two fair coin tosses, then $\{X = 1\}$ consists of the sample outcomes HT and TH, which are the two outcomes to which X assigns the number 1. Since $\{HT, TH\}$ is a subset of the sample space, it is an event. So it makes sense to talk about $P(X = 1)$, or more generally, $P(X = x)$. If $\{X = x\}$ were anything other than an event, it would make no sense to calculate its probability! It does not make sense to write "$P(X)$"; we can only take the probability of an event, not of an r.v.

Let's look at a few examples of PMFs.

Example 3.2.4 (Coin tosses continued). In this example we'll find the PMFs of all the random variables in Example 3.1.2, the example with two fair coin tosses. Here are the r.v.s we defined, along with their PMFs:

- X, the number of Heads. Since X equals 0 if TT occurs, 1 if HT or TH occurs, and 2 if HH occurs, the PMF of X is the function p_X given by

$$p_X(0) = P(X = 0) = 1/4,$$
$$p_X(1) = P(X = 1) = 1/2,$$
$$p_X(2) = P(X = 2) = 1/4,$$

and $p_X(x) = 0$ for all other values of x.

- $Y = 2 - X$, the number of Tails. Reasoning as above or using the fact that

$$P(Y = y) = P(2 - X = y) = P(X = 2 - y) = p_X(2 - y),$$

the PMF of Y is

$$p_Y(0) = P(Y = 0) = 1/4,$$
$$p_Y(1) = P(Y = 1) = 1/2,$$
$$p_Y(2) = P(Y = 2) = 1/4,$$

and $p_Y(y) = 0$ for all other values of y.

Note that X and Y have the same PMF (that is, p_X and p_Y are the same function) even though X and Y are not the same r.v. (that is, X and Y are two different functions from $\{HH, HT, TH, TT\}$ to the real line).

- I, the indicator of the first toss landing Heads. Since I equals 0 if TH or TT occurs and 1 if HH or HT occurs, the PMF of I is

$$p_I(0) = P(I = 0) = 1/2,$$
$$p_I(1) = P(I = 1) = 1/2,$$

and $p_I(i) = 0$ for all other values of i.

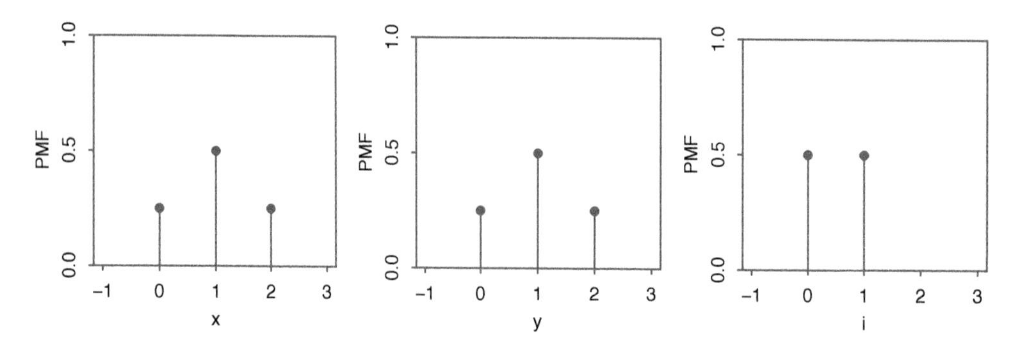

FIGURE 3.3
Left to right: PMFs of X, Y, and I, with X the number of Heads in two fair coin tosses, Y the number of Tails, and I the indicator of Heads on the first toss.

The PMFs of X, Y, and I are plotted in Figure 3.3. Vertical bars are drawn to make it easier to compare the heights of different points. □

Example 3.2.5 (Sum of die rolls). We roll two fair 6-sided dice. Let $T = X + Y$ be the total of the two rolls, where X and Y are the individual rolls. The sample space of this experiment has 36 equally likely outcomes:

$$S = \{(1,1),(1,2),\ldots,(6,5),(6,6)\}.$$

For example, 7 of the 36 outcomes s are shown in the table below, along with the corresponding values of X, Y, and T. After the experiment is performed, we observe values for X and Y, and then the observed value of T is the sum of those values.

s	X	Y	$X+Y$
$(1,2)$	1	2	3
$(1,6)$	1	6	7
$(2,5)$	2	5	7
$(3,1)$	3	1	4
$(4,3)$	4	3	7
$(5,4)$	5	4	9
$(6,6)$	6	6	12

Since the dice are fair, the PMF of X is

$$P(X = j) = 1/6,$$

for $j = 1, 2, \ldots, 6$ (and $P(X = j) = 0$ otherwise); we say that X has a *Discrete Uniform* distribution on $1, 2, \ldots, 6$. Similarly, Y is also Discrete Uniform on $1, 2, \ldots, 6$.

Note that Y has the same *distribution* as X but is not the same *random variable* as X. In fact, we have

$$P(X = Y) = 6/36 = 1/6.$$

Two more r.v.s in this experiment with the same distribution as X are $7 - X$ and $7 - Y$. To see this, we can use the fact that for a standard die, $7 - X$ is the value on the bottom if X is the value on the top. If the top value is equally likely to be any of the numbers $1, 2, \ldots, 6$, then so is the bottom value. Note that even though $7 - X$ has the same distribution as X, it is *never* equal to X in a run of the experiment!

Let's now find the PMF of T. By the naive definition of probability,

$$\begin{aligned}
P(T = 2) &= P(T = 12) = 1/36,\\
P(T = 3) &= P(T = 11) = 2/36,\\
P(T = 4) &= P(T = 10) = 3/36,\\
P(T = 5) &= P(T = 9) = 4/36,\\
P(T = 6) &= P(T = 8) = 5/36,\\
P(T = 7) &= 6/36.
\end{aligned}$$

For all other values of t, $P(T = t) = 0$. We can see directly that the support of T

is $\{2, 3, \ldots, 12\}$ just by looking at the possible totals for two dice, but as a check, note that

$$P(T = 2) + P(T = 3) + \cdots + P(T = 12) = 1,$$

which shows that all possibilities have been accounted for. The symmetry property of T that appears above, $P(T = t) = P(T = 14-t)$, makes sense since each outcome $\{X = x, Y = y\}$ which makes $T = t$ has a corresponding outcome $\{X = 7 - x, Y = 7 - y\}$ of the same probability which makes $T = 14 - t$.

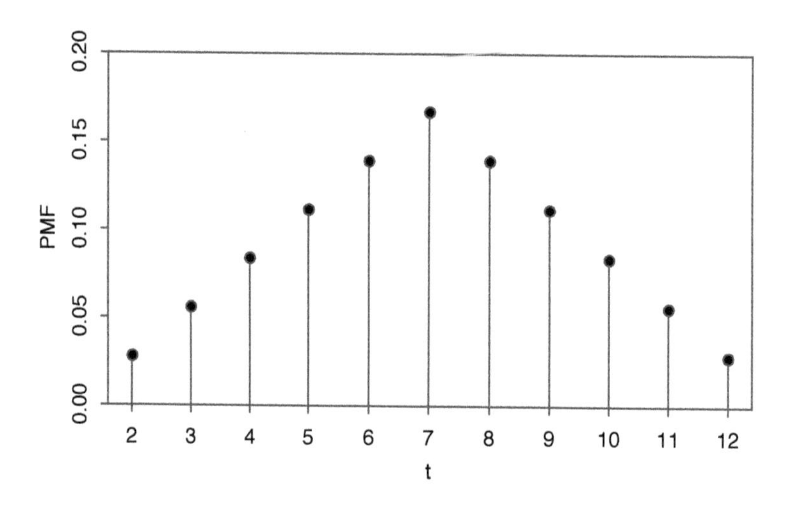

FIGURE 3.4
PMF of the sum of two die rolls.

The PMF of T is plotted in Figure 3.4; it has a triangular shape, and the symmetry noted above is very visible. □

Example 3.2.6 (Children in a U.S. household). Suppose we choose a household in the United States at random. Let X be the number of children in the chosen household. Since X can only take on integer values, it is a discrete r.v. The probability that X takes on the value x is proportional to the number of households in the United States with x children.

Using data from the 2010 General Social Survey [23], we can approximate the proportion of households with 0 children, 1 child, 2 children, etc., and hence approximate the PMF of X, which is plotted in Figure 3.5. □

We will now state the properties of a valid PMF.

Theorem 3.2.7 (Valid PMFs). Let X be a discrete r.v. with support $x_1, x_2, \ldots$ (assume these values are distinct and, for notational simplicity, that the support is countably infinite; the analogous results hold if the support is finite). The PMF p_X of X must satisfy the following two criteria:

- Nonnegative: $p_X(x) > 0$ if $x = x_j$ for some j, and $p_X(x) = 0$ otherwise;
- Sums to 1: $\sum_{j=1}^{\infty} p_X(x_j) = 1$.

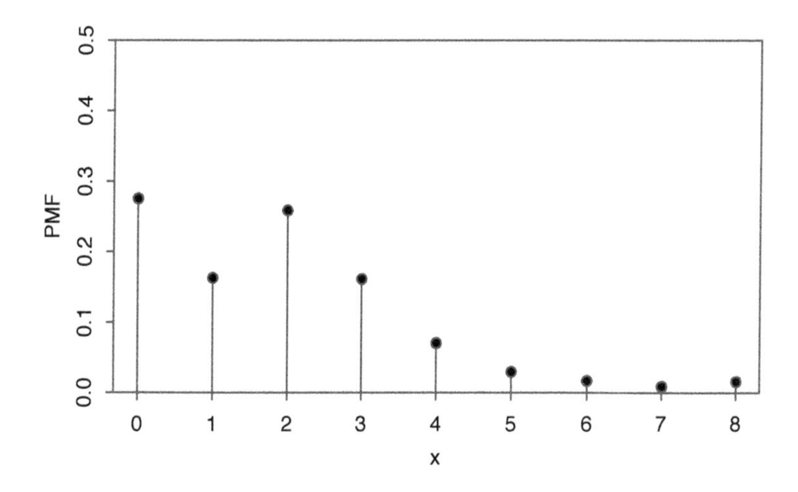

FIGURE 3.5
PMF of the number of children in a randomly selected U.S. household.

Proof. The first criterion is true since probability is nonnegative. The second is true since X must take on *some* value, and the events $\{X = x_j\}$ are disjoint, so

$$\sum_{j=1}^{\infty} P(X = x_j) = P\left(\bigcup_{j=1}^{\infty}\{X = x_j\}\right) = P(X = x_1 \text{ or } X = x_2 \text{ or } \dots) = 1. \quad \blacksquare$$

Conversely, if distinct values $x_1, x_2, \dots$ are specified and we have a function satisfying the two criteria above, then this function *is* the PMF of some r.v.; we will show how to construct such an r.v. in Chapter 5.

We claimed earlier that the PMF is one way of expressing the distribution of a discrete r.v. This is because once we know the PMF of X, we can calculate the probability that X will fall into a given subset of the real numbers by summing over the appropriate values of x, as the next example shows.

Example 3.2.8. Returning to Example 3.2.5, let T be the sum of two fair die rolls. We have already calculated the PMF of T. Now suppose we're interested in the probability that T is in the interval $[1, 4]$. There are only three values in the interval $[1, 4]$ that T can take on, namely, 2, 3, and 4. We know the probability of each of these values from the PMF of T, so

$$P(1 \leq T \leq 4) = P(T = 2) + P(T = 3) + P(T = 4) = 6/36. \quad \square$$

In general, given a discrete r.v. X and a set B of real numbers, if we know the PMF of X we can find $P(X \in B)$, the probability that X is in B, by summing up the heights of the vertical bars at points in B in the plot of the PMF of X. *Knowing the PMF of a discrete r.v. determines its distribution.*

3.3 Bernoulli and Binomial

Some distributions are so ubiquitous in probability and statistics that they have their own names. We will introduce these *named distributions* throughout the book, starting with a very simple but useful case: an r.v. that can take on only two possible values, 0 and 1.

Definition 3.3.1 (Bernoulli distribution). An r.v. X is said to have the *Bernoulli distribution* with parameter p if $P(X = 1) = p$ and $P(X = 0) = 1 - p$, where $0 < p < 1$. We write this as $X \sim \text{Bern}(p)$. The symbol $\sim$ is read "is distributed as".

Any r.v. whose possible values are 0 and 1 has a $\text{Bern}(p)$ distribution, with p the probability of the r.v. equaling 1. This number p in $\text{Bern}(p)$ is called the *parameter* of the distribution; it determines which specific Bernoulli distribution we have. Thus there is not just one Bernoulli distribution, but rather a *family* of Bernoulli distributions, indexed by p. For example, if $X \sim \text{Bern}(1/3)$, it would be correct but incomplete to say "X is Bernoulli"; to fully specify the distribution of X, we should both say its name (Bernoulli) and its parameter value (1/3), which is the point of the notation $X \sim \text{Bern}(1/3)$.

Any event has a Bernoulli r.v. that is naturally associated with it, equal to 1 if the event happens and 0 otherwise. This is called the *indicator random variable* of the event; we will see that such r.v.s are extremely useful.

Definition 3.3.2 (Indicator random variable). The *indicator random variable* of an event A is the r.v. which equals 1 if A occurs and 0 otherwise. We will denote the indicator r.v. of A by I_A or $I(A)$. Note that $I_A \sim \text{Bern}(p)$ with $p = P(A)$.

We often imagine Bernoulli r.v.s using coin tosses, but this is just convenient language for discussing the following general story.

Story 3.3.3 (Bernoulli trial). An experiment that can result in either a "success" or a "failure" (but not both) is called a *Bernoulli trial*. A Bernoulli random variable can be thought of as the *indicator of success* in a Bernoulli trial: it equals 1 if success occurs and 0 if failure occurs in the trial. □

Because of this story, the parameter p is often called the *success probability* of the $\text{Bern}(p)$ distribution. Once we start thinking about Bernoulli trials, it's hard not to start thinking about what happens when we have more than one trial.

Story 3.3.4 (Binomial distribution). Suppose that n *independent* Bernoulli trials are performed, each with the same success probability p. Let X be the number of successes. The distribution of X is called the *Binomial distribution* with parameters n and p. We write $X \sim \text{Bin}(n, p)$ to mean that X has the Binomial distribution with parameters n and p, where n is a positive integer and $0 < p < 1$. □

Notice that we define the Binomial distribution not by its PMF, but by a *story* about the type of experiment that could give rise to a random variable with a Binomial distribution. The most famous distributions in statistics all have stories which explain why they are so often used as models for data, or as the building blocks for more complicated distributions.

Thinking about the named distributions first and foremost in terms of their stories has many benefits. It facilitates pattern recognition, allowing us to see when two problems are essentially identical in structure; it often leads to cleaner solutions that avoid PMF calculations altogether; and it helps us understand how the named distributions are connected to one another. Here it is clear that Bern(p) is the same distribution as Bin($1, p$): the Bernoulli is a special case of the Binomial.

Using the story definition of the Binomial, let's find its PMF.

Theorem 3.3.5 (Binomial PMF). If $X \sim \text{Bin}(n, p)$, then the PMF of X is

$$P(X = k) = \binom{n}{k} p^k (1-p)^{n-k}$$

for $k = 0, 1, \ldots, n$ (and $P(X = k) = 0$ otherwise).

☣ **3.3.6.** To save writing, it is often left implicit that a PMF is zero wherever it is not specified to be nonzero, but in any case it is important to understand what the support of a random variable is, and good practice to check that PMFs are valid. If two discrete r.v.s have the same PMF, then they also must have the same support. So we sometimes refer to the support of a discrete *distribution*; this is the support of any r.v. with that distribution.

Proof. An experiment consisting of n independent Bernoulli trials produces a sequence of successes and failures. The probability of any specific sequence of k successes and $n - k$ failures is $p^k(1-p)^{n-k}$. There are $\binom{n}{k}$ such sequences, since we just need to select where the successes are. Therefore, letting X be the number of successes,

$$P(X = k) = \binom{n}{k} p^k (1-p)^{n-k}$$

for $k = 0, 1, \ldots, n$, and $P(X = k) = 0$ otherwise. This is a valid PMF because it is nonnegative and it sums to 1 by the binomial theorem. ■

Figure 3.6 shows plots of the Binomial PMF for various values of n and p. Note that the PMF of the Bin($10, 1/2$) distribution is symmetric about 5, but when the success probability is not 1/2, the PMF is *skewed.* For a fixed number of trials n, X tends to be larger when the success probability is high and lower when the success probability is low, as we would expect from the story of the Binomial distribution. Also recall that in any PMF plot, the sum of the heights of the vertical bars must be 1.

We've used Story 3.3.4 to find the Bin(n, p) PMF. The story also gives us a straightforward proof of the fact that if X is Binomial, then $n - X$ is also Binomial.

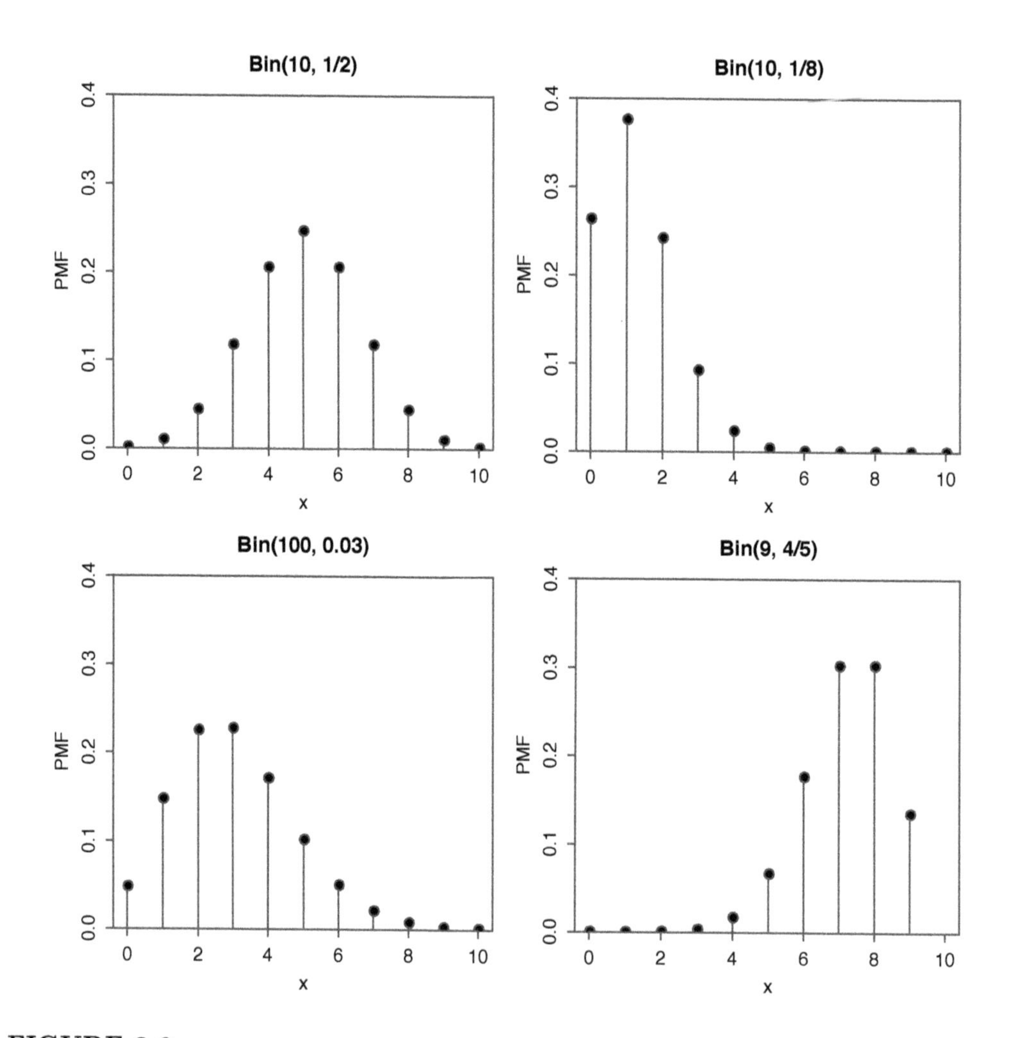

FIGURE 3.6
Some Binomial PMFs. In the lower left, we plot the Bin(100, 0.03) PMF between 0 and 10 only, as the probability of more than 10 successes is close to 0.

Theorem 3.3.7. Let $X \sim \text{Bin}(n, p)$, and $q = 1 - p$ (we often use q to denote the failure probability of a Bernoulli trial). Then $n - X \sim \text{Bin}(n, q)$.

Proof. Using the story of the Binomial, interpret X as the number of successes in n independent Bernoulli trials. Then $n - X$ is the number of failures in those trials. Interchanging the roles of success and failure, we have $n - X \sim \text{Bin}(n, q)$. Alternatively, we can check that $n - X$ has the $\text{Bin}(n, q)$ PMF. Let $Y = n - X$. The PMF of Y is

$$P(Y = k) = P(X = n - k) = \binom{n}{n-k} p^{n-k} q^k = \binom{n}{k} q^k p^{n-k},$$

for $k = 0, 1, \ldots, n$. ■

Corollary 3.3.8. Let $X \sim \text{Bin}(n, p)$ with $p = 1/2$ and n even. Then the distribution of X is symmetric about $n/2$, in the sense that $P(X = n/2 + j) = P(X = n/2 - j)$ for all nonnegative integers j.

Proof. By Theorem 3.3.7, $n - X$ is also $\text{Bin}(n, 1/2)$, so

$$P(X = k) = P(n - X = k) = P(X = n - k)$$

for all nonnegative integers k. Letting $k = n/2 + j$, the desired result follows. This explains why the $\text{Bin}(10, 1/2)$ PMF is symmetric about 5 in Figure 3.6. ■

Example 3.3.9 (Coin tosses continued). Going back to Example 3.1.2, we now know that $X \sim \text{Bin}(2, 1/2)$, $Y \sim \text{Bin}(2, 1/2)$, and $I \sim \text{Bern}(1/2)$. Consistent with Theorem 3.3.7, X and $Y = 2 - X$ have the same distribution, and consistent with Corollary 3.3.8, the distribution of X (and of Y) is symmetric about 1. □

3.4 Hypergeometric

If we have an urn filled with w white and b black balls, then drawing n balls out of the urn *with replacement* yields a $\text{Bin}(n, w/(w + b))$ distribution for the number of white balls obtained in n trials, since the draws are independent Bernoulli trials, each with probability $w/(w+b)$ of success. If we instead sample *without replacement*, as illustrated in Figure 3.7, then the number of white balls follows a *Hypergeometric distribution.*

Story 3.4.1 (Hypergeometric distribution). Consider an urn with w white balls and b black balls. We draw n balls out of the urn at random without replacement, such that all $\binom{w+b}{n}$ samples are equally likely. Let X be the number of white balls in the sample. Then X is said to have the *Hypergeometric distribution* with parameters w, b, and n; we denote this by $X \sim \text{HGeom}(w, b, n)$. □

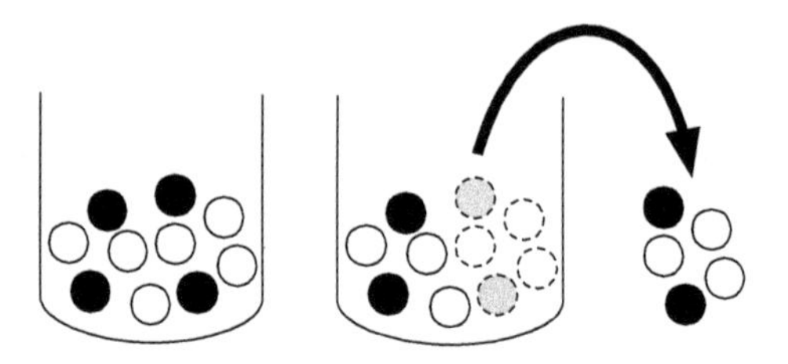

FIGURE 3.7
Hypergeometric story. An urn contains $w = 6$ white balls and $b = 4$ black balls. We sample $n = 5$ without replacement. The number X of white balls in the sample is Hypergeometric; here we observe $X = 3$.

As with the Binomial distribution, we can obtain the PMF of the Hypergeometric distribution from the story.

Theorem 3.4.2 (Hypergeometric PMF). If $X \sim \text{HGeom}(w, b, n)$, then the PMF of X is

$$P(X = k) = \frac{\binom{w}{k}\binom{b}{n-k}}{\binom{w+b}{n}},$$

for integers k satisfying $0 \leq k \leq w$ and $0 \leq n-k \leq b$, and $P(X = k) = 0$ otherwise.

Proof. To get $P(X = k)$, we first count the number of possible ways to draw exactly k white balls and $n - k$ black balls from the urn (without distinguishing between different orderings for getting the same set of balls). If $k > w$ or $n - k > b$, then the draw is impossible. Otherwise, there are $\binom{w}{k}\binom{b}{n-k}$ ways to draw k white and $n - k$ black balls by the multiplication rule, and there are $\binom{w+b}{n}$ total ways to draw n balls. Since all samples are equally likely, the naive definition of probability gives

$$P(X = k) = \frac{\binom{w}{k}\binom{b}{n-k}}{\binom{w+b}{n}}$$

for integers k satisfying $0 \leq k \leq w$ and $0 \leq n-k \leq b$. This PMF is valid because the numerator, summed over all k, equals $\binom{w+b}{n}$ by Vandermonde's identity (Example 1.5.3), so the PMF sums to 1. ■

The Hypergeometric distribution comes up in many scenarios which, on the surface, have little in common with white and black balls in an urn. The essential structure of the Hypergeometric story is that items in a population are classified using two sets of *tags*: in the urn story, each ball is either white or black (this is the first set of tags), and each ball is either sampled or not sampled (this is the second set of tags). Furthermore, at least one of these sets of tags is assigned completely at random (in the urn story, the balls are sampled randomly, with all sets of the correct size equally likely). Then $X \sim \text{HGeom}(w, b, n)$ represents the number of twice-tagged items: in the urn story, balls that are *both* white and sampled.

The next two examples show seemingly dissimilar scenarios that are nonetheless isomorphic to the urn story.

Example 3.4.3 (Elk capture-recapture). A forest has N elk. Today, m of the elk are captured, tagged, and released into the wild. At a later date, n elk are recaptured at random. Assume that the recaptured elk are equally likely to be any set of n of the elk, e.g., an elk that has been captured does not learn how to avoid being captured again.

By the story of the Hypergeometric, the number of tagged elk in the recaptured sample is $\text{HGeom}(m, N-m, n)$. The m tagged elk in this story correspond to the white balls and the $N-m$ untagged elk correspond to the black balls. Instead of sampling n balls from the urn, we recapture n elk from the forest. □

Example 3.4.4 (Aces in a poker hand). In a five-card hand drawn at random from a well-shuffled standard deck, the number of aces in the hand has the $\text{HGeom}(4, 48, 5)$ distribution, which can be seen by thinking of the aces as white balls and the non-aces as black balls. Using the Hypergeometric PMF, the probability that the hand has exactly three aces is

$$\frac{\binom{4}{3}\binom{48}{2}}{\binom{52}{5}} \approx 0.0017.$$ □

The following table summarizes how the above examples can be thought of in terms of two sets of tags. In each example, the r.v. of interest is the number of items falling into both the second and the fourth columns: white and sampled, tagged and recaptured, ace and in one's hand.

Story	First set of tags		Second set of tags	
urn	white	black	sampled	not sampled
elk	tagged	untagged	recaptured	not recaptured
cards	ace	not ace	in hand	not in hand

The next theorem describes a symmetry between two Hypergeometric distributions with different parameters; the proof follows from *swapping* the two sets of tags in the Hypergeometric story.

Theorem 3.4.5. The $\text{HGeom}(w, b, n)$ and $\text{HGeom}(n, w+b-n, w)$ distributions are identical. That is, if $X \sim \text{HGeom}(w, b, n)$ and $Y \sim \text{HGeom}(n, w+b-n, w)$, then X and Y have the same distribution.

Proof. Using the story of the Hypergeometric, imagine an urn with w white balls, b black balls, and a sample of size n made without replacement. Let $X \sim \text{HGeom}(w, b, n)$ be the number of white balls in the sample, thinking of white/black as the first set of tags and sampled/not sampled as the second set of tags. Let $Y \sim \text{HGeom}(n, w+b-n, w)$ be the number of sampled balls among the white balls, thinking of sampled/not sampled as the first set of tags and white/black as the second set of tags. Both X and Y count the number of white sampled balls, so they have the same distribution.

Alternatively, we can check algebraically that X and Y have the same PMF:

$$P(X=k) = \frac{\binom{w}{k}\binom{b}{n-k}}{\binom{w+b}{n}} = \frac{w!b!n!(w+b-n)!}{k!(w+b)!(w-k)!(n-k)!(b-n+k)!},$$

$$P(Y=k) = \frac{\binom{n}{k}\binom{w+b-n}{w-k}}{\binom{w+b}{w}} = \frac{w!b!n!(w+b-n)!}{k!(w+b)!(w-k)!(n-k)!(b-n+k)!}.$$

We prefer the story proof because it is less tedious and more memorable. ■

☣ **3.4.6** (Binomial vs. Hypergeometric). The Binomial and Hypergeometric distributions are often confused. Both are discrete distributions taking on integer values between 0 and n for some n, and both can be interpreted as the number of successes in n Bernoulli trials (for the Hypergeometric, each tagged elk in the recaptured sample can be considered a success and each untagged elk a failure). However, a crucial part of the Binomial story is that the Bernoulli trials involved are *independent*. The Bernoulli trials in the Hypergeometric story are *dependent*, since the sampling is done without replacement: knowing that one elk in our sample is tagged decreases the probability that the second elk will also be tagged.

3.5 Discrete Uniform

A very simple story, closely connected to the naive definition of probability, describes picking a random number from some finite set of possibilities.

Story 3.5.1 (Discrete Uniform distribution). Let C be a finite, nonempty set of numbers. Choose one of these numbers uniformly at random (i.e., all values in C are equally likely). Call the chosen number X. Then X is said to have the *Discrete Uniform distribution* with parameter C; we denote this by $X \sim \text{DUnif}(C)$. □

The PMF of $X \sim \text{DUnif}(C)$ is

$$P(X=x) = \frac{1}{|C|}$$

for $x \in C$ (and 0 otherwise), since a PMF must sum to 1. As with questions based on the naive definition of probability, questions based on a Discrete Uniform distribution reduce to counting problems. Specifically, for $X \sim \text{DUnif}(C)$ and any $A \subseteq C$, we have

$$P(X \in A) = \frac{|A|}{|C|}.$$

Example 3.5.2 (Random slips of paper). There are 100 slips of paper in a hat, each of which has one of the numbers $1, 2, \ldots, 100$ written on it, with no number appearing more than once. Five of the slips are drawn, one at a time.

First consider random sampling with *replacement (with equal probabilities).*

(a) What is the distribution of how many of the drawn slips have a value of at least 80 written on them?

(b) What is the distribution of the value of the jth draw (for $1 \leq j \leq 5$)?

(c) What is the probability that the number 100 is drawn at least once?

Now consider random sampling without *replacement (with all sets of five slips equally likely to be chosen).*

(d) What is the distribution of how many of the drawn slips have a value of at least 80 written on them?

(e) What is the distribution of the value of the jth draw (for $1 \leq j \leq 5$)?

(f) What is the probability that the number 100 is drawn in the sample?

Solution:

(a) By the story of the Binomial, the distribution is $\text{Bin}(5, 0.21)$.

(b) Let X_j be the value of the jth draw. By symmetry, $X_j \sim \text{DUnif}(1, 2, \ldots, 100)$. There aren't certain slips that love being chosen on the jth draw and others that avoid being chosen then; all are equally likely.

(c) Taking complements,

$$P(X_j = 100 \text{ for at least one } j) = 1 - P(X_1 \neq 100, \ldots, X_5 \neq 100).$$

By the naive definition of probability, this is

$$1 - (99/100)^5 \approx 0.049.$$

This solution just uses new notation for concepts from Chapter 1. It is useful to have this new notation since it is compact and flexible. In the above calculation, it is important to see why

$$P(X_1 \neq 100, \ldots, X_5 \neq 100) = P(X_1 \neq 100) \ldots P(X_5 \neq 100).$$

This follows from the naive definition in this case, but a more general way to think about such statements is through *independence* of r.v.s, a concept discussed in detail in Section 3.8.

(d) By the story of the Hypergeometric, the distribution is $\text{HGeom}(21, 79, 5)$.

(e) Let Y_j be the value of the jth draw. By symmetry, $Y_j \sim \text{DUnif}(1, 2, \ldots, 100)$.

Learning any Y_i gives information about the other values (so $Y_1, \ldots, Y_5$ are *not* independent, as defined in Section 3.8), but symmetry still holds since, unconditionally, the jth slip drawn is equally likely to be any of the slips. This is the *unconditional* distribution of Y_j: we are working from a vantage point before drawing any of the slips.

For further insight into why each of $Y_1, \ldots, Y_5$ is Discrete Uniform and how to think about Y_j unconditionally, imagine that instead of one person drawing five slips, one at a time, there are five people who draw one slip each, all reaching into the hat *simultaneously*, with all possibilities equally likely for who gets which slip. This formulation does not change the problem in any important way, and it helps avoid getting distracted by irrelevant chronological details. Label the five people $1, 2, \ldots, 5$ in some way, e.g., from youngest to oldest, and let Z_j be the value drawn by person j. By symmetry, $Z_j \sim \text{DUnif}(1, 2, \ldots, 100)$ for each j; the Z_j's are dependent but, looked at individually, each person is drawing a uniformly random slip.

(f) The events $Y_1 = 100, \ldots, Y_5 = 100$ are disjoint since we are now sampling without replacement, so

$$P(Y_j = 100 \text{ for some } j) = P(Y_1 = 100) + \cdots + P(Y_5 = 100) = 0.05.$$

Sanity check: This answer makes sense intuitively since we can just as well think of first choosing five random slips out of 100 blank slips and then randomly writing the numbers from 1 to 100 on the slips, which gives a $5/100$ chance that the number 100 is on one of the five chosen slips.

It would be bizarre if the answer to (c) were greater than or equal to the answer to (f), since sampling without replacement makes it easier to find the number 100. (For the same reason, when searching for a lost possession it makes more sense to sample locations without replacement than with replacement.) But it makes sense that the answer to (c) is only slightly less than the answer to (f), since it is unlikely in (c) that the same slip will be sampled more than once (though by the birthday problem it's less unlikely than many people would guess).

More generally, if k slips are drawn without replacement, where $0 \leq k \leq 100$, then the same reasoning gives that the probability of drawing the number 100 is $k/100$. Note that this makes sense in the extreme case $k = 100$, since in that case we draw *all* of the slips. □

3.6 Cumulative distribution functions

Another function that describes the distribution of an r.v. is the *cumulative distribution function* (CDF). Unlike the PMF, which only discrete r.v.s possess, the CDF is defined for *all* r.v.s.

Definition 3.6.1. The *cumulative distribution function* (CDF) of an r.v. X is the function F_X given by $F_X(x) = P(X \le x)$. When there is no risk of ambiguity, we sometimes drop the subscript and just write F (or some other letter) for a CDF.

The next example demonstrates that for discrete r.v.s, we can freely convert between CDF and PMF.

Example 3.6.2. Let $X \sim \text{Bin}(4, 1/2)$. Figure 3.8 shows the PMF and CDF of X.

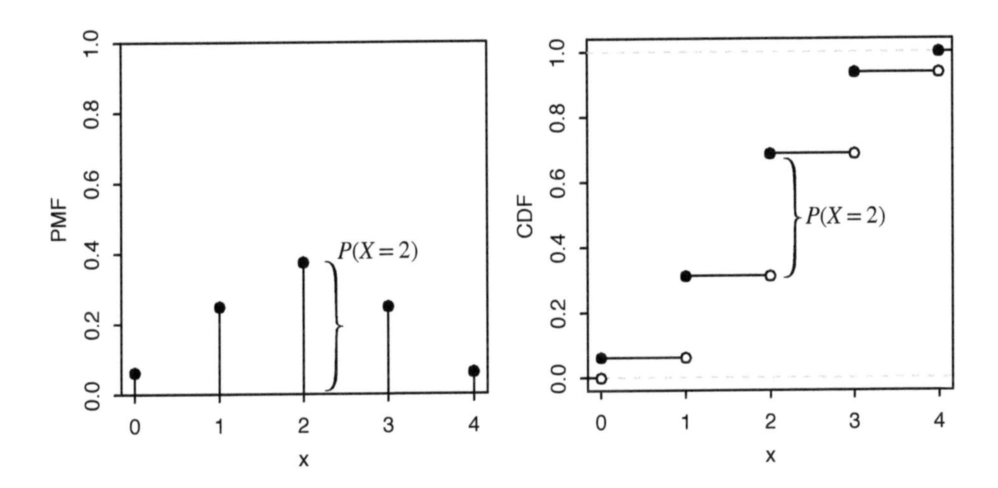

FIGURE 3.8
Bin(4, 1/2) PMF and CDF. The height of the vertical bar $P(X = 2)$ in the PMF is also the height of the jump in the CDF at 2.

- *From PMF to CDF*: To find $P(X \le 1.5)$, which is the CDF evaluated at 1.5, we sum the PMF over all values of the support that are less than or equal to 1.5:

$$P(X \le 1.5) = P(X = 0) + P(X = 1) = \left(\frac{1}{2}\right)^4 + 4\left(\frac{1}{2}\right)^4 = \frac{5}{16}.$$

  Similarly, the value of the CDF at an arbitrary point x is the sum of the heights of the vertical bars of the PMF at values less than or equal to x.

- *From CDF to PMF*: The CDF of a discrete r.v. consists of jumps and flat regions. The height of a jump in the CDF at x is equal to the value of the PMF at x. For example, in Figure 3.8, the height of the jump in the CDF at 2 is the same as the height of the corresponding vertical bar in the PMF; this is indicated in the figure with curly braces. The flat regions of the CDF correspond to values outside the support of X, so the PMF is equal to 0 in those regions. □

Valid CDFs satisfy the following criteria.

Theorem 3.6.3 (Valid CDFs). Any CDF F has the following properties.

- Increasing: If $x_1 \le x_2$, then $F(x_1) \le F(x_2)$.

- Right-continuous: As in Figure 3.8, the CDF is continuous except possibly for having some jumps. Wherever there is a jump, the CDF is continuous from the right. That is, for any a, we have

$$F(a) = \lim_{x \to a^+} F(x).$$

- Convergence to 0 and 1 in the limits:

$$\lim_{x \to -\infty} F(x) = 0 \text{ and } \lim_{x \to \infty} F(x) = 1.$$

Proof. The above criteria are true for *all* CDFs, but for simplicity we will only prove it for the case where F is the CDF of a discrete r.v. X whose possible values are $0, 1, 2, \dots$. As an example of how to visualize the criteria, consider Figure 3.8: the CDF shown there is increasing (with some flat regions), continuous from the right (it is continuous except at jumps, and each jump has an open dot at the bottom and a closed dot at the top), and it converges to 0 as $x \to -\infty$ and to 1 as $x \to \infty$ (in this example, it reaches 0 and 1; in some examples, one or both of these values may be approached but never reached).

The first criterion is true since the event $\{X \leq x_1\}$ is a subset of the event $\{X \leq x_2\}$, so $P(X \leq x_1) \leq P(X \leq x_2)$.

For the second criterion, note that

$$P(X \leq x) = P(X \leq \lfloor x \rfloor),$$

where $\lfloor x \rfloor$ is the greatest integer less than or equal to x. For example, $P(X \leq 4.9) = P(X \leq 4)$ since X is integer-valued. So $F(a + b) = F(a)$ for any $b > 0$ that is small enough so that $a + b < \lfloor a \rfloor + 1$, e.g., for $a = 4.9$, this holds for $0 < b < 0.1$. This implies $F(a) = \lim_{x \to a^+} F(x)$ (in fact, it's much stronger since it says $F(x)$ *equals* $F(a)$ when x is close enough to a and on the right).

For the third criterion, we have $F(x) = 0$ for $x < 0$, and

$$\lim_{x \to \infty} F(x) = \lim_{x \to \infty} P(X \leq \lfloor x \rfloor) = \lim_{x \to \infty} \sum_{n=0}^{\lfloor x \rfloor} P(X = n) = \sum_{n=0}^{\infty} P(X = n) = 1. \quad \blacksquare$$

The converse is true too: we will show in Chapter 5 that given any function F meeting these criteria, we can construct a random variable whose CDF is F.

To recap, we have now seen three equivalent ways of expressing the distribution of a random variable. Two of these are the PMF and the CDF: we know these two functions contain the same information, since we can always figure out the CDF from the PMF and vice versa. Generally the PMF is easier to work with for discrete r.v.s, since evaluating the CDF requires a summation.

A third way to describe a distribution is with a story that explains (in a precise way) how the distribution can arise. We used the stories of the Binomial and Hypergeometric distributions to derive the corresponding PMFs. Thus the story and the PMF also contain the same information, though we can often achieve more intuitive proofs with the story than with PMF calculations.

3.7 Functions of random variables

In this section we will discuss what it means to take a function of a random variable, and we will build understanding for why *a function of a random variable is a random variable*. That is, if X is a random variable, then X^2, e^X, and $\sin(X)$ are also random variables, as is $g(X)$ for any function $g : \mathbb{R} \to \mathbb{R}$.

For example, imagine that two basketball teams (A and B) are playing a seven-game match, and let X be the number of wins for team A (so $X \sim \text{Bin}(7, 1/2)$ if the teams are evenly matched and the games are independent). Let $g(x) = 7 - x$, and let $h(x) = 1$ if $x \geq 4$ and $h(x) = 0$ if $x < 4$. Then $g(X) = 7 - X$ is the number of wins for team B, and $h(X)$ is the indicator of team A winning the majority of the games. Since X is an r.v., both $g(X)$ and $h(X)$ are also r.v.s.

To see how to define functions of an r.v. formally, let's rewind a bit. At the beginning of this chapter, we considered a random variable X defined on a sample space with 6 elements. Figure 3.1 used arrows to illustrate how X maps each pebble in the sample space to a real number, and the left half of Figure 3.2 showed how we can equivalently imagine X writing a real number inside each pebble.

Now we can, if we want, apply the same function g to all the numbers inside the pebbles. Instead of the numbers $X(s_1)$ through $X(s_6)$, we now have the numbers $g(X(s_1))$ through $g(X(s_6))$, which gives a new mapping from sample outcomes to real numbers—we've created a new random variable, $g(X)$.

Definition 3.7.1 (Function of an r.v.). For an experiment with sample space S, an r.v. X, and a function $g : \mathbb{R} \to \mathbb{R}$, $g(X)$ is the r.v. that maps s to $g(X(s))$ for all $s \in S$.

Taking $g(x) = \sqrt{x}$ for concreteness, Figure 3.9 shows that $g(X)$ is the *composition* of the functions X and g, saying "first apply X, then apply g". Figure 3.10 represents $g(X)$ more succinctly by directly labeling the sample outcomes. Both figures show us that $g(X)$ is an r.v.; if X crystallizes to 4, then $g(X)$ crystallizes to 2.

Given a discrete r.v. X with a known PMF, how can we find the PMF of $Y = g(X)$? In the case where g is a one-to-one function, the answer is straightforward: the support of Y is the set of all $g(x)$ with x in the support of X, and

$$P(Y = g(x)) = P(g(X) = g(x)) = P(X = x).$$

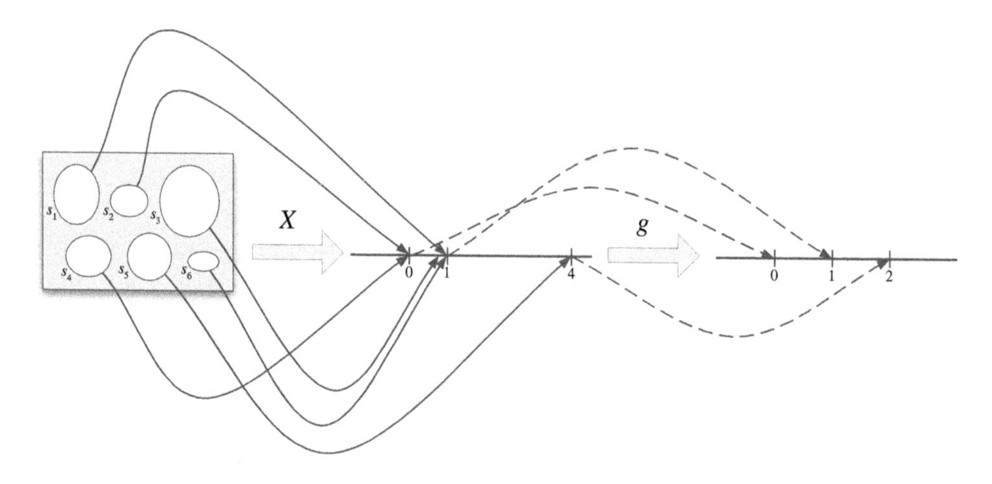

FIGURE 3.9
The r.v. X is defined on a sample space with 6 elements, and has possible values 0, 1, and 4. The function g is the square root function. Composing X and g gives the random variable $g(X) = \sqrt{X}$, which has possible values 0, 1, and 2.

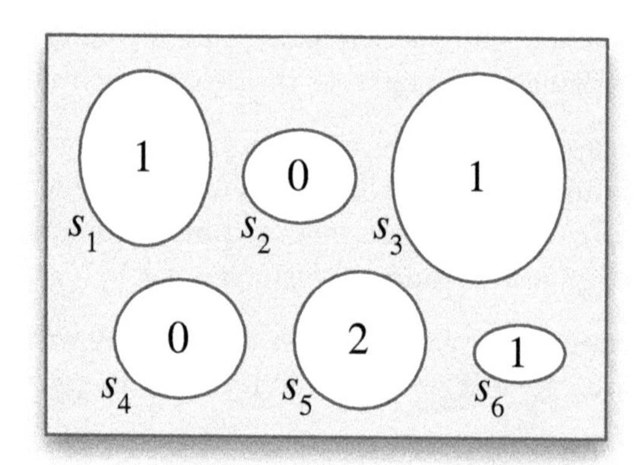

FIGURE 3.10
Since $g(X) = \sqrt{X}$ labels each pebble with a number, it is an r.v.

The case where $Y = g(X)$ with g one-to-one is illustrated in the following tables; the idea is that if the distinct possible values of X are $x_1, x_2, \ldots$ with probabilities $p_1, p_2, \ldots$ (respectively), then the distinct possible values of Y are $g(x_1), g(x_2), \ldots$, with the *same* list $p_1, p_2, \ldots$ of probabilities.

x	$P(X = x)$
x_1	p_1
x_2	p_2
x_3	p_3
$\vdots$	$\vdots$

PMF of X, in table form

y	$P(Y = y)$
$g(x_1)$	p_1
$g(x_2)$	p_2
$g(x_3)$	p_3
$\vdots$	$\vdots$

PMF of Y, in table form

This suggests a strategy for finding the PMF of an r.v. with an unfamiliar distribution: try to express the r.v. as a one-to-one function of an r.v. with a known distribution. The next example illustrates this method.

Example 3.7.2 (Random walk). A particle moves n steps on a number line. The particle starts at 0, and at each step it moves 1 unit to the right or to the left, with equal probabilities. Assume all steps are independent. Let Y be the particle's position after n steps. Find the PMF of Y.

Solution:

Consider each step to be a Bernoulli trial, where right is considered a success and left is considered a failure. Then the number of steps the particle takes to the right is a $\text{Bin}(n, 1/2)$ random variable, which we can name X. If $X = j$, then the particle has taken j steps to the right and $n - j$ steps to the left, giving a final position of $j - (n - j) = 2j - n$. So we can express Y as a one-to-one function of X, namely, $Y = 2X - n$. Since X takes values in $\{0, 1, 2, \ldots, n\}$, Y takes values in $\{-n, 2 - n, 4 - n, \ldots, n\}$.

The PMF of Y can then be found from the PMF of X:

$$P(Y = k) = P(2X - n = k) = P(X = (n + k)/2) = \binom{n}{\frac{n+k}{2}} \left(\frac{1}{2}\right)^n,$$

if k is an integer between $-n$ and n (inclusive) such that $n+k$ is an even number. □

If g is not one-to-one, then for a given y, there may be multiple values of x such that $g(x) = y$. To compute $P(g(X) = y)$, we need to sum up the probabilities of X taking on any of these candidate values of x.

Theorem 3.7.3 (PMF of $g(X)$). Let X be a discrete r.v. and $g : \mathbb{R} \to \mathbb{R}$. Then the support of $g(X)$ is the set of all y such that $g(x) = y$ for at least one x in the support of X, and the PMF of $g(X)$ is

$$P(g(X) = y) = \sum_{x: g(x) = y} P(X = x),$$

for all y in the support of $g(X)$.

Example 3.7.4. Continuing as in the previous example, let D be the particle's distance from the origin after n steps. Assume that n is even. Find the PMF of D.

Solution:

We can write $D = |Y|$; this is a function of Y, but it isn't one-to-one. The event $D = 0$ is the same as the event $Y = 0$. For $k = 2, 4, \ldots, n$, the event $D = k$ is the same as the event $\{Y = k\} \cup \{Y = -k\}$. So the PMF of D is

$$P(D = 0) = \binom{n}{\frac{n}{2}} \left(\frac{1}{2}\right)^n,$$

$$P(D = k) = P(Y = k) + P(Y = -k) = 2\binom{n}{\frac{n+k}{2}} \left(\frac{1}{2}\right)^n,$$

for $k = 2, 4, \ldots, n$. In the final step we used symmetry (imagine a new random walk that moves left each time our random walk moves right, and vice versa) to see that $P(Y = k) = P(Y = -k)$. □

The same reasoning we have used to handle functions of one random variable can be extended to deal with functions of multiple random variables. We have already seen an example of this with the addition function (which maps two numbers x, y to their sum $x + y$): in Example 3.2.5, we saw how to view $T = X + Y$ as an r.v. in its own right, where X and Y are obtained by rolling dice.

Definition 3.7.5 (Function of two r.v.s)**.** Given an experiment with sample space S, if X and Y are r.v.s that map $s \in S$ to $X(s)$ and $Y(s)$ respectively, then $g(X, Y)$ is the r.v. that maps s to $g(X(s), Y(s))$.

Note that we are assuming that X and Y are defined on the same sample space S. Usually we assume that S is chosen to be rich enough to encompass whatever r.v.s we wish to work with. For example, if X is based on a coin flip and Y is based on a die roll, and we initially were using the sample space $S_1 = \{H, T\}$ for X and the sample space $S_2 = \{1, 2, 3, 4, 5, 6\}$ for Y, we can easily redefine X and Y so that both are defined on the richer space $S = S_1 \times S_2 = \{(s_1, s_2) : s_1 \in S_1, s_2 \in S_2\}$.

One way to understand the mapping from S to $\mathbb{R}$ represented by the r.v. $g(X, Y)$ is with a table displaying the values of X, Y, and $g(X, Y)$ under various possible outcomes. Interpreting $X+Y$ as an r.v. is intuitive: if we observe $X = x$ and $Y = y$, then $X+Y$ crystallizes to $x+y$. For a less familiar example like $\max(X, Y)$, students often are unsure how to interpret it as an r.v. But the idea is the same: if we observe $X = x$ and $Y = y$, then $\max(X, Y)$ crystallizes to $\max(x, y)$.

Example 3.7.6 (Maximum of two die rolls)**.** We roll two fair 6-sided dice. Let X be the number on the first die and Y the number on the second die. The following table gives the values of X, Y, and $\max(X, Y)$ under 7 of the 36 outcomes in the sample space, analogously to the table in Example 3.2.5.

s	X	Y	$\max(X,Y)$
(1,2)	1	2	2
(1,6)	1	6	6
(2,5)	2	5	5
(3,1)	3	1	3
(4,3)	4	3	4
(5,4)	5	4	5
(6,6)	6	6	6

So $\max(X,Y)$ assigns a numerical value to each sample outcome. The PMF is

$$\begin{aligned}
P(\max(X,Y) = 1) &= 1/36,\\
P(\max(X,Y) = 2) &= 3/36,\\
P(\max(X,Y) = 3) &= 5/36,\\
P(\max(X,Y) = 4) &= 7/36,\\
P(\max(X,Y) = 5) &= 9/36,\\
P(\max(X,Y) = 6) &= 11/36.
\end{aligned}$$

These probabilities can be obtained by tabulating the values of $\max(x,y)$ in a 6×6 grid and counting how many times each value appears in the grid, or with calculations such as

$$\begin{aligned}
P(\max(X,Y) = 5) &= P(X = 5, Y \leq 4) + P(X \leq 4, Y = 5) + P(X = 5, Y = 5)\\
&= 2P(X = 5, Y \leq 4) + 1/36\\
&= 2(4/36) + 1/36 = 9/36.
\end{aligned}$$

□

☣ **3.7.7** (Category errors and sympathetic magic). Many common mistakes in probability can be traced to confusing two of the following fundamental objects with each other: distributions, random variables, events, and numbers. Such mistakes are examples of *category errors*. In general, a category error is a mistake that doesn't just happen to be wrong, but in fact is necessarily wrong since it is based on the wrong category of object. For example, answering the question "How many people live in Boston?" with "-42" or "π" or "pink elephants" would be a category error—we may not know the population size of a city, but we do know that it is a nonnegative integer at any point in time. To help avoid being categorically wrong, always think about what category an answer should have.

An especially common category error is to confuse a random variable with its distribution. We call this error *sympathetic magic*; this term comes from anthropology, where it is used for the belief that one can influence an object by manipulating a representation of that object. The following saying sheds light on the distinction between a random variable and its distribution:

> The word is not the thing; the map is not the territory. – Alfred Korzybski

We can think of the distribution of a random variable as a map or *blueprint* describing the r.v. Just as different houses can share the same blueprint, different r.v.s can have the same distribution, even if the *experiments* they summarize, and the *sample spaces* they map from, are not the same.

Here are two examples of sympathetic magic:

- Given an r.v. X, trying to get the PMF of $2X$ by multiplying the PMF of X by 2. It does not make sense to multiply a PMF by 2, since the probabilities would no longer sum to 1. As we saw above, if X takes on values x_j with probabilities p_j, then $2X$ takes on values $2x_j$ with probabilities p_j. Therefore the PMF of $2X$ is a horizontal stretch of the PMF of X; it is *not* a vertical stretch, as would result from multiplying the PMF by 2. Figure 3.11 shows the PMF of a discrete r.v. X with support $\{0, 1, 2, 3, 4\}$, along with the PMF of $2X$, which has support $\{0, 2, 4, 6, 8\}$. Note that X can take on odd values, but $2X$ is necessarily even.

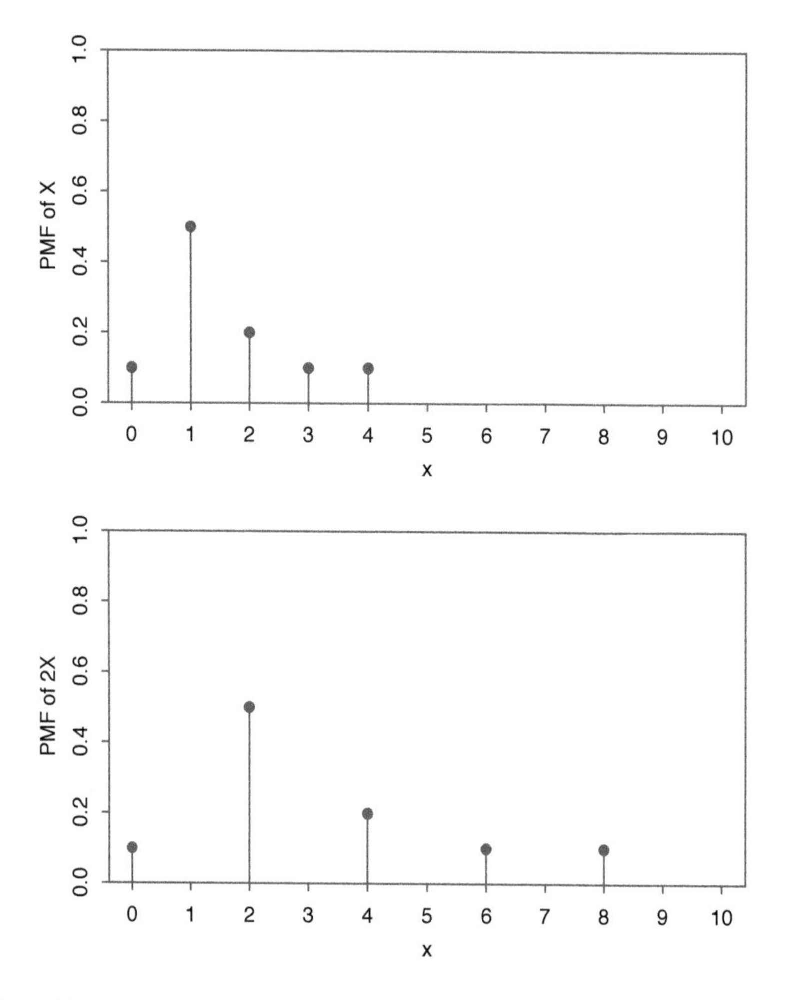

FIGURE 3.11
PMF of X (above) and PMF of $2X$ (below).

- Claiming that because X and Y have the same distribution, X must always equal Y, i.e., $P(X = Y) = 1$. Just because two r.v.s have the same distribution does not mean they are always equal, or *ever* equal. We saw this in Example 3.2.5. As another example, consider flipping a fair coin once. Let X be the indicator of Heads and $Y = 1 - X$ be the indicator of Tails. Both X and Y have the Bern(1/2) distribution, but the event $X = Y$ is impossible. The PMFs of X and Y are the same function, but X and Y are different mappings from the sample space to the real numbers.

  If Z is the indicator of Heads in a second flip (independent of the first flip), then Z is also Bern(1/2), but Z is not the same r.v. as X. Here

$$P(Z = X) = P(HH \text{ or } TT) = 1/2.$$

3.8 Independence of r.v.s

Just as we had the notion of independence of events, we can define independence of random variables. Intuitively, if two r.v.s X and Y are independent, then knowing the value of X gives no information about the value of Y, and vice versa. The definition formalizes this idea.

Definition 3.8.1 (Independence of two r.v.s). Random variables X and Y are said to be *independent* if

$$P(X \leq x, Y \leq y) = P(X \leq x)P(Y \leq y),$$

for all $x, y \in \mathbb{R}$.

In the discrete case, this is equivalent to the condition

$$P(X = x, Y = y) = P(X = x)P(Y = y),$$

for all x, y with x in the support of X and y in the support of Y.

The definition for more than two r.v.s is analogous.

Definition 3.8.2 (Independence of many r.v.s). Random variables $X_1, \dots, X_n$ are *independent* if

$$P(X_1 \leq x_1, \dots, X_n \leq x_n) = P(X_1 \leq x_1) \dots P(X_n \leq x_n),$$

for all $x_1, \dots, x_n \in \mathbb{R}$. For infinitely many r.v.s, we say that they are independent if every finite subset of the r.v.s is independent.

Comparing this to the criteria for independence of n events, it may seem strange that the independence of $X_1, \dots, X_n$ requires just one equality, whereas for events we needed to verify pairwise independence for all $\binom{n}{2}$ pairs, three-way independence for all $\binom{n}{3}$ triplets, and so on. However, upon closer examination of the definition, we see that independence of r.v.s requires the equality to hold for *all* possible $x_1, \dots, x_n$—infinitely many conditions! If we can find even a single list of values $x_1, \dots, x_n$ for which the above equality fails to hold, then $X_1, \dots, X_n$ are not independent.

☣ **3.8.3.** If $X_1, \dots, X_n$ are independent, then they are pairwise independent, i.e., X_i is independent of X_j for $i \neq j$. The idea behind proving that X_i and X_j are independent is to let all the x_k other than x_i, x_j go to ∞ in the definition of independence, since we already know $X_k < \infty$ is true (though it takes some work to give a complete justification for the limit). But pairwise independence does *not* imply independence in general, as we saw in Chapter 2 for events.

Example 3.8.4. In a roll of two fair dice, if X is the number on the first die and Y is the number on the second die, then $X + Y$ is not independent of $X - Y$ since

$$0 = P(X + Y = 12, X - Y = 1) \neq P(X + Y = 12)P(X - Y = 1) = \frac{1}{36} \cdot \frac{5}{36}.$$

Knowing the total is 12 tells us the difference must be 0, so the r.v.s provide information about each other. □

If X and Y are independent then it is also true, e.g., that X^2 is independent of Y^4, since if X^2 provided information about Y^4, then X would give information about Y (using X^2 and Y^4 as intermediaries: X determines X^2, which would give information about Y^4, which in turn would give information about Y). More generally, we have the following result (for which we omit a formal proof).

Theorem 3.8.5 (Functions of independent r.v.s). If X and Y are independent r.v.s, then any function of X is independent of any function of Y.

Definition 3.8.6 (i.i.d.). We will often work with random variables that are independent and have the same distribution. We call such r.v.s *independent and identically distributed*, or *i.i.d.* for short.

☣ **3.8.7** (i. vs. i.d.). "Independent" and "identically distributed" are two often-confused but completely different concepts. Random variables are independent if they provide no information about each other; they are identically distributed if they have the same PMF (or equivalently, the same CDF). Whether two r.v.s are independent has nothing to do with whether they have the same distribution. We can have r.v.s that are:

- independent and identically distributed. Let X be the result of a die roll, and let Y be the result of a second, independent die roll. Then X and Y are i.i.d.
- independent and not identically distributed. Let X be the result of a die roll, and let Y be the closing price of the Dow Jones (a stock market index) a month from now. Then X and Y provide no information about each other (one would fervently hope), and X and Y do not have the same distribution.

- dependent and identically distributed. Let X be the number of Heads in n independent fair coin tosses, and let Y be the number of Tails in those same n tosses. Then X and Y are both distributed $\text{Bin}(n, 1/2)$, but they are highly dependent: if we know X, then we know Y perfectly.

- dependent and not identically distributed. Let X be the indicator of whether the majority party retains control of the House of Representatives in the U.S. after the next election, and let Y be the average favorability rating of the majority party in polls taken within a month of the election. Then X and Y are dependent, and X and Y do not have the same distribution.

By taking a sum of i.i.d. Bernoulli r.v.s, we can write down the story of the Binomial distribution in an algebraic form.

Theorem 3.8.8. If $X \sim \text{Bin}(n, p)$, viewed as the number of successes in n independent Bernoulli trials with success probability p, then we can write $X = X_1 + \cdots + X_n$ where the X_i are i.i.d. $\text{Bern}(p)$.

Proof. Let $X_i = 1$ if the ith trial was a success, and 0 if the ith trial was a failure. It's as though we have a person assigned to each trial, and we ask each person to raise their hand if their trial was a success. If we count the number of raised hands (which is the same as adding up the X_i), we get the total number of successes. ■

An important fact about the Binomial distribution is that the sum of independent Binomial r.v.s with the same success probability is also Binomial.

Theorem 3.8.9. If $X \sim \text{Bin}(n, p)$, $Y \sim \text{Bin}(m, p)$, and X is independent of Y, then $X + Y \sim \text{Bin}(n + m, p)$.

Proof. We present three proofs, since each illustrates a useful technique.

1. LOTP: We can directly find the PMF of $X + Y$ by conditioning on X (or Y, whichever we prefer) and using the law of total probability:

$$\begin{aligned}
P(X+Y=k) &= \sum_{j=0}^{k} P(X+Y=k|X=j)P(X=j) \\
&= \sum_{j=0}^{k} P(Y=k-j)P(X=j) \\
&= \sum_{j=0}^{k} \binom{m}{k-j} p^{k-j} q^{m-k+j} \binom{n}{j} p^j q^{n-j} \\
&= p^k q^{n+m-k} \sum_{j=0}^{k} \binom{m}{k-j}\binom{n}{j} \\
&= \binom{n+m}{k} p^k q^{n+m-k}.
\end{aligned}$$

In the second line, we used the independence of X and Y to justify dropping the conditioning in

$$P(X + Y = k|X = j) = P(Y = k - j|X = j) = P(Y = k - j),$$

and in the last line, we used the fact that

$$\sum_{j=0}^{k} \binom{m}{k-j}\binom{n}{j} = \binom{n+m}{k}$$

by Vandermonde's identity. The resulting expression is the $\text{Bin}(n + m, p)$ PMF, so $X + Y \sim \text{Bin}(n + m, p)$.

2. Representation: A much simpler proof is to represent both X and Y as the sum of i.i.d. $\text{Bern}(p)$ r.v.s: $X = X_1 + \cdots + X_n$ and $Y = Y_1 + \cdots + Y_m$, where the X_i and Y_j are all i.i.d. $\text{Bern}(p)$. Then $X + Y$ is the sum of $n + m$ i.i.d. $\text{Bern}(p)$ r.v.s, so its distribution, by the previous theorem, is $\text{Bin}(n + m, p)$.

3. Story: By the Binomial story, X is the number of successes in n independent trials and Y is the number of successes in m additional independent trials, all with the same success probability, so $X+Y$ is the total number of successes in the $n+m$ trials, which is the story of the $\text{Bin}(n + m, p)$ distribution. ■

Of course, if we have a definition for independence of r.v.s, we should have an analogous definition for conditional independence of r.v.s.

Definition 3.8.10 (Conditional independence of r.v.s). Random variables X and Y are *conditionally independent* given an r.v. Z if for all $x, y \in \mathbb{R}$ and all z in the support of Z,

$$P(X \leq x, Y \leq y|Z = z) = P(X \leq x|Z = z)P(Y \leq y|Z = z).$$

For discrete r.v.s, an equivalent definition is to require

$$P(X = x, Y = y|Z = z) = P(X = x|Z = z)P(Y = y|Z = z).$$

As we might expect from the name, this is the definition of independence, except that we condition on $Z = z$ everywhere, and require the equality to hold for all z in the support of Z.

Definition 3.8.11 (Conditional PMF). For any discrete r.v.s X and Z, the function $P(X = x|Z = z)$, when considered as a function of x for fixed z, is called the *conditional PMF of X given $Z = z$*.

Independence of r.v.s does not imply conditional independence, nor vice versa. First let us show why independence does not imply conditional independence.

Example 3.8.12 (Matching pennies). Consider the simple game called *matching pennies*. Each of two players, A and B, has a fair penny. They flip their pennies independently. If the pennies match, A wins; otherwise, B wins. Let X be 1 if A's penny lands Heads and -1 otherwise, and define Y similarly for B (the r.v.s X and Y are called *random signs*).

Let $Z = XY$, which is 1 if A wins and -1 if B wins. Then X and Y are unconditionally independent, but given $Z = 1$, we know that $X = Y$ (the pennies match). So X and Y are conditionally dependent given Z. □

Example 3.8.13 (Two friends). Consider again the "I have only two friends who ever call me" scenario from Example 2.5.11, except now with r.v. notation. Let X be the indicator of Alice calling me next Friday, Y be the indicator of Bob calling me next Friday, and Z be the indicator of exactly one of them calling me next Friday. Then X and Y are independent (by assumption). But given $Z = 1$, we have that X and Y are completely dependent: given that $Z = 1$, we have $Y = 1 - X$. □

Next let's see why conditional independence does not imply independence.

Example 3.8.14 (Mystery opponent). Suppose that you are going to play two games of tennis against one of two identical twins. Against one of the twins, you are evenly matched, and against the other you have a 3/4 chance of winning. Suppose that you can't tell which twin you are playing against until after the two games. Let Z be the indicator of playing against the twin with whom you're evenly matched, and let X and Y be the indicators of victory in the first and second games, respectively.

Conditional on $Z = 1$, X and Y are i.i.d. Bern(1/2), and conditional on $Z = 0$, X and Y are i.i.d. Bern(3/4). So X and Y are conditionally independent given Z. Unconditionally, X and Y are dependent because observing $X = 1$ makes it more likely that we are playing the twin who is worse. That is,

$$P(Y = 1|X = 1) > P(Y = 1).$$

Past games give us information which helps us infer who our opponent is, which in turn helps us predict future games! Note that this example is isomorphic to the "random coin" scenario from Example 2.3.7. □

3.9 Connections between Binomial and Hypergeometric

The Binomial and Hypergeometric distributions are connected in two important ways. As we will see in this section, we can get from the Binomial to the Hypergeometric by *conditioning*, and we can get from the Hypergeometric to the Binomial by *taking a limit*. We'll start with a motivating example.

Example 3.9.1 (Fisher exact test). A scientist wishes to study whether women or men are more likely to have a certain disease, or whether they are equally likely. A random sample of n women and m men is gathered, and each person is tested for the disease (assume for this problem that the test is completely accurate). The numbers of women and men in the sample who have the disease are X and Y respectively, with $X \sim \text{Bin}(n, p_1)$ and $Y \sim \text{Bin}(m, p_2)$, independently. Here p_1 and p_2 are unknown, and we are interested in testing whether $p_1 = p_2$ (this is known as a *null hypothesis* in statistics).

Consider a 2×2 table with rows corresponding to disease status and columns corresponding to gender. Each entry is the count of how many people have that disease status and gender, so $n + m$ is the sum of all 4 entries. Suppose that it is observed that $X + Y = r$.

The *Fisher exact test* is based on conditioning on both the row and column sums, so n, m, r are all treated as fixed, and then seeing if the observed value of X is "extreme" compared to this conditional distribution. Assuming the null hypothesis, find the conditional PMF of X given $X + Y = r$.

Solution:

First we'll build the 2×2 table, treating n, m, and r as fixed.

	Women	**Men**	**Total**
Disease	x	$r - x$	r
No disease	$n - x$	$m - r + x$	$n + m - r$
Total	n	m	$n + m$

Next, let's compute the conditional PMF $P(X = x|X + Y = r)$. By Bayes' rule,

$$\begin{aligned} P(X = x|X + Y = r) &= \frac{P(X + Y = r|X = x)P(X = x)}{P(X + Y = r)} \\ &= \frac{P(Y = r - x)P(X = x)}{P(X + Y = r)}. \end{aligned}$$

The step $P(X+Y = r|X = x) = P(Y = r-x)$ is justified by the independence of X and Y. Assuming the null hypothesis and letting $p = p_1 = p_2$, we have $X \sim \text{Bin}(n, p)$ and $Y \sim \text{Bin}(m, p)$, independently, so $X + Y \sim \text{Bin}(n + m, p)$. Thus,

$$\begin{aligned} P(X = x|X + Y = r) &= \frac{\binom{m}{r-x} p^{r-x}(1-p)^{m-r+x} \binom{n}{x} p^x (1-p)^{n-x}}{\binom{n+m}{r} p^r (1-p)^{n+m-r}} \\ &= \frac{\binom{n}{x}\binom{m}{r-x}}{\binom{n+m}{r}}. \end{aligned}$$

So the conditional distribution of X is Hypergeometric with parameters n, m, r.

To understand why the Hypergeometric appeared, seemingly out of nowhere, let's connect this problem to the elk story for the Hypergeometric. In the elk story, we are interested in the distribution of the number of tagged elk in the recaptured sample. By analogy, think of women as tagged elk and men as untagged elk. Instead of recapturing r elk at random from the forest, we infect $X + Y = r$ people with the disease; under the null hypothesis, the set of diseased people is equally likely to be any set of r people. Thus, conditional on $X + Y = r$, X represents the number of women among the r diseased individuals. This is exactly analogous to the number of tagged elk in the recaptured sample, which is distributed $\text{HGeom}(n, m, r)$.

An interesting fact, which turns out to be useful in statistics, is that the conditional distribution of X does not depend on p: unconditionally, $X \sim \text{Bin}(n, p)$, but p disappears from the parameters of the conditional distribution! This makes sense upon reflection, since once we know $X + Y = r$, we can work directly with the fact that we have a population with r diseased and $n + m - r$ healthy people, without worrying about the value of p that originally generated the population. □

This motivating example serves as a proof of the following theorem.

Theorem 3.9.2. If $X \sim \text{Bin}(n, p)$, $Y \sim \text{Bin}(m, p)$, and X is independent of Y, then the conditional distribution of X given $X + Y = r$ is $\text{HGeom}(n, m, r)$.

In the other direction, the Binomial is a limiting case of the Hypergeometric.

Theorem 3.9.3. If $X \sim \text{HGeom}(w, b, n)$ and $N = w + b \to \infty$ such that $p = w/(w + b)$ remains fixed, then the PMF of X converges to the $\text{Bin}(n, p)$ PMF.

Proof. We take the stated limit of the $\text{HGeom}(w, b, n)$ PMF:

$$\begin{aligned}
P(X = k) &= \frac{\binom{w}{k}\binom{b}{n-k}}{\binom{w+b}{n}} \\
&= \binom{n}{k}\frac{\binom{w+b-n}{w-k}}{\binom{w+b}{w}} \quad \text{by Theorem 3.4.5} \\
&= \binom{n}{k}\frac{w!}{(w-k)!}\frac{b!}{(b-n+k)!}\frac{(w+b-n)!}{(w+b)!} \\
&= \binom{n}{k}\frac{w(w-1)\dots(w-k+1)b(b-1)\dots(b-n+k+1)}{(w+b)(w+b-1)\dots(w+b-n+1)} \\
&= \binom{n}{k}\frac{p\left(p-\frac{1}{N}\right)\dots\left(p-\frac{k-1}{N}\right)q\left(q-\frac{1}{N}\right)\dots\left(q-\frac{n-k-1}{N}\right)}{\left(1-\frac{1}{N}\right)\left(1-\frac{2}{N}\right)\dots\left(1-\frac{n-1}{N}\right)}.
\end{aligned}$$

As $N \to \infty$, the denominator goes to 1, and the numerator goes to $p^k q^{n-k}$. Thus

$$P(X = k) \to \binom{n}{k}p^k q^{n-k},$$

which is the $\text{Bin}(n, p)$ PMF. ■

The stories of the Binomial and Hypergeometric provide intuition for this result: given an urn with w white balls and b black balls, the Binomial distribution arises from sampling n balls from the urn with replacement, while the Hypergeometric arises from sampling without replacement. As the number of balls in the urn grows very large relative to the number of balls that are drawn, sampling with replacement and sampling without replacement become essentially equivalent. In practical terms, this theorem tells us that if $N = w + b$ is large relative to n, we can approximate the HGeom(w, b, n) PMF by the Bin$(n, w/(w + b))$ PMF.

The birthday problem implies that it is surprisingly likely that some ball will be sampled more than once if sampling with replacement; for example, if 1,200 out of 1,000,000 balls are drawn randomly with replacement, then there is about a 51% chance that some ball will be drawn more than once! But this becomes less and less likely as N grows, and even if it is likely that there will be a few coincidences, the approximation can still be reasonable if it is very likely that the vast majority of balls in the sample are sampled only once each.

3.10 Recap

A random variable (r.v.) is a function assigning a real number to every possible outcome of an experiment. The distribution of an r.v. X is a full specification of the probabilities for the events associated with X, such as $\{X = 3\}$ and $\{1 \leq X \leq 5\}$. The distribution of a discrete r.v. can be defined using a PMF, a CDF, or a story. The PMF of X is the function $P(X = x)$ for $x \in \mathbb{R}$. The CDF of X is the function $P(X \leq x)$ for $x \in \mathbb{R}$. A story for X describes an experiment that could give rise to a random variable with the same distribution as X.

For a PMF to be valid, it must be nonnegative and sum to 1. For a CDF to be valid, it must be increasing, right-continuous, converge to 0 as $x \to -\infty$, and converge to 1 as $x \to \infty$.

It is important to distinguish between a random variable and its distribution: the distribution is a blueprint for building the r.v., but different r.v.s can have the same distribution, just as different houses can be built from the same blueprint.

Four named discrete distributions are the Bernoulli, Binomial, Hypergeometric, and Discrete Uniform. Each of these is actually a *family* of distributions, indexed by parameters; to fully specify one of these distributions, we need to give both the name and the parameter values.

- A Bern(p) r.v. is the indicator of success in a Bernoulli trial with probability of success p.

- A Bin(n, p) r.v. is the number of successes in n independent Bernoulli trials, all with the same probability p of success.

- A HGeom(w, b, n) r.v. is the number of white balls obtained in a sample of size n drawn without replacement from an urn of w white and b black balls.

- A DUnif(C) r.v. is obtained by randomly choosing an element of the finite set C, with equal probabilities for each element.

A function of a random variable is still a random variable. If we know the PMF of X, we can find $P(g(X) = k)$, the PMF of $g(X)$, by translating the event $\{g(X) = k\}$ into an equivalent event involving X, then using the PMF of X.

Two random variables are independent if knowing the value of one r.v. gives no information about the value of the other. This is unrelated to whether the two r.v.s are identically distributed. In Chapter 7, we will learn how to deal with dependent random variables by considering them jointly rather than separately.

We have now seen four fundamental types of objects in probability: distributions, random variables, events, and numbers. Figure 3.12 shows connections between these four fundamental objects. A CDF can be used as a blueprint for generating an r.v., and then there are various events describing the behavior of the r.v., such as the events $X \leq x$ for all x. Knowing the probabilities of these events determines the CDF, taking us full circle. For a discrete r.v. we can also use the PMF as a blueprint, and go from distribution to r.v. to events and back again.

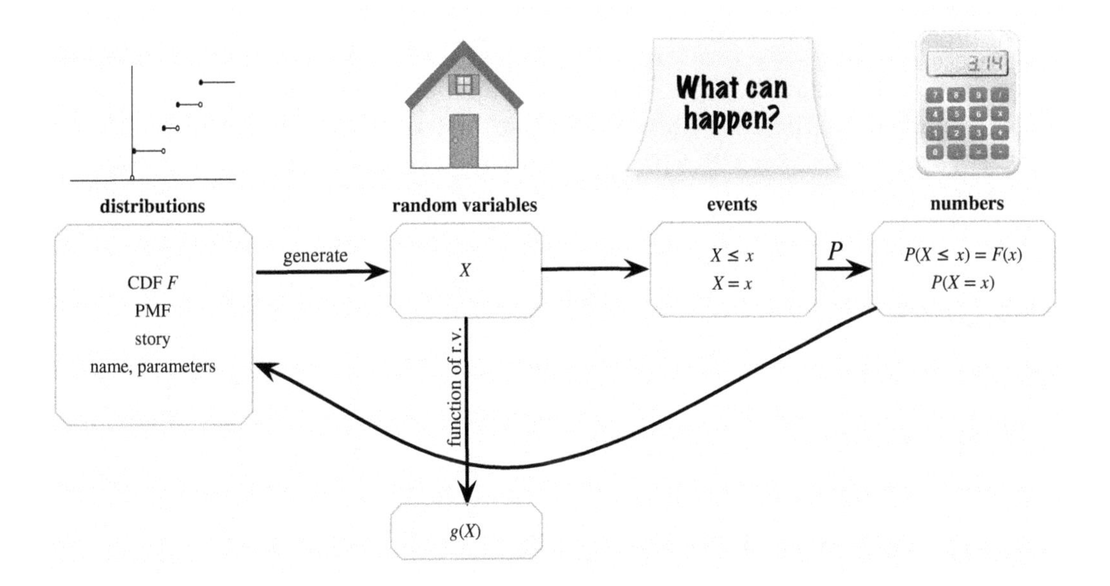

FIGURE 3.12

Four fundamental objects in probability: distributions (blueprints), random variables, events, and numbers. From a CDF F we can generate an r.v. X. From X, we can generate many other r.v.s by taking functions of X. There are various events describing the behavior of X. Most notably, for any constant x the events $X \leq x$ and $X = x$ are of interest. Knowing the probabilities of these events for all x gives us the CDF and (in the discrete case) the PMF, taking us full circle.

3.11 R

Distributions in R

All of the named distributions that we'll encounter in this book have been implemented in R. In this section we'll explain how to work with the Binomial and Hypergeometric distributions in R. We will also explain in general how to generate r.v.s from any discrete distribution with a finite support. Typing `help(distributions)` gives a handy list of built-in distributions; many others are available through R packages that can be loaded.

In general, for many named discrete distributions, three functions starting with `d`, `p`, and `r` will give the PMF, CDF, and random generation, respectively. Note that the function starting with `p` is not the PMF, but rather is the CDF.

Binomial distribution

The Binomial distribution is associated with the following three R functions: `dbinom`, `pbinom`, and `rbinom`. For the Bernoulli distribution we can just use the Binomial functions with $n = 1$.

- `dbinom` is the Binomial PMF. It takes three inputs: the first is the value of x at which to evaluate the PMF, and the second and third are the parameters n and p. For example, `dbinom(3,5,0.2)` returns the probability $P(X = 3)$ where $X \sim \text{Bin}(5, 0.2)$. In other words,

$$\texttt{dbinom(3,5,0.2)} = \binom{5}{3}(0.2)^3(0.8)^2 = 0.0512.$$

- `pbinom` is the Binomial CDF. It takes three inputs: the first is the value of x at which to evaluate the CDF, and the second and third are the parameters. `pbinom(3,5,0.2)` is the probability $P(X \leq 3)$ where $X \sim \text{Bin}(5, 0.2)$. So

$$\texttt{pbinom(3,5,0.2)} = \sum_{k=0}^{3} \binom{5}{k}(0.2)^k(0.8)^{5-k} = 0.9933.$$

- `rbinom` is a function for generating Binomial random variables. For `rbinom`, the first input is *how many* r.v.s we want to generate, and the second and third inputs are still the parameters. Thus the command `rbinom(7,5,0.2)` produces realizations of seven i.i.d. $\text{Bin}(5, 0.2)$ r.v.s. When we ran this command, we got

```
2 1 0 0 1 0 0
```

but you'll probably get something different when you try it!

We can also evaluate PMFs and CDFs at an entire vector of values. For example, recall that `0:n` is a quick way to list the integers from 0 to n. The command `dbinom(0:5,5,0.2)` returns 6 numbers, $P(X=0), P(X=1), \ldots, P(X=5)$, where $X \sim \text{Bin}(5, 0.2)$.

Hypergeometric distribution

The Hypergeometric distribution also has three functions: `dhyper`, `phyper`, and `rhyper`. As one might expect, `dhyper` is the Hypergeometric PMF, `phyper` is the Hypergeometric CDF, and `rhyper` generates Hypergeometric r.v.s. Since the Hypergeometric distribution has three parameters, each of these functions takes *four* inputs. For `dhyper` and `phyper`, the first input is the value at which we wish to evaluate the PMF or CDF, and the remaining inputs are the parameters of the distribution.

Thus `dhyper(k,w,b,n)` returns $P(X = k)$ where $X \sim \text{HGeom}(w, b, n)$, and `phyper(k,w,b,n)` returns $P(X \leq k)$. For `rhyper`, the first input is the number of r.v.s we want to generate, and the remaining inputs are the parameters; `rhyper(100,w,b,n)` generates 100 i.i.d. $\text{HGeom}(w, b, n)$ r.v.s.

Discrete distributions with finite support

We can generate r.v.s from *any* discrete distribution with finite support using the `sample` command. When we first introduced the `sample` command, we said that it can be used in the form `sample(n,k)` or `sample(n,k,replace=TRUE)` to sample k times from the integers 1 through n, either without or with replacement. For example, to generate 5 independent $\text{DUnif}(1, 2, \ldots, 100)$ r.v.s, we can use the command `sample(100,5,replace=TRUE)`.

It turns out that `sample` is far more versatile. If we want to sample from the values $x_1, \ldots, x_n$ with probabilities $p_1, \ldots, p_n$, we simply create a vector `x` containing all the x_i and a vector `p` containing all the p_i, then feed them into `sample`. Suppose we want realizations of i.i.d. r.v.s $X_1, \ldots, X_{100}$ whose PMF is

$$
\begin{aligned}
P(X_j = 0) &= 0.25, \\
P(X_j = 1) &= 0.5, \\
P(X_j = 5) &= 0.1, \\
P(X_j = 10) &= 0.15,
\end{aligned}
$$

and $P(X_j = x) = 0$ for all other values of x. First, we use the `c` function to create vectors with the support of the distribution and the corresponding probabilities.

```
x <- c(0,1,5,10)
p <- c(0.25,0.5,0.1,0.15)
```

Next, we use `sample`. Here's how to get 100 draws from the PMF above:

```
sample(x,100,prob=p,replace=TRUE)
```

The inputs are the vector `x` to sample from, the sample size (100 in this case), the probabilities `p` to use when sampling from `x` (if this is omitted, the probabilities are assumed equal), and whether to sample with replacement.

3.12 Exercises

Exercises marked with Ⓢ have detailed solutions at `http://stat110.net`.

PMFs and CDFs

1. People are arriving at a party one at a time. While waiting for more people to arrive they entertain themselves by comparing their birthdays. Let X be the number of people needed to obtain a birthday match, i.e., before person X arrives no two people have the same birthday, but when person X arrives there is a match. Find the PMF of X.

2. (a) Independent Bernoulli trials are performed, with probability 1/2 of success, until there has been at least one success. Find the PMF of the number of trials performed.

   (b) Independent Bernoulli trials are performed, with probability 1/2 of success, until there has been at least one success and at least one failure. Find the PMF of the number of trials performed.

3. Let X be an r.v. with CDF F, and $Y = \mu + \sigma X$, where μ and σ are real numbers with $\sigma > 0$. (Then Y is called a *location-scale transformation* of X; we will encounter this concept many times in Chapter 5 and beyond.) Find the CDF of Y, in terms of F.

4. Let n be a positive integer and
$$F(x) = \frac{\lfloor x \rfloor}{n}$$
for $0 \leq x \leq n$, $F(x) = 0$ for $x < 0$, and $F(x) = 1$ for $x > n$, where $\lfloor x \rfloor$ is the greatest integer less than or equal to x. Show that F is a CDF, and find the PMF that it corresponds to.

5. (a) Show that $p(n) = \left(\frac{1}{2}\right)^{n+1}$ for $n = 0, 1, 2, \dots$ is a valid PMF for a discrete r.v.

   (b) Find the CDF of a random variable with the PMF from (a).

6. Ⓢ *Benford's law* states that in a very large variety of real-life data sets, the first digit approximately follows a particular distribution with about a 30% chance of a 1, an 18% chance of a 2, and in general
$$P(D = j) = \log_{10}\left(\frac{j+1}{j}\right), \text{ for } j \in \{1, 2, 3, \dots, 9\},$$
where D is the first digit of a randomly chosen element. Check that this is a valid PMF (using properties of logs, not with a calculator).

7. Bob is playing a video game that has 7 levels. He starts at level 1, and has probability p_1 of reaching level 2. In general, given that he reaches level j, he has probability p_j of reaching level $j+1$, for $1 \leq j \leq 6$. Let X be the highest level that he reaches. Find the PMF of X (in terms of $p_1, \dots, p_6$).

8. There are 100 prizes, with one worth \$1, one worth \$2, ..., and one worth \$100. There are 100 boxes, each of which contains one of the prizes. You get 5 prizes by picking random boxes one at a time, *without replacement.* Find the PMF of how much your most valuable prize is worth (as a simple expression in terms of binomial coefficients).

9. Let F_1 and F_2 be CDFs, $0 < p < 1$, and $F(x) = pF_1(x) + (1-p)F_2(x)$ for all x.

   (a) Show directly that F has the properties of a valid CDF (see Theorem 3.6.3). The distribution defined by F is called a *mixture* of the distributions defined by F_1 and F_2.

   (b) Consider creating an r.v. in the following way. Flip a coin with probability p of Heads. If the coin lands Heads, generate an r.v. according to F_1; if the coin lands Tails, generate an r.v. according to F_2. Show that the r.v. obtained in this way has CDF F.

10. (a) Is there a discrete distribution with support $1, 2, 3, \dots,$ such that the value of the PMF at n is proportional to $1/n$?

    Hint: See the math appendix for a review of some facts about series.

    (b) Is there a discrete distribution with support $1, 2, 3, \dots,$ such that the value of the PMF at n is proportional to $1/n^2$?

11. Ⓢ Let X be an r.v. whose possible values are $0, 1, 2, \dots,$ with CDF F. In some countries, rather than using a CDF, the convention is to use the function G defined by $G(x) = P(X < x)$ to specify a distribution. Find a way to convert from F to G, i.e., if F is a known function, show how to obtain $G(x)$ for all real x.

12. (a) Give an example of r.v.s X and Y such that $F_X(x) \leq F_Y(x)$ for all x, where the inequality is strict for some x. Here F_X is the CDF of X and F_Y is the CDF of Y. For the example you gave, sketch the CDFs of both X and Y on the same axes. Then sketch their PMFs on a second set of axes.

    (b) In Part (a), you found an example of two different CDFs where the first is less than or equal to the second everywhere. Is it possible to find two different PMFs where the first is less than or equal to the second everywhere? In other words, find discrete r.v.s X and Y such that $P(X = x) \leq P(Y = x)$ for all x, where the inequality is strict for some x, or show that it is impossible to find such r.v.s.

13. Let X, Y, Z be discrete r.v.s such that X and Y have the same conditional distribution given Z, i.e., for all a and z we have

    $$P(X = a|Z = z) = P(Y = a|Z = z).$$

    Show that X and Y have the same distribution (unconditionally, not just when given Z).

14. Let X be the number of purchases that Fred will make on the online site for a certain company (in some specified time period). Suppose that the PMF of X is $P(X = k) = e^{-\lambda}\lambda^k/k!$ for $k = 0, 1, 2, \dots$. This distribution is called the *Poisson distribution* with parameter λ, and it will be studied extensively in later chapters.

    (a) Find $P(X \geq 1)$ and $P(X \geq 2)$ without summing infinite series.

    (b) Suppose that the company only knows about people who have made at least one purchase on their site (a user sets up an account to make a purchase, but someone who has never made a purchase there doesn't appear in the customer database). If the company computes the number of purchases for everyone in their database, then these data are draws from the *conditional* distribution of the number of purchases, given that at least one purchase is made. Find the conditional PMF of X given $X \geq 1$. (This conditional distribution is called a *truncated Poisson distribution.*)

Named distributions

15. Find the CDF of an r.v. $X \sim \text{DUnif}(1, 2, \dots, n)$.

16. Let $X \sim \text{DUnif}(C)$, and B be a nonempty subset of C. Find the conditional distribution of X, given that X is in B.

17. An airline overbooks a flight, selling more tickets for the flight than there are seats on the plane (figuring that it's likely that some people won't show up). The plane has 100 seats, and 110 people have booked the flight. Each person will show up for the flight with probability 0.9, independently. Find the probability that there will be enough seats for everyone who shows up for the flight.

18. Ⓢ (a) In the World Series of baseball, two teams (call them A and B) play a sequence of games against each other, and the first team to win four games wins the series. Let p be the probability that A wins an individual game, and assume that the games are independent. What is the probability that team A wins the series?

    (b) Give a clear intuitive explanation of whether the answer to (a) depends on whether the teams always play 7 games (and whoever wins the majority wins the series), or the teams stop playing more games as soon as one team has won 4 games (as is actually the case in practice: once the match is decided, the two teams do not keep playing more games).

19. In a chess tournament, n games are being played, independently. Each game ends in a win for one player with probability 0.4 and ends in a draw (tie) with probability 0.6. Find the PMFs of the number of games ending in a draw, and of the number of players whose games end in draws.

20. Suppose that a lottery ticket has probability p of being a winning ticket, independently of other tickets. A gambler buys 3 tickets, hoping this will triple the chance of having at least one winning ticket.

    (a) What is the distribution of how many of the 3 tickets are winning tickets?

    (b) Show that the probability that at least 1 of the 3 tickets is winning is $3p - 3p^2 + p^3$, in two different ways: by using inclusion-exclusion, and by taking the complement of the desired event and then using the PMF of a certain named distribution.

    (c) Show that the gambler's chances of having at least one winning ticket do not quite triple (compared with buying only one ticket), but that they do *approximately* triple if p is small.

21. Ⓢ Let $X \sim \text{Bin}(n, p)$ and $Y \sim \text{Bin}(m, p)$, independent of X. Show that $X - Y$ is *not* Binomial.

22. There are two coins, one with probability p_1 of Heads and the other with probability p_2 of Heads. One of the coins is randomly chosen (with equal probabilities for the two coins). It is then flipped $n \geq 2$ times. Let X be the number of times it lands Heads.

    (a) Find the PMF of X.

    (b) What is the distribution of X if $p_1 = p_2$?

    (c) Give an intuitive explanation of why X is *not* Binomial for $p_1 \neq p_2$ (its distribution is called a *mixture* of two Binomials). You can assume that n is large for your explanation, so that the frequentist interpretation of probability can be applied.

23. There are n people eligible to vote in a certain election. Voting requires registration. Decisions are made independently. Each of the n people will register with probability p_1. Given that a person registers, they will vote with probability p_2. Given that a person votes, they will vote for Kodos (who is one of the candidates) with probability p_3. What is the distribution of the number of votes for Kodos (give the PMF, fully simplified, or the name of the distribution, including its parameters)?

24. Let X be the number of Heads in 10 fair coin tosses.

    (a) Find the conditional PMF of X, given that the first two tosses both land Heads.

    (b) Find the conditional PMF of X, given that at least two tosses land Heads.

25. Ⓢ Alice flips a fair coin n times and Bob flips another fair coin $n+1$ times, resulting in independent $X \sim \text{Bin}(n, \frac{1}{2})$ and $Y \sim \text{Bin}(n+1, \frac{1}{2})$.

    (a) Show that $P(X < Y) = P(n - X < n + 1 - Y)$.

    (b) Compute $P(X < Y)$.

    Hint: Use (a) and the fact that X and Y are integer-valued.

26. If $X \sim \text{HGeom}(w, b, n)$, what is the distribution of $n - X$? Give a short proof.

27. Recall de Montmort's matching problem from Chapter 1: in a deck of n cards labeled 1 through n, a match occurs when the number on the card matches the card's position in the deck. Let X be the number of matching cards. Is X Binomial? Is X Hypergeometric?

28. Ⓢ There are n eggs, each of which hatches a chick with probability p (independently). Each of these chicks survives with probability r, independently. What is the distribution of the number of chicks that hatch? What is the distribution of the number of chicks that survive? (Give the PMFs; also give the names of the distributions and their parameters, if applicable.)

29. Ⓢ A sequence of n independent experiments is performed. Each experiment is a success with probability p and a failure with probability $q = 1 - p$. Show that conditional on the number of successes, all valid possibilities for the list of outcomes of the experiment are equally likely.

30. A certain company has $n + m$ employees, consisting of n women and m men. The company is deciding which employees to promote.

    (a) Suppose for this part that the company decides to promote t employees, where $1 \leq t \leq n + m$, by choosing t random employees (with equal probabilities for each set of t employees). What is the distribution of the number of women who get promoted?

    (b) Now suppose that instead of having a predetermined number of promotions to give, the company decides independently for each employee, promoting the employee with probability p. Find the distributions of the number of women who are promoted, the number of women who are not promoted, and the number of employees who are promoted.

    (c) In the set-up from (b), find the conditional distribution of the number of women who are promoted, given that exactly t employees are promoted.

31. Once upon a time, a famous statistician offered tea to a lady. The lady claimed that she could tell whether milk had been added to the cup before or after the tea. The statistician decided to run some experiments to test her claim.

    (a) The lady is given 6 cups of tea, where it is known in advance that 3 will be milk-first and 3 will be tea-first, in a completely random order. The lady gets to taste each and then guess which 3 were milk-first. Assume for this part that she has no ability whatsoever to distinguish milk-first from tea-first cups of tea. Find the probability that at least 2 of her 3 guesses are correct.

    (b) Now the lady is given one cup of tea, with probability 1/2 of it being milk-first. She needs to say whether she thinks it was milk-first. Let p_1 be the lady's probability of being correct given that it was milk-first, and p_2 be her probability of being correct given that it was tea-first. She claims that the cup was milk-first. Find the *posterior odds* that the cup is milk-first, given this information.

32. In Evan's history class, 10 out of 100 key terms will be randomly selected to appear on the final exam; Evan must then choose 7 of those 10 to define. Since he knows the format of the exam in advance, Evan is trying to decide how many key terms he should study.

    (a) Suppose that Evan decides to study s key terms, where s is an integer between 0 and 100. Let X be the number of key terms appearing on the exam that he has studied. What is the distribution of X? Give the name and parameters, in terms of s.

    (b) Using R or other software, calculate the probability that Evan knows at least 7 of the 10 key terms that appear on the exam, assuming that he studies $s = 75$ key terms.

33. A book has n typos. Two proofreaders, Prue and Frida, independently read the book. Prue catches each typo with probability p_1 and misses it with probability $q_1 = 1 - p_1$, independently, and likewise for Frida, who has probabilities p_2 of catching and $q_2 = 1-p_2$ of missing each typo. Let X_1 be the number of typos caught by Prue, X_2 be the number caught by Frida, and X be the number caught by at least one of the two proofreaders.

    (a) Find the distribution of X.

    (b) For this part only, assume that $p_1 = p_2$. Find the conditional distribution of X_1 given that $X_1 + X_2 = t$.

34. There are n students at a certain school, of whom $X \sim \text{Bin}(n, p)$ are Statistics majors. A simple random sample of size m is drawn ("simple random sample" means sampling without replacement, with all subsets of the given size equally likely).

    (a) Find the PMF of the number of Statistics majors in the sample, using the law of total probability (don't forget to say what the support is). You can leave your answer as a sum (though with some algebra it can be simplified, by writing the binomial coefficients in terms of factorials and using the binomial theorem).

    (b) Give a story proof derivation of the distribution of the number of Statistics majors in the sample; simplify fully.

    Hint: Does it matter whether the students declare their majors before or after the random sample is drawn?

35. Ⓢ Players A and B take turns in answering trivia questions, starting with player A answering the first question. Each time A answers a question, she has probability p_1 of getting it right. Each time B plays, he has probability p_2 of getting it right.

    (a) If A answers m questions, what is the PMF of the number of questions she gets right?

    (b) If A answers m times and B answers n times, what is the PMF of the total number of questions they get right (you can leave your answer as a sum)? Describe exactly when/whether this is a Binomial distribution.

    (c) Suppose that the first player to answer correctly wins the game (with no predetermined maximum number of questions that can be asked). Find the probability that A wins the game.

36. There are n voters in an upcoming election in a certain country, where n is a large, even number. There are two candidates: Candidate A (from the Unite Party) and Candidate B (from the Untie Party). Let X be the number of people who vote for Candidate A. Suppose that each voter chooses randomly whom to vote for, independently and with equal probabilities.

    (a) Find an exact expression for the probability of a tie in the election (so the candidates end up with the same number of votes).

(b) Use Stirling's approximation, which approximates the factorial function as

$$n! \approx \sqrt{2\pi n}\left(\frac{n}{e}\right)^n,$$

to find a simple approximation to the probability of a tie. Your answer should be of the form $1/\sqrt{cn}$, with c a constant (which you should specify).

37. Ⓢ A message is sent over a noisy channel. The message is a sequence $x_1, x_2, \dots, x_n$ of n bits ($x_i \in \{0,1\}$). Since the channel is noisy, there is a chance that any bit might be corrupted, resulting in an error (a 0 becomes a 1 or vice versa). Assume that the error events are independent. Let p be the probability that an individual bit has an error ($0 < p < 1/2$). Let $y_1, y_2, \dots, y_n$ be the received message (so $y_i = x_i$ if there is no error in that bit, but $y_i = 1 - x_i$ if there is an error there).

To help detect errors, the nth bit is reserved for a parity check: x_n is defined to be 0 if $x_1 + x_2 + \dots + x_{n-1}$ is even, and 1 if $x_1 + x_2 + \dots + x_{n-1}$ is odd. When the message is received, the recipient checks whether y_n has the same parity as $y_1 + y_2 + \dots + y_{n-1}$. If the parity is wrong, the recipient knows that at least one error occurred; otherwise, the recipient assumes that there were no errors.

(a) For $n = 5, p = 0.1$, what is the probability that the received message has errors which go undetected?

(b) For general n and p, write down an expression (as a sum) for the probability that the received message has errors which go undetected.

(c) Give a simplified expression, not involving a sum of a large number of terms, for the probability that the received message has errors which go undetected.

Hint for (c): Letting

$$a = \sum_{k \text{ even},\, k \geq 0} \binom{n}{k} p^k (1-p)^{n-k} \text{ and } b = \sum_{k \text{ odd},\, k \geq 1} \binom{n}{k} p^k (1-p)^{n-k},$$

the binomial theorem makes it possible to find simple expressions for $a + b$ and $a - b$, which then makes it possible to obtain a and b.

Independence of r.v.s

38. (a) Give an example of dependent r.v.s X and Y such that $P(X < Y) = 1$.

(b) Give an example of independent r.v.s X and Y such that $P(X < Y) = 1$.

39. Give an example of two discrete random variables X and Y on the same sample space such that X and Y have the same distribution, with support $\{1, 2, \dots, 10\}$, but the event $X = Y$ *never* occurs. If X and Y are independent, is it still possible to construct such an example?

40. Suppose X and Y are discrete r.v.s such that $P(X = Y) = 1$. This means that X and Y always take on the same value.

(a) Do X and Y have the same PMF?

(b) Is it possible for X and Y to be independent?

41. If X, Y, Z are r.v.s such that X and Y are independent and Y and Z are independent, does it follow that X and Z are independent?

Hint: Think about simple and extreme examples.

42. Ⓢ Let X be a random day of the week, coded so that Monday is 1, Tuesday is 2, etc. (so X takes values $1, 2, \dots, 7$, with equal probabilities). Let Y be the next day after X (again represented as an integer between 1 and 7). Do X and Y have the same distribution? What is $P(X < Y)$?

43. (a) Is it possible to have two r.v.s X and Y such that X and Y have the same distribution but $P(X < Y) \geq p$, where:
   - $p = 0.9$?
   - $p = 0.99$?
   - $p = 0.9999999999999$?
   - $p = 1$?

   For each, give an example showing it is possible, or prove it is impossible.
   Hint: Do the previous question first.

   (b) Consider the same question as in Part (a), but now assume that X and Y are independent. Do your answers change?

44. For x and y binary digits (0 or 1), let $x \oplus y$ be 0 if $x = y$ and 1 if $x \neq y$ (this operation is called *exclusive or* (often abbreviated to XOR), or *addition mod 2*).

   (a) Let $X \sim \text{Bern}(p)$ and $Y \sim \text{Bern}(1/2)$, independently. What is the distribution of $X \oplus Y$?

   (b) With notation as in (a), is $X \oplus Y$ independent of X? Is $X \oplus Y$ independent of Y? Be sure to consider both the case $p = 1/2$ and the case $p \neq 1/2$.

   (c) Let $X_1, \dots, X_n$ be i.i.d. Bern(1/2). For each nonempty subset J of $\{1, 2, \dots, n\}$, let

$$Y_J = \bigoplus_{j \in J} X_j,$$

   where the notation means to "add" in the $\oplus$ sense all the elements of J; the order in which this is done doesn't matter since $x \oplus y = y \oplus x$ and $(x \oplus y) \oplus z = x \oplus (y \oplus z)$. Show that $Y_J \sim \text{Bern}(1/2)$ and that these $2^n - 1$ r.v.s are pairwise independent, but not independent. For example, we can use this to simulate 1023 pairwise independent fair coin tosses using only 10 independent fair coin tosses.

   Hint: Apply the previous parts with $p = 1/2$. Show that if J and K are two different nonempty subsets of $\{1, 2, \dots, n\}$, then we can write $Y_J = A \oplus B, Y_K = A \oplus C$, where A consists of the X_i with $i \in J \cap K$, B consists of the X_i with $i \in J \cap K^c$, and C consists of the X_i with $i \in J^c \cap K$. Then A, B, C are independent since they are based on disjoint sets of X_i. Also, at most one of these sets of X_i can be empty. If $J \cap K = \emptyset$, then $Y_J = B, Y_K = C$. Otherwise, compute $P(Y_J = y, Y_K = z)$ by conditioning on whether $A = 1$.

Mixed practice

45. Ⓢ A new treatment for a disease is being tested, to see whether it is better than the standard treatment. The existing treatment is effective on 50% of patients. It is believed initially that there is a 2/3 chance that the new treatment is effective on 60% of patients, and a 1/3 chance that the new treatment is effective on 50% of patients. In a pilot study, the new treatment is given to 20 random patients, and is effective for 15 of them.

   (a) Given this information, what is the probability that the new treatment is better than the standard treatment?

   (b) A second study is done later, giving the new treatment to 20 new random patients. Given the results of the first study, what is the PMF for how many of the new patients the new treatment is effective on? (Letting p be the answer to (a), your answer can be left in terms of p.)

46. Independent Bernoulli trials are performed, with success probability 1/2 for each trial. An important question that often comes up in such settings is how many trials to perform. Many controversies have arisen in statistics over the issue of how to analyze data coming from an experiment where the number of trials can depend on the data collected so far.

    For example, if we can follow the rule "keep performing trials until there are more than twice as many failures as successes, and then stop", then naively looking at the ratio of failures to successes (if and when the process stops) will give more than 2:1 rather than the true theoretical 1:1 ratio; this could be a very misleading result! However, it might *never* happen that there are more than twice as many failures as successes; in this problem, you will find the probability of that happening.

    (a) Two gamblers, A and B, make a series of bets, where each has probability 1/2 of winning a bet, but A gets \$2 for each win and loses \$1 for each loss (a very favorable game for A!). Assume that the gamblers are allowed to borrow money, so they can and do gamble forever. Let p_k be the probability that A, starting with \$$k$, will ever reach \$0, for each $k \geq 0$. Explain how this story relates to the original problem, and how the original problem can be solved if we can find p_k.

    (b) Find p_k.

    Hint: As in the gambler's ruin, set up and solve a difference equation for p_k. We have $p_k \to 0$ as $k \to \infty$ (you don't need to prove this, but it should make sense since the game is so favorable to A, which will result in A's fortune going to ∞; a formal proof, not required here, could be done using the *law of large numbers*, an important theorem from Chapter 10). The solution can be written neatly in terms of the golden ratio.

    (c) Find the probability of ever having more than twice as many failures as successes with independent Bern(1/2) trials, as originally desired.

47. A copy machine is used to make n pages of copies per day. The machine has two trays in which paper gets loaded, and each page used is taken randomly and independently from one of the trays. At the beginning of the day, the trays are refilled so that they each have m pages.

    (a) Let pbinom(x, n, p) be the CDF of the Bin(n, p) distribution, evaluated at x. In terms of pbinom, find a simple expression for the probability that both trays have enough paper on any particular day, when this probability is strictly between 0 and 1 (also specify the values of m for which the probability is 0 and the values for which it is 1).

    Hint: Be careful about whether inequalities are strict, since the Binomial is discrete.

    (b) Using a computer, find the smallest value of m for which there is at least a 95% chance that both trays have enough paper on a particular day, for $n = 10, n = 100, n = 1000$, and $n = 10000$.

    Hint: If you use R, you may find the following commands useful:
    `g <- function(m,n)` *[your answer from (a)]* defines a function g such that $g(m, n)$ is your answer from (a), `g(1:100,100)` gives the vector $(g(1, 100), \ldots, g(100, 100))$, `which(v>0.95)` gives the indices of the components of vector **v** that exceed 0.95, and `min(w)` gives the minimum of a vector **w**.

4

Expectation

4.1 Definition of expectation

In the previous chapter, we introduced the *distribution* of a random variable, which gives us full information about the probability that the r.v. will fall into any particular set. For example, we can say how likely it is that the r.v. will exceed 1000, that it will equal 5, or that it will be in the interval $[0, 7]$. It can be unwieldy to manage so many probabilities though, so often we want just one number summarizing the "average" value of the r.v.

There are several senses in which the word "average" is used, but by far the most commonly used is the *mean* of an r.v., also known as its *expected value*. In addition, much of statistics is about understanding *variability* in the world, so it is often important to know how "spread out" the distribution is; we will formalize this with the concepts of *variance* and *standard deviation*. As we'll see, variance and standard deviation are defined in terms of expected values, so the uses of expected values go far beyond just computing averages.

Given a list of numbers $x_1, x_2, \dots, x_n$, the familiar way to average them is to add them up and divide by n. This is called the *arithmetic mean*, and is defined by

$$\bar{x} = \frac{1}{n} \sum_{j=1}^{n} x_j.$$

More generally, we can define a *weighted mean* of $x_1, \dots, x_n$ as

$$\text{weighted-mean}(x) = \sum_{j=1}^{n} x_j p_j,$$

where the weights $p_1, \dots, p_n$ are pre-specified nonnegative numbers that add up to 1 (so the unweighted mean $\bar{x}$ is obtained when $p_j = 1/n$ for all j).

The definition of expectation for a discrete r.v. is inspired by the weighted mean of a list of numbers, with weights given by probabilities.

Definition 4.1.1 (Expectation of a discrete r.v.). The *expected value* (also called the *expectation* or *mean*) of a discrete r.v. X whose distinct possible values are

$x_1, x_2, \ldots$ is defined by

$$E(X) = \sum_{j=1}^{\infty} x_j P(X = x_j).$$

If the support is finite, then this is replaced by a finite sum. We can also write

$$E(X) = \sum_{x} \underbrace{x}_{\text{value}} \underbrace{P(X = x)}_{\text{PMF at } x},$$

where the sum is over the support of X (in any case, $xP(X = x)$ is 0 for any x not in the support). The expectation is undefined if $\sum_{j=1}^{\infty} |x_j| P(X = x_j)$ diverges, since then the series for $E(X)$ diverges or its value depends on the order in which the x_j are listed.

In words, the expected value of X is a weighted average of the possible values that X can take on, weighted by their probabilities. Let's check that the definition makes sense in a few simple examples:

1. Let X be the result of rolling a fair 6-sided die, so X takes on the values $1, 2, 3, 4, 5, 6$, with equal probabilities. Intuitively, we should be able to get the average by adding up these values and dividing by 6. Using the definition, the expected value is

   $$E(X) = \frac{1}{6}(1 + 2 + \cdots + 6) = 3.5,$$

   as we expected. Note that X *never* equals its mean in this example. This is similar to the fact that the average number of children per household in some country could be 1.8, but that doesn't mean that a typical household has 1.8 children!

2. Let $X \sim \text{Bern}(p)$ and $q = 1 - p$. Then

   $$E(X) = 1p + 0q = p,$$

   which makes sense intuitively since it is between the two possible values of X, compromising between 0 and 1 based on how likely each is. This is illustrated in Figure 4.1 for a case with $p < 1/2$: two pebbles are being balanced on a seesaw. For the seesaw to balance, the fulcrum (shown as a triangle) must be at p, which in physics terms is the *center of mass*.

   The frequentist interpretation would be to consider a large number of independent Bernoulli trials, each with probability p of success. Writing 1 for "success" and 0 for "failure", in the long run we would expect to have data consisting of a list of numbers where the proportion of 1's is very close to p. The average of a list of 0's and 1's *is* the proportion of 1's.

3. Let X have 3 distinct possible values, a_1, a_2, a_3, with probabilities p_1, p_2, p_3, respectively. Imagine running a simulation where n independent draws

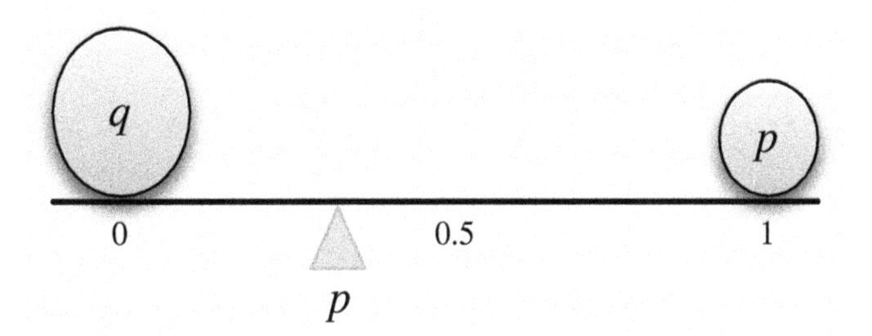

FIGURE 4.1
Center of mass of two pebbles, depicting that $E(X) = p$ for $X \sim \text{Bern}(p)$. Here q and p denote the masses of the two pebbles.

from the distribution of X are generated. For n large, we would expect to have about $p_1 n$ a_1's, $p_2 n$ a_2's, and $p_3 n$ a_3's. (We will look at a more mathematical version of this example when we study the law of large numbers in Chapter 10.) If the simulation results are close to these expected results, then the arithmetic mean of the simulation results is approximately

$$\frac{p_1 n \cdot a_1 + p_2 n \cdot a_2 + p_3 n \cdot a_3}{n} = p_1 a_1 + p_2 a_2 + p_3 a_3 = E(X).$$

Note that $E(X)$ depends only on the *distribution* of X. This follows directly from the definition, but is worth recording since it is fundamental.

Proposition 4.1.2. If X and Y are discrete r.v.s with the same distribution, then $E(X) = E(Y)$ (if either side exists).

Proof. In the definition of $E(X)$, we only need to know the PMF of X. ■

The converse of the above proposition is false since the expected value is just a one-number summary, not nearly enough to specify the entire distribution; it's a measure of where the "center" is but does not determine, for example, how spread out the distribution is or how likely the r.v. is to be positive. Figure 4.2 shows an example of two different PMFs with the same expected value (balancing point).

☣ **4.1.3** (Replacing an r.v. by its expectation)**.** For any discrete r.v. X, the expected value $E(X)$ is a *number* (if it exists). A common mistake is to replace an r.v. by its expectation without justification, which is wrong both mathematically (X is a function, $E(X)$ is a constant) and statistically (it ignores the variability of X), except in the degenerate case where X is a constant.

Notation 4.1.4. We often abbreviate $E(X)$ to EX. Similarly, we often abbreviate $E(X^2)$ to EX^2, and $E(X^n)$ to EX^n.

☣ **4.1.5.** Paying attention to the order of operations is crucial when working with expectation. As stated above, EX^2 is the expectation of the random variable X^2, *not* the square of the number EX. Unless the parentheses explicitly indicate otherwise,

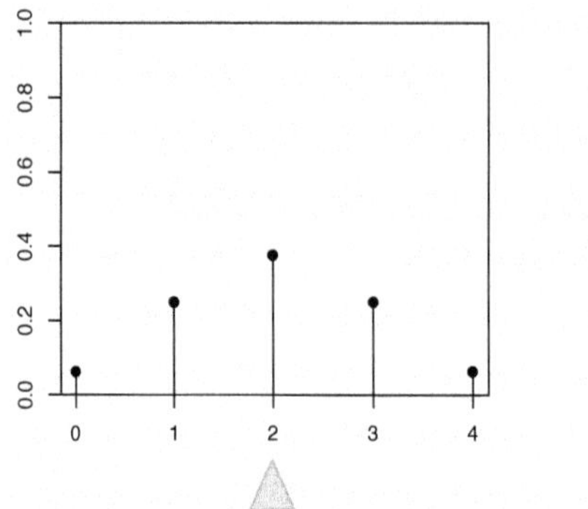

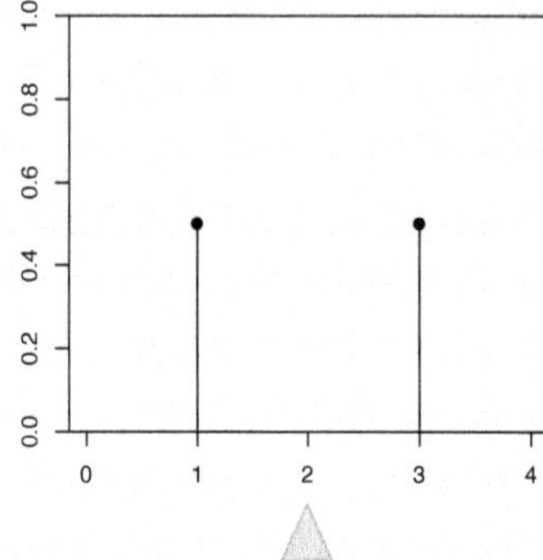

FIGURE 4.2
The expected value does not determine the distribution: different PMFs can have the same balancing point.

for the expectation of an r.v. raised to a power, first we take the power and then we take the expectation. For example, $E(X-1)^4$ is $E\left((X-1)^4\right)$, not $(E(X-1))^4$.

4.2 Linearity of expectation

The most important property of expectation is *linearity*: the expected value of a sum of r.v.s is the sum of the individual expected values.

Theorem 4.2.1 (Linearity of expectation). For any r.v.s X, Y and any constant c,

$$E(X+Y) = E(X) + E(Y),$$
$$E(cX) = cE(X).$$

The second equation says that we can take out constant factors from an expectation; this is both intuitively reasonable and easily verified from the definition. The first equation, $E(X+Y) = E(X) + E(Y)$, also seems reasonable when X and Y are independent. What may be surprising is that it holds even if X and Y are dependent! To build intuition for this, consider the extreme case where X always equals Y. Then $X + Y = 2X$, and both sides of $E(X+Y) = E(X) + E(Y)$ are equal to $2E(X)$, so linearity still holds even in the most extreme case of dependence.

Linearity is true for all r.v.s, not just discrete r.v.s, but in this chapter we will prove it only for discrete r.v.s. Before proving linearity, it is worthwhile to recall some basic facts about averages. If we have a list of numbers, say $(1, 1, 1, 1, 1, 3, 3, 5)$, we can calculate their mean by adding all the values and dividing by the length of the list, so that each element of the list gets a weight of $\frac{1}{8}$:

$$\frac{1}{8}(1+1+1+1+1+3+3+5) = 2.$$

But another way to calculate the mean is to group together all the 1's, all the 3's, and all the 5's, and then take a weighted average, giving appropriate weights to 1's, 3's, and 5's:

$$\frac{5}{8} \cdot 1 + \frac{2}{8} \cdot 3 + \frac{1}{8} \cdot 5 = 2.$$

This insight—that averages can be calculated in two ways, *ungrouped* or *grouped*—is all that is needed to prove linearity! Recall that X is a function which assigns a real number to every outcome s in the sample space. The r.v. X may assign the same value to multiple sample outcomes. When this happens, our definition of expectation groups all these outcomes together into a *super-pebble* whose weight, $P(X = x)$, is the total weight of the constituent pebbles. This grouping process is illustrated in Figure 4.3 for a hypothetical r.v. taking values in $\{0, 1, 2\}$. So our definition of expectation corresponds to the grouped way of taking averages.

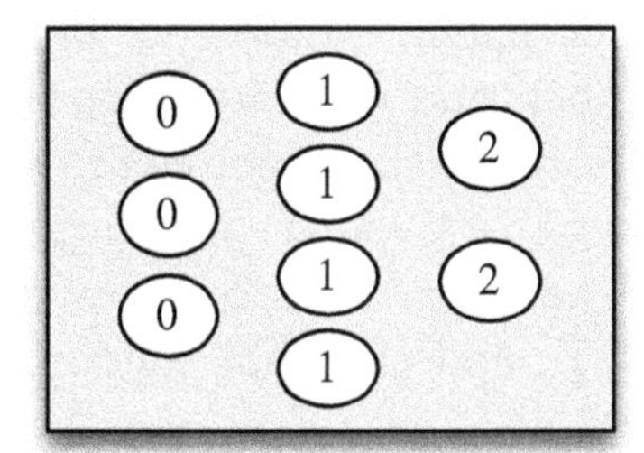

FIGURE 4.3
Left: X assigns a number to each pebble in the sample space. Right: Grouping the pebbles by the value that X assigns to them, the 9 pebbles become 3 super-pebbles. The weight of a super-pebble is the sum of the weights of the constituent pebbles.

The advantage of this definition is that it allows us to work with the distribution of X directly, without returning to the sample space. The disadvantage comes when we have to prove theorems like this one, for if we have another r.v. Y on the same sample space, the super-pebbles created by Y are different from those created from X, with different weights $P(Y = y)$; this makes it difficult to combine $\sum_x xP(X = x)$ and $\sum_y yP(Y = y)$.

Fortunately, we know there's another equally valid way to calculate an average: we can take a weighted average of the values of individual pebbles. In other words, if $X(s)$ is the value that X assigns to pebble s, we can take the weighted average

$$E(X) = \sum_s X(s)P(\{s\}),$$

where $P(\{s\})$ is the weight of pebble s. This corresponds to the ungrouped way of taking averages. The advantage of this definition is that it breaks down the sample space into the smallest possible units, so we are now using the *same* weights $P(\{s\})$ for every random variable defined on this sample space. If Y is another random variable, then

$$E(Y) = \sum_s Y(s)P(\{s\}).$$

We *can* combine $\sum_s X(s)P(\{s\})$ and $\sum_s Y(s)P(\{s\})$, which gives

$$E(X)+E(Y) = \sum_s X(s)P(\{s\})+\sum_s Y(s)P(\{s\}) = \sum_s (X+Y)(s)P(\{s\}) = E(X+Y).$$

Another intuition for linearity of expectation is via the concept of *simulation.* If we simulate many, many times from the distribution of X, the histogram of the simulated values will look very much like the true PMF of X. In particular, the *arithmetic mean* of the simulated values will be very close to the true value of $E(X)$ (the precise nature of this convergence is described by the law of large numbers, an important theorem that we will discuss in detail in Chapter 10).

Let X and Y be r.v.s summarizing a certain experiment. Suppose we perform the experiment n times, where n is a very large number, and we write down the values realized by X and Y each time. For each repetition of the experiment, we obtain an X value, a Y value, and (by adding them) an $X + Y$ value. In Figure 4.4, each row represents a repetition of the experiment. The left column contains the draws of X, the middle column contains the draws of Y, and the right column contains the draws of $X + Y$.

There are two ways to calculate the sum of all the numbers in the last column. The straightforward way is just to add all the numbers in that column. But an equally valid way is to add all the numbers in the first column, add all the numbers in the second column, and then add the two column sums.

Dividing by n everywhere, what we've argued is that the following procedures are equivalent:

- Taking the arithmetic mean of all the numbers in the last column. By the law of large numbers, this is very close to $E(X + Y)$.
- Taking the arithmetic mean of the first column and the arithmetic mean of the second column, then adding the two column means. By the law of large numbers, this is very close to $E(X) + E(Y)$.

Linearity of expectation thus emerges as a simple fact about arithmetic (we're just adding numbers in two different orders)! Notice that nowhere in our argument did we rely on whether X and Y were independent. In fact, in Figure 4.4, X and Y appear to be dependent: Y tends to be large when X is large, and Y tends to be small when X is small (in the language of Chapter 7, we say that X and Y are *positively correlated*). But this dependence is irrelevant: shuffling the draws of Y could completely alter the pattern of dependence between X and Y, but would have no effect on the column sums.

X	Y	$X+Y$
3	4	7
2	2	4
6	8	14
10	23	33
1	−3	−2
1	0	1
5	9	14
4	1	5
$\vdots$	$\vdots$	$\vdots$

$$\frac{1}{n}\sum_{i=1}^{n} x_i \quad + \quad \frac{1}{n}\sum_{i=1}^{n} y_i \quad = \quad \frac{1}{n}\sum_{i=1}^{n}(x_i + y_i)$$

$$E(X) \quad + \quad E(Y) \quad = \quad E(X+Y)$$

FIGURE 4.4
Intuitive view of linearity of expectation. Each row represents a repetition of the experiment; the three columns are the realized values of X, Y, and $X+Y$, respectively. Adding all the numbers in the last column is equivalent to summing the first column and the second column separately, then adding the two column sums. So the mean of the last column is the sum of the first and second column means; this is linearity of expectation.

Linearity is an extremely handy tool for calculating expected values, often allowing us to bypass the definition of expected value altogether. Let's use linearity to find the expectations of the Binomial and Hypergeometric distributions.

Example 4.2.2 (Binomial expectation). For $X \sim \text{Bin}(n, p)$, let's find $E(X)$ in two ways. By definition of expectation,

$$E(X) = \sum_{k=0}^{n} kP(X = k) = \sum_{k=0}^{n} k\binom{n}{k}p^k q^{n-k}.$$

From Example 1.5.2, we know $k\binom{n}{k} = n\binom{n-1}{k-1}$, so

$$\begin{aligned}\sum_{k=0}^{n} k\binom{n}{k}p^k q^{n-k} &= n\sum_{k=0}^{n}\binom{n-1}{k-1}p^k q^{n-k} \\ &= np\sum_{k=1}^{n}\binom{n-1}{k-1}p^{k-1}q^{n-k} \\ &= np\sum_{j=0}^{n-1}\binom{n-1}{j}p^j q^{n-1-j} \\ &= np.\end{aligned}$$

The sum in the penultimate line equals 1 because it is the sum of the $\text{Bin}(n-1, p)$ PMF (or by the binomial theorem). Therefore, $E(X) = np$.

This proof required us to remember combinatorial identities and manipulate binomial coefficients. Using linearity of expectation, we obtain a *much* shorter path to the same result. Let's write X as the sum of n independent $\text{Bern}(p)$ r.v.s:

$$X = I_1 + \cdots + I_n,$$

where each I_j has expectation $E(I_j) = 1p + 0q = p$. By linearity,

$$E(X) = E(I_1) + \cdots + E(I_n) = np. \qquad \square$$

Example 4.2.3 (Hypergeometric expectation). Let $X \sim \text{HGeom}(w, b, n)$, interpreted as the number of white balls in a sample of size n drawn without replacement from an urn with w white and b black balls. As in the Binomial case, we can write X as a sum of Bernoulli random variables,

$$X = I_1 + \cdots + I_n,$$

where I_j equals 1 if the jth ball in the sample is white and 0 otherwise. By symmetry, $I_j \sim \text{Bern}(p)$ with $p = w/(w+b)$, since unconditionally the jth ball drawn is equally likely to be any of the balls.

Unlike in the Binomial case, the I_j are not independent, since the sampling is without replacement: given that a ball in the sample is white, there is a lower chance that another ball in the sample is white. However, linearity still holds for dependent random variables! Thus,

$$E(X) = nw/(w+b).$$ □

As another example of the power of linearity, we can give a quick proof of the intuitive idea that "bigger r.v.s have bigger expectations".

Proposition 4.2.4 (Monotonicity of expectation). Let X and Y be r.v.s such that $X \geq Y$ with probability 1. Then $E(X) \geq E(Y)$, with equality holding if and only if $X = Y$ with probability 1.

Proof. This result holds for all r.v.s, but we will prove it only for discrete r.v.s since this chapter focuses on discrete r.v.s. The r.v. $Z = X - Y$ is nonnegative (with probability 1), so $E(Z) \geq 0$ since $E(Z)$ is defined as a sum of nonnegative terms. By linearity,

$$E(X) - E(Y) = E(X - Y) \geq 0,$$

as desired. If $E(X) = E(Y)$, then by linearity we also have $E(Z) = 0$, which implies that $P(X = Y) = P(Z = 0) = 1$ since if even one term in the sum defining $E(Z)$ is positive, then the whole sum is positive. ■

4.3 Geometric and Negative Binomial

We now introduce two more famous discrete distributions, the Geometric and Negative Binomial, and calculate their expected values.

Story 4.3.1 (Geometric distribution). Consider a sequence of independent Bernoulli trials, each with the same success probability $p \in (0, 1)$, with trials performed until a success occurs. Let X be the number of *failures* before the first successful trial. Then X has the *Geometric distribution* with parameter p; we denote this by $X \sim \text{Geom}(p)$. □

For example, if we flip a fair coin until it lands Heads for the first time, then the number of Tails before the first occurrence of Heads is distributed as $\text{Geom}(1/2)$.

To get the Geometric PMF from the story, imagine the Bernoulli trials as a string of 0's (failures) ending in a single 1 (success). Each 0 has probability $q = 1 - p$ and the final 1 has probability p, so a string of k failures followed by one success has probability $q^k p$.

Theorem 4.3.2 (Geometric PMF). If $X \sim \text{Geom}(p)$, then the PMF of X is

$$P(X = k) = q^k p$$

for $k = 0, 1, 2, \ldots$, where $q = 1 - p$.

This is a valid PMF because, summing a geometric series (see the math appendix for a review of geometric series), we have

$$\sum_{k=0}^{\infty} q^k p = p \sum_{k=0}^{\infty} q^k = p \cdot \frac{1}{1-q} = 1.$$

Just as the binomial theorem shows that the Binomial PMF is valid, a geometric series shows that the Geometric PMF is valid! A geometric series can also be used to obtain the Geometric CDF.

Theorem 4.3.3 (Geometric CDF). If $X \sim \text{Geom}(p)$, then the CDF of X is

$$F(x) = \begin{cases} 1 - q^{\lfloor x \rfloor + 1}, & \text{if } x \geq 0; \\ 0, & \text{if } x < 0, \end{cases}$$

where $q = 1 - p$ and $\lfloor x \rfloor$ is the greatest integer less than or equal to x.

Proof. Let F be the CDF of X. We will find $F(x)$ first for the case $x < 0$, then for the case that x is a nonnegative integer, and lastly for the case that x is a nonnegative real number. For $x < 0$, $F(x) = 0$ since X can't be negative. For n a nonnegative integer,

$$F(n) = \sum_{k=0}^{n} P(X = k) = p \sum_{k=0}^{n} q^k = p \cdot \frac{1 - q^{n+1}}{1 - q} = 1 - q^{n+1}.$$

We can also get the same result from the fact that the event $X \geq n + 1$ means that the first $n + 1$ trials were failures:

$$F(n) = 1 - P(X > n) = 1 - P(X \geq n + 1) = 1 - q^{n+1}.$$

For real $x \geq 0$,

$$F(x) = P(X \leq x) = P(X \leq \lfloor x \rfloor),$$

since X always takes on integer values. For example,

$$P(X \leq 3.7) = P(X \leq 3) + P(3 < X \leq 3.7) = P(X \leq 3).$$

Therefore, F is as claimed. ■

Figure 4.3 displays the Geom(0.5) PMF and CDF from 0 to 6. All Geometric PMFs have a similar shape; the greater the success probability p, the more quickly the PMF decays to 0.

☣ **4.3.4** (Conventions for the Geometric). There are differing conventions for the definition of the Geometric distribution; some sources define the Geometric as the total number of *trials*, including the success. In this book, the Geometric distribution excludes the success, and the *First Success* distribution includes the success.

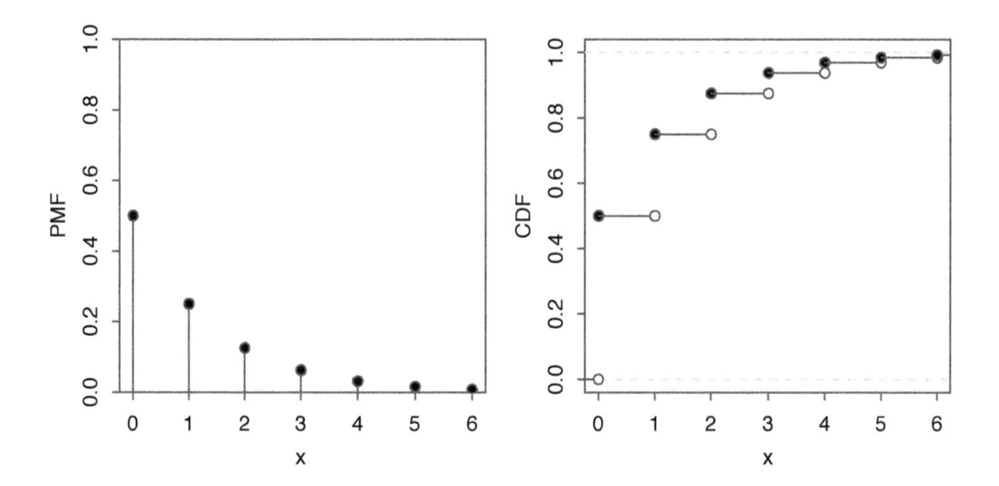

FIGURE 4.5
Geom(0.5) PMF and CDF.

Definition 4.3.5 (First Success distribution). In a sequence of independent Bernoulli trials with success probability p, let Y be the number of *trials* until the first successful trial, including the success. Then Y has the *First Success distribution* with parameter p; we denote this by $Y \sim \text{FS}(p)$.

It is easy to convert back and forth between the two but important to be careful about which convention is being used. If $Y \sim \text{FS}(p)$ then $Y - 1 \sim \text{Geom}(p)$, and we can convert between the PMFs of Y and $Y - 1$ by writing

$$P(Y = k) = P(Y - 1 = k - 1).$$

Conversely, if $X \sim \text{Geom}(p)$, then $X + 1 \sim \text{FS}(p)$.

Example 4.3.6 (Geometric expectation). Let $X \sim \text{Geom}(p)$. By definition,

$$E(X) = \sum_{k=0}^{\infty} kq^k p,$$

where $q = 1 - p$. This sum looks unpleasant; it's not a geometric series because of the extra k multiplying each term. But we notice that each term looks similar to kq^{k-1}, the derivative of q^k (with respect to q), so let's start there:

$$\sum_{k=0}^{\infty} q^k = \frac{1}{1-q}.$$

This geometric series converges since $0 < q < 1$. Differentiating both sides with respect to q, we get

$$\sum_{k=0}^{\infty} kq^{k-1} = \frac{1}{(1-q)^2}.$$

Finally, if we multiply both sides by pq, we recover the original sum we wanted to find:

$$E(X) = \sum_{k=0}^{\infty} kq^k p = pq \sum_{k=0}^{\infty} kq^{k-1} = pq \frac{1}{(1-q)^2} = \frac{q}{p}.$$

In Example 9.1.8, we will give a story proof of the same result, based on first-step analysis: condition on the result of the first trial in the story interpretation of X. If the first trial is a success, we know $X = 0$ and if it's a failure, we have one wasted trial and then are back where we started. □

Example 4.3.7 (First Success expectation). Since we can write $Y \sim \text{FS}(p)$ as $Y = X + 1$ where $X \sim \text{Geom}(p)$, we have

$$E(Y) = E(X + 1) = \frac{q}{p} + 1 = \frac{1}{p}.$$ □

The Negative Binomial distribution generalizes the Geometric distribution: instead of waiting for just one success, we can wait for any predetermined number r of successes.

Story 4.3.8 (Negative Binomial distribution). In a sequence of independent Bernoulli trials with success probability p, if X is the number of *failures* before the rth success, then X is said to have the *Negative Binomial distribution* with parameters r and p, denoted $X \sim \text{NBin}(r, p)$. □

Both the Binomial and the Negative Binomial distributions are based on independent Bernoulli trials; they differ in the *stopping rule* and in what they are counting. The Binomial counts the number of successes in a fixed number of *trials*; the Negative Binomial counts the number of failures until a fixed number of *successes*.

In light of these similarities, it comes as no surprise that the derivation of the Negative Binomial PMF bears a resemblance to the corresponding derivation for the Binomial.

Theorem 4.3.9 (Negative Binomial PMF). If $X \sim \text{NBin}(r, p)$, then the PMF of X is

$$P(X = n) = \binom{n + r - 1}{r - 1} p^r q^n$$

for $n = 0, 1, 2 \dots$, where $q = 1 - p$.

Proof. Imagine a string of 0's and 1's, with 1's representing successes. The probability of any *specific* string of n 0's and r 1's is $p^r q^n$. How many such strings are there? Because we stop as soon as we hit the rth success, the string must terminate in a 1. Among the other $n+r-1$ positions, we choose $r-1$ places for the remaining 1's to go. So the overall probability of exactly n failures before the rth success is

$$P(X = n) = \binom{n + r - 1}{r - 1} p^r q^n, \quad n = 0, 1, 2, \dots.$$ ■

Just as a Binomial r.v. can be represented as a sum of i.i.d. Bernoullis, a Negative Binomial r.v. can be represented as a sum of i.i.d. Geometrics.

Theorem 4.3.10. Let $X \sim \text{NBin}(r, p)$, viewed as the number of failures before the rth success in a sequence of independent Bernoulli trials with success probability p. Then we can write $X = X_1 + \cdots + X_r$ where the X_i are i.i.d. $\text{Geom}(p)$.

Proof. Let X_1 be the number of failures until the first success, X_2 be the number of failures between the first success and the second success, and in general, X_i be the number of failures between the $(i-1)$st success and the ith success.

Then $X_1 \sim \text{Geom}(p)$ by the story of the Geometric distribution. After the first success, the number of additional failures until the next success is still Geometric! So $X_2 \sim \text{Geom}(p)$, and similarly for all the X_i. Furthermore, the X_i are independent because the trials are all independent of each other. Adding the X_i, we get the total number of failures before the rth success, which is X. ■

Using linearity, the expectation of the Negative Binomial now follows without any additional calculations.

Example 4.3.11 (Negative Binomial expectation). Let $X \sim \text{NBin}(r, p)$. By the previous theorem, we can write $X = X_1 + \cdots + X_r$, where the X_i are i.i.d. $\text{Geom}(p)$. By linearity,

$$E(X) = E(X_1) + \cdots + E(X_r) = r \cdot \frac{q}{p}.$$ □

The next example is a famous problem in probability and an instructive application of the Geometric and First Success distributions. It is usually stated as a problem about collecting coupons, hence its name, but we'll use toys instead of coupons.

Example 4.3.12 (Coupon collector). Suppose there are n types of toys, which you are collecting one by one, with the goal of getting a complete set. When collecting toys, the toy types are random (as is sometimes the case, for example, with toys included in cereal boxes or included with kids' meals from a fast food restaurant). Assume that each time you collect a toy, it is equally likely to be any of the n types. What is the expected number of toys needed until you have a complete set?

Solution:

Let N be the number of toys needed; we want to find $E(N)$. Our strategy will be to break up N into a sum of simpler r.v.s so that we can apply linearity. So write

$$N = N_1 + N_2 + \cdots + N_n,$$

where N_1 is the number of toys until the first toy type you haven't seen before (which is always 1, as the first toy is always a new type), N_2 is the additional number of toys until the second toy type you haven't seen before, and so forth. Figure 4.6 illustrates these definitions with $n = 3$ toy types.

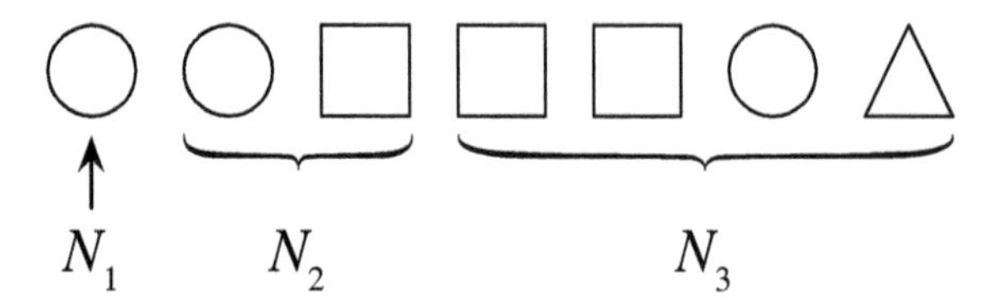

FIGURE 4.6
Coupon collector, $n = 3$. Here N_1 is the time (number of toys collected) until the first new toy type, N_2 is the additional time until the second new type, and N_3 is the additional time until the third new type. The total number of toys for a complete set is $N_1 + N_2 + N_3$.

By the story of the FS distribution, $N_2 \sim \text{FS}((n-1)/n)$: after collecting the first toy type, there's a $1/n$ chance of getting the same toy you already had (failure) and an $(n-1)/n$ chance you'll get something new (success). Similarly, N_3, the additional number of toys until the third new toy type, is distributed $\text{FS}((n-2)/n)$. In general,

$$N_j \sim \text{FS}((n-j+1)/n).$$

By linearity,

$$\begin{aligned}
E(N) &= E(N_1) + E(N_2) + E(N_3) + \cdots + E(N_n) \\
&= 1 + \frac{n}{n-1} + \frac{n}{n-2} + \cdots + n \\
&= n \sum_{j=1}^{n} \frac{1}{j}.
\end{aligned}$$

For large n, this is very close to $n(\log n + 0.577)$.

Before we leave this example, let's take a moment to connect it to our proof of Theorem 4.3.10, the representation of the Negative Binomial as a sum of i.i.d. Geometrics. In both problems, we are waiting for a specified number of successes, and we approach the problem by considering the intervals between successes. There are two major differences:

- In Theorem 4.3.10, we exclude the successes themselves, so the number of failures between two successes is Geometric. In the coupon collector problem, we include the successes because we want to count the total number of toys, so we have First Success r.v.s instead.

- In Theorem 4.3.10, the probability of success in each trial never changes, so the total number of failures is a sum of *i.i.d.* Geometrics. In the coupon collector problem, the probability of success decreases after each success, since it becomes harder and harder to find a new toy type you haven't seen before; so the N_j are not identically distributed, though they are independent. □

☣ **4.3.13** (Expectation of a nonlinear function of an r.v.). Expectation is linear, but in general we do *not* have $E(g(X)) = g(E(X))$ for arbitrary functions g. We must be careful not to move the E around when g is not linear. The next example shows a situation in which $E(g(X))$ is *very* different from $g(E(X))$.

Example 4.3.14 (St. Petersburg paradox)**.** Suppose a wealthy stranger offers to play the following game with you. You will flip a fair coin until it lands Heads for the first time, and you will receive \$2 if the game lasts for 1 round, \$4 if the game lasts for 2 rounds, \$8 if the game lasts for 3 rounds, and in general, $\$2^n$ if the game lasts for n rounds. What is the fair value of this game (the expected payoff)? How much would you be willing to pay to play this game once?

Solution:

Let X be your winnings from playing the game. By definition, $X = 2^N$ where N is the number of rounds that the game lasts. Then X is 2 with probability 1/2, 4 with probability 1/4, 8 with probability 1/8, and so on, so

$$E(X) = \frac{1}{2} \cdot 2 + \frac{1}{4} \cdot 4 + \frac{1}{8} \cdot 8 + \cdots = \infty.$$

The expected winnings are infinite! On the other hand, the number of rounds N that the game lasts is the number of tosses until the first Heads, so $N \sim \text{FS}(1/2)$ and $E(N) = 2$. Thus $E(2^N) = \infty$ while $2^{E(N)} = 4$. Infinity certainly does not equal 4, illustrating the danger of confusing $E(g(X))$ with $g(E(X))$ when g is not linear.

This problem is often considered a paradox because although the game's expected payoff is infinite, most people would not be willing to pay very much to play the game (even if they could afford to lose the money). One explanation is to note that *the amount of money in the real world is finite.* Suppose that if the game lasts longer than 40 rounds, the wealthy stranger flees the country and you get nothing. Since $2^{40} \approx 1.1 \times 10^{12}$, this still gives you the potential to earn over a trillion dollars, and anyway it's incredibly unlikely that the game will last longer than 40 rounds. But in this setting, your expected value is

$$E(X) = \sum_{n=1}^{40} \frac{1}{2^n} \cdot 2^n + \sum_{n=41}^{\infty} \frac{1}{2^n} \cdot 0 = 40.$$

Is this drastic reduction because the wealthy stranger may flee the country? Let's suppose instead that the wealthy stranger caps your winnings at 2^{40}, so if the game lasts more than 40 rounds you will get this amount rather than walking away empty-handed. Now your expected value is

$$E(X) = \sum_{n=1}^{40} \frac{1}{2^n} \cdot 2^n + \sum_{n=41}^{\infty} \frac{1}{2^n} \cdot 2^{40} = 40 + 1 = 41,$$

an increase of only \$1 from the previous scenario. The ∞ in the St. Petersburg paradox is driven by an infinite "tail" of extremely rare events where you get extremely large payoffs. Cutting off this tail at some point, which makes sense in the real world, dramatically reduces the expected value of the game. □

4.4 Indicator r.v.s and the fundamental bridge

This section is devoted to *indicator random variables*, which we already encountered in the previous chapter but will treat in much greater detail here. In particular, we will show that indicator r.v.s are an extremely useful tool for calculating expected values.

Recall from the previous chapter that the indicator r.v. I_A (or $I(A)$) for an event A is defined to be 1 if A occurs and 0 otherwise. So I_A is a Bernoulli random variable, where success is defined as "A occurs" and failure is defined as "A does not occur". Some useful properties of indicator r.v.s are summarized below.

Theorem 4.4.1 (Indicator r.v. properties). Let A and B be events. Then the following properties hold.

1. $(I_A)^k = I_A$ for any positive integer k.
2. $I_{A^c} = 1 - I_A$.
3. $I_{A \cap B} = I_A I_B$.
4. $I_{A \cup B} = I_A + I_B - I_A I_B$.

Proof. Property 1 holds since $0^k = 0$ and $1^k = 1$ for any positive integer k. Property 2 holds since $1 - I_A$ is 1 if A does not occur and 0 if A occurs. Property 3 holds since $I_A I_B$ is 1 if both I_A and I_B are 1, and 0 otherwise. Property 4 holds since

$$I_{A \cup B} = 1 - I_{A^c \cap B^c} = 1 - I_{A^c} I_{B^c} = 1 - (1 - I_A)(1 - I_B) = I_A + I_B - I_A I_B. \quad \blacksquare$$

Indicator r.v.s provide a link between probability and expectation; we call this fact the *fundamental bridge*.

Theorem 4.4.2 (Fundamental bridge between probability and expectation). There is a one-to-one correspondence between events and indicator r.v.s, and the probability of an event A is the expected value of its indicator r.v. I_A:

$$P(A) = E(I_A).$$

Proof. For any event A, we have an indicator r.v. I_A. This is a one-to-one correspondence since A uniquely determines I_A and vice versa (to get from I_A back to A, we can use the fact that $A = \{s \in S : I_A(s) = 1\}$). Since $I_A \sim \text{Bern}(p)$ with $p = P(A)$, we have $E(I_A) = P(A)$. $\blacksquare$

The fundamental bridge connects events to their indicator r.v.s, and allows us to express *any* probability as an expectation. As an example, we give a short proof of inclusion-exclusion and a related inequality known as *Boole's inequality* or *Bonferroni's inequality* using indicator r.v.s.

Example 4.4.3 (Boole, Bonferroni, and inclusion-exclusion). Let $A_1, A_2, \dots, A_n$ be events. Note that

$$I(A_1 \cup \dots \cup A_n) \leq I(A_1) + \dots + I(A_n),$$

since if the left-hand side is 0 this is immediate, and if the left-hand side is 1 then at least one term on the right-hand side must be 1. Taking the expectation of both sides and using linearity and the fundamental bridge, we have

$$P(A_1 \cup \dots \cup A_n) \leq P(A_1) + \dots + P(A_n),$$

which is called *Boole's inequality* or *Bonferroni's inequality*. To prove inclusion-exclusion for $n = 2$, we can take the expectation of both sides in Property 4 of Theorem 4.4.1. For general n, we can use properties of indicator r.v.s as follows:

$$\begin{aligned} 1 - I(A_1 \cup \dots \cup A_n) &= I(A_1^c \cap \dots \cap A_n^c) \\ &= (1 - I(A_1)) \cdots (1 - I(A_n)) \\ &= 1 - \sum_i I(A_i) + \sum_{i<j} I(A_i)I(A_j) - \dots + (-1)^n I(A_1) \cdots I(A_n). \end{aligned}$$

Taking the expectation of both sides, by the fundamental bridge we have proven the inclusion-exclusion theorem. □

Conversely, the fundamental bridge is also extremely useful in many expected value problems. We can often express a complicated discrete r.v. whose distribution we don't know as a sum of indicator r.v.s, which are extremely simple. The fundamental bridge lets us find the expectation of the indicators; then, using linearity, we obtain the expectation of our original r.v. This strategy is extremely useful and versatile—in fact, we already used it when deriving the expectations of the Binomial and Hypergeometric distributions earlier in this chapter!

Recognizing problems that are amenable to this strategy and then defining the indicator r.v.s takes practice, so it is important to study a lot of examples and solve a lot of problems. In applying the strategy to a random variable that counts the number of [noun]s, we should have an indicator for each potential [noun]. This [noun] could be a person, place, or thing; we will see examples of all three types.

We'll start by revisiting two problems from Chapter 1, de Montmort's matching problem and the birthday problem.

Example 4.4.4 (Matching continued). We have a well-shuffled deck of n cards, labeled 1 through n. A card is a *match* if the card's position in the deck matches the card's label. Let X be the number of matches; find $E(X)$.

Solution:

First let's check whether X could have any of the named distributions we have studied. The Binomial and Hypergeometric are the only two candidates since the value of X must be an integer between 0 and n. But neither of these distributions has the right support because X can't take on the value $n-1$: if $n-1$ cards are matches, then the nth card must be a match as well. So X does not follow a named distribution we have studied, but we can readily find its mean using indicator r.v.s: let's write $X = I_1 + I_2 + \cdots + I_n$, where

$$I_j = \begin{cases} 1 & \text{if the } j\text{th card in the deck is a match,} \\ 0 & \text{otherwise.} \end{cases}$$

In other words, I_j is the indicator for A_j, the event that the jth card in the deck is a match. We can imagine that each I_j "raises its hand" to be counted if its card is a match; adding up the raised hands, we get the total number of matches, X.

By the fundamental bridge,

$$E(I_j) = P(A_j) = \frac{1}{n}$$

for all j. So by linearity,

$$E(X) = E(I_1) + \cdots + E(I_n) = n \cdot \frac{1}{n} = 1.$$

The expected number of matched cards is 1, regardless of n. Even though the I_j are dependent in a complicated way that makes the distribution of X neither Binomial nor Hypergeometric, linearity still holds. □

Example 4.4.5 (Distinct birthdays, birthday matches). In a group of n people, under the usual assumptions about birthdays, what is the expected number of distinct birthdays among the n people, i.e., the expected number of days on which at least one of the people was born? What is the expected number of birthday matches, i.e., pairs of people with the same birthday?

Solution:

Let X be the number of distinct birthdays, and write $X = I_1 + \cdots + I_{365}$, where

$$I_j = \begin{cases} 1 & \text{if the } j\text{th day is represented,} \\ 0 & \text{otherwise.} \end{cases}$$

We create an indicator for each *day* of the year because X counts the number of *days* of the year that are represented. By the fundamental bridge,

$$E(I_j) = P(j\text{th day is represented}) = 1 - P(\text{no one born on day } j) = 1 - \left(\frac{364}{365}\right)^n$$

for all j. Then by linearity,

$$E(X) = 365\left(1 - \left(\frac{364}{365}\right)^n\right).$$

Now let Y be the number of birthday matches. Label the people as $1, 2, \ldots, n$, and order the $\binom{n}{2}$ pairs of people in some definite way. Then we can write

$$Y = J_1 + \cdots + J_{\binom{n}{2}},$$

where J_i is the indicator of the ith pair of people having the same birthday. We create an indicator for each *pair of people* since Y counts the number of *pairs of people* with the same birthday. The probability of any two people having the same birthday is $1/365$, so again by the fundamental bridge and linearity,

$$E(Y) = \frac{\binom{n}{2}}{365}. \qquad \square$$

In addition to the fundamental bridge and linearity, the last two examples used a basic form of symmetry to simplify the calculations greatly: within each sum of indicator r.v.s, each indicator had the same expected value. For example, in the matching problem the probability of the jth card being a match does not depend on j, so we can just take n times the expected value of the first indicator r.v.

Other forms of symmetry can also be extremely helpful when available. The next two examples showcase a form of symmetry that stems from having equally likely permutations. Note how symmetry, linearity, and the fundamental bridge are used in tandem to make seemingly very hard problems manageable.

Example 4.4.6 (Putnam problem). A permutation $a_1, a_2, \dots, a_n$ of $1, 2, \dots, n$ has a *local maximum* at j if $a_j > a_{j-1}$ and $a_j > a_{j+1}$ (for $2 \leq j \leq n-1$; for $j = 1$, a local maximum at j means $a_1 > a_2$ while for $j = n$, it means $a_n > a_{n-1}$). For example, $4, 2, 5, 3, 6, 1$ has 3 local maxima, at positions 1, 3, and 5. The Putnam exam (a famous, hard math competition, on which the median score is often a 0) from 2006 posed the following question: for $n \geq 2$, what is the average number of local maxima of a random permutation of $1, 2, \dots, n$, with all $n!$ permutations equally likely?

Solution:

This problem can be solved quickly using indicator r.v.s, symmetry, and the fundamental bridge. Let $I_1, \dots, I_n$ be indicator r.v.s, where I_j is 1 if there is a local maximum at position j, and 0 otherwise. We are interested in the expected value of $\sum_{j=1}^{n} I_j$. For $1 < j < n$, $EI_j = 1/3$ since having a local maximum at j is equivalent to a_j being the largest of a_{j-1}, a_j, a_{j+1}, which has probability $1/3$ since all orders are equally likely. For $j = 1$ or $j = n$, we have $EI_j = 1/2$ since then there is only one neighbor. Thus, by linearity,

$$E\left(\sum_{j=1}^{n} I_j\right) = 2 \cdot \frac{1}{2} + (n-2) \cdot \frac{1}{3} = \frac{n+1}{3}. \qquad \square$$

The next example introduces the *Negative Hypergeometric* distribution, which completes the following table. The table shows the distributions for four sampling schemes: the sampling can be done with or without replacement, and the stopping rule can require a fixed number of draws or a fixed number of successes.

	With replacement	Without replacement
Fixed number of trials	Binomial	Hypergeometric
Fixed number of successes	Negative Binomial	Negative Hypergeometric

Example 4.4.7 (Negative Hypergeometric). An urn contains w white balls and b black balls, which are randomly drawn one by one *without replacement*, until r white balls have been obtained. The number of black balls drawn before drawing the rth white ball has a *Negative Hypergeometric* distribution with parameters w, b, r. We denote this distribution by $\text{NHGeom}(w, b, r)$. Of course, we assume that $r \leq w$. For example, if we shuffle a deck of cards and deal them one at a time, the number of cards dealt before uncovering the first ace is $\text{NHGeom}(4, 48, 1)$.

As another example, suppose a college offers g good courses and b bad courses (for some definition of "good" and "bad"), and a student wants to find 4 good courses to take. Not having any idea which of the courses are good, the student randomly tries out courses one at a time, stopping when they have obtained 4 good courses. Then the number of bad courses the student tries out is $\text{NHGeom}(g, b, 4)$.

We can obtain the PMF of $X \sim \text{NHGeom}(w, b, r)$ by noting that, in the urn context, $X = k$ means that the $(r + k)$th ball chosen is white and exactly $r - 1$ of the first $r + k - 1$ balls chosen are white. This gives

$$P(X = k) = \frac{\binom{w}{r-1}\binom{b}{k}}{\binom{w+b}{r+k-1}} \cdot \frac{w - r + 1}{w + b - r - k + 1}$$

for $k = 0, 1, \ldots, b$ (and 0 otherwise).

Alternatively, we can imagine that we continue drawing balls until the urn has been emptied out; this is valid since whether or not we continue to draw balls after obtaining the rth white ball has no effect on X. Think of the $w + b$ balls as lined up in a random order, the order in which they will be drawn.

Then $X = k$ means that among the first $r + k - 1$ balls there are exactly $r - 1$ white balls, then there is a white ball, and then among the last $w + b - r - k$ balls there are exactly $w - r$ white balls. All $\binom{w+b}{w}$ possibilities for the locations of the white balls in the line are equally likely. So by the naive definition of probability, we have the following slightly simpler expression for the PMF:

$$P(X = k) = \frac{\binom{r+k-1}{r-1}\binom{w+b-r-k}{w-r}}{\binom{w+b}{w}},$$

for $k = 0, 1, \ldots, b$ (and 0 otherwise).

Finding the expected value of a Negative Hypergeometric r.v. directly from the definition of expectation results in complicated sums. But the answer is very simple: for $X \sim \text{NHGeom}(w, b, r)$, we have $E(X) = rb/(w + 1)$.

Let's prove this using indicator r.v.s. As explained above, we can assume that we continue drawing balls until the urn is empty. First consider the case $r = 1$. Label the black balls as $1, 2, \dots, b$, and let I_j be the indicator of black ball j being drawn before any white balls have been drawn. Then $P(I_j = 1) = 1/(w + 1)$ since, listing out the order in which black ball j and the white balls are drawn (ignoring the other balls), all orders are equally likely by symmetry, and $I_j = 1$ is equivalent to black ball j being first in this list. So by linearity,

$$E\left(\sum_{j=1}^{b} I_j\right) = \sum_{j=1}^{b} E(I_j) = b/(w + 1).$$

Sanity check: This answer makes sense since it is increasing in b, decreasing in w, and correct in the extreme cases $b = 0$ (when no black balls will be drawn) and $w = 0$ (when all the black balls will be exhausted before drawing a nonexistent white ball). Moreover, note that $b/(w + 1)$ looks similar to, but is strictly smaller than, b/w, which is the expected value of a $\text{Geom}(w/(w + b))$ r.v. It makes sense that sampling without replacement should give a smaller expected waiting time than sampling with replacement. Similarly, if you are searching for something you lost, it makes more sense to choose locations to check without replacement, rather than wasting time looking over and over again in locations you already ruled out.

For general r, write $X = X_1 + X_2 + \cdots + X_r$, where X_1 is the number of black balls before the first white ball, X_2 is the number of black balls after the first white ball but before the second white ball, etc. By essentially the same argument we used to handle the $r = 1$ case, we have $E(X_j) = b/(w + 1)$ for each j. So by linearity,

$$E(X) = rb/(w + 1). \qquad \square$$

Closely related to indicator r.v.s is an alternative expression for the expectation of a nonnegative integer-valued r.v. X. Rather than summing up values of X times values of the PMF of X, we can sum up probabilities of the form $P(X > n)$ (known as *tail probabilities*), over nonnegative integers n.

Theorem 4.4.8 (Expectation via survival function). Let X be a nonnegative integer-valued r.v. Let F be the CDF of X, and $G(x) = 1 - F(x) = P(X > x)$. The function G is called the *survival function* of X. Then

$$E(X) = \sum_{n=0}^{\infty} G(n).$$

That is, we can obtain the expectation of X by summing up the survival function (or, stated otherwise, summing up *tail probabilities* of the distribution).

Proof. For simplicity, we will prove the result only for the case that X is *bounded*, i.e., there is a nonnegative integer b such that X is always at most b. We can represent X as a sum of indicator r.v.s: $X = I_1 + I_2 + \cdots + I_b$, where $I_n = I(X \geq n)$. For example, if $X = 7$ occurs, then I_1 through I_7 equal 1 while the other indicators equal 0.

By linearity and the fundamental bridge, and the fact that $\{X \geq k\}$ is the same event as $\{X > k-1\}$,

$$E(X) = \sum_{k=1}^{b} E(I_k) = \sum_{k=1}^{b} P(X \geq k) = \sum_{n=0}^{b-1} P(X > n) = \sum_{n=0}^{\infty} G(n). \qquad \blacksquare$$

As a quick example, we use the above result to give another derivation of the mean of a Geometric r.v.

Example 4.4.9 (Geometric expectation redux). Let $X \sim \text{Geom}(p)$, and $q = 1 - p$. Using the Geometric story, $\{X > n\}$ is the event that the first $n+1$ trials are all failures. So by Theorem 4.4.8,

$$E(X) = \sum_{n=0}^{\infty} P(X > n) = \sum_{n=0}^{\infty} q^{n+1} = \frac{q}{1-q} = \frac{q}{p},$$

confirming what we already knew about the mean of a Geometric. □

4.5 Law of the unconscious statistician (LOTUS)

As we saw in the St. Petersburg paradox, $E(g(X))$ does *not* equal $g(E(X))$ in general if g is not linear. So how do we correctly calculate $E(g(X))$? Since $g(X)$ is an r.v., one way is to first find the distribution of $g(X)$ and then use the definition of expectation. Perhaps surprisingly, it turns out that it is possible to find $E(g(X))$ directly using the distribution of X, without first having to find the distribution of $g(X)$. This is done using the *law of the unconscious statistician* (LOTUS).

Theorem 4.5.1 (LOTUS). If X is a discrete r.v. and g is a function from $\mathbb{R}$ to $\mathbb{R}$, then

$$E(g(X)) = \sum_{x} g(x) P(X = x),$$

where the sum is taken over all possible values of X.

This means that we can get the expected value of $g(X)$ knowing only $P(X = x)$, the PMF of X; we don't need to know the PMF of $g(X)$. The name comes from the fact that in going from $E(X)$ to $E(g(X))$ it is tempting just to change x to $g(x)$ in the definition, which can be done very easily and mechanically, perhaps in a state of unconsciousness. On second thought, it may sound too good to be true that finding the distribution of $g(X)$ is not needed for this calculation, but LOTUS says it *is* true.

Before proving LOTUS in general, let's see why it is true in some special cases. Let X have support $0, 1, 2, \ldots$ with probabilities $p_0, p_1, p_2, \ldots$, so the PMF is $P(X = n) = p_n$. Then X^3 has support $0^3, 1^3, 2^3, \ldots$ with probabilities $p_0, p_1, p_2, \ldots$, so

$$E(X) = \sum_{n=0}^{\infty} n p_n,$$

$$E(X^3) = \sum_{n=0}^{\infty} n^3 p_n.$$

As claimed by LOTUS, to edit the expression for $E(X)$ into an expression for $E(X^3)$, we can just change the n in front of the p_n to an n^3. This was an easy example since the function $g(x) = x^3$ is one-to-one. But LOTUS holds much more generally. The key insight needed for the proof of LOTUS for general g is the same as the one we used for the proof of linearity: the expectation of $g(X)$ can be written in ungrouped form as

$$E(g(X)) = \sum_s g(X(s))P(\{s\}),$$

where the sum is over all the pebbles in the sample space, but we can also group the pebbles into super-pebbles according to the value that X assigns to them. Within the super-pebble $X = x$, $g(X)$ always takes on the value $g(x)$. Therefore,

$$\begin{aligned} E(g(X)) &= \sum_s g(X(s))P(\{s\}) \\ &= \sum_x \sum_{s:X(s)=x} g(X(s))P(\{s\}) \\ &= \sum_x g(x) \sum_{s:X(s)=x} P(\{s\}) \\ &= \sum_x g(x)P(X = x). \end{aligned}$$

In the last step, we used the fact that $\sum_{s:X(s)=x} P(\{s\})$ is the weight of the super-pebble $X = x$.

4.6 Variance

One important application of LOTUS is for finding the *variance* of a random variable. Like expected value, variance is a single-number summary of the distribution of a random variable. While the expected value tells us the center of mass of a distribution, the variance tells us how spread out the distribution is.

Definition 4.6.1 (Variance and standard deviation). The *variance* of an r.v. X is

$$\text{Var}(X) = E(X - EX)^2.$$

The square root of the variance is called the *standard deviation* (SD):

$$\text{SD}(X) = \sqrt{\text{Var}(X)}.$$

Recall that when we write $E(X - EX)^2$, we mean the expectation of the random variable $(X - EX)^2$, *not* $(E(X - EX))^2$ (which is 0 by linearity).

The variance of X measures how far X is from its mean on average, but instead of simply taking the average difference between X and its mean EX, we take the average *squared* difference. To see why, note that the average deviation from the mean, $E(X - EX)$, always equals 0 by linearity; positive and negative deviations cancel each other out. By squaring the deviations, we ensure that both positive and negative deviations contribute to the overall variability. However, because variance is an average squared distance, it has the wrong units: if X is in dollars, $\text{Var}(X)$ is in squared dollars. To get back to our original units, we take the square root; this gives us the standard deviation.

One might wonder why variance isn't defined as $E|X-EX|$, which would achieve the goal of counting both positive and negative deviations while maintaining the same units as X. This measure of variability isn't nearly as popular as $E(X - EX)^2$, for a variety of reasons. Most notably, the absolute value function isn't differentiable at 0, whereas the squaring function is differentiable everywhere and is central in various fundamental mathematical results such as the Pythagorean theorem.

An equivalent expression for variance is $\text{Var}(X) = E(X^2) - (EX)^2$. This formula is often easier to work with when doing actual calculations. Since this is the variance formula we will use over and over again, we state it as its own theorem.

Theorem 4.6.2. For any r.v. X,

$$\text{Var}(X) = E(X^2) - (EX)^2.$$

Proof. Let $\mu = EX$. Expanding $(X - \mu)^2$ and using linearity, the variance of X is

$$E(X - \mu)^2 = E(X^2 - 2\mu X + \mu^2) = E(X^2) - 2\mu EX + \mu^2 = E(X^2) - \mu^2. \quad \blacksquare$$

Variance has the following properties. The first two are easily verified from the definition, the third will be addressed in a later chapter, and the last one is proven just after stating it.

- $\text{Var}(X + c) = \text{Var}(X)$ for any constant c. Intuitively, if we shift a distribution to the left or right, that should affect the center of mass of the distribution but not its spread.

- $\text{Var}(cX) = c^2\text{Var}(X)$ for any constant c.

- If X and Y are independent, then $\text{Var}(X + Y) = \text{Var}(X) + \text{Var}(Y)$. We prove this and discuss it more in Chapter 7. This is not true in general if X and Y are dependent. For example, in the extreme case where X always equals Y, we have

$$\text{Var}(X + Y) = \text{Var}(2X) = 4\text{Var}(X) > 2\text{Var}(X) = \text{Var}(X) + \text{Var}(Y)$$

if $\text{Var}(X) > 0$ (which will be true unless X is a constant, as the next property shows).

- $\text{Var}(X) \geq 0$, with equality if and only if $P(X = a) = 1$ for some constant a. In other words, the only random variables that have zero variance are constants (which can be thought of as degenerate r.v.s); all other r.v.s have positive variance.

To prove the last property, note that $\text{Var}(X)$ is the expectation of the *nonnegative* r.v. $(X - EX)^2$, so $\text{Var}(X) \geq 0$. If $P(X = a) = 1$ for some constant a, then $E(X) = a$ and $E(X^2) = a^2$, so $\text{Var}(X) = 0$. Conversely, suppose that $\text{Var}(X) = 0$. Then $E(X - EX)^2 = 0$, which shows that $(X - EX)^2 = 0$ has probability 1, which in turn shows that X equals its mean with probability 1.

☣ **4.6.3** (Variance is not linear). Unlike expectation, variance is *not* linear. The constant comes out *squared* in $\text{Var}(cX) = c^2\text{Var}(X)$, and the variance of the sum of r.v.s may not be the sum of their variances if they are dependent.

Example 4.6.4 (Geometric and Negative Binomial variance). In this example we'll use LOTUS to compute the variance of the Geometric distribution.

Let $X \sim \text{Geom}(p)$. We already know $E(X) = q/p$. By LOTUS,

$$E(X^2) = \sum_{k=0}^{\infty} k^2 P(X = k) = \sum_{k=0}^{\infty} k^2 pq^k = \sum_{k=1}^{\infty} k^2 pq^k.$$

We'll find this using a tactic similar to how we found the expectation, starting from the geometric series

$$\sum_{k=0}^{\infty} q^k = \frac{1}{1-q}$$

and taking derivatives. After differentiating once with respect to q, we have

$$\sum_{k=1}^{\infty} kq^{k-1} = \frac{1}{(1-q)^2}.$$

We start the sum from $k = 1$ since the $k = 0$ term is 0 anyway. If we differentiate again, we'll get $k(k-1)$ instead of k^2 as we want, so let's replenish our supply of q's by multiplying both sides by q. This gives

$$\sum_{k=1}^{\infty} kq^k = \frac{q}{(1-q)^2}.$$

Now we are ready to take another derivative:

$$\sum_{k=1}^{\infty} k^2 q^{k-1} = \frac{1+q}{(1-q)^3},$$

so

$$E(X^2) = \sum_{k=1}^{\infty} k^2 pq^k = pq\frac{1+q}{(1-q)^3} = \frac{q(1+q)}{p^2}.$$

Finally,

$$\text{Var}(X) = E(X^2) - (EX)^2 = \frac{q(1+q)}{p^2} - \left(\frac{q}{p}\right)^2 = \frac{q}{p^2}.$$

This is also the variance of the First Success distribution, since shifting by a constant does not affect the variance.

Since an NBin(r, p) r.v. can be represented as a sum of r i.i.d. Geom(p) r.v.s by Theorem 4.3.10, and since variance is additive for independent random variables, it follows that the variance of the NBin(r, p) distribution is $r \cdot \frac{q}{p^2}$. □

LOTUS is an all-purpose tool for computing $E(g(X))$ for any g, but as it usually leads to complicated sums, it should be used as a last resort. For variance calculations, our trusty indicator r.v.s can sometimes be used in place of LOTUS, as in the next example.

Example 4.6.5 (Binomial variance). Let's find the variance of $X \sim \text{Bin}(n, p)$ using indicator r.v.s to avoid tedious sums. Represent $X = I_1 + I_2 + \cdots + I_n$, where I_j is the indicator of the jth trial being a success. Each I_j has variance

$$\text{Var}(I_j) = E(I_j^2) - (E(I_j))^2 = p - p^2 = p(1-p).$$

(Recall that $I_j^2 = I_j$, so $E(I_j^2) = E(I_j) = p$.)

Since the I_j are independent, we can add their variances to get the variance of their sum:

$$\text{Var}(X) = \text{Var}(I_1) + \cdots + \text{Var}(I_n) = np(1-p).$$

Alternatively, we can find $E(X^2)$ by first finding $E\binom{X}{2}$. The latter sounds more complicated, but actually it is simpler since $\binom{X}{2}$ is the number of *pairs* of successful trials. Creating an indicator r.v. for each pair of trials, we have

$$E\binom{X}{2} = \binom{n}{2}p^2.$$

Thus,

$$n(n-1)p^2 = E(X(X-1)) = E(X^2) - E(X) = E(X^2) - np,$$

which again gives

$$\text{Var}(X) = E(X^2) - (EX)^2 = (n(n-1)p^2 + np) - (np)^2 = np(1-p).$$

Exercise 48 uses this strategy to find the variance of the Hypergeometric. □

4.7 Poisson

The last discrete distribution that we'll introduce in this chapter is the Poisson, which is an extremely popular distribution for modeling discrete data. We'll introduce its PMF, mean, and variance, and then discuss its story in more detail.

Definition 4.7.1 (Poisson distribution). An r.v. X has the *Poisson distribution* with parameter λ, where $\lambda > 0$, if the PMF of X is

$$P(X = k) = \frac{e^{-\lambda}\lambda^k}{k!}, \quad k = 0, 1, 2, \ldots.$$

We write this as $X \sim \text{Pois}(\lambda)$.

This is a valid PMF because of the Taylor series $\sum_{k=0}^{\infty} \frac{\lambda^k}{k!} = e^{\lambda}$.

Example 4.7.2 (Poisson expectation and variance). Let $X \sim \text{Pois}(\lambda)$. We will show that the mean and variance are both equal to λ. For the mean, we have

$$\begin{aligned} E(X) &= e^{-\lambda}\sum_{k=0}^{\infty} k\frac{\lambda^k}{k!} \\ &= e^{-\lambda}\sum_{k=1}^{\infty} k\frac{\lambda^k}{k!} \\ &= \lambda e^{-\lambda}\sum_{k=1}^{\infty} \frac{\lambda^{k-1}}{(k-1)!} \\ &= \lambda e^{-\lambda} e^{\lambda} = \lambda. \end{aligned}$$

First we dropped the $k = 0$ term because it was 0. Then we took a λ out of the sum so that what was left inside was just the Taylor series for e^{λ}.

To get the variance, we first find $E(X^2)$. By LOTUS,

$$E(X^2) = \sum_{k=0}^{\infty} k^2 P(X = k) = e^{-\lambda}\sum_{k=0}^{\infty} k^2\frac{\lambda^k}{k!}.$$

From here, the derivation is very similar to that of the variance of the Geometric. Differentiate the familiar series

$$\sum_{k=0}^{\infty} \frac{\lambda^k}{k!} = e^{\lambda}$$

with respect to λ and replenish:

$$\begin{aligned} \sum_{k=1}^{\infty} k\frac{\lambda^{k-1}}{k!} &= e^{\lambda}, \\ \sum_{k=1}^{\infty} k\frac{\lambda^k}{k!} &= \lambda e^{\lambda}. \end{aligned}$$

Rinse and repeat:

$$\begin{aligned} \sum_{k=1}^{\infty} k^2\frac{\lambda^{k-1}}{k!} &= e^{\lambda} + \lambda e^{\lambda} = e^{\lambda}(1 + \lambda), \\ \sum_{k=1}^{\infty} k^2\frac{\lambda^k}{k!} &= e^{\lambda}\lambda(1 + \lambda). \end{aligned}$$

Finally,

$$E(X^2) = e^{-\lambda} \sum_{k=0}^{\infty} k^2 \frac{\lambda^k}{k!} = e^{-\lambda} e^{\lambda} \lambda(1+\lambda) = \lambda(1+\lambda),$$

so

$$\text{Var}(X) = E(X^2) - (EX)^2 = \lambda(1+\lambda) - \lambda^2 = \lambda.$$

Thus, the mean and variance of a Pois(λ) r.v. are both equal to λ. □

Figure 4.7 shows the PMF and CDF of the Pois(2) and Pois(5) distributions from $k = 0$ to $k = 10$. It appears that the mean of the Pois(2) is around 2 and the mean of the Pois(5) is around 5, consistent with our findings above. The PMF of the Pois(2) is highly skewed, but as λ grows larger, the skewness is reduced and the PMF becomes more bell-shaped.

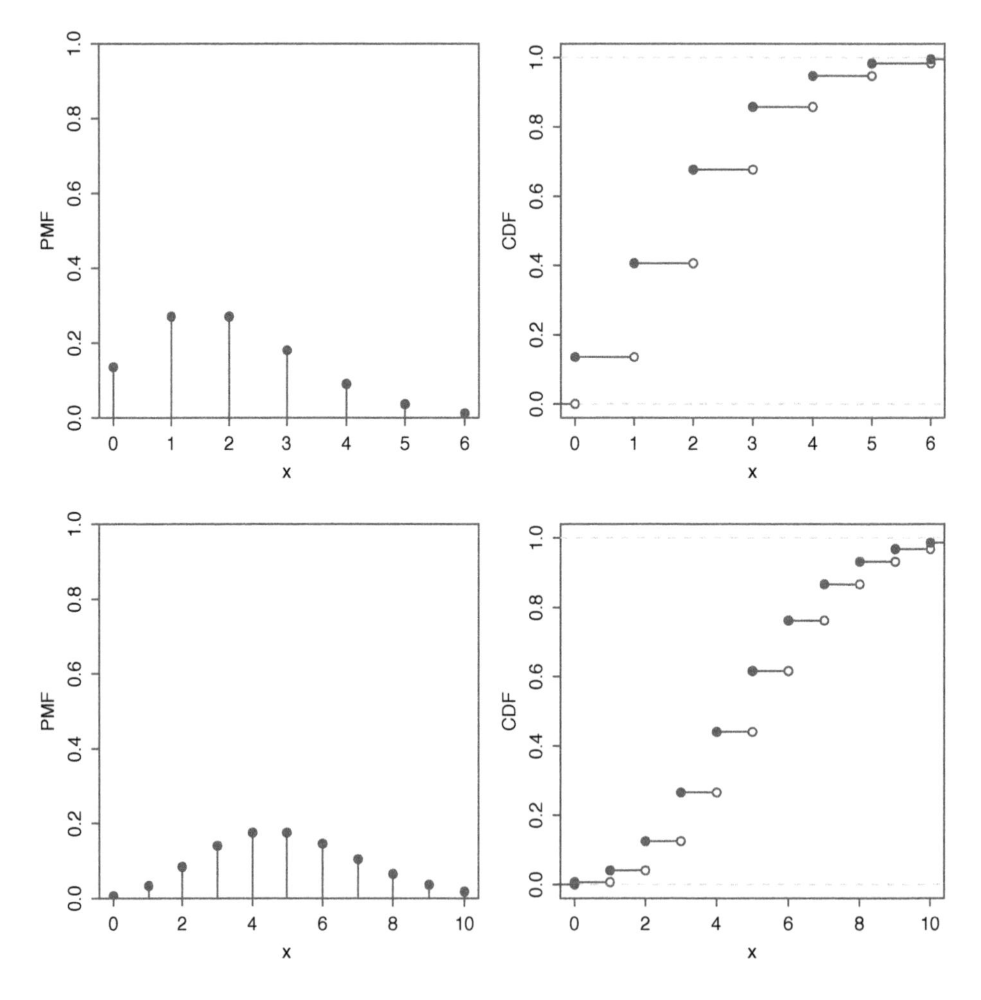

FIGURE 4.7
Top: Pois(2) PMF and CDF. Bottom: Pois(5) PMF and CDF.

The Poisson distribution is often used in situations where we are counting the number of successes in a particular region or interval of time, and there are a large number of trials, each with a small probability of success. For example, the following random variables could follow a distribution that is approximately Poisson.

- The number of emails you receive in an hour. There are a lot of people who could potentially email you in that hour, but it is unlikely that any specific person will actually email you in that hour. Alternatively, imagine subdividing the hour into milliseconds. There are 3.6×10^6 seconds in an hour, but in any specific millisecond it is unlikely that you will get an email.

- The number of chips in a chocolate chip cookie. Imagine subdividing the cookie into small cubes; the probability of getting a chocolate chip in a single cube is small, but the number of cubes is large.

- The number of earthquakes in a year in some region of the world. At any given time and location, the probability of an earthquake is small, but there are a large number of possible times and locations for earthquakes to occur over the course of the year.

The parameter λ is interpreted as the *rate* of occurrence of these rare events; in the examples above, λ could be 20 (emails per hour), 10 (chips per cookie), and 2 (earthquakes per year). The *Poisson paradigm* says that in applications similar to the ones above, we can approximate the distribution of the number of events that occur by a Poisson distribution.

Approximation 4.7.3 (Poisson paradigm). Let $A_1, \ldots, A_n$ be events with $p_j = P(A_j)$, where n is large, the p_j are small, and the A_j are independent or weakly dependent. Let

$$X = \sum_{j=1}^{n} I(A_j)$$

count how many of the A_j occur. Then X is approximately distributed as Pois(λ), with $\lambda = \sum_{j=1}^{n} p_j$.

Proving that the above approximation is good is difficult, and would require first giving precise definitions of weak dependence (there are various ways to measure dependence of r.v.s) and of good approximations (there are various ways to measure how good an approximation is). A remarkable theorem is that if the A_j are independent, $N \sim \text{Pois}(\lambda)$, and B is any set of nonnegative integers, then

$$|P(X \in B) - P(N \in B)| \leq \min\left(1, \frac{1}{\lambda}\right) \sum_{j=1}^{n} p_j^2.$$

This gives an upper bound on how much error is incurred from using a Poisson approximation. It also makes more precise how small the p_j should be: we want $\sum_{j=1}^{n} p_j^2$ to be very small, or at least very small compared to λ. The result can be shown using an advanced technique known as the *Stein-Chen method.*

The Poisson paradigm is also called the *law of rare events*. The interpretation of "rare" is that the p_j are small, not that λ is small. For example, in the email example, the low probability of getting an email from a specific person in a particular hour is offset by the large number of people who could send you an email in that hour.

In the examples we gave above, the number of events that occur isn't *exactly* Poisson because a Poisson random variable has no upper bound, whereas how many of $A_1, \dots, A_n$ occur is at most n, and there is a limit to how many chocolate chips can be crammed into a cookie. But the Poisson distribution often gives good *approximations*. Note that the conditions for the Poisson paradigm to hold are fairly flexible: the n trials can have different success probabilities, and the trials don't have to be independent, though they should not be very dependent. So there are a wide variety of situations that can be cast in terms of the Poisson paradigm. This makes the Poisson a popular model, or at least a starting point, for data whose values are nonnegative integers (called *count data* in statistics).

Example 4.7.4 (Balls in boxes). There are k distinguishable balls and n distinguishable boxes. The balls are randomly placed in the boxes, with all n^k possibilities equally likely. Problems in this setting are called *occupancy problems*, and are at the core of many widely used algorithms in computer science.

(a) Find the expected number of empty boxes (fully simplified, *not* as a sum).

(b) Find the probability that at least one box is empty. Express your answer as a sum of at most n terms.

(c) Now let $n = 1000$, $k = 5806$. The expected number of empty boxes is then approximately 3. Find a good approximation as a decimal for the probability that at least one box is empty. The handy fact $e^3 \approx 20$ may help.

Solution:

(a) Let I_j be the indicator r.v. for the jth box being empty. Then

$$E(I_j) = P(I_j = 1) = \left(1 - \frac{1}{n}\right)^k.$$

By linearity,

$$E\left(\sum_{j=1}^{n} I_j\right) = \sum_{j=1}^{n} E(I_j) = n\left(1 - \frac{1}{n}\right)^k.$$

(b) The probability is 1 for $k < n$. In general, let A_j be the event that box j is empty. By inclusion-exclusion,

$$\begin{aligned} P(A_1 \cup A_2 \cup \cdots \cup A_n) &= \sum_{j=1}^{n} (-1)^{j+1} \binom{n}{j} P(A_1 \cap A_2 \cap \cdots \cap A_j) \\ &= \sum_{j=1}^{n-1} (-1)^{j+1} \binom{n}{j} \left(1 - \frac{j}{n}\right)^k. \end{aligned}$$

(c) The number X of empty boxes is approximately Pois(3), since there are a lot of boxes but each is very unlikely to be empty; the probability that a specific box is empty is $(1-\frac{1}{n})^k = \frac{1}{n} \cdot E(X) \approx 0.003$. So

$$P(X \geq 1) = 1 - P(X = 0) \approx 1 - e^{-3} \approx 1 - \frac{1}{20} = 0.95. \qquad \square$$

Poisson approximation greatly simplifies obtaining a good approximate solution to the birthday problem discussed in Chapter 1, and makes it possible to obtain good approximations to various variations which would be hard to solve exactly.

Example 4.7.5 (Birthday problem continued). If we have m people and make the usual assumptions about birthdays, then each pair of people has probability $p = 1/365$ of having the same birthday, and there are $\binom{m}{2}$ pairs. By the Poisson paradigm the distribution of the number X of birthday matches is approximately Pois(λ), where $\lambda = \binom{m}{2}\frac{1}{365}$. Then the probability of at least one match is

$$P(X \geq 1) = 1 - P(X = 0) \approx 1 - e^{-\lambda}.$$

For $m = 23$, $\lambda = 253/365$ and $1-e^{-\lambda} \approx 0.500002$, which agrees with our finding from Chapter 1 that we need 23 people to have a 50-50 chance of a matching birthday.

Note that even though $m = 23$ is fairly small, the relevant quantity in this problem is actually $\binom{m}{2}$, which is the total number of "trials" for a successful birthday match, so the Poisson approximation still performs well. $\square$

Example 4.7.6 (Near-birthday problem). What if we want to find the number of people required in order to have a 50-50 chance that two people would have birthdays within one day of each other (i.e., on the same day or one day apart)? Unlike the original birthday problem, this is difficult to obtain an exact answer for, but the Poisson paradigm still applies. The probability that any two people have birthdays within one day of each other is 3/365 (choose a birthday for the first person, and then the second person needs to be born on that day, the day before, or the day after). Again there are $\binom{m}{2}$ possible pairs, so the number of within-one-day matches is approximately Pois(λ) where $\lambda = \binom{m}{2}\frac{3}{365}$. Then a calculation similar to the one above tells us that we need $m = 14$ or more. This was a quick approximation, but it turns out that $m = 14$ is the exact answer! $\square$

Example 4.7.7 (Birth-minute and birth-hour). There are 1600 sophomores at a certain college. Throughout this example, make the usual assumptions as in the birthday problem.

(a) Find a Poisson approximation for the probability that there are two sophomores who were born not only on the same day of the year, but also at the same hour *and* the same minute (e.g., both sophomores were born at 8:20 pm on March 31, not necessarily in the same year).

(b) With assumptions as in (a), what is the probability that there are *four* sophomores who were born not only on the same day, but also at the same hour (e.g., all were born between 2 pm and 3 pm on March 31, not necessarily in the same year)?

Give two different Poisson approximations for this value, one based on creating an indicator r.v. for each quadruplet of sophomores, and the other based on creating an indicator r.v. for each possible day-hour. Which do you think is more accurate?

Solution:

(a) This is the birthday problem, with $c = 365 \cdot 24 \cdot 60 = 525600$ categories rather than 365 categories.[1] Let $n = 1600$. Creating an indicator r.v. for each pair of sophomores, by linearity the expected number of pairs born on the same day-hour-minute is

$$\lambda_1 = \binom{n}{2}\frac{1}{c}.$$

By Poisson approximation, the probability of at least one match is approximately

$$1 - \exp(-\lambda_1) \approx 0.9122.$$

This approximation is very accurate: typing `pbirthday(1600, classes=365*24*60)` in R yields 0.9125.

(b) Now there are $b = 365 \cdot 24 = 8760$ categories. Let's explore two different methods of Poisson approximation.

Method 1: Create an indicator for each set of 4 sophomores. By linearity, the expected number of sets of 4 sophomores born on the same day-hour is

$$\lambda_2 = \binom{n}{4}\frac{1}{b^3}.$$

Poisson approximation gives that the desired probability is approximately

$$1 - \exp(-\lambda_2) \approx 0.333.$$

Method 2: Create an indicator for each possible day-hour. Let I_j be the indicator for at least 4 people having been born on the jth day-hour of the year (ordered chronologically), for $1 \leq j \leq b$. Let $p = 1/b$ and $q = 1 - p$. Then

$$\begin{aligned} E(I_j) &= P(I_j = 1) \\ &= 1 - P(\text{at most 3 people born on the } j\text{th day-hour}) \\ &= 1 - q^n - npq^{n-1} - \binom{n}{2}p^2q^{n-2} - \binom{n}{3}p^3q^{n-3}. \end{aligned}$$

The expected number of day-hours on which at least 4 sophomores were born is

$$\lambda_3 = b \cdot E(I_1),$$

with $E(I_1)$ as above. We then have the Poisson approximation

$$1 - \exp(-\lambda_3) \approx 0.295.$$

[1] The song "Seasons of Love" from *Rent* gives a musical interpretation of this fact.

The command `pbirthday(1600, classes = 8760, coincident=4)` in R gives that the correct answer is 0.296. So Method 2 is more accurate than Method 1.

An intuitive explanation for why Method 1 is less accurate is that there is a more substantial dependence in the indicators in that method. For example, being given that sophomores $1, 2, 3, 4$ share the same birth day-hour greatly increases the chance that sophomores $1, 2, 3, 5$ share the same birth day-hour. In contrast, knowing that at least 4 sophomores were born on a specific day-hour provides very little information about whether at least 4 were born on a different specific day-hour. □

4.8 Connections between Poisson and Binomial

The Poisson and Binomial distributions are closely connected, and their relationship is exactly parallel to the relationship between the Binomial and Hypergeometric distributions that we examined in the previous chapter: we can get from the Poisson to the Binomial by *conditioning*, and we can get from the Binomial to the Poisson by *taking a limit*.

Our results will rely on the fact that the sum of independent Poissons is Poisson, just as the sum of independent Binomials is Binomial. We'll prove this result using the law of total probability for now; in Chapter 6 we'll learn a faster method that uses a tool called the moment generating function. Chapter 13 gives further insight into these results.

Theorem 4.8.1 (Sum of independent Poissons). If $X \sim \text{Pois}(\lambda_1)$, $Y \sim \text{Pois}(\lambda_2)$, and X is independent of Y, then $X + Y \sim \text{Pois}(\lambda_1 + \lambda_2)$.

Proof. To get the PMF of $X+Y$, condition on X and use the law of total probability:

$$\begin{aligned}
P(X+Y=k) &= \sum_{j=0}^{k} P(X+Y=k|X=j)P(X=j) \\
&= \sum_{j=0}^{k} P(Y=k-j)P(X=j) \\
&= \sum_{j=0}^{k} \frac{e^{-\lambda_2}\lambda_2^{k-j}}{(k-j)!}\frac{e^{-\lambda_1}\lambda_1^{j}}{j!} \\
&= \frac{e^{-(\lambda_1+\lambda_2)}}{k!}\sum_{j=0}^{k}\binom{k}{j}\lambda_1^{j}\lambda_2^{k-j} \\
&= \frac{e^{-(\lambda_1+\lambda_2)}(\lambda_1+\lambda_2)^k}{k!}.
\end{aligned}$$

The last step used the binomial theorem. Since we've arrived at the Pois$(\lambda_1 + \lambda_2)$ PMF, we have $X + Y \sim \text{Pois}(\lambda_1 + \lambda_2)$.

The story of the Poisson distribution provides intuition for this result. If there are two different types of events occurring at rates λ_1 and λ_2, independently, then the overall event rate is $\lambda_1 + \lambda_2$. ■

Theorem 4.8.2 (Poisson given a sum of Poissons)**.** If $X \sim \text{Pois}(\lambda_1)$, $Y \sim \text{Pois}(\lambda_2)$, and X is independent of Y, then the conditional distribution of X given $X+Y = n$ is Bin $(n, \lambda_1/(\lambda_1 + \lambda_2))$.

Proof. Exactly as in the corresponding proof for the Binomial and Hypergeometric, we use Bayes' rule to compute the conditional PMF $P(X = k|X + Y = n)$:

$$\begin{aligned} P(X = k|X + Y = n) &= \frac{P(X + Y = n|X = k)P(X = k)}{P(X + Y = n)} \\ &= \frac{P(Y = n - k)P(X = k)}{P(X + Y = n)}. \end{aligned}$$

Now we plug in the PMFs of X, Y, and $X + Y$; the last of these is distributed Pois$(\lambda_1 + \lambda_2)$ by the previous theorem. This gives

$$\begin{aligned} P(X = k|X + Y = n) &= \frac{\left(\dfrac{e^{-\lambda_2}\lambda_2^{n-k}}{(n-k)!}\right)\left(\dfrac{e^{-\lambda_1}\lambda_1^{k}}{k!}\right)}{\dfrac{e^{-(\lambda_1+\lambda_2)}\left(\lambda_1+\lambda_2\right)^n}{n!}} \\ &= \binom{n}{k}\frac{\lambda_1^k\lambda_2^{n-k}}{(\lambda_1+\lambda_2)^n} \\ &= \binom{n}{k}\left(\frac{\lambda_1}{\lambda_1+\lambda_2}\right)^k\left(\frac{\lambda_2}{\lambda_1+\lambda_2}\right)^{n-k}, \end{aligned}$$

which is the Bin$(n, \lambda_1/(\lambda_1 + \lambda_2))$ PMF, as desired. ■

Conversely, if we take the limit of the Bin(n, p) distribution as $n \to \infty$ and $p \to 0$ with np fixed, we arrive at a Poisson distribution. This provides the basis for the *Poisson approximation to the Binomial distribution.*

Theorem 4.8.3 (Poisson approximation to Binomial)**.** If $X \sim \text{Bin}(n, p)$ and we let $n \to \infty$ and $p \to 0$ such that $\lambda = np$ remains fixed, then the PMF of X converges to the Pois(λ) PMF. More generally, the same conclusion holds if $n \to \infty$ and $p \to 0$ in such a way that np converges to a constant λ.

This is a special case of the Poisson paradigm, where the A_j are independent with the same probabilities, so that $\sum_{j=1}^{n} I(A_j)$ has a Binomial distribution. In this special case, we can prove that the Poisson approximation makes sense just by taking a limit of the Binomial PMF.

Proof. We will prove this for the case that $\lambda = np$ is fixed while $n \to \infty$ and $p \to 0$, by showing that the Bin(n,p) PMF converges to the Pois(λ) PMF. For $0 \leq k \leq n$,

$$\begin{aligned} P(X = k) &= \binom{n}{k} p^k (1-p)^{n-k} \\ &= \frac{n(n-1)\dots(n-k+1)}{k!} \left(\frac{\lambda}{n}\right)^k \left(1 - \frac{\lambda}{n}\right)^n \left(1 - \frac{\lambda}{n}\right)^{-k} \\ &= \frac{\lambda^k}{k!} \frac{n(n-1)\dots(n-k+1)}{n^k} \left(1 - \frac{\lambda}{n}\right)^n \left(1 - \frac{\lambda}{n}\right)^{-k}. \end{aligned}$$

Letting $n \to \infty$ with k fixed,

$$\begin{aligned} \frac{n(n-1)\dots(n-k+1)}{n^k} &\to 1, \\ \left(1 - \frac{\lambda}{n}\right)^n &\to e^{-\lambda}, \\ \left(1 - \frac{\lambda}{n}\right)^{-k} &\to 1, \end{aligned}$$

where the $e^{-\lambda}$ comes from the compound interest formula from Section A.2.5 of the math appendix. So

$$P(X = k) \to \frac{e^{-\lambda}\lambda^k}{k!},$$

which is the Pois(λ) PMF. ■

This theorem implies that if n is large, p is small, and np is moderate, we can approximate the Bin(n,p) PMF by the Pois(np) PMF. The main thing that matters here is that p should be small; in fact, the result mentioned after the statement of the Poisson paradigm says in this case that the error in approximating $P(X \in B) \approx P(N \in B)$ for $X \sim \text{Bin}(n,p), N \sim \text{Pois}(np)$ is at most $\min(p, np^2)$.

Example 4.8.4 (Visitors to a website). The owner of a certain website is studying the distribution of the number of visitors to the site. Every day, a million people independently decide whether to visit the site, with probability $p = 2 \times 10^{-6}$ of visiting. Give a good approximation for the probability of getting *at least three* visitors on a particular day.

Solution:

Let $X \sim \text{Bin}(n,p)$ be the number of visitors, where $n = 10^6$. It is easy to run into computational difficulties or numerical errors in exact calculations with this distribution since n is so large and p is so small. But since n is large, p is small, and $np = 2$ is moderate, Pois(2) is a good approximation. This gives

$$P(X \geq 3) = 1 - P(X < 3) \approx 1 - e^{-2} - e^{-2} \cdot 2 - e^{-2} \cdot \frac{2^2}{2!} = 1 - 5e^{-2} \approx 0.3233,$$

which turns out to be extremely accurate. □

4.9 *Using probability and expectation to prove existence

An amazing and beautiful fact is that we can use probability and expectation to prove the *existence* of objects with properties we care about. This technique is called the *probabilistic method*, and it is based on two simple but surprisingly powerful ideas. Suppose I want to show that there exists an object in a collection with a certain property. This desire seems at first to have nothing to do with probability; I could simply examine each object in the collection one by one until finding an object with the desired property.

The probabilistic method rejects such painstaking inspection in favor of random selection: our strategy is to pick an object *at random* from the collection and show that there is a positive probability of the random object having the desired property. Note that we are not required to compute the exact probability, but merely to show it is greater than 0. If we can show that the probability of the property holding is positive, then we know that there must exist an object with the property—even if we don't know how to explicitly construct such an object.

Similarly, suppose each object has a score, and I want to show that there exists an object with a "good" score—that is, a score exceeding a particular threshold. Again, we proceed by choosing a *random object* and considering its score, X. We know there is an object in the collection whose score is at least $E(X)$—it's impossible for every object to be below average! If $E(X)$ is already a good score, then there must also be an object in the collection with a good score. Thus we can show the existence of an object with a good score by showing that the average score is already good.

Let's state the two key ideas formally.

- The possibility principle: Let A be the event that a randomly chosen object in a collection has a certain property. If $P(A) > 0$, then there exists an object with the property.
- The good score principle: Let X be the score of a randomly chosen object. If $E(X) \geq c$, then there is an object with a score of at least c.

To see why the possibility principle is true, consider its contrapositive: if there is no object with the desired property, then the probability of a randomly chosen object having the property is 0. Similarly, the contrapositive of the good score principle is "if all objects have a score below c, then the average score is below c", which is true since a weighted average of numbers less than c is a number less than c.

The probabilistic method doesn't tell us *how* to find an object with the desired property; it only assures us that one exists.

Example 4.9.1. A group of 100 people are assigned to 15 committees of size 20, such that each person serves on 3 committees. Show that there exist 2 committees that have at least 3 people in common.

Solution:

A direct approach is inadvisable here: one would have to list all possible committee assignments and compute, for each one, the number of people in common in every pair of committees. The probabilistic method lets us bypass brute-force calculations. To prove the existence of two committees with an overlap of at least three people, we'll calculate the *average* overlap of two *randomly chosen* committees in an arbitrary committee assignment. So choose two committees at random, and let X be the number of people on both committees. We can represent $X = I_1 + I_2 + \cdots + I_{100}$, where $I_j = 1$ if the jth person is on both committees and 0 otherwise. By symmetry, all of the indicators have the same expected value, so $E(X) = 100E(I_1)$, and we just need to find $E(I_1)$.

By the fundamental bridge, $E(I_1)$ is the probability that person 1 (whom we'll name Bob) is on both committees (which we'll call A and B). There are a variety of ways to calculate this probability; one way is to think of Bob's committees as 3 tagged elk in a population of 15. Then A and B are a sample of 2 elk, made without replacement. Using the HGeom(3, 12, 2) PMF, the probability that both of these elk are tagged (i.e., the probability that both committees contain Bob) is $\binom{3}{2}\binom{12}{0}/\binom{15}{2} = 1/35$. Therefore,

$$E(X) = 100/35 = 20/7,$$

which is just shy of the desired "good score" of 3. But hope is not lost! The good score principle says there exist two committees with an overlap of at least 20/7, but since the overlap between two committees must be an integer, an overlap of at least 20/7 implies an overlap of at least 3. Thus, there exist two committees with at least 3 people in common. □

4.9.1 *Communicating over a noisy channel

Another major application of the probabilistic method is in *information theory*, the subject which studies (among other things) how to achieve reliable communication across a noisy channel. Consider the problem of trying to send a message when there is noise. This problem is encountered by millions of people every day, such as when talking on the phone (you may be misheard). Suppose that the message you want to send is represented as a binary vector $x \in \{0, 1\}^k$, and that you want to use a *code* to improve the chance that your message will get through successfully.

Definition 4.9.2 (Codes and rates). Given positive integers k and n, a *code* is a function c that assigns to each input message $x \in \{0, 1\}^k$ a *codeword* $c(x) \in \{0, 1\}^n$. The *rate* of this code is k/n (the number of input bits per output bit). After $c(x)$ is sent, a *decoder* takes the received message, which may be a corrupted version of $c(x)$, and attempts to recover the correct x.

For example, an obvious code would be to repeat yourself a bunch of times, sending x a bunch of times in a row, say m (with m odd); this is called a *repetition code*. The receiver could then *decode* by going with the majority, e.g., decoding the first bit of x as a 1 if that bit was received more times as a 1 than as a 0. But this code may be very inefficient; to get the probability of failure very small, you may need to repeat yourself many times, resulting in a very low rate $1/m$ of communication.

Claude Shannon, the founder of information theory, showed something amazing: even in a very noisy channel, there is a code allowing for very reliable communication at a rate that does not go to 0 as we require the probability of failure to be lower and lower. His proof was even more amazing: he studied the performance of a completely *random* code. Richard Hamming, who worked with Shannon at Bell Labs, described Shannon's approach as follows.

> Courage is another attribute of those who do great things. Shannon is a good example. For some time he would come to work at about 10:00 am, play chess until about 2:00 pm and go home.
>
> The important point is how he played chess. When attacked he seldom, if ever, defended his position, rather he attacked back. Such a method of playing soon produces a very interrelated board. He would then pause a bit, think and advance his queen saying, "I ain't [scared] of nothin'." It took me a while to realize that of course that is why he was able to prove the existence of good coding methods. Who but Shannon would think to average over all random codes and expect to find that the average was close to ideal? I learned from him to say the same to myself when stuck, and on some occasions his approach enabled me to get significant results. [15]

We will prove a version of Shannon's result, for the case of a channel where each transmitted bit gets flipped (from 0 to 1 or from 1 to 0) with probability p, independently. First we need two definitions. A natural measure of distance between binary vectors, named after Hamming, is as follows.

Definition 4.9.3 (Hamming distance). For two binary vectors v and w of the same length, the *Hamming distance* $d(v, w)$ is the number of positions in which they differ. We can write this as

$$d(v, w) = \sum_i |v_i - w_i|.$$

The following function arises very frequently in information theory.

Definition 4.9.4 (Binary entropy function). For $0 < p < 1$, the *binary entropy function* H is given by

$$H(p) = -p \log_2 p - (1-p)\log_2(1-p).$$

We also define $H(0) = H(1) = 0$.

The interpretation of $H(p)$ in information theory is that it is a measure of how much information we get from observing a Bern(p) r.v.; $H(1/2) = 1$ says that a fair coin flip provides 1 bit of information, while $H(1) = 0$ says that with a coin that always lands Heads, there's no information gained from being told the result of the flip, since we already know the result.

Now consider a channel where each transmitted bit gets flipped with probability p, independently. Intuitively, it may seem that smaller p is always better, but note that $p = 1/2$ is actually the worst-case scenario. In that case, technically known as a *useless channel*, it is impossible to send information over the channel: the output will be independent of the input! Analogously, in deciding whether to watch a movie, would you rather hear a review from someone you always disagree with or someone you agree with half the time? We now prove that for $0 < p < 1/2$, it is possible to communicate very reliably with rate very close to $1 - H(p)$.

Theorem 4.9.5 (Shannon). Consider a channel where each transmitted bit gets flipped with probability p, independently. Let $0 < p < 1/2$ and $\epsilon > 0$. There exists a code with rate at least $1 - H(p) - \epsilon$ that can be decoded with probability of error less than ϵ.

Proof. We can assume that $1 - H(p) - \epsilon > 0$, since otherwise there is no constraint on the rate. Let n be a large positive integer (chosen according to conditions given below), and

$$k = \lceil n(1 - H(p) - \epsilon) \rceil + 1.$$

The ceiling function is there since k must be an integer. Choose $p' \in (p, 1/2)$ such that $|H(p') - H(p)| < \epsilon/2$ (this can be done since H is continuous). We will now study the performance of a *random* code C. To generate a random code C, we need to generate a random encoded message $C(x)$ for all possible input messages x.

For each $x \in \{0,1\}^k$, choose $C(x)$ to be a uniformly random vector in $\{0,1\}^n$ (making these choices independently). So we can think of $C(x)$ as a vector consisting of n i.i.d. Bern(1/2) r.v.s. The rate k/n exceeds $1 - H(p) - \epsilon$ by definition, but let's see how well we can decode the received message!

Let $x \in \{0,1\}^k$ be the input message, $C(x)$ be the encoded message, and $Y \in \{0,1\}^n$ be the received message. For now, treat x as deterministic. But $C(x)$ is random since the codewords are chosen randomly, and Y is random since $C(x)$ is random and due to the random noise in the channel. Intuitively, we hope that $C(x)$ will be close to Y (in Hamming distance) and $C(z)$ will be far from Y for all $z \neq x$, in which case it will be clear how to decode Y and the decoding will succeed. To make this precise, decode Y as follows:

If there exists a unique $z \in \{0,1\}^k$ such that $d(C(z), Y) \leq np'$, decode Y to that z; otherwise, declare decoder failure.

We will show that for n large enough, the probability of the decoder failing to recover the correct x is less than ϵ. There are two things that could go wrong:

(a) $d(C(x), Y) > np'$, or

(b) There could be some impostor $z \neq x$ with $d(C(z), Y) \leq np'$.

Note that $d(C(x), Y)$ is an r.v., so $d(C(x), Y) > np'$ is an event. To handle (a), represent

$$d(C(x), Y) = B_1 + \cdots + B_n \sim \text{Bin}(n, p),$$

where B_i is the indicator of the ith bit being flipped. The law of large numbers (see Chapter 10) says that as n grows, the r.v. $d(C(x), Y)/n$ will get very close to p (its expected value), and so will be very unlikely to exceed p':

$$P(d(C(x), Y) > np') = P\left(\frac{B_1 + \cdots + B_n}{n} > p'\right) \to 0 \text{ as } n \to \infty.$$

So by choosing n large enough, we can make

$$P(d(C(x), Y) > np') < \epsilon/4.$$

To handle (b), note that $d(C(z), Y) \sim \text{Bin}(n, 1/2)$ for $z \neq x$, since the n bits in $C(z)$ are i.i.d. Bern(1/2), independent of Y (to show this in more detail, condition on Y using LOTP). Let $B \sim \text{Bin}(n, 1/2)$. By Boole's inequality,

$$P(d(C(z), Y) \leq np' \text{ for some } z \neq x) \leq (2^k - 1)P(B \leq np').$$

To simplify notation, suppose that np' is an integer. A crude way to upper bound a sum of m terms is to use m times the largest term, and a crude way to upper bound a binomial coefficient $\binom{n}{j}$ is to use $r^{-j}(1-r)^{-(n-j)}$ for any $r \in (0, 1)$. Combining these two crudities,

$$P(B \leq np') = \frac{1}{2^n}\sum_{j=0}^{np'}\binom{n}{j} \leq \frac{np'+1}{2^n}\binom{n}{np'} \leq (np'+1)2^{nH(p')-n},$$

using the fact that $(p')^{-np'}(q')^{-nq'} = 2^{nH(p')}$ for $q' = 1 - p'$. Thus,

$$2^k P(B \leq np') \leq (np'+1)2^{n(1-H(p)-\epsilon)+2+n(H(p)+\epsilon/2)-n} = 4(np'+1)2^{-n\epsilon/2} \to 0,$$

so we can choose n to make $P(d(C(z), Y) \leq np' \text{ for some } z \neq x) < \epsilon/4$.

Assume that k and n have been chosen in accordance with the above, and let $F(c, x)$ be the event of failure when code c is used with input message x. Putting together the above results, we have shown that for a *random* C and any *fixed* x,

$$P(F(C, x)) < \epsilon/2.$$

It follows that for each x, there is a code c with $P(F(c, x)) < \epsilon/2$, but this is not good enough: we want *one* code that works well for *all* x! Let X be a uniformly random input message in $\{0, 1\}^k$, independent of C. By LOTP, we have

$$P(F(C, X)) = \sum_x P(F(C, x))P(X = x) < \epsilon/2.$$

Again using LOTP, but this time conditioning on C, we have

$$\sum_c P(F(c, X))P(C = c) = P(F(C, X)) < \epsilon/2.$$

Therefore, there exists a code c such that $P(F(c, X)) < \epsilon/2$, i.e., a code c such that the probability of failure for a random input message X is less than $\epsilon/2$. Lastly, we will improve c, obtaining a code that works well for *all* x, not just a random x. We do this by *expurgating* the worst 50% of the x's. That is, remove as legal input messages the 2^{k-1} values of x with the highest failure probabilities for code c. For all remaining x, we have $P(F(c, x)) < \epsilon$, since otherwise more than half of the $x \in \{0, 1\}^k$ would have more than double the average failure probability (see Markov's inequality in Chapter 10 for more about this kind of argument). By relabeling the remaining x using vectors in $\{0, 1\}^{k-1}$, we obtain a code $c' : \{0, 1\}^{k-1} \to \{0, 1\}^n$ with rate $(k-1)/n \geq 1 - H(p) - \epsilon$ and probability less than ϵ of failure for all input messages in $\{0, 1\}^{k-1}$. ■

There is also a converse to the above theorem, showing that if we require the rate to be at least $1 - H(p) + \epsilon$, it is impossible to find codes that make the probability of error arbitrarily small. This is why $1 - H(p)$ is called the *capacity* of the channel. Shannon also obtained analogous results for much more general channels. These results give theoretical bounds on what can be achieved, without saying explicitly which codes to use. Decades of subsequent work have been devoted to developing specific codes that work well in practice, by coming close to the Shannon bound and allowing for efficient encoding and decoding.

4.10 Recap

The expectation of a discrete r.v. X is

$$E(X) = \sum_x xP(X = x).$$

An equivalent "ungrouped" way of calculating expectation is

$$E(X) = \sum_s X(s)P(\{s\}),$$

where the sum is taken over pebbles in the sample space. Expectation is a single number summarizing the center of mass of a distribution. A single-number summary of the spread of a distribution is the variance, defined by

$$\text{Var}(X) = E(X - EX)^2 = E(X^2) - (EX)^2.$$

The square root of the variance is called the standard deviation.

Expectation is linear:

$$E(cX) = cE(X) \text{ and } E(X+Y) = E(X) + E(Y),$$

regardless of whether X and Y are independent or not. Variance is *not* linear:

$$\text{Var}(cX) = c^2\text{Var}(X),$$

and

$$\text{Var}(X+Y) \neq \text{Var}(X) + \text{Var}(Y)$$

in general (an important exception is when X and Y are independent).

A very important strategy for calculating the expectation of a discrete r.v. X is to express it as a sum of *indicator r.v.s*, and then apply linearity and the fundamental bridge. This technique is especially powerful because the indicator r.v.s need not be independent; linearity holds even for dependent r.v.s. The strategy can be summarized in the following three steps.

1. Represent the r.v. X as a sum of indicator r.v.s. To decide how to define the indicators, think about what X is counting. For example, if X is the number of local maxima, as in the Putnam problem, then we should create an indicator for each local maximum that could occur.

2. Use the fundamental bridge to calculate the expected value of each indicator. When applicable, symmetry may be very helpful at this stage.

3. By linearity of expectation, $E(X)$ can be obtained by adding up the expectations of the indicators.

Another tool for computing expectations is LOTUS, which says we can calculate the expectation of $g(X)$ using only the PMF of X, via

$$E(g(X)) = \sum_x g(x)P(X = x).$$

If g is non-linear, it is a grave mistake to attempt to calculate $E(g(X))$ by swapping the E and the g.

Four new discrete distributions to add to our list are the Geometric, Negative Binomial, Negative Hypergeometric, and Poisson distributions. A Geom(p) r.v. is the number of failures before the first success in a sequence of independent Bernoulli trials with probability p of success, and an NBin(r, p) r.v. is the number of failures before r successes. The Negative Hypergeometric is similar to the Negative Binomial except, in terms of drawing balls from an urn, the Negative Hypergeometric samples *without* replacement and the Negative Binomial samples *with* replacement. (We also introduced the First Success distribution, which is just a Geometric shifted so that the success is included.)

A Poisson r.v. is often used as an approximation for the number of successes that occur when there are many independent or weakly dependent trials, where each trial has a small probability of success. In the Binomial story, all the trials have the same probability p of success, but in the Poisson approximation, different trials can have different (but small) probabilities p_j of success.

The Poisson, Binomial, and Hypergeometric distributions are mutually connected via the operations of conditioning and taking limits, as illustrated in Figure 4.8. In the rest of this book, we'll continue to introduce new named distributions and add them to this family tree, until everything is connected!

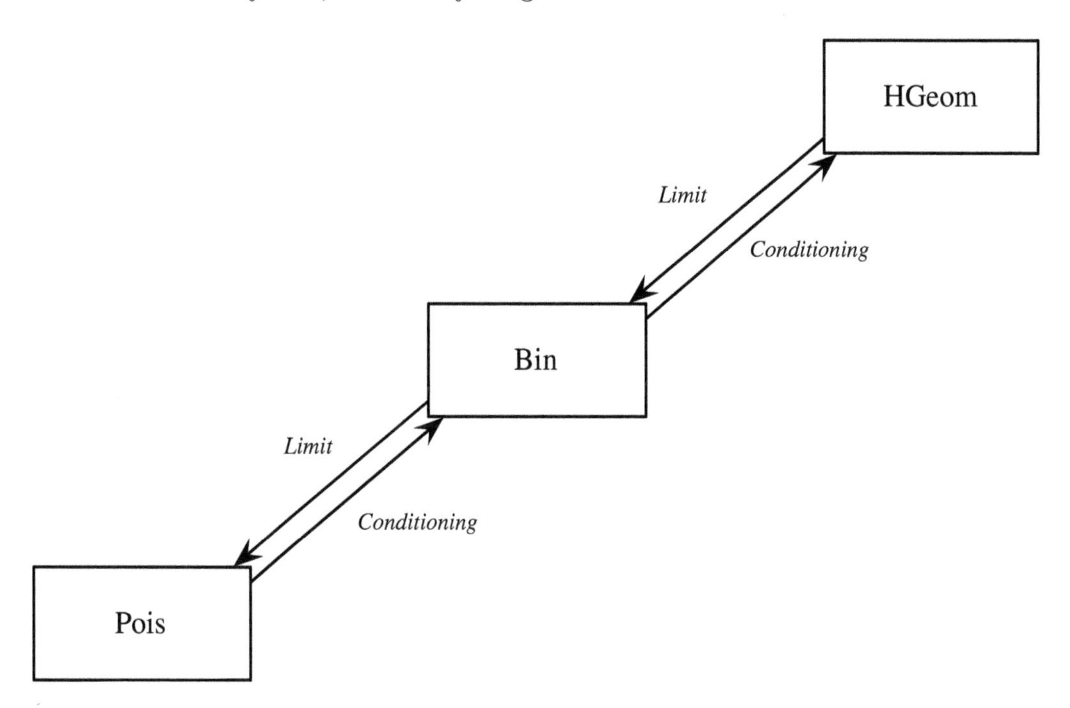

FIGURE 4.8
Relationships between the Poisson, Binomial, and Hypergeometric.

Figure 4.9 expands upon the corresponding figure from the previous chapter, further exploring the connections between the four fundamental objects we have considered: distributions, random variables, events, and numbers.

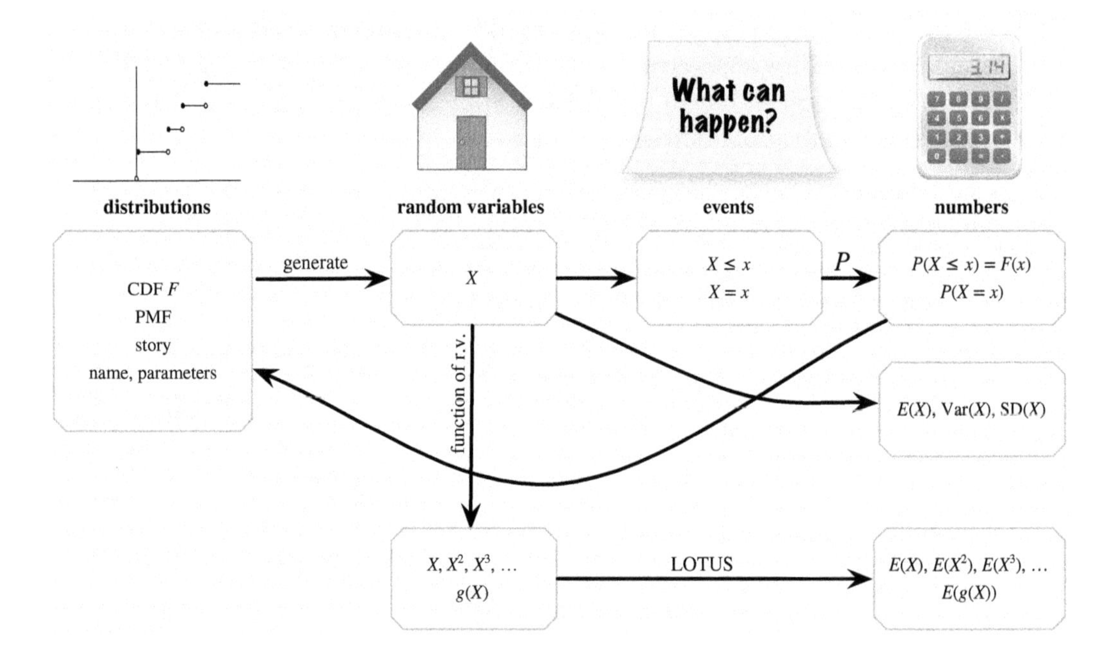

FIGURE 4.9
Four fundamental objects in probability: distributions, random variables, events, and numbers. From an r.v. X, we can generate many other r.v.s by taking functions of X, and we can use LOTUS to find their expected values. The mean, variance, and standard deviation of X express the average and spread of the distribution of X (in particular, they only depend on F, not directly on X itself).

4.11 R

Geometric, Negative Binomial, and Poisson

The three functions for the Geometric distribution in R are `dgeom`, `pgeom`, and `rgeom`, corresponding to the PMF, CDF, and random generation. For `dgeom` and `pgeom`, we need to supply the following as inputs: (1) the value at which to evaluate the PMF or CDF, and (2) the parameter p. For `rgeom`, we need to input (1) the number of random variables to generate and (2) the parameter p.

For example, to calculate $P(X = 3)$ and $P(X \leq 3)$ where $X \sim \text{Geom}(0.5)$, we use `dgeom(3,0.5)` and `pgeom(3,0.5)`, respectively. To generate 100 i.i.d. Geom(0.8) r.v.s, we use `rgeom(100,0.8)`. If instead we want 100 i.i.d. FS(0.8) r.v.s, we just need to add 1 to include the success: `rgeom(100,0.8)+1`.

For the Negative Binomial distribution, we have `dnbinom`, `pnbinom`, and `rnbinom`.

These take three inputs. For example, to calculate the NBin(5, 0.5) PMF at 3, we type `dnbinom(3,5,0.5)`.

Finally, for the Poisson distribution, the three functions are `dpois`, `ppois`, and `rpois`. These take two inputs. For example, to find the Pois(10) CDF at 2, we type `ppois(2,10)`.

Matching simulation

Continuing with Example 4.4.4, let's use simulation to calculate the expected number of matches in a deck of cards. As in Chapter 1, we let `n` be the number of cards in the deck and perform the experiment 10^4 times using `replicate`.

```
n <- 100
r <- replicate(10^4,sum(sample(n)==(1:n)))
```

Now `r` contains the number of matches from each of the 10^4 simulations. But instead of looking at the probability of at least one match, as in Chapter 1, we now want to find the expected number of matches. We can approximate this by the average of all the simulation results, that is, the arithmetic mean of the elements of `r`. This is accomplished with the `mean` function:

```
mean(r)
```

The command `mean(r)` is equivalent to `sum(r)/length(r)`. The result we get is very close to 1, confirming the calculation we did in Example 4.4.4 using indicator r.v.s. You can verify that no matter what value of `n` you choose, `mean(r)` will be very close to 1.

Distinct birthdays simulation

Let's calculate the expected number of distinct birthdays in a group of k people by simulation. We'll let $k = 20$, but you can choose whatever value of k you like.

```
k <- 20
r <- replicate(10^4,{bdays <- sample(365,k,replace=TRUE);
                     length(unique(bdays))})
```

In the second line, `replicate` repeats the expression in the curly braces 10^4 times, so we just need to understand what is inside the curly braces. First, we sample k times with replacement from the numbers 1 through 365 and call these the birthdays of the k people, `bdays`. Then, `unique(bdays)` removes duplicates in the vector `bdays`, and `length(unique(bdays))` calculates the length of the vector after duplicates have been removed. The two commands need to be separated by a semicolon.

Now `r` contains the number of distinct birthdays that we observed in each of the 10^4 simulations. The average number of distinct birthdays across the 10^4 simulations is `mean(r)`. We can compare the simulated value to the theoretical value that we found in Example 4.4.5 using indicator r.v.s:

```
mean(r)
365*(1-(364/365)^k)
```

When we ran the code, both the simulated and theoretical values gave us approximately 19.5.

4.12 Exercises

Exercises marked with Ⓢ have detailed solutions at `http://stat110.net`.

Expectations and variances

1. Bobo, the amoeba from Chapter 2, currently lives alone in a pond. After one minute Bobo will either die, split into two amoebas, or stay the same, with equal probability. Find the expectation and variance for the number of amoebas in the pond after one minute.

2. In the Gregorian calendar, each year has either 365 days (a normal year) or 366 days (a leap year). A year is randomly chosen, with probability $3/4$ of being a normal year and $1/4$ of being a leap year. Find the mean and variance of the number of days in the chosen year.

3. (a) A fair die is rolled. Find the expected value of the roll.

   (b) Four fair dice are rolled. Find the expected total of the rolls.

4. A fair die is rolled some number of times. You can choose whether to stop after $1, 2$, or 3 rolls, and your decision can be based on the values that have appeared so far. You receive the value shown on the last roll of the die, in dollars. What is your optimal strategy (to maximize your expected winnings)? Find the expected winnings for this strategy.

   Hint: Start by considering a simpler version of this problem, where there are at most 2 rolls. For what values of the first roll should you continue for a second roll?

5. Find the mean and variance of a Discrete Uniform r.v. on $1, 2, \dots, n$.

   Hint: See the math appendix for some useful facts about sums.

6. Two teams are going to play a best-of-7 match (the match will end as soon as either team has won 4 games). Each game ends in a win for one team and a loss for the other team. Assume that each team is equally likely to win each game, and that the games played are independent. Find the mean and variance of the number of games played.

7. A certain small town, whose population consists of 100 families, has 30 families with 1 child, 50 families with 2 children, and 20 families with 3 children. The *birth rank* of one of these children is 1 if the child is the firstborn, 2 if the child is the secondborn, and 3 if the child is the thirdborn.

   (a) A random family is chosen (with equal probabilities), and then a random child within that family is chosen (with equal probabilities). Find the PMF, mean, and variance of the child's birth rank.

(b) A random child is chosen in the town (with equal probabilities). Find the PMF, mean, and variance of the child's birth rank.

8. A certain country has four regions: North, East, South, and West. The populations of these regions are 3 million, 4 million, 5 million, and 8 million, respectively. There are 4 cities in the North, 3 in the East, 2 in the South, and there is only 1 city in the West. Each person in the country lives in exactly one of these cities.

   (a) What is the average size of a city in the country? (This is the arithmetic mean of the populations of the cities, and is also the expected value of the population of a city chosen uniformly at random.)

   Hint: Give the cities *names* (labels).

   (b) Show that without further information it is impossible to find the variance of the population of a city chosen uniformly at random. That is, the variance depends on how the people within each region are allocated between the cities in that region.

   (c) A region of the country is chosen uniformly at random, and then a city within that region is chosen uniformly at random. What is the expected population size of this randomly chosen city?

   Hint: First find the selection probability for each city.

   (d) Explain intuitively why the answer to (c) is larger than the answer to (a).

9. Consider the following simplified scenario based on *Who Wants to Be a Millionaire?*, a game show in which the contestant answers multiple-choice questions that have 4 choices per question. The contestant (Fred) has answered 9 questions correctly already, and is now being shown the 10th question. He has no idea what the right answers are to the 10th or 11th questions are. He has one "lifeline" available, which he can apply on any question, and which narrows the number of choices from 4 down to 2. Fred has the following options available.
   (a) Walk away with $16,000.
   (b) Apply his lifeline to the 10th question, and then answer it. If he gets it wrong, he will leave with $1,000. If he gets it right, he moves on to the 11th question. He then leaves with $32,000 if he gets the 11th question wrong, and $64,000 if he gets the 11th question right.
   (c) Same as the previous option, except not using his lifeline on the 10th question, and instead applying it to the 11th question (if he gets the 10th question right).

   Find the expected value of each of these options. Which option has the highest expected value? Which option has the lowest variance?

10. Consider the St. Petersburg paradox (Example 4.3.14), except that you receive $\$n$ rather than $\$2^n$ if the game lasts for n rounds. What is the fair value of this game? What if the payoff is $\$n^2$?

11. Martin has just heard about the following exciting gambling strategy: bet $1 that a fair coin will land Heads. If it does, stop. If it lands Tails, double the bet for the next toss, now betting $2 on Heads. If it does, stop. Otherwise, double the bet for the next toss to $4. Continue in this way, doubling the bet each time and then stopping right after winning a bet. Assume that each individual bet is fair, i.e., has an expected net winnings of 0. The idea is that

$$1 + 2 + 2^2 + 2^3 + \cdots + 2^n = 2^{n+1} - 1,$$

   so the gambler will be $1 ahead after winning a bet, and then can walk away with a profit.

   Martin decides to try out this strategy. However, he only has $31, so he may end up walking away bankrupt rather than continuing to double his bet. On average, how much money will Martin win?

12. Let X be a discrete r.v. with support $-n, -n+1, \ldots, 0, \ldots, n-1, n$ for some positive integer n. Suppose that the PMF of X satisfies the symmetry property $P(X = -k) = P(X = k)$ for all integers k. Find $E(X)$.

13. Ⓢ Are there discrete random variables X and Y such that $E(X) > 100E(Y)$ but Y is greater than X with probability at least 0.99?

14. Let X have PMF
$$P(X = k) = cp^k/k \text{ for } k = 1, 2, \ldots,$$
where p is a parameter with $0 < p < 1$ and c is a normalizing constant. We have $c = -1/\log(1-p)$, as seen from the Taylor series
$$-\log(1-p) = p + \frac{p^2}{2} + \frac{p^3}{3} + \ldots.$$
This distribution is called the *Logarithmic* distribution (because of the log in the above Taylor series), and has often been used in ecology. Find the mean and variance of X.

15. Player A chooses a random integer between 1 and 100, with probability p_j of choosing j (for $j = 1, 2, \ldots, 100$). Player B guesses the number that player A picked, and receives from player A that amount in dollars if the guess is correct (and 0 otherwise).

(a) Suppose for this part that player B knows the values of p_j. What is player B's optimal strategy (to maximize expected earnings)?

(b) Show that if both players choose their numbers so that the probability of picking j is proportional to $1/j$, then neither player has an incentive to change strategies, assuming the opponent's strategy is fixed. (In game theory terminology, this says that we have found a *Nash equilibrium.*)

(c) Find the expected earnings of player B when following the strategy from (b). Express your answer both as a sum of simple terms and as a numerical approximation. Does the value depend on what strategy player A uses?

16. The dean of Blotchville University boasts that the average class size there is 20. But the reality experienced by the majority of students there is quite different: they find themselves in huge courses, held in huge lecture halls, with hardly enough seats or Haribo gummi bears for everyone. The purpose of this problem is to shed light on the situation. For simplicity, suppose that every student at Blotchville University takes only one course per semester.

(a) Suppose that there are 16 seminar courses, which have 10 students each, and 2 large lecture courses, which have 100 students each. Find the dean's-eye-view average class size (the simple average of the class sizes) and the student's-eye-view average class size (the average class size experienced by students, as it would be reflected by surveying students and asking them how big their classes are). Explain the discrepancy intuitively.

(b) Give a short proof that for *any* set of class sizes (not just those given above), the dean's-eye-view average class size will be strictly less than the student's-eye-view average class size, unless all classes have exactly the same size.

Hint: Relate this to the fact that variances are nonnegative.

17. The sociologist Elizabeth Wrigley-Field posed the following puzzle [29]:

> *American fertility fluctuated dramatically in the decades surrounding the Second World War. Parents created the smallest families during the Great Depression, and the largest families during the postwar Baby Boom. Yet children born during the Great Depression came from larger families than those born during the Baby Boom. How can this be?*

(a) For a particular era, let n_k be the number of American families with exactly k children, for each $k \geq 0$. (Assume for simplicity that American history has cleanly been separated into eras, where each era has a well-defined set of families, and each family has a well-defined set of children; we are ignoring the fact that a particular family's size may change over time, that children grow up, etc.) For each $j \geq 0$, let

$$m_j = \sum_{k=0}^{\infty} k^j n_k.$$

For a *family* selected randomly in that era (with all families equally likely), find the expected number of children in the family. Express your answer only in terms of the m_j's.

(b) For a *child* selected randomly in that era (with all children equally likely), find the expected number of children in the child's family, only in terms of the m_j's.

(c) Give an intuitive explanation in words for which of the answers to (a) and (b) is larger, or whether they are equal. Explain how this relates to the Wrigley-Field puzzle.

Named distributions

18. Ⓢ A fair coin is tossed repeatedly, until it has landed Heads at least once and has landed Tails at least once. Find the expected number of tosses.

19. Ⓢ A coin is tossed repeatedly until it lands Heads for the first time. Let X be the number of tosses that are required (including the toss that landed Heads), and let p be the probability of Heads, so that $X \sim \text{FS}(p)$. Find the CDF of X, and for $p = 1/2$ sketch its graph.

20. Let $X \sim \text{Bin}(100, 0.9)$. For each of the following parts, construct an example showing that it is possible, or explain clearly why it is impossible. In this problem, Y is a random variable on the same probability space as X; note that X and Y are not necessarily independent.

    (a) Is it possible to have $Y \sim \text{Pois}(0.01)$ with $P(X \geq Y) = 1$?

    (b) Is it possible to have $Y \sim \text{Bin}(100, 0.5)$ with $P(X \geq Y) = 1$?

    (c) Is it possible to have $Y \sim \text{Bin}(100, 0.5)$ with $P(X \leq Y) = 1$?

21. Ⓢ Let $X \sim \text{Bin}(n, \frac{1}{2})$ and $Y \sim \text{Bin}(n+1, \frac{1}{2})$, independently.

    (a) Let $V = \min(X, Y)$ be the smaller of X and Y, and let $W = \max(X, Y)$ be the larger of X and Y. So if X crystallizes to x and Y crystallizes to y, then V crystallizes to $\min(x, y)$ and W crystallizes to $\max(x, y)$. Find $E(V) + E(W)$.

    (b) Show that $E|X - Y| = E(W) - E(V)$, with notation as in (a).

    (c) Compute $\text{Var}(n - X)$ in two different ways.

22. Ⓢ Raindrops are falling at an average rate of 20 drops per square inch per minute. What would be a reasonable distribution to use for the number of raindrops hitting a particular region measuring 5 inches^2 in t minutes? Why? Using your chosen distribution, compute the probability that the region has no rain drops in a given 3-second time interval.

23. Ⓢ Alice and Bob have just met, and wonder whether they have a mutual friend. Each has 50 friends, out of 1000 other people who live in their town. They think that it's unlikely that they have a friend in common, saying "each of us is only friends with 5% of the people here, so it would be very unlikely that our two 5%'s overlap."

    Assume that Alice's 50 friends are a random sample of the 1000 people (equally likely to be any 50 of the 1000), and similarly for Bob. Also assume that knowing who Alice's friends are gives no information about who Bob's friends are.

(a) Compute the expected number of mutual friends Alice and Bob have.

(b) Let X be the number of mutual friends they have. Find the PMF of X.

(c) Is the distribution of X one of the important distributions we have looked at? If so, which?

24. Let $X \sim \text{Bin}(n, p)$ and $Y \sim \text{NBin}(r, p)$. Using a story about a sequence of Bernoulli trials, prove that $P(X < r) = P(Y > n - r)$.

25. Ⓢ Calvin and Hobbes play a match consisting of a series of games, where Calvin has probability p of winning each game (independently). They play with a "win by two" rule: the first player to win two games more than his opponent wins the match. Find the expected number of games played.

    Hint: Consider the first two games as a pair, then the next two as a pair, etc.

26. Nick and Penny are independently performing independent Bernoulli trials. For concreteness, assume that Nick is flipping a nickel with probability p_1 of Heads and Penny is flipping a penny with probability p_2 of Heads. Let $X_1, X_2, \dots$ be Nick's results and $Y_1, Y_2, \dots$ be Penny's results, with $X_i \sim \text{Bern}(p_1)$ and $Y_j \sim \text{Bern}(p_2)$.

    (a) Find the distribution and expected value of the first time at which they are simultaneously successful, i.e., the smallest n such that $X_n = Y_n = 1$.

    Hint: Define a new sequence of Bernoulli trials and use the story of the Geometric.

    (b) Find the expected time until at least one has a success (including the success).

    Hint: Define a new sequence of Bernoulli trials and use the story of the Geometric.

    (c) For $p_1 = p_2$, find the probability that their first successes are simultaneous, and use this to find the probability that Nick's first success precedes Penny's.

27. Ⓢ Let X and Y be Pois(λ) r.v.s, and $T = X + Y$. Suppose that X and Y are *not* independent, and in fact $X = Y$. Prove or disprove the claim that $T \sim \text{Pois}(2\lambda)$ in this scenario.

28. William is on a treasure hunt. There are t pieces of treasure, each of which is hidden in one of n locations. William searches these locations one by one, without replacement, until he has found all the treasure. (Assume that no location contains more than one piece of treasure, and that William *will* find the treasure piece when he searches a location that does have treasure.) Let X be the number of locations that William searches during his treasure hunt. Find the distribution of X, and find $E(X)$.

29. Let $X \sim \text{Geom}(p)$, and define the function f by $f(x) = P(X = x)$, for all real x. Find $E(f(X))$. (The notation $f(X)$ means first evaluate $f(x)$ in terms of p and x, and then plug in X for x; it is *not* correct to say "$f(X) = P(X = X) = 1$".)

30. (a) Use LOTUS to show that for $X \sim \text{Pois}(\lambda)$ and any function g,

$$E(Xg(X)) = \lambda E(g(X + 1)),$$

    assuming that both sides exist. This is called the *Stein-Chen identity* for the Poisson.

    (b) Find the third moment $E(X^3)$ for $X \sim \text{Pois}(\lambda)$ by using the identity from (a) and a bit of algebra to reduce the calculation to the fact that X has mean λ and variance λ.

31. In many problems about modeling count data, it is found that values of zero in the data are far more common than can be explained well using a Poisson model (we can make $P(X = 0)$ large for $X \sim \text{Pois}(\lambda)$ by making λ small, but that also constrains the mean and variance of X to be small since both are λ). The *Zero-Inflated Poisson* distribution

is a modification of the Poisson to address this issue, making it easier to handle frequent zero values gracefully.

A Zero-Inflated Poisson r.v. X with parameters p and λ can be generated as follows. First flip a coin with probability of p of Heads. Given that the coin lands Heads, $X = 0$. Given that the coin lands Tails, X is distributed Pois(λ). Note that if $X = 0$ occurs, there are two possible explanations: the coin could have landed Heads (in which case the zero is called a *structural zero*), or the coin could have landed Tails but the Poisson r.v. turned out to be zero anyway.

For example, if X is the number of chicken sandwiches consumed by a random person in a week, then $X = 0$ for vegetarians (this is a structural zero), but a chicken-eater could still have $X = 0$ occur by chance (since they might happen not to eat any chicken sandwiches that week).

(a) Find the PMF of a Zero-Inflated Poisson r.v. X.

(b) Explain why X has the same distribution as $(1 - I)Y$, where $I \sim$ Bern(p) is independent of $Y \sim$ Pois(λ).

(c) Find the mean of X in two different ways: directly using the PMF of X, and using the representation from (b). For the latter, you can use the fact (which we prove in Chapter 7) that if r.v.s Z and W are independent, then $E(ZW) = E(Z)E(W)$.

(d) Find the variance of X.

32. Ⓢ A discrete distribution has the *memoryless property* if for X a random variable with that distribution, $P(X \geq j + k | X \geq j) = P(X \geq k)$ for all nonnegative integers j, k.

    (a) If X has a memoryless distribution with CDF F and PMF $p_i = P(X = i)$, find an expression for $P(X \geq j + k)$ in terms of $F(j), F(k), p_j, p_k$.

    (b) Name a discrete distribution which has the memoryless property. Justify your answer with a clear interpretation in words or with a computation.

33. Find values of w, b, r such that the Negative Hypergeometric distribution with parameters w, b, r reduces to a Discrete Uniform on $\{0, 1, \ldots, n\}$. Justify your answer both in terms of the story of the Negative Hypergeometric and in terms of its PMF.

Indicator r.v.s

34. Ⓢ Randomly, k distinguishable balls are placed into n distinguishable boxes, with all possibilities equally likely. Find the expected number of empty boxes.

35. Ⓢ A group of 50 people are comparing their birthdays (as usual, assume their birthdays are independent, are not February 29, etc.). Find the expected number of pairs of people with the same birthday, and the expected number of days in the year on which at least two of these people were born.

36. Ⓢ A group of $n \geq 4$ people are comparing their birthdays (as usual, assume their birthdays are independent, are not February 29, etc.). Let I_{ij} be the indicator r.v. of i and j having the same birthday (for $i < j$). Is I_{12} independent of I_{34}? Is I_{12} independent of I_{13}? Are the I_{ij} independent?

37. Ⓢ A total of 20 bags of Haribo gummi bears are randomly distributed to 20 students. Each bag is obtained by a random student, and the outcomes of who gets which bag are independent. Find the average number of bags of gummi bears that the first three students get in total, and find the average number of students who get at least one bag.

38. Each of $n \geq 2$ people puts their name on a slip of paper (no two have the same name). The slips of paper are shuffled in a hat, and then each person draws one (uniformly at random at each stage, without replacement). Find the average number of people who draw their own names.

39. Two researchers independently select simple random samples from a population of size N, with sample sizes m and n (for each researcher, the sampling is done without replacement, with all samples of the prescribed size equally likely). Find the expected size of the overlap of the two samples.

40. In a sequence of n independent fair coin tosses, what is the expected number of occurrences of the pattern HTH (consecutively)? Note that overlap is allowed, e.g., $HTHTH$ contains two overlapping occurrences of the pattern.

41. You have a well-shuffled 52-card deck. On average, how many pairs of adjacent cards are there such that both cards are red?

42. Suppose there are n types of toys, which you are collecting one by one. Each time you collect a toy, it is equally likely to be any of the n types. What is the expected number of distinct toy types that you have after you have collected t toys? (Assume that you will definitely collect t toys, whether or not you obtain a complete set before then.)

43. A building has n floors, labeled $1, 2, \dots, n$. At the first floor, k people enter the elevator, which is going up and is empty before they enter. Independently, each decides which of floors $2, 3, \dots, n$ to go to and presses that button (unless someone has already pressed it).

    (a) Assume for this part only that the probabilities for floors $2, 3, \dots, n$ are equal. Find the expected number of stops the elevator makes on floors $2, 3, \dots, n$.

    (b) Generalize (a) to the case that floors $2, 3, \dots, n$ have probabilities $p_2, \dots, p_n$ (respectively); you can leave your answer as a finite sum.

44. Ⓢ There are 100 shoelaces in a box. At each stage, you pick two random ends and tie them together. Either this results in a longer shoelace (if the two ends came from different pieces), or it results in a loop (if the two ends came from the same piece). What are the expected number of steps until everything is in loops, and the expected number of loops after everything is in loops? (This is a famous interview problem; leave the latter answer as a sum.)

    Hint: For each step, create an indicator r.v. for whether a loop was created then, and note that the number of free ends goes down by 2 after each step.

45. Show that for any events $A_1, \dots, A_n$,

$$P(A_1 \cap A_2 \cdots \cap A_n) \geq \sum_{j=1}^{n} P(A_j) - n + 1.$$

    Hint: First prove a similar-looking statement about indicator r.v.s, by interpreting what the events $I(A_1 \cap A_2 \cdots \cap A_n) = 1$ and $I(A_1 \cap A_2 \cdots \cap A_n) = 0$ mean.

46. You have a well-shuffled 52-card deck. You turn the cards face up one by one, without replacement. What is the expected number of non-aces that appear before the first ace? What is the expected number between the first ace and the second ace?

47. You are being tested for psychic powers. Suppose that you do not have psychic powers. A standard deck of cards is shuffled, and the cards are dealt face down one by one. Just after each card is dealt, you name any card (as your prediction). Let X be the number of cards you predict correctly. (See Diaconis [5] for much more about the statistics of testing for psychic powers.)

    (a) Suppose that you get no feedback about your predictions. Show that no matter what strategy you follow, the expected value of X stays the same; find this value. (On the other hand, the *variance* may be very different for different strategies. For example, saying "Ace of Spades" every time gives variance 0.)

    Hint: Indicator r.v.s.

(b) Now suppose that you get partial feedback: after each prediction, you are told immediately whether or not it is right (but without the card being revealed). Suppose you use the following strategy: keep saying a specific card's name (e.g., "Ace of Spades") until you hear that you are correct. Then keep saying a different card's name (e.g., "Two of Spades") until you hear that you are correct (if ever). Continue in this way, naming the same card over and over again until you are correct and then switching to a new card, until the deck runs out. Find the expected value of X, and show that it is very close to $e - 1$.

Hint: Indicator r.v.s.

(c) Now suppose that you get complete feedback: just after each prediction, the card is revealed. Call a strategy "stupid" if it allows, e.g., saying "Ace of Spades" as a guess after the Ace of Spades has already been revealed. Show that any non-stupid strategy gives the same expected value for X; find this value.

Hint: Indicator r.v.s.

48. Ⓢ Let X be Hypergeometric with parameters w, b, n.

(a) Find $E\binom{X}{2}$ *by thinking*, without any complicated calculations.

(b) Use (a) to find the variance of X. You should get

$$\text{Var}(X) = \frac{N-n}{N-1}npq,$$

where $N = w + b, p = w/N, q = 1 - p$.

49. There are n prizes, with values \$1, \$2, ..., \$$n$. You get to choose k random prizes, without replacement. What is the expected total value of the prizes you get?

Hint: Express the total value in the form $a_1I_1 + \cdots + a_nI_n$, where the a_j are constants and the I_j are indicator r.v.s. Or find the expected value of the jth prize received directly.

50. Ten random chords of a circle are chosen, independently. To generate each of these chords, two independent uniformly random points are chosen on the circle (intuitively, "uniformly" means that the choice is completely random, with no favoritism toward certain angles; formally, it means that the probability of any arc is proportional to the length of that arc). On average, how many pairs of chords intersect?

Hint: Consider two random chords. An equivalent way to generate them is to pick four independent uniformly random points on the circle, and then pair them up randomly.

51. Ⓢ A hash table is being used to store the phone numbers of k people, storing each person's phone number in a uniformly random location, represented by an integer between 1 and n (see Exercise 27 from Chapter 1 for a description of hash tables). Find the expected number of locations with no phone numbers stored, the expected number with exactly one phone number, and the expected number with more than one phone number (should these quantities add up to n?).

52. A coin with probability p of Heads is flipped n times. The sequence of outcomes can be divided into *runs* (blocks of H's or blocks of T's), e.g., $HHHTTHTTTH$ becomes $\boxed{HHH}\boxed{TT}\boxed{H}\boxed{TTT}\boxed{H}$, which has 5 runs. Find the expected number of runs.

Hint: Start by finding the expected number of tosses (other than the first) where the outcome is different from the previous one.

53. A coin with probability p of Heads is flipped 4 times. Let X be the number of occurrences of HH (for example, $THHT$ has 1 occurrence and $HHHH$ has 3 occurrences). Find $E(X)$ and $\text{Var}(X)$.

54. A population has N people, with ID numbers from 1 to N. Let y_j be the value of some numerical variable for person j, and

$$\bar{y} = \frac{1}{N}\sum_{j=1}^{N} y_j$$

be the population average of the quantity. For example, if y_j is the height of person j then $\bar{y}$ is the average height in the population, and if y_j is 1 if person j holds a certain belief and 0 otherwise, then $\bar{y}$ is the proportion of people in the population who hold that belief. In this problem, $y_1, y_2, \dots, y_n$ are thought of as constants rather than random variables.

A researcher is interested in learning about $\bar{y}$, but it is not feasible to measure y_j for all j. Instead, the researcher gathers a random sample of size n, by choosing people one at a time, with equal probabilities at each stage and without replacement. Let W_j be the value of the numerical variable (e.g., height) for the jth person in the sample. Even though $y_1, \dots, y_n$ are constants, W_j is a random variable because of the random sampling. A natural way to estimate the unknown quantity $\bar{y}$ is using

$$\bar{W} = \frac{1}{n}\sum_{j=1}^{n} W_j.$$

Show that $E(\bar{W}) = \bar{y}$ in two different ways:

(a) by directly evaluating $E(W_j)$ using symmetry;

(b) by showing that $\bar{W}$ can be expressed as a sum over the population by writing

$$\bar{W} = \frac{1}{n}\sum_{j=1}^{N} I_j y_j,$$

where I_j is the indicator of person j being included in the sample, and then using linearity and the fundamental bridge.

55. Ⓢ Consider the following algorithm, known as *bubble sort*, for sorting a list of n distinct numbers into increasing order. Initially they are in a random order, with all orders equally likely. The algorithm compares the numbers in positions 1 and 2, and swaps them if needed, then it compares the new numbers in positions 2 and 3, and swaps them if needed, etc., until it has gone through the whole list. Call this one "sweep" through the list. After the first sweep, the largest number is at the end, so the second sweep (if needed) only needs to work with the first $n-1$ positions. Similarly, the third sweep (if needed) only needs to work with the first $n-2$ positions, etc. Sweeps are performed until $n-1$ sweeps have been completed or there is a swapless sweep.

For example, if the initial list is 53241 (omitting commas), then the following 4 sweeps are performed to sort the list, with a total of 10 comparisons:

$$53241 \to 35241 \to 32541 \to 32451 \to 32415.$$

$$32415 \to 23415 \to 23415 \to 23145.$$

$$23145 \to 23145 \to 21345.$$

$$21345 \to 12345.$$

(a) An *inversion* is a pair of numbers that are out of order (e.g., 12345 has no inversions, while 53241 has 8 inversions). Find the expected number of inversions in the original list.

(b) Show that the expected number of comparisons is between $\frac{1}{2}\binom{n}{2}$ and $\binom{n}{2}$.

Hint: For one bound, think about how many comparisons are made if $n-1$ sweeps are done; for the other bound, use Part (a).

56. A certain basketball player practices shooting free throws over and over again. The shots are independent, with probability p of success.

(a) In n shots, what is the expected number of streaks of 7 consecutive successful shots? (Note that, for example, 9 in a row counts as 3 streaks.)

(b) Now suppose that the player keeps shooting until making 7 shots in a row for the first time. Let X be the number of shots taken. Show that $E(X) \leq 7/p^7$.

Hint: Consider the first 7 trials as a block, then the next 7 as a block, etc.

57. Ⓢ An urn contains red, green, and blue balls. Balls are chosen randomly with replacement (each time, the color is noted and then the ball is put back). Let r, g, b be the probabilities of drawing a red, green, blue ball, respectively ($r + g + b = 1$).

(a) Find the expected number of balls chosen before obtaining the first red ball, not including the red ball itself.

(b) Find the expected number of different *colors* of balls obtained before getting the first red ball.

(c) Find the probability that at least 2 of n balls drawn are red, given that at least 1 is red.

58. Ⓢ Job candidates $C_1, C_2, \ldots$ are interviewed one by one, and the interviewer compares them and keeps an updated list of rankings (if n candidates have been interviewed so far, this is a list of the n candidates, from best to worst). Assume that there is no limit on the number of candidates available, that for any n the candidates $C_1, C_2, \ldots, C_n$ are equally likely to arrive in any order, and that there are no ties in the rankings given by the interview.

Let X be the index of the first candidate to come along who ranks as better than the very first candidate C_1 (so C_X is better than C_1, but the candidates after 1 but prior to X (if any) are worse than C_1. For example, if C_2 and C_3 are worse than C_1 but C_4 is better than C_1, then $X = 4$. All 4! orderings of the first 4 candidates are equally likely, so it could have happened that the first candidate was the best out of the first 4 candidates, in which case $X > 4$.

What is $E(X)$ (which is a measure of how long, on average, the interviewer needs to wait to find someone better than the very first candidate)?

Hint: Find $P(X > n)$ by interpreting what $X > n$ says about how C_1 compares with other candidates, and then apply the result of Theorem 4.4.8.

59. People are arriving at a party one at a time. While waiting for more people to arrive they entertain themselves by comparing their birthdays. Let X be the number of people needed to obtain a birthday match, i.e., before person X arrives there are no two people with the same birthday, but when person X arrives there is a match.

Assume for this problem that there are 365 days in a year, all equally likely. By the result of the birthday problem from Chapter 1, for 23 people there is a 50.7% chance of a birthday match (and for 22 people there is a less than 50% chance). But this has to do with the *median* of X (defined below); we also want to know the *mean* of X, and in this problem we will find it, and see how it compares with 23.

(a) A *median* of a random variable Y is a value m for which $P(Y \leq m) \geq 1/2$ and $P(Y \geq m) \geq 1/2$ (this is also called a median of the *distribution* of Y; note that the notion is completely determined by the CDF of Y). Every distribution has a median, but for some distributions it is not unique. Show that 23 is the *unique* median of X.

(b) Show that $X = I_1 + I_2 + \cdots + I_{366}$, where I_j is the indicator r.v. for the event $X \geq j$. Then find $E(X)$ in terms of p_j's defined by $p_1 = p_2 = 1$ and for $3 \leq j \leq 366$,

$$p_j = \left(1 - \frac{1}{365}\right)\left(1 - \frac{2}{365}\right)\cdots\left(1 - \frac{j-2}{365}\right).$$

(c) Compute $E(X)$ numerically. In R, the pithy command `cumprod(1-(0:364)/365)` produces the vector $(p_2, \ldots, p_{366})$.

(d) Find the variance of X, both in terms of the p_j's and numerically.

Hint: What is I_i^2, and what is $I_i I_j$ for $i < j$? Use this to simplify the expansion

$$X^2 = I_1^2 + \cdots + I_{366}^2 + 2\sum_{j=2}^{366}\sum_{i=1}^{j-1} I_i I_j.$$

Note: In addition to being an entertaining game for parties, the birthday problem has many applications in computer science, such as in a method called the *birthday attack* in cryptography. It can be shown that if there are n days in a year and n is large, then

$$E(X) \approx \sqrt{\frac{\pi n}{2}} + \frac{2}{3}.$$

60. Elk dwell in a certain forest. There are N elk, of which a simple random sample of size n is captured and tagged (so all $\binom{N}{n}$ sets of n elk are equally likely). The captured elk are returned to the population, and then a new sample is drawn. This is an important method that is widely used in ecology, known as *capture-recapture*. If the new sample is also a simple random sample, with some fixed size, then the number of tagged elk in the new sample is Hypergeometric.

    For this problem, assume that instead of having a fixed sample size, elk are sampled one by one without replacement until m tagged elk have been recaptured, where m is specified in advance (of course, assume that $1 \leq m \leq n \leq N$). An advantage of this sampling method is that it can be used to avoid ending up with a very small number of tagged elk (maybe even zero), which would be problematic in many applications of capture-recapture. A disadvantage is not knowing how large the sample will be.

    (a) Find the PMFs of the number of untagged elk in the new sample (call this X) and of the total number of elk in the new sample (call this Y).

    (b) Find the expected sample size EY using symmetry, linearity, and indicator r.v.s.

    (c) Suppose that m, n, N are such that EY is an integer. If the sampling is done with a fixed sample size equal to EY rather than sampling until exactly m tagged elk are obtained, find the expected number of tagged elk in the sample. Is it less than m, equal to m, or greater than m (for $n < N$)?

LOTUS

61. Ⓢ For $X \sim \text{Pois}(\lambda)$, find $E(X!)$ (the average factorial of X), if it is finite.
62. For $X \sim \text{Pois}(\lambda)$, find $E(2^X)$, if it is finite.
63. For $X \sim \text{Geom}(p)$, find $E(2^X)$ (if it is finite) and $E(2^{-X})$ (if it is finite). For each, make sure to clearly state what the values of p are for which it is finite.
64. Ⓢ Let $X \sim \text{Geom}(p)$ and let t be a constant. Find $E(e^{tX})$, as a function of t (this is known as the *moment generating function*; we will see in Chapter 6 how this function is useful).

65. Ⓢ The number of fish in a certain lake is a Pois(λ) random variable. Worried that there might be no fish at all, a statistician adds one fish to the lake. Let Y be the resulting number of fish (so Y is 1 plus a Pois(λ) random variable).

(a) Find $E(Y^2)$.

(b) Find $E(1/Y)$.

66. Ⓢ Let X be a Pois(λ) random variable, where λ is fixed but unknown. Let $\theta = e^{-3\lambda}$, and suppose that we are interested in estimating θ based on the data. Since X is what we observe, our estimator is a function of X, call it $g(X)$. The *bias* of the estimator $g(X)$ is defined to be $E(g(X)) - \theta$, i.e., how far off the estimate is on average; the estimator is *unbiased* if its bias is 0.

(a) For estimating λ, the r.v. X itself is an unbiased estimator. Compute the bias of the estimator $T = e^{-3X}$. Is it unbiased for estimating θ?

(b) Show that $g(X) = (-2)^X$ is an unbiased estimator for θ. (In fact, it turns out to be the only unbiased estimator for θ.)

(c) Explain intuitively why $g(X)$ is a silly choice for estimating θ, despite (b), and show how to improve it by finding an estimator $h(X)$ for θ that is always at least as good as $g(X)$ and sometimes strictly better than $g(X)$. That is,

$$|h(X) - \theta| \leq |g(X) - \theta|,$$

with the inequality sometimes strict.

Poisson approximation

67. Ⓢ Law school courses often have assigned seating to facilitate the Socratic method. Suppose that there are 100 first-year law students, and each takes the same two courses: Torts and Contracts. Both are held in the same lecture hall (which has 100 seats), and the seating is uniformly random and independent for the two courses.

(a) Find the probability that no one has the same seat for both courses (exactly; you should leave your answer as a sum).

(b) Find a simple but accurate approximation to the probability that no one has the same seat for both courses.

(c) Find a simple but accurate approximation to the probability that at least two students have the same seat for both courses.

68. Ⓢ A group of n people play "Secret Santa" as follows: each puts their name on a slip of paper in a hat, picks a name randomly from the hat (without replacement), and then buys a gift for that person. Unfortunately, they overlook the possibility of drawing one's own name, so some may have to buy gifts for themselves (on the bright side, some may like self-selected gifts better). Assume $n \geq 2$.

(a) Find the expected value of the number X of people who pick their own names.

(b) Find the expected number of pairs of people, A and B, such that A picks B's name and B picks A's name (where $A \neq B$ and order doesn't matter).

(c) What is the *approximate* distribution of X if n is large (specify the parameter value or values)? What does $P(X = 0)$ converge to as $n \to \infty$?

69. A survey is being conducted in a city with a million (10^6) people. A sample of size 1000 is collected by choosing people in the city at random, *with* replacement and with equal probabilities for everyone in the city. Find a simple, accurate approximation to the probability that at least one person will get chosen more than once (in contrast, Exercise 26 from Chapter 1 asks for an exact answer).

    Hint: Indicator r.v.s are useful here, but creating 1 indicator for each of the million people is *not* recommended since it leads to a messy calculation. Feel free to use the fact that $999 \approx 1000$.

70. Ⓢ Ten million people enter a certain lottery. For each person, the chance of winning is one in ten million, independently.

    (a) Find a simple, good approximation for the PMF of the number of people who win the lottery.

    (b) Congratulations! You won the lottery. However, there may be other winners. Assume now that the number of winners other than you is $W \sim \text{Pois}(1)$, and that if there is more than one winner, then the prize is awarded to one randomly chosen winner. Given this information, find the probability that you win the prize (simplify).

71. In a group of 90 people, find a simple, good approximation for the probability that there is at least one pair of people such that they share a birthday *and* their biological mothers share a birthday. Assume that no one among the 90 people is the biological mother of another one of the 90 people, nor do two of the 90 people have the same biological mother. Express your answer as a fully simplified fraction in the form a/b, where a and b are positive integers and $b \leq 100$.

    Make the usual assumptions as in the birthday problem. To simplify the calculation, you can use the approximations $365 \approx 360$ and $89 \approx 90$, and the fact that $e^x \approx 1 + x$ for $x \approx 0$.

72. Use Poisson approximations to investigate the following types of coincidences. The usual assumptions of the birthday problem apply.

    (a) How many people are needed to have a 50% chance that at least one of them has the same birthday as *you*?

    (b) How many people are needed to have a 50% chance that there is at least one pair of people who not only were born on the same day of the year, but also were born at the same *hour* (e.g., two people born between 2 pm and 3 pm are considered to have been born at the same hour)?

    (c) Considering that only 1/24 of pairs of people born on the same day were born at the same hour, why isn't the answer to (b) approximately $24 \cdot 23$?

    (d) With 100 people, there is a 64% chance that there is at least one set of 3 people with the same birthday (according to R, using `pbirthday(100,classes=365,coincident=3)` to compute it). Provide two different Poisson approximations for this value, one based on creating an indicator r.v. for each triplet of people, and the other based on creating an indicator r.v. for each day of the year. Which is more accurate?

73. A chess tournament has 100 players. In the first round, they are randomly paired to determine who plays whom (so 50 games are played). In the second round, they are again randomly paired, independently of the first round. In both rounds, all possible pairings are equally likely. Let X be the number of people who play against the same opponent twice.

    (a) Find the expected value of X.

    (b) Explain why X is *not* approximately Poisson.

    (c) Find good approximations to $P(X = 0)$ and $P(X = 2)$, by thinking about games in the second round such that the same pair played each other in the first round.

*Existence

74. Ⓢ Each of 111 people names their 5 favorite movies out of a list of 11 movies.

(a) Alice and Bob are 2 of the 111 people. Assume *for this part only* that Alice's 5 favorite movies out of the 11 are random, with all sets of 5 equally likely, and likewise for Bob, independently. Find the expected number of movies in common to Alice's and Bob's lists of favorite movies.

(b) Show that there are 2 movies such that at least 21 of the people name both of these movies as favorites.

75. Ⓢ The circumference of a circle is colored with red and blue ink such that 2/3 of the circumference is red and 1/3 is blue. Prove that no matter how complicated the coloring scheme is, there is a way to inscribe a square in the circle such that at least three of the four corners of the square touch red ink.

76. Ⓢ A hundred students have taken an exam consisting of 8 problems, and for each problem at least 65 of the students got the right answer. Show that there exist two students who collectively got everything right, in the sense that for each problem, at least one of the two got it right.

77. Ⓢ Ten points in the plane are designated. You have ten circular coins (of the same radius). Show that you can position the coins in the plane (without stacking them) so that all ten points are covered.

Hint: Consider a *honeycomb tiling* of the plane (this is a way to divide the plane into hexagons). You can use the fact from geometry that if a circle is inscribed in a hexagon then the ratio of the area of the circle to the area of the hexagon is $\frac{\pi}{2\sqrt{3}} > 0.9$.

78. Ⓢ Let S be a set of binary strings $a_1 \dots a_n$ of length n (where juxtaposition means concatenation). We call S *k-complete* if for any indices $1 \leq i_1 < \dots < i_k \leq n$ and any binary string $b_1 \dots b_k$ of length k, there is a string $s_1 \dots s_n$ in S such that $s_{i_1} s_{i_2} \dots s_{i_k} = b_1 b_2 \dots b_k$. For example, for $n = 3$, the set $S = \{001, 010, 011, 100, 101, 110\}$ is 2-complete since all 4 patterns of 0's and 1's of length 2 can be found in any 2 positions. Show that if $\binom{n}{k} 2^k (1 - 2^{-k})^m < 1$, then there exists a k-complete set of size at most m.

Mixed practice

79. A hacker is trying to break into a password-protected website by randomly trying to guess the password. Let m be the number of possible passwords.

(a) Suppose for this part that the hacker makes random guesses (with equal probability), *with replacement.* Find the average number of guesses it will take until the hacker guesses the correct password (including the successful guess).

(b) Now suppose that the hacker guesses randomly, *without replacement.* Find the average number of guesses it will take until the hacker guesses the correct password (including the successful guess).
Hint: Use symmetry.

(c) Show that the answer to (a) is greater than the answer to (b) (except in the degenerate case $m = 1$), and explain why this makes sense intuitively.

(d) Now suppose that the website locks out any user after n incorrect password attempts, so the hacker can guess at most n times. Find the PMF of the number of guesses that the hacker makes, both for the case of sampling with replacement and for the case of sampling without replacement.

80. A fair 20-sided die is rolled repeatedly, until a gambler decides to stop. The gambler receives the amount shown on the die when the gambler stops. The gambler decides in advance to roll the die until a value of m or greater is obtained, and then stop (where m is a fixed integer with $1 \leq m \leq 20$).

(a) What is the expected number of rolls (simplify)?

(b) What is the expected square root of the number of rolls (as a sum)?

81. Ⓢ A group of 360 people is going to be split into 120 teams of 3 (where the order of teams and the order within a team don't matter).

(a) How many ways are there to do this?

(b) The group consists of 180 married couples. A random split into teams of 3 is chosen, with all possible splits equally likely. Find the expected number of teams containing married couples.

82. Ⓢ The gambler de Méré asked Pascal whether it is more likely to get at least one six in 4 rolls of a die, or to get at least one double-six in 24 rolls of a pair of dice. Continuing this pattern, suppose that a group of n fair dice is rolled $4 \cdot 6^{n-1}$ times.

(a) Find the expected number of times that "all sixes" is achieved (i.e., how often among the $4 \cdot 6^{n-1}$ rolls it happens that all n dice land 6 simultaneously).

(b) Give a simple but accurate approximation of the probability of having at least one occurrence of "all sixes", for n large (in terms of e but not n).

(c) de Méré finds it tedious to re-roll so many dice. So after one normal roll of the n dice, in going from one roll to the next, with probability 6/7 he leaves the dice in the same configuration and with probability 1/7 he re-rolls. For example, if $n = 3$ and the 7th roll is $(3, 1, 4)$, then 6/7 of the time the 8th roll remains $(3, 1, 4)$ and 1/7 of the time the 8th roll is a new random outcome. Does the expected number of times that "all sixes" is achieved stay the same, increase, or decrease (compared with (a))? Give a short but clear explanation.

83. Ⓢ Five people have just won a \$100 prize, and are deciding how to divide the \$100 up between them. Assume that whole dollars are used, not cents. Also, for example, giving \$50 to the first person and \$10 to the second is different from vice versa.

(a) How many ways are there to divide up the \$100, such that each gets at least \$10?

(b) Assume that the \$100 is randomly divided up, with all of the possible allocations counted in (a) equally likely. Find the expected amount of money that the first person receives.

(c) Let A_j be the event that the jth person receives more than the first person (for $2 \leq j \leq 5$), when the \$100 is randomly allocated as in (b). Are A_2 and A_3 independent?

84. Ⓢ Joe's iPod has 500 different songs, consisting of 50 albums of 10 songs each. He listens to 11 random songs on his iPod, with all songs equally likely and chosen independently (so repetitions may occur).

(a) What is the PMF of how many of the 11 songs are from his favorite album?

(b) What is the probability that there are 2 (or more) songs from the same album among the 11 songs he listens to?

(c) A pair of songs is a *match* if they are from the same album. If, say, the 1st, 3rd, and 7th songs are all from the same album, this counts as 3 matches. Among the 11 songs he listens to, how many matches are there on average?

85. Ⓢ Each day that the Mass Cash lottery is run in Massachusetts, 5 of the integers from 1 to 35 are chosen (randomly and without replacement).

(a) When playing this lottery, find the probability of guessing exactly 3 numbers right, given that you guess at least 1 of the numbers right.

(b) Find an exact expression for the expected number of days needed so that all of the $\binom{35}{5}$ possible lottery outcomes will have occurred.

(c) Approximate the probability that after 50 days of the lottery, every number from 1 to 35 has been picked at least once.

86. A certain country has three political parties, denoted by A, B, and C. Each adult in the country is a member of exactly one of the three parties. There are n adults in the country, consisting of n_A members of party A, n_B members of party B, and n_C members of party C, where n_A, n_B, n_C are positive integers with $n_A + n_B + n_C = n$.

A simple random sample of size m is chosen from the adults in the country (the sampling is done *without* replacement, and all possible samples of size m are equally likely). Let X be the number of members of party A in the sample, Y be the number of members of party B in the sample, and Z be the number of members of party C in the sample.

(a) Find $P(X = x, Y = y, Z = z)$, for x, y, z nonnegative integers with $x + y + z = m$.

(b) Find $E(X)$.

(c) Find $\text{Var}(X)$, and briefly explain why your answer makes sense in the extreme cases $m = 1$ and $m = n$.

87. The U.S. Senate consists of 100 senators, with 2 from each of the 50 states. There are d Democrats in the Senate. A committee of size c is formed, by picking a random set of senators such that all sets of size c are equally likely.

(a) Find the expected number of Democrats on the committee.

(b) Find the expected number of states represented on the committee (by at least one senator).

(c) Find the expected number of states such that both of the state's senators are on the committee.

(d) Each state has a *junior senator* and a *senior senator* (based on which of them has served longer). A committee of size 20 is formed randomly, with all sets of 20 senators equally likely. Find the distribution of the number of junior senators on the committee, and the expected number of junior senators on the committee.

(e) For the committee from (d), find the expected number of states such that both senators from that state are on the committee.

88. A certain college has g good courses and b bad courses, where g and b are positive integers. Alice, who is hoping to find a good course, randomly shops courses one at a time (without replacement) until she finds a good course.

(a) Find the expected number of bad courses that Alice shops before finding a good course (as a simple expression in terms of g and b).

(b) Should the answer to (a) be less than, equal to, or greater than b/g? Explain this using properties of the Geometric distribution.

89. A DNA sequence can be represented as a sequence of letters, where the *alphabet* has 4 letters: A,C,T,G. Suppose such a sequence is generated randomly, where the letters are independent and the probabilities of A,C,T,G are p_A, p_C, p_T, p_G, respectively.

    (a) In a DNA sequence of length 115, what is the variance of the number of occurrences of the letter C?

    (b) In a DNA sequence of length 115, what is the expected number of occurrences of the expression CATCAT? Note that, for example, the expression CATCATCAT counts as 2 occurrences.

    (c) In a DNA sequence of length 6, what is the probability that the expression CAT occurs at least once?

90. Alice is conducting a survey in a town with population size 1000. She selects a simple random sample of size 100 (i.e., sampling without replacement, such that all samples of size 100 are equally likely). Bob is also conducting a survey in this town. Bob selects a simple random sample of size 20, independent of Alice's sample. Let A be the set of people in Alice's sample and B be the set of people in Bob's sample.

    (a) Find the expected number of people in $A \cap B$.

    (b) Find the expected number of people in $A \cup B$.

    (c) The 1000 people consist of 500 married couples. Find the expected number of couples such that both members of the couple are in Bob's sample.

91. The *Wilcoxon rank sum test* is a widely used procedure for assessing whether two groups of observations come from the same distribution. Let group 1 consist of i.i.d. $X_1, \dots, X_m$ with CDF F and group 2 consist of i.i.d. $Y_1, \dots, Y_n$ with CDF G, with all of these r.v.s independent. Assume that the probability of 2 of the observations being equal is 0 (this will be true if the distributions are continuous).

    After the $m + n$ observations are obtained, they are listed in increasing order, and each is assigned a *rank* between 1 and $m + n$: the smallest has rank 1, the second smallest has rank 2, etc. Let R_j be the rank of X_j among all the observations for $1 \leq j \leq m$, and let $R = \sum_{j=1}^{m} R_j$ be the sum of the ranks for group 1.

    Intuitively, the Wilcoxon rank sum test is based on the idea that a very large value of R is evidence that observations from group 1 are usually larger than observations from group 2 (and vice versa if R is very small). But how large is "very large" and how small is "very small"? Answering this precisely requires studying the distribution of the *test statistic* R.

    (a) The *null hypothesis* in this setting is that $F = G$. Show that if the null hypothesis is true, then $E(R) = m(m + n + 1)/2$.

    (b) The *power* of a test is an important measure of how good the test is about saying to reject the null hypothesis if the null hypothesis is false. To study the power of the Wilcoxon rank sum test, we need to study the distribution of R in general. So for this part, we do *not* assume $F = G$. Let $p = P(X_1 > Y_1)$. Find $E(R)$ in terms of m, n, p.

    Hint: Write R_j in terms of indicator r.v.s for X_j being greater than various other r.v.s.

92. The legendary Caltech physicist Richard Feynman and two editors of *The Feynman Lectures on Physics* (Michael Gottlieb and Ralph Leighton) posed the following problem about how to decide what to order at a restaurant. You plan to eat m meals at a certain restaurant, where you have never eaten before. Each time, you will order one dish.

    The restaurant has n dishes on the menu, with $n \geq m$. Assume that if you had tried all the dishes, you would have a definite ranking of them from 1 (your least favorite) to n (your favorite). If you knew which your favorite was, you would be happy to order it always (you never get tired of it).

Before you've eaten at the restaurant, this ranking is completely unknown to you. After you've tried some dishes, you can rank those dishes amongst themselves, but don't know how they compare with the dishes you haven't yet tried. There is thus an *exploration-exploitation tradeoff*: should you try new dishes, or should you order your favorite among the dishes you have tried before?

A natural strategy is to have two phases in your series of visits to the restaurant: an *exploration phase*, where you try different dishes each time, and an *exploitation phase*, where you always order the best dish you obtained in the exploration phase. Let k be the length of the exploration phase (so $m - k$ is the length of the exploitation phase).

Your goal is to maximize the expected sum of the ranks of the dishes you eat there (the rank of a dish is the "true" rank from 1 to n that you would give that dish if you could try all the dishes). Show that the optimal choice is

$$k = \sqrt{2(m+1)} - 1,$$

or this rounded up or down to an integer if needed. Do this in the following steps:

(a) Let X be the rank of the best dish that you find in the exploration phase. Find the expected sum of the ranks of all the dishes you eat (including both phases), in terms of k, n, and $E(X)$.

(b) Find the PMF of X, as a simple expression in terms of binomial coefficients.

(c) Show that

$$E(X) = \frac{k(n+1)}{k+1}.$$

Hint: Use Example 1.5.2 (about the team captain) and Exercise 20 from Chapter 1 (about the hockey stick identity).

(d) Use calculus to find the optimal value of k.

5

Continuous random variables

So far we have been working with discrete random variables, whose possible values can be written down as a list. In this chapter we will discuss *continuous* r.v.s, which can take on any real value in an interval (possibly of infinite length, such as $(0, \infty)$ or the entire real line). First we'll look at properties of continuous r.v.s in general. Then we'll introduce three famous continuous distributions—the Uniform, Normal, and Exponential—which, in addition to having important stories in their own right, serve as building blocks for many other useful continuous distributions.

5.1 Probability density functions

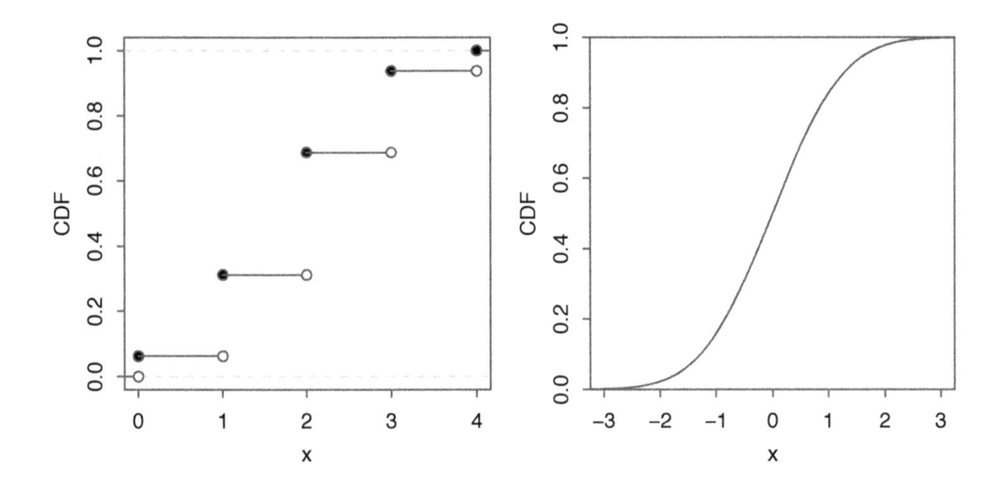

FIGURE 5.1
Discrete vs. continuous r.v.s. Left: The CDF of a discrete r.v. has jumps at each point in the support. Right: The CDF of a continuous r.v. increases smoothly.

Recall that for a discrete r.v., the CDF jumps at every point in the support, and is flat everywhere else. In contrast, for a continuous r.v. the CDF increases smoothly; see Figure 5.1 for a comparison of discrete vs. continuous CDFs.

Definition 5.1.1 (Continuous r.v.). An r.v. has a *continuous distribution* if its

CDF is differentiable. We also allow there to be endpoints (or finitely many points) where the CDF is continuous but not differentiable, as long as the CDF is differentiable everywhere else. A *continuous random variable* is a random variable with a continuous distribution.

For discrete r.v.s, the CDF is awkward to work with because of its jumpiness, and its derivative is almost useless since it's undefined at the jumps and 0 everywhere else. But for continuous r.v.s, the CDF is often convenient to work with, and its derivative is a very useful function, called the *probability density function.*

Definition 5.1.2 (Probability density function). For a continuous r.v. X with CDF F, the *probability density function* (PDF) of X is the derivative f of the CDF, given by $f(x) = F'(x)$. The *support* of X, and of its distribution, is the set of all x where $f(x) > 0$.

An important way in which continuous r.v.s differ from discrete r.v.s is that for a continuous r.v. X, $P(X = x) = 0$ for all x. This is because $P(X = x)$ is the height of a jump in the CDF at x, but the CDF of X has no jumps! Since the PMF of a continuous r.v. would just be 0 everywhere, we work with a PDF instead.

The PDF is analogous to the PMF in many ways, but there is a key difference: for a PDF f, the quantity $f(x)$ is *not* a probability, and in fact it is possible to have $f(x) > 1$ for some values of x. To obtain a probability, we need to *integrate* the PDF. The fundamental theorem of calculus tells us how to get from the PDF back to the CDF.

Proposition 5.1.3 (PDF to CDF). Let X be a continuous r.v. with PDF f. Then the CDF of X is given by

$$F(x) = \int_{-\infty}^{x} f(t)dt.$$

Proof. By definition of PDF, F is an antiderivative of f. So by the fundamental theorem of calculus,

$$\int_{-\infty}^{x} f(t)dt = F(x) - F(-\infty) = F(x).$$ ■

The above result is analogous to how we obtained the value of a discrete CDF at x by summing the PMF over all values less than or equal to x; here we *integrate* the PDF over all values up to x, so the CDF is the *accumulated area* under the PDF. Since we can freely convert between the PDF and the CDF using the inverse operations of integration and differentiation, both the PDF and CDF carry complete information about the distribution of a continuous r.v.

Since the PDF determines the distribution, we should be able to use it to find the probability of X falling into an interval (a, b). A handy fact is that we can include or exclude the endpoints as we wish without altering the probability, since the endpoints have probability 0:

$$P(a < X < b) = P(a < X \leq b) = P(a \leq X < b) = P(a \leq X \leq b).$$

☣ **5.1.4** (Including or excluding endpoints). We can be carefree about including or excluding endpoints as above for continuous r.v.s, but we must not be careless about this for discrete r.v.s.

By definition of CDF and the fundamental theorem of calculus,

$$P(a < X \leq b) = F(b) - F(a) = \int_a^b f(x)dx.$$

Therefore, to find the probability of X falling in the interval $(a, b]$ (or (a, b), $[a, b)$, or $[a, b]$) using the PDF, we simply integrate the PDF from a to b. In general, for an arbitrary region $A \subseteq \mathbb{R}$,

$$P(X \in A) = \int_A f(x)dx.$$

In summary:

To get a desired probability, integrate the PDF over the appropriate range.

Just as a valid PMF must be nonnegative and sum to 1, a valid PDF must be nonnegative and integrate to 1.

Theorem 5.1.5 (Valid PDFs). The PDF f of a continuous r.v. must satisfy the following two criteria:

- Nonnegative: $f(x) \geq 0$;
- Integrates to 1: $\int_{-\infty}^{\infty} f(x)dx = 1$.

Proof. The first criterion is true because probability is nonnegative; if $f(x_0)$ were negative, then we could integrate over a tiny region around x_0 and get a negative probability. Alternatively, note that the PDF at x_0 is the slope of the CDF at x_0, so $f(x_0) < 0$ would imply that the CDF is *decreasing* at x_0, which is not allowed. The second criterion is true since $\int_{-\infty}^{\infty} f(x)dx$ is the probability of X falling somewhere on the real line, which is 1. ■

Conversely, any such function f *is* the PDF of some r.v. This is because if f satisfies these properties, we can integrate it as in Proposition 5.1.3 to get a function F satisfying the properties of a CDF. Then a version of Universality of the Uniform, the main concept in Section 5.3, can be used to create an r.v. with CDF F.

Now let's look at some specific examples of PDFs. The two distributions in the following examples are named the Logistic and Rayleigh distributions, but we won't discuss their stories here; their appearance is intended mainly as a way of getting comfortable with PDFs.

Example 5.1.6 (Logistic). The Logistic distribution has CDF

$$F(x) = \frac{e^x}{1 + e^x}, \quad x \in \mathbb{R}.$$

To get the PDF, we differentiate the CDF, which gives

$$f(x) = \frac{e^x}{(1+e^x)^2}, \quad x \in \mathbb{R}.$$

Let $X \sim$ Logistic. To find $P(-2 < X < 2)$, integrate the PDF from -2 to 2:

$$P(-2 < X < 2) = \int_{-2}^{2} \frac{e^x}{(1+e^x)^2} dx = F(2) - F(-2) \approx 0.76.$$

The integral was easy to evaluate since we already knew that F was an antiderivative for f, and we had a nice expression for F. Otherwise, we could have made the substitution $u = 1 + e^x$, so $du = e^x dx$, giving

$$\int_{-2}^{2} \frac{e^x}{(1+e^x)^2} dx = \int_{1+e^{-2}}^{1+e^2} \frac{1}{u^2} du = \left(-\frac{1}{u}\right)\Big|_{1+e^{-2}}^{1+e^2} \approx 0.76.$$

Figure 5.2 shows the Logistic PDF (left) and CDF (right). On the PDF, the probability $P(-2 < X < 2)$ is represented by the shaded area; on the CDF, it is represented by the height of the curly brace. You can check that the properties of a valid PDF and CDF are satisfied. ☐

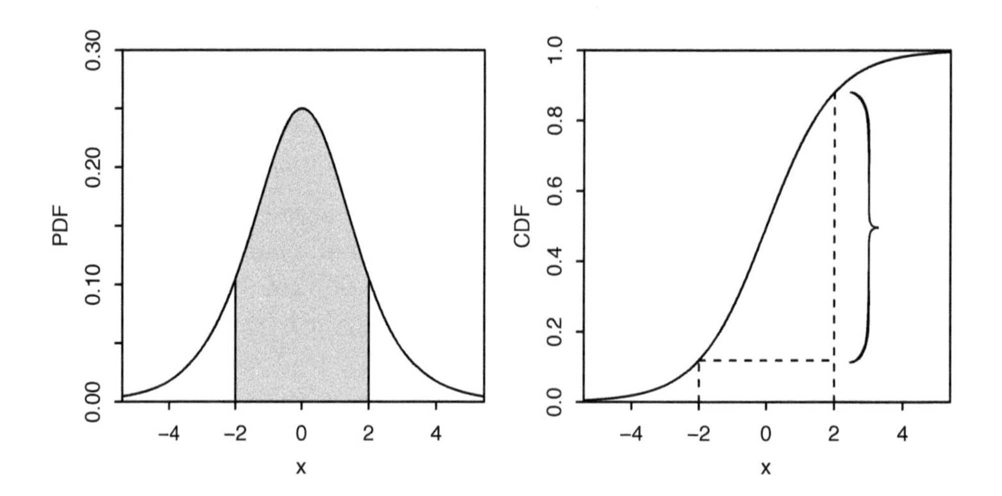

FIGURE 5.2
Logistic PDF and CDF. The probability $P(-2 < X < 2)$ is indicated by the shaded area under the PDF and the height of the curly brace on the CDF.

Example 5.1.7 (Rayleigh). The Rayleigh distribution has CDF

$$F(x) = 1 - e^{-x^2/2}, \quad x > 0.$$

To get the PDF, we differentiate the CDF, which gives

$$f(x) = xc^{-x^2/2}, \quad x > 0.$$

For $x \leq 0$, both the CDF and the PDF are equal to 0.

Let $X \sim$ Rayleigh. To find $P(X > 2)$, we need to integrate the PDF from 2 to ∞. We can do that by making the substitution $u = -x^2/2$, but since we already have the CDF in a nice form we know the integral is $F(\infty) - F(2) = 1 - F(2)$:

$$P(X > 2) = \int_2^\infty x e^{-x^2/2} dx = 1 - F(2) \approx 0.14.$$

The Rayleigh PDF and CDF are plotted in Figure 5.3. Again, probability is represented by a shaded area on the PDF and a vertical height on the CDF. □

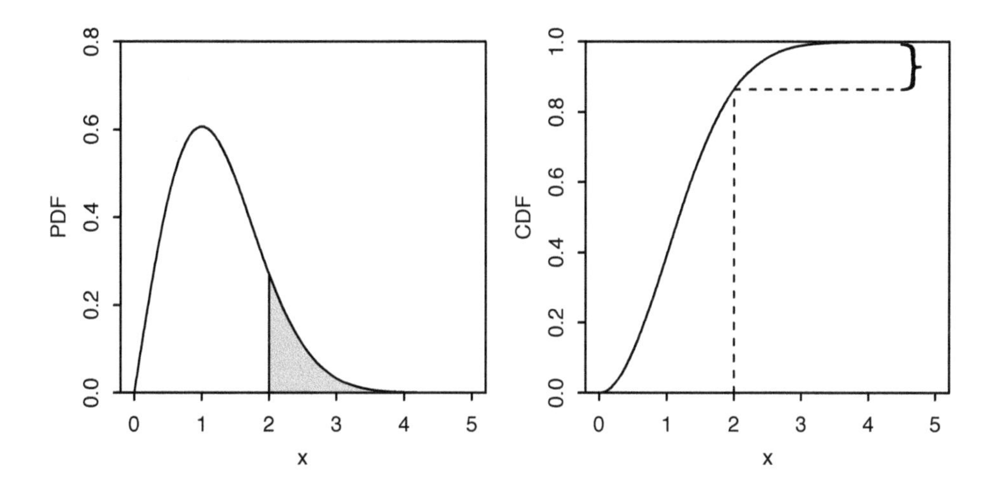

FIGURE 5.3
Rayleigh PDF and CDF. The probability $P(X > 2)$ is indicated by the shaded area under the PDF and the height of the curly brace on the CDF.

Although the height of a PDF at x does not represent a probability, it is closely related to the probability of falling into a tiny interval around x, as the following intuition explains.

Intuition 5.1.8 (Units). Let F be the CDF and f be the PDF of a continuous r.v. X. As mentioned earlier, $f(x)$ is *not* a probability; for example, we could have $f(3) > 1$, and we know $P(X = 3) = 0$. But thinking about the probability of X being *very close* to 3 gives us a way to interpret $f(3)$. Specifically, the probability of X being in a tiny interval of length ϵ, centered at 3, will essentially be $f(3)\epsilon$. This is because

$$P(3 - \epsilon/2 < X < 3 + \epsilon/2) = \int_{3-\epsilon/2}^{3+\epsilon/2} f(x)dx \approx f(3)\epsilon,$$

if the interval is so tiny that f is approximately the constant $f(3)$ on that interval. In general, we can think of $f(x)dx$ as the probability of X being in an infinitesimally small interval containing x, of length dx.

In practice, X often has *units* in some system of measurement, such as units of distance, time, area, or mass. Thinking about the units is not only important in applied problems, but also it often helps in checking that answers make sense.

Suppose for concreteness that X is a length, measured in centimeters (cm). Then $f(x) = dF(x)/dx$ is the probability per cm at x, which explains why $f(x)$ is a probability *density*. Probability is a dimensionless quantity (a number without physical units), so the units of $f(x)$ are cm^{-1}. Therefore, to be able to get a probability again, we need to multiply $f(x)$ by a length. When we do an integral such as $\int_0^5 f(x)dx$, this is achieved by the often-forgotten dx. □

Intuition 5.1.9 (Simulation). For another intuitive way to think about PDFs, consider the following graphical way to *simulate* a draw X from a continuous distribution, based on looking at the graph of the PDF. To generate X, choose a uniformly random point under the PDF curve; this means that the probability of any region under the curve is the *area* of that region. Then let X be the x-coordinate of the random point. This is illustrated in Figure 5.4.

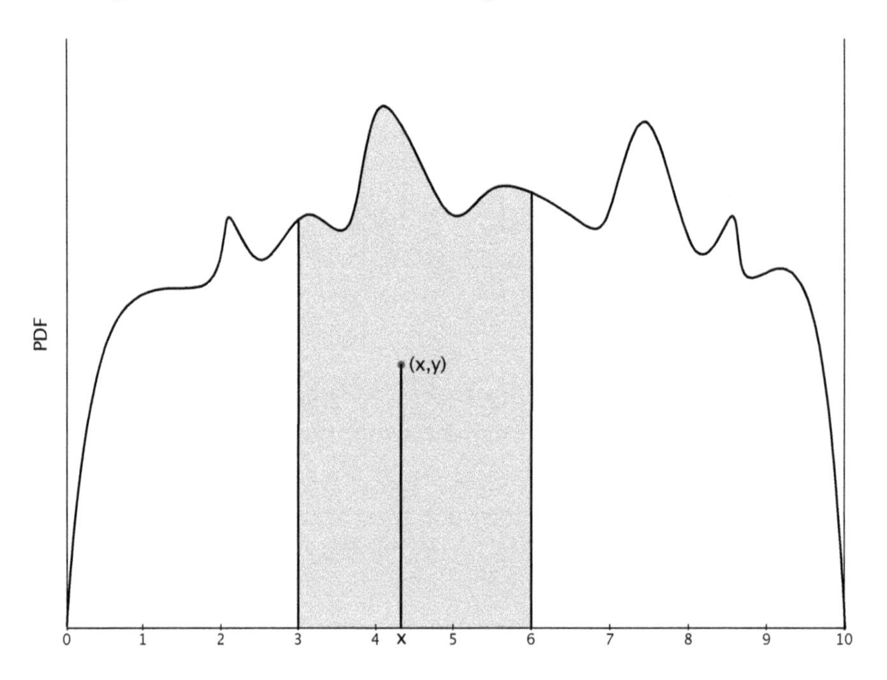

FIGURE 5.4
A complicated PDF (no numbers are shown on the vertical axis since the scale is whatever it needs to be to make the area under the curve 1). To generate an r.v. X with this PDF, choose a uniformly random point (x, y) under the curve and let $X = x$. This method works since, for example, X will be in the interval $[3, 6]$ if and only if the randomly chosen point (x, y) is in the shaded region.

Then X has the desired distribution since, by construction, $P(a \leq X \leq b)$ is the area under the PDF curve between the lines $x = a$ and $x = b$. Thinking about this method helps build intuition for PDFs, by giving us a feel for random variables sampled according to a particular PDF curve. □

The definition of expectation for continuous r.v.s is analogous to the definition for discrete r.v.s: replace the sum with an integral and the PMF with the PDF.

Definition 5.1.10 (Expectation of a continuous r.v.). The *expected value* (also called the *expectation* or *mean*) of a continuous r.v. X with PDF f is

$$E(X) = \int_{-\infty}^{\infty} xf(x)dx.$$

As in the discrete case, the expectation of a continuous r.v. may or may not exist. When discussing expectations, it would be very tedious to have to add "(if it exists)" after every mention of an expectation not yet shown to exist, so we will often leave this implicit.

The integral is taken over the entire real line, but if the support of X is not the entire real line we can just integrate over the support. The units in this definition make sense: if X is measured in centimeters, then so is $E(X)$, since $xf(x)dx$ has units of $\text{cm} \cdot \text{cm}^{-1} \cdot \text{cm} = \text{cm}$.

With this definition, the expected value retains its interpretation as a center of mass. As shown in Figure 5.5, using the Rayleigh PDF for illustrative purposes, the expected value is the balancing point of the PDF, just as it was the balancing point of the PMF in the discrete case.

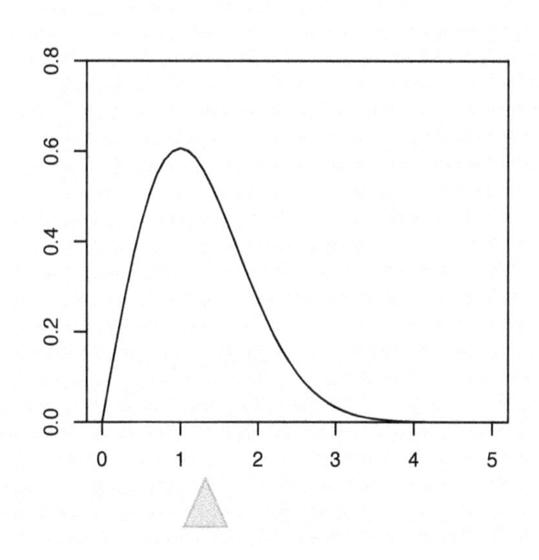

FIGURE 5.5
The expected value of a continuous r.v. is the balancing point of the PDF.

Linearity of expectation holds for continuous r.v.s, as it did for discrete r.v.s (we will show this later in Example 7.2.4). LOTUS also holds for continuous r.v.s, replacing the sum with an integral and the PMF with the PDF:

Theorem 5.1.11 (LOTUS, continuous). If X is a continuous r.v. with PDF f and g is a function from $\mathbb{R}$ to $\mathbb{R}$, then

$$E(g(X)) = \int_{-\infty}^{\infty} g(x)f(x)dx.$$

We now have all the tools we need to tackle the named distributions of this chapter, starting with the Uniform distribution.

5.2 Uniform

Intuitively, a Uniform r.v. on the interval (a, b) is a completely random number between a and b. We formalize the notion of "completely random" on an interval by specifying that the PDF should be *constant* over the interval.

Definition 5.2.1 (Uniform distribution). A continuous r.v. U is said to have the *Uniform distribution* on the interval (a, b) if its PDF is

$$f(x) = \begin{cases} \frac{1}{b-a} & \text{if } a < x < b, \\ 0 & \text{otherwise.} \end{cases}$$

We denote this by $U \sim \text{Unif}(a, b)$.

This is a valid PDF because the area under the curve is just the area of a rectangle with width $b - a$ and height $1/(b - a)$. The CDF is the accumulated area under the PDF:

$$F(x) = \begin{cases} 0 & \text{if } x \leq a, \\ \frac{x-a}{b-a} & \text{if } a < x < b, \\ 1 & \text{if } x \geq b. \end{cases}$$

The Uniform distribution that we will most frequently use is the Unif(0, 1) distribution, also called the standard Uniform. The Unif(0, 1) PDF and CDF are particularly simple: $f(x) = 1$ and $F(x) = x$ for $0 < x < 1$. Figure 5.6 shows the Unif(0, 1) PDF and CDF side by side.

For a general Unif(a, b) distribution, the PDF is constant on (a, b), and the CDF is ramp-shaped, increasing linearly from 0 to 1 as x ranges from a to b.

For Uniform distributions, *probability is proportional to length.*

Proposition 5.2.2. Let $U \sim \text{Unif}(a, b)$, and let (c, d) be a subinterval of (a, b), of length l (so $l = d - c$). Then the probability of U being in (c, d) is proportional to l. For example, a subinterval that is twice as long has twice the probability of containing U, and a subinterval of the same length has the same probability.

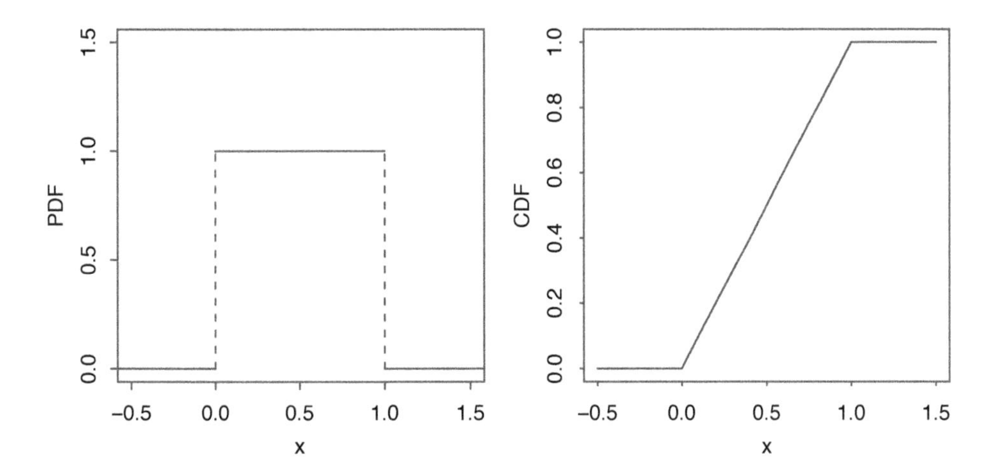

FIGURE 5.6
Unif$(0,1)$ PDF and CDF.

Proof. Since the PDF of U is the constant $\frac{1}{b-a}$ on (a,b), the area under the PDF from c to d is $\frac{l}{b-a}$, which is a constant times l. ■

The above proposition is a very special property of the Uniform; for any other distribution, there are intervals of the same length that have different probabilities. Even after conditioning on a Uniform r.v. being in a certain subinterval, we *still* have a Uniform distribution and thus still have probability proportional to length (within that subinterval); we show this below.

Proposition 5.2.3. Let $U \sim$ Unif(a,b), and let (c,d) be a subinterval of (a,b). Then the conditional distribution of U given $U \in (c,d)$ is Unif(c,d).

Proof. For u in (c,d), the conditional CDF at u is

$$P(U \le u | U \in (c,d)) = \frac{P(U \le u, c < U < d)}{P(U \in (c,d))} = \frac{P(U \in (c,u])}{P(U \in (c,d))} = \frac{u-c}{d-c}.$$

The conditional CDF is 0 for $u \le c$ and 1 for $u \ge d$. So the conditional distribution of U is as claimed. ■

Example 5.2.4. Let's illustrate the above propositions for $U \sim$ Unif$(0,1)$. In this special case, the support has length 1, so probability *is* length: the probability of U falling into the interval $(0, 0.3)$ is 0.3, as is the probability of falling into $(0.3, 0.6)$, $(0.4, 0.7)$, or any other interval of length 0.3 within $(0,1)$.

Now suppose that we learn that $U \in (0.4, 0.7)$. Given this information, the conditional distribution of U is Unif$(0.4, 0.7)$. Then the conditional probability of $U \in (0.4, 0.6)$ is $2/3$, since $(0.4, 0.6)$ provides $2/3$ of the length of $(0.4, 0.7)$. The conditional probability of $U \in (0, 0.6)$ is also $2/3$, since we discard the points to the left of 0.4 when conditioning on $U \in (0.4, 0.7)$. □

Next, let's derive the mean and variance of $U \sim \text{Unif}(a,b)$. The expectation is extremely intuitive: the PDF is constant, so its balancing point should be the midpoint of (a,b). This is exactly what we find by using the definition of expectation for continuous r.v.s:

$$E(U) = \int_a^b x \cdot \frac{1}{b-a} dx = \frac{1}{b-a}\left(\frac{b^2}{2} - \frac{a^2}{2}\right) = \frac{a+b}{2}.$$

For the variance, we first find $E(U^2)$ using the continuous version of LOTUS:

$$E(U^2) = \int_a^b x^2 \frac{1}{b-a} dx = \frac{1}{3} \cdot \frac{b^3 - a^3}{b-a}.$$

Then

$$\text{Var}(U) = E(U^2) - (EU)^2 = \frac{1}{3} \cdot \frac{b^3-a^3}{b-a} - \left(\frac{a+b}{2}\right)^2,$$

which reduces, after factoring $b^3 - a^3 = (b-a)(a^2+ab+b^2)$ and simplifying, to

$$\text{Var}(U) = \frac{(b-a)^2}{12}.$$

The above derivation isn't terribly painful, but there is an easier path, using a technique that is often useful for continuous distributions. The technique is called *location-scale transformation*, and it relies on the observation that shifting and scaling a Uniform r.v. produces another Uniform r.v. Shifting is considered a change of *location* and scaling is a change of *scale*, hence the term location-scale. For example, if X is Uniform on the interval $(1,2)$, then $X+5$ is Uniform on the interval $(6,7)$, $2X$ is Uniform on the interval $(2,4)$, and $2X+5$ is Uniform on $(7,9)$.

Definition 5.2.5 (Location-scale transformation). Let X be a random variable and $Y = \sigma X + \mu$, where σ and μ are constants with $\sigma > 0$. Then we say that Y has been obtained as a *location-scale transformation* of X. Here μ controls how the location is changed and σ controls how the scale is changed.

☣ **5.2.6.** In a location-scale transformation, starting with $X \sim \text{Unif}(a,b)$ and transforming it to $Y = cX + d$ where c and d are constants with $c > 0$, Y is a *linear* function of X and Uniformity is preserved: $Y \sim \text{Unif}(ca+d, cb+d)$. But if Y is defined as a *nonlinear* transformation of X, then Y will *not* be Uniform in general. For example, for $X \sim \text{Unif}(a,b)$ with $0 \leq a < b$, the transformed r.v. $Y = X^2$ has support (a^2, b^2) but is *not* Uniform on that interval. Chapter 8 explores transformations of r.v.s in detail.

In studying Uniform distributions, a useful strategy is to start with an r.v. that has the simplest Uniform distribution, figure things out in the friendly simple case, and then use a location-scale transformation to handle the general case.

Let's see how this works for finding the expectation and variance of the Unif(a,b)

distribution. The location-scale strategy says to start with $U \sim \text{Unif}(0,1)$. Since the PDF of U is just 1 on the interval $(0,1)$, it is easy to see that

$$\begin{aligned} E(U) &= \int_0^1 x dx = \frac{1}{2}, \\ E(U^2) &= \int_0^1 x^2 dx = \frac{1}{3}, \\ \text{Var}(U) &= \frac{1}{3} - \frac{1}{4} = \frac{1}{12}. \end{aligned}$$

Now that we know the answers for U, transforming U into a general $\text{Unif}(a,b)$ r.v. takes just two steps. First we change the support from an interval of length 1 to an interval of length $b-a$, so we multiply U by the scaling factor $b-a$ to obtain a $\text{Unif}(0, b-a)$ r.v. Then we shift everything until the left endpoint of the support is at a. Thus, if $U \sim \text{Unif}(0,1)$, the random variable

$$\tilde{U} = a + (b-a)U$$

is distributed $\text{Unif}(a,b)$. Now the mean and variance of $\tilde{U}$ follow directly from properties of expectation and variance. By linearity of expectation,

$$E(\tilde{U}) = E(a + (b-a)U) = a + (b-a)E(U) = a + \frac{b-a}{2} = \frac{a+b}{2}.$$

By the fact that additive constants don't affect the variance while multiplicative constants come out squared,

$$\text{Var}(\tilde{U}) = \text{Var}(a + (b-a)U) = \text{Var}((b-a)U) = (b-a)^2 \text{Var}(U) = \frac{(b-a)^2}{12}.$$

These agree with our previous answers.

The technique of location-scale transformation will work for any family of distributions such that shifting and scaling an r.v. whose distribution in the family produces another r.v. whose distribution is in the family. This technique does not apply to families of discrete distributions (with a fixed support) since, for example, shifting or scaling $X \sim \text{Bin}(n,p)$ changes the support and produces an r.v. that is no longer Binomial. A Binomial r.v. must be able to take on all integer values between 0 and some upper bound, but $X + 4$ can't take on any value in $\{0,1,2,3\}$ and $2X$ can only take even values, so neither of these r.v.s has a Binomial distribution.

☣ **5.2.7** (Beware of sympathetic magic). When using location-scale transformations, the shifting and scaling should be applied to the *random variables* themselves, not to their PDFs. To confuse these two would be an instance of sympathetic magic (see ☣ 3.7.7), and would result in invalid PDFs. For example, let $U \sim \text{Unif}(0,1)$, so the PDF f has $f(x) = 1$ on $(0,1)$ (and $f(x) = 0$ elsewhere). Then $3U + 1 \sim \text{Unif}(1,4)$, but $3f + 1$ is the function that equals 4 on $(0,1)$ and 1 elsewhere, which is not a valid PDF since it does not integrate to 1.

5.3 Universality of the Uniform

In this section, we will discuss a remarkable property of the Uniform distribution: given a Unif$(0,1)$ r.v., we can construct an r.v. with *any continuous distribution we want.* Conversely, given an r.v. with an arbitrary continuous distribution, we can create a Unif$(0,1)$ r.v. We call this the *universality of the Uniform*, because it tells us the Uniform is a universal starting point for building r.v.s with other distributions. Universality of the Uniform also goes by many other names, such as the *probability integral transform*, *inverse transform sampling*, the *quantile transformation*, and even the *fundamental theorem of simulation*.

To keep the proofs simple, we will state universality of the Uniform for a case where we know the inverse of the desired CDF exists. Similar ideas can be used to simulate a random draw from *any* desired CDF as a function of a Unif$(0,1)$ r.v.

Theorem 5.3.1 (Universality of the Uniform). Let F be a CDF which is a continuous function and strictly increasing on the support of the distribution. This ensures that the inverse function F^{-1} exists, as a function from $(0,1)$ to $\mathbb{R}$. We then have the following results.

1. Let $U \sim$ Unif$(0,1)$ and $X = F^{-1}(U)$. Then X is an r.v. with CDF F.
2. Let X be an r.v. with CDF F. Then $F(X) \sim$ Unif$(0,1)$.

Let's make sure we understand what each part of the theorem is saying. The first part says that if we start with $U \sim$ Unif$(0,1)$ and a CDF F, then we can create an r.v. whose CDF is F by plugging U into the inverse CDF F^{-1}. Since F^{-1} is a function (known as the *quantile function*), U is an r.v., and a function of an r.v. is an r.v., $F^{-1}(U)$ is an r.v.; universality of the Uniform says its CDF is F.

The second part of the theorem goes in the reverse direction, starting from an r.v. X whose CDF is F and then creating a Unif$(0,1)$ r.v. Again, F is a function, X is an r.v., and a function of an r.v. is an r.v., so $F(X)$ is an r.v. Since any CDF is between 0 and 1 everywhere, $F(X)$ must take values between 0 and 1. Universality of the Uniform says that the distribution of $F(X)$ is Uniform on $(0,1)$.

☣ **5.3.2.** The second part of universality of the Uniform involves plugging a random variable X *into its own CDF* F. This may seem strangely self-referential, but it makes sense because F is just a function (that satisfies the properties of a valid CDF), and a function of an r.v. is an r.v. There is a potential notational confusion, however: $F(x) = P(X \leq x)$ by definition, but it would be incorrect to say "$F(X) = P(X \leq X) = 1$". Rather, we should first find an expression for the CDF as a function of x, then replace x with X to obtain an r.v. For example, if the CDF of X is $F(x) = 1 - e^{-x}$ for $x > 0$, then $F(X) = 1 - e^{-X}$.

Understanding the statement of the theorem is the difficult part; the proof is just a couple of lines for each direction.

Proof.

1. Let $U \sim \text{Unif}(0,1)$ and $X = F^{-1}(U)$. For all real x,

$$P(X \leq x) = P(F^{-1}(U) \leq x) = P(U \leq F(x)) = F(x),$$

so the CDF of X is F, as claimed. For the last equality, we used the fact that $P(U \leq u) = u$ for $u \in (0,1)$.

2. Let X have CDF F, and find the CDF of $Y = F(X)$. Since Y takes values in $(0,1)$, $P(Y \leq y)$ equals 0 for $y \leq 0$ and equals 1 for $y \geq 1$. For $y \in (0,1)$,

$$P(Y \leq y) = P(F(X) \leq y) = P(X \leq F^{-1}(y)) = F(F^{-1}(y)) = y.$$

Thus Y has the Unif$(0,1)$ CDF. ■

To gain more insight into what the quantile function F^{-1} and universality of the Uniform mean, let's consider an example that is familiar to millions of students: percentiles on an exam.

Example 5.3.3 (Percentiles). A large number of students take a certain exam, graded on a scale from 0 to 100. Let X be the score of a random student. Continuous distributions are easier to deal with here, so let's approximate the discrete distribution of scores using a continuous distribution. Suppose that X is continuous, with a CDF F that is strictly increasing on $(0,100)$. In reality, there are only finitely many students and only finitely many possible scores, but a continuous distribution may be a good approximation.

Suppose that the median score on the exam is 60, i.e., half of the students score above 60 and the other half score below 60 (a convenient aspect of assuming a continuous distribution is that we don't need to worry about how many students had scores *equal* to 60). That is, $F(60) = 1/2$, or, equivalently, $F^{-1}(1/2) = 60$.

If Fred scores a 72 on the exam, then his *percentile* is the fraction of students who score below a 72. This is $F(72)$, which is some number in $(1/2, 1)$ since 72 is above the median. In general, a student with score x has percentile $F(x)$. Going the other way, if we start with a percentile, say 0.95, then $F^{-1}(0.95)$ is the score that has that percentile. A percentile is also called a *quantile*, which is why F^{-1} is called the quantile function. The function F converts scores to quantiles, and the function F^{-1} converts quantiles to scores.

The strange operation of plugging X into its own CDF now has a natural interpretation: $F(X)$ is the percentile attained by a random student. It often happens that the distribution of scores on an exam looks very non-Uniform. For example, there is no reason to think that 10% of the scores are between 70 and 80, even though $(70,80)$ covers 10% of the range of possible scores.

On the other hand, the distribution of *percentiles* of the students *is* Uniform: the universality property says that $F(X) \sim \text{Unif}(0,1)$. For example, 50% of the students have a percentile of at least 0.5. Universality of the Uniform is expressing the fact that 10% of the students have a percentile between 0 and 0.1, 10% have a percentile between 0.1 and 0.2, 10% have a percentile between 0.2 and 0.3, and so on—a fact that is clear from the definition of percentile. □

To illustrate universality of the Uniform, we will apply it to the two distributions we encountered in the previous section, the Logistic and Rayleigh.

Example 5.3.4 (Universality with Logistic)**.** The Logistic CDF is

$$F(x) = \frac{e^x}{1+e^x}, \quad x \in \mathbb{R}.$$

Suppose we have $U \sim \text{Unif}(0,1)$ and wish to generate a Logistic r.v. Part 1 of the universality property says that $F^{-1}(U) \sim$ Logistic, so we first invert the CDF to get F^{-1}:

$$F^{-1}(u) = \log\left(\frac{u}{1-u}\right).$$

Then we plug in U for u:

$$F^{-1}(U) = \log\left(\frac{U}{1-U}\right).$$

Therefore $\log\left(\frac{U}{1-U}\right) \sim$ Logistic.

We can verify directly that $\log\left(\frac{U}{1-U}\right)$ has the required CDF: start from the definition of CDF, do some algebra to isolate U on one side of the inequality, and then use the CDF of the Uniform distribution. Let's work through these calculations once for practice:

$$\begin{aligned} P\left(\log\left(\frac{U}{1-U}\right) \leq x\right) &= P\left(\frac{U}{1-U} \leq e^x\right) \\ &= P\left(U \leq e^x(1-U)\right) \\ &= P\left(U \leq \frac{e^x}{1+e^x}\right) \\ &= \frac{e^x}{1+e^x}, \end{aligned}$$

which is indeed the Logistic CDF.

We can also use simulation to visualize how universality of the Uniform works. To this end, we generated 1 million $\text{Unif}(0,1)$ random variables. We then transformed each of these values u into $\log\left(\frac{u}{1-u}\right)$; if the universality of the Uniform is correct, the transformed numbers should follow a Logistic distribution.

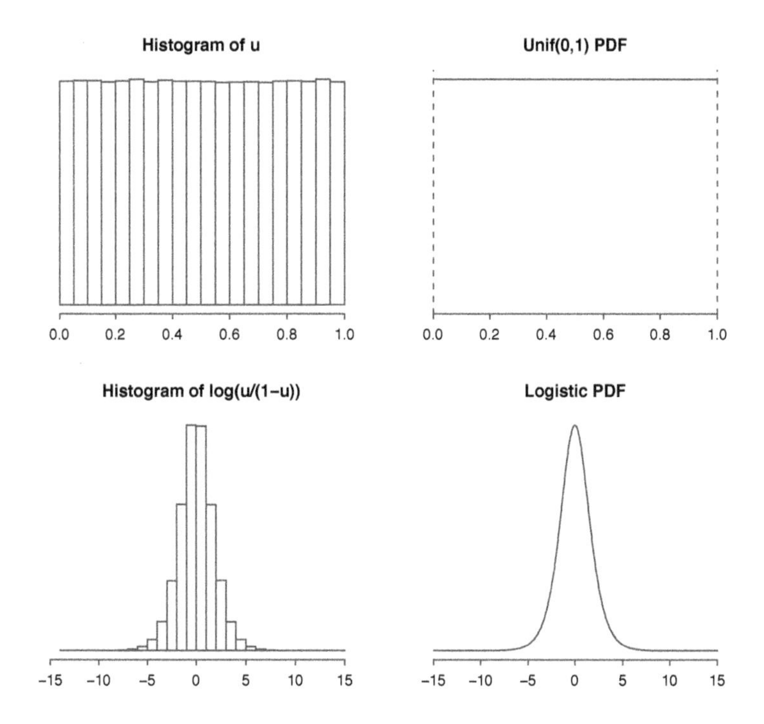

FIGURE 5.7
Top: Histogram of 10^6 draws of $U \sim \text{Unif}(0,1)$, with $\text{Unif}(0,1)$ PDF for comparison. Bottom: Histogram of 10^6 draws of $\log\left(\frac{U}{1-U}\right)$, with Logistic PDF for comparison.

Figure 5.7 displays a histogram of the realizations of U alongside the $\text{Unif}(0,1)$ PDF; below that, we have a histogram of the realizations of $\log\left(\frac{U}{1-U}\right)$ next to the Logistic PDF. As we can see, the second histogram looks very much like the Logistic PDF. Thus, by applying F^{-1}, we were able to transform our Uniform draws into Logistic draws, exactly as claimed by the universality of the Uniform.

Conversely, Part 2 of the universality property states that if $X \sim$ Logistic, then

$$F(X) = \frac{e^X}{1+e^X} \sim \text{Unif}(0,1). \qquad \square$$

Example 5.3.5 (Universality with Rayleigh). The Rayleigh CDF is

$$F(x) = 1 - e^{-x^2/2}, \quad x > 0.$$

The quantile function (the inverse of the CDF) is

$$F^{-1}(u) = \sqrt{-2\log(1-u)},$$

so if $U \sim \text{Unif}(0,1)$, then $F^{-1}(U) = \sqrt{-2\log(1-U)} \sim$ Rayleigh.

We again generated 1 million realizations of $U \sim \text{Unif}(0,1)$ and transformed them

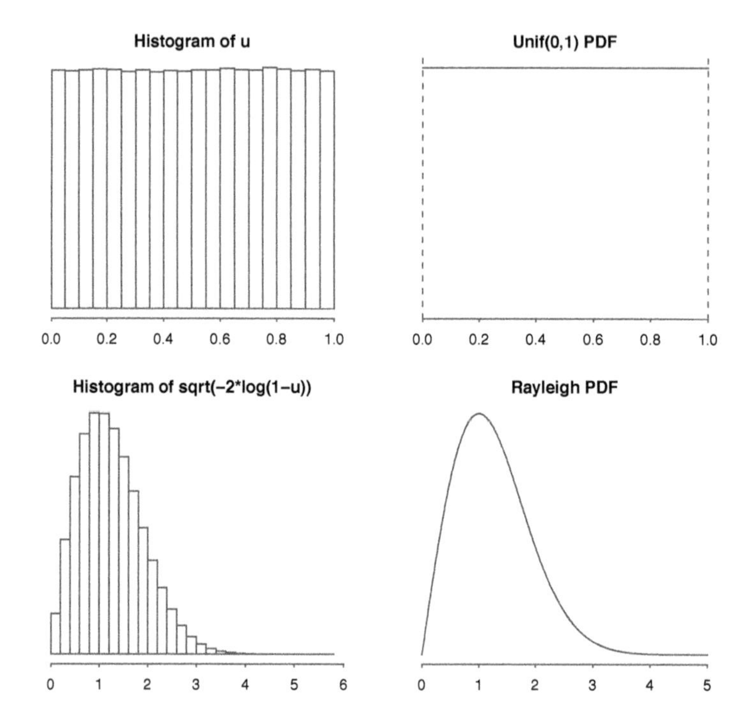

FIGURE 5.8
Top: Histogram of 1 million draws from $U \sim \text{Unif}(0,1)$, with Unif(0, 1) PDF for comparison. Bottom: Histogram of 1 million draws from $\sqrt{-2\log(1-U)}$, with Rayleigh PDF for comparison.

to produce 1 million realizations of $\sqrt{-2\log(1-U)}$. As Figure 5.8 shows, the realizations of $\sqrt{-2\log(1-U)}$ look very similar to the Rayleigh PDF, as predicted by the universality of the Uniform.

Conversely, if $X \sim$ Rayleigh, then $F(X) = 1 - e^{-X^2/2} \sim \text{Unif}(0,1)$. □

Next, let us consider the extent to which universality of the Uniform holds for discrete random variables. The CDF F of a discrete r.v. has jumps and flat regions, so F^{-1} does not exist (in the usual sense). But Part 1 still holds in the sense that given a Uniform random variable, we can construct an r.v. with *any discrete distribution we want.* The difference is that instead of working with the CDF, which is not invertible, it is more straightforward to work with the PMF.

Suppose we want to use $U \sim \text{Unif}(0,1)$ to construct a discrete r.v. X with PMF $p_j = P(X = j)$ for $j = 0, 1, 2, \dots, n$. As illustrated in Figure 5.9, we can chop up the interval $(0,1)$ into pieces of lengths $p_0, p_1, \dots, p_n$. By the properties of a valid PMF, the sum of the p_j's is 1, so this perfectly divides up the interval, without overshooting or undershooting.

Now define X to be the r.v. which equals 0 if U falls into the p_0 interval, 1 if U falls into the p_1 interval, 2 if U falls into the p_2 interval, and so on. Then X is a discrete

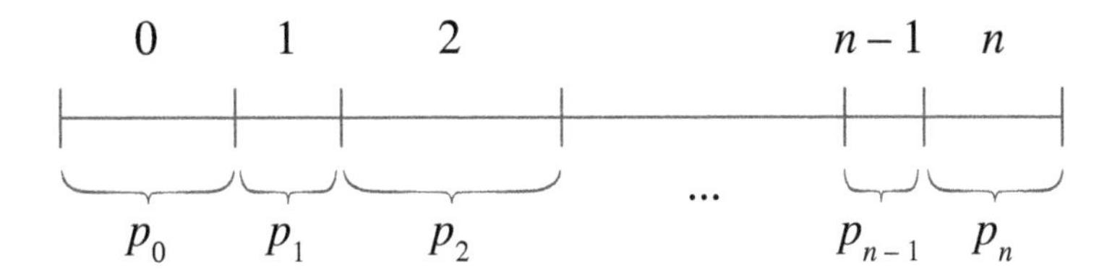

FIGURE 5.9
Given a PMF, chop up the interval $(0, 1)$ into pieces, with lengths given by the PMF values.

r.v. taking on values 0 through n. The probability that $X = j$ is the probability that U falls into the interval of length p_j. But for a Unif$(0, 1)$ r.v., probability *is* length, so $P(X = j)$ is precisely p_j, as desired!

The same trick will work for a discrete r.v. that can take on infinitely many values, such as a Poisson; we'll need to chop $(0, 1)$ into infinitely many pieces, but the total length of the pieces is still 1.

We now know how to take an arbitrary PMF and create an r.v. with that PMF. This fulfills our promise from Chapter 3 that any function with the properties given in Theorem 3.2.7 is the PMF of some r.v.

☣ **5.3.6.** Part 2 of universality of the Uniform, on the other hand, fails for discrete r.v.s. A function of a discrete r.v. is still discrete, so if X is discrete, then $F(X)$ is still discrete. So $F(X)$ doesn't have a Uniform distribution. For example, if $X \sim$ Bern(p), then $F(X)$ has only two possible values: $F(0) = 1 - p$ and $F(1) = 1$.

The upshot of universality is that we can use a Uniform r.v. U to generate r.v.s from both continuous and discrete distributions: in the continuous case, we can plug U into the inverse CDF, and in the discrete case, we can chop up the unit interval according to the desired PMF. Part 1 of universality of the Uniform is often useful in practice when running simulations (since the software being used may know how to generate Uniform r.v.s but not know how to generate r.v.s with the distribution of interest), though the extent to which it is useful depends on how tractable it is to compute the inverse CDF. Part 2 is important for certain widely used techniques in statistical inference, by providing a transformation that converts an r.v. with an *unknown* distribution to an r.v. with a *known*, simple distribution: the Uniform.

Using our analogy of distributions as blueprints and r.v.s as houses, the beauty of the universality property is that the Uniform distribution is a very simple blueprint, and it's easy to create a house from that blueprint; universality of the Uniform then gives us a simple rule for remodeling the Uniform house into a house with any other blueprint, no matter how complicated!

To conclude this section, we give an elegant identity that is often useful for finding the expectation of a nonnegative r.v. The identity also has a neat visual interpretation related to universality of the Uniform, LOTUS, and the relationship between CDFs and quantile functions.

Definition 5.3.7. The *survival function* of an r.v. X with CDF F is the function G given by $G(x) = 1 - F(x) = P(X > x)$.

Theorem 5.3.8 (Expectation by integrating the survival function)**.** Let X be a nonnegative r.v. Its expectation can be found by integrating its survival function:

$$E(X) = \int_0^\infty P(X > x)dx.$$

This result is the continuous analog of Theorem 4.4.8 (note though that it holds for *any* nonnegative r.v., not just for continuous nonnegative r.v.s). Actuaries sometimes call it the *Darth Vader rule*, for obscure reasons; statisticians are more likely to refer to it as finding the expectation by *integrating the survival function.*

Proof. For any number $x \geq 0$, we can write

$$x = \int_0^x dt = \int_0^\infty I(x > t)dt,$$

where $I(x > t)$ is 1 if $x \geq t$ and 0 otherwise. So

$$X(s) = \int_0^\infty I(X(s) > t)dt,$$

for each s in the sample space. We can write this more compactly as

$$X = \int_0^\infty I(X > t)dt.$$

Taking the expectation of both sides and swapping the E with the integral (which can be justified using results from real analysis), we have

$$E(X) = E\left(\int_0^\infty I(X > t)dt\right) = \int_0^\infty E(I(X > t))dt = \int_0^\infty P(X > t)dt. \quad \blacksquare$$

For a visual explanation of this identity, we can graph a CDF and interpret a certain area in two different ways: as the integral of the survival function, and as the integral of the quantile function.

A prototypical CDF of a nonnegative, continuous r.v. with CDF F is shown in Figure 5.10, with the area between the CDF curve and the horizontal line $p = 1$ shaded. This area can be found by integrating $1 - F(x)$, the difference between the line and the curve, from 0 to ∞.

But another way to find this area is to turn your head sideways and integrate with

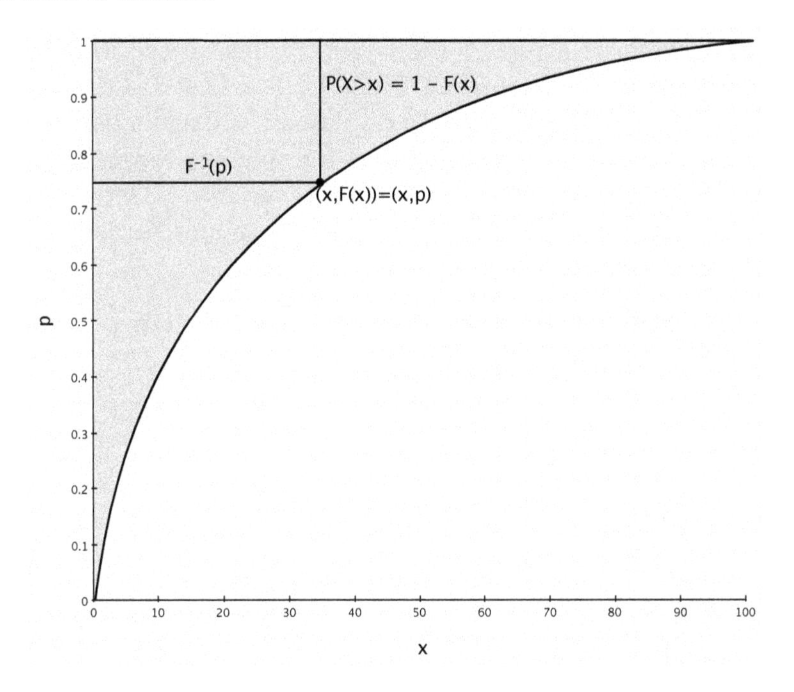

FIGURE 5.10
The area above a certain CDF and below the line $p = 1$ is shaded. This area can be interpreted in two ways: as the integral of the survival function, or as the integral of the quantile function.

respect to the vertical axis variable p rather than the horizontal axis variable x. This gives the integral of the quantile function, which, letting $U \sim \text{Unif}(0, 1)$, is

$$\int_0^1 F^{-1}(p)dp = E(F^{-1}(U)) = E(X),$$

by LOTUS and universality of the Uniform. So again we have

$$\int_0^\infty (1 - F(x))dx = \int_0^1 F^{-1}(p)dp = E(X).$$

5.4 Normal

The Normal distribution is a famous continuous distribution with a bell-shaped PDF. It is extremely widely used in statistics because of a theorem, the *central limit theorem*, which says that under very weak assumptions, the sum of a large number of i.i.d. random variables has an approximately Normal distribution, *regardless* of the distribution of the individual r.v.s. This means we can start with independent r.v.s from almost any distribution, discrete or continuous, but once we add up a bunch of them, the distribution of the resulting r.v. looks like a Normal distribution.

The central limit theorem is a topic for Chapter 10, but in the meantime, we'll introduce the properties of the Normal PDF and CDF and derive the expectation and variance of the Normal distribution. To do this, we will again use the strategy of location-scale transformation by starting with the simplest Normal distribution, the *standard Normal*, which is centered at 0 and has variance 1. After deriving the properties of the standard Normal, we'll be able to get to any Normal distribution we want by shifting and scaling.

Definition 5.4.1 (Standard Normal distribution). A continuous r.v. Z is said to have the *standard Normal distribution* if its PDF φ is given by

$$\varphi(z) = \frac{1}{\sqrt{2\pi}} e^{-z^2/2}, \quad -\infty < z < \infty.$$

We write this as $Z \sim \mathcal{N}(0,1)$ since, as we will show, Z has mean 0 and variance 1.

The constant $\frac{1}{\sqrt{2\pi}}$ in front of the PDF may look surprising (why is something with π needed in front of something with e, when there are no circles in sight?), but it's exactly what is needed to make the PDF integrate to 1. Such constants are called *normalizing constants* because they normalize the total area under the PDF to 1. We'll verify soon that this is a valid PDF.

The standard Normal CDF Φ is the accumulated area under the PDF:

$$\Phi(z) = \int_{-\infty}^{z} \varphi(t)dt = \int_{-\infty}^{z} \frac{1}{\sqrt{2\pi}} e^{-t^2/2} dt.$$

Some people, upon seeing the function Φ for the first time, express dismay that it is left in terms of an integral. Unfortunately, we have little choice in the matter: it turns out to be mathematically impossible to find a closed-form expression for the antiderivative of φ, meaning that we cannot express Φ as a finite sum of more familiar functions like polynomials or exponentials. But closed-form or no, it's still a well-defined function: if we give Φ an input z, it returns the accumulated area under the PDF from $-\infty$ up to z.

Notation 5.4.2. We can tell the Normal distribution must be special because the standard Normal PDF and CDF get their own Greek letters. By convention, we use φ for the standard Normal PDF and Φ for the CDF. We will often use Z to denote a standard Normal random variable.

The standard Normal PDF and CDF are plotted in Figure 5.11. The PDF is bell-shaped and symmetric about 0, and the CDF is S-shaped. These have the same general shape as the Logistic PDF and CDF that we saw Example 5.1.6, but the Normal PDF decays to 0 much more quickly.

There are several important symmetry properties that can be deduced from the standard Normal PDF and CDF.

1. *Symmetry of PDF*: φ satisfies $\varphi(z) = \varphi(-z)$, i.e., φ is an even function.

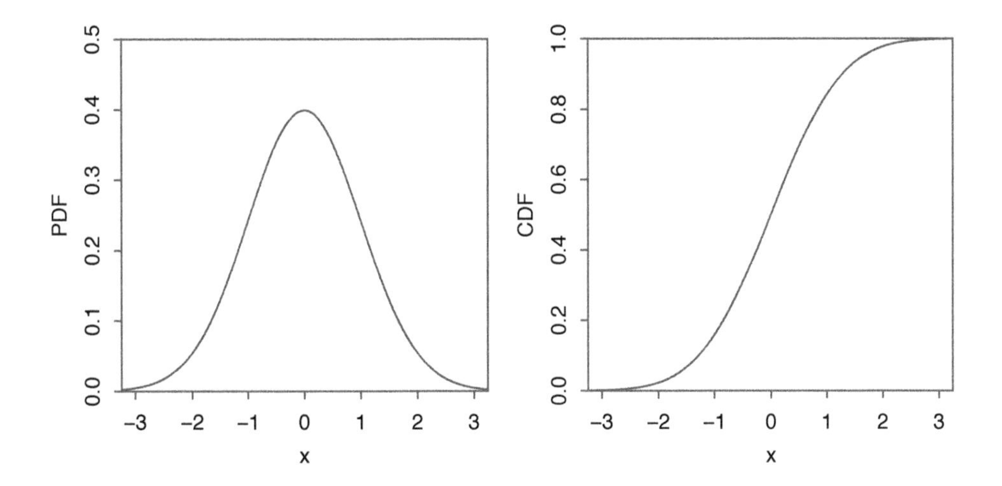

FIGURE 5.11
Standard Normal PDF φ (left) and CDF Φ (right).

2. *Symmetry of tail areas*: The area under the PDF curve to the left of -2, which is $P(Z \leq -2) = \Phi(-2)$ by definition, equals the area to the right of 2, which is $P(Z \geq 2) = 1 - \Phi(2)$. In general, we have

$$\Phi(z) = 1 - \Phi(-z)$$

for all z. This can be seen visually by looking at the PDF curve, and mathematically by substituting $u = -t$ below and using the fact that PDFs integrate to 1:

$$\Phi(-z) = \int_{-\infty}^{-z} \varphi(t)dt = \int_{z}^{\infty} \varphi(u)du = 1 - \int_{-\infty}^{z} \varphi(u)du = 1 - \Phi(z).$$

3. *Symmetry of Z and $-Z$*: If $Z \sim \mathcal{N}(0,1)$, then $-Z \sim \mathcal{N}(0,1)$ as well. To see this, note that the CDF of $-Z$ is

$$P(-Z \leq z) = P(Z \geq -z) = 1 - \Phi(-z),$$

but that is $\Phi(z)$, according to what we just argued. So $-Z$ has CDF Φ.

We need to prove three key facts about the standard Normal, and then we'll be ready to handle general Normal distributions: we need to show that φ is a valid PDF, that $E(Z) = 0$, and that $\text{Var}(Z) = 1$.

To verify the validity of φ, we'll show that the total area under $e^{-z^2/2}$ is $\sqrt{2\pi}$. However, we can't find the antiderivative of $e^{-z^2/2}$ directly, again because of the annoying fact that the antiderivative isn't expressible in closed form. But this doesn't mean we can't do *definite* integrals, with some ingenuity.

An amazing trick saves the day here: write down the integral *twice*. Usually, writing down the same problem repeatedly is more a sign of frustration than a problem-solving strategy. But in this case, it allows a neat conversion to polar coordinates:

$$\left(\int_{-\infty}^{\infty} e^{-z^2/2}dz\right)\left(\int_{-\infty}^{\infty} e^{-z^2/2}dz\right) = \left(\int_{-\infty}^{\infty} e^{-x^2/2}dx\right)\left(\int_{-\infty}^{\infty} e^{-y^2/2}dy\right)$$
$$= \int_{-\infty}^{\infty}\int_{-\infty}^{\infty} e^{-\frac{x^2+y^2}{2}}dxdy$$
$$= \int_{0}^{2\pi}\int_{0}^{\infty} e^{-r^2/2}rdrd\theta.$$

In the first step, we used the fact that z is just a dummy variable in each integral, so we are allowed to give it a different name (or two different names, one for each integral). The extra r that appears in the final step comes from the Jacobian of the transformation to polar coordinates, as explained in Section A.7.2 of the math appendix. That r is also what saves us from the impossibility of the original integral, since we can now use the substitution $u = r^2/2$, $du = rdr$. This gives

$$\int_{0}^{2\pi}\int_{0}^{\infty} e^{-r^2/2}rdrd\theta = \int_{0}^{2\pi}\left(\int_{0}^{\infty} e^{-u}du\right)d\theta$$
$$= \int_{0}^{2\pi} 1d\theta = 2\pi.$$

Therefore,

$$\int_{-\infty}^{\infty} e^{-z^2/2}dz = \sqrt{2\pi},$$

as we wanted to show.

The expectation of the standard Normal has to be 0, by the symmetry of the PDF; no other balancing point would make sense. We can also see this symmetry by looking at the definition of $E(Z)$:

$$E(Z) = \frac{1}{\sqrt{2\pi}}\int_{-\infty}^{\infty} ze^{-z^2/2}dz,$$

and since $g(z) = ze^{-z^2/2}$ is an odd function (see Section A.2.3 of the math appendix for more on even and odd functions), the area under g from $-\infty$ to 0 cancels the area under g from 0 to ∞. Therefore $E(Z) = 0$. In fact, the same argument shows that $E(Z^n) = 0$ for any odd positive integer n.[1]

Getting the mean was easy (one might even say it was EZ), but the variance

[1] A subtlety is that $\infty - \infty$ is undefined, so we also want to check that the area under the curve $z^n e^{-z^2/2}$ from 0 to ∞ is *finite*. But this is true since $e^{-z^2/2}$ goes to 0 extremely quickly (faster than exponential decay), more than offsetting the growth of the polynomial z^n.

calculation is a bit more involved. By LOTUS,

$$\begin{aligned}\mathrm{Var}(Z) &= E(Z^2) - (EZ)^2 = E(Z^2)\\ &= \frac{1}{\sqrt{2\pi}} \int_{-\infty}^{\infty} z^2 e^{-z^2/2} dz\\ &= \frac{2}{\sqrt{2\pi}} \int_{0}^{\infty} z^2 e^{-z^2/2} dz\end{aligned}$$

The last step uses the fact that $z^2 e^{-z^2/2}$ is an even function. Now we use integration by parts with $u = z$ and $dv = ze^{-z^2/2}dz$, so $du = dz$ and $v = -e^{-z^2/2}$:

$$\begin{aligned}\mathrm{Var}(Z) &= \frac{2}{\sqrt{2\pi}} \left(-ze^{-z^2/2} \Big|_0^{\infty} + \int_0^{\infty} e^{-z^2/2} dz \right)\\ &= \frac{2}{\sqrt{2\pi}} \left(0 + \frac{\sqrt{2\pi}}{2} \right)\\ &= 1.\end{aligned}$$

The first term of the integration by parts equals 0 because $e^{-z^2/2}$ decays much faster than z grows, and the second term is $\sqrt{2\pi}/2$ because it's half of the total area under $e^{-z^2/2}$, which we've already proved is $\sqrt{2\pi}$. So indeed, the standard Normal distribution has mean 0 and variance 1.

The general Normal distribution has two parameters, denoted μ and σ^2, which correspond to the mean and variance (so the standard Normal is the special case where $\mu = 0$ and $\sigma^2 = 1$). Starting with a standard Normal r.v. $Z \sim \mathcal{N}(0, 1)$, we can get a Normal r.v. with any mean and variance by a location-scale transformation (shifting and scaling).

Definition 5.4.3 (Normal distribution). If $Z \sim \mathcal{N}(0, 1)$, then

$$X = \mu + \sigma Z$$

is said to have the *Normal distribution* with mean μ and variance σ^2, for any real μ and σ^2 with $\sigma > 0$. We denote this by $X \sim \mathcal{N}(\mu, \sigma^2)$.

It's clear by properties of expectation and variance that X does in fact have mean μ and variance σ^2:

$$\begin{aligned}E(\mu + \sigma Z) &= E(\mu) + \sigma E(Z) = \mu,\\ \mathrm{Var}(\mu + \sigma Z) &= \mathrm{Var}(\sigma Z) = \sigma^2 \mathrm{Var}(Z) = \sigma^2.\end{aligned}$$

Note that we multiply Z by the standard deviation σ, not σ^2; else the units would be wrong and X would have variance σ^4.

Of course, if we can get from Z to X, then we can get from X back to Z. The process of getting a standard Normal from a non-standard Normal is called, appropriately enough, *standardization*. For $X \sim \mathcal{N}(\mu, \sigma^2)$, the *standardized version* of X is

$$\frac{X - \mu}{\sigma} \sim \mathcal{N}(0, 1).$$

We can use standardization to find the CDF and PDF of X in terms of the standard Normal CDF and PDF.

Theorem 5.4.4 (Normal CDF and PDF). Let $X \sim \mathcal{N}(\mu, \sigma^2)$. Then the CDF of X is

$$F(x) = \Phi\left(\frac{x-\mu}{\sigma}\right),$$

and the PDF of X is

$$f(x) = \varphi\left(\frac{x-\mu}{\sigma}\right)\frac{1}{\sigma}.$$

Proof. For the CDF, we start from the definition $F(x) = P(X \leq x)$, standardize, and use the CDF of the standard Normal:

$$F(x) = P(X \leq x) = P\left(\frac{X-\mu}{\sigma} \leq \frac{x-\mu}{\sigma}\right) = \Phi\left(\frac{x-\mu}{\sigma}\right).$$

Then we differentiate to get the PDF, remembering to apply the chain rule:

$$\begin{aligned} f(x) &= \frac{d}{dx}\Phi\left(\frac{x-\mu}{\sigma}\right) \\ &= \varphi\left(\frac{x-\mu}{\sigma}\right)\frac{1}{\sigma}. \end{aligned}$$

We can also write out the PDF as

$$f(x) = \frac{1}{\sqrt{2\pi}\sigma}\exp\left(-\frac{(x-\mu)^2}{2\sigma^2}\right). \qquad \blacksquare$$

Finally, three important benchmarks for the Normal distribution are the probabilities of falling within one, two, and three standard deviations of the mean. The 68-95-99.7% rule tells us that these probabilities are what the name suggests.

Theorem 5.4.5 (68-95-99.7% rule). If $X \sim \mathcal{N}(\mu, \sigma^2)$, then

$$\begin{aligned} P(|X-\mu| < \sigma) &\approx 0.68, \\ P(|X-\mu| < 2\sigma) &\approx 0.95, \\ P(|X-\mu| < 3\sigma) &\approx 0.997. \end{aligned}$$

We can use this rule to get quick approximations of Normal probabilities.[2] Often it is easier to apply the rule after standardizing, in which case we have

$$\begin{aligned} P(|Z| < 1) &\approx 0.68, \\ P(|Z| < 2) &\approx 0.95, \\ P(|Z| < 3) &\approx 0.997. \end{aligned}$$

[2]The 68-95-99.7% rule says that 95% of the time, a Normal random variable will fall within ± 2 standard deviations of its mean. An even more accurate approximation says that 95% of the time, a Normal r.v. is within ± 1.96 SDs of its mean. This explains why the number 1.96 comes up very often in statistics in the context of 95% confidence intervals, which are often created by taking an estimate and putting a buffer zone of 1.96 SDs on either side.

Example 5.4.6 (Practice with the standard Normal CDF)**.** Let $X \sim \mathcal{N}(-1, 4)$. What is $P(|X| < 3)$, exactly (in terms of Φ) and approximately?

Solution:

The event $|X| < 3$ is the same as the event $-3 < X < 3$. We use standardization to express this event in terms of the standard Normal r.v. $Z = (X - (-1))/2$, then apply the 68-95-99.7% rule to get an approximation. The exact answer is

$$P(-3 < X < 3) = P\left(\frac{-3-(-1)}{2} < \frac{X-(-1)}{2} < \frac{3-(-1)}{2}\right) = P(-1 < Z < 2),$$

which is $\Phi(2) - \Phi(-1)$. The 68-95-99.7% rule tells us that $P(-1 < Z < 1) \approx 0.68$ and $P(-2 < Z < 2) \approx 0.95$. In other words, going from ± 1 standard deviation to ± 2 standard deviations adds approximately $0.95 - 0.68 = 0.27$ to the area under the curve. By symmetry, this is evenly divided between the areas $P(-2 < Z < -1)$ and $P(1 < Z < 2)$. Therefore,

$$P(-1 < Z < 2) = P(-1 < Z < 1) + P(1 < Z < 2) \approx 0.68 + \frac{0.27}{2} = 0.815.$$

This is close to the correct value, $\Phi(2) - \Phi(-1) \approx 0.8186$. □

As we will see later in the book, several important distributions can be obtained through transforming Normal r.v.s in natural ways, e.g., squaring or exponentiating. Chapter 8 delves into transformations in depth, but meanwhile there is a lot that we can do just using LOTUS and properties of CDFs.

Example 5.4.7 (Folded Normal)**.** Let $Y = |Z|$ with $Z \sim \mathcal{N}(0, 1)$. The distribution of Y is called a *Folded Normal* with parameters $\mu = 0$ and $\sigma^2 = 1$. In this example, we will derive the mean, variance, and distribution of Y. At first sight, Y may seem tricky to deal with since the absolute value function is not differentiable at 0 (due to its sharp corner), but Y has a perfectly valid continuous distribution.

(a) Find $E(Y)$.

(b) Find $\text{Var}(Y)$.

(c) Find the CDF and PDF of Y.

Solution:

(a) We will derive the PDF of Y later in this example, but to find $E(Y)$, LOTUS says we can work directly with the PDF of Z:

$$E(Y) = E|Z| = \int_{-\infty}^{\infty} |z| \frac{1}{\sqrt{2\pi}} e^{-z^2/2} dz = 2 \int_0^{\infty} z \frac{1}{\sqrt{2\pi}} e^{-z^2/2} dz = \sqrt{\frac{2}{\pi}}.$$

(b) Note that $Y^2 = Z^2$, so we do *not* need to do another integral! We have

$$E(Y^2) = E(Z^2) = \text{Var}(Z) = 1,$$

so

$$\text{Var}(Y) = E(Y^2) - (E(Y))^2 = 1 - \frac{2}{\pi}.$$

(c) For $y \leq 0$, the CDF of Y is $F_Y(y) = P(Y \leq y) = 0$. For $y > 0$, the CDF is

$$F_Y(y) = P(Y \leq y) = P(|Z| \leq y) = P(-y \leq Z \leq y) = \Phi(y) - \Phi(-y) = 2\Phi(y) - 1.$$

So the PDF of Y is $2\varphi(y)$ for $y \geq 0$, and 0 otherwise, where φ is the $\mathcal{N}(0,1)$ PDF.

Sanity check: Note that $2\Phi(y) - 1 \to 2 - 1 = 1$ as $y \to \infty$, as it must, and that the CDF of Y is a continuous function since Φ is continuous and at 0 there is not a jump: $2\Phi(0) - 1 = 0$. Also, the PDF of Y makes sense since taking the absolute value of Z "folds" the probability mass of a negative range of values of Z over to the positive side, e.g., the probability for Z values between -2 and -1 contributes to the probability for Y values between 1 and 2. This results in zero density for negative values and double the density for positive values. □

5.5 Exponential

The Exponential distribution is the continuous counterpart to the Geometric distribution. Recall that a Geometric random variable counts the number of failures before the first success in a sequence of Bernoulli trials. The story of the Exponential distribution is analogous, but we are now waiting for a success in *continuous* time, where successes arrive at a rate of λ successes per unit of time. The average number of successes in a time interval of length t is λt, though the actual number of successes varies randomly. An Exponential random variable represents the waiting time until the first arrival of a success.

Definition 5.5.1 (Exponential distribution). A continuous r.v. X is said to have the *Exponential distribution* with parameter λ, where $\lambda > 0$, if its PDF is

$$f(x) = \lambda e^{-\lambda x}, \quad x > 0.$$

We denote this by $X \sim \text{Expo}(\lambda)$.

The corresponding CDF is

$$F(x) = 1 - e^{-\lambda x}, \quad x > 0.$$

The Expo(1) PDF and CDF are plotted in Figure 5.12. Note the resemblance to the Geometric PMF and CDF pictured in Chapter 4. Exercise 43 explores the sense in which the Geometric converges to the Exponential, in the limit where the Bernoulli trials are performed faster and faster but with smaller and smaller success probabilities.

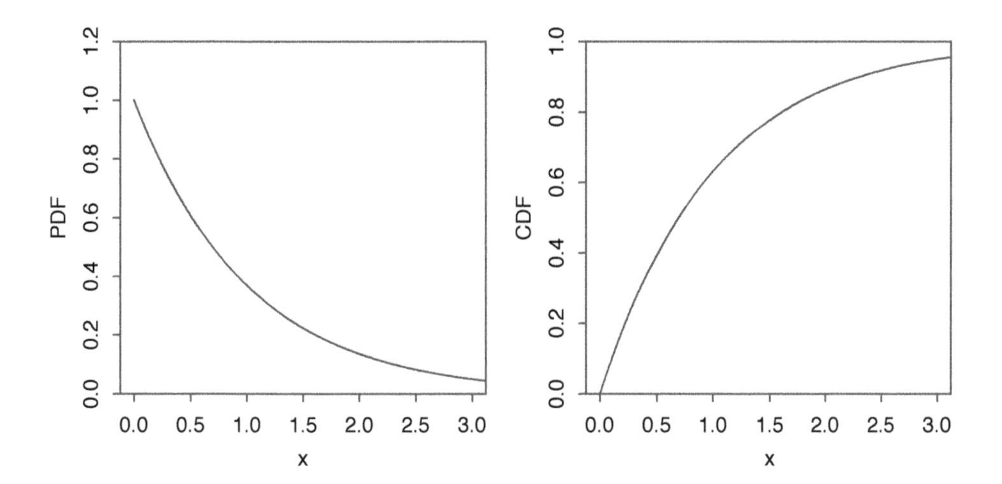

FIGURE 5.12
Expo(1) PDF and CDF.

We've seen how all Uniform and Normal distributions are related to one another via location-scale transformations, and we might wonder whether the Exponential distribution allows this too. Exponential r.v.s are defined to have support $(0, \infty)$, and shifting would change the left endpoint. But scale transformations work nicely, and we can use scaling to get from the simple Expo(1) to the general Expo(λ): if $X \sim \text{Expo}(1)$, then

$$Y = \frac{X}{\lambda} \sim \text{Expo}(\lambda),$$

since

$$P(Y \leq y) = P\left(\frac{X}{\lambda} \leq y\right) = P(X \leq \lambda y) = 1 - e^{-\lambda y}, \quad y > 0.$$

Conversely, if $Y \sim \text{Expo}(\lambda)$, then $\lambda Y \sim \text{Expo}(1)$.

This means that just as we did for the Uniform and the Normal, we can get the mean and variance of the Exponential distribution by starting with $X \sim \text{Expo}(1)$. Both $E(X)$ and $\text{Var}(X)$ are obtained using standard integration by parts calculations. This gives

$$\begin{aligned} E(X) &= \int_0^\infty x e^{-x} dx = 1, \\ E(X^2) &= \int_0^\infty x^2 e^{-x} dx = 2, \\ \text{Var}(X) &= E(X^2) - (EX)^2 = 1. \end{aligned}$$

In the next chapter we'll introduce a new tool called the moment generating function, which will let us get these results without integration.

For $Y = X/\lambda \sim \text{Expo}(\lambda)$ we then have

$$E(Y) = \frac{1}{\lambda}E(X) = \frac{1}{\lambda},$$
$$\text{Var}(Y) = \frac{1}{\lambda^2}\text{Var}(X) = \frac{1}{\lambda^2},$$

so the mean and variance of the Expo(λ) distribution are $1/\lambda$ and $1/\lambda^2$, respectively. As we'd expect intuitively, the faster the rate of arrivals λ, the shorter the average waiting time.

The Exponential distribution has a very special property called the *memoryless property*, which says that even if you've waited for hours or days without success, the success isn't any more likely to arrive soon. In fact, you might as well have just started waiting 10 seconds ago. The definition formalizes this idea.

Definition 5.5.2 (Memoryless property)**.** A continuous distribution is said to have the *memoryless property* if a random variable X from that distribution satisfies

$$P(X \geq s + t | X \geq s) = P(X \geq t)$$

for all $s, t \geq 0$.

Here s represents the time you've already spent waiting; the definition says that after you've waited s minutes, the probability you'll have to wait another t minutes is exactly the same as the probability of having to wait t minutes with no previous waiting time under your belt. Another way to state the memoryless property is that conditional on $X \geq s$, the additional waiting time $X - s$ is still distributed Expo(λ). In particular, this implies

$$E(X | X \geq s) = s + E(X) = s + \frac{1}{\lambda}.$$

(Conditional expectation is explained in detail in Chapter 9, but the meaning should already be clear: for any r.v. X and event A, $E(X|A)$ is the expected value of X given A; this can be defined by replacing the unconditional PMF or PDF of X in the definition of $E(X)$ by the conditional PMF or PDF of X given A.)

Using the definition of conditional probability, we can directly verify that the Exponential distribution has the memoryless property. Let $X \sim \text{Expo}(\lambda)$. Then

$$P(X \geq s + t | X \geq s) = \frac{P(X \geq s + t)}{P(X \geq s)} = \frac{e^{-\lambda(s+t)}}{e^{-\lambda s}} = e^{-\lambda t} = P(X \geq t).$$

What are the implications of the memoryless property? If you're waiting at a bus stop and the time until the bus arrives has an Exponential distribution, then conditional on your having waited 30 minutes, the bus isn't due to arrive soon. The distribution simply *forgets* that you've been waiting for half an hour, and your remaining wait time is the same as if you had just shown up to the bus stop. If the lifetime of a machine has an Exponential distribution, then no matter how long the machine has been functional, conditional on having lived that long, the machine is as good as new: there is no wear-and-tear effect that makes the machine more likely to break down soon. If human lifetimes were Exponential, then conditional on having survived to the age of 80, your remaining lifetime would have the same distribution as that of a newborn baby!

Clearly, the memoryless property is not an appropriate description for human or machine lifetimes. Why then do we care about the Exponential distribution?

1. Some physical phenomena, such as radioactive decay, truly do exhibit the memoryless property, so the Exponential is an important model in its own right.

2. The Exponential distribution is well-connected to other named distributions. In the next section, we'll see how the Exponential and Poisson distributions can be united by a shared story, and we'll discover many more connections in later chapters.

3. The Exponential serves as a building block for more flexible distributions, such as the *Weibull distribution* (introduced in Chapter 6), that allow for a wear-and-tear effect (where older units are due to break down) or a survival-of-the-fittest effect (where the longer you've lived, the stronger you get). To understand these distributions, we first have to understand the Exponential.

The memoryless property is a very special property of the Exponential distribution: no other continuous distribution on $(0, \infty)$ is memoryless! Let's prove this.

Theorem 5.5.3. If X is a positive continuous random variable with the memoryless property, then X has an Exponential distribution.

Proof. Suppose X is a positive continuous r.v. with the memoryless property. Let F be the CDF of X and G be the survival function of X, given by $G(x) = 1 - F(x)$. We will show that $G(x) = e^{-\lambda x}$ for some λ, by first showing that $G(xt) = G(t)^x$ for all real $x > 0$. The memoryless property says that

$$G(s + t) = G(s)G(t)$$

for all $s, t \geq 0$. Putting $s = t$, we have

$$G(2t) = G(t)^2,$$

so

$$G(3t) = G(2t + t) = G(2t)G(t) = G(t)^3, G(4t) = G(t)^4, \dots.$$

We now have

$$G(mt) = G(t)^m$$

for m a positive integer. Let's extend this in stages. Replacing t by $t/2$ in $G(2t) = G(t)^2$, we have $G(t/2) = G(t)^{1/2}$. Similarly,

$$G\left(\frac{t}{n}\right) = G(t)^{1/n}$$

for any positive integer n. It follows that

$$G\left(\frac{m}{n}t\right) = \left(G\left(\frac{t}{n}\right)\right)^m = G(t)^{m/n}$$

for any positive integers m, n, so

$$G(xt) = G(t)^x$$

for all positive *rational* numbers x. Any positive real number can be written as a limit of positive rational numbers so, using the fact that G is a continuous function, the above equation holds for all positive *real* numbers x. Taking $t = 1$, we have

$$G(x) = G(1)^x = e^{-\lambda x},$$

where $\lambda = -\log(G(1)) > 0$. This is exactly the form we wanted for G, so X has an Exponential distribution. ■

The memoryless property is defined analogously for *discrete* distributions: a discrete distribution is memoryless if for X an r.v. with that distribution,

$$P(X \geq j + k | X \geq j) = P(X \geq k)$$

for all nonnegative integers j, k. In view of the analogy between the Geometric and Exponential stories (or if you have solved Exercise 32 from Chapter 4), you might guess that the Geometric distribution is memoryless. If so, you would be correct! If we're waiting for the first Heads in a sequence of fair coin tosses, and in a streak of bad luck we happen to get ten Tails in a row, this has no impact on how many additional tosses we'll need: the coin isn't due for a Heads, nor conspiring against us to perpetually land Tails. The coin is memoryless. The Geometric is the only memoryless discrete distribution taking values in $\{0, 1, 2, \dots\}$, and the Exponential is the only memoryless continuous distribution taking values in $(0, \infty)$.

As practice with the memoryless property, the following example chronicles the adventures of Fred, who experiences firsthand the frustrations of the memoryless property after moving to a town with a memoryless public transportation system.

Example 5.5.4 (Blissville and Blotchville). Fred lives in Blissville, where buses always arrive exactly on time, with the time between successive buses fixed at 10 minutes. Having lost his watch, he arrives at the bus stop at a uniformly random time on a certain day (assume that buses run 24 hours a day, every day, and that the time that Fred arrives is independent of the bus arrival process).

(a) What is the distribution of how long Fred has to wait for the next bus? What is the average time that Fred has to wait?

(b) Given that the bus has not yet arrived after 6 minutes, what is the probability that Fred will have to wait at least 3 more minutes?

(c) Fred moves to Blotchville, a city with inferior urban planning and where buses are much more erratic. Now, when any bus arrives, the time until the next bus arrives is an Exponential random variable with mean 10 minutes. Fred arrives at the bus stop at a random time (assume that Blotchville has followed and will follow this system for all of eternity, and that the time that Fred arrives is independent of the bus arrival process). What is the distribution of Fred's waiting time for the next bus? What is the average time that Fred has to wait?

(d) When Fred complains to a friend how much worse transportation is in Blotchville, the friend says: "Stop whining so much! You arrive at a uniform instant between the previous bus arrival and the next bus arrival. The average length of that interval between buses is 10 minutes, but since you are equally likely to arrive at any time in that interval, your average waiting time is only 5 minutes."

Fred disagrees, both from experience and from solving Part (c) while waiting for the bus. Explain what is wrong with the friend's reasoning.

Solution:

(a) The distribution is Uniform on $(0, 10)$, so the mean is 5 minutes.

(b) Let T be the waiting time. Then

$$P(T \geq 6 + 3|T > 6) = \frac{P(T \geq 9, T > 6)}{P(T > 6)} = \frac{P(T \geq 9)}{P(T > 6)} = \frac{1/10}{4/10} = \frac{1}{4}.$$

In particular, Fred's waiting time in Blissville is not memoryless; conditional on having waited 6 minutes already, there's only a $1/4$ chance that he'll have to wait at least another 3 minutes, whereas if he had just showed up, there would be a $P(T \geq 3) = 7/10$ chance of having to wait at least 3 minutes.

(c) By the memoryless property, the distribution is Exponential with parameter $1/10$ (and mean 10 minutes) regardless of when Fred arrives; how much longer the next bus will take to arrive is independent of how long ago the previous bus arrived. The average time that Fred has to wait is 10 minutes.

(d) Fred's friend is making the mistake, explained in ☣ 4.1.3, of replacing a random variable (the time between buses) by its expectation (10 minutes), thereby ignoring the variability in interarrival times. The *average* length of a time interval between two buses is 10 minutes, but Fred is not equally likely to arrive at any of these intervals: Fred is more likely to arrive during a long interval between buses than to arrive during a short interval between buses. For example, if one interval between buses is 50 minutes and another interval is 5 minutes, then Fred is 10 times more likely to arrive during the 50-minute interval.

This phenomenon is known as *length-biased sampling*, and it comes up in many real-life situations. For example, asking randomly chosen mothers how many children they have yields a different distribution from asking randomly chosen people how many siblings they have, including themselves. Asking students the sizes of their classes and averaging those results may give a much higher value than taking a list of classes and averaging the sizes of each; this is called the *class size paradox*. See exercises 16 and 17 from Chapter 4 for more about the class size paradox and length-biased sampling.

Fred's adventures in Blissville and Blotchville continue in the exercises (see also MacKay [17] for more of Fred's adventures). The bus arrivals in Blotchville follow a *Poisson process*, which is the topic of the next section. □

5.6 Poisson processes

The Exponential distribution is closely connected to the Poisson distribution, as suggested by our use of λ for the parameters of both distributions. In this section we will see that the Exponential and Poisson are linked by a common story, which is the story of the *Poisson process*. A Poisson process is a sequence of arrivals occurring at different points on a timeline, such that the number of arrivals in a particular interval of time has a Poisson distribution. Poisson processes are discussed in much greater detail in Chapter 13, but we already have the tools to understand the definition and basic properties.

Definition 5.6.1 (Poisson process). A process of arrivals in continuous time is called a *Poisson process* with rate λ if the following two conditions hold:

1. The number of arrivals that occur in an interval of length t is a $\text{Pois}(\lambda t)$ random variable.

2. The numbers of arrivals that occur in disjoint intervals are independent of each other. For example, the numbers of arrivals in the intervals $(0, 10), [10, 12)$, and $[15, \infty)$ are independent.

In this section, we will focus on Poisson processes on $(0, \infty)$, but we can also define Poisson processes on $(-\infty, \infty)$ or other intervals, and in Chapter 13 we will introduce Poisson processes in more than one dimension. A sketch of a Poisson process on $(0, \infty)$ is pictured in Figure 5.13. Each X marks the spot of an arrival.

For concreteness, suppose that the arrivals are emails landing in an inbox according to a Poisson process with rate λ. There are several things we might want to know about this process. One question we could ask is: in one hour, *how many* emails will arrive? The answer comes directly from the definition, which tells us that the

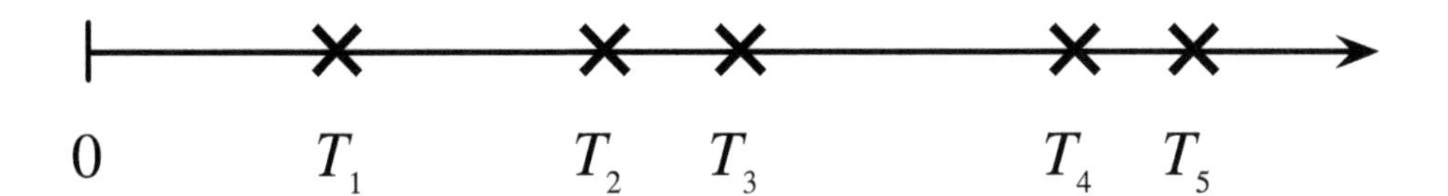

FIGURE 5.13
A Poisson process on $(0, \infty)$. Each X corresponds to an arrival.

number of emails in an hour follows a Pois(λ) distribution. Notice that the number of emails is a nonnegative integer, so a discrete distribution is appropriate.

But we could also flip the question around and ask: *how long* does it take until the first email arrives (measured relative to some fixed starting point)? The waiting time for the first email is a positive real number, so a continuous distribution on $(0, \infty)$ is appropriate. Let T_1 be the time until the first email arrives. To find the distribution of T_1, we just need to understand one crucial fact: saying that the waiting time for the first email is greater than t is the same as saying that *no emails* have arrived between 0 and t. In other words, if N_t is the number of emails that arrive at or before time t, then

$$T_1 > t \text{ is the same event as } N_t = 0.$$

We call this the *count-time duality* because it connects a discrete r.v., N_t, which *counts* the number of arrivals, with a continuous r.v., T_1, which marks the *time* of the first arrival. More generally, the count-time duality says that

$$T_n > t \text{ is the same event as } N_t < n.$$

Saying that the nth arrival has not happened yet as of time t is equivalent to saying that, up until time t, there have been fewer than n arrivals.

If two events are the same, they have the same probability. Since $N_t \sim \text{Pois}(\lambda t)$ by the definition of Poisson process,

$$P(T_1 > t) = P(N_t = 0) = \frac{e^{-\lambda t}(\lambda t)^0}{0!} = e^{-\lambda t}.$$

Therefore $P(T_1 \leq t) = 1 - e^{-\lambda t}$, so $T_1 \sim \text{Expo}(\lambda)$! The time until the first arrival in a Poisson process of rate λ has an Exponential distribution with parameter λ.

What about $T_2 - T_1$, the time between the first and second arrivals? Since disjoint intervals in a Poisson process are independent by definition, the past is irrelevant once the first arrival occurs. Thus $T_2 - T_1$ is independent of the time until the first arrival, and by the same argument as before, $T_2 - T_1$ also has an Exponential distribution with rate λ.

Similarly, $T_3 - T_2 \sim \text{Expo}(\lambda)$ independently of T_1 and $T_2 - T_1$. Continuing in this way, we deduce that all the interarrival times are i.i.d. Expo(λ) random variables. Thus, Poisson processes tie together two important distributions, one discrete and one continuous, and the use of a common symbol λ for both the Poisson and Exponential parameters is felicitous notation, for λ is the arrival rate in the process that unites the two distributions.

☣ **5.6.2.** The total time until the second arrival, T_2, is the sum of two independent Expo(λ) r.v.s, T_1 and $T_2 - T_1$. This does *not* have an Exponential distribution, but rather a Gamma distribution, which is introduced in Chapter 8.

The story of the Poisson process provides intuition for the fact, shown below, that the minimum of independent Exponential r.v.s is another Exponential r.v.

Example 5.6.3 (Minimum of independent Expos). Let $X_1, \dots, X_n$ be independent, with $X_j \sim \text{Expo}(\lambda_j)$. Let $L = \min(X_1, \dots, X_n)$. Show that $L \sim \text{Expo}(\lambda_1 + \dots + \lambda_n)$, and interpret this intuitively.

Solution:

We can find the distribution of L by considering its *survival function* $P(L > t)$, since the survival function is 1 minus the CDF.

$$\begin{aligned} P(L > t) &= P(\min(X_1, \dots, X_n) > t) = P(X_1 > t, \dots, X_n > t) \\ &= P(X_1 > t) \cdots P(X_n > t) = e^{-\lambda_1 t} \cdots e^{-\lambda_n t} = e^{-(\lambda_1 + \dots + \lambda_n)t}. \end{aligned}$$

The second equality holds since saying that the minimum of the X_j is greater than t is the same as saying that all of the X_j are greater than t. The third equality holds by independence of the X_j. Thus, L has the survival function (and the CDF) of an Exponential distribution with parameter $\lambda_1 + \dots + \lambda_n$.

Intuitively, we can interpret the λ_j as the rates of n independent Poisson processes. We can imagine, for example, X_1 as the waiting time for a green car to pass by, X_2 as the waiting time for a blue car to pass by, and so on, assigning a color to each X_j. Then L is the waiting time for a car with any of these colors to pass by, so it makes sense that L has a combined rate of $\lambda_1 + \dots + \lambda_n$. □

☣ **5.6.4.** The minimum of independent Exponentials is Exponential, but the maximum of independent Exponentials is *not* Exponential. However, the result about such a minimum turns out to be useful in studying such a maximum, as illustrated in the next two examples.

Example 5.6.5 (Maximum of 3 independent Exponentials). Three students are working independently on their probability homework. All 3 start at 1 pm on a certain day, and each takes an Exponential time with mean 6 hours to complete the homework. What is the earliest time at which all 3 students will have completed the homework, on average?

Solution: Label the students as 1, 2, 3, and let X_j be how long it takes student j to finish the homework. Let $\lambda = 1/6$, and let T be the time when all 3 students will have completed the homework, so $T = \max(X_1, X_2, X_3)$ with $X_i \sim \text{Expo}(\lambda)$. The CDF of T is

$$P(T \le t) = P(X_1 \le t, X_2 \le t, X_3 \le t) = (1 - e^{-\lambda t})^3.$$

So the PDF of T is

$$f_T(t) = 3\lambda e^{-\lambda t}(1 - e^{-\lambda t})^2.$$

In particular, T is *not* Exponential.

Finding $E(T)$ by integrating $tf_T(t)$ is possible but not especially pleasant. A neater approach is to use the memoryless property and the fact that the minimum of independent Exponentials *is* Exponential. We can decompose

$$T = T_1 + T_2 + T_3,$$

where $T_1 = \min(X_1, X_2, X_3)$ is how long it takes for one student to complete the homework, T_2 is the additional time it takes for a second student to complete the homework, and T_3 is the additional time until all 3 have completed the homework. Then $T_1 \sim \text{Expo}(3\lambda)$, by the result of Example 5.6.3.

By the memoryless property, at the first time when a student completes the homework the other two students are starting from fresh, so $T_2 \sim \text{Expo}(2\lambda)$. Again by the memoryless property, $T_3 \sim \text{Expo}(\lambda)$. The memoryless property also implies that T_1, T_2, T_3 are independent (which would be very useful if we were finding $\text{Var}(T)$). By linearity,

$$E(T) = \frac{1}{3\lambda} + \frac{1}{2\lambda} + \frac{1}{\lambda} = 2 + 3 + 6 = 11,$$

which shows that on average, the 3 students will have all completed the homework at midnight, 11 hours after they started. □

Example 5.6.6 (Machine repair). A certain machine often breaks down and needs to be fixed. At time 0, the machine is working. It works for an Expo(λ) period of time (measured in days), and then breaks down. It then takes an Expo(λ) amount of time to get it fixed, after which it will work for an Expo(λ) time until it breaks down again, after which it will take an Expo(λ) time to get it fixed, etc. Assume that these Expo(λ) r.v.s are i.i.d.

(a) A *transition* occurs when the machine switches from working to being broken, or switches from being broken to working. Find the distribution of the number of transitions that occur in the time interval $(0, t)$.

(b) Hoping to reduce the frequency of breakdowns, the machine is redesigned so that it can continue to function even if one component has failed. The redesigned machine has 5 components, each of which works for an Expo(λ) amount of time and then fails, independently. The machine works properly if and only if at most one component has failed. Currently, all 5 components are working (none have failed). Find the expected time until the machine breaks down.

Solution:

(a) The times between transitions are i.i.d. Expo(λ), so the times at which transitions occur follow a Poisson process of rate λ. So the desired distribution is Pois(λt).

(b) The time until a component fails is Expo(5λ). Then by the memoryless property, the additional time until another component fails is Expo(4λ). So the expected time until the machine breaks down is

$$\frac{1}{5\lambda} + \frac{1}{4\lambda} = \frac{9}{20\lambda}.$$

□

5.7 Symmetry of i.i.d. continuous r.v.s

Continuous r.v.s that are independent and identically distributed have an important symmetry property: all possible orderings are equally likely. Intuitively, this is because if all we are told is that $X_1, \ldots, X_n$ are i.i.d., then they are interchangeable, in the sense that we have been given no information that distinguishes one X_i from another X_j. This is reminiscent of the fact that it is common for someone to say "Do you want the good news first or the bad news first?" but rare for someone to say "I have two pieces of news. Which do you want to hear first?", since in the latter case no distinguishing information has been provided for the two pieces of news.

Proposition 5.7.1. Let $X_1, \ldots, X_n$ be i.i.d. from a continuous distribution. Then

$$P(X_{a_1} < X_{a_2} < \cdots < X_{a_n}) = \frac{1}{n!}$$

for any permutation $a_1, a_2, \ldots, a_n$ of $1, 2, \ldots, n$.

Proof. Let F be the CDF of X_j. By symmetry, all orderings of $X_1, \ldots, X_n$ are equally likely. For example, $P(X_3 < X_2 < X_1) = P(X_1 < X_2 < X_3)$ since both sides have exactly the same structure: they are both of the form $P(A < B < C)$ where A, B, C are i.i.d. draws from F. For any i and j with $i \neq j$, the probability of the *tie* $X_i = X_j$ is 0 since X_i and X_j are independent continuous r.v.s. So the probability of there being at least one tie among $X_1, \ldots, X_n$ is also 0, since

$$P\left(\bigcup_{i \neq j} \{X_i = X_j\}\right) \leq \sum_{i \neq j} P(X_i = X_j) = 0.$$

Thus, $X_1, \ldots, X_n$ are distinct with probability 1, and the probability of any particular ordering is $1/n!$. ■

☣ **5.7.2.** This proposition may fail if the r.v.s are dependent. Let $n = 2$, and consider the extreme case where X_1 and X_2 are so dependent that they are always equal, i.e., $X_1 = X_2$ with probability 1. Then $P(X_1 < X_2) = P(X_2 < X_1) = 0$. For dependent X_1, X_2 we can also make $P(X_1 < X_2) \neq P(X_2 < X_1)$. For an example, see Exercise 42 from Chapter 3.

☣ **5.7.3.** If X and Y are i.i.d. *continuous* r.v.s, then

$$P(X < Y) = P(Y < X) = \frac{1}{2},$$

by symmetry and since the probability of a tie is 0. In contrast, if X and Y are i.i.d. *discrete* r.v.s, it is still true that $P(X < Y) = P(Y < X)$ by symmetry, but this number is less than 1/2 because of the possibility of a tie. For example, if X and Y are i.i.d. nonnegative integer-valued r.v.s with $P(X = j) = c_j$, then

$$1 = P(X < Y) + P(X = Y) + P(Y < X) = 2P(X < Y) + P(X = Y),$$

so

$$P(X < Y) = \frac{1}{2} \cdot (1 - P(X = Y)) = \frac{1}{2} \cdot \left(1 - \sum_{j=0}^{\infty} c_j^2\right) < \frac{1}{2}.$$

The *ranks* of a list of distinct numbers are defined by giving the smallest number a rank of 1, the second smallest a rank of 2, and so on. For example, the ranks for $3.14, 2.72, 1.41, 1.62$ are $4, 3, 1, 2$. Proposition 5.7.1 says that the ranks of i.i.d. continuous $X_1, \ldots, X_n$ are a uniformly random permutation of the numbers $1, \ldots, n$. The next example shows how we can use this symmetry property in conjunction with indicator r.v.s in problems involving *records*, such as the record level of rainfall or the record performance on a high jump.

Example 5.7.4 (Records). Athletes compete one at a time at the high jump. Let X_j be how high the jth jumper jumped, with $X_1, X_2, \ldots$ i.i.d. with a continuous distribution. We say that the jth jumper sets a *record* if X_j is greater than all of $X_{j-1}, \ldots, X_1$.

(a) Is the event "the 110th jumper sets a record" independent of the event "the 111th jumper sets a record"?

(b) Find the mean number of records among the first n jumpers. What happens to the mean as $n \to \infty$?

(c) A *double record* occurs at time j if *both* the jth and $(j-1)$st jumpers set records. Find the mean number of double records among the first n jumpers. What happens to the mean as $n \to \infty$?

Solution:

(a) Let I_j be the indicator r.v. for the jth jumper setting a record. By symmetry, $P(I_j = 1) = 1/j$ (as any of the first j jumps is equally likely to be the highest of those jumps). Also,

$$P(I_{110} = 1, I_{111} = 1) = \frac{109!}{111!} = \frac{1}{110 \cdot 111},$$

since in order for both the 110th and 111th jumps to be records, we need the highest of the first 111 jumps to be in position 111 and the second highest to be in position 110, and the remaining 109 can be in any order. So

$$P(I_{110} = 1, I_{111} = 1) = P(I_{110} = 1)P(I_{111} = 1),$$

which shows that the 110th jumper setting a record is independent of the 111th jumper setting a record. Intuitively, this makes sense since learning that the 111th jumper sets a record gives us no information about the "internal" matter of how the first 110 jumps are arranged amongst themselves.

(b) By linearity, the expected number of records among the first n jumpers is $\sum_{j=1}^{n} \frac{1}{j}$, which goes to ∞ as $n \to \infty$ since the harmonic series diverges.

(c) Let J_j be the indicator r.v. for a double record occurring at time j, for $2 \leq j \leq n$. Then $P(J_j = 1) = \frac{1}{j(j-1)}$, following the logic of Part (a). So the expected number of double records is

$$\sum_{j=2}^{n} \frac{1}{j(j-1)} = \sum_{j=2}^{n} \left(\frac{1}{j-1} - \frac{1}{j} \right) = 1 - \frac{1}{n},$$

since all the other terms cancel out. Thus, the expected number of records goes to ∞ as $n \to \infty$, but the expected number of double records goes to 1. □

5.8 Recap

A continuous r.v. can take on any value in an interval, although the probability that it equals any particular value is 0. The CDF of a continuous r.v. is differentiable, and the derivative is called the probability density function (PDF). Probability is given by area under the PDF curve, *not* by the value of the PDF at a point. We must integrate the PDF to get a probability. The table below summarizes and compares some important concepts in the discrete case and the continuous case.

	Discrete r.v.	**Continuous r.v.**
CDF	$F(x) = P(X \leq x)$	$F(x) = P(X \leq x)$
PMF/PDF	$P(X = x)$ • PMF is height of jump of F at x. • PMF is nonnegative. • PMF sums to 1. • $P(X \in A) = \sum_{x \in A} P(X = x)$.	$f(x) = F'(x)$ • PDF is derivative of F. • PDF is nonnegative. • PDF integrates to 1. • $P(X \in A) = \int_A f(x)dx$.
Expectation	$E(X) = \sum_x xP(X = x)$	$E(X) = \int_{-\infty}^{\infty} xf(x)dx$
LOTUS	$E(g(X)) = \sum_x g(x)P(X = x)$	$E(g(X)) = \int_{-\infty}^{\infty} g(x)f(x)dx$

Three important continuous distributions are the Uniform, Normal, and Exponential. A Unif(a, b) r.v. is a "completely random" number in the interval (a, b), and it has the property that probability is proportional to length. The universality of the Uniform tells us how we can use a Unif$(0, 1)$ r.v. to construct r.v.s from other distributions we may be interested in; it also says that if we plug a continuous r.v. into its own CDF, the resulting r.v. has a Unif$(0, 1)$ distribution.

A $\mathcal{N}(\mu, \sigma^2)$ r.v. has a symmetric bell-shaped PDF centered at μ, with σ controlling how spread out the curve is. The mean is μ and standard deviation is σ. The 68-95-99.7% rule gives important benchmarks for the probability of a Normal r.v. falling within 1, 2, and 3 standard deviations of its mean.

An Expo(λ) r.v. represents the waiting time for the first success in continuous time, analogous to how a Geometric r.v. represents the number of failures before the first success in discrete time; the parameter λ can be interpreted as the rate at which successes arrive. The Exponential distribution has the memoryless property, which says that conditional on our having waited a certain amount of time without success, the distribution of the remaining wait time is exactly the same as if we hadn't waited at all. In fact, the Exponential is the *only* positive continuous distribution with the memoryless property.

A Poisson process is a sequence of arrivals in continuous time such that the number of arrivals in an interval is Poisson (with mean proportional to the length of the interval) and disjoint intervals have independent numbers of arrivals. The interarrival times in a Poisson process of rate λ are i.i.d. Expo(λ) r.v.s.

A useful symmetry property of i.i.d. r.v.s $X_1, X_2, \dots, X_n$ is that all orderings are equally likely. For example, $P(X_1 < X_2 < X_3) = P(X_3 < X_2 < X_1)$. If the X_j are continuous in addition to being i.i.d., then we can also conclude, e.g., that $P(X_1 < X_2 < X_3) = 1/6$, whereas in the discrete case we also have to account for the possibility of ties.

A new strategy that we learned for continuous distributions is location-scale transformation, which says that if shifting and scaling will not take us outside the family of distributions we're studying, then we can start with the simplest member of the family, find the answer for the simple case, then use shifting and scaling to arrive at the general case. For the three main distributions of this chapter, this approach works as follows.

- Uniform: If $U \sim \text{Unif}(0, 1)$, then $\tilde{U} = a + (b - a)U \sim \text{Unif}(a, b)$.
- Normal: If $Z \sim \mathcal{N}(0, 1)$, then $X = \mu + \sigma Z \sim \mathcal{N}(\mu, \sigma^2)$.
- Exponential: If $X \sim \text{Expo}(1)$, then $Y = X/\lambda \sim \text{Expo}(\lambda)$. We do not consider shifts here since a nonzero shift would prevent the support from being $(0, \infty)$.

We can now add the Exponential and Geometric distributions to our diagram of connections between distributions: the Exponential is a continuous limit of the Geometric, and the Poisson and Exponential are connected by the Poisson process.

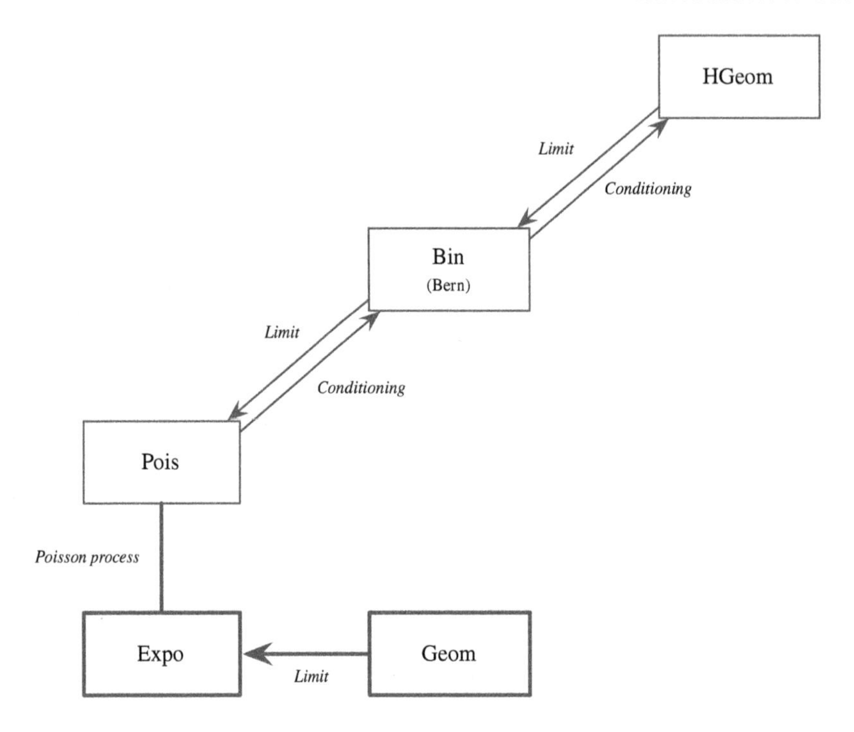

And in our map of the four fundamental objects in probability, we add the PDF as another blueprint for continuous random variables.

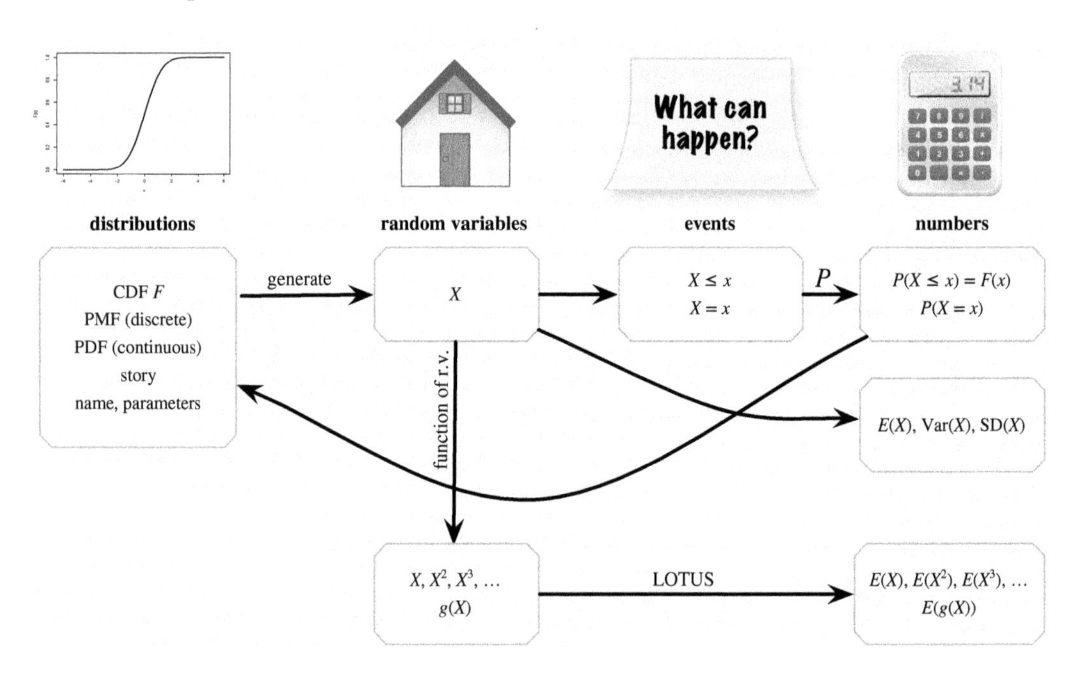

FIGURE 5.14
Four fundamental objects in probability: distributions, random variables, events, and numbers. For a continuous r.v. X, we have $P(X = x) = 0$, so we use the PDF as a blueprint in place of the PMF.

5.9 R

In this section we will introduce continuous distributions in R, learn how to make basic plots, demonstrate the universality of the Uniform by simulation, and simulate arrival times in a Poisson process.

Uniform, Normal, and Exponential distributions

For continuous distributions, the function that starts with `d` is the PDF instead of the PMF. Thus, we have the following functions:

- `dunif`, `punif`, `runif`. To evaluate the Unif(a, b) PDF at x, we use `dunif(x,a,b)`. For the CDF, we use `punif(x,a,b)`. To generate n realizations from the Unif(a, b) distribution, we use `runif(n,a,b)`.
- `dnorm`, `pnorm`, `rnorm`. To evaluate the $\mathcal{N}(\mu, \sigma^2)$ PDF at x, we use `dnorm(x,mu,sigma)`, where `mu` is the mean μ and `sigma` is the *standard deviation* (not variance) σ. For the CDF, we use `pnorm(x,mu,sigma)`. To generate n realizations from the $\mathcal{N}(\mu, \sigma^2)$ distribution, we use `rnorm(n,mu,sigma)`.

  ☣ **5.9.1** (Normal parameters in R). Note that we have to input the standard deviation, not the variance! For example, to get the $\mathcal{N}(10, 3)$ CDF at 12, we use `pnorm(12,10,sqrt(3))`. Ignoring this is a common, disastrous coding error.

- `dexp`, `pexp`, `rexp`. To evaluate the Expo(λ) PDF at x, we use `dexp(x,lambda)`. For the CDF, we use `pexp(x,lambda)`. To generate n realizations from the Expo(λ) distribution, we use `rexp(n,lambda)`.

Due to the importance of location-scale transformations for continuous distributions, R has default parameter settings for each of these three families. The default for the Uniform is Unif$(0, 1)$, the default for the Normal is $\mathcal{N}(0, 1)$, and the default for the Exponential is Expo(1). For example, `dunif(0.5)`, with no additional inputs, evaluates the Unif$(0, 1)$ PDF at 0.5, and `rnorm(10)`, with no additional inputs, generates 10 realizations from the $\mathcal{N}(0, 1)$ distribution. This means there are two ways to generate a $\mathcal{N}(\mu, \sigma^2)$ random variable in R. After choosing our values of μ and σ, such as

```
mu <- 1
sigma <- 2
```

we can do either of the following:

```
rnorm(1,mu,sigma)
mu + sigma*rnorm(1)
```

Either way, we end up generating a draw from the $\mathcal{N}(\mu, \sigma^2)$ distribution.

Plots in R

A simple way to plot a function in R is with the `curve` command. For example,

```
curve(dnorm, from=-3, to=3, n=1000)
```

creates a plot of the standard Normal PDF from -3 to 3. What is actually happening is that R evaluates the function at a finite number of closely spaced points and connects the points with very short lines to create the illusion of a curve. The input `n=1000` tells R to evaluate at 1000 points so that the curve looks very smooth; if we were to choose `n=20`, the piecewise linearity would become very apparent.

Another command that creates plots is called, fittingly, `plot`. This command has many, many possible inputs to customize what the plot looks like; for the sake of demonstration, we'll plot the standard Normal PDF once again, using `plot` instead of `curve`.

The most important inputs to `plot` are a vector of x values and a vector of y values to plot. A useful command for this purpose is `seq`. As introduced in Chapter 1, `seq(a,b,d)` creates the vector of values ranging from `a` to `b`, with successive entries spaced apart by `d`.

```
x <- seq(-3,3,0.01)
y <- dnorm(x)
```

So `x` consists of all numbers from -3 to 3, spaced 0.01 apart, and `y` contains the values of the Normal PDF at each of the points in `x`. Now we simply plot the two with `plot(x,y)`. The default is a scatterplot. For a line plot, we use `plot(x,y,type="l")`. We can also set the axis labels and plot title with `xlab`, `ylab`, and `main`.

```
plot(x,y,type="l",xlab="x",ylab="dnorm(x)",main="N(0,1) PDF")
```

The axis limits can be set manually with `xlim` and `ylim`. If, for example, you wanted the vertical axis to range from 0 to 1, you would add `ylim=c(0,1)` inside the `plot` command.

Finally, to change the color of the plot, add `col="orange"` or `col="green"`, or whatever your favorite color is!

Universality with Logistic

We proved in Example 5.3.4 that for $U \sim \text{Unif}(0,1)$, the r.v. $\log(U/(1-U))$ follows a Logistic distribution. In R, we can simply generate a large number of $\text{Unif}(0,1)$ realizations and transform them.

```
u <- runif(10^4)
x <- log(u/(1-u))
```

Now `x` contains 10^4 realizations from the distribution of $\log(U/(1-U))$. We can visualize them with a histogram, using the command `hist(x)`. The histogram resembles a Logistic PDF, which is reassuring. To control how fine-grained the histogram is, we can set the number of breaks in the histogram: `hist(x,breaks=100)` produces a finer histogram, while `hist(x,breaks=10)` produces a coarser histogram.

Poisson process simulation

To simulate n arrivals in a Poisson process with rate λ, we first generate the interarrival times as i.i.d. Exponentials and store them in a vector:

```
n <- 50
lambda <- 10
x <- rexp(n,lambda)
```

Then we convert the interarrival times into arrival times using the `cumsum` function, which stands for "cumulative sum".

```
t <- cumsum(x)
```

The vector `t` now contains all the simulated arrival times.

5.10 Exercises

Exercises marked with Ⓢ have detailed solutions at `http://stat110.net`.

PDFs and CDFs

1. The Rayleigh distribution from Example 5.1.7 has PDF

$$f(x) = xe^{-x^2/2}, \quad x > 0.$$

Let X have the Rayleigh distribution.

(a) Find $P(1 < X < 3)$.

(b) Find the first quartile, median, and third quartile of X; these are defined to be the values q_1, q_2, q_3 (respectively) such that $P(X \leq q_j) = j/4$ for $j = 1, 2, 3$.

2. (a) Make up a PDF f, with an application for which that PDF would be plausible, where $f(x) > 1$ for all x in a certain interval.

(b) Show that if a PDF f has $f(x) > 1$ for all x in a certain interval, then that interval must have length less than 1.

3. Let F be the CDF of a continuous r.v., and $f = F'$ be the PDF.
(a) Show that g defined by $g(x) = 2F(x)f(x)$ is also a valid PDF.

(b) Show that h defined by $h(x) = \frac{1}{2}f(-x) + \frac{1}{2}f(x)$ is also a valid PDF.

4. Let X be a continuous r.v. with CDF F and PDF f.

   (a) Find the conditional CDF of X given $X > c$, for c a constant with $P(X > c) \neq 0$. That is, find $P(X \leq x|X > a)$ for all c, in terms of F.

   (b) Find the conditional PDF of X given $X > c$ (this is the derivative of the conditional CDF).

   (c) Check that the conditional PDF from (b) is a valid PDF, by showing directly that it is nonnegative and integrates to 1.

5. A circle with a random radius $R \sim \text{Unif}(0, 1)$ is generated. Let A be its area.

   (a) Find the mean and variance of A, without first finding the CDF or PDF of A.

   (b) Find the CDF and PDF of A.

6. The 68-95-99.7% rule gives approximate probabilities of a Normal r.v. being within $1, 2$, and 3 standard deviations of its mean. Derive analogous rules for the following distributions.

   (a) Unif$(0, 1)$.

   (b) Expo(1).

   (c) Expo$(1/2)$. Discuss whether there is one such rule that applies to all Exponential distributions, just as the 68-95-99.7% rule applies to all Normal distributions, not just to the standard Normal.

7. Let
$$F(x) = \frac{2}{\pi}\sin^{-1}\left(\sqrt{x}\right), \text{ for } 0 < x < 1,$$
   $F(x) = 0$ for $x \leq 0$, and $F(x) = 1$ for $x \geq 1$.

   (a) Check that F is a valid CDF, and find the corresponding PDF f. This distribution is called the *Arcsine distribution*, though it also goes by the name Beta$(1/2, 1/2)$ (we will explore the Beta in depth in Chapter 8).

   (b) Explain how it is possible for f to be a valid PDF even though $f(x)$ goes to ∞ as x approaches 0 from the right and as x approaches 1 from the left.

8. The *Beta distribution* with parameters $a = 3, b = 2$ has PDF
$$f(x) = 12x^2(1 - x), \text{ for } 0 < x < 1.$$
   (We will discuss the Beta in detail in Chapter 8.) Let X have this distribution.

   (a) Find the CDF of X.

   (b) Find $P(0 < X < 1/2)$.

   (c) Find the mean and variance of X (without quoting results about the Beta distribution).

9. The *Cauchy distribution* has PDF
$$f(x) = \frac{1}{\pi(1 + x^2)},$$
   for all real x. (We will introduce the Cauchy from another point of view in Chapter 7.) Find the CDF of a random variable with the Cauchy PDF.

   Hint: Recall that the derivative of the inverse tangent function $\arctan(x)$ is $\frac{1}{1+x^2}$.

Uniform and universality of the Uniform

10. Let $U \sim \text{Unif}(0, 8)$.

    (a) Find $P(U \in (0,2) \cup (3,7))$ without using calculus.

    (b) Find the conditional distribution of U given $U \in (3, 7)$.

11. Ⓢ Let U be a Uniform r.v. on the interval $(-1, 1)$ (be careful about minus signs).

    (a) Compute $E(U), \text{Var}(U)$, and $E(U^4)$.

    (b) Find the CDF and PDF of U^2. Is the distribution of U^2 Uniform on $(0, 1)$?

12. Ⓢ A stick is broken into two pieces, at a uniformly random breakpoint. Find the CDF and average of the length of the longer piece.

13. A stick of length 1 is broken at a uniformly random point, yielding two pieces. Let X and Y be the lengths of the shorter and longer pieces, respectively, and let $R = X/Y$ be the ratio of the lengths X and Y.

    (a) Find the CDF and PDF of R.

    (b) Find the expected value of R (if it exists).

    (c) Find the expected value of $1/R$ (if it exists).

14. Let $U_1, \dots, U_n$ be i.i.d. $\text{Unif}(0, 1)$, and $X = \max(U_1, \dots, U_n)$. What is the PDF of X? What is EX?

    Hint: Find the CDF of X first, by translating the event $X \leq x$ into an event involving $U_1, \dots, U_n$.

15. Let $U \sim \text{Unif}(0, 1)$. Using U, construct $X \sim \text{Expo}(\lambda)$.

16. Ⓢ Let $U \sim \text{Unif}(0, 1)$, and

$$X = \log \left(\frac{U}{1-U} \right).$$

    Then X has the Logistic distribution, as defined in Example 5.1.6.

    (a) Write down (but do not compute) an integral giving $E(X^2)$.

    (b) Find $E(X)$ without using calculus.

    Hint: A useful symmetry property here is that $1 - U$ has the same distribution as U.

17. Let $U \sim \text{Unif}(0, 1)$. As a function of U, create an r.v. X with CDF $F(x) = 1 - e^{-x^3}$ for $x > 0$.

18. The *Pareto distribution* with parameter $a > 0$ has PDF $f(x) = a/x^{a+1}$ for $x \geq 1$ (and 0 otherwise). This distribution is often used in statistical modeling.

    (a) Find the CDF of a Pareto r.v. with parameter a; check that it is a valid CDF.

    (b) Suppose that for a simulation you want to run, you need to generate i.i.d. Pareto(a) r.v.s. You have a computer that knows how to generate i.i.d. $\text{Unif}(0, 1)$ r.v.s but does not know how to generate Pareto r.v.s. Show how to do this.

Normal

19. Let $Z \sim \mathcal{N}(0, 1)$. Create an r.v. $Y \sim \mathcal{N}(1, 4)$, as a simple-looking function of Z. Make sure to check that your Y has the correct mean and variance.

20. Engineers sometimes work with the "error function"

$$\text{erf}(z) = \frac{2}{\sqrt{\pi}} \int_0^z e^{-x^2}\,dx,$$

instead of the standard Normal CDF Φ.

(a) Show that the following conversion between Φ and erf holds for all z:

$$\Phi(z) = \frac{1}{2} + \frac{1}{2}\text{erf}\left(\frac{z}{\sqrt{2}}\right).$$

(b) Show that erf is an odd function, i.e., $\text{erf}(-z) = -\text{erf}(z)$.

21. (a) Find the points of inflection of the $\mathcal{N}(0,1)$ PDF φ, i.e., the points where the curve switches from convex (second derivative positive) to concave (second derivative negative) or vice versa.

(b) Use the result of (a) and a location-scale transformation to find the points of inflection of the $\mathcal{N}(\mu,\sigma^2)$ PDF.

22. The distance between two points needs to be measured, in meters. The true distance between the points is 10 meters, but due to measurement error we can't measure the distance exactly. Instead, we will observe a value of $10+\epsilon$, where the error ϵ is distributed $\mathcal{N}(0, 0.04)$. Find the probability that the observed distance is within 0.4 meters of the true distance (10 meters). Give both an exact answer in terms of Φ and an approximate numerical answer.

23. Alice is trying to transmit to Bob the answer to a yes-no question, using a noisy channel. She encodes "yes" as 1 and "no" as 0, and sends the appropriate value. However, the channel adds noise; specifically, Bob receives what Alice sends plus a $\mathcal{N}(0,\sigma^2)$ noise term (the noise is independent of what Alice sends). If Bob receives a value greater than 1/2 he interprets it as "yes"; otherwise, he interprets it as "no".

(a) Find the probability that Bob understands Alice correctly.

(b) What happens to the result from (a) if σ is very small? What about if σ is very large? Explain intuitively why the results in these extreme cases make sense.

24. A woman is pregnant, with a due date of January 10, 2020. Of course, the actual date on which she will give birth is not necessarily the due date. On a timeline, define time 0 to be the instant when January 10, 2020 begins. Suppose that the time T when the woman gives birth has a Normal distribution, centered at 0 and with standard deviation 8 days. What is the probability that she gives birth on her due date? (Your answer should be in terms of Φ, and simplified.)

25. We will show in the next chapter that if X_1 and X_2 are independent with $X_i \sim \mathcal{N}(\mu_i, \sigma_i^2)$, then $X_1 + X_2 \sim \mathcal{N}(\mu_1 + \mu_2, \sigma_1^2 + \sigma_2^2)$. Use this result to find $P(X < Y)$ for $X \sim \mathcal{N}(a,b), Y \sim \mathcal{N}(c,d)$ with X and Y independent.

Hint: Write $P(X < Y) = P(X - Y < 0)$ and then standardize $X - Y$. Check that your answer makes sense in the special case where X and Y are i.i.d.

26. Walter and Carl both often need to travel from Location A to Location B. Walter walks, and his travel time is Normal with mean w minutes and standard deviation σ minutes (travel time can't be negative without using a tachyon beam, but assume that w is so much larger than σ that the chance of a negative travel time is negligible).

Carl drives his car, and his travel time is Normal with mean c minutes and standard deviation 2σ minutes (the standard deviation is larger for Carl due to variability in traffic conditions). Walter's travel time is independent of Carl's. On a certain day, Walter and Carl leave from Location A to Location B at the same time.

(a) Find the probability that Carl arrives first (in terms of Φ and the parameters). For this you can use the important fact, proven in the next chapter, that if X_1 and X_2 are independent with $X_i \sim \mathcal{N}(\mu_i, \sigma_i^2)$, then $X_1 + X_2 \sim \mathcal{N}(\mu_1 + \mu_2, \sigma_1^2 + \sigma_2^2)$.

(b) Give a fully simplified criterion (*not* in terms of Φ), such that Carl has more than a 50% chance of arriving first if and only if the criterion is satisfied.

(c) Walter and Carl want to make it to a meeting at Location B that is scheduled to begin $w+10$ minutes after they depart from Location A. Give a fully simplified criterion (*not* in terms of Φ) such that Carl is more likely than Walter to make it on time for the meeting if and only if the criterion is satisfied.

27. Let $Z \sim \mathcal{N}(0,1)$. We know from the 68-95-99.7% rule that there is a 68% chance of Z being in the interval $(-1,1)$. Give a visual explanation of whether or not there is an interval (a,b) that is shorter than the interval $(-1,1)$, yet which has at least as large a chance as $(-1,1)$ of containing Z.

28. Let $Y \sim \mathcal{N}(\mu, \sigma^2)$. Use the fact that $P(|Y-\mu| < 1.96\sigma) \approx 0.95$ to construct a random interval $(a(Y), b(Y))$ (that is, an interval whose endpoints are r.v.s), such that the probability that μ is in the interval is approximately 0.95. This interval is called a *confidence interval* for μ; such intervals are often desired in statistics when estimating unknown parameters based on data.

29. Let $Y = |X|$, with $X \sim \mathcal{N}(\mu, \sigma^2)$. This is a well-defined continuous r.v., even though the absolute value function is not differentiable at 0 (due to the sharp corner).

(a) Find the CDF of Y in terms of Φ. Be sure to specify the CDF everywhere.

(b) Find the PDF of Y.

(c) Is the PDF of Y continuous at 0? If not, is this a problem as far as using the PDF to find probabilities?

30. Ⓢ Let $Z \sim \mathcal{N}(0,1)$ and let S be a random sign independent of Z, i.e., S is 1 with probability 1/2 and -1 with probability 1/2. Show that $SZ \sim \mathcal{N}(0,1)$.

31. Ⓢ Let $Z \sim \mathcal{N}(0,1)$. Find $E(\Phi(Z))$ *without* using LOTUS, where Φ is the CDF of Z.

32. Ⓢ Let $Z \sim \mathcal{N}(0,1)$ and $X = Z^2$. Then the distribution of X is called *Chi-Square with 1 degree of freedom*. This distribution appears in many statistical methods.

(a) Find a good numerical approximation to $P(1 \le X \le 4)$ using facts about the Normal distribution, without querying a calculator/computer/table about values of the Normal CDF.

(b) Let Φ and φ be the CDF and PDF of Z, respectively. Show that for any $t > 0$, $I(Z > t) \le (Z/t)I(Z > t)$. Using this and LOTUS, derive *Mills' inequality*, which is the following lower bound on Φ:

$$\Phi(t) \ge 1 - \varphi(t)/t.$$

33. Let $Z \sim \mathcal{N}(0,1)$, with CDF Φ. We will show in Chapter 8 that the PDF of Z^2 is the function g given by

$$g(w) = \frac{1}{\sqrt{2\pi w}} e^{-w/2}$$

for $w > 0$, and $g(w) = 0$ for $w \le 0$.

(a) Find expressions for $E(Z^4)$ as integrals in two different ways, one based on the PDF of Z and the other based on the PDF of Z^2.

(b) Find $E(Z^2 + Z + \Phi(Z))$.

34. Ⓢ Let $Z \sim \mathcal{N}(0,1)$. A measuring device is used to observe Z, but the device can only handle positive values, and gives a reading of 0 if $Z \leq 0$; this is an example of *censored data*. So assume that $X = ZI_{Z>0}$ is observed rather than Z, where $I_{Z>0}$ is the indicator of $Z > 0$. Find $E(X)$ and $\text{Var}(X)$.

35. Let $Z \sim \mathcal{N}(0,1)$, and c be a nonnegative constant. Find $E(\max(Z - c, 0))$, in terms of the standard Normal CDF Φ and PDF φ. (This kind of calculation often comes up in quantitative finance.)

    Hint: Use LOTUS, and handle the max symbol by adjusting the limits of integration appropriately. As a check, make sure that your answer reduces to $1/\sqrt{2\pi}$ when $c = 0$; this must be the case since we showed in Example 5.4.7 that $E|Z| = \sqrt{2/\pi}$, and we have $|Z| = \max(Z, 0) + \max(-Z, 0)$ so by symmetry

$$E|Z| = E(\max(Z,0)) + E(\max(-Z,0)) = 2E(\max(Z,0)).$$

Exponential

36. Ⓢ A post office has 2 clerks. Alice enters the post office while 2 other customers, Bob and Claire, are being served by the 2 clerks. She is next in line. Assume that the time a clerk spends serving a customer has an $\text{Expo}(\lambda)$ distribution.

    (a) What is the probability that Alice is the last of the 3 customers to be done being served?

    Hint: No integrals are needed.

    (b) What is the expected total time that Alice needs to spend at the post office?

37. Let T be the time until a radioactive particle decays, and suppose (as is often done in physics and chemistry) that $T \sim \text{Expo}(\lambda)$.

    (a) The *half-life* of the particle is the time at which there is a 50% chance that the particle has decayed (in statistical terminology, this is the *median* of the distribution of T). Find the half-life of the particle.

    (b) Show that for ϵ a small, positive constant, the probability that the particle decays in the time interval $[t, t + \epsilon]$, given that it has survived until time t, does not depend on t and is approximately proportional to ϵ.

    Hint: $e^x \approx 1 + x$ if $x \approx 0$.

    (c) Now consider n radioactive particles, with i.i.d. times until decay $T_1, \ldots, T_n \sim \text{Expo}(\lambda)$. Let L be the first time at which one of the particles decays. Find the CDF of L. Also, find $E(L)$ and $\text{Var}(L)$.

    (d) Continuing (c), find the mean and variance of $M = \max(T_1, \ldots, T_n)$, the *last* time at which one of the particles decays, *without using calculus*.

    Hint: Draw a timeline, apply (c), and remember the memoryless property.

38. Ⓢ Fred wants to sell his car, after moving back to Blissville (where he is happy with the bus system). He decides to sell it to the first person to offer at least \$18,000 for it. Assume that the offers are independent Exponential random variables with mean \$12,000, and that Fred is able to keep getting offers until he obtains one that meets his criterion.

    (a) Find the expected number of offers Fred will have.

    (b) Find the expected amount of money that Fred will get for the car.

39. As in the previous problem, Fred wants to sell his car, and the offers for his car are i.i.d. Exponential r.v.s with mean \$12,000. Assume now though that he will wait until he has 3 offers (no matter how large or small they are), and then accept the largest of the 3 offers. Find the expected amount of money that Fred will get for his car.

40. (a) Fred visits Blotchville again. He finds that the city has installed an electronic display at the bus stop, showing the time when the previous bus arrived. The times between arrivals of buses are still independent Exponentials with mean 10 minutes. Fred waits for the next bus, and then records the time between that bus and the previous bus. On average, what length of time between buses does he see?

    (b) Fred then visits Blunderville, where the times between buses are also 10 minutes on average, and independent. Yet to his dismay, he finds that on average he has to wait more than 1 hour for the next bus when he arrives at the bus stop! How is it possible that the average Fred-to-bus time is greater than the average bus-to-bus time even though Fred arrives at some time between two bus arrivals? Explain this intuitively, and construct a specific discrete distribution for the times between buses showing that this is possible.

41. Fred and Gretchen are waiting at a bus stop in Blotchville. Two bus routes, Route 1 and Route 2, have buses that stop at this bus stop. For Route i, buses arrive according to a Poisson process with rate λ_i buses/minute. The Route 1 process is independent of the Route 2 process. Fred is waiting for a Route 1 bus, and Gretchen is waiting for a Route 2 bus.

    (a) Given that Fred has already waited for 20 minutes, on average how much longer will he have to wait for his bus?

    (b) Find the probability that at least n Route 1 buses will pass by before the first Route 2 bus arrives. The following result from Chapter 7 may be useful here: for independent random variables $X_1 \sim \text{Expo}(\lambda_1), X_2 \sim \text{Expo}(\lambda_2)$, we have $P(X_1 < X_2) = \lambda_1/(\lambda_1+\lambda_2)$.

    (c) For this part only, assume that $\lambda_1 = \lambda_2 = \lambda$. Find the expected time it will take until both Fred and Gretchen have caught their buses.

42. Ⓢ Joe is waiting in continuous time for a book called *The Winds of Winter* to be released. Suppose that the waiting time T until news of the book's release is posted, measured in years relative to some starting point, has an Exponential distribution with $\lambda = 1/5$.

    Joe is not so obsessive as to check multiple times a day; instead, he checks the website *once* at the end of each day. Therefore, he observes the day on which the news was posted, rather than the exact time T. Let X be this measurement, where $X = 0$ means that the news was posted within the first day (after the starting point), $X = 1$ means it was posted on the second day, etc. (assume that there are 365 days in a year). Find the PMF of X. Is this a named distribution that we have studied?

43. The Exponential is the analog of the Geometric in continuous time. This problem explores the connection between Exponential and Geometric in more detail, asking what happens to a Geometric in a limit where the Bernoulli trials are performed faster and faster but with smaller and smaller success probabilities.

    Suppose that Bernoulli trials are being performed in continuous time; rather than only thinking about first trial, second trial, etc., imagine that the trials take place at points on a timeline. Assume that the trials are at regularly spaced times $0, \Delta t, 2\Delta t, \dots$, where Δt is a small positive number. Let the probability of success of each trial be $\lambda \Delta t$, where λ is a positive constant. Let G be the number of failures before the first success (in discrete time), and T be the time of the first success (in continuous time).

    (a) Find a simple equation relating G to T.

    Hint: Draw a timeline and try out a simple example.

(b) Find the CDF of T.

Hint: First find $P(T > t)$.

(c) Show that as $\Delta t \to 0$, the CDF of T converges to the Expo(λ) CDF, evaluating all the CDFs at a fixed $t \geq 0$.

Hint: Use the compound interest limit (see Section A.2.5 of the math appendix).

44. The *Laplace distribution* has PDF

$$f(x) = \frac{1}{2}e^{-|x|}$$

for all real x. The Laplace distribution is also called a *symmetrized Exponential* distribution. Explain this in the following two ways.

(a) Plot the PDFs and explain how they relate.

(b) Let $X \sim \text{Expo}(1)$ and S be a random sign (1 or -1, with equal probabilities), with S and X independent. Find the PDF of SX (by first finding the CDF), and compare the PDF of SX and the Laplace PDF.

45. Emails arrive in an inbox according to a Poisson process with rate 20 emails per hour. Let T be the time at which the 3rd email arrives, measured in hours after a certain fixed starting time. Find $P(T > 0.1)$ without using calculus.

Hint: Apply the count-time duality.

46. Let T be the lifetime of a certain person (how long that person lives), and let T have CDF F and PDF f. The *hazard function* of T is defined by

$$h(t) = \frac{f(t)}{1 - F(t)}.$$

(a) Explain why h is called the hazard function and in particular, why $h(t)$ is the probability density for death at time t, given that the person survived up until then.

(b) Show that an Exponential r.v. has constant hazard function and conversely, if the hazard function of T is a constant then T must be Expo(λ) for some λ.

47. Let T be the lifetime of a person (or animal or gadget), with CDF F and PDF f. Let h be the hazard function, defined as in the previous problem. If we know F then we can calculate f, and then in turn we can calculate h. In this problem, we consider the reverse problem: how to recover F and f from knowing h.

(a) Show that the CDF and hazard function are related by

$$F(t) = 1 - \exp\left(-\int_0^t h(s)ds\right),$$

for all $t > 0$.

Hint: Let $G(t) = 1 - F(t)$ be the survival function, and consider the derivative of $\log G(t)$.

(b) Show that the PDF and hazard function are related by

$$f(t) = h(t)\exp\left(-\int_0^t h(s)ds\right),$$

for all $t > 0$.

Hint: Apply the result of (a).

48. Ⓢ Find $E(X^3)$ for $X \sim \text{Expo}(\lambda)$, using LOTUS and the fact that $E(X) = 1/\lambda$ and $\text{Var}(X) = 1/\lambda^2$, and integration by parts at most once. In the next chapter, we'll learn how to find $E(X^n)$ for all n.

49. Ⓢ The *Gumbel distribution* is the distribution of $-\log X$ with $X \sim \text{Expo}(1)$.

    (a) Find the CDF of the Gumbel distribution.

    (b) Let $X_1, X_2, \dots$ be i.i.d. Expo(1) and let $M_n = \max(X_1, \dots, X_n)$. Show that $M_n - \log n$ converges in distribution to the Gumbel distribution, i.e., as $n \to \infty$ the CDF of $M_n - \log n$ converges to the Gumbel CDF.

Mixed practice

50. Explain intuitively why $P(X < Y) = P(Y < X)$ if X and Y are i.i.d., but equality may not hold if X and Y are not independent or not identically distributed.

51. Let X be an r.v. (discrete or continuous) such that $0 \leq X \leq 1$ always holds. Let $\mu = E(X)$.

    (a) Show that

$$\text{Var}(X) \leq \mu - \mu^2 \leq \frac{1}{4}.$$

    Hint: With probability 1, we have $X^2 \leq X$.

    (b) Show that there is only one possible distribution for X for which $\text{Var}(X) = 1/4$. What is the name of this distribution?

52. The Rayleigh distribution from Example 5.1.7 has PDF

$$f(x) = xe^{-x^2/2}, \quad x > 0.$$

    Let X have the Rayleigh distribution.

    (a) Find $E(X)$ without using much calculus, by interpreting the integral in terms of known results about the Normal distribution.

    (b) Find $E(X^2)$.

    Hint: A nice approach is to use LOTUS and the substitution $u = x^2/2$, and then interpret the resulting integral in terms of known results about the Exponential distribution.

53. Ⓢ Consider an experiment where we observe the value of a random variable X, and estimate the value of an unknown constant θ using some random variable $T = g(X)$ that is a function of X. The r.v. T is called an *estimator*. Think of X as the data observed in the experiment, and θ as an unknown parameter related to the distribution of X.

    For example, consider the experiment of flipping a coin n times, where the coin has an unknown probability θ of Heads. After the experiment is performed, we have observed the value of $X \sim \text{Bin}(n, \theta)$. The most natural estimator for θ is then X/n.

    The *bias* of an estimator T for θ is defined as $b(T) = E(T) - \theta$. The *mean squared error* is the average squared error when using $T(X)$ to estimate θ:

$$\text{MSE}(T) = E(T - \theta)^2.$$

    Show that

$$\text{MSE}(T) = \text{Var}(T) + (b(T))^2.$$

    This implies that for fixed MSE, lower bias can only be attained at the cost of higher variance and vice versa; this is a form of the *bias-variance tradeoff*, a phenomenon which arises throughout statistics.

54. Ⓢ (a) Suppose that we have a list of the populations of every country in the world.

Guess, without looking at data yet, what percentage of the populations have the digit 1 as their first digit (e.g., a country with a population of 1,234,567 has first digit 1 and a country with population 89,012,345 does not).

(b) After having done (a), look through a list of populations and count how many start with a 1. What percentage of countries is this? *Benford's law* states that in a very large variety of real-life data sets, the first digit approximately follows a particular distribution with about a 30% chance of a 1, an 18% chance of a 2, and in general

$$P(D = j) = \log_{10}\left(\frac{j+1}{j}\right), \text{ for } j \in \{1, 2, 3, \ldots, 9\},$$

where D is the first digit of a randomly chosen element. (Exercise 6 from Chapter 3 asks for a proof that this is a valid PMF.) How closely does the percentage found in the data agree with that predicted by Benford's law?

(c) Suppose that we write the random value in some problem (e.g., the population of a random country) in scientific notation as $X \times 10^N$, where N is a nonnegative integer and $1 \leq X < 10$. Assume that X is a continuous r.v. with PDF

$$f(x) = c/x, \text{ for } 1 \leq x \leq 10,$$

and 0 otherwise, with c a constant. What is the value of c (be careful with the bases of logs)? Intuitively, we might hope that the distribution of X does not depend on the choice of units in which X is measured. To see whether this holds, let $Y = aX$ with $a > 0$. What is the PDF of Y (specifying where it is nonzero)?

(d) Show that if we have a random number $X \times 10^N$ (written in scientific notation) and X has the PDF f from (c), then the first digit (which is also the first digit of X) has Benford's law as its PMF.

Hint: What does $D = j$ correspond to in terms of the values of X?

55. Ⓢ (a) Let $X_1, X_2, \ldots$ be independent $\mathcal{N}(0, 4)$ r.v.s., and let J be the smallest value of j such that $X_j > 4$ (i.e., the index of the first X_j exceeding 4). In terms of Φ, find $E(J)$.

(b) Let f and g be PDFs with $f(x) > 0$ and $g(x) > 0$ for all x. Let X be a random variable with PDF f. Find the expected value of the ratio

$$R = \frac{g(X)}{f(X)}.$$

Such ratios come up very often in statistics, when working with a quantity known as a *likelihood ratio* and when using a computational technique known as *importance sampling*.

(c) Define

$$F(x) = e^{-e^{-x}}.$$

This is a CDF and is a continuous, strictly increasing function. Let X have CDF F, and define $W = F(X)$. What are the mean and variance of W?

56. Let $X, Y, Z \sim \mathcal{N}(0, 1)$ be i.i.d.

(a) Find an expression for $E\left(Z^2\, \Phi(Z)\right)$ as an integral.

(b) Find $P(\Phi(Z) < 2/3)$.

(c) Find $P(X < Y < Z)$.

57. Let $X, Y, Z \sim \mathcal{N}(0,1)$ be i.i.d., and $W = (\Phi(Z))^2$.

(a) Find the CDF and PDF of W.

(b) Let f_W be the PDF of W and φ be the PDF of Z. Find unsimplified expressions for $E(W^3)$ as integrals in two different ways, one based on f_W and one based on φ.

(c) Find $P(X + 2Y < 2Z + 3)$, in terms of Φ.

Hint: Move all of the r.v.s to one side of the inequality.

58. Let $Z \sim \mathcal{N}(0,1)$ and $Y = \max(Z, 0)$. So Y is Z if $Z > 0$, and Y is 0 if $Z \leq 0$.

(a) Find an expression for $E(Y)$ as an integral (which can be unsimplified).

(b) Let $Y_1, Y_2, \dots$ be independent r.v.s, each with the same distribution as Y. Let $N = \min\{n : Y_n = 0\}$, i.e., N is the smallest value such that $Y_N = 0$. Find $E(N)$.

(c) Find the CDF of Y in terms of Φ. (Be sure to specify it for all real numbers.)

59. The unit circle $\{(x, y) : x^2 + y^2 = 1\}$ is divided into three arcs by choosing three random points A, B, C on the circle (independently and uniformly), forming arcs between A and B, between A and C, and between B and C. Let L be the length of the arc containing the point $(1, 0)$. What is $E(L)$? Study this by working through the following steps.

(a) Explain what is wrong with the following argument: "The total length of the arcs is 2π, the circumference of the circle. So by symmetry and linearity, each arc has length $2\pi/3$ on average. Referring to the arc containing $(1, 0)$ is just a way to specify one of the arcs (it wouldn't matter if $(1, 0)$ were replaced by $(0, -1)$ or any other specific point on the circle in the statement of the problem). So the expected value of L is $2\pi/3$."

(b) Let the arc containing $(1, 0)$ be divided into two pieces: the piece extending counterclockwise from $(1, 0)$ and the piece extending clockwise from $(1, 0)$. Write $L = L_1 + L_2$, where L_1 and L_2 are the lengths of the counterclockwise and clockwise pieces, respectively. Find the CDF, PDF, and expected value of L_1.

(c) Use (b) to find $E(L)$.

60. Ⓢ As in Example 5.7.4, athletes compete one at a time at the high jump. Let X_j be how high the jth jumper jumped, with $X_1, X_2, \dots$ i.i.d. with a continuous distribution. We say that the jth jumper is "best in recent memory" if they jump higher than the previous 2 jumpers (for $j \geq 3$; the first 2 jumpers don't qualify).

(a) Find the expected number of best in recent memory jumpers among the 3rd through nth jumpers.

(b) Let A_j be the event that the jth jumper is the best in recent memory. Find $P(A_3), P(A_4)$, and $P(A_3 \cap A_4)$. Are A_3 and A_4 independent?

61. Tyrion, Cersei, and n other guests arrive at a party at i.i.d. times drawn from a continuous distribution with support $[0, 1]$, and stay until the end (time 0 is the party's start time and time 1 is the end time). The party will be boring at times when neither Tyrion nor Cersei is there, fun when exactly one of them is there, and awkward when both Tyrion and Cersei are there.

(a) On average, how many of the n other guests will arrive at times when the party is fun?

(b) Jaime and Robert are two of the other guests. By computing both sides in the definition of independence, determine whether the event "Jaime arrives at a fun time" is independent of the event "Robert arrives at a fun time".

(c) Give a clear intuitive explanation of whether the two events from (b) are independent, and whether they are conditionally independent given the arrival times of everyone else, i.e., everyone except Jaime and Robert.

62. Let $X_1, X_2, \ldots$ be the annual rainfalls in Boston (measured in inches) in the years 2101, 2102, ..., respectively. Assume that annual rainfalls are i.i.d. draws from a continuous distribution. A rainfall value is a *record high* if it is greater than those in all previous years (starting with 2101), and a *record low* if it is lower than those in all previous years.

    (a) In the 22nd century (the years 2101 through 2200, inclusive), find the expected number of years that have either a record low or a record high rainfall.

    (b) On average, in how many years in the 22nd century is there a record low followed in the next year by a record high? (Only the record low is required to be in the 22nd century, not the record high.)

    (c) By definition, the year 2101 is a record high (and record low). Let N be the number of years required to get a new record high. Find $P(N > n)$ for all positive integers n, and use this to find the PMF of N.

    Hint: Note that $P(N = n) + P(N > n) = P(N > n - 1)$.

    (d) With notation as above, show that $E(N)$ is infinite.

6

Moments

The nth *moment* of an r.v. X is $E(X^n)$. In this chapter, we explore how the moments of an r.v. shed light on its distribution. We have already seen that the first two moments are useful since they provide the mean $E(X)$ and variance $E(X^2)-(EX)^2$, which are important summaries of the average value of X and how spread out its distribution is. But there is much more to a distribution than its mean and variance. We'll see that the third and fourth moments tell us about the *asymmetry* of a distribution and the behavior of the *tails* or extreme values, two properties that are not captured by the mean and variance. After introducing moments, we'll discuss the *moment generating function* (MGF), which not only helps us compute moments but also provides a useful alternative way to specify a distribution.

6.1 Summaries of a distribution

The mean is called a *measure of central tendency* because it tells us something about the center of a distribution, specifically its center of mass. Other measures of central tendency that are commonly used in statistics are the *median* and the *mode*, which we now define.

Definition 6.1.1 (Median). We say that c is a *median* of a random variable X if $P(X \leq c) \geq 1/2$ and $P(X \geq c) \geq 1/2$. (The simplest way this can happen is if the CDF of X hits $1/2$ exactly at c, but we know that some CDFs have jumps.)

Definition 6.1.2 (Mode). For a discrete r.v. X, we say that c is a *mode* of X if it maximizes the PMF: $P(X = c) \geq P(X = x)$ for all x. For a continuous r.v. X with PDF f, we say that c is a mode if it maximizes the PDF: $f(c) \geq f(x)$ for all x.

As with the mean, the median and mode of an r.v. depend only on its *distribution*, so we can talk about the mean, median, or mode of a distribution without referring to any particular r.v. that has that distribution. For example, if $Z \sim \mathcal{N}(0, 1)$ then the median of Z is 0 (since $\Phi(0) = 1/2$ by symmetry), and we also say that the standard Normal *distribution* has median 0.

Intuitively, the median is a value c such that half the mass of the distribution falls on either side of c (or as close to half as possible, for discrete r.v.s), and the mode is a value that has the greatest mass or density out of all values in the support of

X. If the CDF F is a continuous, strictly increasing function, then $F^{-1}(1/2)$ is the median (and is unique).

Note that a distribution can have multiple medians and multiple modes. Medians have to occur side by side; modes can occur all over the distribution. In Figure 6.1, we show a distribution supported on $[-5, -1] \cup [1, 5]$ that has two modes and infinitely many medians. The PDF is 0 between -1 and 1, so all values between -1 and 1 are medians of the distribution because half of the mass falls on either side. The two modes are at -3 and 3.

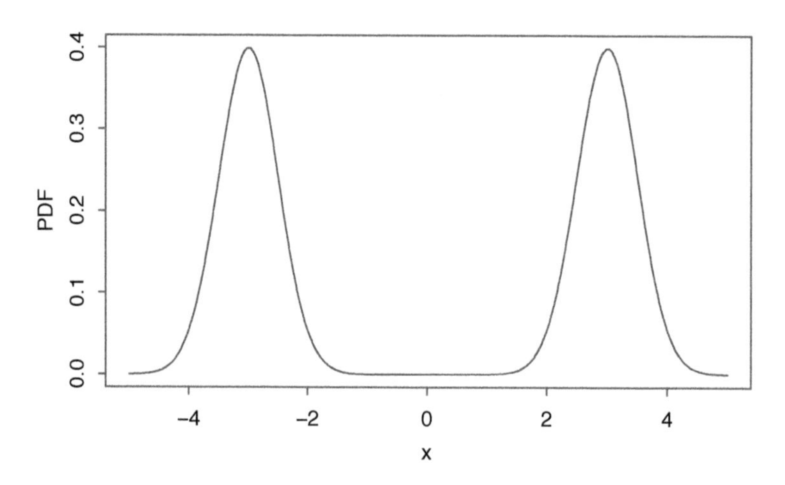

FIGURE 6.1
A distribution with two modes (-3 and 3) and infinitely many medians (all x in the interval $[-1, 1]$).

Example 6.1.3 (Mean, median, and mode for salaries). A certain company has 100 employees. Let $s_1, s_2, \ldots, s_{100}$ be their salaries, sorted in increasing order (we can still do this even if some salaries appear more than once). Let X be the salary of a randomly selected employee (chosen uniformly). The mean, median, and mode for the data set $s_1, s_2, \ldots, s_{100}$ are defined to be the corresponding quantities for X.

What is a typical salary? What is the most useful one-number summary of the salary data? The answer, as is often the case, is *it depends on the goal.* Different summaries reveal different characteristics of the data, so it may be hard to choose just one number—and it is often unnecessary to do so, since usually we can provide several summaries (and plot the data too). Here we briefly compare the mean, median, and mode, though often it makes sense to report all three (and other summaries too).

If the salaries are all different, the mode doesn't give a useful one-number summary since there are 100 modes. If there are only a few possible salaries in the company, the mode becomes more useful. But even then it could happen that, for example, 34 people receive salary a, 33 receive salary b, and 33 receive salary c. Then a is the unique mode, but if we only report a we are ignoring b and c, which just barely missed being modes and which together account for almost 2/3 of the data.

Next, let's consider the median. There are two numbers "in the middle", s_{50} and s_{51}. In fact, *any* number m with $s_{50} \leq m \leq s_{51}$ is a median, since there is at least a 50% chance that the random employee's salary is in $\{s_1, \ldots, s_{50}\}$ (in which case it is at most m) and at least a 50% chance that it is in $\{s_{51}, \ldots, s_{100}\}$ (in which case it is at least m). The usual convention is to choose $m = (s_{50} + s_{51})/2$, the mean of the two numbers in the middle. If the number of employees had been odd, this issue would not have come up; in that case, there is a unique median, the number in the middle when all the salaries are listed in increasing order.

Compared with the mean, the median is much less sensitive to extreme values. For example, if the CEO's salary is changed from being slightly more than anyone else's to vastly more than anyone else's, that could have a large impact on the mean but it has no impact on the median. This robustness is a reason that the median could be a more sensible summary than the mean of what the typical salary is. On the other hand, suppose that we want to know the total cost the company is paying for its employees' salaries. If we only know a mode or a median, we can't extract this information, but if we know the mean we can just multiply it by 100. □

Suppose that we are trying to guess what a not-yet-observed r.v. X will be, by making a prediction c. The mean and the median both seem like natural guesses for c, but which is better? That depends on how "better" is defined. Two natural ways to judge how good c is are the *mean squared error* $E(X-c)^2$ and the *mean absolute error* $E|X-c|$. The following result says what the best guesses are in both cases.

Theorem 6.1.4. Let X be an r.v. with mean μ, and let m be a median of X.

- The value of c that minimizes the mean squared error $E(X-c)^2$ is $c = \mu$.
- A value of c that minimizes the mean absolute error $E|X-c|$ is $c = m$.

Proof. We will first prove a useful identity:

$$E(X-c)^2 = \text{Var}(X) + (\mu - c)^2.$$

This can be checked by expanding everything out, but a faster way is to use the fact that adding a constant doesn't affect the variance:

$$\text{Var}(X) = \text{Var}(X-c) = E(X-c)^2 - (E(X-c))^2 = E(X-c)^2 - (\mu-c)^2.$$

It follows that $c = \mu$ is the unique choice that minimizes $E(X-c)^2$, since that choice makes $(\mu - c)^2 = 0$ and any other choice makes $(\mu-c)^2 > 0$.

Next, let's consider the mean absolute error. Let $a \neq m$. We need to show that $E|X-m| \leq E|X-a|$, which is equivalent to $E(|X-a| - |X-m|) \geq 0$. Assume that $m < a$ (the case $m > a$ can be handled similarly). If $X \leq m$ then

$$|X-a| - |X-m| = a - X - (m - X) = a - m,$$

and if $X > m$ then

$$|X - a| - |X - m| \geq X - a - (X - m) = m - a.$$

Let

$$Y = |X - a| - |X - m|.$$

We can split the definition of $E(Y)$ into 2 parts based on whether $X \leq m$ occurs, using indicator r.v.s. Let I be the indicator r.v. for $X \leq m$, so $1 - I$ is the indicator r.v. for $X > m$. Then

$$\begin{aligned} E(Y) &= E(YI) + E(Y(1 - I)) \\ &\geq (a - m)E(I) + (m - a)E(1 - I) \\ &= (a - m)P(X \leq m) + (m - a)P(X > m) \\ &= (a - m)P(X \leq m) - (a - m)(1 - P(X \leq m)) \\ &= (a - m)(2P(X \leq m) - 1). \end{aligned}$$

By definition of median, we have $2P(X \leq m) - 1 \geq 0$. Thus, $E(Y) \geq 0$, which implies $E(|X - m|) \leq E(|X - a|)$. ■

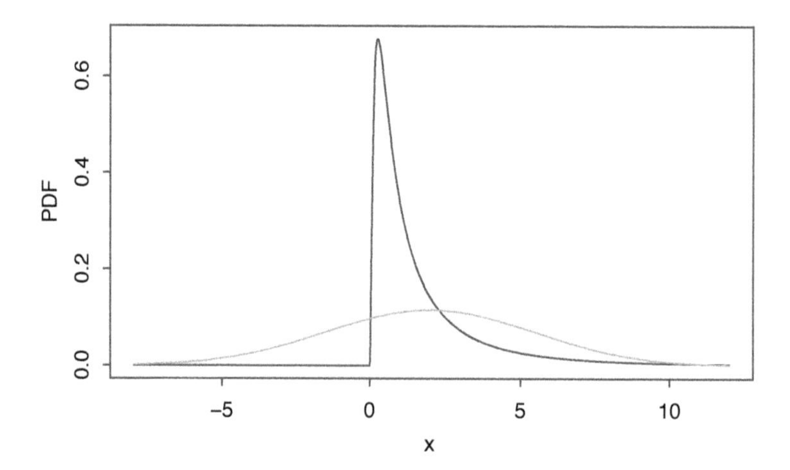

FIGURE 6.2
Two PDFs with mean 2 and variance 12. The light curve is Normal and symmetric; the dark curve is Log-Normal and right-skewed.

Regardless of which measure of central tendency we use in a particular application, it is usually important also to know about the *spread* of the distribution as measured by, for example, the variance. However, there are also major features of a distribution that are not captured by the mean and variance. For example, the two PDFs in Figure 6.2 both have mean 2 and variance 12. The light curve is the $\mathcal{N}(2, 12)$ PDF and the dark curve belongs to the *Log-Normal* family of distributions (the Log-Normal is defined later in this chapter). The Normal curve is symmetric about 2, so its mean, median, and mode are all 2. In contrast, the Log-Normal is heavily *skewed* to the right; this means its right tail is very long compared to its left tail. It has mean 2, but median 1 and mode 0.25. From the mean and variance alone, we would not be able to capture the difference between the asymmetry of the Log-Normal and the symmetry of the Normal.

Now consider Figure 6.3, which shows the PMF of a $\text{Bin}(10, 0.9)$ r.v. on the left and the PMF of 8 plus a $\text{Bin}(10, 0.1)$ r.v. on the right. Both of these distributions have mean, median, *and* mode equal to 9 and variance equal to 0.9, but they look drastically different. We say that the PMF on the left is *left-skewed* and the PMF on the right is *right-skewed*. In this chapter we'll learn that a standard measure of the asymmetry of a distribution is based on the third moment.

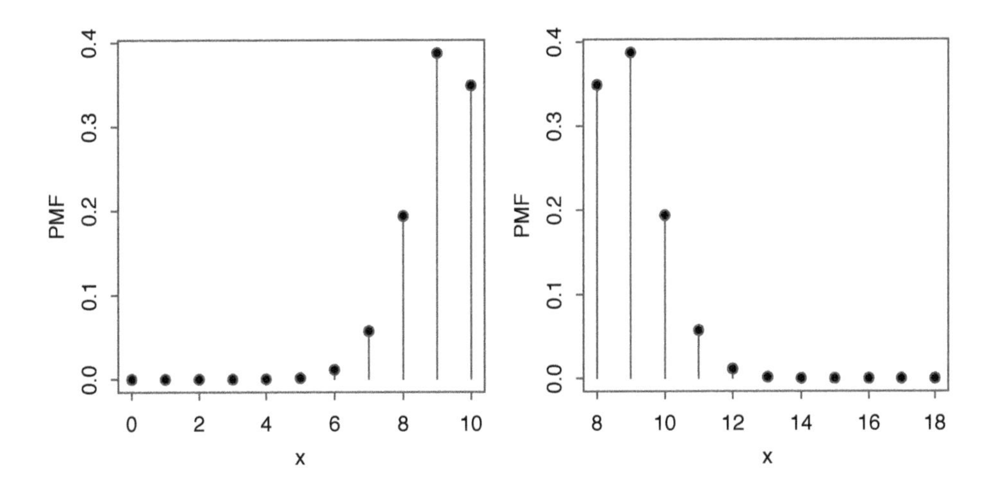

FIGURE 6.3
Left: $\text{Bin}(10, 0.9)$ is left-skewed. Right: $\text{Bin}(10, 0.1)$, shifted to the right by 8, is right-skewed but has the same mean, median, mode, and variance as $\text{Bin}(10, 0.9)$.

The previous two examples considered asymmetric distributions, but symmetric distributions with the same mean and variance can also look very different. The left plot in Figure 6.4 shows two symmetric PDFs with mean 0 and variance 1. The light curve is the $\mathcal{N}(0, 1)$ PDF and the dark curve is the t_3 PDF, scaling it to have variance 1 (we'll define the t_3 distribution in Chapter 10). The dark curve has a sharper peak and heavier tails than the light curve. The tail behavior is magnified in the right plot, where it is easy to see that the dark curve decays to 0 much more gradually, making outcomes far out in the tail much more likely than for the light curve. As we'll learn in this chapter, a standard measure of the heaviness of the tails of a distribution is based on the fourth moment.

In the next section, we will go into further detail of how to interpret moments, especially the first four moments.

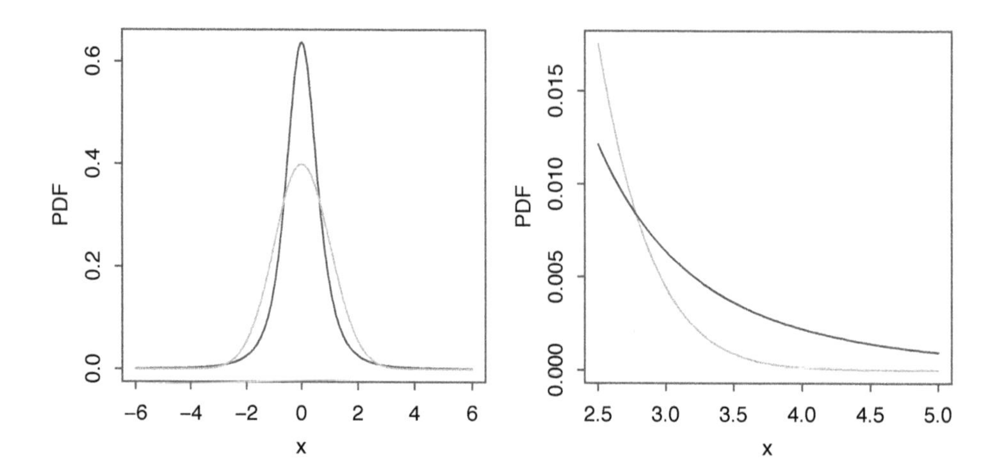

FIGURE 6.4
Left: $\mathcal{N}(0,1)$ PDF (light) and a scaled t_3 PDF (dark). Both have mean 0 and variance 1, but the latter has a sharper peak and heavier tails. Right: magnified version of right tail behavior.

6.2 Interpreting moments

Definition 6.2.1 (Kinds of moments). Let X be an r.v. with mean μ and variance σ^2. For any positive integer n, the nth *moment* of X is $E(X^n)$, the nth *central moment* is $E((X-\mu)^n)$, and the nth *standardized moment* is $E\left(\left(\frac{X-\mu}{\sigma}\right)^n\right)$. Throughout the previous sentence, "if it exists" is left implicit.

In particular, the mean is the first moment and the variance is the second central moment. The term *moment* is borrowed from physics. Let X be a discrete r.v. with distinct possible values $x_1, \ldots, x_n$, and imagine a pebble with mass $m_j = P(X = x_j)$ positioned at x_j on a number line, for each j (as illustrated in Figure 6.5). In physics,

$$E(X) = \sum_{j=1}^{n} m_j x_j$$

is called the *center of mass* of the system, and

$$\text{Var}(X) = \sum_{j=1}^{n} m_j (x_j - E(X))^2$$

is called the *moment of inertia* about the center of mass.

We'll now define skewness, a single-number summary of asymmetry which, as alluded to earlier, is based on the third moment. In fact, skewness is defined to be the third standardized moment.

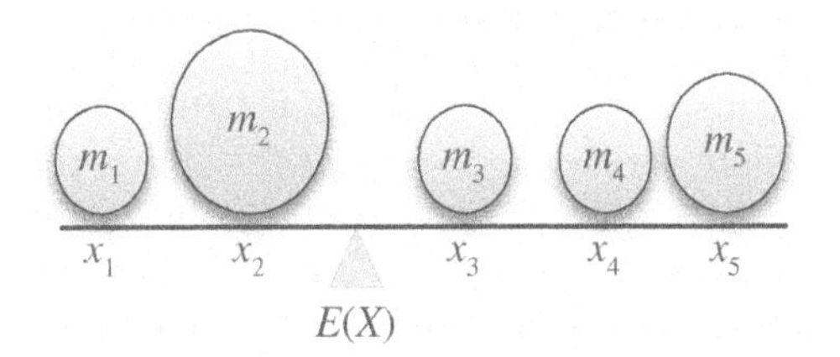

FIGURE 6.5
Physics interpretation of moments. The mean (first moment) of an r.v. corresponds to the center of mass of a collection of pebbles, and the variance (second central moment) corresponds to the moment of inertia about the center of mass.

Definition 6.2.2 (Skewness). The *skewness* of an r.v. X with mean μ and variance σ^2 is the third standardized moment of X:

$$\text{Skew}(X) = E\left(\frac{X-\mu}{\sigma}\right)^3.$$

By standardizing first, we make the definition of Skew(X) not depend on the location or the scale of X, which is reasonable since we already have μ and σ to provide information about location and scale. Also, standardizing first gives the nice property that the units in which X is measured (e.g., inches vs. meters) will not affect the skewness.

To understand how skewness measures asymmetry, we first need to discuss what it means for an r.v. to be symmetric.

Definition 6.2.3 (Symmetry of an r.v.). We say that an r.v. X has a *symmetric distribution about* μ if $X - \mu$ has the same distribution as $\mu - X$. We also say that X is symmetric or that the distribution of X is symmetric; these all have the same meaning.

The number μ in the above definition must be $E(X)$ if the mean exists, since

$$E(X) - \mu = E(X - \mu) = E(\mu - X) = \mu - E(X)$$

simplifies to $E(X) = \mu$. Because of this, it is common to say "X is symmetric" as shorthand for "X is symmetric about its mean" (if the mean exists). The number μ is also a median of the distribution, since if $X - \mu$ has the same distribution as $\mu - X$, then

$$P(X - \mu \leq 0) = P(\mu - X \leq 0),$$

so

$$P(X \leq \mu) = P(X \geq \mu),$$

which implies that

$$P(X \leq \mu) = 1 - P(X > \mu) \geq 1 - P(X \geq \mu) = 1 - P(X \leq \mu),$$

showing that $P(X \leq \mu) \geq 1/2$ and $P(X \geq \mu) \geq 1/2$.

☣ **6.2.4.** Sometimes people say "X is symmetric" to mean "X is symmetric about 0". Note that if X is symmetric about μ, then $X-\mu$ is symmetric about 0. Symmetry about 0 is especially convenient since then $-X$ and X have the same distribution, and the PDF of X (if X is continuous) is an even function, as shown below.

Intuitively, symmetry means that the PDF of X to the left of μ is the mirror image of the PDF of X to the right of μ (for X continuous, and the same holds for the PMF if X is discrete). For example, we have seen before that $X \sim \mathcal{N}(\mu, \sigma^2)$ is symmetric; in terms of the definition, this is because $X - \mu$ and $\mu - X$ are both $\mathcal{N}(0, \sigma^2)$. We have also seen from Corollary 3.3.8 that $X \sim \text{Bin}(n, p)$ is symmetric when $p = 1/2$.

We can also give an algebraic description of what the PDF of a symmetric continuous r.v. looks like.

Proposition 6.2.5 (Symmetry in terms of the PDF)**.** Let X be a continuous r.v. with PDF f. Then X is symmetric about μ if and only if $f(x) = f(2\mu - x)$ for all x.

Proof. Let F be the CDF of X. If symmetry holds, we have

$$F(x) = P(X - \mu \leq x - \mu) = P(\mu - X \leq x - \mu) = P(X \geq 2\mu - x) = 1 - F(2\mu - x).$$

Taking the derivative of both sides yields $f(x) = f(2\mu - x)$. Conversely, suppose that $f(x) = f(2\mu - x)$ holds for all x. Integrating both sides, we have

$$P(X - \mu \leq t) = P(X \leq \mu + t) = \int_{-\infty}^{\mu+t} f(x)dx = \int_{-\infty}^{\mu+t} f(2\mu - x)dx,$$

which, after the substitution $w = 2\mu - x$, becomes

$$\int_{\mu-t}^{\infty} f(w)dw = P(X \geq \mu - t) = P(\mu - X \leq t). \qquad \blacksquare$$

Odd central moments give some] information about symmetry.

Proposition 6.2.6 (Odd central moments of a symmetric distribution)**.** Let X be symmetric about its mean μ. Then for any odd number m, the mth central moment $E(X - \mu)^m$ is 0 if it exists.

Proof. Since $X - \mu$ has the same distribution as $\mu - X$, they have the same mth moment (if it exists):

$$E(X - \mu)^m = E(\mu - X)^m.$$

Let $Y = (X - \mu)^m$. Then $(\mu - X)^m = (-(X - \mu))^m = (-1)^m Y = -Y$, so the above equation just says $E(Y) = -E(Y)$. So $E(Y) = 0$. $\blacksquare$

This leads us to consider using an odd standardized moment as a measure of the skew of a distribution. The first standardized moment is always 0, so the third standardized moment is taken as the definition of skewness. Positive skewness is indicative of having a long right tail relative to the left tail, and negative skewness is indicative of the reverse. (The converse of the above proposition is false, though: there exist asymmetric distributions whose odd central moments are all 0.)

Why not use, say, the fifth standardized moment instead of the third? One reason is that the third standardized moment is usually easier to calculate. Another reason is that we may want to estimate skewness from a data set. It is usually easier to estimate lower moments than higher moments in a stable way since, for example, a large, noisy observation will have a very large, very noisy fifth power. Nevertheless, just as the mean isn't the only useful notion of average and the variance isn't the only useful notion of spread, the third standardized moment isn't the only useful notion of skew.

Another important descriptive feature of a distribution is how heavy its tails are. For a given variance, is the variability explained more by a few rare (extreme) events or by a moderate number of moderate deviations from the mean? This is an important consideration for risk management in finance: for many financial assets, the distribution of returns has a heavy left tail caused by rare but severe crisis events, and failure to account for these rare events can have disastrous consequences, as demonstrated by the 2008 financial crisis.

As with measuring skew, no single measure can perfectly capture the tail behavior, but there is a widely used summary based on the fourth standardized moment.

Definition 6.2.7 (Kurtosis). The *kurtosis* of an r.v. X with mean μ and variance σ^2 is a shifted version of the fourth standardized moment of X:

$$\text{Kurt}(X) = E\left(\frac{X-\mu}{\sigma}\right)^4 - 3.$$

☣ **6.2.8.** The reason for subtracting 3 is that this makes any Normal distribution have kurtosis 0 (as shown in Section 6.5). This provides a convenient basis for comparison. However, some sources define the kurtosis without the 3, in which case they call our version "excess kurtosis".

Figure 6.6 shows three named distributions and lists the skewness and kurtosis of each. The Expo(1) and Pois(4) distributions (left and middle) both have positive skewness and positive kurtosis, indicating that they are right-skewed and their tails are heavier than those of a Normal distribution. The Unif$(0, 1)$ distribution (right) has zero skewness and negative kurtosis: zero skewness because the distribution is symmetric about its mean, and negative kurtosis because it has no tails!

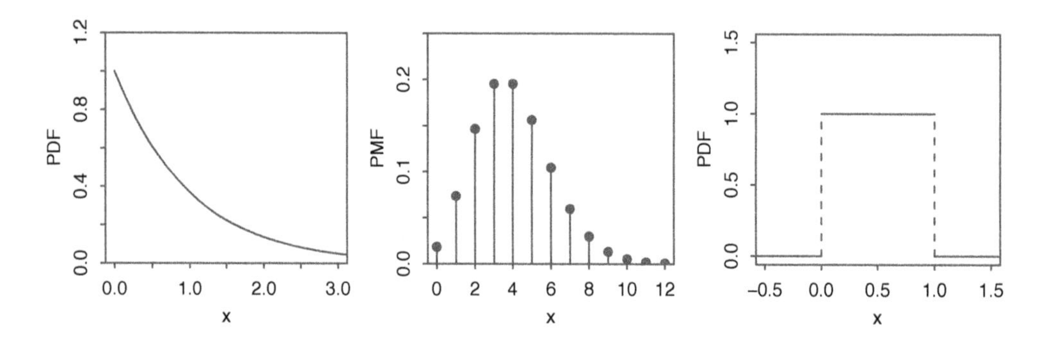

FIGURE 6.6
Skewness and kurtosis of some named distributions. Left: Expo(1) PDF, skewness = 2, kurtosis = 6. Middle: Pois(4) PMF, skewness = 0.5, kurtosis = 0.25. Right: Unif(0, 1) PDF, skewness = 0, kurtosis = −1.2.

6.3 Sample moments

In statistical inference, a central problem is how to use data to estimate unknown parameters of a distribution, or functions of unknown parameters. It is especially common to want to estimate the mean and variance of a distribution. If the data are i.i.d. random variables $X_1, \ldots, X_n$ where the mean $E(X_j)$ is unknown, then the most obvious way to estimate the mean is simply to average the X_j, taking the arithmetic mean.

For example, if the observed data are $3, 1, 1, 5$, then a simple, natural way to estimate the mean of the distribution that generated the data is to use $(3+1+1+5)/4 = 2.5$. This is called the *sample mean*. Similarly, if we want to estimate the second moment of the distribution that generated the data $3, 1, 1, 5$, then a simple, natural way is to use $(3^2 + 1^2 + 1^2 + 5^2)/4 = 9$. This is called the *sample second moment*. In general, sample moments are defined as follows.

Definition 6.3.1 (Sample moments). Let $X_1, \ldots, X_n$ be i.i.d. random variables. The kth *sample moment* is the r.v.

$$M_k = \frac{1}{n}\sum_{j=1}^{n} X_j^k.$$

The *sample mean* $\bar{X}_n$ is the first sample moment:

$$\bar{X}_n = \frac{1}{n}\sum_{j=1}^{n} X_j.$$

In contrast, the *population mean* or *true mean* is $E(X_j)$, the mean of the distribution from which the X_j were drawn.

The law of large numbers, which we prove in Chapter 10, shows that the kth sample moment of i.i.d. random variables $X_1, \dots, X_n$ converges to the kth moment $E(X_1^k)$ as $n \to \infty$. Also, the expected value of the kth sample moment is the kth moment. In statistical terms, we say that the kth sample moment is *unbiased* for estimating the kth moment. It is easy to check this by linearity:

$$E\left(\frac{1}{n}\sum_{j=1}^{n} X_j^k\right) = \frac{1}{n}\left(E(X_1^k) + \cdots + E(X_n^k)\right) = E(X_1^k).$$

The mean and variance of the sample mean are often needed, and have nice expressions that are often needed in statistics.

Theorem 6.3.2 (Mean and variance of sample mean)**.** Let $X_1, \dots, X_n$ be i.i.d. r.v.s with mean μ and variance σ^2. Then the sample mean $\bar{X}_n$ is unbiased for estimating μ. That is,

$$E(\bar{X}_n) = \mu.$$

The variance of $\bar{X}_n$ is given by

$$\text{Var}(\bar{X}_n) = \frac{\sigma^2}{n}.$$

Proof. We have $E(\bar{X}_n) = \mu$ since we showed above that the kth sample moment is unbiased for estimating the kth moment. For the variance, we will use the fact (shown in the next chapter) that the variance of the sum of *independent* r.v.s is the sum of the variances:

$$\text{Var}(\bar{X}_n) = \frac{1}{n^2}\text{Var}(X_1 + \cdots + X_n) = \frac{n}{n^2}\text{Var}(X_1) = \frac{\sigma^2}{n}. \qquad \blacksquare$$

For estimating the variance of the distribution of i.i.d. r.v.s $X_1, \dots, X_n$, a natural approach building on the above concepts is to mimic the formula $\text{Var}(X) = E(X^2) - (EX)^2$ by taking the second sample moment and subtracting the square of the sample mean. There are advantages to this method, but a more common method is as follows.

Definition 6.3.3 (Sample variance and sample standard deviation)**.** Let $X_1, \dots, X_n$ be i.i.d. random variables. The *sample variance* is the r.v.

$$S_n^2 = \frac{1}{n-1}\sum_{j=1}^{n}(X_j - \bar{X}_n)^2.$$

The *sample standard deviation* is the square root of the sample variance.

The idea of the above definition is to mimic the formula $\text{Var}(X) = E(X - E(X))^2$ by averaging the squared distances of the X_j from the sample mean, except with $n-1$ rather than n in the denominator. The motivation for the $n-1$ is that this makes the sample variance S_n^2 *unbiased* for estimating σ^2, i.e., it is correct on average. However, the sample standard deviation S_n is *not* unbiased for estimating σ; we will see in Chapter 10 which way the inequality goes. In any case, unbiasedness is only one of several criteria by which to judge an estimation procedure. For example, in some problems we can get a lower mean squared error in return for allowing a little bit of bias, and this tradeoff may be worthwhile.

Theorem 6.3.4 (Unbiasedness of sample variance). Let $X_1, \dots, X_n$ be i.i.d. r.v.s with mean μ and variance σ^2. Then the sample variance S_n^2 is unbiased for estimating σ^2, i.e.,

$$E(S_n^2) = \sigma^2.$$

Proof. The key to the proof is the handy identity

$$\sum_{j=1}^{n}(X_j - c)^2 = \sum_{j=1}^{n}(X_j - \bar{X}_n)^2 + n(\bar{X}_n - c)^2,$$

which holds for all c. To verify the identity, add and subtract $\bar{X}_n$ in the left-hand sum:

$$\begin{aligned}
\sum_{j=1}^{n}(X_j - c)^2 &= \sum_{j=1}^{n}\left((X_j - \bar{X}_n) + (\bar{X}_n - c)\right)^2 \\
&= \sum_{j=1}^{n}(X_j - \bar{X}_n)^2 + 2\sum_{j=1}^{n}(X_j - \bar{X}_n)(\bar{X}_n - c) + \sum_{j=1}^{n}(\bar{X}_n - c)^2 \\
&= \sum_{j=1}^{n}(X_j - \bar{X}_n)^2 + n(\bar{X}_n - c)^2.
\end{aligned}$$

For the last line, we used the fact that $\bar{X}_n - c$ does not depend on j and the fact that

$$\sum_{j=1}^{n}(X_j - \bar{X}_n) = \sum_{j=1}^{n}X_j - \sum_{j=1}^{n}\bar{X}_n = n\bar{X}_n - n\bar{X}_n = 0.$$

Now let us apply the identity, choosing $c = \mu$. Taking the expectation of both sides,

$$nE(X_1 - \mu)^2 = E\left(\sum_{j=1}^{n}(X_j - \bar{X}_n)^2\right) + nE(\bar{X}_n - \mu)^2.$$

By definition of variance, $E(X_1-\mu)^2 = \text{Var}(X_1) = \sigma^2$, and $E(\bar{X}_n-\mu)^2 = \text{Var}(\bar{X}_n) = \sigma^2/n$. Plugging these results in above and simplifying, we have $E(S_n^2) = \sigma^2$. ■

Similarly, we can define the *sample skewness* to be

$$\frac{\frac{1}{n}\sum_{j=1}^{n}(X_j - \bar{X}_n)^3}{S_n^3},$$

and the *sample kurtosis* to be

$$\frac{\frac{1}{n}\sum_{j=1}^{n}(X_j - \bar{X}_n)^4}{S_n^4} - 3.$$

Beyond the fourth moment, it rapidly gets harder to interpret moments graphically and harder to estimate them well from data if they are unknown. However, in the rest of this chapter we will see that it can still be useful to know *all* the moments of a distribution. We will also study a way of computing moments that is often easier than LOTUS. Both the usefulness and the computation of moments are closely connected to a blueprint called the *moment generating function*, to which we devote most of the rest of this chapter.

6.4 Moment generating functions

> A generating function is a clothesline on which we hang up a sequence of numbers for display. – Herbert Wilf [28]

Generating functions are a powerful tool in combinatorics and probability, bridging between sequences of numbers and the world of calculus. In probability, they are useful for studying both discrete and continuous distributions. The general idea behind a generating function is as follows: starting with a sequence of numbers, create a continuous function—the generating function—that encodes the sequence. We then have all the tools of calculus at our disposal for manipulating the generating function.

A moment generating function, as its name suggests, is a generating function that encodes the *moments* of a distribution. Here is the definition, followed by a few examples.

Definition 6.4.1 (Moment generating function). The *moment generating function* (MGF) of an r.v. X is $M(t) = E(e^{tX})$, as a function of t, if this is finite on some open interval $(-a, a)$ containing 0. Otherwise we say the MGF of X does not exist.

A natural question at this point is "What is the interpretation of t?" The answer is that t has no interpretation in particular; it's just a bookkeeping device that we introduce in order to be able to use calculus instead of working with a discrete sequence of moments.

Note that $M(0) = 1$ for any valid MGF M; whenever you compute an MGF, plug in 0 and see if you get 1, as a quick check!

Example 6.4.2 (Bernoulli MGF). For $X \sim \text{Bern}(p)$, e^{tX} takes on the value e^t with probability p and the value 1 with probability q, so $M(t) = E(e^{tX}) = pe^t + q$. Since this is finite for all values of t, the MGF is defined on the entire real line. □

Example 6.4.3 (Geometric MGF). For $X \sim \text{Geom}(p)$,

$$M(t) = E(e^{tX}) = \sum_{k=0}^{\infty} e^{tk} q^k p = p \sum_{k=0}^{\infty} (qe^t)^k = \frac{p}{1 - qe^t}$$

for $qe^t < 1$, i.e., for t in $(-\infty, \log(1/q))$, which is an open interval containing 0. □

Example 6.4.4 (Uniform MGF). Let $U \sim \text{Unif}(a, b)$. Then the MGF of U is

$$M(t) = E(e^{tU}) = \frac{1}{b-a} \int_a^b e^{tu} du = \frac{e^{tb} - e^{ta}}{t(b-a)}$$

for $t \neq 0$, and $M(0) = 1$. □

The next three theorems give three reasons why the MGF is important. First, the MGF encodes the moments of an r.v. Second, the MGF of an r.v. determines its distribution, like the CDF and PMF/PDF. Third, MGFs make it easy to find the distribution of a sum of independent r.v.s. Let's take these one by one.

Theorem 6.4.5 (Moments via derivatives of the MGF). Given the MGF of X, we can get the nth moment of X by evaluating the nth derivative of the MGF at 0: $E(X^n) = M^{(n)}(0)$.

Proof. This can be seen by noting that the Taylor expansion of $M(t)$ about 0 is

$$M(t) = \sum_{n=0}^{\infty} M^{(n)}(0) \frac{t^n}{n!},$$

while on the other hand, we also have

$$M(t) = E(e^{tX}) = E\left(\sum_{n=0}^{\infty} X^n \frac{t^n}{n!}\right).$$

We are allowed to interchange the expectation and the infinite sum because certain technical conditions are satisfied (this is where we invoke the condition that $E(e^{tX})$ is finite in an interval around 0), so

$$M(t) = \sum_{n=0}^{\infty} E(X^n) \frac{t^n}{n!}.$$

Matching the coefficients of the two expansions, we get $E(X^n) = M^{(n)}(0)$. ■

The above theorem is surprising in that for a continuous r.v. X, to compute moments would seemingly require doing integrals with LOTUS, but with the MGF it is possible to find moments by taking derivatives rather than doing integrals!

Theorem 6.4.6 (MGF determines the distribution). The MGF of a random variable determines its distribution: if two r.v.s have the same MGF, they must have the same distribution. In fact, if there is even a tiny interval $(-a, a)$ containing 0 on which the MGFs are equal, then the r.v.s must have the same distribution.

The above theorem is a difficult result in analysis, so we will not prove it here.

Theorem 6.4.7 (MGF of a sum of independent r.v.s)**.** If X and Y are independent, then the MGF of $X+Y$ is the product of the individual MGFs:

$$M_{X+Y}(t) = M_X(t)M_Y(t).$$

This is true because if X and Y are independent, then $E(e^{t(X+Y)}) = E(e^{tX})E(e^{tY})$ (this follows from results discussed in Chapter 7). Using this fact, we can get the MGFs of the Binomial and Negative Binomial, which are sums of independent Bernoullis and Geometrics, respectively.

Example 6.4.8 (Binomial MGF)**.** The MGF of a Bern(p) r.v. is $pe^t + q$, so the MGF of a Bin(n, p) r.v. is

$$M(t) = (pe^t + q)^n.$$

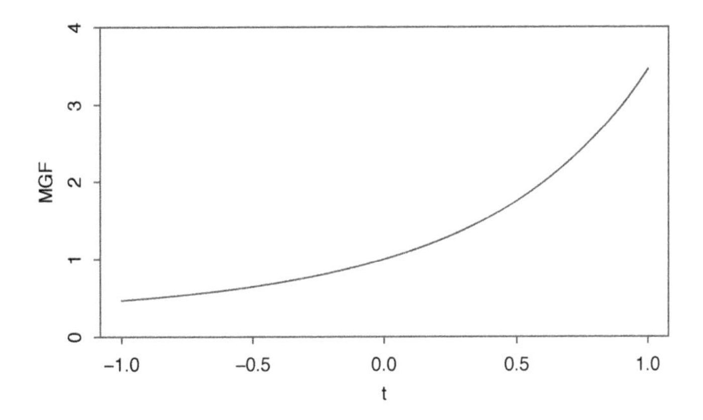

FIGURE 6.7
Bin(2, 1/2) MGF, $M(t) = \left(\frac{1}{2}e^t + \frac{1}{2}\right)^2$. The slope of the MGF at $t = 0$ is 1, so the mean of the distribution is 1. The concavity or second derivative of the MGF at $t = 0$ is 3/2, so the second moment of the distribution is 3/2.

Figure 6.7 plots the MGF of the Bin(2, 1/2) distribution, $M(t) = \left(\frac{1}{2}e^t + \frac{1}{2}\right)^2$, between $t = -1$ and $t = 1$. As with all MGFs, the value of the MGF is 1 at $t = 0$. The first and second moments of the distribution are the first and second derivatives of the MGF, evaluated at $t = 0$; these correspond to the slope and concavity of the plotted curve at $t = 0$. These two derivatives are 1 and 3/2, corresponding to the fact that the Bin(2, 1/2) distribution has mean 1 and variance $3/2 - 1^2 = 1/2$. □

Example 6.4.9 (Negative Binomial MGF)**.** We know the MGF of a Geom(p) r.v. is $\frac{p}{1-qe^t}$ for $qe^t < 1$, so the MGF of $X \sim \text{NBin}(r, p)$ is

$$M(t) = \left(\frac{p}{1-qe^t}\right)^r, \text{ for } qe^t < 1.$$ □

☣ **6.4.10.** Not all r.v.s have an MGF. Some r.v.s X don't even have $E(X)$ exist, or don't have $E(X^n)$ exist for some $n > 1$, in which case the MGF clearly will not exist. But even if all the moments of X exist, the MGF may not exist if the moments grow too quickly. Luckily, there is a way to fix this: inserting an imaginary number! The function $\psi(t) = E(e^{itX})$ with $i = \sqrt{-1}$ is called the *characteristic function* by statisticians and the *Fourier transform* by everyone else. It turns out that the characteristic function *always* exists. In this book we will focus on the MGF rather than the characteristic function, to avoid having to handle imaginary numbers.

As we saw in the previous chapter, location and scale transformations are a fundamental way to build a family of distributions from an initial distribution. For example, starting with $Z \sim \mathcal{N}(0,1)$, we can scale by σ and shift by μ to obtain $X = \mu + \sigma Z \sim \mathcal{N}(\mu, \sigma^2)$. In general, if we have an r.v. X with mean μ and standard deviation $\sigma > 0$, we can create the standardized version $(X - \mu)/\sigma$, and vice versa. Conveniently, it is easy to relate the MGFs of two r.v.s connected by such a transformation.

Proposition 6.4.11 (MGF of location-scale transformation). If X has MGF $M(t)$, then the MGF of $a + bX$ is

$$E(e^{t(a+bX)}) = e^{at}E(e^{btX}) = e^{at}M(bt).$$

For example, let's use this proposition to help obtain the MGFs of the Normal and Exponential distributions.

Example 6.4.12 (Normal MGF). The MGF of a standard Normal r.v. Z is

$$M_Z(t) = E(e^{tZ}) = \int_{-\infty}^{\infty} e^{tz}\frac{1}{\sqrt{2\pi}}e^{-z^2/2}dz.$$

After completing the square, we have

$$M_Z(t) = e^{t^2/2}\int_{-\infty}^{\infty} \frac{1}{\sqrt{2\pi}}e^{-(z-t)^2/2}dz = e^{t^2/2},$$

since the $\mathcal{N}(t,1)$ PDF integrates to 1. Thus, the MGF of $X = \mu + \sigma Z \sim \mathcal{N}(\mu, \sigma^2)$ is

$$M_X(t) = e^{\mu t}M_Z(\sigma t) = e^{\mu t}e^{(\sigma t)^2/2} = e^{\mu t + \frac{1}{2}\sigma^2 t^2}.$$ □

Example 6.4.13 (Exponential MGF). The MGF of $X \sim \text{Expo}(1)$ is

$$M(t) = E(e^{tX}) = \int_0^{\infty} e^{tx}e^{-x}dx = \int_0^{\infty} e^{-x(1-t)}dx = \frac{1}{1-t} \quad \text{for } t < 1.$$

So the MGF of $Y = X/\lambda \sim \text{Expo}(\lambda)$ is

$$M_Y(t) = M_X\left(\frac{t}{\lambda}\right) = \frac{\lambda}{\lambda - t} \quad \text{for } t < \lambda.$$ □

6.5 Generating moments with MGFs

We now give some examples of using an MGF to find moments. . Theorem 6.4.5 shows that we can get moments by differentiating the MGF and evaluating at 0, rather than doing a complicated sum or integral by LOTUS. Better yet, in some cases we can simultaneously find *all* the moments of a distribution via a Taylor expansion, rather than differentiating over and over again.

Example 6.5.1 (Exponential moments). In this example we will show how to use the Exponential MGF to get *all* the moments of the Exponential distribution simultaneously! Let $X \sim \text{Expo}(1)$. The MGF of X is $M(t) = 1/(1-t)$ for $t < 1$.

As shown in Theorem 6.4.5, we could obtain the moments by taking derivatives of the MGF and evaluating at 0. In this case, though, we recognize $1/(1-t)$ as a geometric series, valid in an interval around 0. For $|t| < 1$,

$$M(t) = \frac{1}{1-t} = \sum_{n=0}^{\infty} t^n = \sum_{n=0}^{\infty} n! \frac{t^n}{n!}.$$

On the other hand, we know that $E(X^n)$ is the coefficient of the term involving t^n in the Taylor expansion of $M(t)$:

$$M(t) = \sum_{n=0}^{\infty} E(X^n) \frac{t^n}{n!}.$$

Thus we can match coefficients to conclude that $E(X^n) = n!$ for all n. We not only did not have to do a LOTUS integral, but also we did not, for example, have to take 10 derivatives to get the 10th moment—we got the moments all at once.

To find the moments of $Y \sim \text{Expo}(\lambda)$, use a scale transformation: we can express $Y = X/\lambda$ where $X \sim \text{Expo}(1)$. Therefore $Y^n = X^n/\lambda^n$ and

$$E(Y^n) = \frac{n!}{\lambda^n}.$$

In particular, we have found the mean and variance of Y, making good on our promise from Chapter 5:

$$E(Y) = \frac{1}{\lambda},$$

$$\text{Var}(Y) = E(Y^2) - (EY)^2 = \frac{2}{\lambda^2} - \frac{1}{\lambda^2} = \frac{1}{\lambda^2}.$$

□

Example 6.5.2 (Standard Normal moments). In this example we will find all the moments of the standard Normal distribution. Let $Z \sim \mathcal{N}(0,1)$. We can use the same trick of matching the coefficients of the Taylor expansion.

$$M(t) = e^{t^2/2} = \sum_{n=0}^{\infty} \frac{(t^2/2)^n}{n!} = \sum_{n=0}^{\infty} \frac{t^{2n}}{2^n \cdot n!} = \sum_{n=0}^{\infty} \frac{(2n)!}{2^n \cdot n!} \frac{t^{2n}}{(2n)!}.$$

Therefore

$$E(Z^{2n}) = \frac{(2n)!}{2^n \cdot n!},$$

and the odd moments of Z are equal to 0, which must be true due to the symmetry of the standard Normal. From the story proof about partnerships in Example 1.5.4, we know that $\frac{(2n)!}{2^n \cdot n!}$ is the product of the odd numbers from 1 through $2n - 1$, so

$$E(Z^2) = 1, E(Z^4) = 1 \cdot 3, E(Z^6) = 1 \cdot 3 \cdot 5, \dots.$$

This result also shows that the kurtosis of a Normal r.v. is 0. For $X \sim \mathcal{N}(\mu, \sigma^2)$,

$$\text{Kurt}(X) = E\left(\frac{X-\mu}{\sigma}\right)^4 - 3 = E(Z^4) - 3 = 3 - 3 = 0. \qquad \square$$

Example 6.5.3 (Log-Normal moments). Now let's consider the Log-Normal distribution. We say that Y is *Log-Normal* with parameters μ and σ^2, denoted by $Y \sim \mathcal{LN}(\mu, \sigma^2)$, if $Y = e^X$ where $X \sim \mathcal{N}(\mu, \sigma^2)$.

☣ **6.5.4.** Log-Normal does not mean "log of a Normal", since a Normal can be negative. Rather, Log-Normal means "log *is* Normal". It is important to distinguish between the mean and variance of the Log-Normal and the mean and variance of the underlying Normal. Here we are defining μ and σ^2 to be the mean and variance of the underlying Normal, which is the most common convention.

Interestingly, the Log-Normal MGF does not exist, since $E(e^{tY})$ is infinite for all $t > 0$. Consider the case where $Y = e^Z$ for $Z \sim \mathcal{N}(0, 1)$; by LOTUS,

$$E(e^{tY}) = E(e^{te^Z}) = \int_{-\infty}^{\infty} e^{te^z} \frac{1}{\sqrt{2\pi}} e^{-z^2/2} dz = \int_{-\infty}^{\infty} \frac{1}{\sqrt{2\pi}} e^{te^z - z^2/2} dz.$$

For any $t > 0$, $te^z - z^2/2$ goes to infinity as z grows, so the above integral diverges. Since $E(e^{tY})$ is not finite on an open interval around 0, the MGF of Y does not exist. The same reasoning holds for a general Log-Normal distribution.

However, even though the Log-Normal MGF does not exist, we can still obtain all the moments of the Log-Normal, using the MGF of the *Normal*. For $Y = e^X$ with $X \sim \mathcal{N}(\mu, \sigma^2)$,

$$E(Y^n) = E(e^{nX}) = M_X(n) = e^{n\mu + \frac{1}{2}n^2\sigma^2}.$$

In other words, the nth moment of the Log-Normal is the MGF of the Normal evaluated at $t = n$. Letting

$$m = E(Y) = e^{\mu + \frac{1}{2}\sigma^2},$$

we have, after some algebra,

$$\text{Var}(Y) = E(Y^2) - m^2 = m^2(e^{\sigma^2} - 1).$$

All Log-Normal distributions are right-skewed. For example, Figure 6.2 shows a

Log-Normal PDF in dark, with mean 2 and variance 12. This is the distribution of e^X for $X \sim \mathcal{N}(0, 2\log 2)$, and it is clearly right-skewed. To quantify this, let us compute the skewness of the Log-Normal r.v. $Y = e^X$ for $X \sim \mathcal{N}(0, \sigma^2)$. Letting $m = E(Y) = e^{\frac{1}{2}\sigma^2}$, we have $E(Y^n) = m^{n^2}$ and $\text{Var}(Y) = m^2(m^2 - 1)$, and the third central moment is

$$\begin{aligned} E(Y-m)^3 &= E(Y^3 - 3mY^2 + 3m^2Y - m^3) \\ &= E(Y^3) - 3mE(Y^2) + 2m^3 \\ &= m^9 - 3m^5 + 2m^3. \end{aligned}$$

Thus, the skewness is

$$\text{Skew}(Y) = \frac{E(Y-m)^3}{\text{SD}^3(Y)} = \frac{m^9 - 3m^5 + 2m^3}{m^3(m^2-1)^{3/2}} = (m^2+2)\sqrt{m^2-1},$$

where in the last step we factored $m^6 - 3m^2 + 2 = (m^2+2)(m-1)^2(m+1)^2$. The skewness is positive since $m > 1$, and it increases very quickly as σ grows. □

In the next example, we introduce the Weibull distribution, which is one of the most widely used distributions in *survival analysis* (the study of the duration of time until an event occurs, e.g., modeling how long someone with a particular disease will live).

Example 6.5.5 (Weibull distribution)**.** As you may remember from the previous chapter, the Exponential distribution is memoryless, which makes it unrealistic for, e.g., modeling a human lifetime. Remarkably, simply raising an Exponential r.v. to a power dramatically improves the flexibility and applicability of the distribution.

Let $T = X^{1/\gamma}$, with $X \sim \text{Expo}(\lambda)$ and $\lambda, \gamma > 0$. The distribution of T is called the *Weibull* distribution, and we denote this by $T \sim \text{Wei}(\lambda, \gamma)$. This generalizes the Exponential, with the case $\gamma = 1$ reducing back to the Exponential.

Weibull distributions are widely used in biostatistics, epidemiology, and engineering; there is even an 800-page book devoted to this distribution: *The Weibull Distribution: A Handbook* by Horst Rinne [21].

The PDF of T is

$$f(t) = \gamma\lambda e^{-\lambda t^\gamma} t^{\gamma-1}$$

for $t > 0$, as can be shown by relating the CDF of T to the CDF of X, or by using transformation results from Chapter 8. The PDF looks somewhat complicated, but often when working with a Weibull random variable we can use the strategy of first transforming it back to an Exponential.

For simplicity and concreteness, let's look at a specific Weibull distribution, letting $\lambda = 1$ and $\gamma = 1/3$.

(a) Find $P(T > s + t | T > s)$ for $s, t > 0$. Does T have the memoryless property?

(b) Find the mean and variance of T, and the nth moment $E(T^n)$ for $n = 1, 2, \ldots$.

(c) Determine whether or not the MGF of T exists.

Solution:

(a) The CDF of T is

$$P(T \leq t) = P(X^3 \leq t) = P(X \leq t^{1/3}) = 1 - e^{-t^{1/3}},$$

for $t > 0$. So

$$P(T > s + t | T > s) = \frac{P(T > s + t)}{P(T > s)} = \frac{e^{-(s+t)^{1/3}}}{e^{-s^{1/3}}},$$

which is not the same as $P(T > t) = e^{-t^{1/3}}$. Thus, T does *not* have the memoryless property. Nor could it, since it is not Exponential.

(b) Example 6.5.1 shows that the moments of X are given by $E(X^n) = n!$. This allows us to find the moments of T without doing any additional work! Specifically,

$$E(T^n) = E(X^{3n}) = (3n)!.$$

The mean and variance of T are

$$E(T) = 3! = 6, \text{Var}(T) = 6! - 6^2 = 684.$$

(c) By LOTUS,

$$E(e^{tT}) = E(e^{tX^3}) = \int_0^\infty e^{tx^3 - x} dx.$$

This integral diverges for $t > 0$ since the tx^3 term dominates over the x; more precisely, we have $tx^3 - x > x$ for all x sufficiently large (specifically, for $x > \sqrt{2/t}$), so this integral diverges by comparison with the divergent integral $\int_0^\infty e^x dx$. So the MGF of T does not exist, even though all the moments of T do exist. □

☣ **6.5.6.** Several different parameterizations of the Weibull are commonly used, e.g., we are including a scale in the Exponential and then raising to a power, but it is also common to take an Expo(1) to a power and then rescale. So care is needed when reading various references: always check which convention is being used. Our parameterization here is widely used in medical statistics, and is convenient to work with since we are just raising an Expo(λ) r.v. to a power.

6.6 Sums of independent r.v.s via MGFs

Since the MGF of a sum of independent r.v.s is just the product of the individual MGFs, we now have a new strategy for finding the distribution of a sum of independent r.v.s: multiply the individual MGFs together and see if the product is recognizable as the MGF of a named distribution. The next two examples illustrate this strategy.

Example 6.6.1 (Sum of independent Poissons). Using MGFs, we can easily show that the sum of independent Poissons is Poisson. First let's find the MGF of $X \sim \text{Pois}(\lambda)$:

$$E(e^{tX}) = \sum_{k=0}^{\infty} e^{tk} \frac{e^{-\lambda}\lambda^k}{k!} = e^{-\lambda} \sum_{k=0}^{\infty} \frac{(\lambda e^t)^k}{k!} = e^{-\lambda} e^{\lambda e^t} = e^{\lambda(e^t-1)}.$$

Now let $Y \sim \text{Pois}(\mu)$ be independent of X. The MGF of $X+Y$ is

$$E(e^{tX})E(e^{tY}) = e^{\lambda(e^t-1)} e^{\mu(e^t-1)} = e^{(\lambda+\mu)(e^t-1)},$$

which is the $\text{Pois}(\lambda+\mu)$ MGF. Since the MGF determines the distribution, we have proven that $X+Y \sim \text{Pois}(\lambda+\mu)$. Contrast this with the proof from Chapter 4 (Theorem 4.8.1), which required using the law of total probability and summing over all possible values of X. The proof using MGFs is far less tedious. □

☣ **6.6.2.** It is important that X and Y be independent in the above example. To see why, consider an extreme form of dependence: $X = Y$. In that case, $X + Y = 2X$, which can't possibly be Poisson since its value is always an even number!

Example 6.6.3 (Sum of independent Normals). If we have $X_1 \sim \mathcal{N}(\mu_1, \sigma_1^2)$ and $X_2 \sim \mathcal{N}(\mu_2, \sigma_2^2)$ independently, then the MGF of $X_1 + X_2$ is

$$M_{X_1+X_2}(t) = M_{X_1}(t)M_{X_2}(t) = e^{\mu_1 t + \frac{1}{2}\sigma_1^2 t^2} \cdot e^{\mu_2 t + \frac{1}{2}\sigma_2^2 t^2} = e^{(\mu_1+\mu_2)t + \frac{1}{2}(\sigma_1^2+\sigma_2^2)t^2},$$

which is the $\mathcal{N}(\mu_1+\mu_2, \sigma_1^2+\sigma_2^2)$ MGF. Again, because the MGF determines the distribution, it must be the case that $X_1+X_2 \sim \mathcal{N}(\mu_1+\mu_2, \sigma_1^2+\sigma_2^2)$. Thus the sum of independent Normals is Normal, and the means and variances simply add. □

Example 6.6.4 (Sum is Normal). A converse to the previous example also holds: if X_1 and X_2 are independent and X_1+X_2 is Normal, then X_1 and X_2 must be Normal! This is known as *Cramér's theorem*. Proving this in full generality is difficult, but it becomes much easier if X_1 and X_2 are i.i.d. with MGF $M(t)$. Without loss of generality, we can assume $X_1 + X_2 \sim \mathcal{N}(0, 1)$, and then its MGF is

$$e^{t^2/2} = E(e^{t(X_1+X_2)}) = E(e^{tX_1})E(e^{tX_2}) = (M(t))^2,$$

so $M(t) = e^{t^2/4}$, which is the $\mathcal{N}(0, 1/2)$ MGF. Thus, $X_1, X_2 \sim \mathcal{N}(0, 1/2)$. □

In Chapter 8 we'll discuss a more general technique for finding the distribution of a sum of r.v.s, which applies when the individual MGFs don't exist, or when the product of the individual MGFs is *not* recognizable and we would like to get the PMF/PDF instead.

6.7 *Probability generating functions

In this section we discuss *probability generating functions*, which are similar to MGFs but are guaranteed to exist for nonnegative integer-valued r.v.s. First we'll use PGFs to conquer a seemingly intractable counting problem. Then we'll prove that the PGF of a nonnegative integer-valued r.v. determines its distribution, which we omitted in the more general MGF setting.

Definition 6.7.1 (Probability generating function). The *probability generating function* (PGF) of a nonnegative integer-valued r.v. X with PMF $p_k = P(X = k)$ is the generating function of the PMF. By LOTUS, this is

$$E(t^X) = \sum_{k=0}^{\infty} p_k t^k.$$

The PGF converges to a value in $[-1, 1]$ for all t in $[-1, 1]$ since $\sum_{k=0}^{\infty} p_k = 1$ and $|p_k t^k| \leq p_k$ for $|t| \leq 1$.

The MGF is closely related to the PGF, when both exist: for $t > 0$,

$$E(t^X) = E(e^{X \log t})$$

is the MGF evaluated at $\log t$.

Example 6.7.2 (Generating dice probabilities). Frederick Mosteller, the founder of the Harvard Statistics Department, once recounted the following life-changing moment:

> A key moment in my life occurred in one of those classes during my sophomore year. We had the question: When three dice are rolled what is the chance that the sum of the faces will be 10? The students in this course were very good, but we all got the answer largely by counting on our fingers. When we came to class, I said to the teacher, "That's all very well—we got the answer—but if we had been asked about six dice and the probability of getting 18, we would still be home counting. How do you do problems like that?" He said, "I don't know, but I know a man who probably does and I'll ask him."
>
> One day I was in the library and Professor Edwin G. Olds of the Mathematics Department came in. He shouted at me, "I hear you're interested in the three dice problem." He had a huge voice, and you know how libraries are. I was embarrassed. "Well, come and see me," he said, "and I'll show you about it." "Sure," I said. But I was saying to myself, "I'll never go." Then he said, "What are you doing?" I showed him. "That's nothing important," he said. "Let's go now."
>
> So we went to his office, and he showed me a generating function. It was the most marvelous thing I had ever seen in mathematics. It used mathematics that, up to that time, in my heart of hearts, I had thought was something that mathematicians just did to create homework problems for innocent students in high school and college. I don't know where I had got ideas like that about various parts of mathematics. Anyway, I was stunned when I saw how Olds used this mathematics that I hadn't believed in. He used it in such an unusually outrageous way. It was a total retranslation of the meaning of the numbers. [1]

Let X be the total from rolling 6 fair dice, and let $X_1, \ldots, X_6$ be the individual rolls. What is $P(X = 18)$? It turns out that there are 3431 ways to obtain a sum of 18, so the probability is $3431/6^6 \approx 0.0735$. Listing out all possibilities is extremely tedious, and the tedium would be compounded with the worry of having somehow missed a case. And what if we laboriously listed out all 3431 cases, and then were asked to find $P(X = 19)$?

The PGF of X lets us count the cases in a systematic way. The PGF of X_1 is

$$E(t^{X_1}) = \frac{1}{6}(t + t^2 + \cdots + t^6).$$

Since the X_j are i.i.d., the PGF of X is

$$E\left(t^X\right) = E\left(t^{X_1} \cdots t^{X_6}\right) = E\left(t^{X_1}\right) \cdots E\left(t^{X_6}\right) = \frac{t^6}{6^6}(1 + t + \cdots + t^5)^6.$$

By definition, the coefficient of t^{18} in the PGF is $P(X = 18)$. So the number of ways to get a sum of 18 is the coefficient of t^{18} in $t^6(1 + t + \cdots + t^5)^6$, which is the coefficient of t^{12} in $(1 + t + \cdots + t^5)^6$. Multiplying this out by hand is tedious, but it is vastly easier than listing out 3431 cases, and it can also be done easily on a computer without having to write a special program.

Better yet, we can use the fact that $1 + t + \cdots + t^5$ is a geometric series to write

$$(1 + t + \cdots + t^5)^6 = \frac{(1 - t^6)^6}{(1 - t)^6}.$$

(Assume that $|t| < 1$, which we can do since, as with the MGF, we just need to know how the PGF behaves in an open interval containing 0.) The above equation is just algebra since we have the bookkeeping device t, but would have been hard to fathom if everything were still in sequence notation. By the binomial theorem, the numerator is

$$(1 - t^6)^6 = \sum_{j=0}^{6} \binom{6}{j} (-1)^j t^{6j}.$$

For the denominator, write

$$\frac{1}{(1 - t)^6} = (1 + t + t^2 + \cdots)^6 = \sum_{k=0}^{\infty} a_k t^k.$$

Here a_k is the number of ways to choose one term from each of the six $(1+t+t^2+\ldots)$ factors, such that the degrees add up to k. For example, for $k = 20$ one possibility is to choose the $t^3, 1, t^2, t^{10}, 1, t^5$ terms, respectively, since these choices contribute one t^{20} term when the product is expanded out. So a_k is the number of solutions to $y_1 + y_2 + \cdots + y_6 = k$, with the y_j nonnegative integers. We saw how to count this number of solutions in Chapter 1: a_k is the Bose-Einstein value $\binom{6+k-1}{k} = \binom{k+5}{5}$. So

$$\frac{1}{(1 - t)^6} = \sum_{k=0}^{\infty} \binom{k + 5}{5} t^k.$$

For $0 < t < 1$, another way to see why this equation holds is to write it as

$$\sum_{k=0}^{\infty} \binom{k+5}{5} (1-t)^6 t^k = 1,$$

which we already knew to be true since the NBin$(6, 1-t)$ PMF must sum to 1. (The identity for $(1-t)^{-6}$ is an example of a generalization of the binomial theorem to allow for negative integer powers; this helps explain why the Negative Binomial is called that despite being neither negative nor Binomial!)

Putting together the above results, we just need the coefficient of t^{12} in

$$\left(\sum_{j=0}^{2} \binom{6}{j} (-1)^j t^{6j} \right) \left(\sum_{k=0}^{12} \binom{k+5}{5} t^k \right),$$

where we summed only up to $j = 2$ and $k = 12$ in the two factors since any further terms will not contribute to the coefficient of t^{12}. This lets us reduce the 3431 cases down to just *three* cases: (j, k) is $(0, 12), (1, 6)$, or $(2, 0)$. The coefficient of t^{12} is

$$\binom{17}{5} - 6\binom{11}{5} + \binom{6}{2} = 3431,$$

since, for example, when $j = 1$ and $k = 6$ we get the term

$$-\binom{6}{1} t^6 \cdot \binom{6+5}{5} t^6 = -6\binom{11}{5} t^{12}.$$

Thus,

$$P(X = 18) = \frac{3431}{6^6}.$$ □

Since the PGF is just a handy bookkeeping device for the PMF, it fully determines the distribution (for any nonnegative integer-valued r.v.). The theorem below shows how to pluck the PMF values down from the "clothesline" of the PGF.

Theorem 6.7.3. Let X and Y be nonnegative integer-valued r.v.s, with PGFs g_X and g_Y respectively. Suppose that $g_X(t) = g_Y(t)$ for all t in $(-a, a)$, where $0 < a < 1$. Then X and Y have the same distribution, and their PMF can be recovered by taking derivatives of g_X:

$$P(X = k) = P(Y = k) = \frac{g_X^{(k)}(0)}{k!}.$$

Proof. Write

$$g_X(t) = \sum_{k=0}^{\infty} p_k t^k.$$

Then $g_X(0) = p_0$, so $P(X = 0)$ has been recovered—from knowing the function g_X, we can extract the value of $P(X = 0)$. The derivative is

$$g'_X(t) = \sum_{k=1}^{\infty} kp_k t^{k-1},$$

so $g'_X(0) = p_1$ (swapping the derivative and the infinite sum is justified by results in real analysis). Then $P(X = 1)$ has been recovered. Continuing in this way, we can recover the entire PMF by taking derivatives. ■

Example 6.7.4. Let $X \sim \text{Bin}(n, p)$. The PGF of a Bern(p) r.v. is $pt + q$ (with $q = 1 - p$), so the PGF of X is $g(t) = (pt + q)^n$. The above theorem says that *any* r.v. with this PGF must in fact be Binomial. Furthermore, we can recover the PMF by computing

$$g(0) = q^n, \, g'(0) = npq^{n-1}, \, g''(0)/2! = \binom{n}{2} p^2 q^{n-2}, \ldots.$$

We can avoid having to take derivatives by using the binomial theorem to write

$$g(t) = (pt + q)^n = \sum_{k=0}^{n} \binom{n}{k} p^k q^{n-k} t^k,$$

from which we can directly read off the Binomial PMF.

While we're working with the Binomial PGF, let's see how it can be used to get the moments of a Binomial. Letting $p_k = P(X = k)$, we have

$$g'(t) = np(pt + q)^{n-1} = \sum_{k=1}^{n} kp_k t^{k-1},$$

so

$$g'(1) = np = \sum_{k=1}^{n} kp_k = E(X).$$

Taking the derivative again,

$$g''(t) = n(n-1)p^2(pt + q)^{n-2} = \sum_{k=2}^{n} k(k-1)p_k t^{k-2},$$

so

$$E(X(X-1)) = g''(1) = n(n-1)p^2.$$

Rearranging these results gives another proof that $\text{Var}(X) = npq$. Continuing in this way, we have computed what are called the *factorial moments* of the Binomial:

$$E(X(X-1)\ldots(X-k+1)) = k!\binom{n}{k}p^k.$$

Dividing by $k!$ on both sides, this implies $E\binom{X}{k} = \binom{n}{k}p^k$, which can also be seen with a story proof: $\binom{X}{k}$ is the number of ways to choose k out of the X successful Bernoulli trials, which is the number of ways to choose k out of the n original trials such that all k are successes. Creating an indicator r.v. for each of the $\binom{n}{k}$ subsets of size k and using linearity, the result follows. □

6.8 Recap

A useful way to study a distribution is via its *moments*. The first 4 moments are widely used as a basis for quantitatively describing what the distribution looks like, though many other descriptions are also possible. In particular, the first moment is the mean, the second central moment is the variance, the third standardized moment measures skew (asymmetry), and the fourth standardized moment minus 3 is a measure of how heavy the tails are.

Moments are useful for far more than studying the location and shape of a distribution, especially when the moment generating function (MGF) exists (which is stronger than saying that all the moments exist). The MGF of an r.v. X is the function M defined by

$$M(t) = E(e^{tX}),$$

if this is finite for all t in some open interval containing 0. If the MGF exists, then

$$M(0) = 1,\ M'(0) = E(X),\ M''(0) = E(X^2),\ M'''(0) = E(X^3), \dots.$$

MGFs are useful for three main reasons: for computing moments (as an alternative to LOTUS), for studying sums of independent r.v.s, and since they fully determine the distribution and thus serve as an additional blueprint for a distribution.

We also introduced the Log-Normal and Weibull distributions in this chapter, both of which are widely used in practice. The Log-Normal and Weibull are connected to the Normal and Exponential, respectively, via simple transformations. The log of a Log-Normal r.v. is Normal, and raising a $\text{Wei}(\lambda, \gamma)$ r.v. to the power γ yields an $\text{Expo}(\lambda)$ r.v. Often the best way to study a Log-Normal is to transform it back to Normal, and likewise for Weibull and Exponential. Chapter 8 goes into much more detail about how to work with transformations.

Figure 6.8 augments our map of the connections between fundamental objects in probability. If the MGF of X exists, then the sequence $E(X), E(X^2), E(X^3), \dots$ of moments provides enough information (at least in principle) to determine the distribution of X.

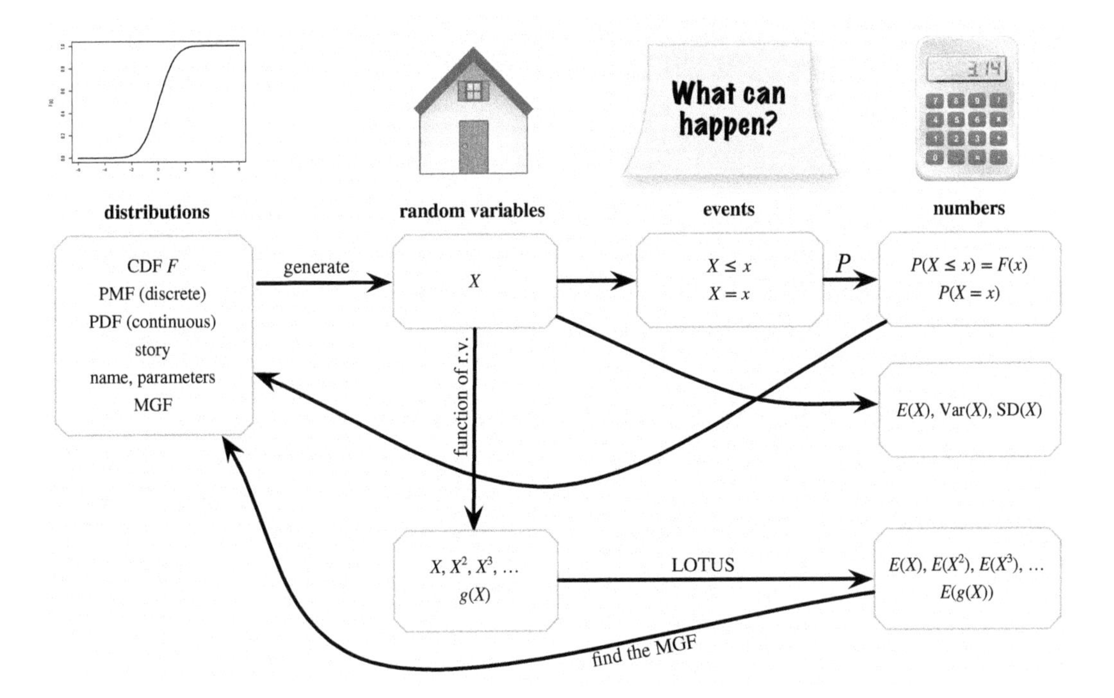

FIGURE 6.8
For an r.v. X, we can study its moments $E(X), E(X^2), \ldots$. These can be computed using LOTUS or by the MGF (if it exists). If the MGF exists, it determines the distribution, taking us full circle and adding to our list of blueprints.

6.9 R

Functions

The MGF of an r.v. is a *function*. As an example of defining and working with functions in R, let's use the $\mathcal{N}(0,1)$ MGF, which is given by $M(t) = e^{t^2/2}$. The code

```
M <- function(t) {exp(t^2/2)}
```

defines M to be this function. The `function(t)` says that we're defining a function of one variable t (called the *argument* of the function). Then, for example, `M(0)` evaluates the function at 0, `M(1:10)` evaluates the function at $1, 2, \ldots, 10$, and `curve(M,from=-3,to=3)` plots the graph of M from -3 to 3. Writing

```
M <- function(x) {exp(x^2/2)}
```

would define the same function M, except that now the argument is named x. Giving the arguments names is helpful for functions of more than one variable, since R then saves us from having to remember the order in which to write the arguments, and allows us to assign default values. For example, the $\mathcal{N}(\mu, \sigma^2)$ MGF is given by $g(t) = \exp(\mu t + \sigma^2 t^2/2)$, which depends on t, μ, and σ. We can define this in R by

```
g <- function(t,mean=0,sd=1) {exp(mean*t + sd^2*t^2/2)}
```

What is `g(1,2,3)`? It's the $\mathcal{N}(2, 3^2)$ MGF evaluated at 1, but it may be hard to remember which argument is which, especially when working with many functions with many arguments over the course of many months. So we can also write `g(t=1,mean=2,sd=3)` or `g(mean=2,sd=3,t=1)` or any of the other 4 permutations to mean the same thing.

Also, when defining g we specified *default values* of 0 for the mean and 1 for the standard deviation, so if we want the $\mathcal{N}(0, 5^2)$ MGF evaluated at 3, we can use `g(t=3,sd=5)` as shorthand. It would be bad here to write `g(3,5)`, since that is ambiguous about which argument is omitted; in fact, R interprets this as `g(t=3,mean=5)`.

Moments

LOTUS makes it easy to *write down* any moment of a continuous r.v. as an integral, and then R can help us *do* the integral numerically, using the `integrate` command. For example, let's approximate the 6th moment of a $\mathcal{N}(0, 1)$ r.v. The code

```
g <- function(x) x^6*dnorm(x)
integrate(g, lower = -Inf, upper = Inf)
```

asks R to compute $\int_{-\infty}^{\infty} g(x)dx$, where $g(x) = x^6\varphi(x)$ with φ the $\mathcal{N}(0, 1)$ PDF. When we ran this, R reported 15 (the correct answer, as we know from this chapter!) and that the absolute error was less than 7.9×10^{-5}. Similarly, to check that the 2nd moment (and variance) of a Unif$(-1, 1)$ r.v. is 1/3, we can use

```
h <- function(x) x^2*dunif(x,-1,1)
integrate(h, lower = -1, upper = 1)
```

☣ **6.9.1.** Numerical integration runs into difficulties for some functions; as usual, checking answers in multiple ways is a good idea. Using `upper = Inf` is preferred to using a large number as the upper limit when integrating up to ∞ (and likewise for a lower limit of $-\infty$). For example, on many systems `integrate(dnorm,0,10^6)` reports 0 while `integrate(dnorm,0,Inf)` reports the correct answer, 0.5.

For moments of a discrete r.v., we can use LOTUS and the `sum` command. For example, to find the 2nd moment of $X \sim$ Pois(7), we can use

```
g <- function(k) k^2*dpois(k,7)
sum(g(0:100))
```

Here we summed up to 100 since it's clear after getting a sense of the terms that the total contribution of all the terms after $k = 100$ is negligible (choosing an upper limit in this way is in contrast to how to use the **integrate** command in the continuous case). The result is extremely close to 56, which is comforting since $E(X^2) = \text{Var}(X) + (EX)^2 = 7 + 49 = 56$.

A sample moment can be found in one line in R. If **x** is a vector of data, then `mean(x)` gives its sample mean and, more generally, `mean(x^n)` gives the nth sample mean for any positive integer n. For example,

```
x <- rnorm(100)
mean(x^6)
```

gives the 6th sample moment of 100 i.i.d. $\mathcal{N}(0, 1)$ r.v.s. How close is it to the true 6th moment? How close are other sample moments to the corresponding true moments?

The sample variance can also be found in one line in R. If **x** is a vector of data, then `var(x)` gives its sample variance. This returns NA (not available) if **x** has length 1, since the $n - 1$ in the denominator is 0 in this case. It makes sense not to return a numerical value in this case, not only because of the definition but also because it would be insane to try to estimate the variability of a population if we only have one observation!

For a simple demonstration of using the sample mean and sample variance to estimate the true mean and true variance of a distribution, we generate 1000 times from a $\mathcal{N}(0, 1)$ distribution and store the values in **z**. We then compute the sample mean and sample variance with **mean** and **var**.

```
z <- rnorm(1000)
mean(z)
var(z)
```

We find that `mean(z)` is close to 0 and `var(z)` is close to 1. You can try this out for a $\mathcal{N}(\mu, \sigma^2)$ distribution (or other distribution) of your choosing; just remember that **rnorm** takes σ and not σ^2 as an input!

The sample standard deviation of **x** can be found using `sd(x)`. This gives the same result as `sqrt(var(x))`.

R does not come with built-in functions for sample skewness or sample kurtosis, but we can define our own functions as follows.

```
skew <- function(x) {
    centralmoment <- mean((x-mean(x))^3)
    centralmoment/(sd(x)^3)
}

kurt <- function(x) {
    centralmoment <- mean((x-mean(x))^4)
    centralmoment/(sd(x)^4) - 3
}
```

Medians and modes

To find the median of a continuous r.v. with CDF F, we need to solve the equation $F(x) = 1/2$ for x, which is equivalent to finding the root (zero) of the function g given by $g(x) = F(x) - 1/2$. This can be done using `uniroot` in R. For example, let's find the median of the Expo(1) distribution. The code

```
g <- function(x) pexp(x) - 1/2
uniroot(g,lower=0,upper=1)
```

asks R to find a root of the desired function between 0 and 1. This returns an answer very close to the true answer of $\log(2) \approx 0.693$. Of course, in this case we can solve $1 - e^{-x} = 1/2$ directly without having to use numerical methods.

☣ **6.9.2.** The `uniroot` command is useful but it only attempts to find *one* root (as the name suggests), and there is no guarantee that it will find a root.

An easier way to find the median of the Expo(1) in R is to use `qexp(1/2)`. The function `qexp` is the quantile function of the Expo(1) distribution, which means that `qexp(p)` is the value of x such that $P(X \leq x) = p$ for $X \sim$ Expo(1).

For finding the mode of a continuous distribution, we can use the `optimize` function in R. For example, let's find the mode of the Gamma$(6, 1)$ distribution, which is an important distribution that we will introduce in the next chapter. Its PDF is proportional to $x^5 e^{-x}$. Using calculus, we can find that the mode is at $x = 5$. Using R, we can find that the mode is very close to $x = 5$ as follows.

```
h <- function(x) x^5*exp(-x)
optimize(h,lower=0,upper=20,maximum=TRUE)
```

If we had wanted to minimize instead of maximize, we could have put `maximum=FALSE`.

Next, let's do a discrete example of median and mode. An interesting fact about the Bin(n, p) distribution is that if the mean np is an integer, then the median and mode are also np (even if the distribution is very skewed). To check this fact about the median for the Bin$(50, 0.2)$ distribution, we can use the following code.

```
n <- 50; p <- 0.2
which.max(pbinom(0:n,n,p)>=0.5)
```

The `which.max` function finds the location of the maximum of a vector, giving the index of the *first* occurrence of a maximum. Since TRUE is encoded as 1 and FALSE is encoded as 0, the first maximum in `pbinom(0:n,n,p)>=0.5` is at the first value for which the CDF is at least 0.5. The output of the above code is 11, but we must be careful to avoid an off-by-one error: the index 11 corresponds to the median being 10, since we started evaluating the CDF at 0. Similarly, `which.max(dbinom(0:n,n,p))` returns 11, showing that the mode is at 10.

The *sample median* of a vector `x` of data can be found using `median(x)`. But `mode(x)` does *not* give the sample mode of `x` (rather, it gives information about what *type* of object `x` is). To find the sample mode (or sample modes, in case there are ties), we can use the following function.

```
datamode <- function(x) {
    t <- table(x)
    m <- max(t)
    as.numeric(names(t[t==m]))
}
```

Log-Normal and Weibull distributions

Analogous to the functions `dnorm`, `pnorm`, and `rnorm` for the Normal distribution, the functions `dlnorm`, `plnorm`, and `rlnorm` give the Log-Normal PDF, the Log-Normal CDF, and random generation of Log-Normal r.v.s, respectively. The parameters used for these functions for the Log-Normal are the mean and standard deviation of the underlying Normal.

For example, `dlnorm(x,1,2)` gives the PDF of the $\mathcal{LN}(1,4)$ distribution (*not* $\mathcal{LN}(1,2)$, nor a Log-Normal whose mean is 1). Because of the relationship between Normal and Log-Normal, `rlnorm(n,mu,sigma)` is equivalent to `exp(rnorm(n,mu,sigma))`.

For the Weibull, we can obtain the PDF, the CDF, and random generation with the functions `dweibull`, `pweibull`, and `rweibull`, respectively. The parametrization in R is different from the one we are using, but it is easy to convert between them: for the $\text{Wei}(\lambda, \gamma)$ distribution, let $a = \gamma$ and $b = \lambda^{-1/\gamma}$.

Then `dweibull(x,a,b)` gives the $\text{Wei}(\lambda, \gamma)$ PDF, `pweibull(x,a,b)` gives the CDF, and `rweibull(n,a,b)` generates n i.i.d. draws from the distribution. Because of the relationship between Exponential and Weibull, another way to generate $\text{Wei}(\lambda, \gamma)$ r.v.s is to generate $\text{Expo}(\lambda)$ r.v.s and then raise each of them to the $1/\gamma$ power.

Dice simulation

In the starred Section 6.7, we showed that in rolling 6 fair dice, the probability of a total of 18 is $3431/6^6 \approx 0.07354$. But the proof was complicated. If we only need an approximate answer, simulation is a much easier approach. And we already know how to do it! Here is the code for a million repetitions:

```
r <- replicate(10^6,sum(sample(6,6,replace=TRUE)))
sum(r==18)/10^6
```

In our simulation this yielded 0.07346, which is very close to 0.07354.

6.10 Exercises

Exercises marked with Ⓢ have detailed solutions at http://stat110.net.

Means, medians, modes, and moments

1. Let $U \sim \text{Unif}(a, b)$. Find the median and mode of U.

2. Let $X \sim \text{Expo}(\lambda)$. Find the median and mode of X.

3. Let X have the *Pareto distribution* with parameter $a > 0$; this means that X has PDF $f(x) = a/x^{a+1}$ for $x \geq 1$ (and 0 otherwise). Find the median and mode of X.

4. Let $X \sim \text{Bin}(n, p)$.

   (a) For $n = 5$, $p = 1/3$, find all medians and all modes of X. How do they compare to the mean?

   (b) For $n = 6$, $p = 1/3$, find all medians and all modes of X. How do they compare to the mean?

5. Let X be Discrete Uniform on $1, 2, \ldots, n$. Find all medians and all modes of X (your answer can depend on whether n is even or odd).

6. Suppose that we have data giving the amount of rainfall in a city each day in a certain year. We want useful, informative summaries of how rainy the city was that year. On the majority of days in that year, it did not rain at all in the city. Discuss and compare the following six summaries: the mean, median, and mode of the rainfall on a randomly chosen day from that year, and the mean, median, and mode of the rainfall on a randomly chosen rainy day from that year (where by "rainy day" we mean that it did rain that day in the city).

7. Let a and b be positive constants. The *Beta distribution* with parameters a and b, which we introduce in detail in Chapter 8, has PDF proportional to $x^{a-1}(1-x)^{b-1}$ for $0 < x < 1$ (and the PDF is 0 outside of this range). Show that for $a > 1, b > 1$, the mode of the distribution is $(a-1)/(a+b-2)$.

   Hint: Take the log of the PDF first (note that this does not affect where the maximum is achieved).

8. Find the median of the Beta distribution with parameters $a = 3$ and $b = 1$ (see the previous problem for information about the Beta distribution).

9. Let Y be Log-Normal with parameters μ and σ^2. So $Y = e^X$ with $X \sim \mathcal{N}(\mu, \sigma^2)$. Three students are discussing the median and the mode of Y. Evaluate and explain whether or not each of the following arguments is correct.

   (a) Student A: The median of Y is e^μ because the median of X is μ and the exponential function is continuous and strictly increasing, so the event $Y \leq e^\mu$ is the same as the event $X \leq \mu$.

   (b) Student B: The mode of Y is e^μ because the mode of X is μ, which corresponds to e^μ for Y since $Y = e^X$.

   (c) Student C: The mode of Y is μ because the mode of X is μ and the exponential function is continuous and strictly increasing, so maximizing the PDF of X is equivalent to maximizing the PDF of $Y = e^X$.

10. A distribution is called *symmetric unimodal* if it is symmetric (about some point) and has a unique mode. For example, any Normal distribution is symmetric unimodal. Let X have a continuous symmetric unimodal distribution for which the mean exists. Show that the mean, median, and mode of X are all equal.

11. Let $X_1, \dots, X_n$ be i.i.d. r.v.s with mean μ, variance σ^2, and skewness γ.

(a) Standardize the X_j by letting

$$Z_j = \frac{X_j - \mu}{\sigma}.$$

Let $\bar{X}_n$ and $\bar{Z}_n$ be the sample means of the X_j and Z_j, respectively. Show that Z_j has the same skewness as X_j, and $\bar{Z}_n$ has the same skewness as $\bar{X}_n$.

(b) Show that the skewness of the sample mean $\bar{X}_n$ is $\gamma/\sqrt{n}$. You can use the fact, shown in Chapter 7, that if X and Y are independent then $E(XY) = E(X)E(Y)$.

Hint: By (a), we can assume $\mu = 0$ and $\sigma^2 = 1$ without loss of generality; if the X_j are not standardized initially, then we can standardize them. If $(X_1 + X_2 + \cdots + X_n)^3$ is expanded out, there are 3 types of terms: terms such as X_1^3, terms such as $3X_1^2X_2$, and terms such as $6X_1X_2X_3$.

(c) What does the result of (b) say about the distribution of $\bar{X}_n$ when n is large?

12. Let c be the speed of light in a vacuum. Suppose that c is unknown, and scientists wish to estimate it. But even more so than that, they wish to estimate c^2, for use in the famous equation $E = mc^2$.

Through careful experiments, they obtain i.i.d. measurements $X_1, \dots, X_n \sim \mathcal{N}(c, \sigma^2)$. Using these data, there are various possible ways to estimate c^2. Two natural ways are: (1) estimate c using the average of the X_j's and then square the estimated c, and (2) average the X_j^2's. So let

$$\bar{X}_n = \frac{1}{n}\sum_{j=1}^{n} X_j,$$

and consider the two estimators

$$T_1 = \bar{X}_n^2 \text{ and } T_2 = \frac{1}{n}\sum_{j=1}^{n} X_j^2.$$

Note that T_1 is the square of the first sample moment and T_2 is the second sample moment.

(a) Find $P(T_1 < T_2)$.

Hint: Start by comparing $(\frac{1}{n}\sum_{j=1}^{n} x_j)^2$ and $\frac{1}{n}\sum_{j=1}^{n} x_j^2$ when $x_1, \dots, x_n$ are *numbers*, by considering a discrete r.v. whose possible values are $x_1, \dots, x_n$.

(b) When an r.v. T is used to estimate an unknown parameter θ, the *bias* of the estimator T is defined to be $E(T) - \theta$. Find the bias of T_1 and the bias of T_2.

Hint: First find the distribution of $\bar{X}_n$. In general, for finding $E(Y^2)$ for an r.v. Y, it is often useful to write it as $E(Y^2) = \text{Var}(Y) + (EY)^2$.

Moment generating functions

13. Ⓢ A fair die is rolled twice, with outcomes X for the first roll and Y for the second roll. Find the moment generating function $M_{X+Y}(t)$ of $X + Y$ (your answer should be a function of t and can contain unsimplified finite sums).

14. Ⓢ Let $U_1, U_2, \dots, U_{60}$ be i.i.d. Unif(0, 1) and $X = U_1 + U_2 + \cdots + U_{60}$. Find the MGF of X.

15. Let $W = X^2+Y^2$, with X, Y i.i.d. $\mathcal{N}(0,1)$. The MGF of X^2 turns out to be $(1-2t)^{-1/2}$ for $t < 1/2$ (you can assume this).

    (a) Find the MGF of W.

    (b) What famous distribution that we have studied so far does W follow (be sure to state the parameters in addition to the name)? In fact, the distribution of W is also a special case of two more famous distributions that we will study in later chapters!

16. Let $X \sim \text{Expo}(\lambda)$. Find the skewness of X, and explain why it is positive and why it does not depend on λ.

    Hint: Recall that $\lambda X \sim \text{Expo}(1)$ and the nth moment of an Expo(1) r.v. is $n!$ for all n.

17. Let $X_1, \dots, X_n$ be i.i.d. with mean μ, variance σ^2, and MGF M. Let

$$Z_n = \sqrt{n}\left(\frac{\bar{X}_n - \mu}{\sigma}\right).$$

    (a) Show that Z_n is a standardized quantity, i.e., it has mean 0 and variance 1.

    (b) Find the MGF of Z_n in terms of M, the MGF of each X_j.

18. Use the MGF of the Geom(p) distribution to give another proof that the mean of this distribution is q/p and the variance is q/p^2, with $q = 1 - p$.

19. Use MGFs to determine whether $X + 2Y$ is Poisson if X and Y are i.i.d. Pois(λ).

20. Ⓢ Let $X \sim \text{Pois}(\lambda)$, and let $M(t)$ be the MGF of X. The *cumulant generating function* is defined to be $g(t) = \log M(t)$. Expanding $g(t)$ as a Taylor series

$$g(t) = \sum_{j=1}^{\infty} \frac{c_j}{j!} t^j$$

    (the sum starts at $j = 1$ because $g(0) = 0$), the coefficient c_j is called the jth *cumulant* of X. Find the jth cumulant of X, for all $j \geq 1$.

21. Ⓢ Let $X_n \sim \text{Bin}(n, p_n)$ for all $n \geq 1$, where np_n is a constant $\lambda > 0$ for all n (so $p_n = \lambda/n$). Let $X \sim \text{Pois}(\lambda)$. Show that the MGF of X_n converges to the MGF of X (this gives another way to see that the Bin(n, p) distribution can be well-approximated by the Pois(λ) when n is large, p is small, and $\lambda = np$ is moderate).

22. Consider a setting where a Poisson approximation should work well: let $A_1, \dots, A_n$ be independent, rare events, with n large and $p_j = P(A_j)$ small for all j. Let $X = I(A_1) + \cdots + I(A_n)$ count how many of the rare events occur, and let $\lambda = E(X)$.

    (a) Find the MGF of X.

    (b) If the approximation $1 + x \approx e^x$ (this is a good approximation when x is very close to 0 but terrible when x is not close to 0) is used to write each factor in the MGF of X as e to a power, what happens to the MGF? Explain why the result makes sense intuitively.

23. Let U_1, U_2 be i.i.d. Unif(0, 1). Example 8.2.5 in Chapter 8 shows that $U_1 + U_2$ has a *Triangle distribution*, with PDF given by

$$f(t) = \begin{cases} t & \text{for } 0 < t \leq 1, \\ 2 - t & \text{for } 1 < t < 2. \end{cases}$$

    The method in Example 8.2.5 is useful but it often leads to difficult integrals, so having alternative methods is important. Show that $U_1 + U_2$ has a Triangle distribution by showing that they have the same MGF.

24. Let X and Y be i.i.d. Expo(1), and $L = X - Y$. The *Laplace distribution* has PDF

$$f(x) = \frac{1}{2}e^{-|x|}$$

for all real x. Use MGFs to show that the distribution of L is Laplace.

25. Let $Z \sim \mathcal{N}(0,1)$, and $Y = |Z|$. So Y has the *Folded Normal* distribution, discussed in Example 5.4.7. Find *two* expressions for the MGF of Y as unsimplified integrals: one integral based on the PDF of Y, and one based on the PDF of Z.

26. Let $X, Y, Z, W \sim \mathcal{N}(0,1)$ be i.i.d.

(a) Find an expression for $E\left(\Phi(Z)e^Z\right)$ as an unsimplified integral.

(b) Find $E(\Phi(Z))$ and $E(e^Z)$ as fully simplified numbers.

7

Joint distributions

When we first introduced random variables and their distributions in Chapter 3, we noted that the individual distributions of two r.v.s do not tell us anything about whether the r.v.s are independent or dependent. For example, two Bern(1/2) r.v.s X and Y could be independent if they indicate Heads on two different coin flips, or dependent if they indicate Heads and Tails, respectively, on the same coin flip. Thus, although the PMF of X is a complete blueprint for X and the PMF of Y is a complete blueprint for Y, these individual PMFs are missing important information about how the two r.v.s are related.

Of course, in real life, we usually care about the relationship between multiple r.v.s in the same experiment. To give just a few examples:

- *Medicine*: To evaluate the effectiveness of a treatment, we may take multiple measurements per patient; an ensemble of blood pressure, heart rate, and cholesterol readings can be more informative than any of these measurements considered separately.

- *Genetics*: To study the relationships between various genetic markers and a particular disease, if we only looked separately at distributions for each genetic marker, we could fail to learn about whether an *interaction* between markers is related to the disease.

- *Time series*: To study how something evolves over time, we can often make a series of measurements over time, and then study the series jointly. There are many applications of such series, such as global temperatures, stock prices, or national unemployment rates. The series of measurements considered jointly can help us deduce trends for the purpose of forecasting future measurements.

This chapter considers *joint distributions*, also called *multivariate distributions*, which capture the previously missing information about how multiple r.v.s interact. We introduce multivariate analogs of the CDF, PMF, and PDF in order to provide a complete specification of the relationship between multiple r.v.s. After this groundwork is in place, we'll study a couple of famous named multivariate distributions, generalizing the Binomial and Normal distributions to higher dimensions.

7.1 Joint, marginal, and conditional

The three key concepts for this section are *joint*, *marginal*, and *conditional* distributions. Recall that the distribution of a single r.v. X provides complete information about the probability of X falling into any subset of the real line. Analogously, the *joint* distribution of two r.v.s X and Y provides complete information about the probability of the vector (X, Y) falling into any subset of the plane. The *marginal* distribution of X is the individual distribution of X, ignoring the value of Y, and the *conditional* distribution of X given $Y = y$ is the updated distribution for X after observing $Y = y$. We'll look at these concepts in the discrete case first, then extend them to the continuous case.

7.1.1 Discrete

The most general description of the joint distribution of two r.v.s is the *joint CDF*, which applies to discrete and continuous r.v.s alike.

Definition 7.1.1 (Joint CDF). The *joint CDF* of r.v.s X and Y is the function $F_{X,Y}$ given by

$$F_{X,Y}(x, y) = P(X \leq x, Y \leq y).$$

The joint CDF of n r.v.s is defined analogously.

Unfortunately, the joint CDF of discrete r.v.s is not a well-behaved function; as in the univariate case, it consists of jumps and flat regions. For this reason, with discrete r.v.s we usually work with the *joint PMF*, which also determines the joint distribution and is much easier to visualize.

Definition 7.1.2 (Joint PMF). The *joint PMF* of discrete r.v.s X and Y is the function $p_{X,Y}$ given by

$$p_{X,Y}(x, y) = P(X = x, Y = y).$$

The joint PMF of n discrete r.v.s is defined analogously.

Just as univariate PMFs must be nonnegative and sum to 1, we require valid joint PMFs to be nonnegative and sum to 1, where the sum is taken over all possible values of X and Y:

$$\sum_x \sum_y P(X = x, Y = y) = 1.$$

The joint PMF determines the distribution because we can use it to find the probability of the event $(X, Y) \in A$ for any set A of points in the support of (X, Y). All we have to do is sum the joint PMF over A:

$$P((X, Y) \in A) = \sum\sum_{(x,y)\in A} P(X = x, Y = y).$$

Figure 7.1 shows a sketch of what the joint PMF of two discrete r.v.s could look like. The height of a vertical bar at (x, y) represents the probability $P(X = x, Y = y)$. For the joint PMF to be valid, the total height of the vertical bars must be 1.

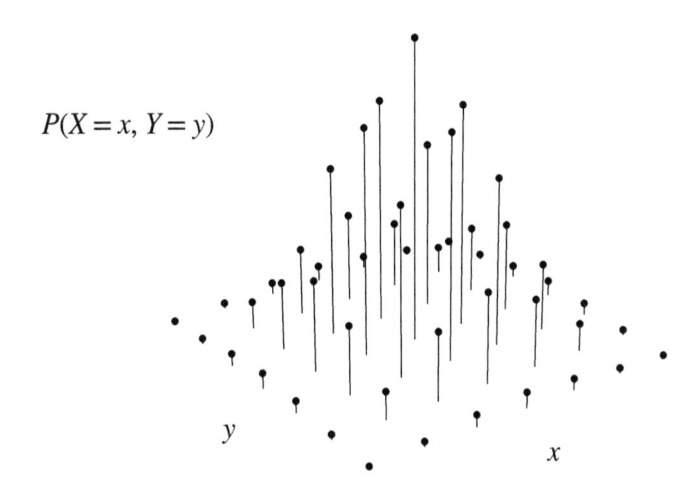

FIGURE 7.1
Joint PMF of discrete r.v.s X and Y.

From the joint distribution of X and Y, we can get the distribution of X alone by summing over the possible values of Y. This gives us the familiar PMF of X that we have seen in previous chapters. In the context of joint distributions, we will call it the *marginal* or unconditional distribution of X, to make it clear that we are referring to the distribution of X alone, without regard for the value of Y.

Definition 7.1.3 (Marginal PMF). For discrete r.v.s X and Y, the *marginal PMF* of X is

$$P(X = x) = \sum_{y} P(X = x, Y = y).$$

The marginal PMF of X *is* the PMF of X, viewing X individually rather than jointly with Y. The above equation follows from the axioms of probability (we are summing over disjoint cases). The operation of summing over the possible values of Y in order to convert the joint PMF into the marginal PMF of X is known as *marginalizing out* Y.

The process of obtaining the marginal PMF from the joint PMF is illustrated in Figure 7.2. Here we take a bird's-eye view of the joint PMF for a clearer perspective; each column of the joint PMF corresponds to a fixed x and each row corresponds to a fixed y. For any x, the probability $P(X = x)$ is the total height of the bars in the corresponding column of the joint PMF: we can imagine taking all the bars in that column and stacking them on top of each other to get the marginal probability. Repeating this for all x, we arrive at the marginal PMF, depicted in bold.

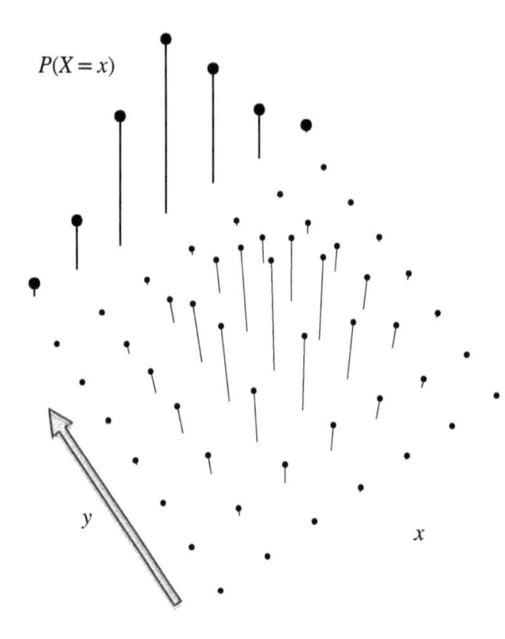

FIGURE 7.2
Bird's-eye view of the joint PMF from Figure 7.1. The marginal PMF $P(X = x)$ is obtained by summing over the joint PMF in the y-direction.

Similarly, the marginal PMF of Y is obtained by summing over all possible values of X. So given the joint PMF, we can marginalize out Y to get the PMF of X, or marginalize out X to get the PMF of Y. But if we only know the marginal PMFs of X and Y, there is no way to recover the joint PMF without further assumptions. It is clear how to stack the bars in Figure 7.2, but very unclear how to unstack the bars after they have been stacked!

Another way to go from joint to marginal distributions is via the joint CDF. In that case, we take a limit rather than a sum: the marginal CDF of X is

$$F_X(x) = P(X \leq x) = \lim_{y \to \infty} P(X \leq x, Y \leq y) = \lim_{y \to \infty} F_{X,Y}(x, y).$$

However, as mentioned above it is usually easier to work with joint PMFs.

Now suppose that we observe the value of X and want to update our distribution of Y to reflect this information. Instead of using the marginal PMF $P(Y = y)$, which does not take into account any information about X, we should use a PMF that conditions on the event $X = x$, where x is the value we observed for X. This naturally leads us to consider *conditional PMFs*.

Definition 7.1.4 (Conditional PMF). For discrete r.v.s X and Y, the *conditional PMF* of Y given $X = x$ is

$$P(Y = y|X = x) = \frac{P(X = x, Y = y)}{P(X = x)}.$$

This is viewed as a function of y for fixed x.

Note that the conditional PMF (for fixed x) *is* a valid PMF. So we can define the conditional expectation of Y given $X = x$, denoted by $E(Y|X = x)$, in the same way that we defined $E(Y)$ except that we replace the PMF of Y with the conditional PMF of Y. Chapter 9 is devoted to conditional expectation.

Figure 7.3 illustrates the definition of conditional PMF. To condition on the event $X = x$, we first take the joint PMF and focus in on the vertical bars where X takes on the value x; in the figure, these are shown in bold. All of the other vertical bars are irrelevant because they are inconsistent with the knowledge that $X = x$ occurred. Since the total height of the bold bars is the marginal probability $P(X = x)$, we then *renormalize* the conditional PMF by dividing by $P(X = x)$; this ensures that the conditional PMF will sum to 1. Therefore conditional PMFs *are* PMFs, just as conditional probabilities are probabilities. Notice that there is a different conditional PMF of Y for every possible value of X; Figure 7.3 highlights just one of these conditional PMFs.

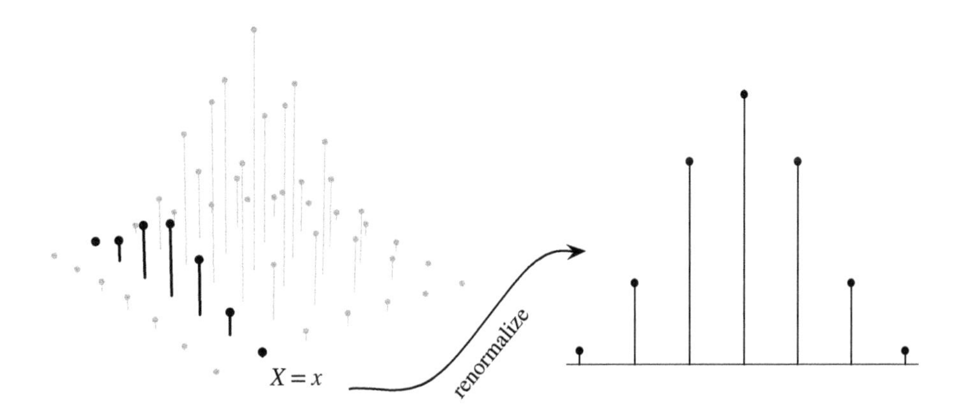

FIGURE 7.3
Conditional PMF of Y given $X = x$. The conditional PMF $P(Y = y|X = x)$ is obtained by renormalizing the column of the joint PMF that is compatible with the event $X = x$.

We can also relate the conditional distribution of Y given $X = x$ to that of X given $Y = y$, using Bayes' rule:

$$P(Y = y|X = x) = \frac{P(X = x|Y = y)P(Y = y)}{P(X = x)}.$$

And using LOTP, we have another way of getting the marginal PMF: the marginal PMF of X is a weighted average of the conditional PMFs $P(X = x|Y = y)$, where the weights are the probabilities $P(Y = y)$:

$$P(X = x) = \sum_y P(X = x|Y = y)P(Y = y).$$

Let's work through a numerical example to complement the graphs we've been looking at.

Example 7.1.5 (2×2 table). The simplest example of a discrete joint distribution is when X and Y are both Bernoulli r.v.s. In this case, the joint PMF is fully specified by the four values $P(X = 1, Y = 1)$, $P(X = 0, Y = 1)$, $P(X = 1, Y = 0)$, and $P(X = 0, Y = 0)$, so we can represent the joint PMF of X and Y using a 2×2 table.

This very simple scenario actually has an important place in statistics, as these so-called *contingency tables* are often used to study whether a treatment is associated with a particular outcome. In such scenarios, X may be the indicator of receiving the treatment, and Y may be the indicator of the outcome of interest.

For example, suppose we randomly sample an adult male from the United States population. Let X be the indicator of the sampled individual being a current smoker, and let Y be the indicator of his developing lung cancer at some point in his life. Suppose the joint PMF is as follows (these numbers are for illustrative purposes; they are not estimated from real data).

	$Y = 1$	$Y = 0$
$X = 1$	$\frac{5}{100}$	$\frac{20}{100}$
$X = 0$	$\frac{3}{100}$	$\frac{72}{100}$

To get the marginal probability $P(Y = 1)$, we add the probabilities in the two cells of the table where $Y = 1$. We do the same for $P(Y = 0)$, $P(X = 1)$, and $P(X = 0)$ and write these probabilities in the margins of the table (making "marginal" an appropriate name!).

	$Y = 1$	$Y = 0$	**Total**
$X = 1$	$\frac{5}{100}$	$\frac{20}{100}$	$\frac{25}{100}$
$X = 0$	$\frac{3}{100}$	$\frac{72}{100}$	$\frac{75}{100}$
Total	$\frac{8}{100}$	$\frac{92}{100}$	$\frac{100}{100}$

This shows that the *marginal distribution* of X is Bern(0.25) and the *marginal distribution* of Y is Bern(0.08). In words, the unconditional probability that the individual is a current smoker is 0.25, and the unconditional probability of his developing lung cancer is 0.08.

Now suppose we observe $X = 1$, i.e., the individual is a current smoker. We can then update our beliefs about his risk for lung cancer.

$$P(Y = 1|X = 1) = \frac{P(X = 1, Y = 1)}{P(X - 1)} = \frac{5/100}{25/100} = 0.2,$$

so the *conditional distribution* of Y given $X = 1$ is Bern(0.2). By a similar calculation, the conditional distribution of Y given $X = 0$ is Bern(0.04). This tells us that the probability of developing lung cancer is 0.2 for current smokers but only 0.04 for non-smokers. □

☣ **7.1.6.** The word "marginal" has opposite meanings in economics and statistics. In economics it refers to a derivative, e.g., marginal revenue is the derivative of revenue with respect to quantity sold. In statistics it refers to an integral or sum, which can be thought of intuitively by writing totals in the margins of a table, as in the above example.

Armed with an understanding of joint, marginal, and conditional distributions, we can revisit the definition of independence that we introduced in Chapter 3.

Definition 7.1.7 (Independence of discrete r.v.s)**.** Random variables X and Y are *independent* if for all x and y,

$$F_{X,Y}(x, y) = F_X(x)F_Y(y).$$

If X and Y are discrete, this is equivalent to the condition

$$P(X = x, Y = y) = P(X = x)P(Y = y)$$

for all x, y, and it is also equivalent to the condition

$$P(Y = y|X = x) = P(Y = y)$$

for all x, y such that $P(X = x) > 0$.

Using the terminology from this chapter, the definition says that for independent r.v.s, the *joint CDF* factors into the product of the *marginal CDFs*, or that the *joint PMF* factors into the product of the *marginal PMFs*. Remember that in general, the marginal distributions do *not* determine the joint distribution: this is the reason why we wanted to study joint distributions in the first place! But in the special case of independence, the marginal distributions are all we need in order to specify the joint distribution; we can get the joint PMF by multiplying the marginal PMFs.

Another way of looking at independence is that *all the conditional PMFs* are the same as the *marginal PMF*. That is, starting with the marginal PMF of Y, no updating is necessary when we condition on $X = x$, regardless of what x is.

Example 7.1.8 (Independence in the 2×2 table)**.** Returning to the table from the previous example, we can use these two views of independence to see why X and Y are not independent.

	$Y = 1$	$Y = 0$	**Total**
$X = 1$	$\frac{5}{100}$	$\frac{20}{100}$	$\frac{25}{100}$
$X = 0$	$\frac{3}{100}$	$\frac{72}{100}$	$\frac{75}{100}$
Total	$\frac{8}{100}$	$\frac{92}{100}$	$\frac{100}{100}$

First, the joint PMF is not the product of the marginal PMFs. For example,

$$P(X = 1, Y = 1) \neq P(X = 1)P(Y = 1).$$

Finding even one pair of values x and y such that

$$P(X = x, Y = y) \neq P(X = x)P(Y = y)$$

is enough to rule out independence.

Second, we found that the marginal distribution of Y is Bern(0.08), whereas the conditional distribution of Y given $X = 1$ is Bern(0.2) and the conditional distribution of Y given $X = 0$ is Bern(0.04). Since conditioning on the value of X alters the distribution of Y, X and Y are not independent: learning whether or not the sampled individual is a current smoker gives us information about the probability that he will develop lung cancer.

Although we have found that X and Y are dependent, we cannot make conclusions about whether smoking *causes* lung cancer based on this association alone. As we learned from Simpson's paradox, misleading associations can arise when we fail to account for confounding variables. □

We'll do one more example of a discrete joint distribution to round out this section. We've named it the *chicken-egg story*; in it, we use wishful thinking to find a joint PMF, and our efforts land us a surprising independence result.

Story 7.1.9 (Chicken-egg). Suppose a chicken lays a random number of eggs, N, where $N \sim \text{Pois}(\lambda)$. Each egg independently hatches with probability p and fails to hatch with probability $q = 1 - p$. Let X be the number of eggs that hatch and Y the number that do not hatch, so $X + Y = N$. What is the joint PMF of X and Y?

Solution:

We seek the joint PMF $P(X = i, Y = j)$ for nonnegative integers i and j. Conditional on the total number of eggs N, the eggs are independent Bernoulli trials with probability of success p, so by the story of the Binomial, the conditional distributions of X and Y are $X|N = n \sim \text{Bin}(n, p)$ and $Y|N = n \sim \text{Bin}(n, q)$. Since our lives would be easier if only we knew the total number of eggs, let's use wishful thinking: condition on N and apply the law of total probability. This gives

$$P(X = i, Y = j) = \sum_{n=0}^{\infty} P(X = i, Y = j|N = n)P(N = n).$$

The sum is over all possible values of n, holding i and j fixed. But unless $n = i + j$, it is impossible for X to equal i and Y to equal j. For example, the only way there can be 5 hatched eggs and 6 unhatched eggs is if there are 11 eggs in total. So

$$P(X = i, Y = j|N = n) = 0$$

unless $n = i + j$, which means all other terms in the sum can be dropped:

$$P(X = i, Y = j) = P(X = i, Y = j|N = i + j)P(N = i + j).$$

Conditional on $N = i + j$, the events $X = i$ and $Y = j$ are exactly the same event, so keeping both is redundant. We'll keep $X = i$; the rest is a matter of plugging in the Binomial PMF to get $P(X = i|N = i + j)$ and the Poisson PMF to get $P(N = i + j)$. Thus,

$$\begin{aligned} P(X = i, Y = j) &= P(X = i|N = i + j)P(N = i + j) \\ &= \binom{i+j}{i} p^i q^j \cdot \frac{e^{-\lambda}\lambda^{i+j}}{(i+j)!} \\ &= \frac{e^{-\lambda p}(\lambda p)^i}{i!} \cdot \frac{e^{-\lambda q}(\lambda q)^j}{j!}. \end{aligned}$$

The joint PMF factors into the product of the Pois(λp) PMF (as a function of i) and the Pois(λq) PMF (as a function of j). This tells us two elegant facts: (1) X and Y are independent, since their joint PMF is the product of their marginal PMFs, and (2) $X \sim \text{Pois}(\lambda p)$ and $Y \sim \text{Pois}(\lambda q)$.

At first it may seem deeply counterintuitive that X is independent of Y. Doesn't knowing that a lot of eggs hatched mean that there are probably not so many that didn't hatch? For a *fixed* number of eggs, this independence would be impossible: knowing the number of hatched eggs would perfectly determine the number of unhatched eggs. But in this example, the number of eggs is *random*, following a Poisson distribution, and this happens to be the right kind of randomness to make X and Y unconditionally independent. This is a very special property of the Poisson. □

The chicken-egg story supplements the following result from Chapter 4:

Theorem 7.1.10. If $X \sim \text{Pois}(\lambda p)$, $Y \sim \text{Pois}(\lambda q)$, and X and Y are independent, then $N = X + Y \sim \text{Pois}(\lambda)$ and $X|N = n \sim \text{Bin}(n, p)$.

By the chicken-egg story, we now have the converse to this theorem.

Theorem 7.1.11. If $N \sim \text{Pois}(\lambda)$ and $X|N = n \sim \text{Bin}(n, p)$, then $X \sim \text{Pois}(\lambda p)$, $Y = N - X \sim \text{Pois}(\lambda q)$, and X and Y are independent.

☣ **7.1.12.** In the chicken-egg story, it is *not* valid to say "$P(X = x|N = n) = P(X = x|X + Y = n) = P(X = x|Y = n - x)$", since in $P(X = x|N = n)$ we are *not* conditioning on $X = x$. Indeed, $P(X = x|N = n)$ is the Bin(n, p) PMF, whereas $P(X = x|Y = n - x) = P(X = x)$ is the Pois(λp) PMF. This blunder of plugging in x for X on the right side of the conditioning bar illustrates how crucial it is to carefully distinguish between X and x, and between the left side and the right side of the conditioning bar.

7.1.2 Continuous

Once we have a handle on discrete joint distributions, it isn't much harder to consider continuous joint distributions. We simply make the now-familiar substitutions of integrals for sums and PDFs for PMFs, remembering that the probability of any individual point is now 0.

Formally, in order for X and Y to have a continuous joint distribution, we require that the joint CDF

$$F_{X,Y}(x, y) = P(X \leq x, Y \leq y)$$

be differentiable with respect to x and y. The partial derivative with respect to x and y is called the *joint PDF*. The joint PDF determines the joint distribution, as does the joint CDF.

Definition 7.1.13 (Joint PDF). If X and Y are continuous with joint CDF $F_{X,Y}$, their *joint PDF* is the derivative of the joint CDF with respect to x and y:

$$f_{X,Y}(x, y) = \frac{\partial^2}{\partial x \partial y} F_{X,Y}(x, y).$$

We require valid joint PDFs to be nonnegative and integrate to 1:

$$f_{X,Y}(x, y) \geq 0, \text{ and } \int_{-\infty}^{\infty} \int_{-\infty}^{\infty} f_{X,Y}(x, y) dx dy = 1.$$

In the univariate case, the PDF was the function we integrated to get the probability of an interval. Similarly, the joint PDF of two r.v.s is the function we integrate to get the probability of a two-dimensional region. For example,

$$P(X < 3,\ 1 < Y < 4) = \int_{1}^{4} \int_{-\infty}^{3} f_{X,Y}(x, y) dx dy.$$

For a general region $A \subseteq \mathbb{R}^2$,

$$P((X, Y) \in A) = \iint\limits_{A} f_{X,Y}(x, y) dx dy.$$

Figure 7.4 shows a sketch of what a joint PDF of two r.v.s could look like. As usual with continuous r.v.s, we need to keep in mind that the height of the surface $f_{X,Y}(x, y)$ at a single point does *not* represent a probability. The probability of any specific point in the plane is 0. Now that we've gone up a dimension, the probability of any line or curve in the plane is also 0. The only way we can get nonzero probability is by integrating over a region of *positive area* in the xy-plane.

When we integrate the joint PDF over a region A, we are calculating the volume under the surface of the joint PDF and above A. Thus, probability is represented by *volume under the joint PDF*. The total volume under a valid joint PDF is 1.

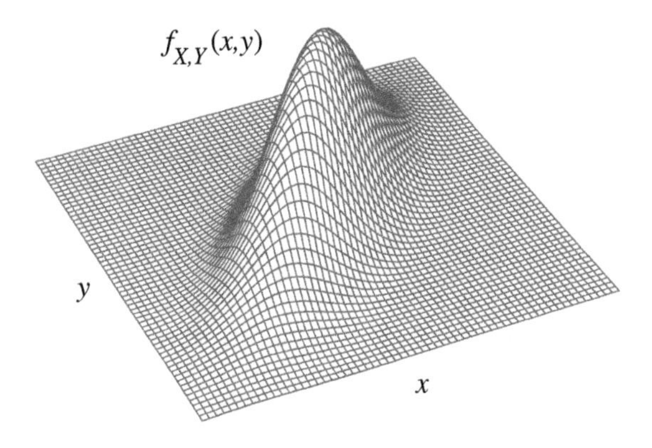

FIGURE 7.4
Joint PDF of continuous r.v.s X and Y.

In the discrete case, we get the marginal PMF of X by summing over all possible values of Y in the joint PMF. In the continuous case, we get the *marginal PDF* of X by integrating over all possible values of Y in the joint PDF.

Definition 7.1.14 (Marginal PDF). For continuous r.v.s X and Y with joint PDF $f_{X,Y}$, the *marginal PDF* of X is

$$f_X(x) = \int_{-\infty}^{\infty} f_{X,Y}(x,y)dy.$$

This *is* the PDF of X, viewing X individually rather than jointly with Y.

To simplify notation, we have mainly been looking at the joint distribution of two r.v.s rather than n r.v.s, but marginalization works analogously with any number of variables. For example, if we have the joint PDF of X, Y, Z, W but want the joint PDF of X, W, we just have to integrate over all possible values of Y and Z:

$$f_{X,W}(x,w) = \int_{-\infty}^{\infty} \int_{-\infty}^{\infty} f_{X,Y,Z,W}(x,y,z,w)dydz.$$

Conceptually this is easy—just integrate over the unwanted variables to get the joint PDF of the wanted variables—but computing it may or may not be easy.

Returning to the case of the joint distribution of two r.v.s X and Y, let's consider how to update our distribution for Y after observing the value of X, using the *conditional PDF*.

Definition 7.1.15 (Conditional PDF). For continuous r.v.s X and Y with joint PDF $f_{X,Y}$, the *conditional PDF* of Y given $X = x$ is

$$f_{Y|X}(y|x) = \frac{f_{X,Y}(x,y)}{f_X(x)},$$

for all x with $f_X(x) > 0$. This is considered as a function of y for fixed x. As a convention, in order to make $f_{Y|X}(y|x)$ well-defined for all real x, let $f_{Y|X}(y|x) = 0$ for all x with $f_X(x) = 0$.

Notation 7.1.16. The subscripts that we place on all the f's are just to remind us that we have three different functions on our plate. We could just as well write $g(y|x) = f(x,y)/h(x)$, where f is the joint PDF, h is the marginal PDF of X, and g is the conditional PDF of Y given $X = x$, but that makes it more difficult to remember which letter stands for which function.

Figure 7.5 illustrates the definition of conditional PDF. We take a vertical slice of the joint PDF corresponding to the observed value of X. Since the total area under this slice is $f_X(x)$, we then divide by $f_X(x)$ to ensure that the conditional PDF will have an area of 1. So the conditional PDF of Y given $X = x$ satisfies the properties of a valid PDF, for any x in the support of X.

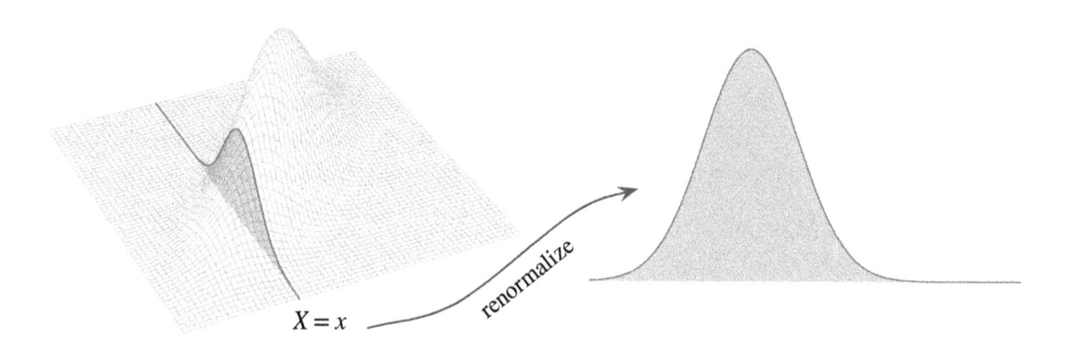

FIGURE 7.5
Conditional PDF of Y given $X = x$. The conditional PDF $f_{Y|X}(y|x)$ is obtained by renormalizing the slice of the joint PDF at the fixed value x.

☣ **7.1.17.** How can we speak of conditioning on $X = x$ for X a continuous r.v., considering that this event has probability 0? Rigorously speaking, we are actually conditioning on the event that X falls within a small interval containing x, say $X \in (x - \epsilon, x + \epsilon)$, and then taking a limit as ϵ approaches 0 from the right. We will not fuss over this technicality; fortunately, many important results such as Bayes' rule work in the continuous case exactly as one would hope.

Note that we can recover the joint PDF $f_{X,Y}$ if we have the conditional PDF $f_{Y|X}$ and the corresponding marginal f_X:

$$f_{X,Y}(x,y) = f_{Y|X}(y|x)f_X(x).$$

Similarly, we can recover the joint PDF if we have $f_{X|Y}$ and f_Y:

$$f_{X,Y}(x,y) = f_{X|Y}(x|y)f_Y(y).$$

This allows us to develop continuous versions of Bayes' rule and LOTP. The continuous versions are analogous to the discrete versions, with probability density functions in place of probabilities and integrals in place of sums.

Theorem 7.1.18 (Continuous form of Bayes' rule and LOTP)**.** For continuous r.v.s X and Y, we have the following continuous form of Bayes' rule:

$$f_{Y|X}(y|x) = \frac{f_{X|Y}(x|y)f_Y(y)}{f_X(x)}, \text{ for } f_X(x) > 0.$$

And we have the following continuous form of the law of total probability:

$$f_X(x) = \int_{-\infty}^{\infty} f_{X|Y}(x|y)f_Y(y)dy.$$

Proof. By definition of conditional PDFs, we have

$$f_{Y|X}(y|x)f_X(x) = f_{X,Y}(x,y) = f_{X|Y}(x|y)f_Y(y).$$

The continuous version of Bayes' rule follows immediately from dividing by $f_X(x)$. The continuous version of LOTP follows immediately from integrating with respect to y:

$$f_X(x) = \int_{-\infty}^{\infty} f_{X,Y}(x,y)dy = \int_{-\infty}^{\infty} f_{X|Y}(x|y)f_Y(y)dy. \quad \blacksquare$$

Out of curiosity, let's see what would have happened if we had plugged in the other expression for $f_{X,Y}(x,y)$ instead in the proof of LOTP:

$$f_X(x) = \int_{-\infty}^{\infty} f_{X,Y}(x,y)dy = \int_{-\infty}^{\infty} f_{Y|X}(y|x)f_X(x)dy = f_X(x)\int_{-\infty}^{\infty} f_{Y|X}(y|x)dy.$$

This just says that, for any x with $f_X(x) > 0$,

$$\int_{-\infty}^{\infty} f_{Y|X}(y|x)dy = 1,$$

confirming the fact that conditional PDFs must integrate to 1.

We now have versions of Bayes' rule and LOTP for two discrete r.v.s and for two continuous r.v.s. Better yet, there are also versions when we have one discrete r.v. and one continuous r.v. After understanding the discrete versions, it is easy to remember and use the other versions since they are analogous, replacing probabilities by PDFs when appropriate. For example, for X discrete and Y continuous, we have the following version of LOTP:

$$P(X = x) = \int_{-\infty}^{\infty} P(X = x|Y = y)f_Y(y)dy.$$

Taking X to be the indicator r.v. of an event A and $x = 1$, we have an expression for a general probability $P(A)$ based on conditioning on a continuous r.v. Y:

$$P(A) = \int_{-\infty}^{\infty} P(A|Y = y)f_Y(y)dy.$$

Here are the four versions of Bayes' rule, summarized in a table.

	Y **discrete**	Y **continuous**
X **discrete**	$P(Y=y\|X=x)=\frac{P(X=x\|Y=y)P(Y=y)}{P(X=x)}$	$f_Y(y\|X=x)=\frac{P(X=x\|Y=y)f_Y(y)}{P(X=x)}$
X **continuous**	$P(Y=y\|X=x)=\frac{f_X(x\|Y=y)P(Y=y)}{f_X(x)}$	$f_{Y\|X}(y\|x)=\frac{f_{X\|Y}(x\|y)f_Y(y)}{f_X(x)}$

And here are the four versions of LOTP, summarized in a table. The top row gives expressions for $P(X = x)$, while the bottom row gives expressions for $f_X(x)$.

	Y **discrete**	Y **continuous**
X **discrete**	$\sum_y P(X=x\|Y=y)P(Y=y)$	$\int_{-\infty}^{\infty} P(X=x\|Y=y)f_Y(y)dy$
X **continuous**	$\sum_y f_X(x\|Y=y)P(Y=y)$	$\int_{-\infty}^{\infty} f_{X\|Y}(x\|y)f_Y(y)dy$

Finally, let's discuss the definition of independence for continuous r.v.s; then we'll turn to concrete examples. As in the discrete case, we can view independence of continuous r.v.s in two ways. One is that the joint CDF factors into the product of the marginal CDFs, or the joint PDF factors into the product of the marginal PDFs. The other is that the conditional PDF of Y given $X = x$ is the same as the marginal PDF of Y, so conditioning on X provides no information about Y.

Definition 7.1.19 (Independence of continuous r.v.s). Random variables X and Y are *independent* if for all x and y,

$$F_{X,Y}(x,y) = F_X(x)F_Y(y).$$

If X and Y are continuous with joint PDF $f_{X,Y}$, this is equivalent to the condition

$$f_{X,Y}(x,y) = f_X(x)f_Y(y)$$

for all x, y, and it is also equivalent to the condition

$$f_{Y|X}(y|x) = f_Y(y)$$

for all x, y such that $f_X(x) > 0$.

☣ **7.1.20.** The marginal PDF of Y, $f_Y(y)$, is a function of y only; it cannot depend on x in any way. The conditional PDF $f_{Y|X}(y|x)$ *can* depend on x in general. Only in the special case of independence is $f_{Y|X}(y|x)$ free of x.

Sometimes we have a joint PDF for X and Y that factors as a function of x times a function of y, without knowing in advance whether these functions are the marginal PDFs, or even whether they *are* valid PDFs. The next result addresses this situation.

Proposition 7.1.21. Suppose that the joint PDF $f_{X,Y}$ of X and Y factors as

$$f_{X,Y}(x, y) = g(x)h(y)$$

for all x and y, where g and h are nonnegative functions. Then X and Y are independent. Also, if either g or h is a valid PDF, then the other one is a valid PDF too and g and h are the marginal PDFs of X and Y, respectively. (The analogous result in the discrete case also holds.)

Proof. Let $c = \int_{-\infty}^{\infty} h(y)dy$. Multiplying and dividing by c, we can write

$$f_{X,Y}(x, y) = cg(x) \cdot \frac{h(y)}{c}.$$

(The point of this is that $h(y)/c$ is a valid PDF.) Then the marginal PDF of X is

$$f_X(x) = \int_{-\infty}^{\infty} f_{X,Y}(x, y)dy = cg(x) \int_{-\infty}^{\infty} \frac{h(y)}{c} dy = cg(x).$$

It follows that $\int_{-\infty}^{\infty} cg(x)dx = 1$ since a marginal PDF *is* a valid PDF (knowing the integral of h gave us the integral of g for free!). Then the marginal PDF of Y is

$$f_Y(y) = \int_{-\infty}^{\infty} f_{X,Y}(x, y)dx = \frac{h(y)}{c} \int_{-\infty}^{\infty} cg(x)dx = \frac{h(y)}{c}.$$

Thus, X and Y are independent with PDFs $cg(x)$ and $h(y)/c$, respectively. If g or h is already a valid PDF, then $c = 1$, so the other one is also a valid PDF. ■

☣ **7.1.22.** In the above proposition, we need the joint PDF to factor as a function of x times a function of y for *all* (x, y) in the plane $\mathbb{R}^2$, not just for (x, y) with $f_{X,Y}(x, y) > 0$. The reason for this is illustrated in the next example.

A simple case of a continuous joint distribution is when the joint PDF is constant over some region in the plane. In the following example, we'll compare a joint PDF that is constant on a square to a joint PDF that is constant on a disk.

Example 7.1.23 (Uniform on a region in the plane). Let (X, Y) be a completely random point in the square $\{(x, y) : x, y \in [0, 1]\}$, in the sense that the joint PDF of X and Y is constant over the square and 0 outside of it:

$$f_{X,Y}(x, y) = \begin{cases} 1 & \text{if } x, y \in [0, 1], \\ 0 & \text{otherwise.} \end{cases}$$

The constant 1 is chosen so that the joint PDF will integrate to 1. This distribution is called the *Uniform distribution* on the square.

Intuitively, it makes sense that X and Y should be Unif$(0, 1)$ marginally. We can check this by computing

$$f_X(x) = \int_0^1 f_{X,Y}(x, y)dy = \int_0^1 1dy = 1,$$

and similarly for f_Y. Furthermore, X and Y are independent, since the joint PDF factors into the product of the marginal PDFs (this just reduces to $1 = 1 \cdot 1$, but it's important to note that the value of X does not constrain the possible values of Y). So the conditional distribution of Y given $X = x$ is Unif(0, 1), regardless of x.

Now let (X, Y) be a completely random point in the unit disk $\{(x, y) : x^2 + y^2 \leq 1\}$, with joint PDF

$$f_{X,Y}(x, y) = \begin{cases} \frac{1}{\pi} & \text{if } x^2 + y^2 \leq 1, \\ 0 & \text{otherwise.} \end{cases}$$

Again, the constant $1/\pi$ is chosen to make the joint PDF integrate to 1; the value follows from the fact that the integral of 1 over some region in the plane is the area of that region.

Note that X and Y are *not* independent, since in general, knowing the value of X constrains the possible values of Y: larger values of $|X|$ restrict Y to be in a smaller range. It would fall into ☣ 7.1.22 disastrously to conclude independence from the fact that $f_{X,Y}(x, y) = g(x)h(y)$ for all (x, y) in the disk, where $g(x) = 1/\pi$ and $h(y) = 1$ are constant functions. To see from the definition that X and Y are not independent, note that, for example, $f_{X,Y}(0.9, 0.9) = 0$ since $(0.9, 0.9)$ is not in the unit disk, but $f_X(0.9)f_Y(0.9) \neq 0$ since 0.9 is in the supports of both X and Y.

The marginal distribution of X is now

$$f_X(x) = \int_{-\sqrt{1-x^2}}^{\sqrt{1-x^2}} \frac{1}{\pi} dy = \frac{2}{\pi}\sqrt{1 - x^2}, \quad -1 \leq x \leq 1.$$

By symmetry, $f_Y(y) = \frac{2}{\pi}\sqrt{1 - y^2}$. Note that the marginal distributions of X and Y are *not* Uniform on $[-1, 1]$; rather, X and Y are more likely to fall near 0 than near ± 1.

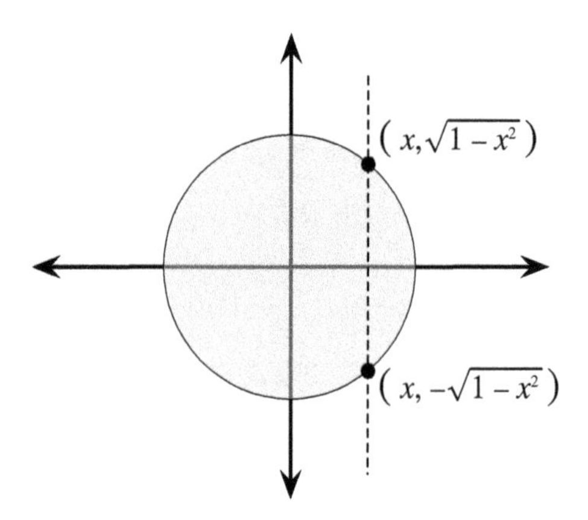

FIGURE 7.6
Bird's-eye view of the Uniform joint PDF on the unit disk. Conditional on $X = x$, Y is restricted to the interval $[-\sqrt{1 - x^2}, \sqrt{1 - x^2}]$.

Suppose we observe $X = x$. As illustrated in Figure 7.6, this constrains Y to lie in the interval $[-\sqrt{1-x^2}, \sqrt{1-x^2}]$. Specifically, the conditional distribution of Y given $X = x$ is

$$f_{Y|X}(y|x) = \frac{f_{X,Y}(x,y)}{f_X(x)} = \frac{\frac{1}{\pi}}{\frac{2}{\pi}\sqrt{1-x^2}} = \frac{1}{2\sqrt{1-x^2}}$$

for $-\sqrt{1-x^2} \leq y \leq \sqrt{1-x^2}$, and 0 otherwise. This conditional PDF is constant as a function of y, which tells us that the conditional distribution of Y is Uniform on the interval $[-\sqrt{1-x^2}, \sqrt{1-x^2}]$. The fact that the conditional PDF is not free of x confirms the fact that X and Y are not independent.

In general, for a region R in the plane, the Uniform distribution on R is defined to have joint PDF that is constant inside R and 0 outside R. The constant is the reciprocal of the area of R. If R is the rectangle $\{(x,y) : a \leq x \leq b, c \leq y \leq d\}$, then X and Y will be independent; unlike for a disk, the vertical slices of a rectangle all look the same. But for any region where the value of X constrains the possible values of Y or vice versa, X and Y will *not* be independent. □

As another example of working with joint PDFs, let's consider a question that comes up often when dealing with Exponentials of different rates.

Example 7.1.24 (Comparing Exponentials of different rates). Let $T_1 \sim \text{Expo}(\lambda_1)$ and $T_2 \sim \text{Expo}(\lambda_2)$ be independent. Find $P(T_1 < T_2)$. For example, T_1 could be the lifetime of a refrigerator and T_2 could be the lifetime of a stove (if we are willing to assume Exponential distributions for these), and then $P(T_1 < T_2)$ is the probability that the refrigerator fails before the stove. We know from Chapter 5 that $\min(T_1, T_2) \sim \text{Expo}(\lambda_1 + \lambda_2)$, which tells us about *when* the first appliance failure will occur, but we may also want to know about *which* appliance will fail first.

Solution:

We just need to integrate the joint PDF of T_1 and T_2 over the appropriate region, which is all (t_1, t_2) with $t_1 > 0, t_2 > 0$, and $t_1 < t_2$. This yields

$$\begin{aligned}
P(T_1 < T_2) &= \int_0^\infty \int_0^{t_2} \lambda_1 e^{-\lambda_1 t_1} \lambda_2 e^{-\lambda_2 t_2} dt_1 dt_2 \\
&= \int_0^\infty \left(\int_0^{t_2} \lambda_1 e^{-\lambda_1 t_1} dt_1 \right) \lambda_2 e^{-\lambda_2 t_2} dt_2 \\
&= \int_0^\infty (1 - e^{-\lambda_1 t_2}) \lambda_2 e^{-\lambda_2 t_2} dt_2 \\
&= 1 - \int_0^\infty \lambda_2 e^{-(\lambda_1+\lambda_2) t_2} dt_2 \\
&= 1 - \frac{\lambda_2}{\lambda_1 + \lambda_2} \\
&= \frac{\lambda_1}{\lambda_1 + \lambda_2}.
\end{aligned}$$

This result makes sense intuitively if we interpret λ_1 and λ_2 as rates. For example, if refrigerators have twice the failure rate of stoves, then it says that the odds are 2 to 1 in favor of the refrigerator failing first. As a simple check, note that the answer reduces to 1/2 when $\lambda_1 = \lambda_2$, which must be true by symmetry.

An alternative method of getting the same result is to use LOTP to condition on T_1 (or condition on T_2). A third approach, using a story about Poisson processes, is in Chapter 13. □

Our last example in this section demonstrates how we can use the joint distribution of X and Y to derive the distribution of a function of X and Y.

Example 7.1.25 (Cauchy PDF). Let X and Y be i.i.d. $\mathcal{N}(0,1)$, and let $T = X/Y$. (We can define T arbitrarily in the case $Y = 0$; the choice of how to define T in that case has no effect on the distribution of T, since $P(Y = 0) = 0$.) The distribution of T is a famous named distribution called the *Cauchy* distribution, and we will encounter it again in later chapters. Meanwhile, find the PDF of T.

Solution:

We'll find an expression for the CDF of T first, and then differentiate to get the PDF. We can write

$$F_T(t) = P(T \leq t) = P\left(\frac{X}{Y} \leq t\right) = P\left(\frac{X}{|Y|} \leq t\right),$$

since the r.v.s $\frac{X}{Y}$ and $\frac{X}{|Y|}$ are identically distributed by the symmetry of the standard Normal distribution. Now $|Y|$ is nonnegative, so we can multiply by it on both sides without reversing the direction of the inequality. Thus, we are interested in finding

$$F_T(t) = P(X \leq t|Y|).$$

We calculate this probability by integrating the joint PDF of X and Y over the region where $X \leq t|Y|$ holds. The joint PDF of X and Y is just the product of the marginal PDFs, by independence. So

$$\begin{aligned} F_T(t) &= P(X \leq t|Y|) \\ &= \int_{-\infty}^{\infty} \int_{-\infty}^{t|y|} \frac{1}{\sqrt{2\pi}} e^{-x^2/2} \frac{1}{\sqrt{2\pi}} e^{-y^2/2} dx dy. \end{aligned}$$

Note that the inner limits of integration (the limits for x) depend on y, whereas the outer limits of integration (the limits for y) can't depend on x (see the math appendix for more about limits of integration with multiple integrals). With some manipulations, we can get the double integral down to a single integral:

$$\begin{aligned} F_T(t) &= \int_{-\infty}^{\infty} \frac{1}{\sqrt{2\pi}} e^{-y^2/2} \left(\int_{-\infty}^{t|y|} \frac{1}{\sqrt{2\pi}} e^{-x^2/2} dx \right) dy \\ &= \int_{-\infty}^{\infty} \frac{1}{\sqrt{2\pi}} e^{-y^2/2} \Phi(t|y|) dy \\ &= \sqrt{\frac{2}{\pi}} \int_0^{\infty} e^{-y^2/2} \Phi(ty) dy. \end{aligned}$$

Alternatively, we can get the same result without doing a double integral by using one of the versions of LOTP. Letting I be the indicator r.v. for the event $X \leq t|Y|$, we again have

$$\begin{aligned} P(I=1) &= \int_{-\infty}^{\infty} P(I=1|Y=y) f_Y(y)dy \\ &= \int_{-\infty}^{\infty} \frac{1}{\sqrt{2\pi}} e^{-y^2/2} \Phi(t|y|)dy. \end{aligned}$$

At this point we seem to be stuck with an integral we don't know how to do. Fortunately, we were asked to find the PDF, not the CDF, so instead of evaluating the integral, we can just differentiate it with respect to t (not with respect to y, which is a dummy variable). We are permitted to interchange the order of integration and differentiation under mild technical conditions, which are met here. (This technique is known as *differentiation under the integral sign*, or *DUThIS* for short.) Then

$$\begin{aligned} f_T(t) = F_T'(t) &= \sqrt{\frac{2}{\pi}} \int_0^{\infty} \frac{\partial}{\partial t} \left(e^{-y^2/2} \Phi(ty) \right) dy \\ &= \sqrt{\frac{2}{\pi}} \int_0^{\infty} y e^{-y^2/2} \varphi(ty) dy \\ &= \frac{1}{\pi} \int_0^{\infty} y e^{-\frac{(1+t^2)y^2}{2}} dy \\ &= \frac{1}{\pi(1+t^2)}, \end{aligned}$$

using the substitution $u = (1+t^2)y^2/2$, $du = (1+t^2)ydy$ for the final step. So the PDF of T is

$$f_T(t) = \frac{1}{\pi(1+t^2)}, \quad t \in \mathbb{R}.$$

Since

$$\int_{-\infty}^{\infty} \frac{1}{1+t^2} dt = \arctan(\infty) - \arctan(-\infty) = \pi,$$

we have obtained a valid PDF. If we want to obtain the CDF too, we can integrate the PDF over the appropriate interval:

$$F_T(t) = \int_{-\infty}^{t} \frac{1}{\pi(1+u^2)} du = \frac{1}{\pi} \arctan(t) + \frac{1}{2}.$$

As we mentioned, the distribution of T is called the *Cauchy* distribution. The Cauchy PDF is similar in shape to the Normal bell curve, but with tails that decay less quickly to 0. Figure 7.7 superimposes the Cauchy and standard Normal PDFs; the heavier tails of the Cauchy PDF are evident.

An interesting fact about the Cauchy distribution is that although the PDF is symmetric about 0, its expected value does not exist, since the integral $\int_{-\infty}^{\infty} \frac{t}{\pi(1+t^2)} dt$ diverges: note that for large t, $\frac{t}{1+t^2} \approx \frac{1}{t}$, and $\int_1^{\infty} \frac{1}{t} dt = \infty$. It would be a blunder to write "$E(\frac{X}{Y}) = E(X)E(\frac{1}{Y}) = 0 \cdot E(\frac{1}{Y}) = 0$", since $E(\frac{1}{Y})$ also does not exist. □

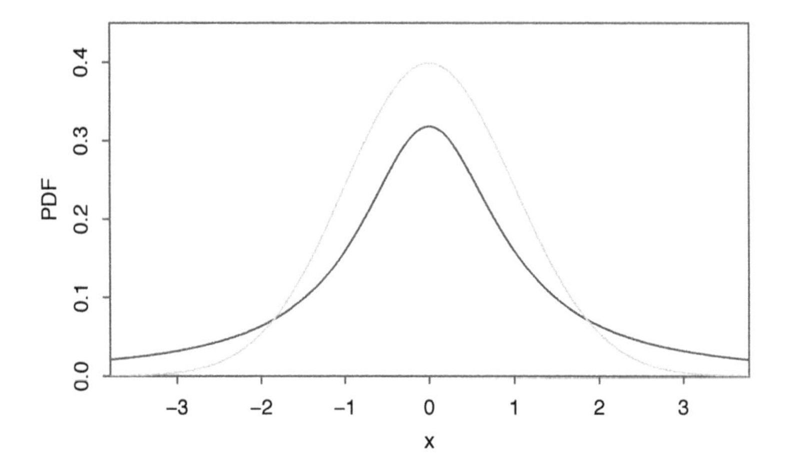

FIGURE 7.7
Cauchy PDF (dark) and $\mathcal{N}(0, 1)$ PDF (light). The Cauchy distribution has much heavier tails than the Normal distribution.

7.1.3 Hybrid

It is also possible that we could be interested in the joint distribution of a discrete r.v. and a continuous r.v. This case was mentioned when discussing the four forms of Bayes' rule and LOTP. Conceptually it is analogous to the other cases, but since the notation can be tricky, we'll work through an example that sheds light on it.

Example 7.1.26 (Which company made the lightbulb?). A lightbulb was manufactured by one of two companies. Bulbs that are made by Company 0 last an Expo(λ_0) amount of time, and bulbs made by Company 1 last an Expo(λ_1) amount of time, with $\lambda_0 < \lambda_1$. The bulb of interest here was made by Company 0 with probability p_0 and by Company 1 with probability $p_1 = 1 - p_0$, but from inspecting the bulb we don't know which company made it.

Let T be how long the bulb lasts, and I be the indicator of it having been made by Company 1.

(a) Find the CDF and PDF of T.

(b) Does T have the memoryless property?

(c) Find the conditional distribution of I given $T = t$. What happens to this as $t \to \infty$?

Solution:

Since T is a continuous r.v. and I is discrete, the joint distribution of T and I is a hybrid, as illustrated in Figure 7.8. In a joint PDF of two continuous r.v.s, there are infinitely many vertical slices of the joint PDF that we can take, each corresponding to a different conditional PDF. Here there are only two conditional PDFs of T, one for $I = 0$ and one for $I = 1$. As stated in the problem, the conditional distribution of T given $I = 0$ is Expo(λ_0) and given $I = 1$ is Expo(λ_1). The marginal distribution of I is Bern(p_1).

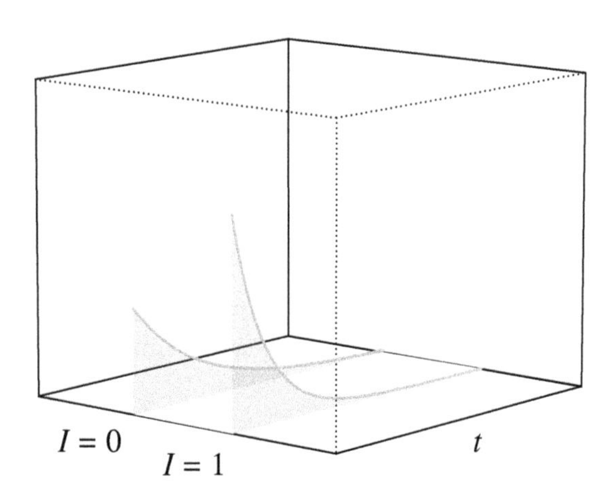

FIGURE 7.8
Hybrid joint distribution of T and I.

Thus, we are given the joint distribution of T and I in terms of (1) the marginal distribution of I and (2) the conditional distribution of T given I. The problem then asks us to flip it around and find (1) the marginal distribution of T and (2) the conditional distribution of I given T. Phrased in this way, it becomes clear that LOTP and Bayes' rule will be our friends.

(a) In this part we are asked to derive the marginal distribution of T. For the CDF, we use the law of total probability, conditioning on I:

$$\begin{aligned} F_T(t) = P(T \leq t) &= P(T \leq t|I = 0)p_0 + P(T \leq t|I = 1)p_1 \\ &= (1 - e^{-\lambda_0 t})p_0 + (1 - e^{-\lambda_1 t})p_1 \\ &= 1 - p_0 e^{-\lambda_0 t} - p_1 e^{-\lambda_1 t} \end{aligned}$$

for $t > 0$. The marginal PDF is the derivative of the CDF:

$$f_T(t) = p_0 \lambda_0 e^{-\lambda_0 t} + p_1 \lambda_1 e^{-\lambda_1 t}, \text{ for } t > 0.$$

We could also have gotten this directly from the "X continuous, Y discrete" version of LOTP, but we did not write out a proof of that version of LOTP, and working through this example helps show *why* that version of LOTP works.

(b) Since $\lambda_0 \neq \lambda_1$, the above expression for the PDF does not reduce to the form $\lambda e^{-\lambda t}$. So the distribution of T is *not* Exponential, which implies that it does not have the memoryless property. (The distribution of T is called a *mixture* of two Exponentials.)

(c) Using a hybrid form of Bayes' rule, which is the "X continuous, Y discrete" version in the table of versions of Bayes' rule, we have

$$P(I = 1|T = t) = \frac{f_T(t|I = 1)P(I = 1)}{f_T(t)},$$

where $f_T(t|I = 1)$ is the conditional PDF of T given $I = 1$, evaluated at t. Using the fact that $T|I = 1 \sim \text{Expo}(\lambda_1)$ and the marginal PDF derived in (a),

$$P(I = 1|T = t) = \frac{p_1\lambda_1 e^{-\lambda_1 t}}{p_0\lambda_0 e^{-\lambda_0 t} + p_1\lambda_1 e^{-\lambda_1 t}} = \frac{p_1\lambda_1}{p_0\lambda_0 e^{(\lambda_1-\lambda_0)t} + p_1\lambda_1}.$$

Thus the conditional distribution of I given $T = t$ is Bernoulli with this probability of success. This probability goes to 0 as $t \to \infty$, which makes sense intuitively: the longer the bulb lasts, the more confident we will be that it was made by Company 0, as their bulbs have a lower failure rate λ and a higher life expectancy $\frac{1}{\lambda}$. □

7.2 2D LOTUS

The two-dimensional version of LOTUS lets us calculate the expectation of a random variable that is a function of two random variables X and Y, using the joint distribution of X and Y.

Theorem 7.2.1 (2D LOTUS). Let g be a function from $\mathbb{R}^2$ to $\mathbb{R}$. If X and Y are discrete, then

$$E(g(X,Y)) = \sum_x \sum_y g(x,y)P(X = x, Y = y).$$

If X and Y are continuous with joint PDF $f_{X,Y}$, then

$$E(g(X,Y)) = \int_{-\infty}^{\infty}\int_{-\infty}^{\infty} g(x,y)f_{X,Y}(x,y)dxdy.$$

Like its 1D counterpart, 2D LOTUS saves us from having to find the distribution of $g(X,Y)$ in order to calculate its expectation. Instead, having the joint PMF or joint PDF of X and Y is enough.

One use of 2D LOTUS is to find the expected distance between two r.v.s.

Example 7.2.2 (Expected distance between two Uniforms). Let X and Y be i.i.d. Unif(0, 1) r.v.s. Find $E(|X - Y|)$.

Solution:

Since the joint PDF is 1 on the unit square $\{(x, y) : x, y \in [0, 1]\}$, 2D LOTUS gives

$$\begin{aligned} E(|X - Y|) &= \int_0^1 \int_0^1 |x - y| dx dy \\ &= \int_0^1 \int_y^1 (x - y) dx dy + \int_0^1 \int_0^y (y - x) dx dy \\ &= 2 \int_0^1 \int_y^1 (x - y) dx dy = 1/3. \end{aligned}$$

First we broke up the integral into two parts so we could eliminate the absolute value; then we used symmetry.

Incidentally, by solving this problem, we have also figured out the expected value of $M = \max(X, Y)$ and $L = \min(X, Y)$. Since $M + L$ is the same r.v. as $X + Y$ and $M - L$ is the same r.v. as $|X - Y|$,

$$\begin{aligned} E(M + L) &= E(X + Y) = 1, \\ E(M - L) &= E(|X - Y|) = 1/3. \end{aligned}$$

This is a system of two equations and two unknowns, which we can solve to get $E(M) = 2/3$ and $E(L) = 1/3$. As a check, $E(M)$ exceeds $E(L)$, as it should, and $E(M)$ and $E(L)$ are equidistant from $1/2$, as should be the case by symmetry. ☐

Example 7.2.3 (Expected distance between two Normals). For $X, Y \stackrel{\text{i.i.d.}}{\sim} \mathcal{N}(0, 1)$, find $E(|X - Y|)$.

Solution:

We could again use 2D LOTUS, giving

$$E(|X - Y|) = \int_{-\infty}^{\infty} \int_{-\infty}^{\infty} |x - y| \frac{1}{\sqrt{2\pi}} e^{-x^2/2} \frac{1}{\sqrt{2\pi}} e^{-y^2/2} dx dy,$$

but an easier solution uses the fact that the sum or difference of independent Normals is Normal, as we proved using MGFs in Chapter 6. Then $X - Y \sim \mathcal{N}(0, 2)$, so we can write $X - Y = \sqrt{2} Z$ where $Z \sim \mathcal{N}(0, 1)$, and $E(|X - Y|) = \sqrt{2} E|Z|$. Thus, we have reduced a 2D LOTUS to a 1D LOTUS! It was shown in Example 5.4.7 that

$$E|Z| = \sqrt{\frac{2}{\pi}},$$

so

$$E(|X - Y|) = \frac{2}{\sqrt{\pi}}.$$

☐

We can also use 2D LOTUS to give another proof of linearity of expectation.

Example 7.2.4 (Linearity via 2D LOTUS). Let X and Y be continuous r.v.s (the analogous method also works in the discrete case). By 2D LOTUS,

$$\begin{aligned} E(X+Y) &= \int_{-\infty}^{\infty}\int_{-\infty}^{\infty}(x+y)f_{X,Y}(x,y)dxdy \\ &= \int_{-\infty}^{\infty}\int_{-\infty}^{\infty}xf_{X,Y}(x,y)dxdy + \int_{-\infty}^{\infty}\int_{-\infty}^{\infty}yf_{X,Y}(x,y)dxdy \\ &= E(X)+E(Y). \end{aligned}$$

This is a short proof of linearity of expectation. For the last step, we used 2D LOTUS and the fact that X is a function of X and Y (that happens to be degenerate in the sense that it doesn't involve Y), and similarly for Y. Another way to get the last step is to write

$$\int_{-\infty}^{\infty}\int_{-\infty}^{\infty}yf_{X,Y}(x,y)dxdy = \int_{-\infty}^{\infty}y\int_{-\infty}^{\infty}f_{X,Y}(x,y)dxdy = \int_{-\infty}^{\infty}yf_Y(y)dy = E(Y),$$

where we took y out from the inner integral (since y is held constant when integrating with respect to x) and then recognized the marginal PDF of Y. For the $E(X)$ term we can first swap the order of integration, from $dxdy$ to $dydx$, and then the same argument that we used for the $E(Y)$ term can be applied. □

7.3 Covariance and correlation

Just as the mean and variance provided single-number summaries of the distribution of a single r.v., covariance is a single-number summary of the joint distribution of two r.v.s. Roughly speaking, covariance measures a tendency of two r.v.s to go up or down together, relative to their means: positive covariance between X and Y indicates that when X goes up, Y also tends to go up, and negative covariance indicates that when X goes up, Y tends to go down. Here is the precise definition.

Definition 7.3.1 (Covariance). The *covariance* between r.v.s X and Y is

$$\text{Cov}(X,Y) = E((X-EX)(Y-EY)).$$

Multiplying this out and using linearity, we have an equivalent expression:

$$\text{Cov}(X,Y) = E(XY) - E(X)E(Y).$$

Let's think about the definition intuitively. If X and Y tend to move in the same direction, then $X - EX$ and $Y - EY$ will tend to be either both positive or both negative, so $(X - EX)(Y - EY)$ will be positive on average, giving a positive covariance. If X and Y tend to move in opposite directions, then $X - EX$ and $Y - EY$ will tend to have opposite signs, giving a negative covariance.

If X and Y are independent, then their covariance is zero. We say that r.v.s with zero covariance are *uncorrelated.*

Theorem 7.3.2. If X and Y are independent, then they are uncorrelated.

Proof. We'll show this in the case where X and Y are continuous, with PDFs f_X and f_Y. Since X and Y are independent, their joint PDF is the product of the marginal PDFs. By 2D LOTUS,

$$\begin{aligned} E(XY) &= \int_{-\infty}^{\infty}\int_{-\infty}^{\infty} xyf_X(x)f_Y(y)dxdy \\ &= \int_{-\infty}^{\infty} yf_Y(y)\left(\int_{-\infty}^{\infty} xf_X(x)dx\right)dy \\ &= \int_{-\infty}^{\infty} xf_X(x)dx \int_{-\infty}^{\infty} yf_Y(y)dy \\ &= E(X)E(Y). \end{aligned}$$

The proof in the discrete case is the same, with PMFs instead of PDFs. ■

The converse of this theorem is false: just because X and Y are uncorrelated does not mean they are independent. For example, let $X \sim \mathcal{N}(0,1)$, and let $Y = X^2$. Then $E(XY) = E(X^3) = 0$ because the odd moments of the standard Normal distribution are equal to 0 by symmetry. Thus X and Y are uncorrelated,

$$\text{Cov}(X,Y) = E(XY) - E(X)E(Y) = 0 - 0 = 0,$$

but they are certainly not independent: Y is a function of X, so knowing X gives us perfect information about Y. Covariance is a measure of *linear* association, so r.v.s can be dependent in nonlinear ways and still have zero covariance, as this example demonstrates. The bottom right plot of Figure 7.9 shows draws from the joint distribution of X and Y in this example. The other three plots illustrate positive correlation, negative correlation, and independence.

Covariance has the following key properties.

1. $\text{Cov}(X,X) = \text{Var}(X)$.
2. $\text{Cov}(X,Y) = \text{Cov}(Y,X)$.
3. $\text{Cov}(X,c) = 0$ for any constant c.
4. $\text{Cov}(aX,Y) = a\text{Cov}(X,Y)$ for any constant a.
5. $\text{Cov}(X+Y,Z) = \text{Cov}(X,Z) + \text{Cov}(Y,Z)$.
6. $\text{Cov}(X+Y,Z+W) = \text{Cov}(X,Z) + \text{Cov}(X,W) + \text{Cov}(Y,Z) + \text{Cov}(Y,W)$.
7. $\text{Var}(X+Y) = \text{Var}(X) + \text{Var}(Y) + 2\text{Cov}(X,Y)$. For n r.v.s $X_1,\dots,X_n$,

$$\text{Var}(X_1 + \dots + X_n) = \text{Var}(X_1) + \dots + \text{Var}(X_n) + 2\sum_{i<j}\text{Cov}(X_i,X_j).$$

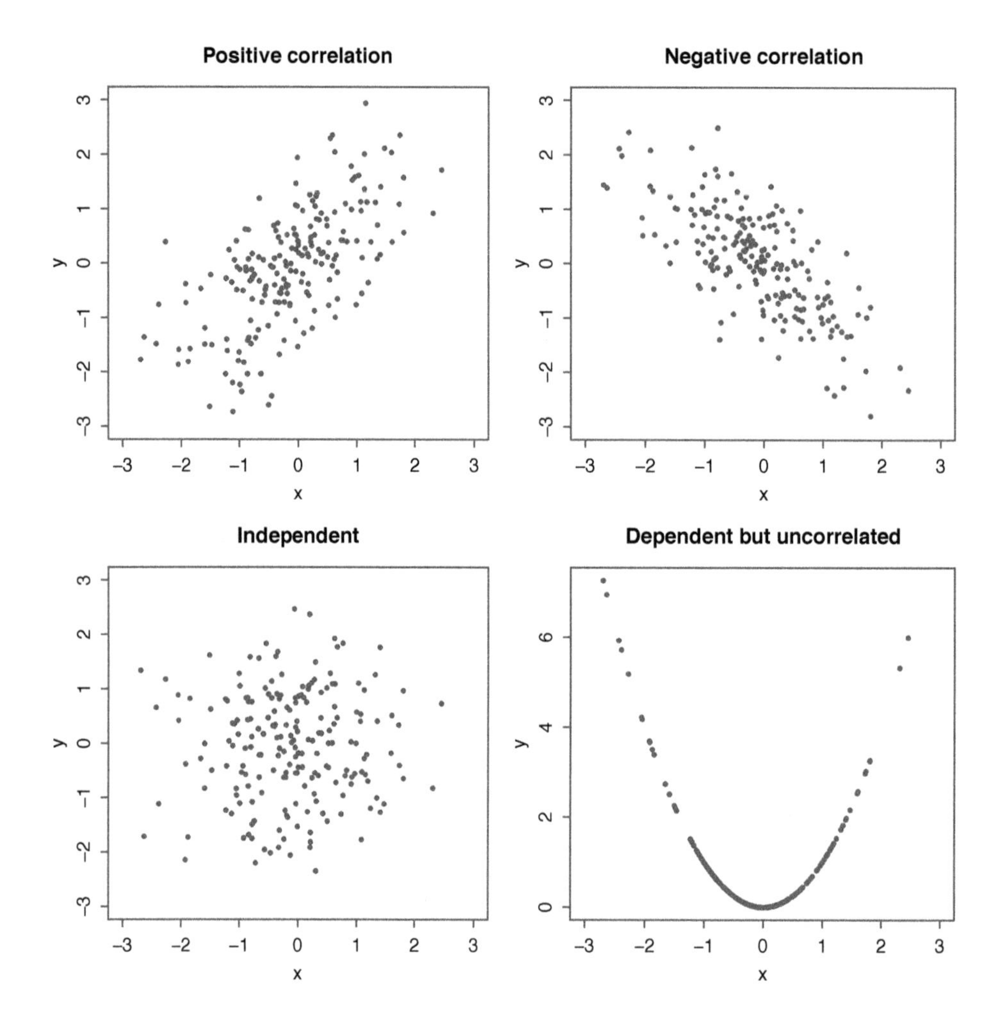

FIGURE 7.9
Draws from the joint distribution of (X, Y) under various dependence structures. Top left: X and Y are positively correlated. Top right: X and Y are negatively correlated. Bottom left: X and Y are independent, hence uncorrelated. Bottom right: Y is a deterministic function of X, but X and Y are uncorrelated.

The first five properties follow readily from the definition and basic properties of expectation. Property 6 follows from Property 2 and Property 5, by expanding

$$\begin{aligned}\mathrm{Cov}(X+Y,Z+W) &= \mathrm{Cov}(X,Z+W)+\mathrm{Cov}(Y,Z+W)\\ &= \mathrm{Cov}(Z+W,X)+\mathrm{Cov}(Z+W,Y)\\ &= \mathrm{Cov}(Z,X)+\mathrm{Cov}(W,X)+\mathrm{Cov}(Z,Y)+\mathrm{Cov}(W,Y)\\ &= \mathrm{Cov}(X,Z)+\mathrm{Cov}(X,W)+\mathrm{Cov}(Y,Z)+\mathrm{Cov}(Y,W).\end{aligned}$$

Property 7 follows from writing the variance of an r.v. as its covariance with itself (by Property 1) and then using Property 6 repeatedly.

We have now fulfilled our promise from Chapter 4 that for independent r.v.s, the variance of the sum is the sum of the variances:

$$\mathrm{Var}\left(\sum_{j=1}^{n} X_j\right) = \sum_{j=1}^{n}\mathrm{Var}(X_j) \text{ if } X_1,\dots,X_n \text{ are independent.}$$

By Theorem 7.3.2, independent r.v.s are uncorrelated, so in that case all the covariance terms drop out of the expression in Property 7.

☣ **7.3.3.** If X and Y are independent, then the properties of covariance give

$$\mathrm{Var}(X-Y) = \mathrm{Var}(X)+\mathrm{Var}(-Y) = \mathrm{Var}(X)+\mathrm{Var}(Y).$$

It is a common mistake to claim "$\mathrm{Var}(X-Y) = \mathrm{Var}(X)-\mathrm{Var}(Y)$"; this is a category error since $\mathrm{Var}(X)-\mathrm{Var}(Y)$ could be negative. For general X and Y, we have

$$\mathrm{Var}(X-Y) = \mathrm{Var}(X)+\mathrm{Var}(Y)-2\mathrm{Cov}(X,Y).$$

Since covariance depends on the units in which X and Y are measured—if we decide to measure X in centimeters rather than meters, the covariance is multiplied by 100—it is easier to interpret a unitless version of covariance called correlation.

Definition 7.3.4 (Correlation). The *correlation* between r.v.s X and Y is

$$\mathrm{Corr}(X,Y) = \frac{\mathrm{Cov}(X,Y)}{\sqrt{\mathrm{Var}(X)\mathrm{Var}(Y)}}.$$

(This is undefined in the degenerate cases $\mathrm{Var}(X)=0$ or $\mathrm{Var}(Y)=0$.)

Notice that shifting and scaling X and Y has no effect on their correlation. Shifting does not affect $\mathrm{Cov}(X,Y)$, $\mathrm{Var}(X)$, or $\mathrm{Var}(Y)$, so the correlation is unchanged. As for scaling, the fact that we divide by the standard deviations of X and Y ensures that the scale factor cancels out:

$$\mathrm{Corr}(cX,Y) = \frac{\mathrm{Cov}(cX,Y)}{\sqrt{\mathrm{Var}(cX)\mathrm{Var}(Y)}} = \frac{c\mathrm{Cov}(X,Y)}{\sqrt{c^2\mathrm{Var}(X)\mathrm{Var}(Y)}} = \mathrm{Corr}(X,Y).$$

Correlation is convenient to interpret because it does not depend on the units of measurement and is always between -1 and 1.

Theorem 7.3.5 (Correlation bounds)**.** For any r.v.s X and Y,

$$-1 \leq \text{Corr}(X, Y) \leq 1.$$

Proof. Without loss of generality we can assume X and Y have variance 1, since scaling does not change the correlation. Let $\rho = \text{Corr}(X, Y) = \text{Cov}(X, Y)$. Using the fact that variance is nonnegative, along with Property 7 of covariance, we have

$$\text{Var}(X + Y) = \text{Var}(X) + \text{Var}(Y) + 2\text{Cov}(X, Y) = 2 + 2\rho \geq 0,$$

$$\text{Var}(X - Y) = \text{Var}(X) + \text{Var}(Y) - 2\text{Cov}(X, Y) = 2 - 2\rho \geq 0.$$

Thus, $-1 \leq \rho \leq 1$. ■

Here is an example of how to calculate covariance and correlation.

Example 7.3.6 (Exponential max and min)**.** Let X and Y be i.i.d. Expo(1) r.v.s. Find the correlation between $\max(X, Y)$ and $\min(X, Y)$.

Solution:

Let $M = \max(X, Y)$ and $L = \min(X, Y)$. By the memoryless property and results from Chapter 5, we know that $L \sim \text{Expo}(2)$, $M - L \sim \text{Expo}(1)$, and $M - L$ is independent of L (see Example 5.6.5). Therefore,

$$\text{Cov}(M, L) = \text{Cov}(M - L + L, L) = \text{Cov}(M - L, L) + \text{Cov}(L, L) = 0 + \text{Var}(L) = \frac{1}{4},$$

$$\text{Var}(M) = \text{Var}(M - L + L) = \text{Var}(M - L) + \text{Var}(L) = 1 + \frac{1}{4} = \frac{5}{4},$$

and

$$\text{Corr}(M, L) = \frac{\text{Cov}(M, L)}{\sqrt{\text{Var}(M)\text{Var}(L)}} = \frac{\frac{1}{4}}{\sqrt{\frac{5}{4} \cdot \frac{1}{4}}} = \frac{1}{\sqrt{5}}.$$ □

☣ **7.3.7.** In the above example, it makes sense that the correlation is positive because M is constrained to be at least as large as L. The following argument would be a blunder: "Either $M = X, L = Y$ or $M = Y, L = X$, so either $\text{Cov}(M, L) = \text{Cov}(X, Y)$ or $\text{Cov}(M, L) = \text{Cov}(Y, X)$. But $\text{Cov}(Y, X) = \text{Cov}(X, Y)$ so we always have $\text{Cov}(M, L) = \text{Cov}(X, Y) = 0$." It is true that either $M = X, L = Y$ will occur or $M = Y, L = X$ will occur, but these are *events*, not deterministic cases. The argument would fall apart if if were written carefully using conditional probability.

Covariance properties can also be a helpful tool for finding *variances*, especially when the r.v. of interest is a sum of dependent random variables. The next example uses properties of covariance to derive the variance of the Hypergeometric distribution. If you did Exercise 48 from Chapter 4, you can compare the two derivations.

Example 7.3.8 (Hypergeometric variance). Find $\text{Var}(X)$ for $X \sim \text{HGeom}(w, b, n)$.

Solution:

Interpret X as the number of white balls in a sample of size n from an urn with w white and b black balls. We can represent X as a sum of indicator random variables, $X = I_1 + \cdots + I_n$, where I_j is the indicator of the jth ball in the sample being white. Each I_j has mean $p = w/(w+b)$ and variance $p(1-p)$, but because the I_j are dependent, we cannot simply add their variances. Instead, we apply properties of covariance:

$$\begin{aligned}\text{Var}(X) &= \text{Var}\left(\sum_{j=1}^{n} I_j\right)\\ &= \text{Var}(I_1) + \cdots + \text{Var}(I_n) + 2\sum_{i<j} \text{Cov}(I_i, I_j)\\ &= np(1-p) + 2\binom{n}{2}\text{Cov}(I_1, I_2),\end{aligned}$$

since all $\binom{n}{2}$ pairs of indicators have the same covariance by symmetry. Now we just need to find $\text{Cov}(I_1, I_2)$. By the fundamental bridge,

$$\begin{aligned}\text{Cov}(I_1, I_2) &= E(I_1 I_2) - E(I_1)E(I_2)\\ &= P(\text{1st and 2nd balls both white}) - P(\text{1st ball white})P(\text{2nd ball white})\\ &= \frac{w}{w+b} \cdot \frac{w-1}{w+b-1} - p^2.\end{aligned}$$

Plugging this into the above formula and simplifying, we eventually obtain

$$\text{Var}(X) = \frac{N-n}{N-1} np(1-p),$$

where $N = w + b$. This differs from the Binomial variance of $np(1-p)$ by a factor of $\frac{N-n}{N-1}$, which is known as the *finite population correction*. The discrepancy arises from the fact that in the Binomial story, we sample with replacement, so the same ball can be drawn multiple times; in the Hypergeometric story, we sample without replacement, so each ball appears in the sample at most once.

If we consider N to be the "population size" of the urn, then as N grows very large relative to the sample size n, it becomes extremely unlikely that in sampling with replacement, we would draw the same ball more than once. Thus sampling with replacement and sampling without replacement become equivalent in the limit as $N \to \infty$ with n fixed, and the finite population correction approaches 1.

The other case where sampling with and without replacement are equivalent is the simple case where we only draw one ball from the urn, and indeed, the finite population correction also equals 1 when $n = 1$. $\square$

The last two sections in this chapter introduce the Multinomial and Multivariate Normal distributions. The Multinomial is the most famous discrete multivariate distribution, and the Multivariate Normal is the most famous continuous multivariate distribution.

7.4 Multinomial

The Multinomial distribution is a generalization of the Binomial. Whereas the Binomial distribution counts the successes in a fixed number of trials that can only be categorized as success or failure, the Multinomial distribution keeps track of trials whose outcomes can fall into multiple categories, such as excellent, adequate, poor; or red, yellow, green, blue.

Story 7.4.1 (Multinomial distribution). Each of n objects is independently placed into one of k categories. An object is placed into category j with probability p_j, where the p_j are nonnegative and $\sum_{j=1}^{k} p_j = 1$. Let X_1 be the number of objects in category 1, X_2 the number of objects in category 2, etc., so that $X_1 + \cdots + X_k = n$. Then $\mathbf{X} = (X_1, \ldots, X_k)$ is said to have the *Multinomial distribution* with parameters n and $\mathbf{p} = (p_1, \ldots, p_k)$. We write this as $\mathbf{X} \sim \text{Mult}_k(n, \mathbf{p})$. □

We call $\mathbf{X}$ a *random vector* because it is a vector of random variables. The joint PMF of $\mathbf{X}$ can be derived from the story.

Theorem 7.4.2 (Multinomial joint PMF). If $\mathbf{X} \sim \text{Mult}_k(n, \mathbf{p})$, then the joint PMF of $\mathbf{X}$ is

$$P(X_1 = n_1, \ldots, X_k = n_k) = \frac{n!}{n_1! n_2! \ldots n_k!} \cdot p_1^{n_1} p_2^{n_2} \ldots p_k^{n_k},$$

for $n_1, \ldots, n_k$ satisfying $n_1 + \cdots + n_k = n$.

Proof. If $n_1, \ldots, n_k$ don't add up to n, then the event $\{X_1 = n_1, \ldots, X_k = n_k\}$ is impossible: every object has to go somewhere, and new objects can't appear out of nowhere. If $n_1, \ldots, n_k$ do add up to n, then any particular way of putting n_1 objects into category 1, n_2 objects into category 2, etc., has probability $p_1^{n_1} p_2^{n_2} \ldots p_k^{n_k}$, and there are

$$\frac{n!}{n_1! n_2! \ldots n_k!}$$

ways to do this, as discussed in Example 1.4.18 in the context of rearranging the letters in STATISTICS. So the joint PMF is as claimed. ■

Since we've specified the joint distribution of $\mathbf{X}$, we have enough information to determine the marginal and conditional distributions, as well as the covariance between any two components of $\mathbf{X}$.

Let's take these one by one, starting with the marginal distribution of X_j, which is the jth component of $\mathbf{X}$. Were we to blindly apply the definition, we would have to sum the joint PMF over all components of $\mathbf{X}$ other than X_j. The prospect of $k-1$ summations is an unpleasant one, to say the least. Fortunately, we can avoid tedious calculations if we instead use the story of the Multinomial distribution: X_j is the number of objects in category j, where each of the n objects independently belongs to category j with probability p_j. Define success as landing in category j. Then we just have n independent Bernoulli trials, so the marginal distribution of X_j is $\text{Bin}(n, p_j)$.

Theorem 7.4.3 (Multinomial marginals)**.** The marginals of a Multinomial are Binomial. Specifically, if $\mathbf{X} \sim \text{Mult}_k(n, \mathbf{p})$, then $X_j \sim \text{Bin}(n, p_j)$.

More generally, whenever we merge multiple categories together in a Multinomial random vector, we get another Multinomial random vector. For example, suppose we randomly sample n people in a country with 5 political parties. (If the sampling is done without replacement, the n trials are not independent, but independence is a good approximation as long as the population is large relative to the sample, as we discussed in Theorem 3.9.3 and Example 7.3.8.) Let

$$\mathbf{X} = (X_1, \dots, X_5) \sim \text{Mult}_5(n, (p_1, \dots, p_5))$$

represent the political party affiliations of the sample, i.e., X_j is the number of people in the sample who support party j.

Suppose that parties 1 and 2 are the dominant parties, while parties 3 through 5 are minor third parties. If we decide that instead of keeping track of all 5 parties, we only want to count the number of people in party 1, party 2, or "other", then we can define a new random vector that lumps all the third parties into one category:

$$\mathbf{Y} = (X_1, X_2, X_3 + X_4 + X_5).$$

By the story of the Multinomial,

$$\mathbf{Y} \sim \text{Mult}_3(n, (p_1, p_2, p_3 + p_4 + p_5)).$$

Of course, this idea applies to merging categories in any Multinomial, not just in the context of political parties.

Theorem 7.4.4 (Multinomial lumping)**.** If $\mathbf{X} \sim \text{Mult}_k(n, \mathbf{p})$, then for any distinct i and j, $X_i + X_j \sim \text{Bin}(n, p_i + p_j)$. The random vector of counts obtained from merging categories i and j is still Multinomial. For example, merging categories 1 and 2 gives

$$(X_1 + X_2, X_3, \dots, X_k) \sim \text{Mult}_{k-1}(n, (p_1 + p_2, p_3, \dots, p_k)).$$

Now for the conditional distributions. Suppose we get to observe X_1, the number of objects in category 1, and we wish to update our distribution for the other categories $(X_2, \dots, X_k)$. One way to do this is with the definition of conditional PMF:

$$P(X_2 = n_2, \dots, X_k = n_k | X_1 = n_1) = \frac{P(X_1 = n_1, X_2 = n_2, \dots, X_k = n_k)}{P(X_1 = n_1)}.$$

The numerator is the joint PMF of the Multinomial, and the denominator is the marginal PMF of X_1, both of which we have already derived. However, we prefer to use the Multinomial story to deduce the conditional distribution of $(X_2, \dots, X_k)$ without algebra. Given that there are n_1 objects in category 1, the remaining $n - n_1$ objects fall into categories 2 through k, independently of one another. By Bayes' rule, the conditional probability of falling into category j is

$$P(\text{in category } j | \text{not in category } 1) = \frac{P(\text{in category } j)}{P(\text{not in category } 1)} = \frac{p_j}{p_2 + \dots + p_k},$$

for $j = 2, \dots, k$. This makes intuitive sense: the updated probabilities are proportional to the original probabilities $(p_2, \dots, p_k)$, but these must be renormalized to yield a valid probability vector. Putting it all together, we have the following result.

Theorem 7.4.5 (Multinomial conditioning). If $\mathbf{X} \sim \text{Mult}_k(n, \mathbf{p})$, then

$$(X_2, \dots, X_k) \big| X_1 = n_1 \sim \text{Mult}_{k-1}(n - n_1, (p'_2, \dots, p'_k)),$$

where $p'_j = p_j / (p_2 + \dots + p_k)$.

Finally, we know that components within a Multinomial random vector are dependent since they are constrained by $X_1 + \dots + X_k = n$. To find the covariance between X_i and X_j, we can use the marginal and lumping properties we have just discussed.

Theorem 7.4.6 (Covariance in a Multinomial). Let $(X_1, \dots, X_k) \sim \text{Mult}_k(n, \mathbf{p})$, where $\mathbf{p} = (p_1, \dots, p_k)$. For $i \neq j$, $\text{Cov}(X_i, X_j) = -np_ip_j$.

Proof. For concreteness, let $i = 1$ and $j = 2$. Using the lumping property and the marginal distributions of a Multinomial, we know $X_1 + X_2 \sim \text{Bin}(n, p_1 + p_2)$, $X_1 \sim \text{Bin}(n, p_1)$, and $X_2 \sim \text{Bin}(n, p_2)$. Therefore

$$\text{Var}(X_1 + X_2) = \text{Var}(X_1) + \text{Var}(X_2) + 2\text{Cov}(X_1, X_2)$$

becomes

$$n(p_1 + p_2)(1 - (p_1 + p_2)) = np_1(1 - p_1) + np_2(1 - p_2) + 2\text{Cov}(X_1, X_2).$$

Solving for $\text{Cov}(X_1, X_2)$ gives $\text{Cov}(X_1, X_2) = -np_1p_2$. By the same logic, for $i \neq j$, we have $\text{Cov}(X_i, X_j) = -np_ip_j$.

The components are negatively correlated, as we would expect: if we know there are a lot of objects in category i, then there aren't as many objects left over that could possibly be in category j. Exercise 65 asks for a different proof of this result, using indicators. ■

☢ **7.4.7** (Independent trials but dependent components). The k *components* of a Multinomial are dependent, but the n objects in the story of the Multinomial are independently categorized. In the extreme case $k = 2$, a $\text{Mult}_k(n, \mathbf{p})$ random vector looks like $(X, n - X)$ with $X \sim \text{Bin}(n, p_1)$, which we can think of as (number of successes, number of failures), where "success" is defined as getting category 1. The number of successes is perfectly negatively correlated with the number of failures, even though the trials are independent.

We conclude this section with a heroic example that ties together many of the most important concepts for the Multinomial.

Example 7.4.8 (Statwoman). The superhero Statwoman uses probability and statistics to fight crime. She has battled with countless foes, sometimes even fighting several at the same time. For simplicity though, assume that each of her battles is with exactly one of the following adversaries: the Confounder, the Extrapolator, and the Overfitter.

Suppose that Statwoman will have n battles next year (with n a positive integer), and that each battle is with the Confounder with probability p_1, the Extrapolator with probability p_2, and the Overfitter with probability p_3, independently. Here p_1, p_2, p_3 are nonnegative and sum to 1. Let X_1, X_2, X_3 be the numbers of battles Statwoman will have with the Confounder, the Extrapolator, and the Overfitter next year, respectively.

(a) Find the joint distribution of X_1, X_2, X_3.

(b) Find the correlation between X_1 and X_2.

(c) Suppose for this part only that it turns out that the Extrapolator and the Overfitter have been devising evil plots together, so it is of interest to study their combined number of skirmishes with Statwoman. Let $X_{23} = X_2 + X_3$. Find the joint distribution of X_1, X_{23}.

(d) Suppose for this part only that the parameters p_1, p_2, p_3 are *unknown*, $n = 360$, and it is observed that exactly 36 of Statwoman's battles are with the Overfitter. A natural way to estimate p_3 would be to use $36/360 = 0.1$. The *maximum likelihood estimate* (MLE) of p_3 is the value of p_3 that makes the observed data, $X_3 = 36$, as likely as possible. That is, the MLE is the value of p_3 that maximizes $P(X_3 = 36)$. Show that the MLE is the natural estimate, 0.1.

(e) Assume for this part only that the Overfitter has been captured, through heroic efforts! So assume that all of Statwoman's battles next year will be with one of her other two adversaries. Find the joint PMF of X_1, X_2, given that $X_3 = 0$.

(f) Now suppose that, instead of the number of battles being a constant n, the number of battles is $N \sim \text{Pois}(\lambda)$. Find the joint distribution of X_1, X_2, X_3.

Solution:

(a) By the story of the Multinomial, $(X_1, X_2, X_3) \sim \text{Mult}_3(n, (p_1, p_2, p_3))$.

(b) We have $X_1 \sim \text{Bin}(n, p_1), X_2 \sim \text{Bin}(n, p_2)$, and $\text{Cov}(X_1, X_2) = -np_1p_2$, so

$$\text{Corr}(X_1, X_2) = \frac{\text{Cov}(X_1, X_2)}{\text{SD}(X_1)\text{SD}(X_2)} = \frac{-np_1p_2}{\sqrt{np_1(1-p_1)np_2(1-p_2)}} = -\sqrt{\frac{p_1p_2}{(1-p_1)(1-p_2)}}.$$

(c) By the lumping property of the Multinomial,

$$(X_1, X_{23}) \sim \text{Mult}_2(n, (p_1, p_2 + p_3)).$$

(d) In general for $X \sim \text{Bin}(n, p)$, if $X = x$ is observed then the MLE of p is the value of p that maximizes the function

$$L(p) = \binom{n}{x} p^x (1-p)^{n-x}.$$

(The function L is called the *likelihood function* in statistics. It is the probability of the data, regarded as a function of the parameter with the data treated as fixed.)

As if often the case when dealing with a product of positive numbers, it is helpful to take the log. It is equivalent to find the value $\hat{p}$ that maximizes

$$\log L(p) = \log \binom{n}{x} + x \log p + (n - x) \log(1 - p),$$

since log is a continuous, strictly increasing function. Setting the derivative of $\log L(p)$ (with respect to p, holding x constant) equal to 0, we have

$$\frac{x}{\hat{p}} - \frac{n-x}{1-\hat{p}} = 0,$$

which rearranges to $\hat{p} = x/n$. We have found a maximum since the second derivative of $\log L(p)$ is

$$-\frac{x}{p^2} - \frac{n-x}{(1-p)^2} < 0.$$

So the MLE of p_3 is $\hat{p}_3 = 36/360 = 0.1$.

(e) By the result on Multinomial conditioning,

$$(X_1, X_2)|(X_3 = 0) \sim \text{Mult}_2\left(n, \left(\frac{p_1}{p_1 + p_2}, \frac{p_2}{p_1 + p_2}\right)\right).$$

(f) Reasoning as in the chicken-egg story,

$$P(X_1 = x_1, X_2 = x_2, X_3 = x_3) = \sum_{n=0}^{\infty} P(X_1 = x_1, X_2 = x_2, X_3 = x_3 | N = n) P(N = n),$$

where all terms of the sum are 0 except the term with $n = x_1 + x_2 + x_3$. For this value of n,

$$\begin{aligned} P(X_1 = x_1, X_2 = x_2, X_3 = x_3) &= P(X_1 = x_1, X_2 = x_2, X_3 = x_3 | N = n)P(N = n) \\ &= \frac{n!}{x_1!x_2!x_3!} p_1^{x_1} p_2^{x_2} p_3^{x_3} \cdot e^{-\lambda} \frac{\lambda^n}{n!} \\ &= \frac{e^{-\lambda p_1}(\lambda p_1)^{x_1}}{x_1!} \cdot \frac{e^{-\lambda p_2}(\lambda p_2)^{x_2}}{x_2!} \cdot \frac{e^{-\lambda p_3}(\lambda p_3)^{x_3}}{x_3!}, \end{aligned}$$

for all nonnegative integers x_1, x_2, x_3. Therefore, X_1, X_2, X_3 are independent, with $X_j \sim \text{Pois}(\lambda p_j)$. This result is a Multinomial extension of the chicken-egg story. □

7.5 Multivariate Normal

The Multivariate Normal is a continuous multivariate distribution that generalizes the Normal distribution into higher dimensions. We will not work with the rather unwieldy joint PDF of the Multivariate Normal. Instead we define the Multivariate Normal by its relationship to the ordinary Normal.

Definition 7.5.1 (Multivariate Normal distribution). A k-dimensional random vector $\mathbf{X} = (X_1, \dots, X_k)$ is said to have a *Multivariate Normal* (MVN) distribution if every linear combination of the X_j has a Normal distribution. That is, we require

$$t_1 X_1 + \cdots + t_k X_k$$

to have a Normal distribution for any constants $t_1, \dots, t_k$. If $t_1 X_1 + \cdots + t_k X_k$ is a constant (such as when all $t_i = 0$), we consider it to have a Normal distribution, albeit a degenerate Normal with variance 0. An important special case is $k = 2$; this distribution is called the *Bivariate Normal* (BVN).

If $(X_1, \dots, X_k)$ is MVN, then the marginal distribution of X_1 is Normal, since we can take t_1 to be 1 and all other t_j to be 0. Similarly, the marginal distribution of each X_j is Normal. However, the converse is false: it is possible to have Normally distributed r.v.s $X_1, \dots, X_k$ such that $(X_1, \dots, X_k)$ is not Multivariate Normal.

Example 7.5.2 (Non-example of MVN). Here is an example of two r.v.s whose marginal distributions are Normal but whose joint distribution is not Bivariate Normal. Let $X \sim \mathcal{N}(0, 1)$, and let

$$S = \begin{cases} 1 & \text{with probability } 1/2 \\ -1 & \text{with probability } 1/2 \end{cases}$$

be a *random sign* independent of X. Then $Y = SX$ is a standard Normal r.v., due to the symmetry of the Normal distribution (see Exercise 30 from Chapter 5). However,

(X, Y) is not Bivariate Normal because $P(X + Y = 0) = P(S = -1) = 1/2$, which implies that $X + Y$ can't be Normal (or, for that matter, have any continuous distribution). Since $X + Y$ is a linear combination of X and Y that is not Normally distributed, (X, Y) is not Bivariate Normal. □

Example 7.5.3 (Actual MVN). For $Z, W \stackrel{\text{i.i.d.}}{\sim} \mathcal{N}(0, 1)$, (Z, W) is Bivariate Normal because the sum of independent Normals is Normal. Also, $(Z + 2W, 3Z + 5W)$ is Bivariate Normal, since an arbitrary linear combination

$$t_1(Z + 2W) + t_2(3Z + 5W)$$

can also be written as a linear combination of Z and W,

$$(t_1 + 3t_2)Z + (2t_1 + 5t_2)W,$$

which is Normal. □

The above example showed that if we start with a Multivariate Normal and take linear combinations of the components, we form a new Multivariate Normal. The next two theorems state that we can also produce new MVNs from old MVNs with the operations of subsetting and concatenation.

Theorem 7.5.4. If (X_1, X_2, X_3) is Multivariate Normal, then so is the subvector (X_1, X_2).

Proof. Any linear combination $t_1X_1 + t_2X_2$ can be thought of as a linear combination of X_1, X_2, X_3 where the coefficient of X_3 is 0. So $t_1X_1 + t_2X_2$ is Normal for all t_1, t_2, which shows that (X_1, X_2) is MVN. ■

Theorem 7.5.5. If $\mathbf{X} = (X_1, \dots, X_n)$ and $\mathbf{Y} = (Y_1, \dots, Y_m)$ are Multivariate Normal random vectors with $\mathbf{X}$ independent of $\mathbf{Y}$, then the concatenated random vector $\mathbf{W} = (X_1, \dots, X_n, Y_1, \dots, Y_m)$ is Multivariate Normal.

Proof. Any linear combination $s_1X_1 + \dots + s_nX_n + t_1Y_1 + \dots + t_mY_m$ is Normal since $s_1X_1 + \dots + s_nX_n$ and $t_1Y_1 + \dots + t_mY_m$ are Normal (by definition of MVN) and are independent, so their sum is Normal (as shown in Chapter 6 using MGFs). ■

A Multivariate Normal distribution is fully specified by knowing the mean of each component, the variance of each component, and the covariance or correlation between any two components. Another way to say this is that the parameters of an MVN random vector $(X_1, \dots, X_k)$ are as follows:

- the *mean vector* $(\mu_1, \dots, \mu_k)$, where $E(X_j) = \mu_j$;
- the *covariance matrix*, which is the $k \times k$ matrix of covariances between components, arranged so that the row i, column j entry is $\text{Cov}(X_i, X_j)$.

For example, in order to fully specify a Bivariate Normal distribution for (X, Y), we need to know five parameters:

- the means $E(X)$, $E(Y)$;
- the variances $\text{Var}(X)$, $\text{Var}(Y)$;
- the correlation $\text{Corr}(X, Y)$.

We will show in Example 8.1.10 that the joint PDF of a Bivariate Normal (X, Y) with $\mathcal{N}(0, 1)$ marginal distributions and correlation $\rho \in (-1, 1)$ is

$$f_{X,Y}(x, y) = \frac{1}{2\pi\tau} \exp\left(-\frac{1}{2\tau^2}(x^2 + y^2 - 2\rho xy)\right),$$

with $\tau = \sqrt{1 - \rho^2}$. Figure 7.10 plots the joint PDFs for two different Bivariate Normal distributions with $\mathcal{N}(0, 1)$ marginals, along with the corresponding contour plots. On the left, X and Y are uncorrelated, so the level curves of the joint PDF are circles. On the right, X and Y have a correlation of 0.75, so the level curves are ellipsoidal, reflecting the fact that Y tends to be large when X is large.

Just as the distribution of an r.v. is determined by its CDF, PMF/PDF, or MGF, the joint distribution of a random vector is determined by its joint CDF, joint PMF/PDF, or *joint MGF*, which we now define.

Definition 7.5.6 (Joint MGF). The *joint moment generating function* (joint MGF) of a random vector $\mathbf{X} = (X_1, \dots, X_k)$ is the function M defined by

$$M(\mathbf{t}) = E(e^{\mathbf{t}'\mathbf{X}}) = E\left(e^{t_1X_1 + \cdots + t_kX_k}\right),$$

for $\mathbf{t} = (t_1, \dots, t_k) \in \mathbb{R}^k$. We require this expectation to be finite in a box containing the origin in $\mathbb{R}^k$; otherwise we say the joint MGF does not exist.

For a Multivariate Normal random vector, the joint MGF is particularly nice because the term in the exponent, $t_1X_1 + \cdots + t_kX_k$, is a Normal r.v. by definition. This means we can use what we know about the univariate Normal MGF to find the Multivariate Normal joint MGF! Recall that for any Normal r.v. W,

$$E(e^W) = e^{E(W) + \frac{1}{2}\text{Var}(W)}.$$

Therefore the joint MGF of a Multivariate Normal $(X_1, \dots, X_k)$ is

$$E(e^{t_1X_1 + \cdots + t_kX_k}) = \exp\left(t_1E(X_1) + \cdots + t_kE(X_k) + \frac{1}{2}\text{Var}(t_1X_1 + \cdots + t_kX_k)\right).$$

The variance term can be expanded using properties of covariance.

We know that in general, independence is a stronger condition than zero correlation; r.v.s can be uncorrelated but not independent. A special property of the Multivariate Normal distribution is that for r.v.s whose joint distribution is MVN, independence and zero correlation are equivalent conditions.

Theorem 7.5.7. Within an MVN random vector, uncorrelated implies independent. That is, if $\mathbf{X} \sim \text{MVN}$ can be written as $\mathbf{X} = (\mathbf{X}_1, \mathbf{X}_2)$, where $\mathbf{X}_1$ and $\mathbf{X}_2$ are

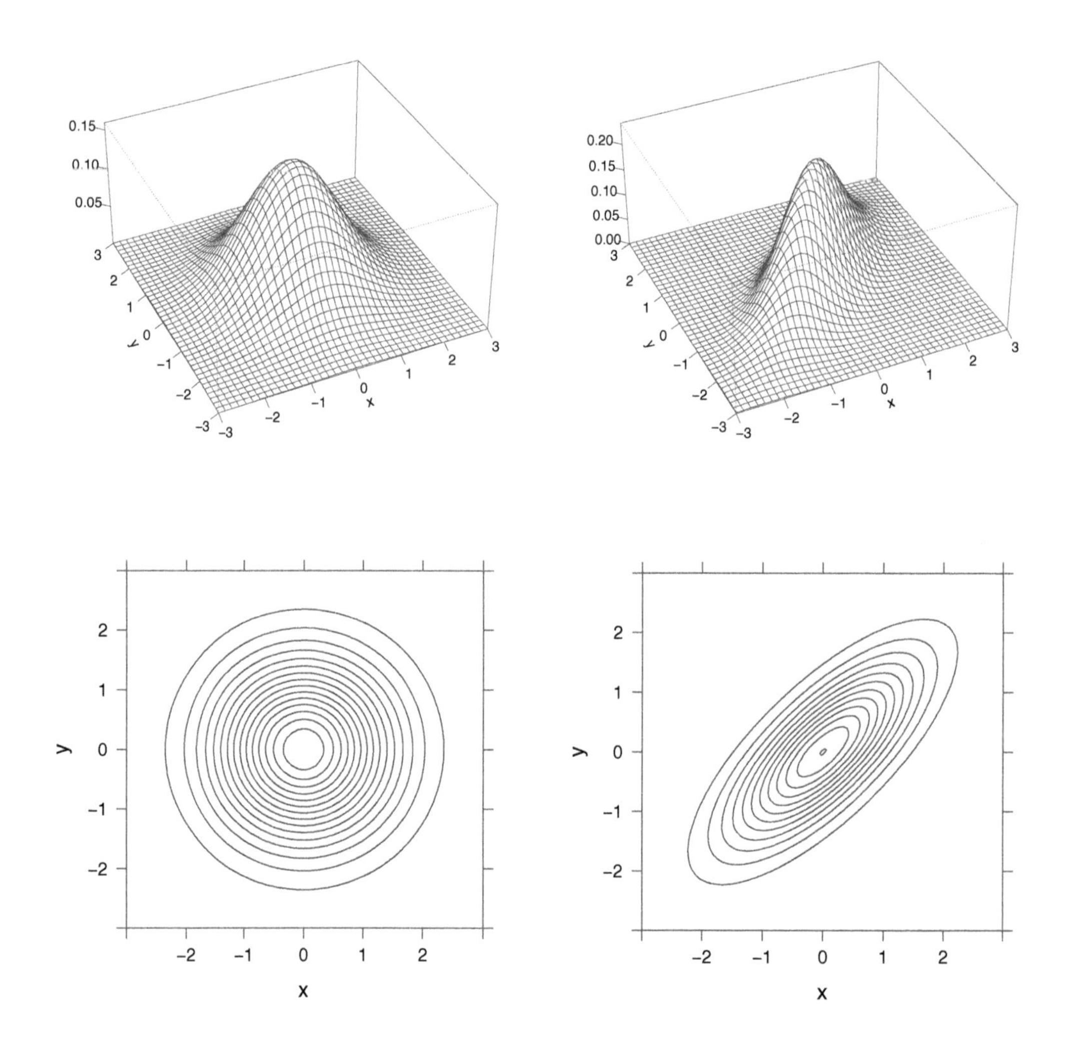

FIGURE 7.10
Joint PDFs of two Bivariate Normal distributions, and the corresponding contour plots. On the left, X and Y are marginally $\mathcal{N}(0,1)$ and have zero correlation. On the right, X and Y are marginally $\mathcal{N}(0,1)$ and have correlation 0.75.

subvectors, and every component of $\mathbf{X}_1$ is uncorrelated with every component of $\mathbf{X}_2$, then $\mathbf{X}_1$ and $\mathbf{X}_2$ are independent.[1]

In particular, if (X, Y) is Bivariate Normal and $\text{Corr}(X, Y) = 0$, then X and Y are independent.

Proof. We will prove this theorem in the case of a Bivariate Normal; the proof in higher dimensions is analogous. Let (X, Y) be Bivariate Normal with $E(X) = \mu_1$, $E(Y) = \mu_2$, $\text{Var}(X) = \sigma_1^2$, $\text{Var}(Y) = \sigma_2^2$, and $\text{Corr}(X, Y) = \rho$. The joint MGF is

$$\begin{aligned} M_{X,Y}(s,t) = E(e^{sX+tY}) &= \exp\left(s\mu_1 + t\mu_2 + \frac{1}{2}\text{Var}(sX+tY)\right) \\ &= \exp\left(s\mu_1 + t\mu_2 + \frac{1}{2}(s^2\sigma_1^2 + t^2\sigma_2^2 + 2st\sigma_1\sigma_2\rho)\right). \end{aligned}$$

If $\rho = 0$, the joint MGF reduces to

$$M_{X,Y}(s,t) = \exp\left(s\mu_1 + t\mu_2 + \frac{1}{2}(s^2\sigma_1^2 + t^2\sigma_2^2)\right).$$

But this is also the joint MGF of (Z, W) where $Z \sim \mathcal{N}(\mu_1, \sigma_1^2)$ and $W \sim \mathcal{N}(\mu_2, \sigma_2^2)$ and Z is independent of W. Since the joint MGF determines the joint distribution, it must be that (X, Y) has the same joint distribution as (Z, W). Therefore X and Y are independent. ■

This theorem does not apply to Example 7.5.2. In that example, as you can verify, X and Y are uncorrelated and not independent, but this does not contradict the theorem because (X, Y) is not BVN. The next two examples show situations where the theorem *does* apply.

Example 7.5.8 (Independence of sum and difference). Let $X, Y \overset{\text{i.i.d.}}{\sim} \mathcal{N}(0, 1)$. Find the joint distribution of $(X + Y, X - Y)$.

Solution:

Since $(X + Y, X - Y)$ is Bivariate Normal and

$$\text{Cov}(X+Y, X-Y) = \text{Var}(X) - \text{Cov}(X,Y) + \text{Cov}(Y,X) - \text{Var}(Y) = 0,$$

$X + Y$ is independent of $X - Y$. Furthermore, they are i.i.d. $\mathcal{N}(0, 2)$. By the same method, we have that if $X \sim \mathcal{N}(\mu_1, \sigma^2)$ and $Y \sim \mathcal{N}(\mu_2, \sigma^2)$ are independent (with the same variance), then $X + Y$ is independent of $X - Y$.

It can be shown that the independence of the sum and difference is a unique characteristic of the Normal! That is, if X and Y are i.i.d. and $X + Y$ is independent of $X - Y$, then X and Y must have Normal distributions.

[1]Independence of random vectors is defined analogously to independence of random variables. In particular, if $\mathbf{X}_i$ has joint PDF $f_{\mathbf{X}_i}$, then it says that $f_{\mathbf{X}}(\mathbf{x}_1, \mathbf{x}_2) = f_{\mathbf{X}_1}(\mathbf{x}_1) f_{\mathbf{X}_2}(\mathbf{x}_2)$ for all $\mathbf{x}_1, \mathbf{x}_2$.

In Exercise 72, you'll extend this example to the case where X and Y are Bivariate Normal with general correlation ρ. □

Example 7.5.9 (Independence of sample mean and sample variance). Let $X_1, \dots, X_n$ be i.i.d. $\mathcal{N}(\mu, \sigma^2)$, with $n \geq 2$. Define

$$\bar{X}_n = \frac{1}{n}(X_1 + \cdots + X_n),$$

$$S_n^2 = \frac{1}{n-1}\sum_{j=1}^{n}(X_j - \bar{X}_n)^2.$$

As shown in Chapter 6, the sample mean $\bar{X}_n$ has expectation μ (the true mean) and the sample variance S_n^2 has expectation σ^2 (the true variance). Show that $\bar{X}_n$ and S_n^2 are independent by applying MVN ideas to $(\bar{X}_n, X_1 - \bar{X}_n, \dots, X_n - \bar{X}_n)$.

Solution:

The vector $(\bar{X}_n, X_1 - \bar{X}_n, \dots, X_n - \bar{X}_n)$ is MVN since any linear combination of its components can be written as a linear combination of $X_1, \dots, X_n$. Furthermore, $E(X_j - \bar{X}_n) = 0$ by linearity. Now we compute the covariance of $\bar{X}_n$ with $X_j - \bar{X}_n$:

$$\text{Cov}(\bar{X}_n, X_j - \bar{X}_n) = \text{Cov}(\bar{X}_n, X_j) - \text{Cov}(\bar{X}_n, \bar{X}_n).$$

For $\text{Cov}(\bar{X}_n, X_j)$, we can expand out $\bar{X}_n$, and most of the terms cancel due to independence:

$$\text{Cov}(\bar{X}_n, X_j) = \text{Cov}\left(\frac{1}{n}X_1 + \cdots + \frac{1}{n}X_n, X_j\right) = \text{Cov}\left(\frac{1}{n}X_j, X_j\right) = \frac{1}{n}\text{Var}(X_j) = \frac{\sigma^2}{n}.$$

For $\text{Cov}(\bar{X}_n, \bar{X}_n)$, we use properties of variance:

$$\text{Cov}(\bar{X}_n, \bar{X}_n) = \text{Var}(\bar{X}_n) = \frac{1}{n^2}\left(\text{Var}(X_1) + \cdots + \text{Var}(X_n)\right) = \frac{\sigma^2}{n}.$$

Therefore $\text{Cov}(\bar{X}_n, X_j - \bar{X}_n) = 0$, which means $\bar{X}_n$ is uncorrelated with every component of $(X_1 - \bar{X}_n, \dots, X_n - \bar{X}_n)$. Since uncorrelated implies independent within an MVN vector, we have that $\bar{X}_n$ is independent of the vector $(X_1 - \bar{X}_n, \dots, X_n - \bar{X}_n)$. But S_n^2 is a function of $(X_1 - \bar{X}_n, \dots, X_n - \bar{X}_n)$, so $\bar{X}_n$ is also independent of S_n^2.

It can be shown that the independence of the sample mean and variance is another unique characteristic of the Normal! If the X_j followed any other distribution, then $\bar{X}_n$ and S_n^2 would be dependent. □

Example 7.5.10 (Bivariate Normal generation). Suppose that we have access to i.i.d. random variables $X, Y \sim \mathcal{N}(0,1)$, but want to generate a Bivariate Normal random vector (Z, W) with $\text{Corr}(Z, W) = \rho$ and Z, W marginally $\mathcal{N}(0,1)$, for the purpose of running a simulation. How can we construct Z and W from linear combinations of X and Y?

Solution:

By definition of Multivariate Normal, any (Z, W) of the form

$$Z = aX + bY$$
$$W = cX + dY$$

will be Bivariate Normal. So let's try to find suitable a, b, c, d. The means are already 0. Setting the variances equal to 1 gives

$$a^2 + b^2 = 1, c^2 + d^2 = 1.$$

Setting the covariance of Z and W equal to ρ gives

$$ac + bd = \rho.$$

There are more unknowns than equations here, and we just need *one* solution. To simplify, let's look for a solution with $b = 0$. Then $a^2 = 1$, so let's take $a = 1$. Now $ac + bd = \rho$ reduces to $c = \rho$, and then we can use $c^2 + d^2 = 1$ to find a suitable d. Putting everything together, we can generate (Z, W) as

$$Z = X$$
$$W = \rho X + \sqrt{1 - \rho^2} Y.$$

Note that in the extreme case $\rho = 1$ (known as *perfect positive correlation*) this says to let $W = Z \sim \mathcal{N}(0, 1)$, in the extreme case $\rho = -1$ (known as *perfect negative correlation*) it says to let $W = -Z$ with $Z \sim \mathcal{N}(0, 1)$, and in the simple case $\rho = 0$ it says to just let $(Z, W) = (X, Y)$. □

7.6 Recap

Joint distributions allow us to describe the behavior of multiple random variables that arise from the same experiment. Important functions associated with a joint distribution are the joint CDF, joint PMF/PDF, marginal PMF/PDF, and conditional PMF/PDF. The table on the next page summarizes these definitions for two discrete r.v.s and two continuous r.v.s. Joint distributions can also be a hybrid of discrete and continuous, in which case we mix and match PMFs and PDFs.

	Two discrete r.v.s	Two continuous r.v.s
Joint CDF	$F_{X,Y}(x,y) = P(X \leq x, Y \leq y)$	$F_{X,Y}(x,y) = P(X \leq x, Y \leq y)$
Joint PMF/PDF	$P(X = x, Y = y)$ • Joint PMF is nonnegative. • Joint PMF sums to 1. • $P((X,Y) \in A) = \sum\sum_{(x,y)\in A} P(X = x, Y = y)$.	$f_{X,Y}(x,y) = \frac{\partial^2}{\partial x \partial y} F_{X,Y}(x,y)$ • Joint PDF is nonnegative. • Joint PDF integrates to 1. • $P((X,Y) \in A) = \iint_A f_{X,Y}(x,y)dxdy$.
Marginal PMF/PDF	$P(X = x) = \sum_y P(X = x, Y = y)$ $= \sum_y P(X = x \| Y = y)P(Y = y)$	$f_X(x) = \int_{-\infty}^{\infty} f_{X,Y}(x,y)dy$ $= \int_{-\infty}^{\infty} f_{X\|Y}(x\|y)f_Y(y)dy$
Conditional PMF/PDF	$P(Y = y \| X = x) = \frac{P(X = x, Y = y)}{P(X = x)}$ $= \frac{P(X = x \| Y = y)P(Y = y)}{P(X = x)}$	$f_{Y\|X}(y\|x) = \frac{f_{X,Y}(x,y)}{f_X(x)}$ $= \frac{f_{X\|Y}(x\|y)f_Y(y)}{f_X(x)}$
Independence	$P(X \leq x, Y \leq y) = P(X \leq x)P(Y \leq y)$ $P(X = x, Y = y) = P(X = x)P(Y = y)$ for all x and y.	$P(X \leq x, Y \leq y) = P(X \leq x)P(Y \leq y)$ $f_{X,Y}(x,y) = f_X(x)f_Y(y)$ for all x and y.
LOTUS	$E(g(X,Y)) = \sum_y \sum_x g(x,y)P(X = x, Y = y)$	$E(g(X,Y)) = \int_{-\infty}^{\infty}\int_{-\infty}^{\infty} g(x,y)f_{X,Y}(x,y)dxdy$

Covariance is a single-number summary of the tendency of two r.v.s to move in the same direction. If two r.v.s are independent, then they are uncorrelated, but the converse does not hold. Correlation is a unitless, standardized version of covariance that is always between -1 and 1.

Two important named multivariate distributions are the Multinomial and Multivariate Normal. The Multinomial distribution is a generalization of the Binomial; a $\text{Mult}_k(n, \mathbf{p})$ random vector counts the number of objects, out of n, that fall into each of k categories, where $\mathbf{p}$ is the vector of probabilities for the k categories.

The Multivariate Normal distribution is a generalization of the Normal; a random vector is defined to be MVN if any linear combination of its components has a Normal distribution. A key property of the MVN distribution is that within an MVN random vector, uncorrelated implies independent.

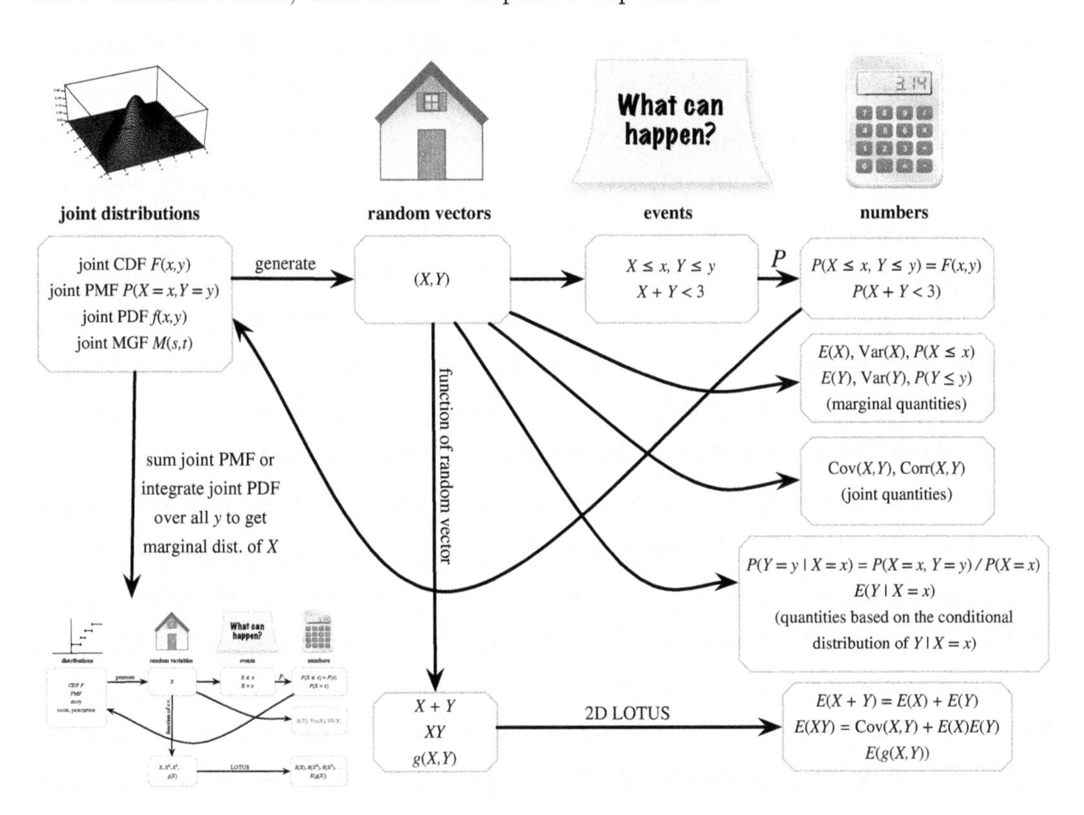

FIGURE 7.11

Fundamental objects of probability for multivariate distributions. A joint distribution is determined by a joint CDF, joint PMF/PDF, or joint MGF. A random vector (X, Y) gives rise to many useful joint, marginal, and conditional quantities. Using 2D LOTUS, we can find the expected value of a function of X and Y. Summing the joint PMF or integrating the joint PDF over all y gives the marginal distribution of X, bringing us back to the case of a one-dimensional distribution.

Figure 7.11 extends our diagram of the fundamental objects of probability to the multivariate setting (taken as bivariate to simplify notation). A joint distribution can be used to generate random vectors (X, Y). Various joint, marginal, and conditional quantities can then be studied. Summing the joint PMF or integrating the joint PDF over all y gives the marginal distribution of X, bringing us back to the one-dimensional realm.

7.7 R

Multinomial

The functions for the Multinomial distribution are `dmultinom` (which is the joint PMF of the Multinomial distribution) and `rmultinom` (which generates realizations of Multinomial random vectors). The joint CDF of the Multinomial is a pain to work with, so it is not built into R.

To use `dmultinom`, we have to input the value at which to evaluate the joint PMF, as well as the parameters of the distribution. For example,

```
x <- c(2,0,3)
n <- 5
p <- c(1/3,1/3,1/3)
dmultinom(x,n,p)
```

returns the probability $P(X_1 = 2, X_2 = 0, X_3 = 3)$, where

$$\mathbf{X} = (X_1, X_2, X_3) \sim \text{Mult}_3(5, (1/3, 1/3, 1/3)).$$

Of course, `n` has to equal `sum(x)`; if we attempted to do `dmultinom(x,7,p)`, R would report an error.

For `rmultinom`, the first input is the number of Multinomial random vectors to generate, and the other inputs are the same. When we typed `rmultinom(10,n,p)` with `n` and `p` as above, R gave us the following matrix:

```
0   2   1   3   2   3   1   2   3   4
2   2   2   2   3   0   1   2   0   0
3   1   2   0   0   2   3   1   2   1
```

Each column of the matrix corresponds to a draw from the $\text{Mult}_3(5, (1/3, 1/3, 1/3))$ distribution. In particular, the sum of each column is 5.

Multivariate Normal

Functions for the Multivariate Normal distribution are located in the package `mvtnorm`. Online resources can teach you how to install packages in R for your system, but for many systems an easy way is to use the `install.packages` command, e.g., by typing `install.packages("mvtnorm")` to install the `mvtnorm` package. After installing it, load the package with `library(mvtnorm)`. Then `dmvnorm` can be used for calculating the joint PDF, and `rmvnorm` can be used for generating random vectors. For example, suppose that we want to generate 1000 independent Bivariate Normal pairs (Z, W), with correlation $\rho = 0.7$ and $\mathcal{N}(0,1)$ marginals. To do this, we can enter the following:

```
meanvector <- c(0,0)
rho <- 0.7
covmatrix <- matrix(c(1,rho,rho,1), nrow = 2, ncol = 2)
r <- rmvnorm(n = 10^3, mean = meanvector, sigma = covmatrix)
```

The covariance matrix here is

$$\begin{pmatrix} 1 & \rho \\ \rho & 1 \end{pmatrix}$$

because

- $\text{Cov}(Z, Z) = \text{Var}(Z) = 1$ (this is the upper left entry),
- $\text{Cov}(W, W) = \text{Var}(W) = 1$ (this is the lower right entry),
- $\text{Cov}(Z, W) = \text{Corr}(Z, W)\text{SD}(Z)\text{SD}(W) = \rho$ (this is the other two entries).

Now `r` is a 1000×2 matrix, with each row a BVN random vector. To see these as points in the plane, we can use `plot(r)` to make a scatter plot, from which the strong positive correlation should be clear. To estimate the covariance of Z and W, we can use `cov(r)`, which the true covariance matrix.

Example 7.5.10 gives another approach to the BVN generation problem:

```
rho <- 0.7
tau <- sqrt(1-rho^2)
x <- rnorm(10^3)
y <- rnorm(10^3)
z <- x
w <- rho*x + tau*y
```

This gives the Z-coordinates in a vector `z` and the W-coordinates in a vector `w`. If we want to put them into one 1000×2 matrix as we had above, we can type `cbind(z,w)` to bind the vectors together as columns.

Cauchy

We can work with the Cauchy distribution introduced in Example 7.1.25 using the three functions `dcauchy`, `pcauchy`, and `rcauchy`. Only one input is needed; for example, `dcauchy(0)` is the Cauchy PDF evaluated at 0.

For an amusing demonstration of the very heavy tails of the Cauchy distribution, try creating a histogram of 1000 simulated values of the Cauchy distribution:

```
hist(rcauchy(1000))
```

Due to extreme values in the tails of the distribution, this histogram looks nothing like the PDF of the distribution from which it was generated.

7.8 Exercises

Exercises marked with Ⓢ have detailed solutions at `http://stat110.net`.

Joint, marginal, and conditional distributions

1. Alice and Bob arrange to meet for lunch on a certain day at noon. However, neither is known for punctuality. They both arrive independently at uniformly distributed times between noon and 1 pm on that day. Each is willing to wait up to 15 minutes for the other to show up. What is the probability they will meet for lunch that day?

2. Alice, Bob, and Carl arrange to meet for lunch on a certain day. They arrive independently at uniformly distributed times between 1 pm and 1:30 pm on that day.

   (a) What is the probability that Carl arrives first?

   For the rest of this problem, assume that Carl arrives first at 1:10 pm, and condition on this fact.

   (b) What is the probability that Carl will be waiting alone for more than 10 minutes?

   (c) What is the probability that Carl will have to wait more than 10 minutes until his party is complete?

   (d) What is the probability that the person who arrives second will have to wait more than 5 minutes for the third person to show up?

3. One of two doctors, Dr. Hibbert and Dr. Nick, is called upon to perform a series of n surgeries. Let H be the indicator r.v. for Dr. Hibbert performing the surgeries, and suppose that $E(H) = p$. Given that Dr. Hibbert is performing the surgeries, each surgery is successful with probability a, independently. Given that Dr. Nick is performing the surgeries, each surgery is successful with probability b, independently. Let X be the number of successful surgeries.

   (a) Find the joint PMF of H and X.

   (b) Find the marginal PMF of X.

   (c) Find the conditional PMF of H given $X = k$.

4. A fair coin is flipped twice. Let X be the number of Heads in the two tosses, and Y be the indicator r.v for the tosses landing the same way.

   (a) Find the joint PMF of X and Y.

   (b) Find the marginal PMFs of X and Y.

(c) Are X and Y independent?

(d) Find the conditional PMFs of Y given $X = x$ and of X given $Y = y$.

5. A fair die is rolled, and then a coin with probability p of Heads is flipped as many times as the die roll says, e.g., if the result of the die roll is a 3, then the coin is flipped 3 times. Let X be the result of the die roll and Y be the number of times the coin lands Heads.

(a) Find the joint PMF of X and Y. Are they independent?

(b) Find the marginal PMFs of X and Y.

(c) Find the conditional PMFs of Y given $X = x$ and of Y given $X = x$.

6. A committee of size k is chosen from a group of n women and m men. All possible committees of size k are equally likely. Let X and Y be the numbers of women and men on the committee, respectively.

(a) Find the joint PMF of X and Y. Be sure to specify the support.

(b) Find the marginal PMF of X in two different ways: by doing a computation using the joint PMF, and using a story.

(c) Find the conditional PMF of Y given that $X = x$.

7. A stick of length L (a positive constant) is broken at a uniformly random point X. Given that $X = x$, another breakpoint Y is chosen uniformly on the interval $[0, x]$.

(a) Find the joint PDF of X and Y. Be sure to specify the support.

(b) We already know that the marginal distribution of X is $\text{Unif}(0, L)$. Check that marginalizing out Y from the joint PDF agrees that this is the marginal distribution of X.

(c) We already know that the conditional distribution of Y given $X = x$ is $\text{Unif}(0, x)$. Check that using the definition of conditional PDFs (in terms of joint and marginal PDFs) agrees that this is the conditional distribution of Y given $X = x$.

(d) Find the marginal PDF of Y.

(e) Find the conditional PDF of X given $Y = y$.

8. (a) Five cards are randomly chosen from a standard deck, one at a time *with replacement*. Let X, Y, Z be the numbers of chosen queens, kings, and other cards. Find the joint PMF of X, Y, Z.

(b) Find the joint PMF of X and Y.

Hint: In summing the joint PMF of X, Y, Z over the possible values of Z, note that most terms are 0 because of the constraint that the number of chosen cards is five.

(c) Now assume instead that the sampling is without replacement (all 5-card hands are equally likely). Find the joint PMF of X, Y, Z.

Hint: Use the naive definition of probability.

9. Let X and Y be i.i.d. $\text{Geom}(p)$, and $N = X + Y$.

(a) Find the joint PMF of X, Y, N.

(b) Find the joint PMF of X and N.

(c) Find the conditional PMF of X given $N = n$, and give a simple description in words of what the result says.

10. Let X and Y be i.i.d. Expo(λ), and $T = X + Y$.

(a) Find the conditional CDF of T given $X = x$. Be sure to specify where it is zero.

(b) Find the conditional PDF $f_{T|X}(t|x)$, and verify that it is a valid PDF.

(c) Find the conditional PDF $f_{X|T}(x|t)$, and verify that it is a valid PDF.

Hint: This can be done using Bayes' rule without having to know the marginal PDF of T, by recognizing what the conditional PDF is up to a normalizing constant—then the normalizing constant must be whatever is needed to make the conditional PDF valid.

(d) In Example 8.2.4, we will show that the marginal PDF of T is $f_T(t) = \lambda^2 te^{-\lambda t}$, for $t > 0$. Give a short alternative proof of this fact, based on the previous parts and Bayes' rule.

11. Let X, Y, Z be r.v.s such that $X \sim \mathcal{N}(0, 1)$ and conditional on $X = x$, Y and Z are i.i.d. $\mathcal{N}(x, 1)$.

(a) Find the joint PDF of X, Y, Z.

(b) By definition, Y and Z are conditionally independent given X. Discuss intuitively whether or not Y and Z are also unconditionally independent.

(c) Find the joint PDF of Y and Z. You can leave your answer as an integral, though the integral can be done with some algebra (such as completing the square) and facts about the Normal distribution.

12. Let $X \sim$ Expo(λ), and let c be a positive constant.

(a) If you remember the memoryless property, you already know that the conditional distribution of X given $X > c$ is the same as the distribution of $c + X$ (think of waiting c minutes for a "success" and then having a fresh Expo(λ) additional waiting time). Derive this in another way, by finding the conditional CDF of X given $X > c$ and the conditional PDF of X given $X > c$.

(b) Find the conditional CDF of X given $X < c$ and the conditional PDF of X given $X < c$.

13. Let X and Y be i.i.d. Expo(λ). Find the conditional distribution of X given $X < Y$ in two different ways:

(a) by using calculus to find the conditional PDF.

(b) without using calculus, by arguing that the conditional distribution of X given $X < Y$ is the same distribution as the unconditional distribution of $\min(X, Y)$, and then applying an earlier result about the minimum of independent Exponentials.

14. Ⓢ (a) A stick is broken into three pieces by picking two points independently and uniformly along the stick, and breaking the stick at those two points. What is the probability that the three pieces can be assembled into a triangle?

Hint: A triangle can be formed from 3 line segments of lengths a, b, c if and only if $a, b, c \in (0, 1/2)$. The probability can be interpreted geometrically as proportional to an area in the plane, avoiding all calculus, but make sure for that approach that the distribution of the random point in the plane is Uniform over some region.

(b) Three legs are positioned uniformly and independently on the perimeter of a round table. What is the probability that the table will stand?

15. Let X and Y be continuous r.v.s., with joint CDF $F(x, y)$. Show that the probability that (X, Y) falls into the rectangle $[a_1, a_2] \times [b_1, b_2]$ is

$$F(a_2, b_2) - F(a_1, b_2) + F(a_1, b_1) - F(a_2, b_1).$$

(a) What is the probability that the Blotchville company bus arrives first?

Hint: One good way is to use the continuous law of total probability.

(b) What is the CDF of Fred's waiting time for a bus?

27. A longevity study is being conducted on n married hobbit couples. Let p be the probability that an individual hobbit lives at least until their eleventy-first birthday, and assume that the lifespans of different hobbits are independent. Let N_0, N_1, N_2 be the number of couples in which neither hobbit reaches age eleventy-one, one hobbit does but not the other, and both hobbits reach eleventy-one, respectively.

(a) Find the joint PMF of N_0, N_1, N_2.

For the rest of this problem, suppose that it is observed that exactly h of the cohort of hobbits reach their eleventy-first birthdays.

(b) Using (a) and the definition of conditional probability, find the conditional PMF of N_2 given this information, up to a normalizing constant (that is, you do not need to find the normalizing constant in this part, but just to give a simplified expression that is proportional to the conditional PMF). For simplicity, you can and should ignore multiplicative constants in this part; this includes multiplicative factors that are functions of h, since h is now being treated as a known constant.

(c) Now obtain the conditional PMF of N_2 using a direct counting argument, now including any normalizing constants needed in order to have a valid conditional PMF.

(d) Discuss intuitively whether or not p should appear in the answer to (c).

(e) What is the conditional expectation of N_2, given the above information (simplify fully)? This can be done without doing any messy sums.

28. There are n stores in a shopping center, labeled from 1 to n. Let X_i be the number of customers who visit store i in a particular month, and suppose that $X_1, X_2, \dots, X_n$ are i.i.d. with PMF $p(x) = P(X_i = x)$. Let $I \sim \text{DUnif}(1, 2, \dots, n)$ be the label of a randomly chosen store, so X_I is the number of customers at a randomly chosen store.

(a) For $i \neq j$, find $P(X_i = X_j)$ in terms of a sum involving the PMF $p(x)$.

(b) Find the joint PMF of I and X_I. Are they independent?

(c) Does X_I, the number of customers for a random store, have the same marginal distribution as X_1, the number of customers for store 1?

(d) Let $J \sim \text{DUnif}(1, 2, \dots, n)$ also be the label of a randomly chosen store, with I and J independent. Find $P(X_I = X_J)$ in terms of a sum involving the PMF $p(x)$. How does $P(X_I = X_J)$ compare to $P(X_i = X_j)$ for fixed i, j with $i \neq j$?

29. Let X and Y be i.i.d. $\text{Geom}(p)$, $L = \min(X, Y)$, and $M = \max(X, Y)$.

(a) Find the joint PMF of L and M. Are they independent?

(b) Find the marginal distribution of L in two ways: using the joint PMF, and using a story.

(c) Find EM.

Hint: A quick way is to use (b) and the fact that $L + M = X + Y$.

(d) Find the joint PMF of L and $M - L$. Are they independent?

30. Let X, Y have the joint CDF

$$F(x, y) = 1 - e^{-x} - e^{-y} + e^{-(x+y+\theta xy)},$$

for $x > 0, y > 0$ (and $F(x, y) = 0$ otherwise), where the parameter θ is in $[0, 1]$.

(a) Find the joint PDF of X, Y. For which values of θ (if any) are they independent?

(b) Explain why we require θ to be in $[0, 1]$.

(c) Find the marginal PDFs of X and Y by working directly from the joint PDF from (a). When integrating, do *not* use integration by parts or computer assistance; rather, *pattern-match* to facts we know about moments of famous distributions.

(d) Find the marginal CDFs of X and Y by working directly from the joint CDF.

2D LOTUS

31. Ⓢ Let X and Y be i.i.d. Unif$(0, 1)$. Find the standard deviation of the distance between X and Y.

32. Ⓢ Let X, Y be i.i.d. Expo(λ). Find $E|X - Y|$ in two different ways: (a) using 2D LOTUS and (b) using the memoryless property without any calculus.

33. Alice walks into a post office with 2 clerks. Both clerks are in the midst of serving customers, but Alice is next in line. The clerk on the left takes an Expo(λ_1) time to serve a customer, and the clerk on the right takes an Expo(λ_2) time to serve a customer. Let T be the amount of time Alice has to wait until it is her turn.

(a) Write down expressions for the mean and variance of T, in terms of double integrals (which you do not need to evaluate).

(b) Find the distribution, mean, and variance of T, *without using calculus.*

34. Let (X, Y) be a uniformly random point in the triangle in the plane with vertices $(0, 0), (0, 1), (1, 0)$. Find Cov(X, Y). (Exercise 18 is about joint, marginal, and conditional PDFs in this setting.)

35. A random point is chosen uniformly in the unit disk $\{(x, y) : x^2 + y^2 \leq 1\}$. Let R be its distance from the origin.

(a) Find $E(R)$ using 2D LOTUS.

Hint: To do the integral, convert to polar coordinates (see the math appendix).

(b) Find the CDFs of R^2 and of R *without using calculus*, using the fact that for a Uniform distribution on a region, probability within that region is proportional to area. Then get the PDFs of R^2 and of R, and find $E(R)$ in two more ways: using the definition of expectation, and using a 1D LOTUS by thinking of R as a function of R^2.

36. Let X and Y be discrete r.v.s.

(a) Use 2D LOTUS (without assuming linearity) to show that $E(X+Y) = E(X)+E(Y)$.

(b) Now suppose that X and Y are independent. Use 2D LOTUS to show that they are uncorrelated, i.e., $E(XY) = E(X)E(Y)$.

37. Let X and Y be i.i.d. continuous random variables with PDF f, mean μ, and variance σ^2. We know that the expected squared distance of X from its mean is σ^2, and likewise for Y; this problem is about the expected squared distance of X from Y.

(a) Use 2D LOTUS to express $E(X - Y)^2$ as a double integral.

(b) By expanding $(x-y)^2 = x^2 - 2xy + y^2$ and evaluating the double integral from (a), show that

$$E(X-Y)^2 = 2\sigma^2.$$

(c) Give an alternative proof of the result from (b), based on the trick of adding and subtracting μ:

$$(X-Y)^2 = (X-\mu+\mu-Y)^2 = (X-\mu)^2 - 2(X-\mu)(Y-\mu) + (Y-\mu)^2.$$

Covariance

38. Ⓢ Let X and Y be r.v.s. Is it correct to say "$\max(X,Y) + \min(X,Y) = X + Y$"? Is it correct to say "$\text{Cov}(\max(X,Y), \min(X,Y)) = \text{Cov}(X,Y)$ since either the max is X and the min is Y or vice versa, and covariance is symmetric"? Explain.

39. Ⓢ Two fair, six-sided dice are rolled (one green and one orange), with outcomes X and Y for the green die and the orange die, respectively.

    (a) Compute the covariance of $X+Y$ and $X-Y$.

    (b) Are $X+Y$ and $X-Y$ independent?

40. Let X and Y be i.i.d. Unif$(0,1)$.

    (a) Compute the covariance of $X+Y$ and $X-Y$.

    (b) Are $X+Y$ and $X-Y$ independent?

41. Ⓢ Let X and Y be standardized r.v.s (i.e., marginally they each have mean 0 and variance 1) with correlation $\rho \in (-1,1)$. Find a, b, c, d (in terms of ρ) such that $Z = aX + bY$ and $W = cX + dY$ are uncorrelated but still standardized.

42. Ⓢ Let X be the number of distinct birthdays in a group of 110 people (i.e., the number of days in a year such that at least one person in the group has that birthday). Under the usual assumptions (no February 29, all the other 365 days of the year are equally likely, and the day when one person is born is independent of the days when the other people are born), find the mean and variance of X.

43. (a) Let X and Y be Bernoulli r.v.s, possibly with different parameters. Show that if X and Y are uncorrelated, then they are independent.

    (b) Give an example of three Bernoulli r.v.s such that each pair of them is uncorrelated, yet the three r.v.s are dependent.

44. Find the variance of the number of toys needed until you have a complete set in Example 4.3.12 (the coupon collector problem), as a sum.

45. A random triangle is formed in some way, such that all pairs of angles have the same joint distribution. What is the correlation between two of the angles (assuming that the variance of the angles is nonzero)?

46. Each of $n \geq 2$ people puts their name on a slip of paper (no two have the same name). The slips of paper are shuffled in a hat, and then each person draws one (uniformly at random at each stage, without replacement). Find the standard deviation of the number of people who draw their own names.

47. As in Example 4.4.7, an urn contains w white balls and b black balls. The balls are randomly drawn one by one *without replacement* until r white balls have been drawn. Let $X \sim \text{NHGeom}(w, b, r)$ be the number of black balls drawn before drawing the rth white ball. In this exercise, you will derive Var(X).

    As explained in Example 4.4.7, we can assume that we continue drawing balls until the urn has been emptied out. Label the black balls as $1, 2, \dots, b$, and write $X = \sum_{j=1}^{b} I_j$, where I_j is the indicator of black ball j being drawn before the rth white ball is drawn.

(a) Show that

$$E(I_j) = \frac{r}{w+1}.$$

(b) Give an intuitive explanation of whether I_i and I_j are positively correlated, uncorrelated, or negatively correlated, for $i \neq j$.

(c) Show that for $i \neq j$,

$$E(I_i I_j) = \frac{\binom{r+1}{2}}{\binom{w+2}{2}} = \frac{(r+1)r}{(w+2)(w+1)}.$$

Hint: Imagine $w + 2$ slots, into which black balls i and j and the w white balls will be placed. All orderings for these $w + 2$ balls are equally likely.

(d) Find an expression for $\text{Var}(X)$. With some algebra (which you don't have to do), your expression should simplify to

$$\text{Var}(X) = \frac{rb(w+b+1)(w-r+1)}{(w+1)^2(w+2)}.$$

48. Ⓢ Athletes compete one at a time at the high jump. Let X_j be how high the jth jumper jumped, with $X_1, X_2, \dots$ i.i.d. with a continuous distribution. We say that the jth jumper sets a *record* if X_j is greater than all of $X_{j-1}, \dots, X_1$.

    Find the variance of the number of records among the first n jumpers (as a sum). What happens to the variance as $n \to \infty$?

49. Ⓢ A chicken lays a $\text{Pois}(\lambda)$ number N of eggs. Each egg hatches a chick with probability p, independently. Let X be the number which hatch, so $X|N = n \sim \text{Bin}(n, p)$.
    Find the correlation between N (the number of eggs) and X (the number of eggs which hatch). Simplify; your final answer should work out to a simple function of p (the λ should cancel out).

50. Let $X_1, \dots, X_n$ be random variables such that $\text{Corr}(X_i, X_j) = \rho$ for all $i \neq j$. Show that $\rho \geq -\frac{1}{n-1}$. This is a bound on how negatively correlated a collection of r.v.s can all be with each other.
    Hint: Assume $\text{Var}(X_i) = 1$ for all i; this can be done without loss of generality, since rescaling two r.v.s does not affect the correlation between them. Then use the fact that $\text{Var}(X_1 + \dots + X_n) \geq 0$.

51. Let X and Y be independent r.v.s. Show that

    $$\text{Var}(XY) = \text{Var}(X)\text{Var}(Y) + (EX)^2\text{Var}(Y) + (EY)^2\text{Var}(X).$$

    Hint: It is often useful when working with a second moment $E(T^2)$ to express it as $\text{Var}(T) + (ET)^2$.

52. Stat 110 shirts come in 3 sizes: small, medium, and large. There are n shirts of each size (where $n \geq 2$). There are $3n$ students. For each size, n of the students have that size as the best fit. This seems ideal. But suppose that instead of giving each student the right size shirt, each student is given a shirt completely randomly (all allocations of the shirts to the students, with one shirt per student, are equally likely). Let X be the number of students who get their right size shirt.

    (a) Find $E(X)$.

    (b) Give each student an ID number from 1 to $3n$, such that the right size shirt is small for students 1 through n, medium for students $n+1$ through $2n$, and large for students

$2n+1$ through $3n$. Let A_j be the event that student j gets their right size shirt. Find $P(A_1, A_2)$ and $P(A_1, A_{n+1})$.

(c) Find $\text{Var}(X)$.

53. Ⓢ A drunken man wanders around randomly in a large space. At each step, he moves one unit of distance North, South, East, or West, with equal probabilities. Choose coordinates such that his initial position is $(0,0)$ and if he is at (x,y) at some time, then one step later he is at $(x, y+1), (x, y-1), (x+1, y)$, or $(x-1, y)$. Let (X_n, Y_n) and R_n be his position and distance from the origin after n steps, respectively.

General hint: Note that X_n is a sum of r.v.s with possible values $-1, 0, 1$, and likewise for Y_n, but be careful throughout the problem about independence.

(a) Determine whether or not X_n is independent of Y_n.
(b) Find $\text{Cov}(X_n, Y_n)$.
(c) Find $E(R_n^2)$.

54. Ⓢ A scientist makes two measurements, considered to be independent standard Normal r.v.s. Find the correlation between the larger and smaller of the values.

Hint: Note that $\max(x,y) + \min(x,y) = x + y$ and $\max(x,y) - \min(x,y) = |x - y|$.

55. Let $U \sim \text{Unif}(-1, 1)$ and $V = 2|U| - 1$.

(a) Find the distribution of V (give the PDF and, if it is a named distribution we have studied, its name and parameters).

Hint: Find the support of V, and then find the CDF of V by reducing $P(V \leq v)$ to probability calculations about U.

(b) Show that U and V are uncorrelated, but not independent. This is also another example illustrating the fact that knowing the marginal distributions of two r.v.s does not determine the joint distribution.

56. Ⓢ Consider the following method for creating a *bivariate Poisson* (a joint distribution for two r.v.s such that both marginals are Poissons). Let $X = V + W, Y = V + Z$ where V, W, Z are i.i.d. $\text{Pois}(\lambda)$ (the idea is to have something borrowed and something new but not something old or something blue).

(a) Find $\text{Cov}(X, Y)$.

(b) Are X and Y independent? Are they conditionally independent given V?

(c) Find the joint PMF of X, Y (as a sum).

57. You are playing an exciting game of Battleship. Your opponent secretly positions ships on a 10 by 10 grid and you try to guess where the ships are. Each of your guesses is a *hit* if there is a ship there and a *miss* otherwise.

The game has just started and your opponent has 3 ships: a battleship (length 4), a submarine (length 3), and a destroyer (length 2). (Usually there are 5 ships to start, but to simplify the calculations we are considering 3 here.) You are playing a variation in which you unleash a *salvo*, making 5 simultaneous guesses. Assume that your 5 guesses are a simple random sample drawn from the 100 grid positions, i.e., all sets of 5 grid positions are equally likely.

Find the mean and variance of the number of distinct ships you will hit in your salvo. (Give exact answers in terms of binomial coefficients or factorials, and also numerical values computed using a computer.)

Hint: First work in terms of the number of ships *missed*, expressing this as a sum of indicator r.v.s. Then use the fundamental bridge and naive definition of probability.

58. This problem explores a visual interpretation of covariance. Data are collected for n individuals, where for each individual two variables are measured (e.g., height and weight). Assume independence *across* individuals (e.g., person 1's variables gives no information about the other people), but not *within* individuals (e.g., a person's height and weight may be correlated).

    Let $(x_1, y_1), \ldots, (x_n, y_n)$ be the n data points, with $n \geq 2$. The data are considered here as fixed, known numbers—they are the observed values after performing an experiment. Imagine plotting all the points (x_i, y_i) in the plane, and drawing the rectangle determined by each pair of points. For example, the points $(1, 3)$ and $(4, 6)$ determine the rectangle with vertices $(1, 3), (1, 6), (4, 6), (4, 3)$.

    The *signed area* contributed by (x_i, y_i) and (x_j, y_j) is the area of the rectangle they determine if the slope of the line between them is positive, and is the negative of the area of the rectangle they determine if the slope of the line between them is negative. (Define the signed area to be 0 if $x_i = x_j$ or $y_i = y_j$, since then the rectangle is degenerate.) So the signed area is positive if a higher x value goes with a higher y value for the pair of points, and negative otherwise. Assume that the x_i are all distinct and the y_i are all distinct.

    (a) The *sample covariance* of the data is defined to be

$$r = \frac{1}{n} \sum_{i=1}^{n} (x_i - \bar{x})(y_i - \bar{y}),$$

    where

$$\bar{x} = \frac{1}{n} \sum_{i=1}^{n} x_i \text{ and } \bar{y} = \frac{1}{n} \sum_{i=1}^{n} y_i$$

    are the sample means. (There are differing conventions about whether to divide by $n-1$ or n in the definition of sample covariance, but that need not concern us here.)

    Let (X, Y) be one of the (x_i, y_i) pairs, chosen uniformly at random. Determine precisely how $\text{Cov}(X, Y)$ is related to the sample covariance.

    (b) Let (X, Y) be as in (a), and $(\tilde{X}, \tilde{Y})$ be an independent draw from the same distribution. That is, (X, Y) and $(\tilde{X}, \tilde{Y})$ are randomly chosen from the n points, independently (so it is possible for the same point to be chosen twice).

    Express the total signed area of the rectangles as a constant times $E((X - \tilde{X})(Y - \tilde{Y}))$. Then show that the sample covariance of the data is a constant times the total signed area of the rectangles.

    Hint: Consider $E((X-\tilde{X})(Y-\tilde{Y}))$ in two ways: as the average signed area of the random rectangle formed by (X, Y) and $(\tilde{X}, \tilde{Y})$, and using properties of expectation to relate it to $\text{Cov}(X, Y)$. For the former, consider the n^2 possibilities for which point (X, Y) is and which point $(\tilde{X}, \tilde{Y})$; note that n such choices result in degenerate rectangles.

    (c) Based on the interpretation from (b), give intuitive explanations of why for any r.v.s W_1, W_2, W_3 and constants a_1, a_2, covariance has the following properties:

    (i) $\text{Cov}(W_1, W_2) = \text{Cov}(W_2, W_1)$;
    (ii) $\text{Cov}(a_1 W_1, a_2 W_2) = a_1 a_2 \text{Cov}(W_1, W_2)$;
    (iii) $\text{Cov}(W_1 + a_1, W_2 + a_2) = \text{Cov}(W_1, W_2)$;
    (iv) $\text{Cov}(W_1, W_2 + W_3) = \text{Cov}(W_1, W_2) + \text{Cov}(W_1, W_3)$.

59. A statistician is trying to estimate an unknown parameter θ based on some data. She has available two independent estimators $\hat{\theta}_1$ and $\hat{\theta}_2$ (an estimator is a function of the data, used to estimate a parameter). For example, $\hat{\theta}_1$ could be the sample mean of a subset of the data and $\hat{\theta}_2$ could be the sample mean of another subset of the data, disjoint from the subset used to calculate $\hat{\theta}_1$. Assume that both of these estimators are unbiased, i.e., $E(\hat{\theta}_j) = \theta$.

Rather than having a bunch of separate estimators, the statistician wants one combined estimator. It may not make sense to give equal weights to $\hat{\theta}_1$ and $\hat{\theta}_2$ since one could be much more reliable than the other, so she decides to consider combined estimators of the form

$$\hat{\theta} = w_1\hat{\theta}_1 + w_2\hat{\theta}_2,$$

a weighted combination of the two estimators. The weights w_1 and w_2 are nonnegative and satisfy $w_1 + w_2 = 1$.

(a) Check that $\hat{\theta}$ is also unbiased, i.e., $E(\hat{\theta}) = \theta$.

(b) Determine the optimal weights w_1, w_2, in the sense of minimizing the mean squared error $E(\hat{\theta} - \theta)^2$. Express your answer in terms of the variances of $\hat{\theta}_1$ and $\hat{\theta}_2$. The optimal weights are known as *Fisher weights.*

Hint: As discussed in Exercise 53 from Chapter 5, mean squared error is variance plus squared bias, so in this case the mean squared error of $\hat{\theta}$ is $\text{Var}(\hat{\theta})$. Note that there is no need for multivariable calculus here, since $w_2 = 1 - w_1$.

(c) Give a simple description of what the estimator found in (b) amounts to if the data are i.i.d. random variables $X_1, \dots, X_n, Y_1, \dots, Y_m$, $\hat{\theta}_1$ is the sample mean of $X_1, \dots, X_n$, and $\hat{\theta}_2$ is the sample mean of $Y_1, \dots, Y_m$.

Chicken-egg

60. Ⓢ A Pois(λ) number of people vote in a certain election. Each voter votes for candidate A with probability p and for candidate B with probability $q = 1 - p$, independently of all the other voters. Let V be the difference in votes, defined as the number of votes for A minus the number for B.

(a) Find $E(V)$.

(b) Find $\text{Var}(V)$.

61. A traveler gets lost $N \sim \text{Pois}(\lambda)$ times on a long journey. When lost, the traveler asks someone for directions with probability p. Let X be the number of times that the traveler is lost and asks for directions, and Y be the number of times that the traveler is lost and does not ask for directions.

(a) Find the joint PMF of N, X, Y. Are they independent?

(b) Find the joint PMF of N, X. Are they independent?

(c) Find the joint PMF of X, Y. Are they independent?

62. The number of people who visit the Leftorium store in a day is Pois(100). Suppose that 10% of customers are *sinister* (left-handed), and 90% are *dexterous* (right-handed). Half of the sinister customers make purchases, but only a third of the dexterous customers make purchases. The characteristics and behavior of people are independent, with probabilities as described in the previous two sentences. On a certain day, there are 42 people who arrive at the store but leave without making a purchase. Given this information, what is the conditional PMF of the number of customers on that day who make a purchase?

63. A chicken lays n eggs. Each egg independently does or doesn't hatch, with probability p of hatching. For each egg that hatches, the chick does or doesn't survive (independently of the other eggs), with probability s of survival. Let $N \sim \text{Bin}(n, p)$ be the number of eggs which hatch, X be the number of chicks which survive, and Y be the number of chicks which hatch but don't survive (so $X + Y = N$). Find the marginal PMF of X, and the joint PMF of X and Y. Are X and Y independent?

64. There will be $X \sim \text{Pois}(\lambda)$ courses offered at a certain school next year.

(a) Find the expected number of choices of 4 courses (in terms of λ, fully simplified), assuming that simultaneous enrollment is allowed if there are time conflicts.

(b) Now suppose that simultaneous enrollment is not allowed. Suppose that most faculty only want to teach on Tuesdays and Thursdays, and most students only want to take courses that start at 10 am or later, and as a result there are only four possible time slots: 10 am, 11:30 am, 1 pm, 2:30 pm (each course meets Tuesday-Thursday for an hour and a half, starting at one of these times). Rather than trying to avoid major conflicts, the school schedules the courses completely randomly: after the list of courses for next year is determined, they randomly get assigned to time slots, independently and with probability 1/4 for each time slot.

Let X_{am} and X_{pm} be the number of morning and afternoon courses for next year, respectively (where "morning" means starting before noon). Find the joint PMF of X_{am} and X_{pm}, i.e., find $P(X_{\text{am}} = a, X_{\text{pm}} = b)$ for all a, b.

(c) Continuing as in (b), let X_1, X_2, X_3, X_4 be the number of 10 am, 11:30 am, 1 pm, 2:30 pm courses for next year, respectively. What is the joint distribution of X_1, X_2, X_3, X_4? (The result is completely analogous to that of $X_{\text{am}}, X_{\text{pm}}$; you can derive it by thinking conditionally, but for this part you are also allowed to just use the fact that the result is analogous to that of (b).) Use this to find the expected number of choices of 4 non-conflicting courses (in terms of λ, fully simplified). What is the ratio of the expected value from (a) to this expected value?

Multinomial

65. Ⓢ Let $(X_1, \dots, X_k)$ be Multinomial with parameters n and $(p_1, \dots, p_k)$. Use indicator r.v.s to show that $\text{Cov}(X_i, X_j) = -np_ip_j$ for $i \neq j$.

66. Ⓢ Consider the birthdays of 100 people. Assume people's birthdays are independent, and the 365 days of the year (exclude the possibility of February 29) are equally likely. Find the covariance and correlation between how many of the people were born on January 1 and how many were born on January 2.

67. A certain course has a freshmen, b sophomores, c juniors, and d seniors. Let X be the number of freshmen and sophomores (total), Y be the number of juniors, and Z be the number of seniors in a random sample of size n, where for Part (a) the sampling is *with* replacement and for Part (b) the sampling is *without* replacement (for both parts, at each stage the allowed choices have equal probabilities).

(a) Find the joint PMF of X, Y, Z, for sampling with replacement.

(b) Find the joint PMF of X, Y, Z, for sampling without replacement.

68. Ⓢ A group of $n \geq 2$ people decide to play an exciting game of Rock-Paper-Scissors. As you may recall, Rock smashes Scissors, Scissors cuts Paper, and Paper covers Rock (despite Bart Simpson once saying "Good old rock, nothing beats that!").

Usually this game is played with 2 players, but it can be extended to more players as follows. If exactly 2 of the 3 choices appear when everyone reveals their choice, say $a, b \in \{\text{Rock}, \text{Paper}, \text{Scissors}\}$ where a beats b, the game is decisive: the players who chose a win, and the players who chose b lose. Otherwise, the game is indecisive and the players play again.

For example, with 5 players, if one player picks Rock, two pick Scissors, and two pick Paper, the round is indecisive and they play again. But if 3 pick Rock and 2 pick Scissors, then the Rock players win and the Scissors players lose the game.

Assume that the n players independently and randomly choose between Rock, Scissors, and Paper, with equal probabilities. Let X, Y, Z be the number of players who pick Rock, Scissors, Paper, respectively in one game.

(a) Find the joint PMF of X, Y, Z.

(b) Find the probability that the game is decisive. Simplify your answer.

(c) What is the probability that the game is decisive for $n = 5$? What is the limiting probability that a game is decisive as $n \to \infty$? Explain briefly why your answer makes sense.

69. Ⓢ Emails arrive in an inbox according to a Poisson process with rate λ (so the number of emails in a time interval of length t is distributed as $\text{Pois}(\lambda t)$, and the numbers of emails arriving in disjoint time intervals are independent). Let X, Y, Z be the numbers of emails that arrive from 9 am to noon, noon to 6 pm, and 6 pm to midnight (respectively) on a certain day.

(a) Find the joint PMF of X, Y, Z.

(b) Find the conditional joint PMF of X, Y, Z given that $X + Y + Z = 36$.

(c) Find the conditional PMF of $X + Y$ given that $X + Y + Z = 36$.

(d) Find $E(X + Y|X + Y + Z = 36)$ and $\text{Var}(X + Y|X + Y + Z = 36)$. (Conditional expectation and conditional variance given an event are defined in the same way as expectation and variance, using the conditional distribution given the event in place of the unconditional distribution; these concepts are explored more in Chapter 9.)

70. Let X be the number of statistics majors in a certain college in the Class of 2030, viewed as an r.v. Each statistics major chooses between two tracks: a general track in statistical principles and methods, and a track in quantitative finance. Suppose that each statistics major chooses randomly which of these two tracks to follow, independently, with probability p of choosing the general track. Let Y be the number of statistics majors who choose the general track, and Z be the number of statistics majors who choose the quantitative finance track.

(a) Suppose that $X \sim \text{Pois}(\lambda)$. (This isn't the exact distribution in reality since a Poisson is unbounded, but it may be a very good approximation.) Find the correlation between X and Y.

(b) Let n be the size of the Class of 2030, where n is a known constant. For this part and the next, instead of assuming that X is Poisson, assume that each of the n students chooses to be a statistics major with probability r, independently. Find the joint distribution of Y, Z, and the number of non-statistics majors, and their marginal distributions.

(c) Continuing as in (b), find the correlation between X and Y.

71. In humans (and many other organisms), genes come in pairs. Consider a gene of interest, which comes in two types (*alleles*): type a and type A. The *genotype* of a person for that gene is the types of the two genes in the pair: AA, Aa, or aa (aA is equivalent to Aa). According to the Hardy-Weinberg law, for a population in equilibrium, the frequencies of AA, Aa, aa will be $p^2, 2p(1-p), (1-p)^2$, respectively, for some p with $0 < p < 1$. Suppose that the Hardy-Weinberg law holds, and that n people are drawn randomly from the population, independently. Let X_1, X_2, X_3 be the number of people in the sample with genotypes AA, Aa, aa, respectively.

(a) What is the joint PMF of X_1, X_2, X_3?

(b) What is the distribution of the number of people in the sample who have an A?

(c) What is the distribution of how many of the $2n$ genes among the people are A's?

(d) Now suppose that p is unknown, and must be estimated using the observed data X_1, X_2, X_3. The *maximum likelihood estimator* (MLE) of p is the value of p for which the observed data are as likely as possible. Find the MLE of p.

(e) Now suppose that p is unknown, and that our observations can't distinguish between AA and Aa. So for each person in the sample, we just know whether or not that person is an aa (in genetics terms, AA and Aa have the same *phenotype*, and we only get to observe the phenotypes, not the genotypes). Find the MLE of p.

Multivariate Normal

72. Ⓢ Let (X, Y) be Bivariate Normal, with X and Y marginally $\mathcal{N}(0,1)$ and with correlation ρ between X and Y.

    (a) Show that $(X+Y, X-Y)$ is also Bivariate Normal.

    (b) Find the joint PDF of $X+Y$ and $X-Y$ (without using calculus), assuming that $-1 < \rho < 1$.

73. Let the joint PDF of X and Y be

$$f_{X,Y}(x,y) = c\exp\left(-\frac{x^2}{2} - \frac{y^2}{2}\right) \text{ for all } x \text{ and } y,$$

    where c is a constant.

    (a) Find c to make this a valid joint PDF.

    (b) What are the marginal distributions of X and Y? Are X and Y independent?

    (c) Is (X, Y) Bivariate Normal?

74. Let the joint PDF of X and Y be

$$f_{X,Y}(x,y) = c\exp\left(-\frac{x^2}{2} - \frac{y^2}{2}\right) \text{ for } xy > 0,$$

    where c is a constant (the joint PDF is 0 for $xy \le 0$).

    (a) Find c to make this a valid joint PDF.

    (b) What are the marginal distributions of X and Y? Are X and Y independent?

    (c) Is (X, Y) Bivariate Normal?

75. Let X, Y, Z be i.i.d. $\mathcal{N}(0,1)$. Find the joint MGF of $(X+2Y, 3X+4Z, 5Y+6Z)$.

76. Let X and Y be i.i.d. $\mathcal{N}(0,1)$, and let S be a random sign (1 or -1, with equal probabilities) independent of (X, Y).

    (a) Determine whether or not $(X, Y, X+Y)$ is Multivariate Normal.

    (b) Determine whether or not $(X, Y, SX+SY)$ is Multivariate Normal.

    (c) Determine whether or not (SX, SY) is Multivariate Normal.

77. Let (X, Y) be Bivariate Normal with $X \sim \mathcal{N}(0, \sigma_1^2)$ and $Y \sim \mathcal{N}(0, \sigma_2^2)$ marginally and with $\text{Corr}(X, Y) = \rho$. Find a constant c such that $Y - cX$ is independent of X.

    Hint: First find c (in terms of ρ, σ_1, σ_2) such that $Y - cX$ and X are uncorrelated.

78. A mother and a father have 6 children. The 8 heights in the family (in inches) are $\mathcal{N}(\mu, \sigma^2)$ r.v.s (with the same distribution, but not necessarily independent).

    (a) Assume for this part that the heights are all independent. On average, how many of the children are taller than *both* parents?

    (b) Let X_1 be the height of the mother, X_2 be the height of the father, and $Y_1, \dots, Y_6$ be the heights of the children. Suppose that $(X_1, X_2, Y_1, \dots, Y_6)$ is Multivariate Normal, with $\mathcal{N}(\mu, \sigma^2)$ marginals and $\text{Corr}(X_1, Y_j) = \rho$ for $1 \le j \le 6$, with $\rho < 1$. On average, how many of the children are more than 1 inch taller than their mother?

Mixed practice

79. Cars pass by a certain point on a road according to a Poisson process with rate λ cars/minute. Let $N_t \sim \text{Pois}(\lambda t)$ be the number of cars that pass by that point in the time interval $[0, t]$, with t measured in minutes.

    (a) A certain device is able to count cars as they pass by, but it does not record the arrival times. At time 0, the counter on the device is reset to 0. At time 3 minutes, the device is observed and it is found that exactly 1 car had passed by. Given this information, find the conditional CDF of when that car arrived. Also describe in words what the result says.

    (b) In the late afternoon, you are counting blue cars. Each car that passes by is blue with probability b, independently of all other cars. Find the joint PMF and marginal PMFs of the number of blue cars and number of non-blue cars that pass by the point in 10 minutes.

80. In a U.S. election, there will be $V \sim \text{Pois}(\lambda)$ registered voters. Suppose each registered voter is a registered Democrat with probability p and a registered Republican with probability $1 - p$, independent of other voters. Also, each registered voter shows up to the polls with probability s and stays home with probability $1 - s$, independent of other voters and independent of their own party affiliation. In this problem, we are interested in X, the number of registered Democrats who actually vote.

    (a) What is the distribution of X, before we know anything about the number of registered voters?

    (b) Suppose we learn that $V = v$; that is, v people registered to vote. What is the conditional distribution of X given this information?

    (c) Suppose we learn there were d registered Democrats and r registered Republicans (where $d + r = v$). What is the conditional distribution of X given this information?

    (d) Finally, we learn in addition to all of the above information that n people showed up at the polls on election day. What is the conditional distribution of X given this information?

81. A certain college has m freshmen, m sophomores, m juniors, and m seniors. A certain class at the college consists of a simple random sample of size n students, i.e., all sets of n of the $4m$ students are equally likely. Let $X_1, \dots, X_4$ be the numbers of freshmen, $\dots$, seniors in the class.

    (a) Find the joint PMF of X_1, X_2, X_3, X_4.

    (b) Give both an intuitive explanation and a mathematical justification for whether or not the distribution from (a) is Multinomial.

    (c) Find $\text{Cov}(X_1, X_3)$, fully simplified.

    Hint: Take the variance of both sides of $X_1 + X_2 + X_3 + X_4 = n$.

82. Let $X \sim \text{Expo}(\lambda)$ and let Y be a nonnegative random variable, discrete or continuous, whose MGF M is finite everywhere. Show that $P(Y < X) = M(c)$ for a certain value of c (which you should specify).

83. A publishing company employs two proofreaders, Prue and Frida. When Prue is proofreading a book, for each typo she has probability p of catching it and $q = 1-p$ of missing it, independently. When Frida is proofreading a book, for each typo she has probability f of catching it and $g = 1 - f$ of missing it, independently.

    (a) A certain book draft has n typos. The company randomly assigns it to one of the two proofreaders, with equal probabilities. Find the distribution of the number of typos that get detected.

(b) Another book is being written. When a draft of the book is complete, it will have a Pois(λ) number of typos, and will be assigned to Prue to proofread. Find the probability that Prue catches exactly k typos, given that she misses exactly m typos.

84. Two authors, Bob and Martha, are about to begin writing an epic co-authored book, *The Adventures of Aaron the Aardwolf.* It will take them A years to write. When they finish this book, they will immediately begin work on new, individually authored books. Bob will spend X years writing *The Bilinear Bonanza of Bonnie the Butterfly*, and Martha will spend Y years writing *Memoirs of Maude the Magnificent Mangabey*, independently. Suppose that A, X, Y are i.i.d. Expo(λ). On a timeline where time 0 is defined as the time when they begin their collaboration, consider the following quantities.

    A: time at which *The Adventures of Aaron the Aardwolf* is completed;
    B: time at which *The Bilinear Bonanza of Bonnie the Butterfly* is completed;
    M: time at which *Memoirs of Maude the Magnificent Mangabey* is completed;
    T: time at which the last to be completed of these three books is completed.

    (a) Find the distribution of B (which is also the distribution of M).

    (b) Find Cov(A, B).

    (c) Find $E(T)$.

85. A DNA sequence can be represented as a sequence of letters, where the *alphabet* has 4 letters: A,C,G,T. Suppose that a random DNA sequence of length $n \geq 4$ is formed by independently generating letters one at a time, with p_A, p_C, p_G, p_T the probabilities of A,C,G,T, respectively, where $p_A + p_C + p_G + p_T = 1$.

    (a) Find the covariance between the number of A's and the number of C's in the sequence.

    (b) It is observed that the sequence contains exactly a A's, c C's, g G's, and t T's, where $a + c + g + t = n$ and $a \geq 2$. Given this information, find the probability that the first A in the sequence is followed immediately by another A.

    Hint: How does this part relate to Exercise 74 in Chapter 2?

    (c) Given the information from (b) about how many times each letter occurs, find the expected number of occurrences of the expression CAT in the sequence.

86. To test for a certain disease, the level of a certain substance in the blood is measured. Let T be this measurement, considered as a continuous r.v. The patient tests positive (i.e., is declared to have the disease) if $T > t_0$ and tests negative if $T \leq t_0$, where t_0 is a threshold decided upon in advance. Let D be the indicator of having the disease. As discussed in Example 2.3.9, the *sensitivity* of the test is the probability of testing positive given that the patient has the disease, and the *specificity* of the test is the probability of testing negative given that the patient does not have the disease.

    (a) The *ROC (receiver operator characteristic) curve* of the test is the plot of sensitivity vs. 1 minus specificity, where sensitivity (the vertical axis) and 1 minus specificity (the horizontal axis) are viewed as functions of the threshold t_0. ROC curves are widely used in medicine and engineering as a way to study the performance of procedures for classifying individuals into two groups (in this case, the two groups are "diseased people" and "non-diseased people").

    Given that $D = 1$, T has CDF G and PDF g; given that $D = 0$, T has CDF H and PDF h. Here g and h are positive on an interval $[a, b]$ and 0 outside this interval. Show that the area under the ROC curve is the probability that a randomly selected diseased person has a higher T value than a randomly selected non-diseased person.

(b) Explain why the result of (a) makes sense in two extreme cases: when $g = h$, and when there is a threshold t_0 such that $P(T > t_0|D = 1)$ and $P(T \leq t_0|D = 0)$ are very close to 1.

87. Let J be Discrete Uniform on $\{1, 2, \dots, n\}$.

(a) Find $E(J)$ and $\text{Var}(J)$, fully simplified, using results from Section A.8 of the math appendix.

(b) Discuss intuitively whether the results in (a) should be approximately the same as the mean and variance (respectively) of a Uniform distribution on a certain interval.

(c) Let $X_1, \dots, X_n$ be i.i.d. $\mathcal{N}(0, 1)$ r.v.s, and let $R_1, \dots, R_n$ be their ranks (the smallest X_i has rank 1, the next has rank 2, ..., and the largest has rank n). Explain why

$$R_n = 1 + \sum_{j=1}^{n-1} I_j,$$

where $I_j = I(X_n > X_j)$. Then use this to find $E(R_n)$ and $\text{Var}(R_n)$ directly using symmetry, linearity, the fundamental bridge, and properties of covariance.

(d) Explain how the results of (a) and (c) relate. Then prove the identities

$$\sum_{j=1}^{n} j = \frac{n(n+1)}{2} \text{ and } \sum_{j=1}^{n} j^2 = \frac{n(n+1)(2n+1)}{6},$$

by giving them probabilistic interpretations.

88. Ⓢ A *network* consists of n *nodes*, each pair of which may or may not have an *edge* joining them. For example, a social network can be modeled as a group of n nodes (representing people), where an edge between i and j means they know each other. Assume the network is undirected and does not have edges from a node to itself (for a social network, this says that if i knows j, then j knows i and that, contrary to Socrates' advice, a person does not know himself or herself). A *clique* of size k is a set of k nodes where every node has an edge to every other node (i.e., within the clique, everyone knows everyone). An *anticlique* of size k is a set of k nodes where there are no edges between them (i.e., within the anticlique, no one knows anyone else). For example, the picture below shows a network with nodes labeled $1, 2, \dots, 7$, where $\{1, 2, 3, 4\}$ is a clique of size 4, and $\{3, 5, 7\}$ is an anticlique of size 3.

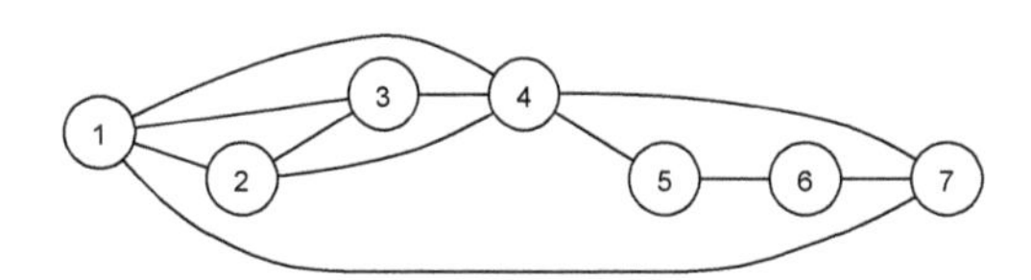

(a) Form a random network with n nodes by independently flipping fair coins to decide for each pair $\{x, y\}$ whether there is an edge joining them. Find the expected number of cliques of size k (in terms of n and k).

(b) A *triangle* is a clique of size 3. For a random network as in (a), find the variance of the number of triangles (in terms of n).

Hint: Find the covariances of the indicator random variables for each possible clique. There are $\binom{n}{3}$ such indicator r.v.s, some pairs of which are dependent.

*(c) Suppose that $\binom{n}{k} < 2^{\binom{k}{2}-1}$. Show that there is a network with n nodes containing no cliques of size k or anticliques of size k.

Hint: Explain why it is enough to show that for a random network with n nodes, the probability of the desired property is positive; then consider the complement.

89. Ⓢ Shakespeare wrote a total of 884647 words in his known works. Of course, many words are used more than once, and the number of distinct words in Shakespeare's known writings is 31534 (according to one computation). This puts a lower bound on the size of Shakespeare's vocabulary, but it is likely that Shakespeare knew words which he did not use in these known writings.

More specifically, suppose that a new poem of Shakespeare were uncovered, and consider the following (seemingly impossible) problem: give a good prediction of the number of words in the new poem that do not appear anywhere in Shakespeare's previously known works.

Ronald Thisted and Bradley Efron studied this problem in the papers [8] and [9], developing theory and methods and then applying the methods to try to determine whether Shakespeare was the author of a poem discovered by a Shakespearean scholar in 1985. A simplified version of their method is developed in the problem below. The method was originally invented by Alan Turing (the founder of computer science) and I.J. Good as part of the effort to break the German Enigma code during World War II.

Let N be the number of distinct words that Shakespeare knew, and assume these words are numbered from 1 to N. Suppose for simplicity that Shakespeare wrote only two plays, A and B. The plays are reasonably long and they are of the same length. Let X_j be the number of times that word j appears in play A, and Y_j be the number of times it appears in play B, for $1 \leq j \leq N$.

(a) Explain why it is reasonable to model X_j as being Poisson, and Y_j as being Poisson with the same parameter as X_j.

(b) Let the numbers of occurrences of the word "eyeball" (which was coined by Shakespeare) in the two plays be independent Pois(λ) r.v.s. Show that the probability that "eyeball" is used in play B but not in play A is

$$e^{-\lambda}(\lambda - \lambda^2/2! + \lambda^3/3! - \lambda^4/4! + \dots).$$

(c) Now assume that λ from (b) is unknown and is itself taken to be a random variable to reflect this uncertainty. So let λ have a PDF f_0. Let X be the number of times the word "eyeball" appears in play A and Y be the corresponding value for play B. Assume that the conditional distribution of X, Y given λ is that they are independent Pois(λ) r.v.s. Show that the probability that "eyeball" is used in play B but not in play A is the alternating series

$$P(X = 1) - P(X = 2) + P(X = 3) - P(X = 4) + \dots.$$

Hint: Condition on λ and use Part (b).

(d) Assume that every word's numbers of occurrences in A and B are distributed as in Part (c), where λ may be different for different words but f_0 is fixed. Let W_j be the number of words that appear exactly j times in play A. Show that the expected number of distinct words appearing in play B but not in play A is

$$E(W_1) - E(W_2) + E(W_3) - E(W_4) + \dots.$$

(This shows that $W_1 - W_2 + W_3 - W_4 + \dots$ is an *unbiased* predictor of the number of distinct words appearing in play B but not in play A: on average it is correct. Moreover, it can be computed just from having seen play A, without needing to know f_0 or any of the λ_j. This method can be extended in various ways to give predictions for unobserved plays based on observed plays.)

8

Transformations

The topic for this chapter is *transformations* of random variables and random vectors. After applying a function to a random variable X or random vector $\mathbf{X}$, the goal is to find the distribution of the transformed random variable or joint distribution of the transformed random vector.

Transformations of random variables appear all over the place in statistics. Here are a few examples, to preview the kinds of transformations we'll be looking at in this chapter.

- *Unit conversion*: In one dimension, we've already seen how standardization and location-scale transformations can be useful tools for learning about an entire family of distributions. A location-scale change is *linear*, converting an r.v. X to the r.v. $Y = aX + b$ where a and b are constants (with $a > 0$).

  There are also many situations in which we may be interested in *nonlinear* transformations, e.g., converting from the dollar-yen exchange rate to the yen-dollar exchange rate, or converting information like "Janet's waking hours yesterday consisted of 8 hours of work, 4 hours visiting friends, and 4 hours surfing the web" to the format "Janet was awake for 16 hours yesterday; she spent $\frac{1}{2}$ of that time working, $\frac{1}{4}$ of that time visiting friends, and $\frac{1}{4}$ of that time surfing the web". The *change of variables formula*, which is the first result in this chapter, shows what happens to the distribution when a random vector is transformed.

- *Sums and averages as summaries*: It is common in statistics to summarize n observations by their sum or sample average. Turning $X_1, \dots, X_n$ into the sum $T = X_1 + \dots + X_n$ or sample mean $\bar{X}_n = T/n$ is a transformation from $\mathbb{R}^n$ to $\mathbb{R}$.

  The term for a sum of independent random variables is *convolution*. We have already encountered stories and MGFs as two techniques for dealing with convolutions. In this chapter, *convolution sums and integrals*, which are based on the law of total probability, will give us another way of obtaining the distribution of a sum of r.v.s.

- *Extreme values*: In many contexts, we may be interested in the distribution of the most extreme observations. For disaster preparedness, government agencies may be concerned about the most extreme flood or earthquake in a 100-year period; in finance, a portfolio manager with an eye toward risk management will want to know the worst 1% or 5% of portfolio returns. In these applications, we are concerned with the maximum or minimum of a set of observations. The transformation that *sorts* observations, turning $X_1, \dots, X_n$ into the *order statistics* $\min(X_1, \dots, X_n), \dots, \max(X_1, \dots, X_n)$, is a transformation from $\mathbb{R}^n$ to $\mathbb{R}^n$ that is not invertible. Order statistics are addressed in the last section in this chapter.

Furthermore, it is especially important to us to understand transformations because of the approach we've taken to learning about the named distributions. Starting from a few basic distributions, we have defined other distributions as transformations of these elementary building blocks, in order to understand how the named distributions are related to one another. We'll continue in that spirit here as we introduce two new distributions, the Beta and Gamma, which generalize the Uniform and Exponential.

We already have quite a few tools in our toolbox for dealing with transformations, so let's review those briefly. First, if we are only looking for the expectation of $g(X)$, LOTUS shows us the way: it tells us that the PMF or PDF of X is enough for calculating $E(g(X))$. LOTUS also applies to functions of several r.v.s, as we learned in the previous chapter.

If we need the full distribution of $g(X)$, not just its expectation, our approach depends on whether X is discrete or continuous.

- In the discrete case, we get the PMF of $g(X)$ by translating the event $g(X) = y$ into an equivalent event involving X. To do so, we look for all values x such that $g(x) = y$; as long as X equals any of these x's, the event $g(X) = y$ will occur. This gives the formula

$$P(g(X) = y) = \sum_{x:g(x)=y} P(X = x).$$

  For a one-to-one g, the situation is particularly simple, because there is only one value of x such that $g(x) = y$, namely $g^{-1}(y)$. Then we can use

$$P(g(X) = y) = P(X = g^{-1}(y))$$

  to convert between the PMFs of X and $g(X)$, as also discussed in Section 3.7. For example, it is extremely easy to convert between the Geometric and First Success distributions.

- In the continuous case, a universal approach is to start from the CDF of $g(X)$, and translate the event $g(X) \leq y$ into an equivalent event involving X. For general g, we may have to think carefully about how to express $g(X) \leq y$ in terms of X, and there is no easy formula we can plug into. But when g is continuous and strictly increasing, the translation is easy: $g(X) \leq y$ is the same as $X \leq g^{-1}(y)$, so

$$F_{g(X)}(y) = P(g(X) \leq y) = P(X \leq g^{-1}(y)) = F_X(g^{-1}(y)).$$

  We can then differentiate with respect to y to get the PDF of $g(X)$. This gives a one-dimensional version of the *change of variables formula*, which generalizes to invertible transformations in multiple dimensions.

8.1 Change of variables

Theorem 8.1.1 (Change of variables in one dimension). Let X be a continuous r.v. with PDF f_X, and let $Y = g(X)$, where g is differentiable and strictly increasing (or strictly decreasing). Then the PDF of Y is given by

$$f_Y(y) = f_X(x)\left|\frac{dx}{dy}\right|,$$

where $x = g^{-1}(y)$. The support of Y is all $g(x)$ with x in the support of X.

Proof. Let g be strictly increasing. The CDF of Y is

$$F_Y(y) = P(Y \leq y) = P(g(X) \leq y) = P(X \leq g^{-1}(y)) = F_X(g^{-1}(y)) = F_X(x),$$

so by the chain rule, the PDF of Y is

$$f_Y(y) = f_X(x)\frac{dx}{dy}.$$

The proof for g strictly decreasing is analogous. In that case the PDF ends up as $-f_X(x)\frac{dx}{dy}$, which is nonnegative since $\frac{dx}{dy} < 0$ if g is strictly decreasing. Using $|\frac{dx}{dy}|$, as in the statement of the theorem, covers both cases. ■

When applying the change of variables formula, we can choose whether to compute $\frac{dx}{dy}$, or compute $\frac{dy}{dx}$ and take the reciprocal. By the chain rule, these give the same result, so we can do whichever is easier.

☣ **8.1.2.** When finding the distribution of Y, be sure to:

- Check the assumptions of the change of variables theorem carefully if you wish to apply it (if it doesn't apply, a good strategy is to start with the CDF of Y).
- Express your final answer for the PDF of Y as a function of y.
- Specify the support of Y.

The change of variables formula (in the strictly increasing g case) is easy to remember when written in the form

$$f_Y(y)dy = f_X(x)dx,$$

which has an aesthetically pleasing symmetry to it. This formula also makes sense if we think about *units*. For example, let X be a measurement in inches and $Y = 2.54X$ be the conversion into centimeters (cm). Then the units of $f_X(x)$ are inches^{-1} and the units of $f_Y(y)$ are cm^{-1}, so it would be absurd to say something like "$f_Y(y) = f_X(x)$". But dx is measured in inches and dy is measured in cm, so $f_Y(y)dy$ and $f_X(x)dx$ are unitless quantities, and it makes sense to equate them. Better yet,

$f_X(x)dx$ and $f_Y(y)dy$ have probability interpretations (recall from Chapter 5 that $f_X(x)dx$ is essentially the probability that X is in a tiny interval of length dx, centered at x), which makes it easier to think intuitively about what the change of variables formula is saying.

The next two examples derive the PDFs of two r.v.s that are defined as transformations of a standard Normal r.v. In the first example the change of variables formula applies; in the second example it does not.

Example 8.1.3 (Log-Normal PDF)**.** Let $X \sim \mathcal{N}(0,1)$, $Y = e^X$. In Chapter 6 we named the distribution of Y the Log-Normal, and we found all of its moments using the MGF of the Normal distribution. Now we can use the change of variables formula to find the PDF of Y, since $g(x) = e^x$ is strictly increasing. Let $y = e^x$, so $x = \log y$ and $dy/dx = e^x$. Then

$$f_Y(y) = f_X(x)\left|\frac{dx}{dy}\right| = \varphi(x)\frac{1}{e^x} = \varphi(\log y)\frac{1}{y}, \quad y > 0.$$

Note that after applying the change of variables formula, we write everything on the right-hand side in terms of y, and we specify the support of the distribution. To determine the support, we just observe that as x ranges from $-\infty$ to ∞, e^x ranges from 0 to ∞.

We can get the same result by working from the definition of the CDF, translating the event $Y \leq y$ into an equivalent event involving X. For $y > 0$,

$$F_Y(y) = P(Y \leq y) = P(e^X \leq y) = P(X \leq \log y) = \Phi(\log y),$$

so the PDF is again

$$f_Y(y) = \frac{d}{dy}\Phi(\log y) = \varphi(\log y)\frac{1}{y}, \quad y > 0. \qquad \square$$

Example 8.1.4 (Chi-Square PDF)**.** Let $X \sim \mathcal{N}(0,1)$, $Y = X^2$. The distribution of Y is an example of a *Chi-Square* distribution, which is formally introduced in Chapter 10. To find the PDF of Y, we can no longer apply the change of variables formula because $g(x) = x^2$ is not one-to-one; instead we start from the CDF.

By drawing the graph of $y = x^2$, we can see that the event $X^2 \leq y$ is equivalent to the event $-\sqrt{y} \leq X \leq \sqrt{y}$. Then

$$F_Y(y) = P(X^2 \leq y) = P(-\sqrt{y} \leq X \leq \sqrt{y}) = \Phi(\sqrt{y}) - \Phi(-\sqrt{y}) = 2\Phi(\sqrt{y}) - 1,$$

so

$$f_Y(y) = 2\varphi(\sqrt{y}) \cdot \frac{1}{2}y^{-1/2} = \varphi(\sqrt{y})y^{-1/2}, \quad y > 0. \qquad \square$$

The following example sheds light on an unexpected appearance of a decidedly non-Normal distribution.

Example 8.1.5 (Lighthouse). A lighthouse on a shore is shining light toward the ocean at a random angle U (measured in radians), where

$$U \sim \text{Unif}\left(\frac{-\pi}{2}, \frac{\pi}{2}\right).$$

Consider a line which is parallel to the shore and 1 mile away from the shore, as illustrated in Figure 8.1. An angle of 0 would mean the ray of light is perpendicular to the shore, while an angle of $\pi/2$ would mean the ray is along the shore, shining to the right from the perspective of the figure.

Let X be the point that the light hits on the line, where the line's origin is the point on the line that is closest to the lighthouse. Find the distribution of X.

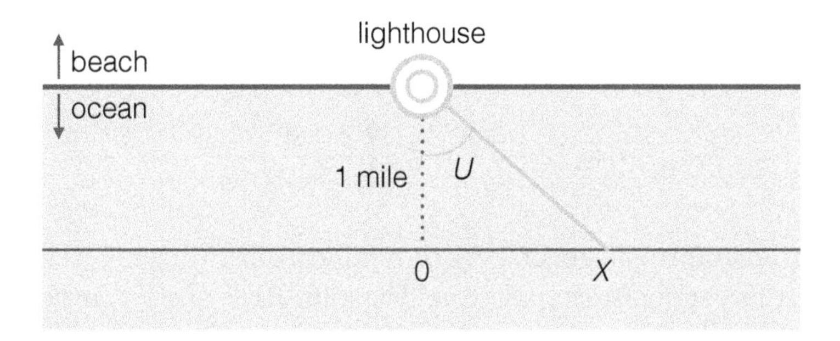

FIGURE 8.1
A lighthouse shining light at a random angle U, viewed from above.

Solution: Looking at the right triangle in Figure 8.1, the length of the opposite side of U divided by the length of the adjacent side of U is $X/1 = X$, so

$$X = \tan(U).$$

(The figure illustrates a case where $U > 0$ and, correspondingly, $X > 0$, but the same relationship holds when $U \leq 0$.) Let x be a possible value of X and u be the corresponding possible value of U, so

$$x = \tan(u) \text{ and } u = \arctan(x).$$

By the change of variables formula, which applies since tan is a differentiable, strictly increasing function on $(-\pi/2, \pi/2)$,

$$f_X(x) = f_U(u)\frac{du}{dx} = \frac{1}{\pi} \cdot \frac{1}{1+x^2},$$

which shows that X is Cauchy. In particular, this implies that $E|X|$ is infinite (since the expected value of a Cauchy does not exist), so on average X is infinitely far from the origin of the line!

The fact that X is Cauchy also makes sense in light of universality of the Uniform. As shown in Example 7.1.25, the Cauchy CDF is

$$F(x) = \frac{1}{\pi}\arctan(x) + 0.5.$$

The inverse is $F^{-1}(v) = \tan\left(\pi\left(v - 0.5\right)\right)$, so for $V \sim \text{Unif}(0,1)$ we have

$$F^{-1}(V) = \tan\left(\pi\left(V - 0.5\right)\right) \sim \text{Cauchy}.$$

This agrees with our earlier result since $\pi\left(V - 0.5\right) \sim \text{Unif}(-\pi/2, \pi/2)$. □

We can also use the change of variables formula to find the PDF of a location-scale transformation.

Example 8.1.6 (PDF of a location-scale transformation). Let X have PDF f_X, and let $Y = a + bX$, with $b \neq 0$. Let $y = a + bx$, to mirror the relationship between Y and X. Then $\frac{dy}{dx} = b$, so the PDF of Y is

$$f_Y(y) = f_X(x)\left|\frac{dx}{dy}\right| = f_X\left(\frac{y-a}{b}\right)\frac{1}{|b|}.$$ □

The change of variables formula generalizes to n dimensions, where it tells us how to use the joint PDF of a random vector $\mathbf{X}$ to get the joint PDF of the transformed random vector $\mathbf{Y} = g(\mathbf{X})$. The formula is analogous to the one-dimensional version, but it involves a multivariate generalization of the derivative called a *Jacobian matrix*; see sections A.6 and A.7 of the math appendix for more about Jacobians.

Theorem 8.1.7 (Change of variables). Let $\mathbf{X} = (X_1, \dots, X_n)$ be a continuous random vector with joint PDF $f_{\mathbf{X}}$. Let $g : A_0 \to B_0$ be an invertible function, where A_0 and B_0 are open[1] subsets of $\mathbb{R}^n$, A_0 contains the support of $\mathbf{X}$, and B_0 is the range of g.

Let $\mathbf{Y} = g(\mathbf{X})$, and mirror this by letting $\mathbf{y} = g(\mathbf{x})$. Since g is invertible, we also have $\mathbf{X} = g^{-1}(\mathbf{Y})$ and $\mathbf{x} = g^{-1}(\mathbf{y})$.

Suppose that all the partial derivatives $\frac{\partial x_i}{\partial y_j}$ exist and are continuous, so we can form the *Jacobian matrix*

$$\frac{\partial \mathbf{x}}{\partial \mathbf{y}} = \begin{pmatrix} \frac{\partial x_1}{\partial y_1} & \frac{\partial x_1}{\partial y_2} & \cdots & \frac{\partial x_1}{\partial y_n} \\ \vdots & & & \vdots \\ \frac{\partial x_n}{\partial y_1} & \frac{\partial x_n}{\partial y_2} & \cdots & \frac{\partial x_n}{\partial y_n} \end{pmatrix}.$$

Also assume that the determinant of this Jacobian matrix is never 0. Then the joint PDF of $\mathbf{Y}$ is

$$f_{\mathbf{Y}}(\mathbf{y}) = f_{\mathbf{X}}\left(\mathbf{g}^{-1}(\mathbf{y})\right) \cdot \left|\left|\frac{\partial \mathbf{x}}{\partial \mathbf{y}}\right|\right| \quad \text{for } \mathbf{y} \in B_0,$$

[1]A set $C \subset \mathbb{R}^n$ is *open* if for each $\mathbf{x} \in C$, there exists $\epsilon > 0$ such that all points with distance less than ϵ from $\mathbf{x}$ are contained in C. Sometimes we take $A_0 = B_0 = \mathbb{R}^n$, but often we would like more flexibility for the domain and range of g. For example, if $n = 2$, and X_1 and X_2 have support $(0, \infty)$, we may want to work with the open set $A_0 = (0,\infty) \times (0,\infty)$ rather than all of $\mathbb{R}^2$.

and 0 otherwise. (The inner bars around the Jacobian say to take the determinant and the outer bars say to take the absolute value.)

That is, to convert $f_{\mathbf{X}}(\mathbf{x})$ to $f_{\mathbf{Y}}(\mathbf{y})$ we express the $\mathbf{x}$ in $f_{\mathbf{X}}(\mathbf{x})$ in terms of $\mathbf{y}$ and then multiply by the absolute value of the determinant of the Jacobian $\partial\mathbf{x}/\partial\mathbf{y}$.

As in the 1D case,

$$\left|\frac{\partial \mathbf{x}}{\partial \mathbf{y}}\right| = \left|\frac{\partial \mathbf{y}}{\partial \mathbf{x}}\right|^{-1},$$

so we can compute whichever of the two Jacobians is easier, and then at the end express the joint PDF of $\mathbf{Y}$ as a function of $\mathbf{y}$.

We will not prove the change of variables formula here, but the idea is to apply the change of variables formula from multivariable calculus and the fact that if A is a region in A_0 and $B = \{g(\mathbf{x}) : \mathbf{x} \in A\}$ is the corresponding region in B_0, then $\mathbf{X} \in A$ is equivalent to $\mathbf{Y} \in B$—they are the same event. So $P(\mathbf{X} \in A) = P(\mathbf{Y} \in B)$, which shows that

$$\int_A f_{\mathbf{X}}(\mathbf{x})d\mathbf{x} = \int_B f_{\mathbf{Y}}(\mathbf{y})d\mathbf{y}.$$

The change of variables formula from multivariable calculus (which is reviewed in the math appendix) can then be applied to the integral on the left-hand side, with the substitution $\mathbf{x} = g^{-1}(\mathbf{y})$.

☣ **8.1.8.** A crucial conceptual difference between transformations of discrete r.v.s and transformations of continuous r.v.s is that with discrete r.v.s we don't need a Jacobian, while with continuous r.v.s we do need a Jacobian. For example, let X be a positive r.v. and $Y = X^3$. If X is discrete, then

$$P(Y = y) = P(X = y^{1/3})$$

converts between the PMFs. But if X is continuous, we need a Jacobian (which in one dimension is just a derivative) to convert between the PDFs:

$$f_Y(y) = f_X(x)\frac{dx}{dy} = f_X(y^{1/3})\frac{1}{3y^{2/3}}.$$

Exercise 23 is a cautionary tale about someone who failed to use a Jacobian when it was needed.

The next two examples apply the 2D change of variables formula.

Example 8.1.9 (Box-Muller). Let $U \sim \text{Unif}(0, 2\pi)$, and let $T \sim \text{Expo}(1)$ be independent of U. Define

$$X = \sqrt{2T}\cos U \text{ and } Y = \sqrt{2T}\sin U.$$

Find the joint PDF of (X, Y). Are they independent? What are their marginal distributions?

Solution:

The joint PDF of U and T is

$$f_{U,T}(u,t) = \frac{1}{2\pi}e^{-t},$$

for $u \in (0, 2\pi)$ and $t > 0$. Viewing (X, Y) as a point in the plane,

$$X^2 + Y^2 = 2T(\cos^2 U + \sin^2 U) = 2T$$

is the squared distance from the origin and U is the angle; that is, $(\sqrt{2T}, U)$ expresses (X, Y) in polar coordinates.

Since we can recover (U, T) from (X, Y), the transformation is invertible. The Jacobian matrix

$$\frac{\partial(x,y)}{\partial(u,t)} = \begin{pmatrix} -\sqrt{2t}\sin u & \frac{1}{\sqrt{2t}}\cos u \\ \sqrt{2t}\cos u & \frac{1}{\sqrt{2t}}\sin u \end{pmatrix}$$

exists, has continuous entries, and has absolute determinant

$$|-\sin^2 u - \cos^2 u| = 1$$

(which is never 0). Then letting $x = \sqrt{2t}\cos u$, $y = \sqrt{2t}\sin u$ to mirror the transformation from (U, T) to (X, Y), we have

$$\begin{aligned} f_{X,Y}(x,y) &= f_{U,T}(u,t) \cdot \left| \left| \frac{\partial(u,t)}{\partial(x,y)} \right| \right| \\ &= \frac{1}{2\pi}e^{-t} \cdot 1 \\ &= \frac{1}{2\pi}e^{-\frac{1}{2}(x^2+y^2)} \\ &= \frac{1}{\sqrt{2\pi}}e^{-x^2/2} \cdot \frac{1}{\sqrt{2\pi}}e^{-y^2/2}, \end{aligned}$$

for all real x and y.

The joint PDF $f_{X,Y}$ factors into a function of x times a function of y, so X and Y are independent. Furthermore, we recognize the joint PDF as the product of two standard Normal PDFs, so X and Y are i.i.d. $\mathcal{N}(0, 1)$ r.v.s! This result is called the *Box-Muller* method for generating Normal r.v.s. □

Example 8.1.10 (Bivariate Normal joint PDF). In Chapter 7, we saw some properties of the Bivariate Normal distribution and found its joint MGF. Now let's find its joint PDF.

Let (Z, W) be BVN with $\mathcal{N}(0, 1)$ marginals and $\text{Corr}(Z, W) = \rho$. (If we want the joint PDF when the marginals are not standard Normal, we can standardize both components separately and use the result below.) Assume that $-1 < \rho < 1$ since otherwise the distribution is degenerate (with Z and W perfectly correlated).

As shown in Example 7.5.10, we can construct (Z, W) as

$$Z = X$$
$$W = \rho X + \tau Y,$$

with $\tau = \sqrt{1-\rho^2}$ and X, Y i.i.d. $\mathcal{N}(0,1)$. We also need the inverse transformation. Solving $Z = X$ for X, we have $X = Z$. Plugging this into $W = \rho X + \tau Y$ and solving for Y, we have

$$X = Z$$
$$Y = -\frac{\rho}{\tau} Z + \frac{1}{\tau} W.$$

The Jacobian is

$$\frac{\partial(x,y)}{\partial(z,w)} = \begin{pmatrix} 1 & 0 \\ -\frac{\rho}{\tau} & \frac{1}{\tau} \end{pmatrix},$$

which has absolute determinant $1/\tau$. So by the change of variables formula,

$$\begin{aligned} f_{Z,W}(z,w) &= f_{X,Y}(x,y) \cdot \left| \left| \frac{\partial(x,y)}{\partial(z,w)} \right| \right| \\ &= \frac{1}{2\pi\tau} \exp\left(-\frac{1}{2}(x^2+y^2)\right) \\ &= \frac{1}{2\pi\tau} \exp\left(-\frac{1}{2}(z^2 + (-\frac{\rho}{\tau}z + \frac{1}{\tau}w)^2)\right) \\ &= \frac{1}{2\pi\tau} \exp\left(-\frac{1}{2\tau^2}(z^2 + w^2 - 2\rho z w)\right), \text{ for all real } z, w. \end{aligned}$$

In the last step we multiplied things out and used the fact that $\rho^2 + \tau^2 = 1$. □

8.2 Convolutions

A *convolution* is a sum of independent random variables. As we mentioned earlier, we often add independent r.v.s because the sum is a useful summary of an experiment (in n Bernoulli trials, we may only care about the total number of successes), and because sums lead to averages, which are also useful (in n Bernoulli trials, the *proportion* of successes).

The main task in this section is to determine the distribution of $T = X + Y$, where X and Y are independent r.v.s whose distributions are known. In previous chapters, we've already seen how stories and MGFs can help us accomplish this task. For example, we used stories to show that the sum of independent Binomials with the same success probability is Binomial, and that the sum of i.i.d. Geometrics is Negative Binomial. We used MGFs to show that a sum of independent Normals is Normal.

A third method for obtaining the distribution of T is by using a *convolution sum or integral.* The formulas are given in the following theorem. As we'll see, a convolution sum is nothing more than the law of total probability, conditioning on the value of either X or Y; a convolution integral is analogous.

Theorem 8.2.1 (Convolution sums and integrals)**.** Let X and Y be independent r.v.s and $T = X + Y$ be their sum. If X and Y are discrete, then the PMF of T is

$$\begin{aligned} P(T = t) &= \sum_x P(Y = t - x)P(X = x) \\ &= \sum_y P(X = t - y)P(Y = y). \end{aligned}$$

If X and Y are continuous, then the PDF of T is

$$\begin{aligned} f_T(t) &= \int_{-\infty}^{\infty} f_Y(t - x)f_X(x)dx \\ &= \int_{-\infty}^{\infty} f_X(t - y)f_Y(y)dy. \end{aligned}$$

Proof. For the discrete case, we use LOTP, conditioning on X:

$$\begin{aligned} P(T = t) &= \sum_x P(X + Y = t|X = x)P(X = x) \\ &= \sum_x P(Y = t - x|X = x)P(X = x) \\ &= \sum_x P(Y = t - x)P(X = x). \end{aligned}$$

Conditioning on Y instead, we obtain the second formula for the PMF of T.

☣ **8.2.2.** We use the assumption that X and Y are independent in order to get from $P(Y = t - x|X = x)$ to $P(Y = t - x)$ in the last step. We are only justified in dropping the condition $X = x$ if the conditional distribution of Y given $X = x$ is the same as the marginal distribution of Y, i.e., X and Y are independent. A common mistake is to assume that after plugging in x for X, we've "already used the information" that $X = x$, when in fact we need an independence assumption to drop the condition. Otherwise we destroy information without justification.

In the continuous case, since the value of a PDF at a point is not a probability, we first find the CDF, and then differentiate to get the PDF. By LOTP,

$$\begin{aligned}F_T(t) = P(X+Y \le t) &= \int_{-\infty}^{\infty} P(X+Y \le t|X=x) f_X(x)dx \\ &= \int_{-\infty}^{\infty} P(Y \le t-x) f_X(x)dx \\ &= \int_{-\infty}^{\infty} F_Y(t-x) f_X(x)dx.\end{aligned}$$

Again, we need independence to drop the condition $X = x$. To get the PDF, we then differentiate with respect to t, interchanging the order of integration and differentiation. This gives

$$f_T(t) = \int_{-\infty}^{\infty} f_Y(t-x) f_X(x)dx.$$

Conditioning on Y instead, we get the second formula for f_T.

An alternative derivation uses the change of variables formula in two dimensions. The only snag is that the change of variables formula requires an *invertible* transformation from $\mathbb{R}^2$ to $\mathbb{R}^2$, but $(X, Y) \mapsto X+Y$ maps $\mathbb{R}^2$ to $\mathbb{R}$ and is not invertible. We can get around this by adding a redundant component to the transformation, in order to make it invertible. Accordingly, we consider the invertible transformation $(X, Y) \mapsto (X+Y, X)$ (using $(X, Y) \mapsto (X+Y, Y)$ would be equally valid). Once we have the joint PDF of $X+Y$ and X, we integrate out X to get the marginal PDF of $X+Y$.

Let $T = X+Y$, $W = X$, and let $t = x+y$, $w = x$. It may seem redundant to give X the new name "W", but doing this makes it easier to distinguish between pre-transformation variables and post-transformation variables: we are transforming $(X, Y) \mapsto (T, W)$. Then

$$\frac{\partial(t, w)}{\partial(x, y)} = \begin{pmatrix} 1 & 1 \\ 1 & 0 \end{pmatrix}$$

has absolute determinant equal to 1, so $\left|\left|\frac{\partial(x,y)}{\partial(t,w)}\right|\right|$ is also 1. Thus, the joint PDF of T and W is

$$f_{T,W}(t, w) = f_{X,Y}(x, y) = f_X(x) f_Y(y) = f_X(w) f_Y(t-w),$$

and the marginal PDF of T is

$$f_T(t) = \int_{-\infty}^{\infty} f_{T,W}(t, w)dw = \int_{-\infty}^{\infty} f_X(x) f_Y(t-x)dx,$$

in agreement with our result above. ■

☣ **8.2.3.** It is not hard to remember the convolution integral formula by reasoning by analogy from

$$P(T = t) = \sum_x P(Y = t-x) P(X = x)$$

to

$$f_T(t) = \int_{-\infty}^{\infty} f_Y(t-x) f_X(x) dx.$$

But care is still needed. For example, Exercise 23 shows that an analogous-looking formula for the PDF of the product of two independent continuous r.v.s is wrong: a Jacobian is needed (for convolutions, the absolute Jacobian determinant is 1 so it isn't noticeable in the convolution integral formula).

Since convolution sums are just the law of total probability, we have already used them in previous chapters without mentioning the word convolution; see, for example, the first and most tedious proof of Theorem 3.8.9 (sum of independent Binomials), as well as the proof of Theorem 4.8.1 (sum of independent Poissons). In the following examples, we find the distribution of a sum of Exponentials and a sum of Uniforms using a convolution integral.

Example 8.2.4 (Exponential convolution). Let $X, Y \stackrel{\text{i.i.d.}}{\sim} \text{Expo}(\lambda)$. Find the distribution of $T = X + Y$.

Solution:

For $t > 0$, the convolution formula gives

$$f_T(t) = \int_{-\infty}^{\infty} f_Y(t-x) f_X(x) dx = \int_0^t \lambda e^{-\lambda(t-x)} \lambda e^{-\lambda x} dx,$$

where we restricted the integral to be from 0 to t since we need $t - x > 0$ and $x > 0$ for the PDFs inside the integral to be nonzero. Simplifying, we have

$$f_T(t) = \lambda^2 \int_0^t e^{-\lambda t} dx = \lambda^2 t e^{-\lambda t}, \text{ for } t > 0.$$

This is known as the Gamma$(2, \lambda)$ distribution. We will introduce the Gamma distribution in detail in Section 8.4. □

Example 8.2.5 (Uniform convolution). Let $X, Y \stackrel{\text{i.i.d.}}{\sim} \text{Unif}(0,1)$. Find the distribution of $T = X + Y$.

Solution:

The PDF of X (and of Y) is

$$g(x) = \begin{cases} 1, & x \in (0,1), \\ 0, & \text{otherwise.} \end{cases}$$

The convolution formula gives

$$f_T(t) = \int_{-\infty}^{\infty} f_Y(t-x) f_X(x) dx = \int_{-\infty}^{\infty} g(t-x) g(x) dx.$$

The integrand is 1 if and only if $0 < t - x < 1$ and $0 < x < 1$; this is a parallelogram-shaped constraint. Equivalently, the constraint is $\max(0, t-1) < x < \min(t, 1)$.

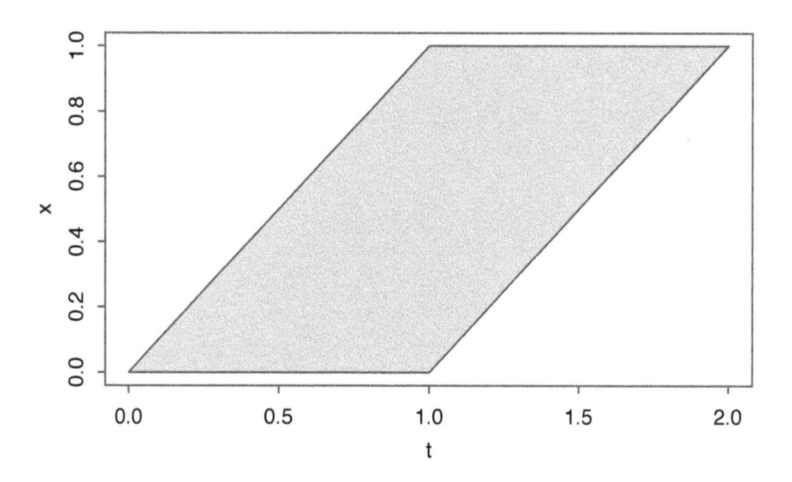

FIGURE 8.2
Region in the (t, x)-plane where $g(t-x)g(x)$ is 1.

From Figure 8.2, we see that for $0 < t \leq 1$, x is constrained to be in $(0, t)$, and for $1 < t < 2$, x is constrained to be in $(t-1, 1)$. Therefore, the PDF of T is a piecewise linear function:

$$f_T(t) = \begin{cases} \displaystyle\int_0^t dx = t & \text{for } 0 < t \leq 1, \\ \displaystyle\int_{t-1}^1 dx = 2 - t & \text{for } 1 < t < 2. \end{cases}$$

Figure 8.3 plots the PDF of T. It is shaped like a triangle with vertices at 0, 1, and 2, so it is called the Triangle$(0, 1, 2)$ distribution.

Heuristically, it makes sense that T is more likely to take on values near the middle than near the extremes: a value near 1 can be obtained if both X and Y are moderate, if X is large but Y is small, or if Y is large but X is small. In contrast, a value near 2 is only possible if both X and Y are large. Thinking back to Example 3.2.5, the PMF of the sum of two die rolls was also shaped like a triangle. A single die roll has a *Discrete Uniform* distribution on the integers 1 through 6, so in that problem we were looking at a convolution of two Discrete Uniforms. It makes sense that the PDF we obtained here is similar in shape. □

8.3 Beta

In this section and the next, we will introduce two continuous distributions, the Beta and Gamma, which are related to several named distributions we have already

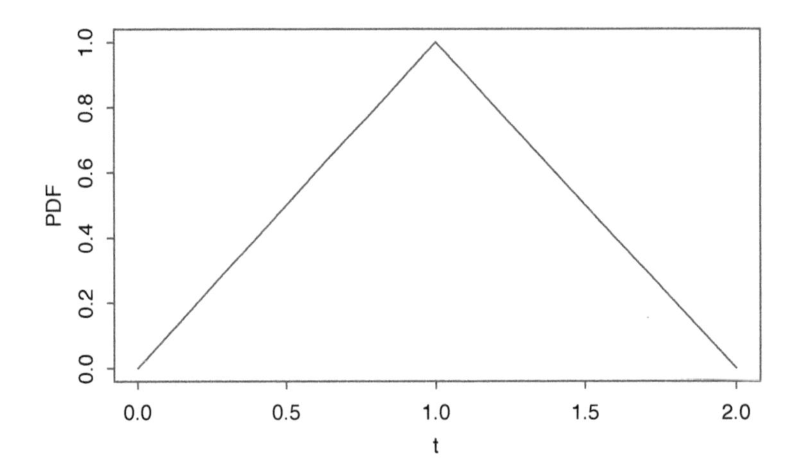

FIGURE 8.3
PDF of $T = X + Y$, where X and Y are i.i.d. Unif$(0, 1)$.

studied and are also related to each other via a shared story. This is an interlude from the subject of transformations, but we'll eventually need to use a change of variables to tie the Beta and Gamma distributions together.

The Beta distribution is a continuous distribution on the interval $(0, 1)$. It is a generalization of the Unif$(0, 1)$ distribution, allowing the PDF to be non-constant on $(0, 1)$.

Definition 8.3.1 (Beta distribution). An r.v. X is said to have the *Beta distribution* with parameters a and b, where $a > 0$ and $b > 0$, if its PDF is

$$f(x) = \frac{1}{\beta(a, b)} x^{a-1}(1 - x)^{b-1}, \quad 0 < x < 1,$$

where the constant $\beta(a, b)$ is chosen to make the PDF integrate to 1. We write this as $X \sim \text{Beta}(a, b)$.

Taking $a = b = 1$, the Beta$(1, 1)$ PDF is constant on $(0, 1)$, so the Beta$(1, 1)$ and Unif$(0, 1)$ distributions are the same. By varying the values of a and b, we get PDFs with a variety of shapes; Figure 8.4 shows four examples. Here are a couple of general patterns:

- If $a < 1$ and $b < 1$, the PDF is U-shaped and opens upward. If $a > 1$ and $b > 1$, the PDF opens down.
- If $a = b$, the PDF is symmetric about $1/2$. If $a > b$, the PDF favors values larger than $1/2$; if $a < b$, the PDF favors values smaller than $1/2$.

By definition, the constant $\beta(a, b)$ satisfies

$$\beta(a, b) = \int_0^1 x^{a-1}(1 - x)^{b-1} dx.$$

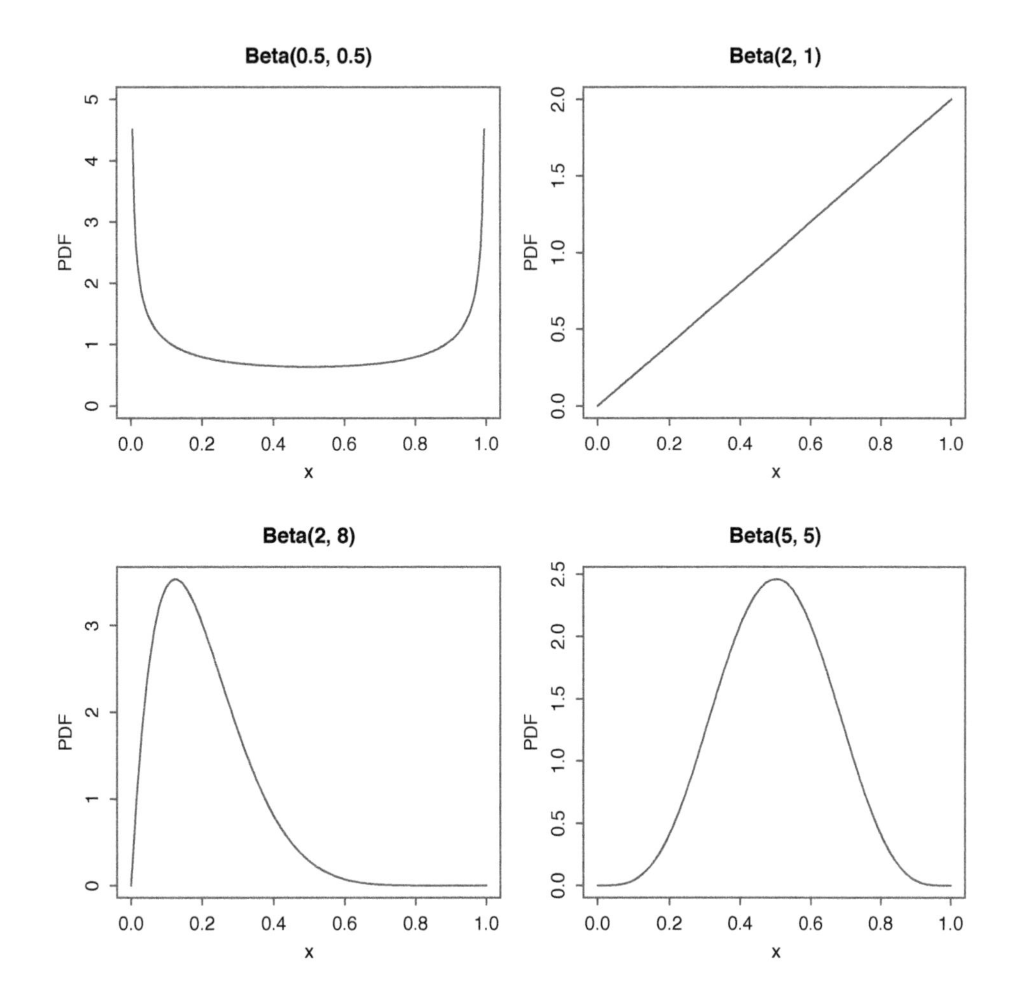

FIGURE 8.4
Beta PDFs for various values of a and b. Clockwise from top left: Beta$(0.5, 0.5)$, Beta$(2, 1)$, Beta$(5, 5)$, Beta$(2, 8)$.

An integral of this form is called a *beta integral*, and we will need to do some calculus to derive a general formula for $\beta(a, b)$. But in the special case where a and b are positive integers, Thomas Bayes figured out how to do the integral using a story proof rather than calculus!

Story 8.3.2 (Bayes' billiards). Show *without using calculus* that for any integers k and n with $0 \leq k \leq n$,

$$\int_0^1 \binom{n}{k} x^k (1-x)^{n-k} dx = \frac{1}{n+1}.$$

Solution:

By telling two stories, we will show that the left-hand and right-hand sides are both equal to $P(X = k)$, where X is an r.v. that we will construct.

Story 1: Start with $n+1$ balls, n white and 1 gray. Randomly throw each ball onto the unit interval $[0, 1]$, such that the positions of the balls are i.i.d. Unif$(0, 1)$. Let X be the number of white balls to the left of the gray ball; X is a discrete r.v. with possible values $0, 1, \ldots, n$. Figure 8.5 illustrates the setup of our experiment.

To get the probability of the event $X = k$, we use LOTP, conditioning on the position of the gray ball, which we'll call B. Conditional on $B = p$, the number of white balls landing to the left of p has a Bin(n, p) distribution, since we can consider each of the white balls to be an independent Bernoulli trial, where success is defined as landing to the left of p. Let f be the PDF of B; $f(p) = 1$ since $B \sim$ Unif$(0, 1)$. So

$$P(X = k) = \int_0^1 P(X = k | B = p) f(p) dp = \int_0^1 \binom{n}{k} p^k (1-p)^{n-k} dp.$$

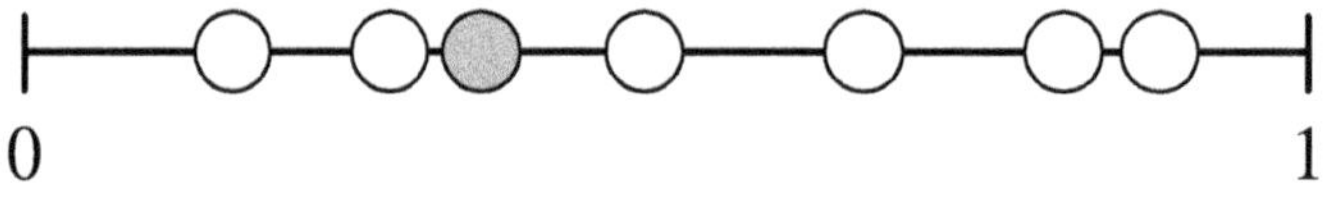

0 1

FIGURE 8.5
Bayes' billiards. Here we throw $n = 6$ white balls and one gray ball onto the unit interval, and we observe $X = 2$ balls to the left of the gray ball.

Story 2: Start with $n + 1$ balls, all white. Randomly throw each ball onto the unit interval; then choose one ball at random and paint it gray. Again, let X be the number of white balls to the left of the gray ball. By symmetry, any one of the $n + 1$ balls is equally likely to be painted gray, so

$$P(X = k) = \frac{1}{n+1}$$

for $k = 0, 1, \ldots, n$.

Here's the crux: X has the same distribution in the two stories! It does not matter whether we paint the gray ball first and then throw, or whether we throw first and then paint the gray ball. So $P(X = k)$ is the same in Story 1 and Story 2, and

$$\int_0^1 \binom{n}{k} p^k (1-p)^{n-k} dp = \frac{1}{n+1}$$

for $k = 0, 1, \ldots, n$. Despite the k's in the integrand, the value of the integral doesn't depend on k. Substituting $a-1$ for k and $b-1$ for $n-k$, this shows that for positive integer values of a and b,

$$\beta(a, b) = \frac{1}{(a+b-1)\binom{a+b-2}{a-1}} = \frac{(a-1)!(b-1)!}{(a+b-1)!}.$$

Later in this chapter, we'll learn what $\beta(a, b)$ is for general a and b. □

The Beta is a flexible family of distributions on $(0, 1)$, and has many stories. One of these stories is that a Beta r.v. is often used to represent an *unknown probability*. That is, we can use the Beta to put probabilities on unknown probabilities!

Story 8.3.3 (Beta-Binomial conjugacy). We have a coin that lands Heads with probability p, but we don't know what p is. Our goal is to *infer* the value of p after observing the outcomes of n tosses of the coin. The larger that n is, the more accurately we should be able to estimate p.

There are several ways to go about doing this. One major approach is *Bayesian inference*, which treats all unknown quantities as random variables. In the Bayesian approach, we would treat the unknown probability p as a random variable and give p a distribution. This is called a *prior* distribution, and it reflects our uncertainty about the true value of p before observing the coin tosses. After the experiment is performed and the data are gathered, the prior distribution is updated using Bayes' rule; this yields the *posterior* distribution, which reflects our new beliefs about p.

Let's see what happens if the prior distribution on p is a Beta distribution. Let $p \sim \text{Beta}(a, b)$ for known constants a and b, and let X be the number of Heads in n tosses of the coin. Conditional on knowing the true value of p, the tosses would just be independent Bernoulli trials with probability p of success, so

$$X|p \sim \text{Bin}(n, p).$$

Note that X is *not* marginally Binomial; it is *conditionally* Binomial, given p. The marginal distribution of X is called the *Beta-Binomial distribution*. To get the posterior distribution of p, we use Bayes' rule (in a hybrid form, since X is discrete and p is continuous). Letting $f(p)$ be the prior distribution and $f(p|X = k)$ be the posterior distribution after observing k Heads,

$$f(p|X=k) = \frac{P(X=k|p)f(p)}{P(X=k)}$$
$$= \frac{\binom{n}{k}p^k(1-p)^{n-k} \cdot \frac{1}{\beta(a,b)}p^{a-1}(1-p)^{b-1}}{P(X=k)}.$$

The denominator, which is the marginal PMF of X, is given by

$$P(X=k) = \int_0^1 P(X=k|p)f(p)dp = \int_0^1 \binom{n}{k}p^k(1-p)^{n-k}f(p)dp.$$

For $a = b = 1$ (which gives a Unif(0, 1) prior on p), we showed in the Bayes' billiards story that $P(X = k) = 1/(n + 1)$, i.e., X is Discrete Uniform on $\{0, 1, \ldots, n\}$. For a and b any positive integers, we can again use Bayes' billiards to find $P(X = k)$; interestingly, X then turns out to have a Negative Hypergeometric distribution (the story and PMF of which we saw in Example 4.4.7).

But it does not seem easy to find $P(X = k)$ in general for real a, b. Relatedly, we still have not evaluated $\beta(a, b)$ in general. Are we stuck?

Actually, the calculation is much easier than it appears at first. The conditional PDF $f(p|X = k)$ is a *function of* p, which means everything that doesn't depend on p is just a constant. We can drop all these constants and find the PDF up to a multiplicative constant (and then the normalizing constant is whatever it needs to be to make the PDF integrate to 1). This gives

$$f(p|X=k) \propto p^{a+k-1}(1-p)^{b+n-k-1},$$

which is the Beta$(a + k, b + n - k)$ PDF, up to a multiplicative constant. Therefore, the posterior distribution of p is

$$p|X = k \sim \text{Beta}(a+k, b+n-k).$$

The posterior distribution of p after observing $X = k$ is still a Beta distribution! This is a special relationship between the Beta and Binomial distributions called *conjugacy*: if we have a Beta prior distribution on p and data that are conditionally Binomial given p, then when going from prior to posterior, we don't leave the family of Beta distributions. We say that *the Beta is the conjugate prior of the Binomial.*

Furthermore, notice the very simple formula for updating the distribution of p. We just add the number of observed successes, k, to the first parameter of the Beta distribution, and the number of observed failures, $n - k$, to the second parameter. So a and b have a concrete interpretation in this context, at least when a and b are positive integers: think of $a - 1$ as the number of prior successes and $b - 1$ as the number of prior failures in earlier experiments (these prior experiments could be real or imagined). Adding on k successes and $n - k$ failures from the current experiment, we have $a+k-1$ successes and $b+n-k-1$ failures, and the Beta(a,b) prior gets updated to a Beta$(a+k, b+n-k)$ posterior.

As in Section 2.6, we can sequentially update our beliefs as we get more and more evidence: we start with a prior distribution and update it to get a posterior distribution, which becomes the new prior distribution, which we update to get a new posterior distribution, etc. The beauty here is that all of this can be done within the Beta family of distributions, with easy updates to the parameters based on tallying the observed successes and failures.

For concreteness, Figure 8.6 shows the case where the prior is Beta$(1,1)$ (which is equivalent to Unif$(0,1)$, as noted earlier), and we observe $n = 5$ coin tosses, all of which happen to land Heads. Then the posterior is Beta$(6,1)$, which is plotted on the right half of Figure 8.6. Notice how the posterior distribution incorporates the evidence from the coin tosses: larger values of p have higher density, consistent with the fact that we observed all Heads.

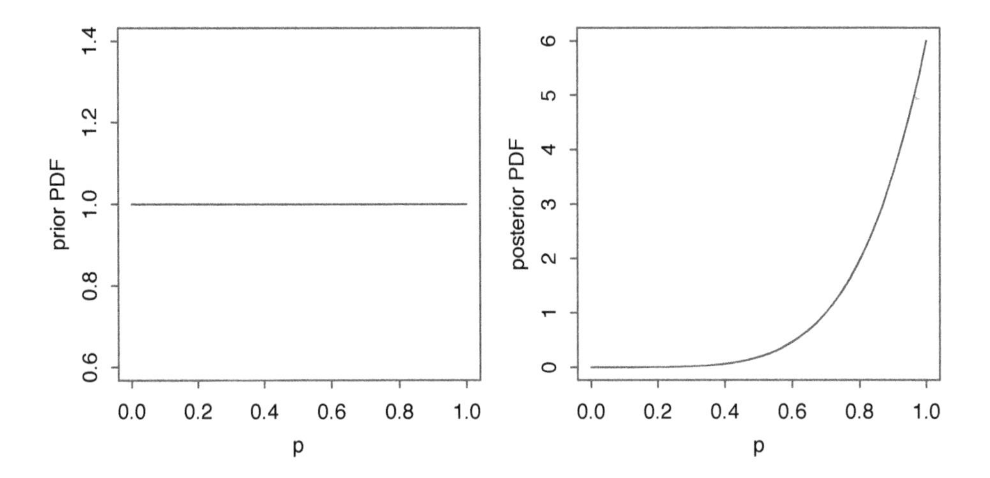

FIGURE 8.6
Beta is conjugate prior of Binomial. Left: prior is Unif$(0,1)$. Right: after observing 5 Heads in 5 tosses, posterior is Beta$(6,1)$.

This model is a continuous analog of Example 2.3.7, our very first example of Bayes' rule. In that example, we also had a coin whose probability of Heads p was unknown, but our prior information led us to believe that p could only take on one of two possible values, $1/2$ or $3/4$. For this reason, our prior distribution on p—though we didn't call it that at the time!—was discrete. In particular, our prior PMF was

$$P(p = 1/2) = 1/2,$$
$$P(p = 3/4) = 1/2.$$

After observing three Heads in a row, we updated this PMF to obtain the posterior PMF, which assigned a probability of 0.23 to $p = 1/2$ and a probability of 0.77

to $p = 3/4$. The same logic applies to the example in this chapter, except that we now give p a continuous prior distribution, which is appropriate if we believe that p could possibly take on any value between 0 and 1. □

To conclude the section, here are two examples that illustrate how Beta-Binomial conjugacy and Bayes' billiards can make some seemingly complicated problems much easier.

Example 8.3.4 (Bayes' serum). A new treatment, Bayes' serum, has just been developed for the disease conditionitis. A clinical trial is about to be conducted, to study how effective the treatment is. Bayes' serum will be applied to n patients who have conditionitis. Given p, the patients' outcomes are independent, with each patient having probability p of being cured by the treatment. But p is unknown. To quantify our uncertainty about p, we model p as a random variable, with prior distribution $p \sim \text{Unif}(0, 1)$.

(a) Find the probability that exactly k out of the n patients will be cured by the treatment (unconditionally, *not* given p).

(b) Now suppose that the treatment is extremely effective in the clinical trial: all n patients are cured! Given this information, find the probability that p exceeds $1/2$. Your answer should be fully simplified, and expressed only in terms of n.

Solution:

(a) By LOTP followed by Bayes' billiards or pattern-matching to a Beta PDF, the probability is

$$\int_0^1 \binom{n}{k} p^k (1-p)^{n-k} dp = \frac{1}{n+1}.$$

(b) By Beta-Binomial conjugacy, the posterior distribution of p given the data is $\text{Beta}(1+n, 1)$, which has PDF $(n+1)p^n$ for $0 < p < 1$. Thus,

$$P\left(p > \frac{1}{2}\right) = \int_{1/2}^1 (n+1)p^n dp = p^{n+1}\Big|_{1/2}^1 = 1 - \frac{1}{2^{n+1}}.$$ □

Example 8.3.5. A basketball player will shoot $N \sim \text{Pois}(\lambda)$ free throws in a game tomorrow. Let X_j be the indicator of her making their jth free throw, and $X = X_1 + \cdots + X_N$ be the total number of free throws she makes in the game (so $X = 0$ if $N = 0$). To model our uncertainty about how good a free throw shooter she is, let $p \sim \text{Beta}(a, b)$. Given p, the player has probability p of making a free throw and probability $q = 1 - p$ of missing it. Assume that $X_1, X_2, \ldots$ are conditionally independent given p, and that N is independent of $p, X_1, X_2, \ldots$.

(a) Find the conditional distribution of X given N, p.

(b) Find the conditional distribution of X given p.

(c) Find the conditional distribution of X given N, for the case $a = b = 1$.

(d) Find the conditional distribution of p given X, N.

Solution:

(a) By the story of the Binomial, $X|(N,p) \sim \text{Bin}(N,p)$.

(b) By the chicken-egg story, $X|p \sim \text{Pois}(\lambda p)$.

(c) For $k = 0, 1, \dots, n$, by LOTP and Bayes' billiards we have

$$P(X = k|N = n) = \int_0^1 \binom{n}{k} p^k (1-p)^{n-k} dp = \frac{1}{n+1}.$$

So the conditional distribution of X given N is Discrete Uniform on $\{0, 1, \dots, N\}$.

(d) By Beta-Binomial conjugacy, $p|(X,N) \sim \text{Beta}(a+X, b+N-X)$. □

8.4 Gamma

The Gamma distribution is a continuous distribution on the positive real line; it is a generalization of the Exponential distribution. While an Exponential r.v. represents the waiting time for the first success under conditions of memorylessness, we shall see that a Gamma r.v. represents the total waiting time for multiple successes.

Before writing down the PDF, we first introduce the *gamma function*, a very famous function in mathematics that extends the factorial function beyond the realm of nonnegative integers.

Definition 8.4.1 (Gamma function). The *gamma function* Γ is defined by

$$\Gamma(a) = \int_0^\infty x^a e^{-x} \frac{dx}{x},$$

for real numbers $a > 0$.

We could also cancel an x and write the integrand as $x^{a-1}e^{-x}$, but it turns out to be convenient having the $\frac{dx}{x}$ since it is common to make a transformation of the form $u = cx$, and then we have the handy fact that $\frac{du}{u} = \frac{dx}{x}$. Here are two important properties of the gamma function.

- $\Gamma(a+1) = a\Gamma(a)$ for all $a > 0$. This follows from integration by parts:

$$\Gamma(a+1) = \int_0^\infty x^a e^{-x} dx = -x^a e^{-x}\Big|_0^\infty + a\int_0^\infty x^{a-1}e^{-x}dx = 0 + a\Gamma(a).$$

- $\Gamma(n) = (n-1)!$ if n is a positive integer. This can be proved by induction, starting with $n = 1$ and using the recursive relation $\Gamma(a+1) = a\Gamma(a)$. Thus, if we evaluate the gamma function at positive integer values, we recover the factorial function (albeit shifted by 1).

Now let's suppose that on a whim, we decide to divide both sides of the above definition by $\Gamma(a)$. We have

$$1 = \int_0^\infty \frac{1}{\Gamma(a)} x^a e^{-x} \frac{dx}{x},$$

so the function under the integral is a valid PDF supported on $(0, \infty)$. This is the definition of the PDF of the Gamma distribution. Specifically, we say that X has the *Gamma distribution* with parameters a and 1, denoted $X \sim \text{Gamma}(a, 1)$, if its PDF is

$$f_X(x) = \frac{1}{\Gamma(a)} x^a e^{-x} \frac{1}{x}, \quad x > 0.$$

From the Gamma$(a, 1)$ distribution, we obtain the general Gamma distribution by a scale transformation: if $X \sim \text{Gamma}(a, 1)$ and $\lambda > 0$, then the distribution of $Y = X/\lambda$ is called the Gamma(a, λ) distribution. By the change of variables formula with $x = \lambda y$ and $dx/dy = \lambda$, the PDF of Y is

$$f_Y(y) = f_X(x) \left| \frac{dx}{dy} \right| = \frac{1}{\Gamma(a)} (\lambda y)^a e^{-\lambda y} \frac{1}{\lambda y} \lambda = \frac{1}{\Gamma(a)} (\lambda y)^a e^{-\lambda y} \frac{1}{y}, \quad y > 0.$$

This is summarized in the following definition.

Definition 8.4.2 (Gamma distribution). An r.v. Y is said to have the *Gamma distribution* with parameters a and λ, where $a > 0$ and $\lambda > 0$, if its PDF is

$$f(y) = \frac{1}{\Gamma(a)} (\lambda y)^a e^{-\lambda y} \frac{1}{y}, \quad y > 0.$$

We write $Y \sim \text{Gamma}(a, \lambda)$.

Taking $a = 1$, the Gamma$(1, \lambda)$ PDF is $f(y) = \lambda e^{-\lambda y}$ for $y > 0$, so the Gamma$(1, \lambda)$ and Expo(λ) distributions are the same. The extra parameter a allows Gamma PDFs to have a greater variety of shapes. Figure 8.7 shows four Gamma PDFs. For small values of a, the PDF is skewed, but as a increases, the PDF starts to look more symmetrical and bell-shaped; we will learn the reason for this in Chapter 10. Increasing λ compresses the PDF toward smaller values, as we can see by comparing the Gamma$(3, 1)$ and Gamma$(3, 0.5)$ PDFs.

Let's find the mean, variance, and other moments of the Gamma distribution, starting with $X \sim \text{Gamma}(a, 1)$. We'll use properties of the gamma function as well as the technique of *doing integrals by pattern recognition*. For the mean, we write down the definition of $E(X)$,

$$E(X) = \int_0^\infty \frac{1}{\Gamma(a)} x^{a+1} e^{-x} \frac{dx}{x},$$

but instead of attempting a gruesome integration by parts, we *recognize* that after taking out $1/\Gamma(a)$, what's left is precisely the gamma function evaluated at $a + 1$. Therefore

$$E(X) = \frac{\Gamma(a+1)}{\Gamma(a)} = \frac{a\Gamma(a)}{\Gamma(a)} = a.$$

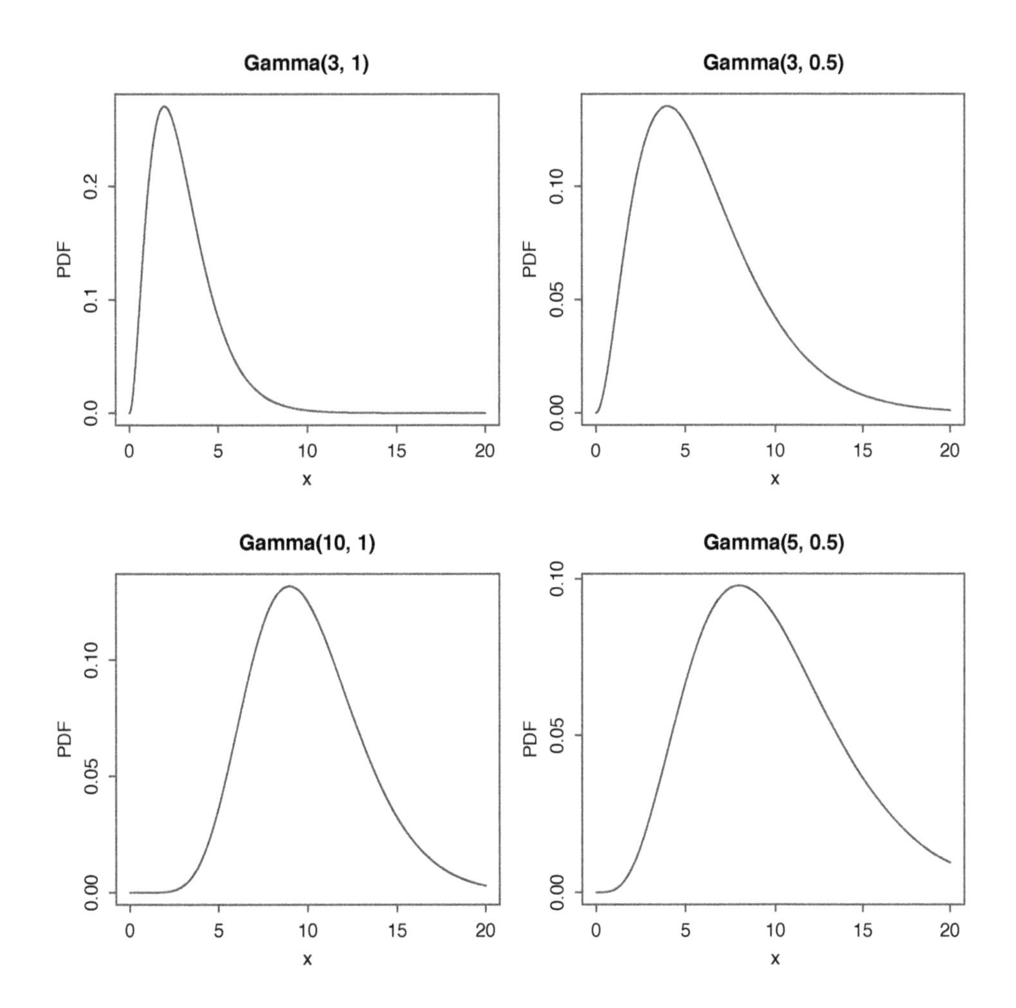

FIGURE 8.7
Gamma PDFs for various values of a and λ. Clockwise from top left: Gamma$(3, 1)$, Gamma$(3, 0.5)$, Gamma$(5, 0.5)$, Gamma$(10, 1)$.

Similarly, for the variance, LOTUS gives us an integral expression for the second moment,

$$E(X^2) = \int_0^\infty \frac{1}{\Gamma(a)} x^{a+2} e^{-x} \frac{dx}{x},$$

and we recognize the gamma function evaluated at $a + 2$. Therefore

$$E(X^2) = \frac{\Gamma(a+2)}{\Gamma(a)} = \frac{(a+1)a\Gamma(a)}{\Gamma(a)} = (a+1)a$$

and

$$\text{Var}(X) = (a+1)a - a^2 = a.$$

So for $X \sim \text{Gamma}(a, 1)$, $E(X) = \text{Var}(X) = a$.

The cth moment is no harder than the second moment; we just use LOTUS and recognize the definition of $\Gamma(a + c)$. This gives

$$E(X^c) = \int_0^\infty \frac{1}{\Gamma(a)} x^{a+c} e^{-x} \frac{dx}{x} = \frac{\Gamma(a+c)}{\Gamma(a)}$$

for all real c such that the integral converges, i.e., for $c > -a$.

We can now transform to $Y = X/\lambda \sim \text{Gamma}(a, \lambda)$ to get

$$\begin{aligned}
E(Y) &= \frac{1}{\lambda} E(X) = \frac{a}{\lambda}, \\
\text{Var}(Y) &= \frac{1}{\lambda^2} \text{Var}(X) = \frac{a}{\lambda^2}, \\
E(Y^c) &= \frac{1}{\lambda^c} E(X^c) = \frac{1}{\lambda^c} \cdot \frac{\Gamma(a+c)}{\Gamma(a)}, \quad c > -a.
\end{aligned}$$

Looking back at the Gamma PDF plots, they are consistent with our finding that the mean and variance are increasing in a and decreasing in λ.

So far, we've been learning about the Gamma distribution using the PDF, which allowed us to discern general patterns from PDF plots and to derive the mean and variance. But the PDF doesn't provide much insight about why we'd ever *use* the Gamma distribution, and it doesn't give us much of an interpretation for the parameters a and λ. For this, we need to connect the Gamma to other named distributions through stories. The rest of this section is devoted to stories for the Gamma distribution.

In the special case where a is an integer, we can represent a $\text{Gamma}(a, \lambda)$ r.v. as a sum (convolution) of i.i.d. $\text{Expo}(\lambda)$ r.v.s.

Theorem 8.4.3. Let $X_1, \ldots, X_n$ be i.i.d. $\text{Expo}(\lambda)$. Then

$$X_1 + \cdots + X_n \sim \text{Gamma}(n, \lambda).$$

Proof. The Expo(λ) MGF is $\frac{\lambda}{\lambda-t}$ for $t < \lambda$, so the MGF of $X_1 + \cdots + X_n$ is

$$M_n(t) = \left(\frac{\lambda}{\lambda - t}\right)^n$$

for $t < \lambda$. Let $Y \sim \text{Gamma}(n, \lambda)$; we'll show that the MGF of Y is the same as that of $X_1 + \cdots + X_n$. By LOTUS,

$$E(e^{tY}) = \int_0^\infty e^{ty} \frac{1}{\Gamma(n)} (\lambda y)^n e^{-\lambda y} \frac{dy}{y}.$$

Again, we'll use pattern recognition to do the integral. We just need to do algebraic manipulations until what's left inside the integral is a recognizable Gamma PDF:

$$\begin{aligned} E(e^{tY}) &= \int_0^\infty e^{ty} \frac{1}{\Gamma(n)} (\lambda y)^n e^{-\lambda y} \frac{dy}{y} \\ &= \frac{\lambda^n}{(\lambda - t)^n} \int_0^\infty \frac{1}{\Gamma(n)} e^{-(\lambda - t)y} ((\lambda - t)y)^n \frac{dy}{y}. \end{aligned}$$

We pulled λ^n out of the integral, then multiplied by $(\lambda - t)^n$ on the inside while dividing by it on the outside. Now the expression inside the integral is the Gamma($n, \lambda - t$) PDF, assuming $t < \lambda$. Since PDFs integrate to 1, we have

$$E(e^{tY}) = \left(\frac{\lambda}{\lambda - t}\right)^n$$

for $t < \lambda$; if $t \geq \lambda$ the integral fails to converge.

We have shown that $X_1 + \cdots + X_n$ and $Y \sim \text{Gamma}(n, \lambda)$ have the same MGF. Since the MGF determines the distribution, $X_1 + \cdots + X_n \sim \text{Gamma}(n, \lambda)$.

Alternatively, we can compute the convolution integral inductively. It suffices to consider the case $\lambda = 1$, since after having done that case we can rescale the r.v.s to prove the result for general λ. Let $T_n = X_1 + \cdots + X_n$. We will show by induction that $T_n \sim \text{Gamma}(n, 1)$ for all $n \geq 1$. For $n = 1$, this is true since the Gamma(1, 1) PDF is e^{-x} (for $x > 0$), which is the same as the Expo(1) PDF. Now assume that $T_n \sim \text{Gamma}(n, 1)$, and show that $T_{n+1} \sim \text{Gamma}(n + 1, 1)$. The PDF of $T_{n+1} = T_n + X_{n+1}$ is given by the convolution integral

$$\int_0^\infty f_{T_n}(x) f_{X_{n+1}}(t-x) dx = \frac{1}{\Gamma(n)} \int_0^t x^{n-1} e^{-x} e^{-(t-x)} dx = \frac{e^{-t}}{\Gamma(n)} \int_0^t x^{n-1} dx = \frac{t^n e^{-t}}{\Gamma(n+1)}$$

for $t > 0$, which completes the induction. ■

Thus, if $Y \sim \text{Gamma}(a, \lambda)$ with a an integer, we can represent Y as a sum of i.i.d. Expo(λ) r.v.s, $X_1 + \cdots + X_a$, and get the mean and variance right away:

$$\begin{aligned} E(Y) &= E(X_1 + \cdots + X_a) = aE(X_1) = \frac{a}{\lambda}, \\ \text{Var}(Y) &= \text{Var}(X_1 + \cdots + X_a) = a\text{Var}(X_1) = \frac{a}{\lambda^2}, \end{aligned}$$

in agreement with the results we derived earlier for general a.

Theorem 8.4.3 also allows us to connect the Gamma distribution to the story of the Poisson process. We showed in Chapter 5 that in a Poisson process of rate λ, the interarrival times are i.i.d. Expo(λ) r.v.s. But the total waiting time T_n for the nth arrival is the sum of the first n interarrival times; for instance, Figure 8.8 illustrates how T_3 is the sum of the 3 interarrival times X_1, X_2, X_3. Therefore, by the theorem, $T_n \sim \text{Gamma}(n, \lambda)$. The interarrival times in a Poisson process are Exponential r.v.s, while the raw arrival times are Gamma r.v.s.

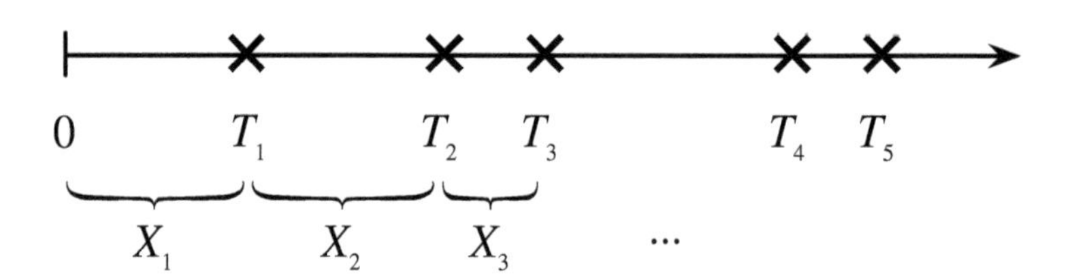

FIGURE 8.8
Poisson process. The interarrival times X_j are i.i.d. Expo(λ), while the raw arrival times T_j are Gamma(j, λ).

☣ **8.4.4.** Unlike the X_j, the T_j are *not* independent, since they are constrained to be increasing; nor are they identically distributed.

At last, we have an interpretation for the parameters of the Gamma(a, λ) distribution. In the Poisson process story, a is the *number of successes* we are waiting for, and λ is the *rate* at which successes arrive; $Y \sim \text{Gamma}(a, \lambda)$ is the total waiting time for the ath arrival in a Poisson process of rate λ.

A consequence of this story is that a convolution of Gammas with the same λ is still Gamma. Exercise 29 explores this fact from several perspectives.

When we introduced the Exponential distribution, we viewed it as the continuous analog of the Geometric distribution: the Geometric waits for the first success in discrete time, and the Exponential waits for the first success in continuous time. Likewise, we can now say that the Gamma distribution is the continuous analog of the Negative Binomial distribution: the Negative Binomial is a sum of Geometric waiting times, and the Gamma is a sum of Exponential waiting times. In Exercise 54 you will use MGFs to show that the Gamma distribution can be obtained as a continuous limit of the Negative Binomial distribution.

A final story about the Gamma is that it shares the same special relationship with the Poisson that the Beta shares with the Binomial: the Gamma is the conjugate prior of the Poisson. Earlier we saw that the Beta distribution can represent an unknown probability of success because its support is $(0, 1)$. The Gamma distribution, on the other hand, can represent an unknown *rate* in a Poisson process because its support is $(0, \infty)$.

To investigate, we'll return to Blotchville, where buses arrive in a Poisson process of rate λ. Previously it was assumed that $\lambda = 1/10$, so that the times between buses were i.i.d. Exponentials with mean 10 minutes, but now we'll assume Fred doesn't know the rate λ at which buses arrive and needs to figure it out. Fred will follow the Bayesian approach and treat the unknown rate as a random variable.

Story 8.4.5 (Gamma-Poisson conjugacy). In Blotchville, buses arrive at a certain bus stop according to a Poisson process with rate λ buses per hour, where λ is unknown. Based on his very memorable adventures in Blotchville, Fred quantifies his uncertainty about λ using the prior $\lambda \sim \text{Gamma}(r_0, b_0)$, where r_0 and b_0 are known, positive constants with r_0 an integer.

To better understand the bus system, Fred wants to learn more about λ. He is a very patient person, and decides that he will sit at the bus stop for t hours and count how many buses arrive in this time interval. Let Y be the number of buses in this time interval, and suppose Fred observes that $Y = y$.

(a) Find Fred's hybrid joint distribution for Y and λ.

(b) Find Fred's marginal distribution for Y.

(c) Find Fred's posterior distribution for λ, i.e., his conditional distribution of λ given the data y.

(d) Find Fred's posterior mean $E(\lambda|Y = y)$ and posterior variance $\text{Var}(\lambda|Y = y)$.

Solution: Notice the similarities between the structure of this problem and that of the hybrid joint distribution from Example 7.1.26. We know that $\lambda \sim \text{Gamma}(r_0, b_0)$ marginally, and by definition of Poisson process, conditional on knowing the true rate λ, the number of buses in an interval of length t is distributed $\text{Pois}(\lambda t)$. In other words, what we're given is

$$\begin{aligned} \lambda &\sim \text{Gamma}(r_0, b_0) \\ Y|\lambda &\sim \text{Pois}(\lambda t). \end{aligned}$$

Then we are asked to flip it around: find the marginal distribution of Y and the conditional distribution of λ given $Y = y$, which is the posterior distribution. This is characteristic of Bayesian inference: we have a *prior* distribution for the unknown parameters (in this case, a Gamma distribution for λ) and a *model for the data* conditional on the unknown parameters (in this case, a Poisson distribution for Y given λ), and we use Bayes' rule to get the distribution of the unknowns conditional on the observed data. So let's get started.

(a) Let f_0 be the prior PDF of λ. The hybrid joint distribution of Y and λ is

$$f(y, \lambda) = P(Y = y|\lambda) f_0(\lambda) = \frac{e^{-\lambda t}(\lambda t)^y}{y!} \frac{(b_0\lambda)^{r_0} e^{-b_0\lambda}}{\lambda \Gamma(r_0)},$$

for $y = 0, 1, 2, \ldots$ and $\lambda > 0$. The hybrid joint distribution is plotted in Figure 8.9;

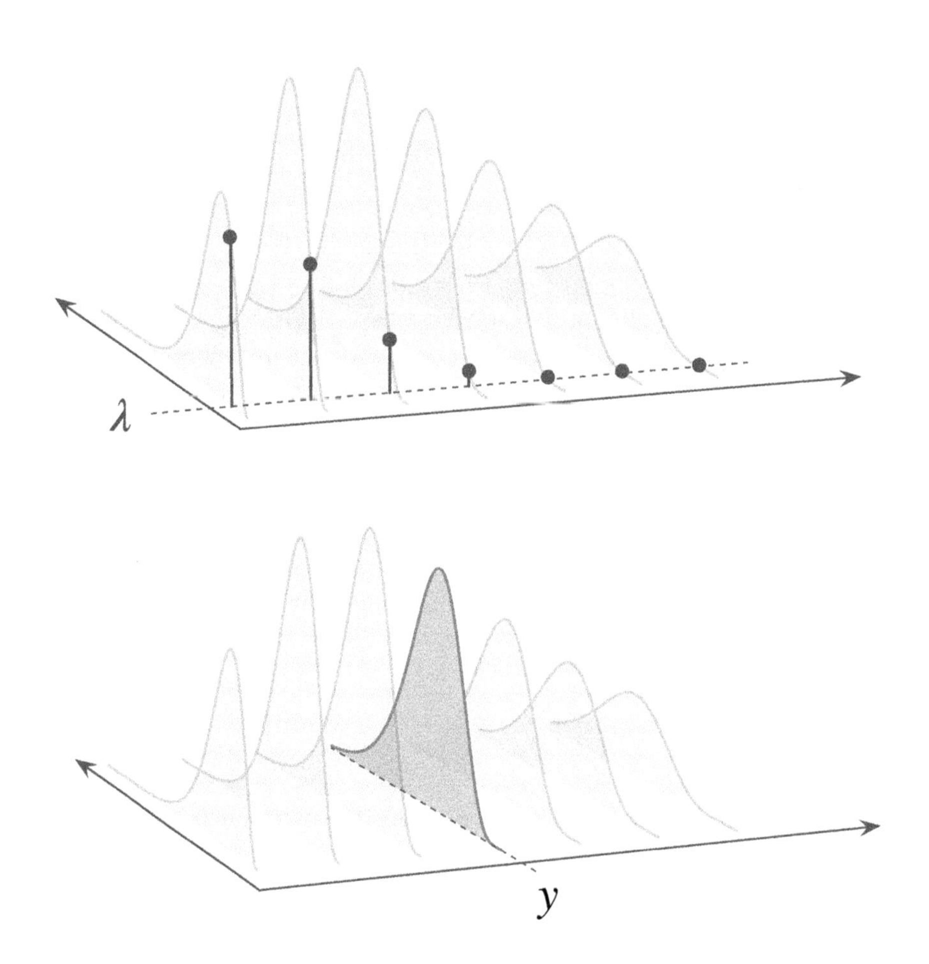

FIGURE 8.9
Hybrid joint distribution of Y and λ. (a) Conditioning on a particular value of λ, the relative heights form a Poisson PMF. (b) In the other direction, conditioning on $Y = y$ gives the posterior distribution of λ.

there is a conditional PMF of Y for every value of λ and a conditional PDF of λ for every value of Y.

(b) To get the marginal PMF of Y, we integrate out λ from the hybrid joint distribution; this is also a form of LOTP. This gives

$$\begin{aligned} P(Y = y) &= \int_0^\infty P(Y = y|\lambda) f_0(\lambda) d\lambda \\ &= \int_0^\infty \frac{e^{-\lambda t}(\lambda t)^y}{y!} \frac{(b_0 \lambda)^{r_0} e^{-b_0 \lambda}}{\Gamma(r_0)} \frac{d\lambda}{\lambda}. \end{aligned}$$

Let's do the integral by pattern recognition, focusing in on the terms involving λ. We spot λ^{r_0+y} and $e^{-(b_0+t)\lambda}$ lurking in the integrand, which suggests pattern-matching to a Gamma(r_0+y, b_0+t) PDF. Pull out all the terms that don't depend on λ, then multiply by whatever it takes to get the desired PDF inside the integral, remembering to multiply by the reciprocal on the outside:

$$\begin{aligned}P(Y=y) &= \frac{t^y b_0^{r_0}}{y!\Gamma(r_0)}\int_0^\infty e^{-(b_0+t)\lambda}\lambda^{r_0+y}\frac{d\lambda}{\lambda}\\ &= \frac{\Gamma(r_0+y)}{y!\Gamma(r_0)}\frac{t^y b_0^{r_0}}{(b_0+t)^{r_0+y}}\int_0^\infty \frac{1}{\Gamma(r_0+y)}e^{-(b_0+t)\lambda}((b_0+t)\lambda)^{r_0+y}\frac{d\lambda}{\lambda}\\ &= \frac{(r_0+y-1)!}{(r_0-1)!y!}\left(\frac{t}{b_0+t}\right)^y\left(\frac{b_0}{b_0+t}\right)^{r_0}.\end{aligned}$$

In the last step, we used the property $\Gamma(n)=(n-1)!$, which is applicable because r_0 is an integer. This is the $\text{NBin}(r_0, b_0/(b_0+t))$ PMF, so the marginal distribution of Y is Negative Binomial with parameters r_0 and $b_0/(b_0+t)$.

(c) By Bayes' rule, the posterior PDF of λ is given by

$$f_1(\lambda|y) = \frac{P(Y=y|\lambda)f_0(\lambda)}{P(Y=y)}.$$

We found $P(Y=y)$ in the previous part, but since it does not depend on λ, we can just treat it as part of the normalizing constant. Absorbing this and other multiplicative factors that don't depend on λ into the normalizing constant,

$$f_1(\lambda|y) \propto e^{-\lambda t}\lambda^y\lambda^{r_0}e^{-b_0\lambda}\frac{1}{\lambda} = e^{-(b_0+t)\lambda}\lambda^{r_0+y}\frac{1}{\lambda},$$

which shows that the posterior distribution of λ is $\text{Gamma}(r_0+y, b_0+t)$.

When going from prior to posterior, the distribution of λ stays in the Gamma family, so the Gamma is indeed the conjugate prior for the Poisson.

Now that we have the posterior PDF of λ, we have a more elegant approach to solving (b). Rearranging Bayes' rule, the marginal PMF of Y is

$$P(Y=y) = \frac{P(Y=y|\lambda)f_0(\lambda)}{f_1(\lambda|y)},$$

where we know the numerator from the statement of the problem and the denominator from the calculation we just did. Plugging in these ingredients and simplifying again yields

$$Y \sim \text{NBin}(r_0, b_0/(b_0+t)).$$

(d) Since conditional PDFs *are* PDFs, it is perfectly fine to calculate the expectation and variance of λ with respect to the posterior distribution. The mean and variance of the $\text{Gamma}(r_0+y, b_0+t)$ distribution give us

$$E(\lambda|Y=y) = \frac{r_0+y}{b_0+t} \quad \text{and} \quad \text{Var}(\lambda|Y=y) = \frac{r_0+y}{(b_0+t)^2}.$$

This example gives another interpretation for the parameters in a Gamma when it is being used as a conjugate prior. Fred's Gamma(r_0, b_0) prior got updated to a Gamma(r_0+y, b_0+t) posterior after observing y arrivals in t hours. We can imagine that in the past, Fred observed r_0 buses arrive in b_0 hours; then after the new data, he has observed $r_0 + y$ buses in $b_0 + t$ hours. So we can interpret r_0 as the number of prior arrivals and b_0 as the total time required for those prior arrivals. □

8.5 Beta-Gamma connections

In this section, we will unite the Beta and Gamma distributions with a common story. As an added bonus, the story will give us an expression for the normalizing constant of the Beta(a, b) PDF in terms of gamma functions, and it will allow us to easily find the expectation of the Beta(a, b) distribution.

Story 8.5.1 (Bank–post office)**.** While running errands, you need to go to the bank, then to the post office. Let $X \sim$ Gamma(a, λ) be your waiting time in line at the bank, and let $Y \sim$ Gamma(b, λ) be your waiting time in line at the post office (with the same λ for both). Assume X and Y are independent. What is the joint distribution of $T = X + Y$ (your total wait at the bank and post office) and $W = \frac{X}{X+Y}$ (the fraction of your waiting time spent at the bank)?

Solution:

We'll do a change of variables in two dimensions to get the joint PDF of T and W. Let $t = x + y$, $w = \frac{x}{x+y}$. Then $x = tw, y = t(1-w)$, and

$$\frac{\partial(x,y)}{\partial(t,w)} = \begin{pmatrix} w & t \\ 1-w & -t \end{pmatrix},$$

which has an absolute determinant of t. Therefore

$$\begin{aligned} f_{T,W}(t,w) &= f_{X,Y}(x,y) \cdot \left|\left|\frac{\partial(x,y)}{\partial(t,w)}\right|\right| \\ &= f_X(x) f_Y(y) \cdot t \\ &= \frac{1}{\Gamma(a)}(\lambda x)^a e^{-\lambda x}\frac{1}{x} \cdot \frac{1}{\Gamma(b)}(\lambda y)^b e^{-\lambda y}\frac{1}{y} \cdot t \\ &= \frac{1}{\Gamma(a)}(\lambda t w)^a e^{-\lambda t w}\frac{1}{tw} \cdot \frac{1}{\Gamma(b)}(\lambda t(1-w))^b e^{-\lambda t(1-w)}\frac{1}{t(1-w)} \cdot t. \end{aligned}$$

Let's group all the terms involving w together, and all the terms involving t together:

$$\begin{aligned} f_{T,W}(t,w) &= \frac{1}{\Gamma(a)\Gamma(b)} w^{a-1}(1-w)^{b-1}(\lambda t)^{a+b} e^{-\lambda t}\frac{1}{t} \\ &= \left(\frac{\Gamma(a+b)}{\Gamma(a)\Gamma(b)} w^{a-1}(1-w)^{b-1}\right)\left(\frac{1}{\Gamma(a+b)}(\lambda t)^{a+b} e^{-\lambda t}\frac{1}{t}\right), \end{aligned}$$

for $0 < w < 1$ and $t > 0$. The form of the joint PDF, together with Proposition 7.1.21, tells us several things:

1. Since the joint PDF factors into a function of t times a function of w, we have that T and W are independent: the total waiting time is independent of the fraction of time spent at the bank.

2. We recognize the marginal PDF of T and deduce that $T \sim \text{Gamma}(a + b, \lambda)$.

3. The PDF of W is

$$f_W(w) = \frac{\Gamma(a+b)}{\Gamma(a)\Gamma(b)} w^{a-1}(1-w)^{b-1}, \quad 0 < w < 1,$$

by Proposition 7.1.21 or just by integrating out T from the joint PDF of T and W. This PDF is proportional to the Beta(a, b) PDF, so it *is* the Beta(a, b) PDF! Note that as a byproduct of the calculation we have just done, we have found the normalizing constant of the Beta distribution:

$$\frac{1}{\beta(a,b)} = \frac{\Gamma(a+b)}{\Gamma(a)\Gamma(b)}$$

is the constant that goes in front of the Beta(a, b) PDF. □

To summarize, the bank–post office story tells us that when we add independent Gamma r.v.s X and Y with the same rate λ, the total $X+Y$ has a Gamma distribution, the fraction $X/(X + Y)$ has a Beta distribution, and the total is independent of the fraction.

We can use this result to find the mean of $W \sim \text{Beta}(a, b)$ without the slightest trace of calculus.

Example 8.5.2 (Beta expectation)**.** With notation as above, note that since T and W are independent, they are uncorrelated: $E(TW) = E(T)E(W)$. Writing this in terms of X and Y, we have

$$E\left((X+Y)\cdot\frac{X}{X+Y}\right) = E(X+Y)E\left(\frac{X}{X+Y}\right),$$
$$E(X) = E(X+Y)E\left(\frac{X}{X+Y}\right),$$
$$\frac{E(X)}{E(X+Y)} = E\left(\frac{X}{X+Y}\right).$$

Ordinarily, the last equality would be a horrendous blunder: faced with an expectation like $E(X/(X + Y))$, we are not generally permitted to move the E into the numerator and denominator as we please. In this case, however, the bank–post office story justifies the move, so finding the expectation of W happily reduces to finding the expectations of X and $X + Y$:

$$E(W) = E\left(\frac{X}{X+Y}\right) = \frac{E(X)}{E(X+Y)} = \frac{a/\lambda}{a/\lambda + b/\lambda} = \frac{a}{a+b}.$$

Another approach is to proceed from the definition of expectation:

$$E(W) = \int_0^1 \frac{\Gamma(a+b)}{\Gamma(a)\Gamma(b)} w^a (1-w)^{b-1} dw.$$

By pattern recognition, the integrand is a Beta$(a+1, b)$ PDF, up to a normalizing constant. After obtaining an exact match for the PDF, we apply properties of the gamma function:

$$\begin{aligned}
E(W) &= \frac{\Gamma(a+b)}{\Gamma(a)} \frac{\Gamma(a+1)}{\Gamma(a+b+1)} \int_0^1 \frac{\Gamma(a+b+1)}{\Gamma(a+1)\Gamma(b)} w^a (1-w)^{b-1} dw \\
&= \frac{\Gamma(a+b)}{\Gamma(a)} \frac{a\Gamma(a)}{(a+b)\Gamma(a+b)} \\
&= \frac{a}{a+b}.
\end{aligned}$$

In Exercise 30, you will use this approach to find the variance and the other moments of the Beta distribution. □

Now that we know the Beta normalizing constant, we can also quickly obtain the PMF of the Beta-Binomial distribution.

Example 8.5.3 (Beta-Binomial PMF). Let $X|p \sim \text{Bin}(n, p)$, with $p \sim \text{Beta}(a, b)$. As mentioned in Story 8.3.3, X has a *Beta-Binomial* distribution. Find the marginal distribution of X.

Solution: Let $f(p)$ be the Beta(a, b) PDF. Then

$$\begin{aligned}
P(X = k) &= \int_0^1 P(X = k|p) f(p) dp \\
&= \frac{1}{\beta(a,b)} \int_0^1 \binom{n}{k} p^k (1-p)^{n-k} p^{a-1} (1-p)^{b-1} dp \\
&= \frac{\binom{n}{k}}{\beta(a,b)} \int_0^1 p^{a+k-1} (1-p)^{b+n-k-1} dp \\
&= \binom{n}{k} \frac{\beta(a+k, b+n-k)}{\beta(a,b)},
\end{aligned}$$

for $k = 0, 1, \ldots, n$. □

8.6 Order statistics

The final transformation we will consider in this chapter is the transformation that takes n random variables $X_1, \ldots, X_n$ and sorts them in order, producing the transformed r.v.s $\min(X_1, \dots, X_n), \dots, \max(X_1, \dots, X_n)$. The transformed r.v.s are called the *order statistics*,[2] and they are often useful when we are concerned with the distribution of extreme values, as we alluded to earlier.

Furthermore, like the sample mean $\bar{X}_n$, the order statistics serve as useful summaries of an experiment, since we can use them to determine the cutoffs for the worst 5% of observations, the worst 25%, the best 25%, and so forth (such cutoffs are called the *quantiles* of the sample).

Definition 8.6.1 (Order statistics). For r.v.s $X_1, X_2, \dots, X_n$, the *order statistics* are the random variables $X_{(1)}, X_{(2)}, \dots, X_{(n)}$, where

$$
\begin{aligned}
&X_{(1)} = \min(X_1, \dots, X_n),\\
&X_{(2)} \text{ is the second-smallest of } X_1, \dots, X_n,\\
&\vdots\\
&X_{(n-1)} \text{ is the second-largest of } X_1, \dots, X_n,\\
&X_{(n)} = \max(X_1, \dots, X_n).
\end{aligned}
$$

Note that $X_{(1)} \leq X_{(2)} \leq \dots \leq X_{(n)}$ by definition. We call $X_{(j)}$ the *jth order statistic*. If n is odd, $X_{((n+1)/2)}$ is called the *sample median* of $X_1, \dots, X_n$.

☣ **8.6.2.** The order statistics $X_{(1)}, \dots, X_{(n)}$ are r.v.s, and each $X_{(j)}$ is a function of $X_1, \dots, X_n$. Even if the original r.v.s are independent, the order statistics are *dependent*: if we know that $X_{(1)} = 100$, then $X_{(n)}$ is forced to be at least 100.

We will focus our attention on the case where $X_1, \dots, X_n$ are i.i.d. continuous r.v.s. The reason is that with discrete r.v.s, there is a positive probability of tied values; with continuous r.v.s, the probability of a tie is exactly 0, which makes matters much easier. Thus, for the rest of this section, assume $X_1, \dots, X_n$ are i.i.d. and continuous, with CDF F and PDF f. We will derive the marginal CDF and PDF of each individual order statistic $X_{(j)}$, as well as the joint PDF of $(X_{(1)}, \dots, X_{(n)})$.

A complication we run into right away is that the transformation to order statistics is not invertible: starting with $\min(X, Y) = 3$ and $\max(X, Y) = 5$, we can't tell whether the original values of X and Y were 3 and 5, respectively, or 5 and 3. Therefore the change of variables formula from $\mathbb{R}^n$ to $\mathbb{R}^n$ does not apply. Instead we will take a direct approach, using pictures to guide us when necessary.

Let's start with the CDF of $X_{(n)} = \max(X_1, \dots X_n)$. Since $X_{(n)}$ is less than x if

[2]This term sometimes causes confusion. In statistics (the field of study), any function of the data is called a *statistic*. If $X_1, \dots, X_n$ are the data, then $\min(X_1, \dots, X_n)$ is a statistic, and so is $\max(X_1, \dots, X_n)$. They are called order statistics because we get them by sorting the data in order (from smallest to largest).

and only if all of the X_j are less than x, the CDF of $X_{(n)}$ is

$$\begin{aligned} F_{X_{(n)}}(x) &= P(\max(X_1, \dots, X_n) \leq x) \\ &= P(X_1 \leq x, \dots, X_n \leq x) \\ &= P(X_1 \leq x) \dots P(X_n \leq x) \\ &= (F(x))^n, \end{aligned}$$

where F is the CDF of the individual X_i. Similarly, $X_{(1)} = \min(X_1, \dots, X_n)$ exceeds x if and only if all of the X_j exceed x, so the CDF of $X_{(1)}$ is

$$\begin{aligned} F_{X_{(1)}}(x) &= 1 - P(\min(X_1, \dots, X_n) > x) \\ &= 1 - P(X_1 > x, \dots, X_n > x) \\ &= 1 - (1 - F(x))^n. \end{aligned}$$

The same logic lets us find the CDF of $X_{(j)}$. For the event $X_{(j)} \leq x$ to occur, we need at least j of the X_i to fall to the left of x. This is illustrated in Figure 8.10.

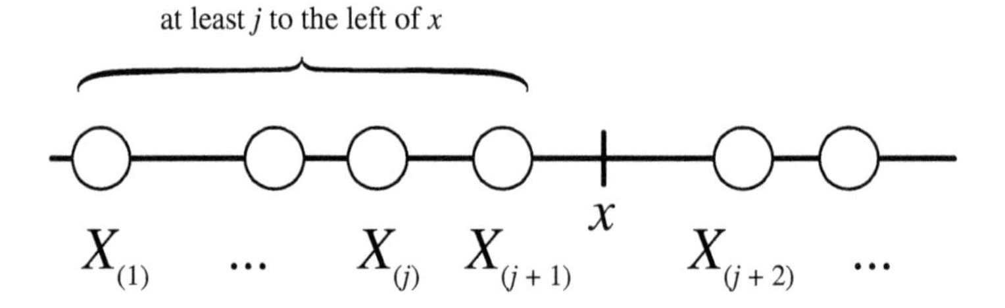

FIGURE 8.10
The event $X_{(j)} \leq x$ is equivalent to the event "at least j X_i's fall to the left of x".

Since it appears that the number of X_i to the left of x will be important to us, let's define a new random variable, N, to keep track of just that: define N to be the number of X_i that land to the left of x. Each X_i lands to the left of x with probability $F(x)$, independently. If we define success as landing to the left of x, we have n independent Bernoulli trials with probability $F(x)$ of success, so $N \sim \text{Bin}(n, F(x))$. Then, by the Binomial PMF,

$$\begin{aligned} P(X_{(j)} \leq x) &= P(\text{at least } j \text{ of the } X_i \text{ are to the left of } x) \\ &= P(N \geq j) \\ &= \sum_{k=j}^{n} \binom{n}{k} F(x)^k (1 - F(x))^{n-k}. \end{aligned}$$

We thus have the following result for the CDF of $X_{(j)}$.

Theorem 8.6.3 (CDF of order statistic). Let $X_1, \dots, X_n$ be i.i.d. continuous r.v.s with CDF F. Then the CDF of the jth order statistic $X_{(j)}$ is

$$P(X_{(j)} \leq x) = \sum_{k=j}^{n} \binom{n}{k} F(x)^k (1 - F(x))^{n-k}.$$

To get the PDF of $X_{(j)}$, we can differentiate the CDF with respect to x, but the resulting expression is ugly (though it can be simplified). Instead we will take a more direct approach. Consider $f_{X_{(j)}}(x)dx$, the probability that the jth order statistic falls into an infinitesimal interval of length dx around x. The only way this can happen is illustrated in Figure 8.11. We need one of the X_i to fall into the infinitesimal interval around x, *and* we need exactly $j-1$ of the X_i to fall to the left of x, leaving the remaining $n-j$ to fall to the right of x.

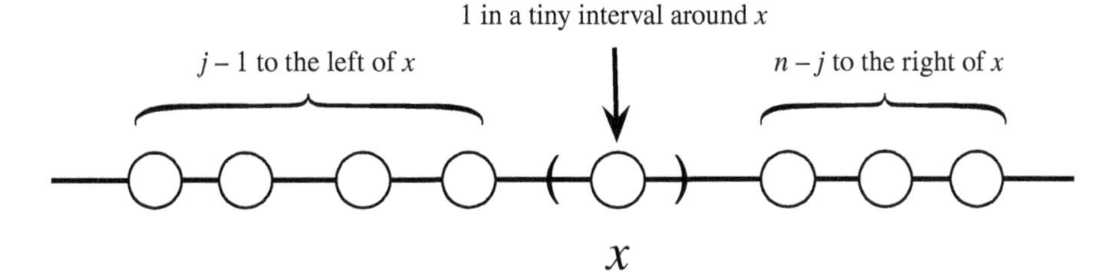

FIGURE 8.11
In order for $X_{(j)}$ to fall within a small interval of x, we require that one of the X_i fall within the small interval and that exactly $j-1$ fall to the left of x.

What is the probability of this extremely specific event? Let's break up the experiment into stages.

- First, we choose which one of the X_i will fall into the infinitesimal interval around x. There are n such choices, each of which occurs with probability $f(x)dx$, where f is the PDF of the X_i.
- Next, we choose exactly $j-1$ out of the remaining $n-1$ to fall to the left of x. There are $\binom{n-1}{j-1}$ such choices, each with probability $F(x)^{j-1}(1-F(x))^{n-j}$ by the $\text{Bin}(n, F(x))$ PMF.

We multiply the probabilities of the two stages to get

$$f_{X_{(j)}}(x)dx = nf(x)dx\binom{n-1}{j-1}F(x)^{j-1}(1-F(x))^{n-j}.$$

Dropping the dx's from both sides gives us the PDF we desire.

Theorem 8.6.4 (PDF of order statistic). Let $X_1, \ldots, X_n$ be i.i.d. continuous r.v.s with CDF F and PDF f. Then the marginal PDF of the jth order statistic $X_{(j)}$ is

$$f_{X_{(j)}}(x) = n\binom{n-1}{j-1}f(x)F(x)^{j-1}(1-F(x))^{n-j}.$$

In general, the order statistics of $X_1, \ldots, X_n$ will not follow a named distribution, but the order statistics of the standard Uniform distribution are an exception.

Example 8.6.5 (Order statistics of Uniforms). Let $U_1, \dots, U_n$ be i.i.d. Unif(0, 1). Then for $0 \le x \le 1$, $f(x) = 1$ and $F(x) = x$, so the PDF of $U_{(j)}$ is

$$f_{U_{(j)}}(x) = n\binom{n-1}{j-1}x^{j-1}(1-x)^{n-j}.$$

This is the Beta$(j, n-j+1)$ PDF! So $U_{(j)} \sim \text{Beta}(j, n-j+1)$, and $E(U_{(j)}) = \frac{j}{n+1}$.

The simple case $n = 2$ is consistent with Example 7.2.2, where we used 2D LOTUS to show that for i.i.d. $U_1, U_2 \sim \text{Unif}(0, 1)$,

$$E(\max(U_1, U_2)) = 2/3,\ E(\min(U_1, U_2)) = 1/3.$$

Now that we know $\max(U_1, U_2)$ and $\min(U_1, U_2)$ follow Beta distributions, the expectation of the Beta distribution confirms our earlier findings. □

8.7 Recap

In this chapter we discussed three broad classes of transformations:

- invertible transformations $\mathbf{Y} = g(\mathbf{X})$ of continuous random vectors, which can be handled with the change of variables formula (under some technical assumptions, most notably that the partial derivatives $\partial x_i/\partial y_j$ exist and are continuous);
- convolutions, for which we can determine the distribution using (in decreasing order of preference) stories, MGFs, or convolution sums/integrals;
- the transformation of i.i.d. continuous r.v.s to their order statistics.

Figure 8.12 illustrates connections between the original random vector (X, Y) and the transformed random vector $(Z, W) = g(X, Y)$, where g is an invertible transformation satisfying certain technical assumptions. The change of variables formula uses Jacobians to take us back and forth between the joint PDF of (X, Y) and the joint PDF of (Z, W).

Let A be a region in the support of (X, Y), and $B = \{g(x, y) : (x, y) \in A\}$ be the corresponding region in the support of (Z, W). Then $(X, Y) \in A$ is the same event as $(Z, W) \in B$, so

$$P((X, Y) \in A) = P((Z, W) \in B).$$

To find this probability, we can either integrate the joint PDF of (X, Y) over A or integrate the joint PDF of (Z, W) over B.

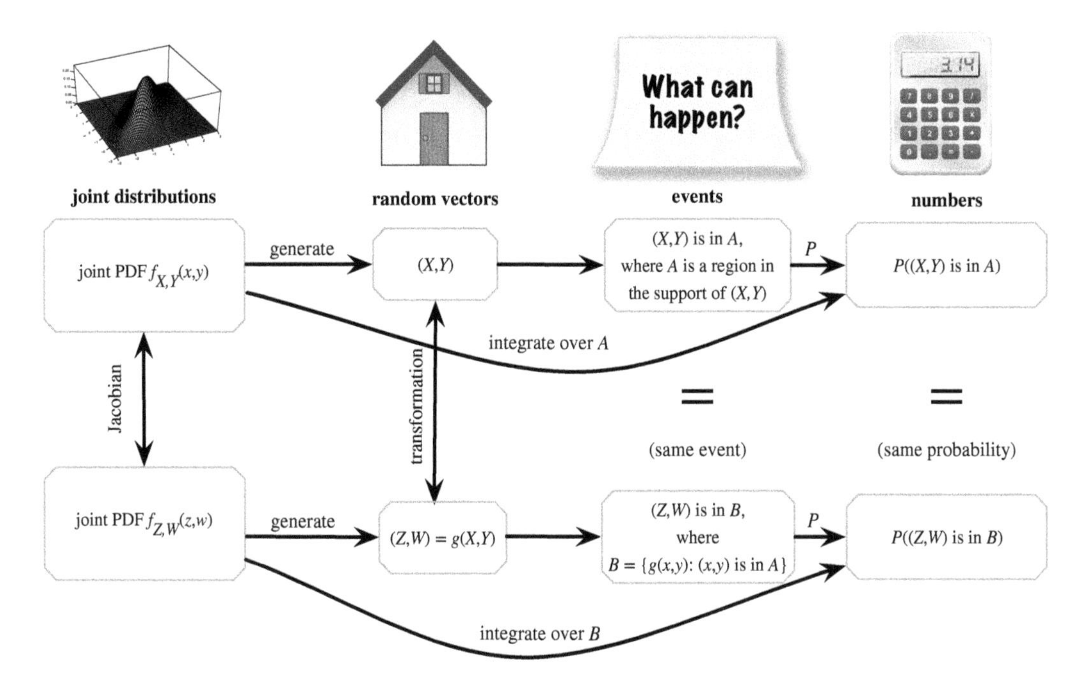

FIGURE 8.12
Let $(Z, W) = g(X, Y)$, where g is an invertible transformation satisfying certain technical assumptions. The change of variables formula lets us go back and forth between the joint PDFs of (X, Y) and (Z, W).

In this chapter, as in many others, we made extensive use of Bayes' rule and LOTP, especially in continuous or hybrid forms. And we often used the strategy of integration by pattern recognition. Since any valid PDF must integrate to 1, and by now we know lots of valid PDFs, we can often use probability to help us do calculus in addition to using calculus to help us do probability!

The two new distributions we introduced are the Beta and Gamma, which are laden with stories and connections to other distributions. The Beta is a generalization of the Unif(0, 1) distribution, and it has the following stories.

- *Order statistics of the Uniform*: The jth order statistic of n i.i.d. Unif(0, 1) r.v.s is distributed Beta$(j, n - j + 1)$.
- *Unknown probability, conjugate prior of the Binomial*: If $p \sim$ Beta(a, b) and $X|p \sim$ Bin(n, p), then $p|X = k \sim$ Beta$(a + k, b + n - k)$. The posterior distribution of p stays within the Beta family of distributions after updating based on Binomial data, a property known as conjugacy. The parameters a and b can be interpreted as the prior number of successes and failures, respectively.

The Gamma is a generalization of the Exponential distribution, and it has the following stories.

- *Poisson process*: In a Poisson process of rate λ, the total waiting time for n arrivals is distributed Gamma(n, λ). Thus the Gamma is the continuous analog of the Negative Binomial distribution.

- *Unknown rate, conjugate prior of the Poisson*: If $\lambda \sim \text{Gamma}(r_0, b_0)$ and $Y|\lambda \sim \text{Pois}(\lambda t)$, then $\lambda|Y = y \sim \text{Gamma}(r_0 + y, b_0 + t)$. The posterior distribution of λ stays within the Gamma family of distributions after updating based on Poisson data. The parameters r_0 and b_0 can be interpreted as the prior number of observed successes and the total waiting time for those successes, respectively.

The Beta and Gamma distributions are related by the bank–post office story, which says that if $X \sim \text{Gamma}(a, \lambda)$, $Y \sim \text{Gamma}(b, \lambda)$ are independent, then $X + Y \sim \text{Gamma}(a + b, \lambda)$, $\frac{X}{X+Y} \sim \text{Beta}(a, b)$, with $X + Y$ and $\frac{X}{X+Y}$ independent.

The diagram of connections, which we last saw in Chapter 5, is hereby updated to include the Beta and Gamma distributions. Distributions listed in parentheses are special cases of the ones not in parentheses.

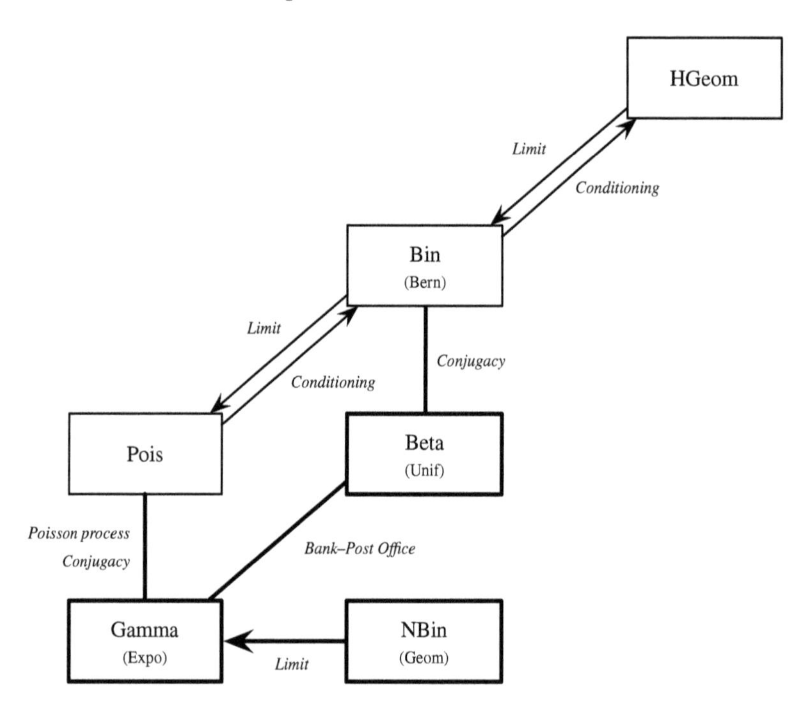

8.8 R

Beta and Gamma distributions

The Beta and Gamma distributions are programmed into R.

- `dbeta`, `pbeta`, `rbeta`: To evaluate the Beta(a, b) PDF or CDF at x, we use `dbeta(x,a,b)` and `pbeta(x,a,b)`. To generate n realizations from the Beta(a, b) distribution, we use `rbeta(n,a,b)`.

- `dgamma`, `pgamma`, `rgamma`: To evaluate the Gamma(a, λ) PDF or CDF at x, we use `dgamma(x,a,lambda)` or `pgamma(x,a,lambda)`. To generate n realizations from the Gamma(a, λ) distribution, we use `rgamma(n,a,lambda)`.

For example, we can check that the Gamma$(3, 2)$ distribution has mean $3/2$ and variance $3/4$. To do this, we generate a large number of Gamma$(3, 2)$ random variables using `rgamma`, then compute their `mean` and `var`:

```
y <- rgamma(10^5,3,2)
mean(y)
var(y)
```

Did you get values that were close to 1.5 and 0.75, respectively?

Convolution of Uniforms

Using R, we can quickly verify that for $X, Y \overset{\text{i.i.d.}}{\sim} \text{Unif}(0, 1)$, the distribution of $T = X + Y$ is triangular in shape:

```
x <- runif(10^5)
y <- runif(10^5)
hist(x+y)
```

The histogram looks like an ascending and then descending staircase, a discrete approximation to a triangle.

Bayes' billiards

In the Bayes' billiards story, we have n white balls and 1 gray ball, throw them onto the unit interval completely at random, and count the number of white balls to the left of the gray ball. Letting p be the position of the gray ball and X be the number of white balls to the left of the gray ball, we have

$$\begin{aligned} p &\sim \text{Unif}(0, 1) \\ X|p &\sim \text{Bin}(n, p). \end{aligned}$$

By performing this experiment a large number of times, we can verify the results we derived in this chapter about the marginal PMF of X and the posterior PDF of p given $X = x$. We'll let the number of simulations be called `nsim`, to avoid a name conflict with the number of white balls, `n`, which we set equal to 10:

```
nsim <- 10^5
n <- 10
```

We simulate 10^5 values of p, then simulate 10^5 values from the conditional distribution of X given p:

```
p <- runif(nsim)
x <- rbinom(nsim,n,p)
```

Notice that we feed the entire vector `p` into `rbinom`. This means that the first element of `x` is generated using the first element of `p`, the second element of `x` is generated using the second element of `p`, and so forth. Thus, conditional on a particular element of `p`, the corresponding element of `x` is Binomial, but the elements of `p` are themselves Uniform, exactly as the model specifies.

According to the Bayes' billiards argument, the marginal distribution of X should be Discrete Uniform on the integers 0 through n. Is this in fact the case? We can make a histogram of `x` to check! Because the distribution of X is discrete, we tell R to make the histogram breaks at $-0.5, 0.5, 1.5, \dots$ so that each bar is centered at an integer value:

```
hist(x,breaks=seq(-0.5,n+0.5,1))
```

Indeed, all the histogram bars are approximately equal in height, consistent with a Discrete Uniform distribution.

Now for the posterior distribution of p given $X = x$. Conditioning is very simple in R. To consider only the simulated values of p where the value of X was 3, we use square brackets, like this: `p[x==3]`. In particular, we can create a histogram of these values to see what the posterior distribution of p given $X = 3$ looks like; try `hist(p[x==3])`.

According to the Beta-Binomial conjugacy result, the true posterior distribution is $p|X = 3 \sim \text{Beta}(4, 8)$. We can plot the histogram of `p[x==3]` next to a histogram of simulated values from the Beta$(4, 8)$ distribution to confirm that they look similar:

```
par(mfrow=c(1,2))
hist(p[x==3])
hist(rbeta(10^4,4,8))
```

The first line tells R we want two side-by-side plots, and the second and third lines create the histograms.

Simulating order statistics

Simulating order statistics in R is easy: we simply simulate i.i.d. r.v.s and use `sort` to sort them in increasing order. For example,

```
sort(rnorm(10))
```

produces one realization of $X_{(1)}, \dots, X_{(10)}$, where $X_1, \dots, X_{10}$ are i.i.d. $\mathcal{N}(0, 1)$.

If we want to plot a histogram of realizations of, say, $X_{(9)}$, we'll need to use `replicate`:

```
order_stats <- replicate(10^4, sort(rnorm(10)))
```

This creates a matrix, `order_stats`, with 10 rows. The ith row of the matrix contains 10^4 realizations of $X_{(i)}$. Now we can create a histogram of $X_{(9)}$, simply by selecting row 9 of the matrix:

```
x9 <- order_stats[9,]
hist(x9)
```

We can also compute summaries like `mean(x9)` and `var(x9)`.

8.9 Exercises

Exercises marked with Ⓢ have detailed solutions at `http://stat110.net`.

Change of variables

1. Find the PDF of e^{-X} for $X \sim \text{Expo}(1)$.
2. Find the PDF of X^7 for $X \sim \text{Expo}(\lambda)$.
3. Find the PDF of Z^3 for $Z \sim \mathcal{N}(0,1)$.
4. Ⓢ Find the PDF of Z^4 for $Z \sim \mathcal{N}(0,1)$.
5. Find the PDF of $|Z|$ for $Z \sim \mathcal{N}(0,1)$.
6. Ⓢ Let $U \sim \text{Unif}(0,1)$. Find the PDFs of U^2 and $\sqrt{U}$.
7. Let $U \sim \text{Unif}(0, \frac{\pi}{2})$. Find the PDF of $\sin(U)$.
8. (a) Find the distribution of X^2 for $X \sim \text{DUnif}(0,1,\dots,n)$.

   (b) Find the distribution of X^2 for $X \sim \text{DUnif}(-n,-n+1,\dots,0,1,\dots,n)$.
9. Let $X \sim \text{Bern}(p)$ and let a and b be constants with $a < b$. Find a simple transformation of X that yields an r.v. that equals a with probability $1-p$ and equals b with probability p.
10. Let $X \sim \text{Pois}(\lambda)$ and Y be the indicator of X being odd. Find the PMF of Y.

    Hint: Find $P(Y=0) - P(Y=1)$ by writing $P(Y=0)$ and $P(Y=1)$ as series and then using the fact that $(-1)^k$ is 1 if k is even and -1 if k is odd.
11. Let T be a continuous r.v. and $V = 1/T$. Show that their CDFs are related as follows:

$$F_V(v) = \begin{cases} F_T(0) + 1 - F_T(\frac{1}{v}) \text{ for } v > 0, \\ F_T(0) \text{ for } v = 0, \\ F_T(0) - F_T(\frac{1}{v}) \text{ for } v < 0. \end{cases}$$

12. Let T be the ratio X/Y of two i.i.d. $\mathcal{N}(0,1)$ r.v.s. X, Y. This is the *Cauchy* distribution and, as shown in Example 7.1.25, it has PDF

$$f_T(t) = \frac{1}{\pi(1+t^2)}.$$

(a) Use the result of the previous problem to find the CDF of $1/T$. Then use calculus to find the PDF of $1/T$. (Note that the one-dimensional change of variables formula does not apply directly, since the function $g(t) = 1/t$, even though it has $g'(t) < 0$ for all $t \neq 0$, is undefined at $t = 0$ and is not a strictly decreasing function on its domain.)

(b) Show that $1/T$ has the same distribution as T without using calculus, in 140 characters or fewer.

13. Let X and Y be i.i.d. Expo(λ), and $T = \log(X/Y)$. Find the CDF and PDF of T.

14. Let X and Y have joint PDF $f_{X,Y}(x,y)$, and transform $(X,Y) \mapsto (T,W)$ linearly by letting

$$T = aX + bY \text{ and } W = cX + dY,$$

where a, b, c, d are constants such that $ad - bc \neq 0$.

(a) Find the joint PDF $f_{T,W}(t,w)$ (in terms of $f_{X,Y}$, though your answer should be written as a function of t and w).

(b) For the case where $T = X + Y, W = X - Y$, show that

$$f_{T,W}(t,w) = \frac{1}{2} f_{X,Y}\left(\frac{t+w}{2}, \frac{t-w}{2}\right).$$

15. Ⓢ Let X, Y be continuous r.v.s with a *spherically symmetric* joint distribution, which means that the joint PDF is of the form $f(x,y) = g(x^2+y^2)$ for some function g. Let (R,θ) be the polar coordinates of (X,Y), so $R^2 = X^2 + Y^2$ is the squared distance from the origin and θ is the angle (in $[0, 2\pi)$), with $X = R\cos\theta, Y = R\sin\theta$.

(a) Explain intuitively why R and θ are independent. Then prove this by finding the joint PDF of (R,θ).

(b) What is the joint PDF of (R,θ) when (X,Y) is Uniform in the unit disk $\{(x,y) : x^2+y^2 \leq 1\}$?

(c) What is the joint PDF of (R,θ) when X and Y are i.i.d. $\mathcal{N}(0,1)$?

16. Let X and Y be i.i.d. $\mathcal{N}(0,1)$ r.v.s, $T = X + Y$, and $W = X - Y$. We know from Example 7.5.8 that T and W are independent $\mathcal{N}(0,2)$ r.v.s (note that (T,W) is Multivariate Normal with $\text{Cov}(T,W) = 0$). Give another proof of this fact, using the change of variables theorem.

17. Let X and Y be i.i.d. $\mathcal{N}(0,1)$ r.v.s, and (R,θ) be the polar coordinates for the point (X,Y), so $X = R\cos\theta$ and $Y = R\sin\theta$ with $R \geq 0$ and $\theta \in [0, 2\pi)$. Find the joint PDF of R^2 and θ. Also find the marginal distributions of R^2 and θ, giving their names (and parameters) if they are distributions we have studied before.

18. Let X and Y be independent positive r.v.s, with PDFs f_X and f_Y, respectively. Let T be the ratio X/Y.

(a) Find the joint PDF of T and X, using a Jacobian.

(b) Find the marginal PDF of T, as a single integral.

19. Let X and Y be i.i.d. Expo(λ), and transform them to $T = X + Y, W = X/Y$.

(a) Find the joint PDF of T and W. Are they independent?

(b) Find the marginal PDFs of T and W.

Convolutions

20. Let $U \sim \text{Unif}(0,1)$ and $X \sim \text{Expo}(1)$, independently. Find the PDF of $U + X$.

21. Let X and Y be i.i.d. Expo(1). Use a convolution integral to show that the PDF of $L = X - Y$ is $f(t) = \frac{1}{2}e^{-|t|}$ for all real t; this is known as the *Laplace distribution.*

22. Use a convolution integral to show that if $X \sim \mathcal{N}(\mu_1, \sigma^2)$ and $Y \sim \mathcal{N}(\mu_2, \sigma^2)$ are independent, then $T = X + Y \sim \mathcal{N}(\mu_1 + \mu_2, 2\sigma^2)$ (to simplify the calculation, we are assuming that the variances are equal). You can use a standardization (location-scale) idea to reduce to the standard Normal case before setting up the integral.

    Hint: Complete the square.

23. Ⓢ Let X and Y be independent positive r.v.s, with PDFs f_X and f_Y, respectively, and consider the product $T = XY$. When asked to find the PDF of T, Jacobno argues that:

    "It's like a convolution, with a product instead of a sum. To have $T = t$ we need $X = x$ and $Y = t/x$ for some x; that has probability $f_X(x)f_Y(t/x)$, so summing up these possibilities we get that the PDF of T is $\int_0^\infty f_X(x)f_Y(t/x)dx$."

    Evaluate Jacobno's argument, while getting the PDF of T (as an integral) in 2 ways:

    (a) using the continuous version of the law of total probability to get the CDF, and then taking the derivative (you can assume that swapping the derivative and integral is valid);

    (b) by taking the log of both sides of $T = XY$ and doing a convolution (and then converting back to get the PDF of T).

24. Let X and Y be i.i.d. Discrete Uniform r.v.s on $\{0, 1, \ldots, n\}$, where n is a positive integer. Find the PMF of $T = X + Y$.

    Hint: In finding $P(T = k)$, it helps to consider the cases $0 \leq k \leq n$ and $n + 1 \leq k \leq 2n$ separately. Be careful about the range of summation in the convolution sum.

25. Let X and Y be i.i.d. $\text{Unif}(0,1)$, and let $W = X - Y$.

    (a) Find the mean and variance of W, without yet deriving the PDF.

    (b) Show that the distribution of W is symmetric about 0, without yet deriving the PDF.

    (c) Find the PDF of W.

    (d) Use the PDF of W to verify your results from (a) and (b).

    (e) How does the distribution of W relate to the distribution of $X + Y$, the Triangle distribution derived in Example 8.2.5? Give a precise description, e.g., using the concepts of location and scale.

26. Let X and Y be i.i.d. $\text{Unif}(0,1)$, and $T = X + Y$. We derived the distribution of T (a Triangle distribution) in Example 8.2.5, using a convolution integral. Since (X, Y) is Uniform in the unit square $\{(x, y) : 0 < x < 1, 0 < y < 1\}$, we can also interpret $P((X, Y) \in A)$ as the *area* of A, for any region A within the unit square. Use this idea to find the CDF of T, by interpreting the CDF (evaluated at some point) as an area.

27. Let X, Y, Z be i.i.d. $\text{Unif}(0,1)$, and $W = X + Y + Z$. Find the PDF of W.

    Hint: We already know the PDF of $X + Y$. Be careful about limits of integration in the convolution integral; there are 3 cases that should be considered separately.

Beta and Gamma

28. Ⓢ Let $B \sim \text{Beta}(a, b)$. Find the distribution of $1 - B$ in two ways: (a) using a change of variables and (b) using a story proof. Also explain why the result makes sense in terms of Beta being the conjugate prior for the Binomial.

29. Ⓢ Let $X \sim \text{Gamma}(a, \lambda)$ and $Y \sim \text{Gamma}(b, \lambda)$ be independent, with a and b integers. Show that $X + Y \sim \text{Gamma}(a + b, \lambda)$ in three ways: (a) with a convolution integral; (b) with MGFs; (c) with a story proof.

30. Let $B \sim \text{Beta}(a, b)$. Use integration by pattern recognition to find $E(B^k)$ for positive integers k. In particular, show that

$$\text{Var}(B) = \frac{ab}{(a+b)^2(a+b+1)}.$$

31. Ⓢ Fred waits $X \sim \text{Gamma}(a, \lambda)$ minutes for the bus to work, and then waits $Y \sim \text{Gamma}(b, \lambda)$ for the bus going home, with X and Y independent. Is the ratio X/Y independent of the total wait time $X + Y$?

32. Ⓢ The F-test is a very widely used statistical test based on the $F(m, n)$ distribution, which is the distribution of $\frac{X/m}{Y/n}$ with $X \sim \text{Gamma}(\frac{m}{2}, \frac{1}{2}), Y \sim \text{Gamma}(\frac{n}{2}, \frac{1}{2})$. Find the distribution of $mV/(n + mV)$ for $V \sim F(m, n)$.

33. Ⓢ Customers arrive at the Leftorium store according to a Poisson process with rate λ customers per hour. The true value of λ is unknown, so we treat it as a random variable. Suppose that our prior beliefs about λ can be expressed as $\lambda \sim \text{Expo}(3)$. Let X be the number of customers who arrive at the Leftorium between 1 pm and 3 pm tomorrow. Given that $X = 2$ is observed, find the posterior PDF of λ.

34. Ⓢ Let X and Y be independent, positive r.v.s. with finite expected values.

(a) Give an example where $E(\frac{X}{X+Y}) \neq \frac{E(X)}{E(X+Y)}$, computing both sides exactly.

Hint: Start by thinking about the simplest examples you can think of!

(b) If X and Y are i.i.d., then is it necessarily true that $E(\frac{X}{X+Y}) = \frac{E(X)}{E(X+Y)}$?

(c) Now let $X \sim \text{Gamma}(a, \lambda)$ and $Y \sim \text{Gamma}(b, \lambda)$. Show *without using calculus* that

$$E\left(\frac{X^c}{(X+Y)^c}\right) = \frac{E(X^c)}{E((X+Y)^c)}$$

for every real $c > 0$.

35. Let $T = X^{1/\gamma}$, with $X \sim \text{Expo}(\lambda)$ and $\lambda, \gamma > 0$. So T has a Weibull distribution, as discussed in Example 6.5.5. Using LOTUS and the definition of the gamma function, we showed in this chapter that for $Y \sim \text{Gamma}(a, \lambda)$,

$$E(Y^c) = \frac{1}{\lambda^c} \cdot \frac{\Gamma(a+c)}{\Gamma(a)},$$

for all real $c > -a$. Use this result to show that

$$E(T) = \frac{\Gamma(1 + 1/\gamma)}{\lambda^{1/\gamma}} \text{ and } \text{Var}(T) = \frac{\Gamma(1 + 2/\gamma) - (\Gamma(1 + 2/\gamma))^2}{\lambda^{2/\gamma}}.$$

36. Alice walks into a post office with 2 clerks. Both clerks are in the midst of serving customers, but Alice is next in line. The clerk on the left takes an $\text{Expo}(\lambda_1)$ time to serve a customer, and the clerk on the right takes an $\text{Expo}(\lambda_2)$ time to serve a customer. Let T_1 be the time until the clerk on the left is done serving their current customer, and define T_2 likewise for the clerk on the right.

(a) If $\lambda_1 = \lambda_2$, is T_1/T_2 independent of $T_1 + T_2$?

Hint: Note that $T_1/T_2 = (T_1/(T_1 + T_2))/(T_2/(T_1 + T_2))$.

(b) Find $P(T_1 < T_2)$ (do not assume $\lambda_1 = \lambda_2$ here or in the next part, but do check that your answers make sense in that special case).

(c) Find the expected total amount of time that Alice spends in the post office (assuming that she leaves immediately after she is done being served).

37. Let $X \sim \text{Pois}(\lambda t)$ and $Y \sim \text{Gamma}(j, \lambda)$, where j is a positive integer. Show using a story about a Poisson process that

$$P(X \geq j) = P(Y \leq t).$$

38. Visitors arrive at a certain scenic park according to a Poisson process with rate λ visitors per hour. Fred has just arrived (independent of anyone else), and will stay for an $\text{Expo}(\lambda_2)$ number of hours. Find the distribution of the number of other visitors who arrive at the park while Fred is there.

39. (a) Let $p \sim \text{Beta}(a, b)$, where a and b are positive real numbers. Find $E(p^2(1-p)^2)$, fully simplified (Γ should not appear in your final answer).

Two teams, A and B, have an upcoming match. They will play five games and the winner will be declared to be the team that wins the majority of games. Given p, the outcomes of games are independent, with probability p of team A winning and $1 - p$ of team B winning. But you don't know p, so you decide to model it as an r.v., with $p \sim \text{Unif}(0, 1)$ a priori (before you have observed any data).

To learn more about p, you look through the historical records of previous games between these two teams, and find that the previous outcomes were, in chronological order, $AAABBAABAB$. (Assume that the true value of p has not been changing over time and will be the same for the match, though your *beliefs* about p may change over time.)

(b) Does your posterior distribution for p, given the historical record of games between A and B, depend on the specific order of outcomes or only on the fact that A won exactly 6 of the 10 games on record? Explain.

(c) Find the posterior distribution for p, given the historical data.

The posterior distribution for p from (c) becomes your new prior distribution, and the match is about to begin!

(d) Conditional on p, is the indicator of A winning the first game of the match positively correlated with, uncorrelated with, or negatively correlated with the indicator of A winning the second game? What about if we only condition on the historical data?

(e) Given the historical data, what is the expected value for the probability that the match is not yet decided when going into the fifth game (viewing this probability as an r.v. rather than a number, to reflect our uncertainty about it)?

40. An engineer is studying the reliability of a product by performing a sequence of n trials. Reliability is defined as the probability of success. In each trial, the product succeeds with probability p and fails with probability $1-p$. The trials are conditionally independent given p. Here p is unknown (else the study would be unnecessary!). The engineer takes a Bayesian approach, with $p \sim \text{Unif}(0, 1)$ as prior.

Let r be a desired reliability level and c be the corresponding confidence level, in the sense that, given the data, the probability is c that the true reliability p is at least r. For example, if $r = 0.9, c = 0.95$, we can be 95% sure, given the data, that the product is at least 90% reliable. Suppose that it is observed that the product succeeds all n times. Find a simple equation for c as a function of r.

Order statistics

41. Ⓢ Let $X \sim \text{Bin}(n, p)$ and $B \sim \text{Beta}(j, n-j+1)$, where n is a positive integer and j is a positive integer with $j \leq n$. Show using a story about order statistics that
$$P(X \geq j) = P(B \leq p).$$
This shows that the CDF of the continuous r.v. B is closely related to the CDF of the discrete r.v. X, and is another connection between the Beta and Binomial.

42. Show that for i.i.d. continuous r.v.s X, Y, Z,
$$P(X < \min(Y, Z)) + P(Y < \min(X, Z)) + P(Z < \min(X, Y)) = 1.$$

43. Show that
$$\int_0^x \frac{n!}{(j-1)!(n-j)!} t^{j-1}(1-t)^{n-j}\,dt = \sum_{k=j}^{n} \binom{n}{k} x^k (1-x)^{n-k},$$
without using calculus, for all $x \in [0, 1]$ and j, n positive integers with $j \leq n$.

44. Let $X_1, \ldots, X_n$ be i.i.d. continuous r.v.s with PDF f and a strictly increasing CDF F. Suppose that we know that the jth order statistic of n i.i.d. Unif(0, 1) r.v.s is a Beta$(j, n-j+1)$, but we have forgotten the formula and derivation for the distribution of the jth order statistic of $X_1, \ldots, X_n$. Show how we can recover the PDF of $X_{(j)}$ quickly using a change of variables.

45. Ⓢ Let X and Y be independent Expo(λ) r.v.s and $M = \max(X, Y)$. Show that M has the same distribution as $X + \frac{1}{2}Y$, in two ways: (a) using calculus and (b) by remembering the memoryless property and other properties of the Exponential.

46. Ⓢ (a) If X and Y are i.i.d. continuous r.v.s with CDF $F(x)$ and PDF $f(x)$, then $M = \max(X, Y)$ has PDF $2F(x)f(x)$. Now let X and Y be discrete and i.i.d., with CDF $F(x)$ and PMF $f(x)$. Explain in words why the PMF of M is *not* $2F(x)f(x)$.

(b) Let X and Y be i.i.d. Bern(1/2) r.v.s, $M = \max(X, Y)$ and $L = \min(X, Y)$. Find the joint PMF of M and L, i.e., $P(M = a, L = b)$, and the marginal PMFs of M and L.

47. Let $X_1, X_2, \ldots$ be i.i.d. r.v.s with CDF F, and let $M_n = \max(X_1, X_2, \ldots, X_n)$. Find the joint distribution of M_n and M_{n+1}, for each $n \geq 1$.

48. Ⓢ Let $X_1, X_2, \ldots, X_n$ be i.i.d. r.v.s with CDF F and PDF f. Find the joint PDF of the order statistics $X_{(i)}$ and $X_{(j)}$ for $1 \leq i < j \leq n$, by drawing and thinking about a picture.

49. Ⓢ Two women are pregnant, both with the same due date. On a timeline, define time 0 to be the instant when the due date begins. Suppose that the time when the woman gives birth has a Normal distribution, centered at 0 and with standard deviation 8 days. The two birth times are i.i.d. Let T be the time of the first of the two births (in days).

(a) Show that
$$E(T) = \frac{-8}{\sqrt{\pi}}.$$
Hint: For any two random variables X and Y, we have $\max(X, Y) + \min(X, Y) = X + Y$ and $\max(X, Y) - \min(X, Y) = |X - Y|$. Example 7.2.3 derives the expected distance between two i.i.d. $\mathcal{N}(0, 1)$ r.v.s.

(b) Find Var(T), in terms of integrals. You can leave your answers unsimplified for this part, but it can be shown that the answer works out to
$$\text{Var}(T) = 64\left(1 - \frac{1}{\pi}\right).$$

50. We are about to observe random variables $Y_1, Y_2, \ldots, Y_n$, i.i.d. from a continuous distribution. We will need to predict an independent future observation Y_{new}, which will also have the same distribution. The distribution is unknown, so we will construct our prediction using $Y_1, Y_2, \ldots, Y_n$ rather than the distribution of Y_{new}. In forming a prediction, we do not want to report only a single number; rather, we want to give a *predictive interval* with "high confidence" of containing Y_{new}. One approach to this is via order statistics.

    (a) For fixed j and k with $1 \leq j < k \leq n$, find $P(Y_{\text{new}} \in [Y_{(j)}, Y_{(k)}])$.

    Hint: By symmetry, all orderings of $Y_1, \ldots, Y_n, Y_{\text{new}}$ are equally likely.

    (b) Let $n = 99$. Construct a predictive interval, as a function of $Y_1, \ldots, Y_n$, such that the probability of the interval containing Y_{new} is 0.95.

51. Let $X_1, \ldots, X_n$ be i.i.d. continuous r.v.s with n odd. Show that the median of the distribution of the sample median of the X_i's is the median of the distribution of the X_i's.

    Hint: Start by reading the problem carefully; it is crucial to distinguish between the median of a distribution (as defined in Chapter 6) and the sample median of a collection of r.v.s (as defined in this chapter). Of course they are closely related: the sample median of i.i.d. r.v.s is a very natural way to estimate the true median of the distribution that the r.v.s are drawn from. Two approaches to evaluating a sum that might come up are (i) use the first story proof example and first story proof exercise from Chapter 1, or (ii) use the fact that, by the story of the Binomial, $Y \sim \text{Bin}(n, 1/2)$ implies $n - Y \sim \text{Bin}(n, 1/2)$.

Mixed practice

52. Let $U_1, U_2, \ldots, U_n$ be i.i.d. Unif(0, 1), and let $X_j = -\log(U_j)$ for all j.

    (a) Find the distribution of X_j. What is its name?

    (b) Find the distribution of the product $U_1 U_2 \ldots U_n$.

    Hint: First take the log.

53. Ⓢ A DNA sequence can be represented as a sequence of letters, where the *alphabet* has 4 letters: A,C,T,G. Suppose such a sequence is generated randomly, where the letters are independent and the probabilities of A,C,T,G are p_1, p_2, p_3, p_4, respectively.

    (a) In a DNA sequence of length 115, what is the expected number of occurrences of the expression "CATCAT" (in terms of the p_j)? (Note that, for example, the expression "CATCATCAT" counts as 2 occurrences.)

    (b) What is the probability that the first A appears earlier than the first C appears, as letters are generated one by one (in terms of the p_j)?

    (c) For this part, assume that the p_j are unknown. Suppose we treat p_2 as a Unif(0, 1) r.v. before observing any data, and that then the first 3 letters observed are "CAT". Given this information, what is the probability that the next letter is C?

54. Ⓢ Consider independent Bernoulli trials with probability p of success for each. Let X be the number of failures incurred before getting a total of r successes.

    (a) Determine what happens to the distribution of $\frac{p}{1-p}X$ as $p \to 0$, using MGFs; what is the PDF of the limiting distribution, and its name and parameters if it is one we have studied?

    Hint: Start by finding the Geom(p) MGF. Then find the MGF of $\frac{p}{1-p}X$, and use the fact that if the MGFs of Y_n converges to the MGF of Y, then the CDF of Y_n converges to the CDF of Y.

    (b) Explain intuitively why the result of (a) makes sense.

9

Conditional expectation

Given that you've read the earlier chapters, you already know what conditional expectation is: expectation, but using *conditional* probabilities. This is an essential concept, for reasons analogous to why we need conditional probability:

- Conditional expectation is a powerful tool for calculating expectations. Using strategies such as conditioning on what we wish we knew and first-step analysis, we can often decompose complicated expectation problems into simpler pieces.
- Conditional expectation is a relevant quantity in its own right, allowing us to predict or estimate unknowns based on whatever evidence is currently available. For example, in statistics we often want to predict a response variable (such as test scores or earnings) based on explanatory variables (such as number of practice problems solved or enrollment in a job training program).

There are two different but closely linked notions of conditional expectation:

- *Conditional expectation $E(Y|A)$ given an event*: let Y be an r.v., and A be an event. If we learn that A occurred, our updated expectation for Y is denoted by $E(Y|A)$ and is computed analogously to $E(Y)$, except using conditional probabilities given A.
- *Conditional expectation $E(Y|X)$ given a random variable*: a more subtle question is how to define $E(Y|X)$, where X and Y are both r.v.s. Intuitively, $E(Y|X)$ is the r.v. that best predicts Y using only the information available from X.

In this chapter, we explore the definitions, properties, intuitions, and applications of both forms of conditional expectation.

9.1 Conditional expectation given an event

Recall that the expectation $E(Y)$ of a discrete r.v. Y is a weighted average of its possible values, where the weights are the PMF values $P(Y = y)$. After learning that an event A occurred, we want to use weights that have been updated to reflect this new information. The definition of $E(Y|A)$ simply replaces the probability $P(Y = y)$ with the conditional probability $P(Y = y|A)$.

Similarly, if Y is continuous, $E(Y)$ is still a weighted average of the possible values of Y, with an integral in place of a sum and the PDF value $f(y)$ in place of a PMF value. If we learn that A occurred, we update the expectation for Y by replacing $f(y)$ with the conditional PDF $f(y|A)$.

Definition 9.1.1 (Conditional expectation given an event). Let A be an event with positive probability. If Y is a discrete r.v., then the *conditional expectation of Y given A* is

$$E(Y|A) = \sum_y yP(Y = y|A),$$

where the sum is over the support of Y. If Y is a continuous r.v. with PDF f, then

$$E(Y|A) = \int_{-\infty}^{\infty} yf(y|A)dy,$$

where the conditional PDF $f(y|A)$ is defined as the derivative of the conditional CDF $F(y|A) = P(Y \leq y|A)$, and can also be computed by a hybrid version of Bayes' rule:

$$f(y|A) = \frac{P(A|Y = y)f(y)}{P(A)}.$$

Intuition 9.1.2. To gain intuition for $E(Y|A)$, let's consider approximating it via simulation (or via the frequentist perspective, based on repeating the same experiment many times). Imagine generating a large number n of replications of the experiment for which Y is a numerical summary. We then have Y-values $y_1, \dots, y_n$, and we can approximate

$$E(Y) \approx \frac{1}{n}\sum_{j=1}^{n} y_j.$$

To approximate $E(Y|A)$, we restrict to the replications where A occurred, and average only *those* Y-values. This can be written as

$$E(Y|A) \approx \frac{\sum_{j=1}^{n} y_j I_j,}{\sum_{j=1}^{n} I_j},$$

where I_j is the indicator of A occurring in the jth replication. This is undefined if A never occurred in the simulation, which makes sense since then there is no simulation data about what the "A occurred" scenario is like. We would like to have n large enough so that there are many occurrences of A (if A is a rare event, more sophisticated techniques for approximating $E(Y|A)$ may be needed).

The principle is simple though: $E(Y|A)$ is approximately the average of Y in a large number of simulation runs in which A occurred. □

☣ **9.1.3.** Confusing conditional expectation and unconditional expectation is a dangerous mistake. More generally, not keeping careful track of what you *should be* conditioning on and what you *are* conditioning on is a recipe for disaster.

For a life-or-death example of the previous biohazard, consider life expectancy.

Example 9.1.4 (Life expectancy). Fred is 30 years old, and he hears that the average life expectancy in his country is 80 years. Should he conclude that, on average, he has 50 years of life left? No, there is a crucial piece of information that he must condition on: the fact that he has lived to age 30 already. Letting T be Fred's lifespan, we have the cheerful news that

$$E(T) < E(T|T \geq 30).$$

The left-hand side is Fred's life expectancy at birth (it implicitly conditions on the fact that he is born), and the right-hand side is Fred's life expectancy given that he reaches age 30.

A harder question is how to decide on an appropriate estimate to use for $E(T)$. Is it just 80, the overall average for his country? In almost every country, women have a longer average life expectancy than men, so it makes sense to condition on Fred being a man. But should we also condition on what city he was born in? Should we condition on racial and financial information about his parents, or the time of day when he was born? Intuitively, we would like estimates that are both accurate and relevant for Fred, but there is a tradeoff since if we condition on more characteristics of Fred, then there are fewer people who match those characteristics to use as data for estimating the life expectancy.

Now consider some specific numbers for the United States. A Social Security Administration study estimated that between 1900 and 2000, the average life expectancy at birth in the U.S. for men increased from 46 to 74, and for women increased from 49 to 79. Tremendous gains! But much of the gain is due to decreases in child mortality. For a 30-year-old person in 1900, the average number of years remaining was 35 for a man and 36 for a woman; in 2000, the corresponding numbers were 46 for a man and 50 for a woman.

There are some subtle statistical issues in obtaining these estimates. For example, how were estimates for life expectancy for someone born in 2000 obtained without waiting at least until the year 2100? Estimating survival distributions is a very important topic in biostatistics and actuarial science. □

The law of total probability allows us to get unconditional probabilities by slicing up the sample space and computing conditional probabilities in each slice. The same idea works for computing unconditional expectations.

Theorem 9.1.5 (Law of total expectation). Let $A_1, \ldots, A_n$ be a partition of a sample space, with $P(A_i) > 0$ for all i, and let Y be a random variable on this sample space. Then

$$E(Y) = \sum_{i=1}^{n} E(Y|A_i)P(A_i).$$

In fact, since all probabilities are expectations by the fundamental bridge, the law of total probability is a special case of the law of total expectation. To see this, let $Y = I_B$ for an event B; then the above theorem says

$$P(B) = E(I_B) = \sum_{i=1}^{n} E(I_B|A_i)P(A_i) = \sum_{i=1}^{n} P(B|A_i)P(A_i),$$

which is exactly LOTP. The law of total expectation is, in turn, a special case of a major result called *Adam's law* (Theorem 9.3.7), so we will not prove it yet.

There are many interesting examples of using wishful thinking to break up an unconditional expectation into conditional expectations. We begin with two cautionary tales about the importance of conditioning carefully and not destroying information without justification.

Example 9.1.6 (Two-envelope paradox). A stranger presents you with two identical-looking, sealed envelopes, each of which contains a check for some positive amount of money. You are informed that one of the envelopes contains exactly twice as much money as the other. You can choose either envelope. Which do you prefer: the one on the left or the one on the right? (Assume that the expected amount of money in each envelope is finite—certainly a good assumption in the real world!)

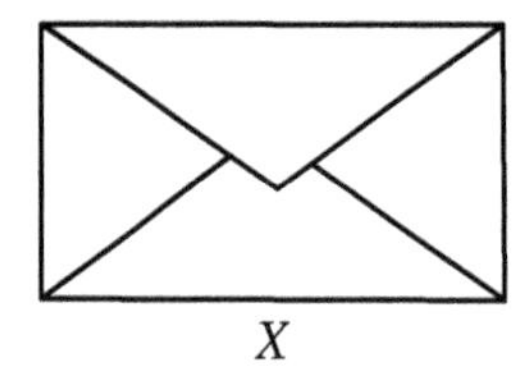

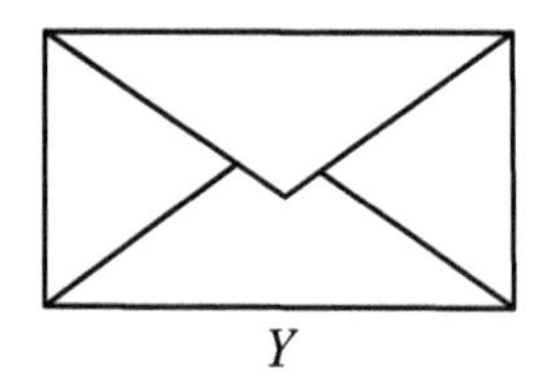

FIGURE 9.1
Two envelopes, where one contains twice as much money as the other. Either $Y = 2X$ or $Y = X/2$, with equal probabilities. Which would you prefer?

Solution:

Let X and Y be the amounts in the left and right envelopes, respectively. By symmetry, there is no reason to prefer one envelope over the other (we are assuming there is no prior information that the stranger is left-handed and left-handed people prefer putting more money on the left). Concluding by symmetry that $E(X) = E(Y)$, it seems that you should not care which envelope you get.

But as you daydream about what's inside the envelopes, another argument occurs to you: suppose that the left envelope has \$100. Then the right envelope either has \$50 or \$200. The average of \$50 and \$200 is \$125, so it seems then that the right envelope is better. But there was nothing special about \$100 here; for any value x for the left envelope, the average of $2x$ and $x/2$ is greater than x, suggesting that the right envelope is better. This is bizarre though, since not only does it contradict the symmetry argument, but also the same reasoning could be applied starting with the right envelope, leading to switching back and forth forever!

Let us try to formalize this argument to see what's going on. We have $Y = 2X$ or $Y = X/2$, with equal probabilities. By Theorem 9.1.5,

$$E(Y) = E(Y|Y = 2X) \cdot \frac{1}{2} + E(Y|Y = X/2) \cdot \frac{1}{2}.$$

One might then think that this is

$$E(2X) \cdot \frac{1}{2} + E(X/2) \cdot \frac{1}{2} = \frac{5}{4}E(X),$$

suggesting a 25% gain from switching from the left to the right envelope. But there is a blunder in that calculation: $E(Y|Y = 2X) = E(2X|Y = 2X)$, but there is no justification for dropping the $Y = 2X$ condition after plugging in $2X$ for Y.

To put it another way, let I be the indicator of the event $Y = 2X$, so that $E(Y|Y = 2X) = E(2X|I = 1)$. If we know that X is independent of I, then we can drop the condition $I = 1$. But in fact we have just *proven* that X and I can't be independent: if they were, we'd have a paradox! Surprisingly, *observing X gives information about whether X is the bigger value or the smaller value.* If we learn that X is very large, we might guess that X is larger than Y, but what is considered very large? Is 10^{12} very large, even though it is tiny compared with 10^{100}? The two-envelope paradox says that no matter what the distribution of X is, there are reasonable ways to define "very large" relative to that distribution.

In Exercise 8 you will look at a related problem, in which the amounts of money in the two envelopes are i.i.d. random variables. You'll show that if you are allowed to look inside one of the envelopes and then decide whether to switch, there is a strategy that allows you to get the better envelope more than 50% of the time! □

The next example vividly illustrates the importance of conditioning on *all* the information. The phenomenon revealed here arises in many real-life decisions about what to buy and what investments to make.

Example 9.1.7 (Mystery prize). You are approached by another stranger, who gives you an opportunity to bid on a mystery box containing a mystery prize! The value of the prize is completely unknown, except that it is worth at least nothing, and at most a million dollars. So the true value V of the prize is considered to be Uniform on [0,1] (measured in millions of dollars).

You can choose to bid any amount b (in millions of dollars). You have the chance to get the prize for considerably less than it is worth, but you could also lose money if you bid too much. Specifically, if $b < 2V/3$, then the bid is rejected and nothing is gained or lost. If $b \geq 2V/3$, then the bid is accepted and your net payoff is $V - b$ (since you pay b to get a prize worth V). What is your optimal bid b, to maximize the expected payoff?

Solution:

Your bid $b \geq 0$ must be a predetermined constant (not based on V, since V is

FIGURE 9.2
When bidding on an unknown asset, beware the winner's curse, and condition on the relevant information.

unknown!). To find the expected payoff W, condition on whether the bid is accepted. The payoff is $V - b$ if the bid is accepted and 0 if the bid is rejected. So

$$\begin{aligned} E(W) &= E(W|b \geq 2V/3)P(b \geq 2V/3) + E(W|b < 2V/3)P(b < 2V/3) \\ &= E(V - b|b \geq 2V/3)P(b \geq 2V/3) + 0 \\ &= \left(E(V|V \leq 3b/2) - b\right) P(V \leq 3b/2). \end{aligned}$$

For $b \geq 2/3$, the event $V \leq 3b/2$ has probability 1, so the right-hand side is $1/2 - b$, which is negative. Now assume $b < 2/3$. Then $V \leq 3b/2$ has probability $3b/2$. Given that $V \leq 3b/2$, the conditional distribution of V is Uniform on $[0, 3b/2]$. Therefore,

$$E(W) = \left(E(V|V \leq 3b/2) - b\right) P(V \leq 3b/2) = (3b/4 - b)\,(3b/2) = -3b^2/8.$$

The above expression is negative except at $b = 0$, so the optimal bid is 0: you shouldn't play this game!

Alternatively, condition on which of the following events occurs: $A = \{V < b/2\}$, $B = \{b/2 \leq V \leq 3b/2\}$, $C = \{V > 3b/2\}$. We have

$$E(W|A) = E(V - b|A) < E(b/2 - b|A) = -b/2 \leq 0,$$

$$E(W|B) = E\left(\frac{b/2 + 3b/2}{2} - b|B\right) = 0,$$

$$E(W|C) = 0,$$

so we should just set $b = 0$ and walk away.

The moral of this story is to *condition on all the information.* It is crucial in the above calculation to use $E(V|V \leq 3b/2)$ rather than $E(V) = 1/2$; knowing that the bid was accepted gives information about how much the mystery prize is worth, so we shouldn't destroy that information. This problem is related to the so-called *winner's curse*, which says that the winner in an auction with incomplete information tends to profit less than they expect (unless they understand probability!). This is because in many settings, the expected value of the item that they bid on *given that they won the bid* is less than the unconditional expected value they originally had in mind. For $b \geq 2/3$, conditioning on $V \leq 3b/2$ does nothing since we know in advance that $V \leq 1$, but such a bid is ludicrously high. For any $b < 2/3$, finding out that your bid is accepted lowers your expectation:

$$E(V|V \leq 3b/2) < E(V).$$

□

The remaining examples use first-step analysis to calculate unconditional expectations. First, as promised in Chapter 4, we derive the expectation of the Geometric distribution using first-step analysis.

Example 9.1.8 (Geometric expectation redux)**.** Let $X \sim \text{Geom}(p)$. Interpret X as the number of Tails before the first Heads in a sequence of coin flips with probability p of Heads. To get $E(X)$, we condition on the outcome of the first toss: if it lands Heads, then X is 0 and we're done; if it lands Tails, then we've wasted one toss and are back to where we started, by memorylessness. Therefore,

$$\begin{aligned} E(X) &= E(X|\text{first toss } H) \cdot p + E(X|\text{first toss } T) \cdot q \\ &= 0 \cdot p + (1 + E(X)) \cdot q, \end{aligned}$$

which gives $E(X) = q/p$. □

The next example derives expected waiting times for some more complicated patterns, using two steps of conditioning.

Example 9.1.9 (Time until *HH* vs. *HT*)**.** You toss a fair coin repeatedly. What is the expected number of tosses until the pattern *HT* appears for the first time? What about the expected number of tosses until *HH* appears for the first time?

Solution:

Let W_{HT} be the number of tosses until *HT* appears. As we can see from Figure 9.3, W_{HT} is the waiting time for the first Heads, which we'll call W_1, plus the additional waiting time for the first Tails after the first Heads, which we'll call W_2. By the story of the First Success distribution, W_1 and W_2 are i.i.d. FS(1/2), so $E(W_1) = E(W_2) = 2$ and $E(W_{HT}) = 4$.

Finding the expected waiting time for *HH*, $E(W_{HH})$, is more complicated. We can't apply the same logic as for $E(W_{HT})$: as shown in Figure 9.4, if the first Heads is immediately followed by Tails, our progress is destroyed and we must start from scratch. But this *is* progress for us in solving the problem, since the fact that the

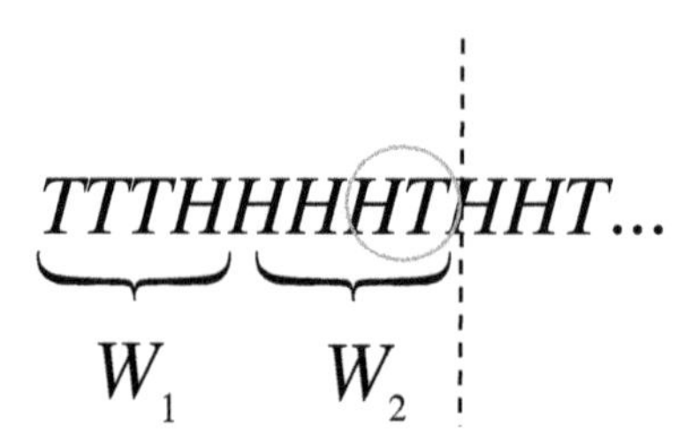

FIGURE 9.3
Waiting time for HT is the waiting time for the first Heads, W_1, plus the additional waiting time for the next Tails, W_2. Durable partial progress is possible!

system can get reset suggests the strategy of first-step analysis. Let's condition on the outcome of the first toss:

$$E(W_{HH}) = E(W_{HH}|\text{first toss } H)\frac{1}{2} + E(W_{HH}|\text{first toss } T)\frac{1}{2}.$$

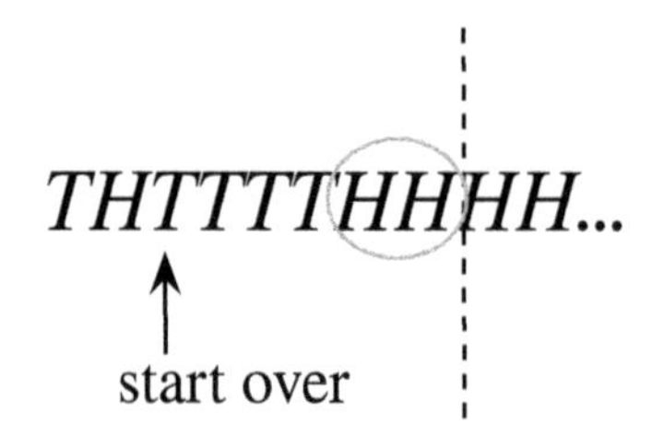

FIGURE 9.4
When waiting for *HH*, partial progress can easily be destroyed.

For the second term, $E(W_{HH}|\text{first toss T}) = 1 + E(W_{HH})$ by memorylessness. For the first term, we compute $E(W_{HH}|\text{1st toss H})$ by further conditioning on the outcome of the second toss. If the second toss is Heads, we have obtained *HH* in two tosses. If the second toss is Tails, we've wasted two tosses and have to start all over! This gives

$$E(W_{HH}|\text{first toss } H) = 2 \cdot \frac{1}{2} + (2 + E(W_{HH})) \cdot \frac{1}{2}.$$

Therefore,

$$E(W_{HH}) = \left(2 \cdot \frac{1}{2} + (2 + E(W_{HH})) \cdot \frac{1}{2}\right)\frac{1}{2} + (1 + E(W_{HH}))\frac{1}{2}.$$

Solving for $E(W_{HH})$, we get $E(W_{HH}) = 6$.

It might seem surprising at first that the expected waiting time for *HH* is greater than the expected waiting time for *HT*. How do we reconcile this with the fact that in two tosses of the coin, *HH* and *HT* both have a $1/4$ chance of appearing? Why aren't the average waiting times the same by symmetry?

As we solved this problem, we in fact noticed an important *asymmetry*. When waiting for HT, once we get the first Heads, we've achieved partial progress that cannot be destroyed: if the Heads is followed by another Heads, we're in the same position as before, and if the Heads is followed by a Tails, we're done. By contrast, when waiting for HH, even after getting the first Heads, we could be sent back to square one if the Heads is followed by a Tails. This suggests the average waiting time for HH should be longer. Symmetry implies that the average waiting time for HH is the same as that for TT, and that for HT is the same as that for TH, but it does not imply that the average waiting times for HH and HT are the same.

More intuition into what's going on can be obtained by considering a long string of coin flips, as in Figure 9.5. We notice right away that appearances of HH can overlap, while appearances of HT must be disjoint. For example, $HHHHHH$ has 5 occurrences of HH, but $HTHTHT$ has only 3 occurrences of HT. Since there are the same average number of HHs and HTs, but HHs sometimes clump together, the average waiting time for HH must be larger than that of HT to compensate.

FIGURE 9.5
Clumping. (a) Appearances of HH can overlap. (b) Appearances of HT must be disjoint.

Related problems occur in information theory when compressing a message, and in genetics when looking for recurring patterns (called *motifs*) in DNA sequences. □

Our final example in this section uses wishful thinking for *both* probabilities and expectations to study a question about a random walk.

Example 9.1.10 (Random walk on the integers). An immortal drunk man wanders around randomly on the integers. He starts at the origin, and at each step he moves 1 unit to the right or 1 unit to the left, with equal probabilities, independently of all his previous steps. Let b be a googolplex (this is 10^g, where $g = 10^{100}$ is a googol).

(a) Find a simple expression for the probability that the immortal drunk visits b before returning to the origin for the first time.

(b) Find the expected number of times that the immortal drunk visits b before returning to the origin for the first time.

Solution:

(a) Let B be the event that the drunk man visits b before returning to the origin for the first time and let L be the event that his first move is to the left. Then

$P(B|L) = 0$ since any path from -1 to b must pass through 0. For $P(B|L^c)$, we are exactly in the setting of the gambler's ruin problem, where player A starts with \$1, player B starts with \$$(b-1)$, and the rounds are fair. Applying that result, we have

$$P(B) = P(B|L)P(L) + P(B|L^c)P(L^c) = \frac{1}{b} \cdot \frac{1}{2} = \frac{1}{2b}.$$

(b) Let N be the number of visits to b before returning to the origin for the first time, and let $p = 1/(2b)$ be the probability found in (a). Then

$$E(N) = E(N|N=0)P(N=0) + E(N|N \geq 1)P(N \geq 1) = pE(N|N \geq 1).$$

The conditional distribution of N given $N \geq 1$ is $\text{FS}(p)$: given that the man reaches b, by symmetry there is probability p of returning to the origin before visiting b again (call this "success") and probability $1-p$ of returning to b again before returning to the origin (call this "failure"). Note that the trials are independent since the situation is the same each time he is at b, independent of the past history. Thus $E(N|N \geq 1) = 1/p$, and

$$E(N) = pE(N|N \geq 1) = p \cdot \frac{1}{p} = 1.$$

Surprisingly, the result doesn't depend on the value of b, and our proof didn't require knowing the value of p. □

9.2 Conditional expectation given an r.v.

In this section we introduce conditional expectation given a random variable. That is, we want to understand what it means to write $E(Y|X)$ for an r.v. X. We will see that $E(Y|X)$ is a *random variable* that is, in a certain sense, our best prediction of Y, assuming we get to know X.

The key to understanding $E(Y|X)$ is first to understand $E(Y|X=x)$. Since $X = x$ is an event, $E(Y|X=x)$ is just the conditional expectation of Y given this event, and it can be computed using the conditional distribution of Y given $X = x$.

If Y is discrete, we use the conditional PMF $P(Y = y|X = x)$ in place of the unconditional PMF $P(Y = y)$:

$$E(Y|X=x) = \sum_y yP(Y=y|X=x).$$

Analogously, if Y is continuous, we use the conditional PDF $f_{Y|X}(y|x)$ in place of the unconditional PDF:

$$E(Y|X=x) = \int_{-\infty}^{\infty} y f_{Y|X}(y|x)dy.$$

Notice that because we sum or integrate over y, $E(Y|X = x)$ is a function of x only. We can give this function a name, like g: let $g(x) = E(Y|X = x)$. We define $E(Y|X)$ as the random variable obtained by finding the form of the function $g(x)$, then *plugging in X for x*.

Definition 9.2.1 (Conditional expectation given an r.v.). Let $g(x) = E(Y|X = x)$. Then the *conditional expectation of Y given X*, denoted $E(Y|X)$, is defined to be the random variable $g(X)$. In other words, if after doing the experiment X crystallizes into x, then $E(Y|X)$ crystallizes into $g(x)$.

☣ **9.2.2.** The notation in this definition sometimes causes confusion. It does *not* say "$g(x) = E(Y|X = x)$, so $g(X) = E(Y|X = X)$, which equals $E(Y)$ because $X = X$ is always true". Rather, we should first compute the function $g(x)$, *then* plug in X for x. For example, if $g(x) = x^2$, then $g(X) = X^2$. A similar biohazard is ☣ 5.3.2, about the meaning of $F(X)$ in the universality of the Uniform.

☣ **9.2.3.** By definition, $E(Y|X)$ is a function of X, so it is a random variable. (This does *not* mean there are no examples where $E(Y|X)$ is a constant. A constant is a degenerate r.v., and a constant function of X. For example, if X and Y are independent then $E(Y|X) = E(Y)$, which is a constant.) Thus it makes sense to compute quantities like $E(E(Y|X))$ and $\text{Var}(E(Y|X))$, the mean and variance of the r.v. $E(Y|X)$. It is easy to be ensnared by category errors when working with conditional expectation, so it is important to keep in mind that conditional expectations of the form $E(Y|A)$ are numbers, while those of the form $E(Y|X)$ are random variables.

Here are some quick examples of how to calculate conditional expectation. In both examples, we don't need to do a sum or integral to get $E(Y|X = x)$ because a more direct approach is available.

Example 9.2.4. A stick of length 1 is broken at a point X chosen uniformly at random. Given that $X = x$, we then choose another breakpoint Y uniformly on the interval $[0, x]$. Find $E(Y|X)$, and its mean and variance.

Solution:

From the description of the experiment, $X \sim \text{Unif}(0, 1)$ and $Y|X = x \sim \text{Unif}(0, x)$. Then $E(Y|X = x) = x/2$, so by plugging in X for x, we have

$$E(Y|X) = X/2.$$

The expected value of $E(Y|X)$ is

$$E(E(Y|X)) = E(X/2) = 1/4.$$

(We will show in the next section that a general property of conditional expectation is that $E(E(Y|X)) = E(Y)$, so it also follows that $E(Y) = 1/4$.) The variance of $E(Y|X)$ is

$$\text{Var}(E(Y|X)) = \text{Var}(X/2) = 1/48. \qquad \square$$

Example 9.2.5. For $X, Y \stackrel{\text{i.i.d.}}{\sim} \text{Expo}(\lambda)$, find $E(\max(X,Y)|\min(X,Y))$.

Solution:

Let $M = \max(X,Y)$ and $L = \min(X,Y)$. By the memoryless property, $M - L$ is independent of L, and $M - L \sim \text{Expo}(\lambda)$ (see Example 7.3.6). Therefore

$$E(M|L=l) = E(L|L=l) + E(M-L|L=l) = l + E(M-L) = l + \frac{1}{\lambda},$$

and $E(M|L) = L + \frac{1}{\lambda}$. □

9.3 Properties of conditional expectation

Conditional expectation has some very useful properties.

- Dropping what's independent: If X and Y are independent, then $E(Y|X) = E(Y)$.
- Taking out what's known: For any function h, $E(h(X)Y|X) = h(X)E(Y|X)$.
- Linearity: $E(Y_1 + Y_2|X) = E(Y_1|X) + E(Y_2|X)$, and $E(cY|X) = cE(Y|X)$ for c a constant (the latter is a special case of taking out what's known).
- Adam's law: $E(E(Y|X)) = E(Y)$.
- Projection interpretation: The r.v. $Y - E(Y|X)$, which is called the *residual* from using X to predict Y, is uncorrelated with $h(X)$ for any function h.

Let's discuss each property individually.

Theorem 9.3.1 (Dropping what's independent)**.** If X and Y are independent, then $E(Y|X) = E(Y)$.

This is true because independence implies $E(Y|X = x) = E(Y)$ for all x, hence $E(Y|X) = E(Y)$. Intuitively, if X provides no information about Y, then our best guess for Y, even if we get to know X, is still the unconditional mean $E(Y)$. However, the converse is false: a counterexample is given in Example 9.3.3 below.

Theorem 9.3.2 (Taking out what's known)**.** For any function h,

$$E(h(X)Y|X) = h(X)E(Y|X).$$

Intuitively, when we take expectations given X, we are treating X as if it has crystallized into a known constant. Then any function of X, say $h(X)$, also acts like a known constant while we are conditioning on X. Taking out what's known is the conditional version of the unconditional fact that $E(cY) = cE(Y)$. The difference is that $E(cY) = cE(Y)$ asserts that two *numbers* are equal, while taking out what's known asserts that two *random variables* are equal.

Example 9.3.3. Let $Z \sim \mathcal{N}(0,1)$ and $Y = Z^2$. Find $E(Y|Z)$ and $E(Z|Y)$.

Solution: Since Y is a function of Z, $E(Y|Z) = E(Z^2|Z) = Z^2$ by taking out what's known. To get $E(Z|Y)$, notice that conditional on $Y = y$, Z equals $\sqrt{y}$ or $-\sqrt{y}$ with equal probabilities by the symmetry of the standard Normal, so $E(Z|Y = y) = 0$ and $E(Z|Y) = 0$.

In this case, although Y provides a lot of information about Z, narrowing down the possible values of Z to just two values, Y only tells us about the magnitude of Z and not its sign. For this reason, $E(Z|Y) = E(Z)$ despite the dependence between Z and Y. This example illustrates that the converse of Theorem 9.3.1 is false. □

Theorem 9.3.4 (Linearity). $E(Y_1 + Y_2|X) = E(Y_1|X) + E(Y_2|X)$.

This result is the conditional version of the unconditional fact that $E(Y_1 + Y_2) = E(Y_1) + E(Y_2)$, and is true since conditional probabilities *are* probabilities.

☣ **9.3.5.** It is incorrect to write "$E(Y|X_1 + X_2) = E(Y|X_1) + E(Y|X_2)$"; linearity applies on the left side of the conditioning bar, not on the right side!

Example 9.3.6. Let $X_1, \dots, X_n$ be i.i.d., and $S_n = X_1 + \dots + X_n$. Find $E(X_1|S_n)$.

Solution:

By symmetry,

$$E(X_1|S_n) = E(X_2|S_n) = \dots = E(X_n|S_n),$$

and by linearity,

$$E(X_1|S_n) + \dots + E(X_n|S_n) = E(S_n|S_n) = S_n.$$

Therefore,

$$E(X_1|S_n) = S_n/n = \bar{X}_n,$$

the sample mean of the X_j's. This is an intuitive result: if we have 2 i.i.d. r.v.s X_1, X_2 and learn that $X_1 + X_2 = 10$, it makes sense to guess that X_1 is 5 (accounting for half of the total). Similarly, if we have n i.i.d. r.v.s and get to know their sum, our best guess for any one of them is the sample mean. □

The next theorem connects conditional expectation to unconditional expectation. It goes by many names, including the law of total expectation, the law of iterated expectation (which has a terrible acronym for something glowing with truth), and the tower property. We call it *Adam's law* because it is used so frequently that it deserves a pithy name, and since it is often used in conjunction with another law we'll encounter soon, which has a complementary name.

Theorem 9.3.7 (Adam's law). For any r.v.s X and Y,

$$E(E(Y|X)) = E(Y).$$

Proof. We present the proof in the case where X and Y are both discrete (the proofs for other cases are analogous). Let $E(Y|X) = g(X)$. We proceed by applying

LOTUS, expanding the definition of $g(x)$ to get a double sum, and then swapping the order of summation:

$$\begin{aligned} E(g(X)) &= \sum_x g(x)P(X=x) \\ &= \sum_x \left(\sum_y yP(Y=y|X=x) \right) P(X=x) \\ &= \sum_x \sum_y yP(X=x)P(Y=y|X=x) \\ &= \sum_y y \sum_x P(X=x, Y=y) \\ &= \sum_y yP(Y=y) = E(Y). \end{aligned}$$

■

Adam's law is a more compact, more general version of the law of total expectation (Theorem 9.1.5). For X discrete, the statements

$$E(Y) = \sum_x E(Y|X=x)P(X=x)$$

and

$$E(Y) = E(E(Y|X))$$

mean the same thing, since if we let $E(Y|X=x) = g(x)$, then

$$E(E(Y|X)) = E(g(X)) = \sum_x g(x)P(X=x) = \sum_x E(Y|X=x)P(X=x).$$

Armed with Adam's law, we have a powerful strategy for finding an expectation $E(Y)$, by conditioning on an r.v. X that we wish we knew. First obtain $E(Y|X)$ by treating X as known, and then take the expectation of $E(Y|X)$. We will see various examples of this later in the chapter.

Just as there are forms of Bayes' rule and LOTP with extra conditioning, as discussed in Chapter 2, there is a version of Adam's law with extra conditioning.

Theorem 9.3.8 (Adam's law with extra conditioning)**.** For any r.v.s X, Y, Z,

$$E(E(Y|X,Z)|Z) = E(Y|Z).$$

The above equation is Adam's law, except with extra conditioning on Z inserted everywhere. It is true because conditional probabilities *are* probabilities. So we are free to use Adam's law to help us find both unconditional expectations and conditional expectations.

Using Adam's law, we can also prove the last item on our list of properties of conditional expectation.

Theorem 9.3.9 (Projection interpretation)**.** For any function h, the random variable $Y - E(Y|X)$ is uncorrelated with $h(X)$. Equivalently,

$$E((Y - E(Y|X))h(X)) = 0.$$

(This is equivalent since $E(Y - E(Y|X)) = 0$, by linearity and Adam's law.)

Proof. We have

$$\begin{aligned} E((Y - E(Y|X))h(X)) &= E(h(X)Y) - E(h(X)E(Y|X)) \\ &= E(h(X)Y) - E(E(h(X)Y|X)) \end{aligned}$$

by Theorem 9.3.2 (here we're "putting back what's known" in the inner expectation). By Adam's law, the second term is equal to $E(h(X)Y)$. ■

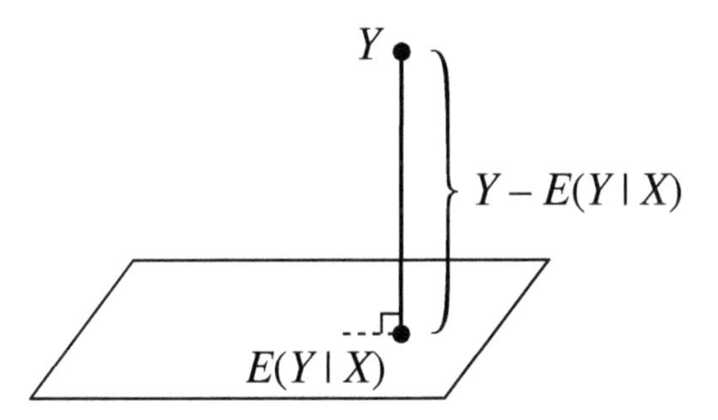

FIGURE 9.6
The conditional expectation $E(Y|X)$ is the *projection* of Y onto the space of all functions of X, shown here as a plane. The residual $Y - E(Y|X)$ is orthogonal to the plane: it's perpendicular to (uncorrelated with) any function of X.

From a geometric perspective, we can visualize Theorem 9.3.9 as in Figure 9.6. In a certain sense (described below), $E(Y|X)$ is the function of X that is *closest* to Y; we say that $E(Y|X)$ is the *projection* of Y into the space of all functions of X. The "line" from Y to $E(Y|X)$ in the figure is orthogonal (perpendicular) to the "plane", since any other route from Y to $E(Y|X)$ would be longer. This orthogonality turns out to be the geometric interpretation of Theorem 9.3.9.

The details of this perspective are given in the next section, which is starred since it requires knowledge of linear algebra. But even without delving into the linear algebra, the projection picture gives some useful intuition. As mentioned earlier, we can think of $E(Y|X)$ as a prediction for Y based on X. This is an extremely common problem in statistics: predict or estimate the future observations or unknown parameters based on data. The projection interpretation of conditional expectation implies that $E(Y|X)$ is the *best predictor* of Y based on X, in the sense that it is the function of X with the lowest *mean squared error* (expected squared difference between Y and the prediction of Y).

Example 9.3.10 (Linear regression). An extremely widely used method for data analysis in statistics is *linear regression*. In its most basic form, the linear regression model uses a single explanatory variable X to predict a response variable Y, and it assumes that the conditional expectation of Y is *linear* in X:

$$E(Y|X) = a + bX.$$

(a) Show that an equivalent way to express this is to write

$$Y = a + bX + \epsilon,$$

where ϵ is an r.v. (called the *error*) with $E(\epsilon|X) = 0$.

(b) Solve for the constants a and b in terms of $E(X)$, $E(Y)$, $\text{Cov}(X, Y)$, and $\text{Var}(X)$.

Solution:

(a) Let $Y = a + bX + \epsilon$, with $E(\epsilon|X) = 0$. Then by linearity,

$$E(Y|X) = E(a|X) + E(bX|X) + E(\epsilon|X) = a + bX.$$

Conversely, suppose that $E(Y|X) = a + bX$, and define

$$\epsilon = Y - (a + bX).$$

Then $Y = a + bX + \epsilon$, with

$$E(\epsilon|X) = E(Y|X) - E(a + bX|X) = E(Y|X) - (a + bX) = 0.$$

(b) First, by Adam's law, taking the expectation of both sides gives

$$E(Y) = a + bE(X).$$

Note that ϵ has mean 0 and X and ϵ are uncorrelated, since

$$E(\epsilon) = E(E(\epsilon|X)) = E(0) = 0$$

and

$$E(\epsilon X) = E(E(\epsilon X|X)) = E(XE(\epsilon|X)) = E(0) = 0.$$

Taking the covariance with X of both sides in $Y = a + bX + \epsilon$, we have

$$\text{Cov}(X, Y) = \text{Cov}(X, a) + b\,\text{Cov}(X, X) + \text{Cov}(X, \epsilon) = b\,\text{Var}(X).$$

Thus,

$$b = \frac{\text{Cov}(X, Y)}{\text{Var}(X)},$$

$$a = E(Y) - b\,E(X) = E(Y) - \frac{\text{Cov}(X, Y)}{\text{Var}(X)} \cdot E(X).$$

□

9.4 *Geometric interpretation of conditional expectation

This section explains in more detail the geometric perspective shown in Figure 9.6, using some concepts from linear algebra. Consider the vector space consisting of all random variables on a certain probability space, such that the random variables all have finite variance. Each vector or point in the space is a random variable (here we are using "vector" in the linear algebra sense, not in the sense of a random vector from Chapter 7). Define the inner product of two r.v.s U and V to be

$$\langle U, V \rangle = E(UV).$$

(For this definition to satisfy the axioms for an inner product, we need the convention that two r.v.s are considered the same if they are equal with probability 1.)

The squared length of an r.v. X is

$$||X||^2 = \langle X, X \rangle = EX^2,$$

and the squared distance between two r.v.s U and V is

$$||U - V||^2 = E(U - V)^2.$$

The interpretations become especially nice if $E(U) = E(V) = 0$, since then:

- $||U||^2 = \text{Var}(U)$, and $||U|| = \text{SD}(U)$.
- $\langle U, V \rangle = \text{Cov}(U, V)$, and the cosine of the "angle" between U and V is $\text{Corr}(U, V)$.
- U and V are orthogonal (i.e., $\langle U, V \rangle = 0$) if and only if they are uncorrelated.

To interpret $E(Y|X)$ geometrically, consider the space of all random variables (with finite variance) that can be expressed as functions of X. This is a subspace of the vector space. In Figure 9.6, the subspace of random variables of the form $h(X)$ is represented by a plane. To get $E(Y|X)$, we *project* Y onto the plane. Then the residual $Y - E(Y|X)$ is orthogonal to $h(X)$ for all functions h, and $E(Y|X)$ is the function of X that best predicts Y, where "best" here means that the mean squared error $E(Y - g(X))^2$ is minimized by choosing $g(X) = E(Y|X)$.

The projection interpretation is a helpful way to think about many of the properties of conditional expectation. For example, if $Y = h(X)$ is a function of X, then Y itself is already in the plane, so it is its own projection; this explains why

$$E(h(X)|X) = h(X).$$

We can think of *unconditional* expectation as a projection too: $E(Y)$ can be thought of as $E(Y|0)$, the projection of Y onto the space of all constants (and indeed, $E(Y)$ is the constant c that minimizes $E(Y - c)^2$, as we proved in Theorem 6.1.4).

We can now also give a geometric interpretation for Adam's law: $E(Y)$ says to project Y in one step onto the space of all constants; $E(E(Y|X))$ says to do it in two steps, by first projecting onto the plane and then projecting $E(Y|X)$ onto the space of all constants, which is a line within that plane. Adam's law says that the one-step and two-step methods yield the same result.

In the next section we will introduce *Eve's law*, which serves the same purpose for variance as Adam's law does for expectation. As a preview and to further explore the geometric interpretation of conditional expectation, let's look at $\text{Var}(Y)$ from the perspective of this section. Assume that $E(Y) = 0$ (if $E(Y) \neq 0$, we can *center* Y by subtracting $E(Y)$; doing so has no effect on the variance of Y).

We can decompose Y into two orthogonal terms, the residual $Y - E(Y|X)$ and the conditional expectation $E(Y|X)$:

$$Y = (Y - E(Y|X)) + E(Y|X).$$

The two terms are orthogonal since $Y - E(Y|X)$ is uncorrelated with any function of X, and $E(Y|X)$ *is* a function of X. So by the Pythagorean theorem,

$$||Y||^2 = ||Y - E(Y|X)||^2 + ||E(Y|X)||^2.$$

That is,

$$\text{Var}(Y) = \text{Var}(Y - E(Y|X)) + \text{Var}(E(Y|X)).$$

As we will see in the next section, this identity is a form of Eve's law. So it turns out that Eve's law, which may look cryptic at first glance, can be interpreted as just being the Pythagorean theorem for a "triangle" whose sides are the vectors $Y - E(Y|X), E(Y|X)$, and Y.

9.5 Conditional variance

Once we've defined conditional expectation given an r.v., we have a natural way to define conditional variance given a random variable: replace all instances of $E(\cdot)$ in the definition of unconditional variance with $E(\cdot|X)$.

Definition 9.5.1 (Conditional variance). The *conditional variance of* Y *given* X is

$$\text{Var}(Y|X) = E((Y - E(Y|X))^2|X).$$

This is equivalent to

$$\text{Var}(Y|X) = E(Y^2|X) - (E(Y|X))^2.$$

☣ **9.5.2.** Like $E(Y|X)$, $\text{Var}(Y|X)$ is a random variable, and it is a function of X.

Since conditional variance is defined in terms of conditional expectations, we can use results about conditional expectation to help us calculate conditional variance. Here's an example.

Example 9.5.3. Let $Z \sim \mathcal{N}(0,1)$ and $Y = Z^2$. Find $\text{Var}(Y|Z)$ and $\text{Var}(Z|Y)$.

Solution:

Without any calculations we can see that $\text{Var}(Y|Z) = 0$: conditional on Z, Y is a known constant, and the variance of a constant is 0. By the same reasoning, $\text{Var}(h(Z)|Z) = 0$ for any function h.

To get $\text{Var}(Z|Y)$, apply the definition:

$$\text{Var}(Z|Z^2) = E(Z^2|Z^2) - (E(Z|Z^2))^2.$$

The first term equals Z^2. The second term equals 0 by symmetry, as we found in Example 9.3.3. Thus $\text{Var}(Z|Z^2) = Z^2$, which we can write as $\text{Var}(Z|Y) = Y$. □

In the next example, we will practice working with conditional expectation and conditional variance in the context of the Bivariate Normal.

Example 9.5.4 (Conditional expectation and conditional variance in a BVN)**.** Let (Z, W) be Bivariate Normal, with $\text{Corr}(Z, W) = \rho$ and Z, W marginally $\mathcal{N}(0,1)$. Find $E(W|Z)$ and $\text{Var}(W|Z)$.

Solution: We can assume that (Z, W) has been constructed as in Example 7.5.10, since $E(W|Z)$ and $\text{Var}(W|Z)$ depend only on the joint distribution of (Z, W), not on the specific method that was used to create (Z, W). So let

$$Z = X$$
$$W = \rho X + \sqrt{1-\rho^2}Y,$$

with X, Y i.i.d. $\mathcal{N}(0,1)$. We can then solve the problem very neatly, without having to resort to messy integrals based on the Bivariate Normal joint PDF. The conditional expectation is

$$E(W|Z) = E(W|X) = \rho X + \sqrt{1-\rho^2}E(Y|X) = \rho X + \sqrt{1-\rho^2}E(Y) = \rho Z,$$

since X and Y are independent. And the conditional variance is

$$\text{Var}(W|Z) = \text{Var}(W|X) = \text{Var}(\sqrt{1-\rho^2}Y|X) = (1-\rho^2)\text{Var}(Y) = 1-\rho^2,$$

since ρX acts as a constant if we are given X, and Y is independent of X.

Interestingly, the same argument with the roles of Z and W reversed shows that

$$E(Z|W) = \rho W, \text{ and } \text{Var}(Z|W) = 1-\rho^2.$$

One might have guessed that if we should multiply by ρ to go from an observed value of Z to a predicted value of W, then we should *divide* by ρ to go from an observed value of W to a predicted value of Z. But the above results say to multiply by the same ρ, regardless of whether using Z to predict W or vice versa! This is closely related to the fact that correlation is symmetric ($\text{Corr}(Z, W) = \rho = \text{Corr}(W, Z)$) and to an important concept in statistics known as *regression toward the mean*. □

We learned in the previous section that Adam's law relates conditional expectation to unconditional expectation. A companion result for Adam's law is *Eve's law*, which relates conditional variance to unconditional variance.

Theorem 9.5.5 (Eve's law). For any r.v.s X and Y,

$$\text{Var}(Y) = E(\text{Var}(Y|X)) + \text{Var}(E(Y|X)).$$

The ordering of E's and Var's on the right-hand side spells EVVE, whence the name Eve's law. Eve's law is also known as the *law of total variance* or the *variance decomposition formula*.

Proof. Let $g(X) = E(Y|X)$. By Adam's law, $E(g(X)) = E(Y)$. Then

$$E(\text{Var}(Y|X)) = E(E(Y^2|X) - g(X)^2) = E(Y^2) - E(g(X)^2),$$

$$\text{Var}(E(Y|X)) = E(g(X)^2) - (Eg(X))^2 = E(g(X)^2) - (EY)^2.$$

Adding these equations, we have Eve's law. ■

To visualize Eve's law, imagine a population where each person has a value of X and a value of Y. We can divide this population into subpopulations, one for each possible value of X. For example, if X represents age and Y represents height, we can group people based on age. Then there are two sources contributing to the variation in people's heights in the overall population. First, within each age group, people have different heights. The average amount of variation in height within each age group is the *within-group variation*, $E(\text{Var}(Y|X))$. Second, across age groups, the average heights are different. The variance of average heights across age groups is the *between-group variation*, $\text{Var}(E(Y|X))$. Eve's law says that to get the total variance of Y, we simply add these two sources of variation.

Figure 9.7 illustrates Eve's law in the simple case where we have three age groups. The average amount of scatter within each of the groups $X = 1$, $X = 2$, and $X = 3$ is the within-group variation, $E(\text{Var}(Y|X))$. The variance of the group means $E(Y|X = 1)$, $E(Y|X = 2)$, and $E(Y|X = 3)$ is the between-group variation, $\text{Var}(E(Y|X))$.

Another way to think about Eve's law is in terms of *prediction*. If we wanted to predict someone's height based on their age alone, the ideal scenario would be if everyone within an age group had exactly the same height, while different age groups had different heights. Then, given someone's age, we would be able to predict their height perfectly. In other words, the ideal scenario for prediction is *no* within-group

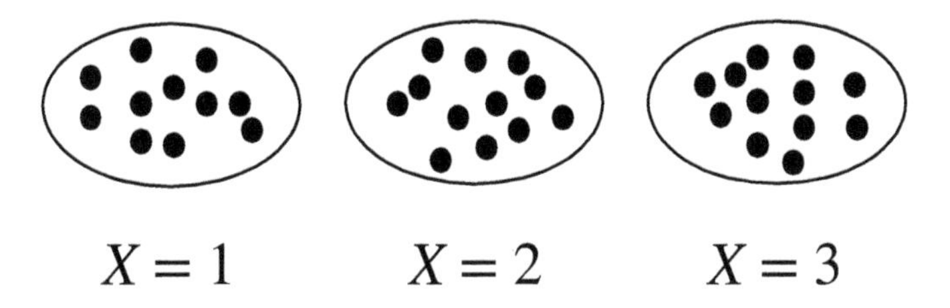

FIGURE 9.7
Eve's law says that total variance is the sum of within-group and between-group variation.

variation in height, since the within-group variation cannot be explained by age differences. For this reason, within-group variation is also called *unexplained variation*, and between-group variation is also called *explained variation*. Eve's law says that the overall variance of Y is the sum of unexplained and explained variation.

We can also write Eve's law in the form

$$\text{Var}(Y) = \text{Var}(Y - E(Y|X)) + \text{Var}(E(Y|X)),$$

since, letting W be the residual $Y - E(Y|X)$,

$$\text{Var}(Y - E(Y|X)) = E(W^2) = E(E(W^2|X)) = E(\text{Var}(Y|X)).$$

Again this says that we can decompose variance into within-group variation plus between-group variation.

☣ **9.5.6.** Let Y be an r.v. and A be an event. It is wrong to say "$\text{Var}(Y) = \text{Var}(Y|A)P(A) + \text{Var}(Y|A^c)P(A^c)$", even though this looks analogous to the law of total expectation. (For a simple counterexample, let $Y \sim \text{Bern}(1/2)$ and A be the event $Y = 0$. Then $\text{Var}(Y|A)$ and $\text{Var}(Y|A^c)$ are both 0, but $\text{Var}(Y) = 1/4$.)

Instead, we should use Eve's law if we want to condition on whether or not A occurred: letting I be the indicator of A,

$$\text{Var}(Y) = E(\text{Var}(Y|I)) + \text{Var}(E(Y|I)).$$

To see how this expression relates to the "wrong expression", let

$$p = P(A),\ q = P(A^c),\ a = E(Y|A),\ b = E(Y|A^c),\ v = \text{Var}(Y|A),\ w = \text{Var}(Y|A^c).$$

Then $E(Y|I)$ is a with probability p and b with probability q, and $\text{Var}(Y|I)$ is v with probability p and w with probability q. So

$$E(\text{Var}(Y|I)) = vp + wq = \text{Var}(Y|A)P(A) + \text{Var}(Y|A^c)P(A^c),$$

which is exactly the "wrong expression", and $\text{Var}(Y)$ consists of this plus the term

$$\text{Var}(E(Y|I)) = a^2p + b^2q - (ap + bq)^2.$$

It is crucial to account for *both* within-group and between-group variation.

9.6 Adam and Eve examples

We conclude this chapter with several examples showing how Adam's law and Eve's law allow us to find the mean and variance of complicated r.v.s, especially in situations that involve multiple levels of randomness.

In our first example, the r.v. of interest is a *random sum*: the sum of a random number of random variables. There are thus two levels of randomness: first, each term in the sum is a random variable; second, the number of terms in the sum is also a random variable.

Example 9.6.1 (Random sum). A store receives N customers in a day, where N is an r.v. with finite mean and variance. Let X_j be the amount spent by the jth customer at the store. Assume that each X_j has mean μ and variance σ^2, and that N and all the X_j are independent of one another. Find the mean and variance of the random sum $X = \sum_{j=1}^{N} X_j$, which is the store's total revenue in a day, in terms of μ, σ^2, $E(N)$, and $\text{Var}(N)$.

Solution:

Since X is a sum, our first impulse might be to claim "$E(X) = N\mu$ by linearity". Alas, this would be a category error, since $E(X)$ is a number and $N\mu$ is a random variable. The key is that X is not merely a sum, but a random sum; the number of terms we are adding up is itself random, whereas linearity applies to sums with a *fixed* number of terms.

Yet this category error actually suggests the correct strategy: if only we were allowed to treat N as a constant, then linearity would apply. So let's condition on N. By linearity of *conditional* expectation,

$$E(X|N) = E\left(\sum_{j=1}^{N} X_j|N\right) = \sum_{j=1}^{N} E(X_j|N) = \sum_{j=1}^{N} E(X_j) = N\mu.$$

We used the independence of the X_j and N to assert $E(X_j|N) = E(X_j)$ for all j. Note that the statement "$E(X|N) = N\mu$" is not a category error because both sides of the equality are r.v.s that are functions of N. Finally, by Adam's law,

$$E(X) = E(E(X|N)) = E(N\mu) = \mu E(N).$$

This is a pleasing result: the average total revenue is the average amount spent per customer, multiplied by the average number of customers.

For $\text{Var}(X)$, we again condition on N to get $\text{Var}(X|N)$:

$$\text{Var}(X|N) = \text{Var}\left(\sum_{j=1}^{N} X_j|N\right) = \sum_{j=1}^{N} \text{Var}(X_j|N) = \sum_{j=1}^{N} \text{Var}(X_j) = N\sigma^2.$$

Eve's law then tells us how to obtain the unconditional variance of X:

$$\begin{aligned}\mathrm{Var}(X) &= E(\mathrm{Var}(X|N)) + \mathrm{Var}(E(X|N)) \\ &= E(N\sigma^2) + \mathrm{Var}(N\mu) \\ &= \sigma^2 E(N) + \mu^2 \mathrm{Var}(N).\end{aligned}$$ □

In the next example, two levels of randomness arise because our experiment takes place in two stages. We sample a city from a group of cities, then sample citizens within the city. This is an example of a *multilevel model.*

Example 9.6.2 (Random sample from a random city). To study the prevalence of a disease in several cities of interest within a certain county, we pick a city at random, then pick a random sample of n people from that city. This is a form of a widely used survey technique known as *cluster sampling.*

Let Q be the proportion of diseased people in the chosen city, and let X be the number of diseased people in the sample. As illustrated in Figure 9.8 (where white dots represent healthy individuals and black dots represent diseased individuals), different cities may have very different prevalences. Since each city has its own disease prevalence, Q is a random variable. Suppose that $Q \sim \mathrm{Unif}(0, 1)$. Also assume that conditional on Q, each individual in the sample independently has probability Q of having the disease; this is true if we sample with replacement from the chosen city, and is approximately true if we sample without replacement but the population size is large. Find $E(X)$ and $\mathrm{Var}(X)$.

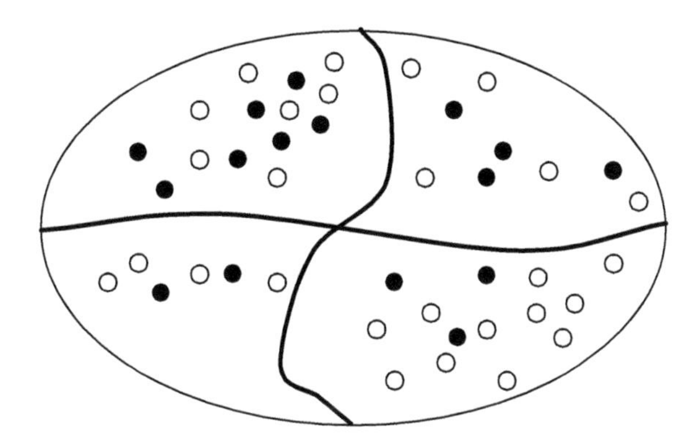

FIGURE 9.8
A certain oval-shaped county has 4 cities. Each city has healthy people (represented as white dots) and diseased people (represented as black dots). A random city is chosen, and then a random sample of n people is chosen from within that city. There are two components to the variability in the number of diseased people in the sample: variation due to different cities having different disease prevalence, and variation due to the randomness of the sample within the chosen city.

Solution:

With our assumptions, $X|Q \sim \mathrm{Bin}(n, Q)$; this notation says that conditional on knowing the disease prevalence in the chosen city, we can treat Q as a constant, and each sampled individual is an independent Bernoulli trial with probability Q of success. Using the mean and variance of the Binomial distribution, $E(X|Q) = nQ$ and $\text{Var}(X|Q) = nQ(1-Q)$. Furthermore, using the moments of the standard Uniform distribution, $E(Q) = 1/2$, $E(Q^2) = 1/3$, and $\text{Var}(Q) = 1/12$. Now we can apply Adam's law and Eve's law to get the unconditional mean and variance of X:

$$E(X) = E(E(X|Q)) = E(nQ) = \frac{n}{2},$$

$$\begin{aligned}\text{Var}(X) &= E(\text{Var}(X|Q)) + \text{Var}(E(X|Q)) \\ &= E(nQ(1-Q)) + \text{Var}(nQ) \\ &= nE(Q) - nE(Q^2) + n^2\text{Var}(Q) \\ &= \frac{n}{6} + \frac{n^2}{12}.\end{aligned}$$

Note that the structure of this problem is identical to that in the story of Bayes' billiards. Therefore, we actually know the distribution of X, not just its mean and variance: X is Discrete Uniform on $\{0, 1, 2, \ldots, n\}$. But the Adam-and-Eve approach can be applied when Q has a more complicated distribution, or with more levels in the multilevel model, whether or not it is feasible to work out the distribution of X. For example, we could have people within cities within counties within states within countries. □

Last but not least, we revisit Story 8.4.5, the Gamma-Poisson problem from the previous chapter.

Example 9.6.3 (Gamma-Poisson revisited). Recall that Fred decided to find out about the rate of Blotchville's Poisson process of buses by waiting at the bus stop for t hours and counting the number of buses Y. He then used the data to update his prior distribution $\lambda \sim \text{Gamma}(r_0, b_0)$. Thus, Fred was using the *two-level model*

$$\begin{aligned}\lambda &\sim \text{Gamma}(r_0, b_0) \\ Y|\lambda &\sim \text{Pois}(\lambda t).\end{aligned}$$

We found that under Fred's model, the marginal distribution of Y is Negative Binomial with parameters $r = r_0$ and $p = b_0/(b_0 + t)$. In particular,

$$\begin{aligned}E(Y) &= \frac{rq}{p} = \frac{r_0 t}{b_0}, \\ \text{Var}(Y) &= \frac{rq}{p^2} = \frac{r_0 t(b_0 + t)}{b_0^2}.\end{aligned}$$

Let's independently verify this with Adam's law and Eve's law. Using results about the Poisson distribution, the conditional mean and variance of Y given λ are $E(Y|\lambda) = \text{Var}(Y|\lambda) = \lambda t$. Using results about the Gamma distribution, the marginal mean and variance of λ are $E(\lambda) = r_0/b_0$ and $\text{Var}(\lambda) = r_0/b_0^2$. For Adam and Eve, this is all that is required:

$$\begin{aligned} E(Y) &= E(E(Y|\lambda)) = E(\lambda t) = \frac{r_0 t}{b_0}, \\ \text{Var}(Y) &= E(\text{Var}(Y|\lambda)) + \text{Var}(E(Y|\lambda)) \\ &= E(\lambda t) + \text{Var}(\lambda t) \\ &= \frac{r_0 t}{b_0} + \frac{r_0 t^2}{b_0^2} = \frac{r_0 t(b_0 + t)}{b_0^2}, \end{aligned}$$

which is consistent with our earlier answers. The difference is that when using Adam and Eve, we don't need to know that Y is Negative Binomial! If we had been too lazy to derive the marginal distribution of Y, or if we weren't so lucky as to have a named distribution for Y, Adam and Eve would still deliver the mean and variance of Y (though not the PMF).

Lastly, let's compare the mean and variance of Y under the two-level model to the mean and variance we would get if Fred were absolutely sure of the true value of λ. In other words, suppose we replaced λ by its mean, $E(\lambda) = r_0/b_0$, making λ a constant instead of an r.v. Then the marginal distribution of the number of buses (which we'll call $\tilde{Y}$ under the new assumptions) would just be Poisson with parameter $r_0 t/b_0$. Then we would have

$$\begin{aligned} E(\tilde{Y}) &= \frac{r_0 t}{b_0}, \\ \text{Var}(\tilde{Y}) &= \frac{r_0 t}{b_0}. \end{aligned}$$

Notice that $E(\tilde{Y}) = E(Y)$, but $\text{Var}(\tilde{Y}) < \text{Var}(Y)$: the extra term $r_0 t^2/b_0^2$ from Eve's law is missing. Intuitively, when we fix λ at its mean, we are eliminating a level of uncertainty in the model, and this causes a reduction in the unconditional variance.

Figure 9.9 overlays the plots of two PMFs, that of $Y \sim \text{NBin}(r_0, b_0/(b_0+t))$ in gray and that of $\tilde{Y} \sim \text{Pois}(r_0 t/b_0)$ in black. The values of the parameters are arbitrarily chosen to be $r_0 = 5$, $b_0 = 1$, $t = 2$. These two PMFs have the same center of mass, but the PMF of Y is noticeably more dispersed. □

9.7 Recap

To calculate an unconditional expectation, we can divide up the sample space and use the law of total expectation

$$E(Y) = \sum_{i=1}^{n} E(Y|A_i)P(A_i),$$

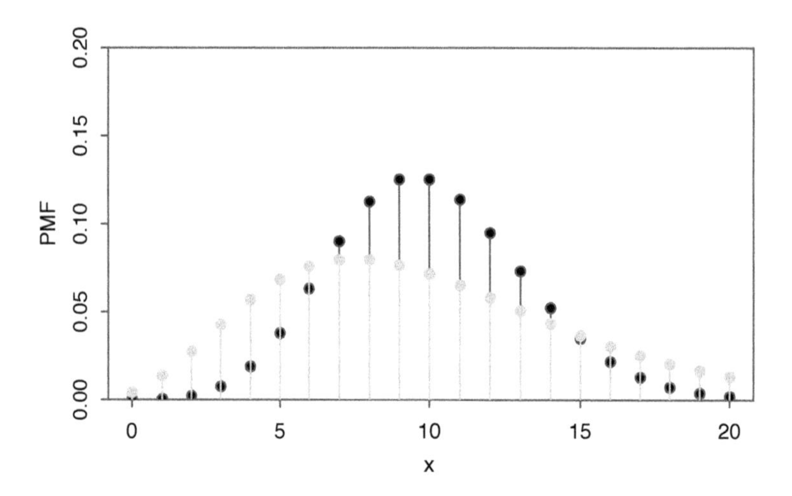

FIGURE 9.9
PMF of $Y \sim \text{NBin}(r_0, b_0/(b_0 + t))$ in gray and $\tilde{Y} \sim \text{Pois}(r_0 t/b_0)$ in black, where $r_0 = 5$, $b_0 = 1$, $t = 2$.

but we must be careful not to destroy information in subsequent steps (such as by forgetting in the midst of a long calculation to condition on something that needs to be conditioned on). In problems with a recursive structure, we can also use first-step analysis for expectations.

The conditional expectation $E(Y|X)$ and conditional variance $\text{Var}(Y|X)$ are random variables that are functions of X; they are obtained by treating X as if it were a known constant. If X and Y are independent, then $E(Y|X) = E(Y)$ and $\text{Var}(Y|X) = \text{Var}(Y)$. Conditional expectation has the properties

$$E(h(X)Y|X) = h(X)E(Y|X)$$
$$E(Y_1 + Y_2|X) = E(Y_1|X) + E(Y_2|X),$$

analogous to the properties $E(cY) = cE(Y)$ and $E(Y_1 + Y_2) = E(Y_1) + E(Y_2)$ for unconditional expectation. The conditional expectation $E(Y|X)$ is also the random variable that makes the residual $Y - E(Y|X)$ uncorrelated with any function of X, which means we can interpret it geometrically as a projection.

Finally, Adam's law and Eve's law,

$$E(Y) = E(E(Y|X))$$
$$\text{Var}(Y) = E(\text{Var}(Y|X)) + \text{Var}(E(Y|X)),$$

often help us calculate $E(Y)$ and $\text{Var}(Y)$ in problems that feature multiple forms or levels of randomness.

Figure 9.10 illustrates how the number $E(Y|X = x)$ connects with the r.v. $E(Y|X)$, whose expectation is $E(Y)$ by Adam's law. Additionally, it shows how the ingredients in Eve's law are formed and come together to give a useful decomposition of $\text{Var}(Y)$ in terms of quantities that condition on X.

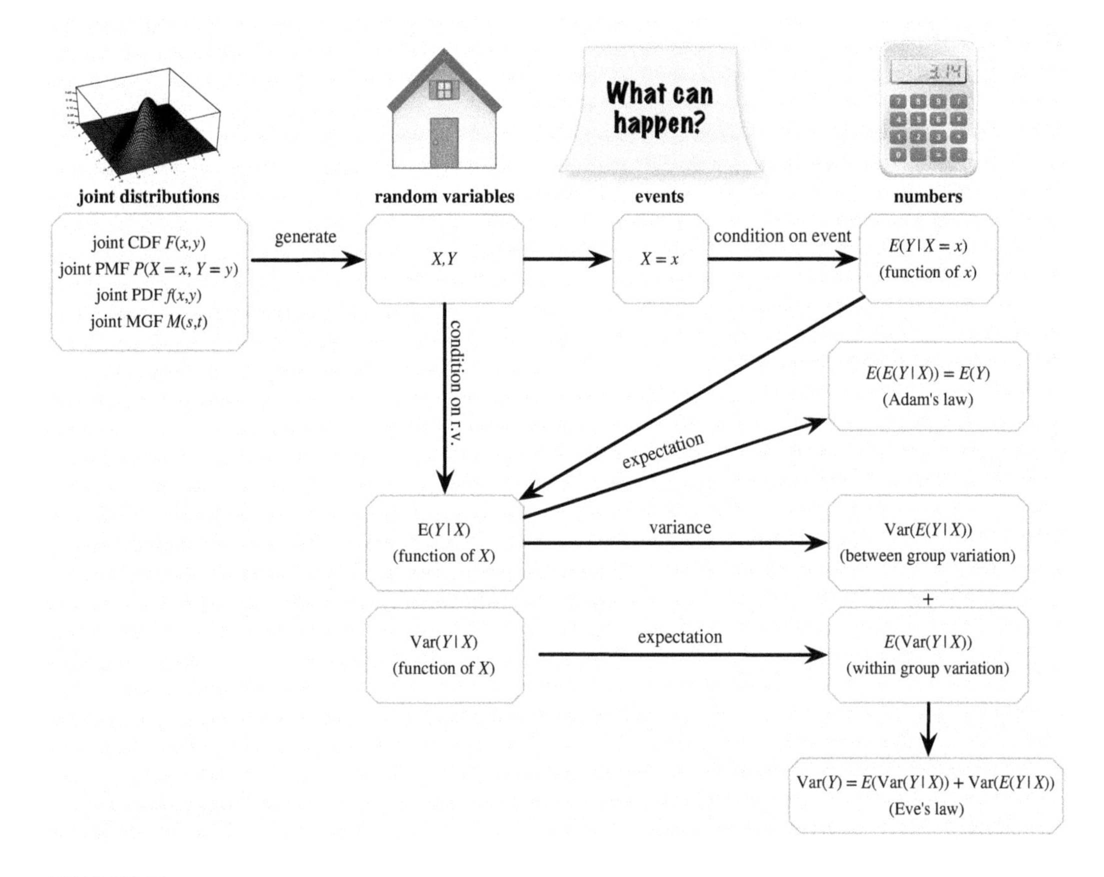

FIGURE 9.10
We often observe an r.v. X and want to predict another r.v. Y based on the information about X. If we observe that $X = x$, then we can condition on that event and use $E(Y|X = x)$ as our prediction. The conditional expectation $E(Y|X)$ is the r.v. that takes the value $E(Y|X = x)$ when $X = x$. Adam's law lets us compute the unconditional expectation $E(Y)$ by starting with the conditional expectation $E(Y|X)$. Similarly, Eve's law lets us compute $\text{Var}(Y)$ in terms of quantities that condition on X.

9.8 R

Mystery prize simulation

We can use simulation to show that in Example 9.1.7, the example of bidding on a mystery prize with unknown value, any bid will lead to a negative payout on average. First choose a bid `b` (we chose 0.6); then simulate a large number of hypothetical mystery prizes and store them in `v`:

```
b <- 0.6
```

```
nsim <- 10^5
v <- runif(nsim)
```

The bid is accepted if `b > (2/3)*v`. To get the average profit conditional on an accepted bid, we use square brackets to keep only those values of `v` satisfying the condition:

```
mean(v[b > (2/3)*v]) - b
```

This value is negative regardless of `b`, as you can check by experimenting with different values of `b`.

Time until *HH* vs. *HT*

To verify the results of Example 9.1.9, we can start by generating a long sequence of fair coin tosses. This is done with the `sample` command. We use `paste` with the `collapse=""` argument to turn these tosses into a single string of H's and T's:

```
paste(sample(c("H","T"),100,replace=TRUE),collapse="")
```

A sequence of length 100 is enough to virtually guarantee that both *HH* and *HT* will have appeared at least once.

To determine how many tosses are required on average to see *HH* and *HT*, we need to generate many sequences of coin tosses. For this, we use our familiar friend `replicate`:

```
r <- replicate(10^3,paste(sample(c("H","T"),100,replace=T),
                          collapse=""))
```

Now `r` contains a thousand sequences of coin tosses, each of length 100. To find the first appearance of *HH* in each of these sequences, you can use the `str_locate` command from the `stringr` package. After you've installed and loaded the package,

```
t <- str_locate(r,"HH")
```

creates a two-column table `t`, whose columns contain the starting and ending positions of the first appearance of *HH* in each sequence of coin tosses. (Use `head(t)` to display the first few rows of the table and get an idea of what your results look like.) What we want are the ending positions, given by the second column. In particular, we want the average value of the second column, which is an approximation of the average waiting time for *HH*:

```
mean(t[,2])
```

Is your answer around 6? Trying again with `"HT"` instead of `"HH"`, is your answer around 4?

Linear regression

In Example 9.3.10, we derived formulas for the slope and intercept of a linear regression model, which can be used to predict a response variable using an explanatory variable. Let's try to apply these formulas to a simulated dataset:

```
x <- rnorm(100)
y <- 3 + 5*x + rnorm(100)
```

The vector `x` contains 100 realizations of the random variable $X \sim \mathcal{N}(0,1)$, and the vector `y` contains 100 realizations of the random variable $Y = a + bX + \epsilon$ where $\epsilon \sim \mathcal{N}(0,1)$. As we can see, the true values of a and b for this dataset are 3 and 5, respectively. We can visualize the data as a scatterplot with `plot(x,y)`.

Now let's see if we can get good estimates of the true a and b, using the formulas in Example 9.3.10:

```
b <- cov(x,y) / var(x)
a <- mean(y) - b*mean(x)
```

Here `cov(x,y)`, `var(x)`, and `mean(x)` provide the sample covariance, sample variance, and sample mean, estimating the quantities $\text{Cov}(X,Y)$, $\text{Var}(X)$, and $E(X)$, respectively. (We have discussed sample mean and sample variance in detail in earlier chapters. Sample covariance is defined analogously, and is a natural way to estimate the true covariance.)

You should find that `b` is close to 5 and `a` is close to 3. These estimated values define the *line of best fit*. The `abline` command lets us plot the line of best fit on top of our scatterplot:

```
plot(x,y)
abline(a=a,b=b)
```

The first argument to `abline` is the intercept of the line, and the second argument is the slope.

9.9 Exercises

Exercises marked with Ⓢ have detailed solutions at `http://stat110.net`.

Conditional expectation given an event

1. Fred wants to travel from Blotchville to Blissville, and is deciding between 3 options (involving different routes or different forms of transportation). The jth option would take an average of μ_j hours, with a standard deviation of σ_j hours. Fred randomly

chooses between the 3 options, with equal probabilities. Let T be how long it takes for him to get from Blotchville to Blissville.

(a) Find $E(T)$. Is it simply $(\mu_1 + \mu_2 + \mu_3)/3$, the average of the expectations?

(b) Find $\text{Var}(T)$. Is it simply $(\sigma_1^2 + \sigma_2^2 + \sigma_3^2)/3$, the average of the variances?

2. While Fred is sleeping one night, X legitimate emails and Y spam emails are sent to him. Suppose that X and Y are independent, with $X \sim \text{Pois}(10)$ and $Y \sim \text{Pois}(40)$. When he wakes up, he observes that he has 30 new emails in his inbox. Given this information, what is the expected value of how many new legitimate emails he has?

3. A group of 21 women and 14 men are enrolled in a medical study. Each of them has a certain disease with probability p, independently. It is then found (through extremely reliable testing) that exactly 5 of the people have the disease. Given this information, what is the expected number of women who have the disease?

4. A researcher studying crime is interested in how often people have gotten arrested. Let $X \sim \text{Pois}(\lambda)$ be the number of times that a random person got arrested in the last 10 years. However, data from police records are being used for the researcher's study, and people who were never arrested in the last 10 years do not appear in the records. In other words, the police records have a *selection bias*: they only contain information on people who *have* been arrested in the last 10 years.

   So averaging the numbers of arrests for people in the police records does not directly estimate $E(X)$; it makes more sense to think of the police records as giving us information about the *conditional* distribution of how many times a person was arrest, given that the person was arrested at least once in the last 10 years. The conditional distribution of X, given that $X \geq 1$, is called a *truncated Poisson distribution* (see Exercise 14 from Chapter 3 for another example of this distribution).

   (a) Find $E(X|X \geq 1)$

   (b) Find $\text{Var}(X|X \geq 1)$.

5. A fair 20-sided die is rolled repeatedly, until a gambler decides to stop. The gambler pays \$1 per roll, and receives the amount shown on the die when the gambler stops (e.g., if the die is rolled 7 times and the gambler decides to stop then, with an 18 as the value of the last roll, then the net payoff is \$18 − \$7 = \$11). Suppose the gambler uses the following strategy: keep rolling until a value of m or greater is obtained, and then stop (where m is a fixed integer between 1 and 20).

   (a) What is the expected net payoff?

   Hint: The average of consecutive integers $a, a+1, \dots, a+n$ is the same as the average of the first and last of these. See the math appendix for more information about series.

   (b) Use R or other software to find the optimal value of m.

6. Let $X \sim \text{Expo}(\lambda)$. Find $E(X|X < 1)$ in two different ways:

   (a) by calculus, working with the conditional PDF of X given $X < 1$.

   (b) without calculus, by expanding $E(X)$ using the law of total expectation.

7. You are given an opportunity to bid on a mystery box containing a mystery prize! The value of the prize is completely unknown, except that it is worth at least nothing, and at most a million dollars. So the true value V of the prize is considered to be Uniform on [0,1] (measured in millions of dollars).

   You can choose to bid any nonnegative amount b (in millions of dollars). If $b < \frac{1}{4}V$, then your bid is rejected and nothing is gained or lost. If $b \geq \frac{1}{4}V$, then your bid is accepted and your net payoff is $V - b$ (since you pay b to get a prize worth V).

Find your expected payoff as a function of b (be sure to specify it for all $b \geq 0$). Then find the optimal bid b, to maximize your expected payoff.

8. Ⓢ You get to choose between two envelopes, each of which contains a check for some positive amount of money. Unlike in the two-envelope paradox, it is not given that one envelope contains twice as much money as the other envelope. Instead, assume that the two values were generated independently from some distribution on the positive real numbers, with no information given about what that distribution is.

   After picking an envelope, you can open it and see how much money is inside (call this value x), and then you have the option of switching. As no information has been given about the distribution, it may seem impossible to have better than a 50% chance of picking the better envelope. Intuitively, we may want to switch if x is "small" and not switch if x is "large", but how do we define "small" and "large" in the grand scheme of all possible distributions? [The last sentence was a rhetorical question.]

   Consider the following strategy for deciding whether to switch. Generate a threshold $T \sim \text{Expo}(1)$, and switch envelopes if and only if the observed value x is less than the value of T. Show that this strategy succeeds in picking the envelope with more money with probability strictly greater than $1/2$.

   Hint: Let t be the value of T (generated by a random draw from the Expo(1) distribution). First explain why the strategy works very well if t happens to be in between the two envelope values, and does no harm in any case (i.e., there is no case in which the strategy succeeds with probability strictly less than $1/2$).

9. There are two envelopes, each of which has a check for a Unif$(0, 1)$ amount of money, measured in thousands of dollars. The amounts in the two envelopes are independent. You get to choose an envelope and open it, and then you can either keep that amount or switch to the other envelope and get whatever amount is in that envelope.

   Suppose that you use the following strategy: choose an envelope and open it. If you observe U, then stick with that envelope with probability U, and switch to the other envelope with probability $1 - U$.

   (a) Find the probability that you get the larger of the two amounts.

   (b) Find the expected value of what you will receive.

10. Suppose n people are bidding on a mystery prize that is up for auction. The bids are to be submitted in secret, and the individual who submits the highest bid wins the prize. The ith bidder receives a signal X_i, with $X_1, \dots, X_n$ i.i.d. The value of the prize, V, is defined to be the sum of the individual bidders' signals:

$$V = X_1 + \cdots + X_n.$$

   This is known in economics as the *wallet game*: we can imagine that the n people are bidding on the total amount of money in their wallets, and each person's signal is the amount of money in their own wallet. Of course, the wallet is a metaphor; the game can also be used to model company takeovers, where each of two companies bids to take over the other, and a company knows its own value but not the value of the other company.

   For this problem, assume the X_i are i.i.d. Unif$(0, 1)$.

   (a) Before receiving her signal, what is bidder 1's unconditional expectation for V?

   (b) Conditional on receiving the signal $X_1 = x_1$, what is bidder 1's expectation for V?

   (c) Suppose each bidder submits a bid equal to their conditional expectation for V, i.e., bidder i bids $E(V|X_i = x_i)$. Conditional on receiving the signal $X_1 = x_1$ and *winning the auction*, what is bidder 1's expectation for V? Explain intuitively why this quantity is always less than the quantity calculated in (b).

11. Ⓢ A coin with probability p of Heads is flipped repeatedly. For (a) and (b), suppose that p is a known constant, with $0 < p < 1$.

(a) What is the expected number of flips until the pattern HT is observed?

(b) What is the expected number of flips until the pattern HH is observed?

(c) Now suppose that p is unknown, and that we use a Beta(a, b) prior to reflect our uncertainty about p (where a and b are known constants and are greater than 2). In terms of a and b, find the corresponding answers to (a) and (b) in this setting.

12. A coin with probability p of Heads is flipped repeatedly, where $0 < p < 1$. The sequence of outcomes can be divided into *runs* (blocks of H's or blocks of T's), e.g., $HHHTTTTHTTTHH$ becomes $\boxed{HHH}\boxed{TTTT}\boxed{H}\boxed{TTT}\boxed{HH}$, which has 5 runs, with lengths $3, 4, 1, 3, 2$, respectively. Assume that the coin is flipped at least until the start of the third run.

(a) Find the expected length of the first run.

(b) Find the expected length of the second run.

13. A fair 6-sided die is rolled once. Find the expected number of additional rolls needed to obtain a value at least as large as that of the first roll.

14. A fair 6-sided die is rolled repeatedly.

(a) Find the expected number of rolls needed to get a 1 followed right away by a 2.

Hint: Start by conditioning on whether or not the first roll is a 1.

(b) Find the expected number of rolls needed to get two consecutive 1's.

(c) Let a_n be the expected number of rolls needed to get the same value n times in a row (i.e., to obtain a streak of n consecutive j's for some not-specified-in-advance value of j). Find a recursive formula for a_{n+1} in terms of a_n.

Hint: Divide the time until there are $n + 1$ consecutive appearances of the same value into two pieces: the time until there are n consecutive appearances, and the rest.

(d) Find a simple, explicit formula for a_n for all $n \geq 1$. What is a_7 (numerically)?

Conditional expectation given a random variable

15. Ⓢ Let X_1, X_2 be i.i.d., and let $\bar{X} = \frac{1}{2}(X_1 + X_2)$ be the sample mean. In many statistics problems, it is useful or important to obtain a conditional expectation given $\bar{X}$. As an example of this, find $E(w_1X_1 + w_2X_2|\bar{X})$, where w_1, w_2 are constants with $w_1 + w_2 = 1$.

16. Let $X_1, X_2, \ldots$ be i.i.d. r.v.s with mean 0, and let $S_n = X_1 + \cdots + X_n$. As shown in Example 9.3.6, the expected value of the first term given the sum of the first n terms is

$$E(X_1|S_n) = \frac{S_n}{n}.$$

Generalize this result by finding $E(S_k|S_n)$ for all positive integers k and n.

17. Ⓢ Consider a group of n roommate pairs at a college (so there are $2n$ students). Each of these $2n$ students independently decides randomly whether to take a certain course, with probability p of success (where "success" is defined as taking the course).

Let N be the number of students among these $2n$ who take the course, and let X be the number of roommate pairs where both roommates in the pair take the course. Find $E(X)$ and $E(X|N)$.

18. Ⓢ Show that $E((Y - E(Y|X))^2|X) = E(Y^2|X) - (E(Y|X))^2$, so these two expressions for $\text{Var}(Y|X)$ agree.

    Hint for the variance: Adding a constant (or something acting as a constant) does not affect variance.

19. Let X be the height of a randomly chosen adult man, and Y be his father's height, where X and Y have been standardized to have mean 0 and standard deviation 1. Suppose that (X, Y) is Bivariate Normal, with $X, Y \sim \mathcal{N}(0, 1)$ and $\text{Corr}(X, Y) = \rho$.

    (a) Let $y = ax + b$ be the equation of the best line for predicting Y from X (in the sense of minimizing the mean squared error), e.g., if we were to observe $X = 1.3$ then we would predict that Y is $1.3a + b$. Now suppose that we want to use Y to predict X, rather than using X to predict Y. Give and explain an *intuitive guess* for what the slope is of the best line for predicting X from Y.

    (b) Find a constant c (in terms of ρ) and an r.v. V such that $Y = cX + V$, with V independent of X.

    Hint: Start by finding c such that $\text{Cov}(X, Y - cX) = 0$.

    (c) Find a constant d (in terms of ρ) and an r.v. W such that $X = dY + W$, with W independent of Y.

    (d) Find $E(Y|X)$ and $E(X|Y)$.

    (e) Reconcile (a) and (d), if your intuitive guess in (a) differed from what the results of (d) implied. Give a clear and correct intuitive explanation of the relationship between the slope of the best line for predicting Y from X and the slope of the best line for predicting X from Y.

20. Let $\mathbf{X} \sim \text{Mult}_5(n, \mathbf{p})$.

    (a) Find $E(X_1|X_2)$ and $\text{Var}(X_1|X_2)$.

    (b) Find $E(X_1|X_2 + X_3)$.

21. Let Y be a discrete r.v., A be an event with $0 < P(A) < 1$, and I_A be the indicator r.v. for A.

    (a) Explain precisely how the r.v. $E(Y|I_A)$ relates to the numbers $E(Y|A)$ and $E(Y|A^c)$.

    (b) Show that $E(Y|A) = E(YI_A)/P(A)$, directly from the definitions of expectation and conditional expectation.

    Hint: Let $X = YI_A$, and then find an expression for the PMF of X.

    (c) Use (b) to give a short proof of the fact that $E(Y) = E(Y|A)P(A) + E(Y|A^c)P(A^c)$.

22. Show that the following version of LOTP, which we encountered in Section 7.1, is also a consequence of Adam's law: for any event A and continuous r.v. X with PDF f_X,

$$P(A) = \int_{-\infty}^{\infty} P(A|X = x) f_X(x) dx.$$

    Hint: Consider $E(I(A)|X = x)$.

23. Ⓢ Let X and Y be random variables with finite variances, and let $W = Y - E(Y|X)$. This is a *residual*: the difference between the true value of Y and the predicted value of Y based on X.

    (a) Compute $E(W)$ and $E(W|X)$.

    (b) Compute $\text{Var}(W)$, for the case that $W|X \sim \mathcal{N}(0, X^2)$ with $X \sim \mathcal{N}(0, 1)$.

24. Ⓢ One of two identical-looking coins is picked from a hat randomly, where one coin has probability p_1 of Heads and the other has probability p_2 of Heads. Let X be the number of Heads after flipping the chosen coin n times. Find the mean and variance of X.

25. Kelly makes a series of n bets, each of which she has probability p of winning, independently. Initially, she has x_0 dollars. Let X_j be the amount she has immediately after her jth bet is settled. Let f be a constant in $(0, 1)$, called the *betting fraction*. On each bet, Kelly wagers a fraction f of her wealth, and then she either wins or loses that amount. For example, if her current wealth is \$100 and $f = 0.25$, then she bets \$25 and either gains or loses that amount. (A famous choice when $p > 1/2$ is $f = 2p - 1$, which is known as the *Kelly criterion*.) Find $E(X_n)$ (in terms of n, p, f, x_0).

    Hint: First find $E(X_{j+1}|X_j)$.

26. Let $N \sim \text{Pois}(\lambda_1)$ be the number of movies that will be released next year. Suppose that for each movie the number of tickets sold is $\text{Pois}(\lambda_2)$, independent of other movies and of N. Find the mean and variance of the number of movie tickets that will be sold next year.

27. A party is being held from 8:00 pm to midnight on a certain night, and $N \sim \text{Pois}(\lambda)$ people are going to show up. They will all arrive at uniformly random times while the party is going on, independently of each other and of N.

    (a) Find the expected time at which the first person arrives, given that at least one person shows up. Give both an exact answer in terms of λ, measured in minutes after 8:00 pm, and an answer rounded to the nearest minute for $\lambda = 20$, expressed in time notation (e.g., 8:20 pm).

    (b) Find the expected time at which the last person arrives, given that at least one person shows up. As in (a), give both an exact answer and an answer rounded to the nearest minute for $\lambda = 20$.

28. Ⓢ We wish to estimate an unknown parameter θ, based on an r.v. X we will get to observe. As in the Bayesian perspective, assume that X and θ have a joint distribution. Let $\hat{\theta}$ be the estimator (which is a function of X). Then $\hat{\theta}$ is said to be *unbiased* if $E(\hat{\theta}|\theta) = \theta$, and $\hat{\theta}$ is said to be the *Bayes procedure* if $E(\theta|X) = \hat{\theta}$.

    (a) Let $\hat{\theta}$ be unbiased. Find $E(\hat{\theta} - \theta)^2$ (the average squared difference between the estimator and the true value of θ), in terms of marginal moments of $\hat{\theta}$ and θ.

    Hint: Condition on θ.

    (b) Repeat (a), except in this part suppose that $\hat{\theta}$ is the *Bayes procedure* rather than assuming that it is unbiased.

    Hint: Condition on X.

    (c) Show that it is *impossible* for $\hat{\theta}$ to be both the Bayes procedure and unbiased, except in silly problems where we get to know θ perfectly by observing X.

    Hint: If Y is a nonnegative r.v. with mean 0, then $P(Y = 0) = 1$.

29. Show that if $E(Y|X) = c$ is a constant, then X and Y are uncorrelated.

    Hint: Use Adam's law to find $E(Y)$ and $E(XY)$.

30. Show by example that it is possible to have uncorrelated X and Y such that $E(Y|X)$ is not a constant.

    Hint: Consider a standard Normal and its square.

31. Ⓢ Emails arrive one at a time in an inbox. Let T_n be the time at which the nth email arrives (measured on a continuous scale from some starting point in time). Suppose that the waiting times between emails are i.i.d. $\text{Expo}(\lambda)$, i.e., $T_1, T_2 - T_1, T_3 - T_2, \ldots$ are i.i.d. $\text{Expo}(\lambda)$.

Each email is non-spam with probability p, and spam with probability $q = 1 - p$ (independently of the other emails and of the waiting times). Let X be the time at which the first non-spam email arrives (so X is a continuous r.v., with $X = T_1$ if the 1st email is non-spam, $X = T_2$ if the 1st email is spam but the 2nd one isn't, etc.).

(a) Find the mean and variance of X.

(b) Find the MGF of X. What famous distribution does this imply that X has (be sure to state its parameter values)?

Hint for both parts: Let N be the number of emails until the first non-spam (including that one), and write X as a sum of N terms; then condition on N.

32. Customers arrive at a store according to a Poisson process of rate λ customers per hour. Each makes a purchase with probability p, independently. Given that a customer makes a purchase, the amount spent has mean μ (in dollars) and variance σ^2.

(a) Find the mean and variance of how much a random customer spends (note that the customer may spend nothing).

(b) Find the mean and variance of the revenue the store obtains in an 8-hour time interval, using (a) and results from this chapter.

(c) Find the mean and variance of the revenue the store obtains in an 8-hour time interval, using the chicken-egg story and results from this chapter.

33. Fred's beloved computer will last an Expo(λ) amount of time until it has a malfunction. When that happens, Fred will try to get it fixed. With probability p, he will be able to get it fixed. If he is able to get it fixed, the computer is good as new again and will last an additional, independent Expo(λ) amount of time until the next malfunction (when again he is able to get it fixed with probability p, and so on). If after any malfunction Fred is unable to get it fixed, he will buy a new computer. Find the expected amount of time until Fred buys a new computer. (Assume that the time spent on computer diagnosis, repair, and shopping is negligible.)

34. A green die is rolled until it lands 1 for the first time. An orange die is rolled until it lands 6 for the first time. The dice are fair, six-sided dice. Let T_1 be the sum of the values of the rolls of the green die (including the 1 at the end) and T_6 be the sum of the values of the rolls of the orange die (including the 6 at the end). Two students are debating whether $E(T_1) = E(T_6)$ or $E(T_1) < E(T_6)$. They kindly gave permission to quote their arguments here.

Student A: We have $E(T_1) = E(T_6)$. By Adam's law, the expected sum of the rolls of a die is the expected number of rolls times the expected value of one roll, and each of these factors is the same for the two dice. In more detail, let N_1 be the number of rolls of the green die and N_6 be the number of rolls of the orange die. By Adam's law and linearity,

$$E(T_1) = E(E(T_1|N_1)) = E(3.5N_1) = 3.5E(N_1),$$

and the same method applied to the orange die gives $3.5E(N_6)$, which equals $3.5E(N_1)$.

Student B: Actually, $E(T_1) < E(T_6)$. I agree that the expected number of rolls is the same for the two dice, but the key difference is that we *know* the last roll is a 1 for the green die and a 6 for the orange die. The expected totals are the same for the two dice *excluding* the last roll of each, and then including the last roll makes $E(T_1) < E(T_6)$.

(a) Discuss in words the extent to which Student A's argument is convincing and correct.

(b) Discuss in words the extent to which Student B's argument is convincing and correct.

(c) Give careful derivations of $E(T_1)$ and $E(T_6)$.

35. Ⓢ Judit plays in a total of $N \sim \text{Geom}(s)$ chess tournaments in her career. Suppose that in each tournament she has probability p of winning the tournament, independently. Let T be the number of tournaments she wins in her career.

(a) Find the mean and variance of T.

(b) Find the MGF of T. What is the name of this distribution (with its parameters)?

36. In Story 8.4.5, we showed (among other things) that if $\lambda \sim \text{Gamma}(r_0, b_0)$ and $Y|\lambda \sim \text{Pois}(\lambda)$, then the marginal distribution of Y is $\text{NBin}(r_0, b_0/(b_0+1))$. Derive this result using Adam's law and MGFs.

Hint: Consider the conditional MGF of $Y|\lambda$.

37. Let $X_1, \dots, X_n$ be i.i.d. r.v.s with mean μ and variance σ^2, and $n \geq 2$. A *bootstrap sample* of $X_1, \dots, X_n$ is a sample of n r.v.s $X_1^*, \dots, X_n^*$ formed from the X_j by sampling with replacement with equal probabilities. Let $\bar{X}^*$ denote the sample mean of the bootstrap sample:

$$\bar{X}^* = \frac{1}{n}\left(X_1^* + \cdots + X_n^*\right).$$

(a) Calculate $E(X_j^*)$ and $\text{Var}(X_j^*)$ for each j.

(b) Calculate $E(\bar{X}^*|X_1, \dots, X_n)$ and $\text{Var}(\bar{X}^*|X_1, \dots, X_n)$.
Hint: Conditional on $X_1, \dots, X_n$, the X_j^* are independent, with a PMF that puts probability $1/n$ at each of the points $X_1, \dots, X_n$. As a check, your answers should be random variables that are functions of $X_1, \dots, X_n$.

(c) Calculate $E(\bar{X}^*)$ and $\text{Var}(\bar{X}^*)$.

(d) Explain intuitively why $\text{Var}(\bar{X}) < \text{Var}(\bar{X}^*)$.

38. An insurance company covers disasters in two neighboring regions, R_1 and R_2. Let I_1 and I_2 be the indicator r.v.s for whether R_1 and R_2 are hit by the insured disaster, respectively. The indicators I_1 and I_2 may be dependent. Let $p_j = E(I_j)$ for $j = 1, 2$, and $p_{12} = E(I_1 I_2)$.

The company reimburses a total cost of

$$C = I_1 \cdot T_1 + I_2 \cdot T_2$$

to these regions, where T_j has mean μ_j and variance σ_j^2. Assume that T_1 and T_2 are independent of each other and that (T_1, T_2) is independent of (I_1, I_2).

(a) Find $E(C)$.

(b) Find $\text{Var}(C)$.

39. Ⓢ A certain stock has low volatility on some days and high volatility on other days. Suppose that the probability of a low volatility day is p and of a high volatility day is $q = 1 - p$, and that on low volatility days the percent change in the stock price is $\mathcal{N}(0, \sigma_1^2)$, while on high volatility days the percent change is $\mathcal{N}(0, \sigma_2^2)$, with $\sigma_1 < \sigma_2$.

Let X be the percent change of the stock on a certain day. The distribution is said to be a *mixture* of two Normal distributions, and a convenient way to represent X is as $X = I_1 X_1 + I_2 X_2$ where I_1 is the indicator r.v. of having a low volatility day, $I_2 = 1 - I_1$, $X_j \sim \mathcal{N}(0, \sigma_j^2)$, and I_1, X_1, X_2 are independent.

(a) Find $\text{Var}(X)$ in two ways: using Eve's law, and by using properties of covariance to calculate $\text{Cov}(I_1 X_1 + I_2 X_2, I_1 X_1 + I_2 X_2)$.

(b) Recall from Chapter 6 that the *kurtosis* of an r.v. Y with mean μ and standard deviation σ is defined by

$$\text{Kurt}(Y) = \frac{E(Y-\mu)^4}{\sigma^4} - 3.$$

Find the kurtosis of X (in terms of $p, q, \sigma_1^2, \sigma_2^2$, fully simplified). The result will show that even though the kurtosis of any Normal distribution is 0, the kurtosis of X is positive and in fact can be very large depending on the parameter values.

40. Let X_1, X_2, and Y be random variables, such that Y has finite variance. Let

$$A = E(Y|X_1) \text{ and } B = E(Y|X_1, X_2).$$

Show that

$$\text{Var}(A) \leq \text{Var}(B).$$

Also, check that this make sense in the extreme cases where Y is independent of X_1 and where $Y = h(X_2)$ for some function h.

Hint: Use Eve's law on B.

41. Show that for any r.v.s X and Y,

$$E(Y|E(Y|X)) = E(Y|X).$$

This has a nice intuitive interpretation if we think of $E(Y|X)$ as the prediction we would make for Y based on X: given the prediction we would use for predicting Y from X, we no longer need to know X to predict Y—we can just use the prediction we have! For example, letting $E(Y|X) = g(X)$, if we observe $g(X) = 7$, then we may or may not know what X is (since g may not be one-to-one). But even without knowing X, we know that the prediction for Y based on X is 7.

Hint: Use Adam's law with extra conditioning.

42. A researcher wishes to know whether a new treatment for the disease conditionitis is more effective than the standard treatment. It is unfortunately not feasible to do a randomized experiment, but the researcher does have the medical records of patients who received the new treatment and those who received the standard treatment. She is worried, though, that doctors tend to give the new treatment to younger, healthier patients. If this is the case, then naively comparing the outcomes of patients in the two groups would be like comparing apples and oranges.

Suppose each patient has background variables $\mathbf{X}$, which might be age, height and weight, and measurements relating to previous health status. Let Z be the indicator of receiving the new treatment. The researcher fears that Z is dependent on $\mathbf{X}$, i.e., that the distribution of $\mathbf{X}$ given $Z = 1$ is different from the distribution of $\mathbf{X}$ given $Z = 0$.

In order to compare apples to apples, the researcher wants to match every patient who received the new treatment to a patient with similar background variables who received the standard treatment. But $\mathbf{X}$ could be a high-dimensional random vector, which often makes it very difficult to find a match with a similar value of $\mathbf{X}$.

The *propensity score* reduces the possibly high-dimensional vector of background variables down to a single number (then it is much easier to match someone to a person with a similar propensity score than to match someone to a person with a similar value of $\mathbf{X}$). The propensity score of a person with background characteristics $\mathbf{X}$ is defined as

$$S = E(Z|\mathbf{X}).$$

By the fundamental bridge, a person's propensity score is their probability of receiving the treatment, given their background characteristics. Show that conditional on S, the treatment indicator Z is independent of the background variables $\mathbf{X}$.

Hint: This problem relates to the previous one. Show that $P(Z = 1|S, \mathbf{X}) = P(Z = 1|S)$, which is equivalent to showing $E(Z|S, \mathbf{X}) = E(Z|S)$.

43. This exercise develops a useful identity for covariance, similar in spirit to Adam's law for expectation and Eve's law for variance. First define *conditional covariance* in a manner analogous to how we defined conditional variance:

$$\text{Cov}(X, Y|Z) = E\big((X - E(X|Z))(Y - E(Y|Z))|Z\big).$$

(a) Show that

$$\text{Cov}(X, Y|Z) = E(XY|Z) - E(X|Z)E(Y|Z).$$

This should be true since it is the conditional version of the fact that

$$\text{Cov}(X, Y) = E(XY) - E(X)E(Y)$$

and conditional probabilities *are* probabilities, but for this problem you should prove it directly using properties of expectation and conditional expectation.

(b) *ECCE*, or the *law of total covariance*, says that

$$\text{Cov}(X, Y) = E(\text{Cov}(X, Y|Z)) + \text{Cov}(E(X|Z), E(Y|Z)).$$

That is, the covariance of X and Y is the expected value of their conditional covariance plus the covariance of their conditional expectations, where all these conditional quantities are conditional on Z. Prove this identity.

Hint: We can assume without loss of generality that $E(X) = E(Y) = 0$, since adding a constant to an r.v. has no effect on its covariance with any r.v. Then expand out the covariances on the right-hand side of the identity and apply Adam's law.

Mixed practice

44. A group of n friends often go out for dinner together. At their dinners, they play "credit card roulette" to decide who pays the bill. This means that at each dinner, one person is chosen uniformly at random to pay the entire bill (independently of what happens at the other dinners).

    (a) Find the probability that in k dinners, no one will have to pay the bill more than once (do not simplify for the case $k \leq n$, but do simplify fully for the case $k > n$).

    (b) Find the expected number of dinners it takes in order for everyone to have paid at least once (you can leave your answer as a finite sum of simple-looking terms).

    (c) Alice and Bob are two of the friends. Find the covariance between how many times Alice pays and how many times Bob pays in k dinners.

45. As in the previous problem, a group of n friends play "credit card roulette" at their dinners. In this problem, let the number of dinners be a Pois(λ) r.v.

    (a) Alice is one of the friends. Find the correlation between how many dinners Alice pays for and how many free dinners Alice gets.

    (b) The costs of the dinners are i.i.d. Gamma(a, b) r.v.s, independent of the number of dinners. Find the mean and variance of the total cost.

46. Joe will read $N \sim \text{Pois}(\lambda)$ books next year. Each book has a Pois(μ) number of pages, with book lengths independent of each other and independent of N.

    (a) Find the expected number of book pages that Joe will read next year.

    (b) Find the variance of the number of book pages Joe will read next year.

    (c) For each of the N books, Joe likes it with probability p and dislikes it with probability $1-p$, independently. Find the conditional distribution of how many of the N books Joe likes, given that he dislikes exactly d of the books.

47. Buses arrive at a certain bus stop according to a Poisson process of rate λ. Each bus has n seats and, at the instant when it arrives at the stop, has a Bin(n, p) number of passengers. Assume that the numbers of passengers on different buses are independent of each other, and independent of the arrival times of the buses.

Let N_t be the number of buses that arrive in the time interval $[0, t]$, and X_t be the total number of passengers on the buses that arrive in the time interval $[0, t]$.

(a) Find the mean and variance of N_t.

(b) Find the mean and variance of X_t.

(c) A bus is *full* if it has exactly n passengers when it arrives at the stop. Find the probability that exactly $a + b$ buses arrive in $[0, t]$, of which a are full and b are not full.

48. Paul and n other runners compete in a marathon. Their times are independent continuous r.v.s with CDF F.

(a) For $j = 1, 2, \ldots, n$, let A_j be the event that anonymous runner j completes the race faster than Paul. Explain whether the events A_j are independent, and whether they are conditionally independent given Paul's time to finish the race.

(b) For the rest of this problem, let N be the number of runners who finish faster than Paul. Find $E(N)$. (Your answer should depend only on n, since Paul's time is an r.v.)

(c) Find the conditional distribution of N, given that Paul's time to finish the marathon is t.

(d) Find $\text{Var}(N)$. (Your answer should depend only on n, since Paul's time is an r.v.)

Hint: Let T be Paul's time, and use Eve's law to condition on T. Alternatively, use indicator r.v.s.

49. Emails arrive in an inbox according to a Poisson process of rate λ emails per hour.

(a) Find the name and parameters of the conditional distribution of the number of emails that arrive in the first 2 hours of an 8-hour time period, given that exactly n emails arrive in that time period.

(b) Each email is legitimate with probability p and spam with probability $q = 1 - p$, independently. Find the name and parameters of the conditional distribution of the number of legitimate emails that arrive in an 8-hour time period, given that exactly s spams arrived in that time period.

(c) Reading an email takes a random amount of time, with mean μ hours and standard deviation σ hours. These reading times are i.i.d. and independent of the email arrival process. Find the (unconditional) mean and variance of the total time it takes to read all the emails that arrive in an 8-hour time period.

50. An actuary wishes to estimate various quantities related to the number of insurance claims and the dollar amounts of those claims for someone named Fred. Suppose that Fred will make N claims next year, where $N|\lambda \sim \text{Pois}(\lambda)$. But λ is unknown, so the actuary, taking a Bayesian approach, gives λ a prior distribution based on past experience. Specifically, the prior is $\lambda \sim \text{Expo}(1)$. The dollar amount of a claim is Log-Normal with parameters μ and σ^2 (here μ and σ^2 are the mean and variance of the underlying Normal), with μ and σ^2 known. The dollar amounts of the claims are i.i.d. and independent of N.

(a) Find $E(N)$ and $\text{Var}(N)$ using properties of conditional expectation (your answers should not depend on λ, since λ is unknown and being treated as an r.v.!).

(b) Find the mean and variance of the total dollar amount of all the claims.

(c) Find the distribution of N. If it is a named distribution we have studied, give its name and parameters.

(d) Find the posterior distribution of λ, given that it is observed that Fred makes $N = n$ claims next year. If it is a named distribution we have studied, give its name and parameters.

51. Ⓢ Let X_1, X_2, X_3 be independent with $X_i \sim \text{Expo}(\lambda_i)$ (so with possibly different rates). Recall from Chapter 7 that

$$P(X_1 < X_2) = \frac{\lambda_1}{\lambda_1 + \lambda_2}.$$

(a) Find $E(X_1 + X_2 + X_3 | X_1 > 1, X_2 > 2, X_3 > 3)$ in terms of $\lambda_1, \lambda_2, \lambda_3$.

(b) Find $P(X_1 = \min(X_1, X_2, X_3))$, the probability that the first of the three Exponentials is the smallest.

Hint: Restate this in terms of X_1 and $\min(X_2, X_3)$.

(c) For the case $\lambda_1 = \lambda_2 = \lambda_3 = 1$, find the PDF of $\max(X_1, X_2, X_3)$. Is this one of the important distributions we have studied?

52. Ⓢ A task is randomly assigned to one of two people (with probability 1/2 for each person). If assigned to the first person, the task takes an $\text{Expo}(\lambda_1)$ length of time to complete (measured in hours), while if assigned to the second person it takes an $\text{Expo}(\lambda_2)$ length of time to complete (independent of how long the first person would have taken). Let T be the time taken to complete the task.

(a) Find the mean and variance of T.

(b) Suppose instead that the task is assigned to *both* people, and let X be the time taken to complete it (by whoever completes it first, with the two people working independently). It is observed that after 24 hours, the task has not yet been completed. Conditional on this information, what is the expected value of X?

53. Suppose for this problem that "true IQ" is a meaningful concept rather than a reified social construct. Suppose that in the U.S. population, the distribution of true IQs is Normal with mean 100 and SD 15. A person is chosen at random from this population to take an IQ test. The test is a noisy measure of true ability: it's correct on average but has a Normal measurement error with SD 5.

Let μ be the person's true IQ, viewed as a random variable, and let Y be her score on the IQ test. Then we have

$$Y|\mu \sim \mathcal{N}(\mu, 5^2)$$
$$\mu \sim \mathcal{N}(100, 15^2).$$

(a) Find the unconditional mean and variance of Y.

(b) Find the marginal distribution of Y. One way is via the MGF.

(c) Find $\text{Cov}(\mu, Y)$.

54. Ⓢ A certain genetic characteristic is of interest. It can be measured numerically. Let X_1 and X_2 be the values of the genetic characteristic for two twin boys. Given that they are identical twins, $X_1 = X_2$ and X_1 has mean 0 and variance σ^2; given that they are fraternal twins, X_1 and X_2 have mean 0, variance σ^2, and correlation ρ. The probability that the twins are identical is 1/2. Find $\text{Cov}(X_1, X_2)$ in terms of ρ, σ^2.

55. Ⓢ The Mass Cash lottery randomly chooses 5 of the numbers from $1, 2, \ldots, 35$ each day (without repetitions within the choice of 5 numbers). Suppose that we want to know how long it will take until all numbers have been chosen. Let a_j be the average number of additional days needed if we are missing j numbers (so $a_0 = 0$ and a_{35} is the average number of days needed to collect all 35 numbers). Find a recursive formula for the a_j.

56. Two chess players, Vishy and Magnus, play a series of games. Given p, the game results are i.i.d. with probability p of Vishy winning, and probability $q = 1 - p$ of Magnus winning (assume that each game ends in a win for one of the two players). But p is

unknown, so we will treat it as an r.v. To reflect our uncertainty about p, we use the prior $p \sim \text{Beta}(a, b)$, where a and b are known positive *integers* and $a \geq 2$.

(a) Find the expected number of games needed in order for Vishy to win a game (including the win). Simplify fully; your final answer should not use factorials or Γ.

(b) Explain in terms of independence vs. conditional independence the direction of the inequality between the answer to (a) and $1 + E(G)$ for $G \sim \text{Geom}(\frac{a}{a+b})$.

(c) Find the conditional distribution of p given that Vishy wins exactly 7 out of the first 10 games.

57. *Laplace's law of succession* says that if $X_1, X_2, \ldots, X_{n+1}$ are conditionally independent Bern(p) r.v.s given p, but p is given a Unif(0, 1) prior to reflect ignorance about its value, then

$$P(X_{n+1} = 1 | X_1 + \cdots + X_n = k) = \frac{k+1}{n+2}.$$

As an example, Laplace discussed the problem of predicting whether the sun will rise tomorrow, given that the sun did rise every time for all n days of recorded history; the above formula then gives $(n+1)/(n+2)$ as the probability of the sun rising tomorrow (of course, assuming independent trials with p unchanging over time may be a very unreasonable model for the sunrise problem).

(a) Find the posterior distribution of p given $X_1 = x_1, X_2 = x_2, \ldots, X_n = x_n$, and show that it only depends on the sum of the x_j (so we only need the one-dimensional quantity $x_1 + x_2 + \cdots + x_n$ to obtain the posterior distribution, rather than needing all n data points).

(b) Prove Laplace's law of succession, using a form of the law of total probability to find $P(X_{n+1} = 1 | X_1 + \cdots + X_n = k)$ by conditioning on p. (The next exercise, which is closely related, involves an equivalent Adam's law proof.)

58. Two basketball teams, A and B, play an n game match. Let X_j be the indicator of team A winning the jth game. Given p, the r.v.s $X_1, \ldots, X_n$ are i.i.d. with $X_j | p \sim \text{Bern}(p)$. But p is unknown, so we will treat it as an r.v. Let the prior distribution be $p \sim \text{Unif}(0, 1)$, and let X be the number of wins for team A.

(a) Find $E(X)$ and $\text{Var}(X)$.

(b) Use Adam's law to find the probability that team A will win game $j+1$, given that they win exactly a of the first j games. (The previous exercise, which is closely related, involves an equivalent LOTP proof.)

Hint: Letting C be the event that team A wins exactly a of the first j games,

$$P(X_{j+1} = 1 | C) = E(X_{j+1} | C) = E(E(X_{j+1} | C, p) | C) = E(p | C).$$

(c) Find the PMF of X. (There are various ways to do this, including a very fast way to see it based on results from earlier chapters.)

(d) The Putnam exam from 2002 posed the following problem:

Shanille O'Keal shoots free throws on a basketball court. She hits the first and misses the second, and thereafter the [conditional] probability that she hits the next shot is equal to the proportion of shots she has hit so far. What is the probability she hits exactly 50 of her first 100 shots?

Solve this Putnam problem by applying the result of Part (c). Be sure to explain why it is valid to apply that result, despite the fact that the Putnam problem does not seem to be using the same model, e.g., it does not mention a prior distribution, let alone mention a Unif(0, 1) prior.

59. Let $X|p \sim \text{Bin}(n, p)$, with $p \sim \text{Beta}(a, b)$. So X has a *Beta-Binomial distribution*, as mentioned in Story 8.3.3 and Example 8.5.3. Find $E(X)$ and $\text{Var}(X)$.

60. An election is being held. There are two candidates, A and B, and there are n voters. The probability of voting for Candidate A varies by city. There are m cities, labeled $1, 2, \ldots, m$. The jth city has n_j voters, so $n_1 + n_2 + \cdots + n_m = n$. Let X_j be the number of people in the jth city who vote for Candidate A, with $X_j|p_j \sim \text{Bin}(n_j, p_j)$. To reflect our uncertainty about the probability of voting in each city, we treat $p_1, \ldots, p_m$ as r.v.s, with prior distribution asserting that they are i.i.d. Unif$(0, 1)$. Assume that $X_1, \ldots, X_m$ are independent, both unconditionally and conditional on $p_1, \ldots, p_m$. Let X be the total number of votes for Candidate A.

(a) Find the marginal distribution of X_1 and the posterior distribution of $p_1|(X_1 = k_1)$.

(b) Find $E(X)$ and $\text{Var}(X)$ in terms of n and s, where $s = n_1^2 + n_2^2 + \cdots + n_m^2$.

10

Inequalities and limit theorems

"What should I do if I can't calculate a probability or expectation exactly?" Almost everyone who uses probability has to deal with this sometimes. Don't panic. There are powerful strategies available: *simulate* it, *bound* it, or *approximate* it.

- *Simulations using Monte Carlo*: We have already seen many examples of simulations in this book; the R sections give numerous examples where a few lines of code and a few seconds on a computer suffice to get good approximate answers. "Monte Carlo" just means that the simulations use random numbers (the term originated from the Monte Carlo Casino in Monaco).

  Monte Carlo simulation is an extremely powerful technique, and there are many problems where it is the *only* reasonable approach currently available. So why not always just do a simulation? Here are a few reasons:

  1. The simulation may need to run for an extremely long time, even on a fast computer. A major extension known as *Markov chain Monte Carlo*, introduced in Chapter 12, greatly increases the range of problems for which Monte Carlo simulation is feasible. But even then, the simulation may need to run for a vast, unknown amount of time to get decent answers to the problem.

  2. We may hope to get a good answer *for all values of the parameters of the problem*. For example, in the coupon collector problem (Example 4.3.12) we saw that with n toy types, it takes about $n \log n$ toys on average to get a complete set. This is a simple, memorable answer. It would be easy to simulate this for any *specific* n, e.g., we could run a coupon collector process for $n = 20$ and get an answer of around 60. But this would not make it easy to see the *general* $n \log n$ result.

  3. A simulation result is easy to criticize: how do you know you ran it long enough? How do you know your result is close to the truth? How close is "close"? We may want *provable guarantees* instead.

- *Bounds using inequalities*: A bound on a probability gives a provable guarantee that the probability is in a certain range. In the first part of this chapter, we'll introduce several important inequalities in probability. These inequalities will often allow us to narrow down the range of possible values for the exact answer, that is, to determine an upper bound and/or lower bound. A bound may not provide a good approximation—if we have bounds of [0.2, 0.6] for a probability we're trying to find, the exact answer could be anywhere within that range—but at least we know the exact answer is *guaranteed* to be inside the bounds.

- *Approximations using limit theorems*: Later in the chapter, we will discuss the two most famous theorems in probability: the *law of large numbers* and the *central limit theorem*. Both tell us what happens to the sample mean as we obtain more and more data. Limit theorems let us make approximations which are likely to work well when we have a large number of data points. We conclude the chapter by using limit theorems to study a couple of important named distributions in statistics.

10.1 Inequalities

10.1.1 Cauchy-Schwarz: a marginal bound on a joint expectation

The Cauchy-Schwarz inequality is one of the most famous inequalities in all of mathematics. In probability, it takes the following form.

Theorem 10.1.1 (Cauchy-Schwarz). For any r.v.s X and Y with finite variances,

$$|E(XY)| \leq \sqrt{E(X^2)E(Y^2)}.$$

Proof. For any t,

$$0 \leq E(Y - tX)^2 = E(Y^2) - 2tE(XY) + t^2E(X^2).$$

Where did t come from? The idea is to introduce t so that we have *infinitely* many inequalities, one for each value of t, and then we can use calculus to find the value of t that gives us the *best* inequality. Differentiating the right-hand side with respect to t and setting it equal to 0, we get that $t = E(XY)/E(X^2)$ minimizes the right-hand side, resulting in the tightest bound. Plugging in this value of t, we have the Cauchy-Schwarz inequality. ■

If X and Y are uncorrelated, then $E(XY) = E(X)E(Y)$, which depends only on the *marginal* expectations $E(X)$ and $E(Y)$. But in general, calculating $E(XY)$ exactly requires knowing the *joint* distribution of X and Y (and being able to work with it). The Cauchy-Schwarz inequality lets us bound $E(XY)$ in terms of the *marginal* second moments $E(X^2)$ and $E(Y^2)$.

If X and Y have mean 0, then Cauchy-Schwarz already has a very familiar statistical interpretation: it says that their correlation is between -1 and 1.

Example 10.1.2. Let $E(X) = E(Y) = 0$. Then

$$E(XY) = \text{Cov}(X, Y),\ E(X^2) = \text{Var}(X),\ E(Y^2) = \text{Var}(Y),$$

so Cauchy-Schwarz reduces to the statement $|\text{Corr}(X, Y)| \leq 1$. Of course, we already knew this from Theorem 7.3.5. Now let's see what happens if we drop the assumption that the means are 0. Applying Cauchy-Schwarz to the *centered* r.v.s $X - E(X)$ and $Y - E(Y)$, we again have that $|\text{Corr}(X, Y)| \leq 1$. □

Cauchy-Schwarz can often be applied in creative ways. For example, if we write $X = X \cdot 1$, then Cauchy-Schwarz gives $|E(X \cdot 1)| \leq \sqrt{E(X^2)E(1^2)}$, which reduces to $E(X^2) \geq (EX)^2$. This gives a quick new proof that variances are nonnegative. As another example, we will obtain an upper bound on the probability of a nonnegative r.v. equaling 0.

Example 10.1.3 (Second moment method). Let X be a nonnegative r.v., and suppose that we want an upper bound on $P(X = 0)$. For example, X could be the number of questions that Fred gets wrong on an exam (then $P(X = 0)$ is the probability of Fred getting a perfect score), or X could be the number of pairs of people at a party with the same birthday (then $P(X = 0)$ is the probability of no birthday matches). Note that

$$X = XI(X > 0),$$

where $I(X > 0)$ is the indicator of $X > 0$. This is true since if $X = 0$ then both sides are 0, while if $X > 0$ then both sides are X. By Cauchy-Schwarz,

$$E(X) = E(XI(X > 0)) \leq \sqrt{E(X^2)E(I(X > 0))}.$$

Rearranging this and using the fundamental bridge, we have

$$P(X > 0) \geq \frac{(EX)^2}{E(X^2)},$$

or equivalently,

$$P(X = 0) \leq \frac{\text{Var}(X)}{E(X^2)}.$$

Applying this bound is sometimes called the *second moment method*. For example, let's apply the bound in the case that

$$X = I_1 + \cdots + I_n,$$

where the I_j are uncorrelated indicator r.v.s. Let $p_j = E(I_j)$. Then

$$\text{Var}(X) = \sum_{j=1}^{n} \text{Var}(I_j) = \sum_{j=1}^{n} (p_j - p_j^2) = \sum_{j=1}^{n} p_j - \sum_{j=1}^{n} p_j^2 = \mu - c,$$

where $\mu = E(X), c = \sum_{j=1}^{n} p_j^2$. Also, $E(X^2) = \text{Var}(X) + (EX)^2 = \mu^2 + \mu - c$. So

$$P(X = 0) \leq \frac{\text{Var}(X)}{E(X^2)} = \frac{\mu - c}{\mu^2 + \mu - c} \leq \frac{1}{\mu + 1},$$

where the last inequality is easily checked by cross-multiplying. In general, it is wrong to say "if X has a high mean, then it has a small chance of being 0", since it could be that X is usually 0 but has a small chance of being extremely large. But in our current setting, we have a simple, quantitative way to say that X having a high mean *does* mean that X is unlikely to be 0.

For example, suppose there are 14 people in a room. How likely is it that there are two people with the same birthday or birthdays one day apart? This is much harder to solve exactly than the birthday problem, so in Example 4.7.6 we used a Poisson approximation. But we may want a *guarantee* from a bound rather than worrying about whether the Poisson approximation is good enough. Let X be the number of "near birthday" pairs. Using indicator r.v.s, we have $E(X) = \binom{14}{2}\frac{3}{365}$. So

$$P(X=0) \leq \frac{1}{E(X)+1} < 0.573.$$

The true answer for $P(X = 0)$ turns out to be 0.46 (to two decimal places), which is consistent with the bound. □

Cauchy-Schwarz also allows us to deduce the existence of a joint MGF from the existence of marginal MGFs; this is another example of the benefit of being able to bound a joint distribution quantity by marginal distribution quantities.

Example 10.1.4 (Existence of joint MGF). Let X_1 and X_2 be jointly distributed r.v.s, not necessarily independent or identically distributed. Show that if X_1 and X_2 both have MGFs marginally, then the random vector (X_1, X_2) has a joint MGF.

Solution:

Recall from Chapter 7 that the joint MGF is defined as $M(s,t) = E(e^{sX_1+tX_2})$, and exists if the expectation is finite in a box around the origin. The marginal MGFs are $E(e^{sX_1})$ and $E(e^{tX_2})$; each is required to be finite in an interval around the origin.

Suppose the MGFs of X_1 and X_2 are finite on $(-a, a)$. Fix s and t in $(-a/2, a/2)$. By Cauchy-Schwarz,

$$E(e^{sX_1+tX_2}) \leq \sqrt{E(e^{2sX_1})E(e^{2tX_2})}.$$

The right-hand side is finite by assumption, so $E(e^{sX_1+tX_2})$ is finite in the box $\{(s,t) : s, t \in (-a/2, a/2)\}$. Hence the joint MGF of (X_1, X_2) exists. □

10.1.2 Jensen: an inequality for convexity

In ☣ 4.3.13, we discussed that for nonlinear functions g, $E(g(X))$ may be very different from $g(E(X))$. If g is either a convex function or a concave function, Jensen's inequality tells us exactly which of $E(g(X))$ and $g(E(X))$ is greater. See the math appendix for information about convex and concave functions. Often we can take the second derivative to test for convexity or concavity: assuming that g'' exists, g being convex is equivalent to $g''(x) \geq 0$ everywhere in the domain, and g being concave is equivalent to $g''(x) \leq 0$ everywhere in the domain.

Theorem 10.1.5 (Jensen). Let X be a random variable. If g is a convex function, then $E(g(X)) \geq g(E(X))$. If g is a concave function, then $E(g(X)) \leq g(E(X))$. In both cases, the only way that equality can hold is if there are constants a and b such that $g(X) = a + bX$ with probability 1.

Proof. If g is convex, then all lines that are tangent to g lie below g (see Figure 10.1). In particular, let $\mu = E(X)$, and consider the tangent line at the point $(\mu, g(\mu))$. (If g is differentiable at μ then the tangent line is unique; otherwise, choose any tangent line at μ.) Denoting this tangent line by $a + bx$, we have $g(x) \geq a + bx$ for all x by convexity, so $g(X) \geq a + bX$. Taking the expectation of both sides,

$$E(g(X)) \geq E(a + bX) = a + bE(X) = a + b\mu = g(\mu) = g(E(X)),$$

as desired. If g is concave, then $h = -g$ is convex, so we can apply what we just proved to h to see that the inequality for g is reversed from the convex case.

Lastly, assume that equality holds in the convex case. Let $Y = g(X) - a - bX$. Then Y is a nonnegative r.v. with $E(Y) = 0$, so $P(Y = 0) = 1$ (even a tiny nonzero chance of $Y > 0$ occurring would make $E(Y) > 0$). So equality holds if and only if $P(g(X) = a + bX) = 1$. For the concave case, we can use the same argument with $Y = a + bX - g(X)$. ■

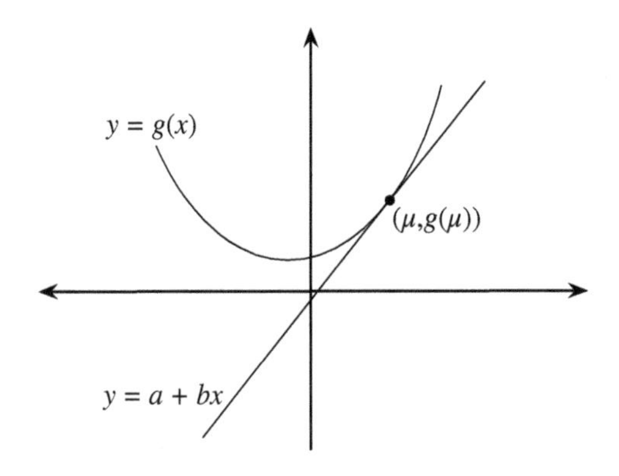

FIGURE 10.1
Since g is convex, the tangent lines lie below the curve. In particular, the tangent line at the point $(\mu, g(\mu))$ lies below the curve.

Let's check Jensen's inequality in a couple of simple known cases.

- Since $g(x) = x^2$ is convex (its second derivative is 2), Jensen's inequality says $E(X^2) \geq (EX)^2$, which we already knew to be true since variances are nonnegative (or by Cauchy-Schwarz).
- In the St. Petersburg paradox from Chapter 4, we found that $E(2^N) > 2^{EN}$, where $N \sim \text{FS}(1/2)$. Jensen's inequality concurs since $g(x) = 2^x$ is convex (to find $g''(x)$, write $2^x = e^{x \log 2}$). Moreover, it tells us that the direction of the inequality

would be the same no matter what distribution N has! The inequality will be strict unless N is constant (with probability 1).

If we ever forget the direction of Jensen's inequality, these simple cases make it easy to recover the correct direction. Here are a few more quick examples of Jensen's inequality:

- $E|X| \geq |EX|$,
- $E(1/X) \geq 1/(EX)$, for positive r.v.s X,
- $E(\log(X)) \leq \log(EX)$, for positive r.v.s X.

As another example, we can use Jensen's inequality to see the direction of the bias if we estimate an unknown standard deviation using the sample standard deviation of a data set.

Example 10.1.6 (Bias of sample standard deviation). Let $X_1, \dots, X_n$ be i.i.d. random variables with variance σ^2. Recall from Theorem 6.3.4 that the sample variance S_n^2 is unbiased for estimating σ^2. That is, $E(S_n^2) = \sigma^2$. However, we are often more interested in estimating the standard deviation σ. A natural estimator for σ is the sample standard deviation, S_n.

Jensen's inequality shows us that S_n is biased for estimating σ. Moreover, it tells us which way the inequality goes:

$$E(S_n) = E(\sqrt{S_n^2}) \leq \sqrt{E(S_n^2)} = \sigma,$$

so the sample standard deviation tends to underestimate the true standard deviation. How biased it is depends on the distribution (so there is no universal way to fix the bias, in contrast to the fact that defining sample variance with $n-1$ in the denominator makes it unbiased for all distributions). Fortunately, the bias is typically minor if the sample size is reasonably large. □

One area in which Jensen's inequality is important is in *information theory*, the study of how to quantify information. The principles of information theory have become essential for communication and compression (e.g., for MP3s and cell phones). Here are a few examples of applications of Jensen's inequality.

Example 10.1.7 (Entropy). The *surprise* of learning that an event with probability p happened is defined as $\log_2(1/p)$, measured in a unit called *bits*. Low-probability events have high surprise, while an event with probability 1 has zero surprise. The log is there so that if we observe two independent events A and B, the total surprise is the same as the surprise from observing $A \cap B$. The log is base 2 so that if we learn that an event with probability 1/2 happened, the surprise is 1, which corresponds to having received 1 bit of information.

Let X be a discrete r.v. whose distinct possible values are $a_1, a_2, \dots, a_n$, with probabilities $p_1, p_2 \dots, p_n$ respectively (so $p_1 + p_2 + \cdots + p_n = 1$). The *entropy* of X is defined to be the average surprise of learning the value of X:

$$H(X) = \sum_{j=1}^{n} p_j \log_2(1/p_j).$$

Note that the entropy of X depends only on the probabilities p_j, not on the values a_j. So for example, $H(X^3) = H(X)$, since X^3 has distinct possible values $a_1^3, a_2^3, \dots, a_n^3$, with probabilities $p_1, p_2, \dots, p_n$—the same list of p_j's as for X.

Using Jensen's inequality, show that the maximum possible entropy for X is when its distribution is uniform over $a_1, a_2, \dots, a_n$, i.e., $p_j = 1/n$ for all j. This makes sense intuitively, since learning the value of X conveys the most information on average when X is equally likely to take any of its values, and the least possible information if X is a constant.

Solution:

Let $X \sim \text{DUnif}(a_1, \dots, a_n)$, so that

$$H(X) = \sum_{j=1}^{n} \frac{1}{n} \log_2(n) = \log_2(n).$$

Let Y be an r.v. that takes on values $1/p_1, \dots, 1/p_n$ with probabilities $p_1, \dots, p_n$, respectively (with the natural modification if the $1/p_j$ have some repeated values, e.g., if $1/p_1 = 1/p_2$ but none of the others are this value, then it gets $p_1 + p_2 = 2p_1$ as its probability). Then $H(Y) = E(\log_2(Y))$ by LOTUS, and $E(Y) = n$. So by Jensen's inequality,

$$H(Y) = E(\log_2(Y)) \leq \log_2(E(Y)) = \log_2(n) = H(X).$$

Since the entropy of an r.v. depends only on the probabilities p_j and not on the specific values that the r.v. takes on, the entropy of Y is unchanged if we alter the support from $1/p_1, \dots, 1/p_n$ to $a_1, \dots, a_n$. Therefore X, which is uniform on $a_1, \dots, a_n$, has entropy at least as large as that of any other r.v. with support $a_1, \dots, a_n$. □

Example 10.1.8 (Kullback-Leibler divergence). Let $\mathbf{p} = (p_1, \dots, p_n)$ and $\mathbf{r} = (r_1, \dots, r_n)$ be probability vectors (so each is nonnegative and sums to 1). Think of each as a possible PMF for a random variable whose support consists of n distinct values. The *Kullback-Leibler divergence* between $\mathbf{p}$ and $\mathbf{r}$ is defined as

$$D(\mathbf{p}, \mathbf{r}) = \sum_{j=1}^{n} p_j \log_2(1/r_j) - \sum_{j=1}^{n} p_j \log_2(1/p_j).$$

This is the difference between the average surprise we will experience when the actual probabilities are $\mathbf{p}$ but we are instead working with $\mathbf{r}$ (for example, if $\mathbf{p}$ is unknown and $\mathbf{r}$ is our current guess for $\mathbf{p}$), and our average surprise when we work with $\mathbf{p}$. Show that the Kullback-Leibler divergence is nonnegative.

Solution:

Using properties of logs, we have

$$D(\mathbf{p}, \mathbf{r}) = -\sum_{j=1}^{n} p_j \log_2 \left(\frac{r_j}{p_j}\right).$$

Let Y be a random variable that takes on values r_j/p_j with probabilities p_j, so that $D(\mathbf{p}, \mathbf{r}) = -E(\log_2(Y))$ by LOTUS. Then by Jensen's inequality,

$$D(\mathbf{p}, \mathbf{r}) = -E(\log_2(Y)) \geq -\log_2(E(Y)) = -\log_2(1) = 0,$$

with equality if and only if $\mathbf{p} = \mathbf{r}$. This result tells us that we're more surprised on average when we work with the wrong probabilities than when we work with the correct probabilities. □

Example 10.1.9 (Log probability scoring). Imagine that on a multiple-choice exam, instead of circling just one of the answer choices, you are asked to assign a probability of correctness to each choice. Your score on a particular question is the log of the probability that you assign to the correct answer. The maximum score for a particular question is 0, and the minimum score is $-\infty$, attained if you assign zero probability to the correct answer.[1]

Suppose your personal probabilities of correctness for each of the n answer choices are $p_1, \dots, p_n$, where the p_j are positive and sum to 1. Show that your expected score on a question is maximized if you report your true probabilities p_j, not any other probabilities. In other words, under log probability scoring, you have no incentive to lie about your beliefs and pretend to be more or less confident than you really are (assuming that your goal is to maximize your expected score).

Solution:

This example is isomorphic to the previous one! Your expected score on a question is $\sum_{j=1}^{n} p_j \log p_j$ if you report your true probabilities $\mathbf{p}$, and $\sum_{j=1}^{n} p_j \log r_j$ if you report false probabilities $\mathbf{r}$. The difference between these two is precisely the Kullback-Leibler divergence between $\mathbf{p}$ and $\mathbf{r}$. This is always nonnegative, as we proved in the previous example. Therefore your expected score is maximized when you report your true probabilities. □

10.1.3 Markov, Chebyshev, Chernoff: bounds on tail probabilities

The inequalities in this section provide bounds on the probability of an r.v. taking on an "extreme" value in the right or left tail of a distribution.

[1]Joe's philosophy professor as an undergraduate at Caltech, Alan Hájek, used precisely this system. He warned the class never to put a probability of zero, since a score of $-\infty$ would not only give a $-\infty$ on that exam, but also it would spill through and result in a $-\infty$ for the whole semester, since a weighted average with even a tiny positive weight on a $-\infty$ yields $-\infty$. Despite this warning, some students did put probability zero on the correct answers.

Theorem 10.1.10 (Markov). For any r.v. X and constant $a > 0$,

$$P(|X| \geq a) \leq \frac{E|X|}{a}.$$

Proof. Let $Y = \frac{|X|}{a}$. We need to show that $P(Y \geq 1) \leq E(Y)$. Note that

$$I(Y \geq 1) \leq Y,$$

since if $I(Y \geq 1) = 0$ then the inequality reduces to $Y \geq 0$, and if $I(Y \geq 1) = 1$ then $Y \geq 1$ (because the indicator says so). Taking the expectation of both sides, we have Markov's inequality. ■

For an intuitive interpretation, let X be the income of a randomly selected individual from a population. Taking $a = 2E(X)$, Markov's inequality says that $P(X \geq 2E(X)) \leq 1/2$, i.e., it is impossible for more than half the population to make at least twice the average income. This is clearly true, since if over half the population were earning at least twice the average income, the average income would be higher! Similarly, $P(X \geq 3E(X)) \leq 1/3$: you can't have more than 1/3 of the population making at least three times the average income, since those people would already drive the average above what it is.

Markov's inequality is a very crude bound because it requires absolutely no assumptions about X. The right-hand side of the inequality could be greater than 1, or even infinite; this is not very helpful when trying to bound a number that we already know to be between 0 and 1. Surprisingly, the following two inequalities, which can be derived from Markov's inequality with almost no additional work, can often give us bounds that are much better than Markov's.

Theorem 10.1.11 (Chebyshev). Let X have mean μ and variance σ^2. Then for any $a > 0$,

$$P(|X - \mu| \geq a) \leq \frac{\sigma^2}{a^2}.$$

Proof. By Markov's inequality,

$$P(|X - \mu| \geq a) = P((X - \mu)^2 \geq a^2) \leq \frac{E(X - \mu)^2}{a^2} = \frac{\sigma^2}{a^2}.$$ ■

Substituting $c\sigma$ for a, for $c > 0$, we have the following equivalent form of Chebyshev's inequality:

$$P(|X - \mu| \geq c\sigma) \leq \frac{1}{c^2}.$$

This gives us an upper bound on the probability of an r.v. being more than c standard deviations away from its mean, e.g., there can't be more than a 25% chance of being 2 or more standard deviations from the mean.

The idea for proving Chebyshev from Markov was to square $|X - \mu|$ and *then* apply Markov. Similarly, it is often fruitful to perform other transformations before applying Markov. *Chernoff's bound*, which is widely used in engineering, uses this idea with an exponential function.

Theorem 10.1.12 (Chernoff). For any r.v. X and constants $a > 0$ and $t > 0$,

$$P(X \geq a) \leq \frac{E(e^{tX})}{e^{ta}}.$$

Proof. The transformation g with $g(x) = e^{tx}$ is invertible and strictly increasing. So by Markov's inequality, we have

$$P(X \geq a) = P(e^{tX} \geq e^{ta}) \leq \frac{E(e^{tX})}{e^{ta}}. \qquad \blacksquare$$

At first it may not be clear what Chernoff's bound has to offer that Markov's inequality doesn't, but it has two very nice features:

1. The right-hand side can be optimized over t to give the tightest upper bound, as in the proof of Cauchy-Schwarz.
2. If the MGF of X exists, then the numerator in the bound *is* the MGF, and some of the useful properties of MGFs can come into play.

Let's now compare the three bounds just discussed by applying them to a simple example where the true probability is known.

Example 10.1.13 (Bounds on a Normal tail probability). Let $Z \sim \mathcal{N}(0, 1)$. By the 68-95-99.7% rule, we know that $P(|Z| > 3)$ is approximately 0.003; the exact value is $2 \cdot \Phi(-3)$. Let's see what upper bounds are obtained from Markov's, Chebyshev's, and Chernoff's inequalities.

- Markov: In Chapter 5, we found that $E|Z| = \sqrt{2/\pi}$. Then

$$P(|Z| > 3) \leq \frac{E|Z|}{3} = \frac{1}{3} \cdot \sqrt{\frac{2}{\pi}} \approx 0.27.$$

- Chebyshev:

$$P(|Z| > 3) \leq \frac{1}{9} \approx 0.11.$$

- Chernoff (after using symmetry of the Normal):

$$P(|Z| > 3) = 2P(Z > 3) \leq 2e^{-3t}E(e^{tZ}) = 2e^{-3t} \cdot e^{t^2/2},$$

  using the MGF of the standard Normal distribution.

  The right-hand side is minimized at $t = 3$, as found by setting the derivative equal to 0, possibly after taking the log first (which is a good idea since it doesn't affect where the minimum occurs and it means we just have to minimize a quadratic polynomial). Plugging in $t = 3$, we have

$$P(|Z| > 3) \leq 2e^{-9/2} \approx 0.022.$$

All of these upper bounds are correct, but Chernoff's bound is the best by far. This example also illustrates the distinction between a bound and an approximation, as we explained in the introduction to this chapter. Markov's inequality tells us that the tail probability $P(|Z| > 3)$ is *at most* 0.27, but it would be a blunder to say that $P(|Z| > 3)$ is *approximately* 0.27—we'd be off by a factor of about 100. ☐

10.2 Law of large numbers

We turn next to two theorems, the law of large numbers and the central limit theorem, which describe the behavior of the sample mean of i.i.d. r.v.s as the sample size grows. Throughout this section and the next, assume we have i.i.d. $X_1, X_2, X_3, \dots$ with finite mean μ and finite variance σ^2. For all positive integers n, let

$$\bar{X}_n = \frac{X_1 + \cdots + X_n}{n}$$

be the sample mean of X_1 through X_n. The sample mean is itself an r.v., with mean μ and variance σ^2/n:

$$E(\bar{X}_n) = \frac{1}{n}E(X_1 + \cdots + X_n) = \frac{1}{n}(E(X_1) + \cdots + E(X_n)) = \mu,$$

$$\text{Var}(\bar{X}_n) = \frac{1}{n^2}\text{Var}(X_1 + \cdots + X_n) = \frac{1}{n^2}(\text{Var}(X_1) + \cdots + \text{Var}(X_n)) = \frac{\sigma^2}{n}.$$

The *law of large numbers* (LLN) says that as n grows, the sample mean $\bar{X}_n$ converges to the true mean μ (in a sense that is explained below). The LLN comes in two versions, "strong" (SLLN) and "weak" (WLLN), which use slightly different definitions of what it means for a sequence of r.v.s to converge to a number. We will state both versions, and prove the second using Chebyshev's inequality.

Theorem 10.2.1 (Strong law of large numbers)**.** The sample mean $\bar{X}_n$ converges to the true mean μ pointwise, with probability 1. Recalling that r.v.s are functions from the sample space S to $\mathbb{R}$, this form of convergence says that $\bar{X}_n(s) \to \mu$ for each point $s \in S$, except that the convergence is allowed to fail on some set B_0 of exceptions, as long as $P(B_0) = 0$. In short, $P(\bar{X}_n \to \mu) = 1$.

Theorem 10.2.2 (Weak law of large numbers)**.** For all $\epsilon > 0$, $P(|\bar{X}_n - \mu| > \epsilon) \to 0$ as $n \to \infty$. (This form of convergence is called *convergence in probability*.)

Proof. Fix $\epsilon > 0$. By Chebyshev's inequality,

$$P(|\bar{X}_n - \mu| > \epsilon) \leq \frac{\sigma^2}{n\epsilon^2}.$$

As $n \to \infty$, the right-hand side goes to 0, and so must the left-hand side. ■

The law of large numbers is essential for simulations, statistics, and science. Consider generating "data" from a large number of independent replications of an experiment, performed either by computer simulation or in the real world. Every time we use the average value in the replications of somc quantity to approximate its theoretical average, we are implicitly appealing to the LLN.

Example 10.2.3 (Running proportion of Heads). Let $X_1, X_2, \ldots$ be i.i.d. Bern(1/2). Interpreting the X_j as indicators of Heads in a string of fair coin tosses, $\bar{X}_n$ is the proportion of Heads after n tosses. The SLLN says that with probability 1, when the sequence of r.v.s $\bar{X}_1, \bar{X}_2, \bar{X}_3, \ldots$ crystallizes into a sequence of numbers, the sequence of numbers will converge to 1/2. Mathematically, there are bizarre outcomes such as *HHHHHH*... and *HHTHHTHHTHHT*..., but collectively they have zero probability of occurring. The WLLN says that for any $\epsilon > 0$, the probability of $\bar{X}_n$ being more than ϵ away from 1/2 can be made as small as we like by letting n grow.

As an illustration, we simulated six sequences of fair coin tosses and, for each sequence, computed $\bar{X}_n$ as a function of n. Of course, in real life we cannot simulate infinitely many coin tosses, so we stopped after 300 tosses. Figure 10.2 plots $\bar{X}_n$ as a function of n for each sequence.

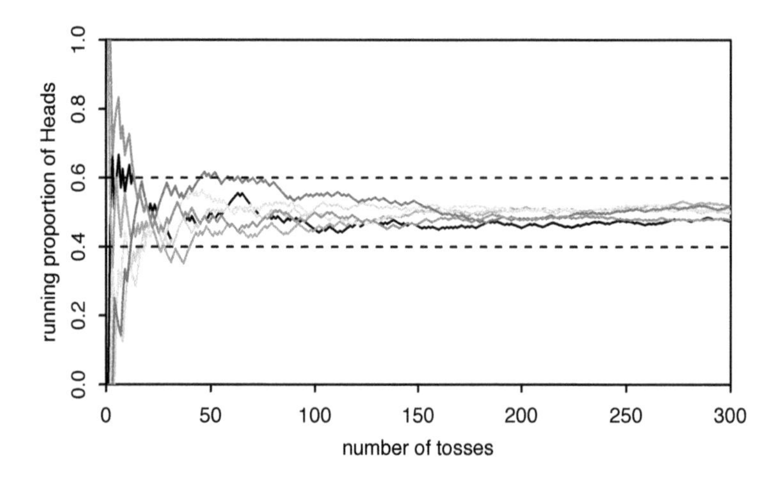

FIGURE 10.2
Running proportion of Heads in 6 sequences of fair coin tosses. Dashed lines at 0.6 and 0.4 are plotted for reference. As the number of tosses increases, the proportion of Heads approaches 1/2.

At the beginning, we can see that there is quite a bit of fluctuation in the running proportion of Heads. As the number of coin tosses increases, however, $\text{Var}(\bar{X}_n)$ gets smaller and smaller, and $\bar{X}_n$ approaches 1/2. □

☣ **10.2.4** (LLN does not contradict the fact that a coin is memoryless)**.** In the above example, the law of large numbers states that the proportion of Heads converges to 1/2, but this does *not* imply that after a long string of Heads, the coin is "due" for a Tails to balance things out. Rather, the convergence takes place through *swamping*: past tosses are swamped by the infinitely many tosses that are yet to come.

A sequence of i.i.d. Bernoullis is the simplest possible example of the LLN, but this simple case forms the basis for extremely useful methods in statistics, as the following examples illustrate.

Example 10.2.5 (Monte Carlo integration)**.** Let f be a complicated function whose integral $\int_a^b f(x)dx$ we want to approximate. Assume that $0 \leq f(x) \leq c$ so that we know the integral is finite. On the surface, this problem doesn't involve probability, as $\int_a^b f(x)dx$ is just a number. But where there is no randomness, we can create our own! The technique of *Monte Carlo integration* uses random samples to obtain approximations of definite integrals when exact integration methods are unavailable.

Let A be the rectangle in the (x, y)-plane given by $a \leq x \leq b$ and $0 \leq y \leq c$. Let B be the region under the curve $y = f(x)$ (and above the x-axis) for $a \leq x \leq b$, so the desired integral is the area of region B. Our strategy will be to take random samples from A, then calculate the proportion of the samples that also fall into the area B. This is depicted in Figure 10.3: points in B are in black, and points not in B are in white.

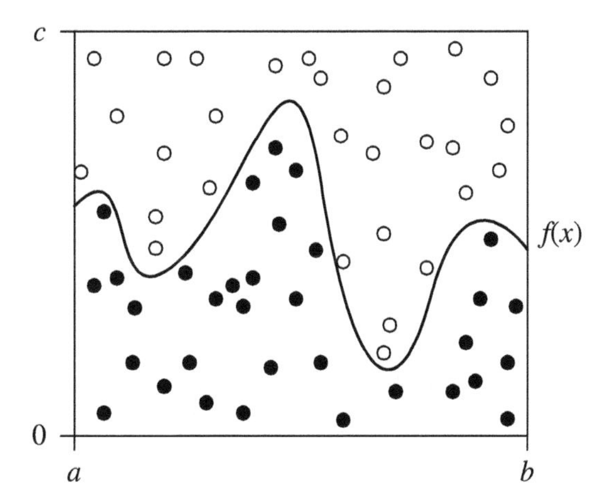

FIGURE 10.3
Monte Carlo integration. To approximate the area under $f(x)$ from $x = a$ to $x = b$, generate random points in the rectangle $[a, b] \times [0, c]$, and approximate the area under $f(x)$ by the proportion of points falling underneath the curve, multiplied by the total area of the rectangle.

To see why this works, suppose we pick i.i.d. points $(X_1, Y_1), (X_2, Y_2), \dots, (X_n, Y_n)$ uniformly in the rectangle A. Define indicator r.v.s $I_1, \dots, I_n$ by letting $I_j = 1$ if (X_j, Y_j) is in B and $I_j = 0$ otherwise. Then the I_j are Bernoulli r.v.s whose success probability is precisely the ratio of the area of B to the area of A. Letting $p = E(I_j)$,

$$p = E(I_j) = P(I_j = 1) = \frac{\int_a^b f(x)dx}{c(b-a)}.$$

We can estimate p using $\frac{1}{n}\sum_{j=1}^n I_j$, and then estimate the desired integral by

$$\int_a^b f(x)dx \approx c(b-a)\frac{1}{n}\sum_{j=1}^n I_j.$$

Since the I_j are i.i.d. with mean p, it follows from the law of large numbers that with probability 1, the estimate converges to the true value of the integral as the number of points approaches infinity. □

Example 10.2.6 (Convergence of empirical CDF). Let $X_1, \dots, X_n$ be i.i.d. random variables with CDF F. For every number x, let $R_n(x)$ count how many of $X_1, \dots, X_n$ are less than or equal to x; that is,

$$R_n(x) = \sum_{j=1}^n I(X_j \leq x).$$

Since the indicators $I(X_j \leq x)$ are i.i.d. with probability of success $F(x)$, we know $R_n(x)$ is Binomial with parameters n and $F(x)$.

The *empirical CDF* of $X_1, \dots, X_n$ is defined as

$$\hat{F}_n(x) = \frac{R_n(x)}{n},$$

considered as a function of x. Before we observe $X_1, \dots, X_n$, $\hat{F}_n(x)$ is a random variable for each x. After we observe $X_1, \dots, X_n$, $\hat{F}_n(x)$ crystallizes into a particular value at each x, so $\hat{F}_n$ crystallizes into a particular CDF, which can be used to estimate the true CDF F if the latter is unknown.

For example, suppose $n = 4$ and we observe $X_1 = x_1, X_2 = x_2, X_3 = x_3, X_4 = x_4$. Then the graph of $\frac{R_4(x)}{4}$ starts at 0 and then jumps by 1/4 every time one of the x_j's is reached. In other words, $\frac{R_4(x)}{4}$ is the CDF of a discrete random variable taking on values $x_1, \dots, x_4$, each with probability 1/4. This is illustrated in Figure 10.4.

Now we can ask, what happens to $\hat{F}_n$ as $n \to \infty$? This is a natural question if we are using $\hat{F}_n$ as an estimate of the true F; does the approximation fare well in the limit? The law of large numbers provides the answer: for every x, $R_n(x)$ is the sum of n i.i.d. Bern(p) r.v.s, where $p = F(x)$. So by the SLLN, $\hat{F}_n(x) \to F(x)$ with probability 1 as $n \to \infty$.

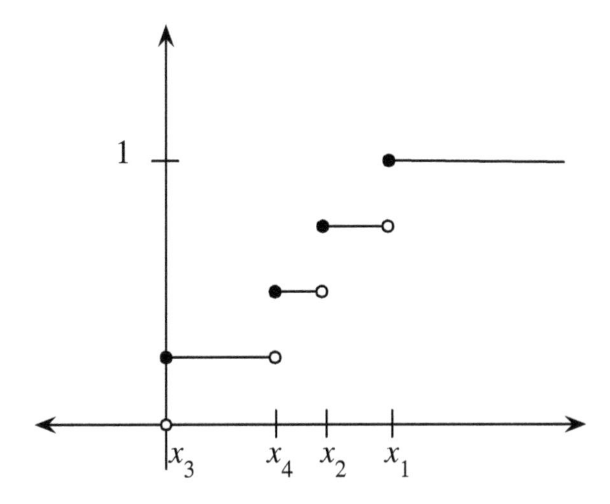

FIGURE 10.4
Empirical CDF after observing $X_1 = x_1, X_2 = x_2, X_3 = x_3, X_4 = x_4$. The graph jumps by $1/4$ every time one of the x_j's is reached.

The empirical CDF is commonly used in *nonparametric statistics*, a branch of statistics that tries to understand a random sample without making strong assumptions about the family of distributions from which it originated. For example, instead of assuming $X_1, \ldots, X_n \sim \mathcal{N}(\mu, \sigma^2)$, a nonparametric method would allow $X_1, \ldots, X_n \sim F$ for an arbitrary CDF F, then use the empirical CDF as an approximation for F. The law of large numbers is what assures us that this approximation is valid in the limit as we collect more and more samples: at every value of x, the empirical CDF converges to the true CDF. □

10.3 Central limit theorem

As in the previous section, let $X_1, X_2, X_3, \ldots$ be i.i.d. with mean μ and variance σ^2. The law of large numbers says that as $n \to \infty$, $\bar{X}_n$ converges to the constant μ (with probability 1). But what is its *distribution* along the way to becoming a constant? This is addressed by the *central limit theorem* (CLT), which, as its name suggests, is a limit theorem of central importance in statistics.

The CLT states that for large n, the distribution of $\bar{X}_n$ after standardization approaches a standard Normal distribution. By standardization, we mean that we subtract μ, the mean of $\bar{X}_n$, and divide by $\sigma/\sqrt{n}$, the standard deviation of $\bar{X}_n$.

Theorem 10.3.1 (Central limit theorem). As $n \to \infty$,

$$\sqrt{n}\left(\frac{\bar{X}_n - \mu}{\sigma}\right) \to \mathcal{N}(0, 1) \text{ in distribution.}$$

In words, this means that the CDF of the left-hand side converges to Φ, the CDF of the standard Normal distribution.

Proof. We will prove the CLT assuming that the MGF of the X_j exists, though the theorem holds much more generally. Let $M(t) = E(e^{tX_j})$, and without loss of generality let $\mu = 0$, $\sigma^2 = 1$ (since we end up standardizing $\bar{X}_n$ for the theorem, we might as well standardize the X_j in the first place). Then $M(0) = 1$, $M'(0) = \mu = 0$, and $M''(0) = \sigma^2 = 1$.

We wish to show that the MGF of $\sqrt{n}\bar{X}_n = (X_1 + \cdots + X_n)/\sqrt{n}$ converges to the MGF of the $\mathcal{N}(0,1)$ distribution, which is $e^{t^2/2}$. This is a valid strategy because of a theorem that says that if $Z_1, Z_2, \ldots$ are r.v.s whose MGFs converge to the MGF of a continuous r.v. Z, then the CDF of Z_n converges to the CDF of Z. (We omit the proof of this result since it requires some difficult analysis. But it should at least seem plausible, in view of the fact that the MGF of an r.v. determines its distribution.)

By properties of MGFs,

$$\begin{aligned} E(e^{t(X_1+\cdots+X_n)/\sqrt{n}}) &= E(e^{tX_1/\sqrt{n}})E(e^{tX_2/\sqrt{n}})\ldots E(e^{tX_n/\sqrt{n}}) \\ &= \left(M\left(\frac{t}{\sqrt{n}}\right)\right)^n. \end{aligned}$$

Letting $n \to \infty$, we get the indeterminate form 1^∞, so instead we should take the limit of the logarithm, $n \log M(\frac{t}{\sqrt{n}})$, and then exponentiate at the end. This gives

$$\begin{aligned} \lim_{n\to\infty} n \log M\left(\frac{t}{\sqrt{n}}\right) &= \lim_{y\to 0} \frac{\log M(yt)}{y^2} && \text{where } y = 1/\sqrt{n} \\ &= \lim_{y\to 0} \frac{tM'(yt)}{2yM(yt)} && \text{by L'Hôpital's rule} \\ &= \frac{t}{2} \lim_{y\to 0} \frac{M'(yt)}{y} && \text{since } M(yt) \to 1 \\ &= \frac{t^2}{2} \lim_{y\to 0} M''(yt) && \text{by L'Hôpital's rule} \\ &= \frac{t^2}{2}. \end{aligned}$$

Therefore $\left(M(\frac{t}{\sqrt{n}})\right)^n$, the MGF of $\sqrt{n}\bar{X}_n$, approaches $e^{t^2/2}$, the $\mathcal{N}(0,1)$ MGF. ■

The CLT is an *asymptotic* result, telling us about the limiting distribution of $\bar{X}_n$ as $n \to \infty$, but it also suggests an *approximation* for the distribution of $\bar{X}_n$ when n is large but finite.

Approximation 10.3.2 (Central limit theorem, approximation form)**.** For large n, the distribution of $\bar{X}_n$ is approximately $\mathcal{N}(\mu, \sigma^2/n)$.

Proof. Change the arrow in the CLT to $\dot{\sim}$, an approximate distribution sign:

$$\sqrt{n}\left(\frac{\bar{X}_n - \mu}{\sigma}\right) \dot{\sim} \mathcal{N}(0,1).$$

Then by a location-scale transformation,

$$\bar{X}_n \dot{\sim} \mathcal{N}(\mu, \sigma^2/n).$$

■

Of course, we already knew from properties of expectation and variance that $\bar{X}_n$ has mean μ and variance σ^2/n; the central limit theorem gives us the additional information that $\bar{X}_n$ is approximately *Normal* with said mean and variance.

Let's take a moment to admire the generality of this result. The distribution of the individual X_j can be *anything in the world*, as long as the mean and variance are finite. We could have a discrete distribution like the Binomial, a bounded distribution like the Beta, a skewed distribution like the Log-Normal, or a distribution with multiple peaks and valleys. No matter what, the act of averaging will cause Normality to emerge. In Figure 10.5 we show histograms of the distribution of $\bar{X}_n$ for 4 different starting distributions and for $n = 1, 5, 30, 100$. As n increases, the distribution of $\bar{X}_n$ starts to look Normal, regardless of the distribution of the X_j.

This does not mean that the distribution of the X_j is irrelevant, however. If the X_j have a highly skewed or multimodal distribution, we may need n to be very large before the Normal approximation becomes accurate; at the other extreme, if the X_j are already i.i.d. Normals, the distribution of $\bar{X}_n$ is exactly $\mathcal{N}(\mu, \sigma^2/n)$ for all n. Since there are no infinite datasets in the real world, the quality of the Normal approximation for finite n is an important consideration.

Example 10.3.3 (Running proportion of Heads, revisited)**.** As in Example 10.2.3, let $X_1, X_2, \ldots$ be i.i.d. Bern(1/2). Before, we used the law of large numbers to conclude that $\bar{X}_n \to 1/2$ as $n \to \infty$. Now, using the central limit theorem, we can say more: $E(\bar{X}_n) = 1/2$ and $\text{Var}(\bar{X}_n) = 1/(4n)$, so for large n,

$$\bar{X}_n \dot{\sim} \mathcal{N}\left(\frac{1}{2}, \frac{1}{4n}\right).$$

This additional information allows us to quantify what kind of deviation from the mean is typical for a given n. For example, when $n = 100$, $\text{SD}(\bar{X}_n) = 1/20 = 0.05$, so if the Normal approximation is valid, then by the 68-95-99.7% rule there's a 95% chance that $\bar{X}_n$ will be in the interval $[0.40, 0.60]$. □

The CLT says that the sample mean $\bar{X}_n$ is approximately Normal, but since the sum $W_n = X_1 + \cdots + X_n = n\bar{X}_n$ is just a scaled version of $\bar{X}_n$, the CLT also implies W_n is approximately Normal. If the X_j have mean μ and variance σ^2, W_n has mean $n\mu$ and variance $n\sigma^2$. The CLT then states that for large n,

$$W_n \dot{\sim} \mathcal{N}(n\mu, n\sigma^2).$$

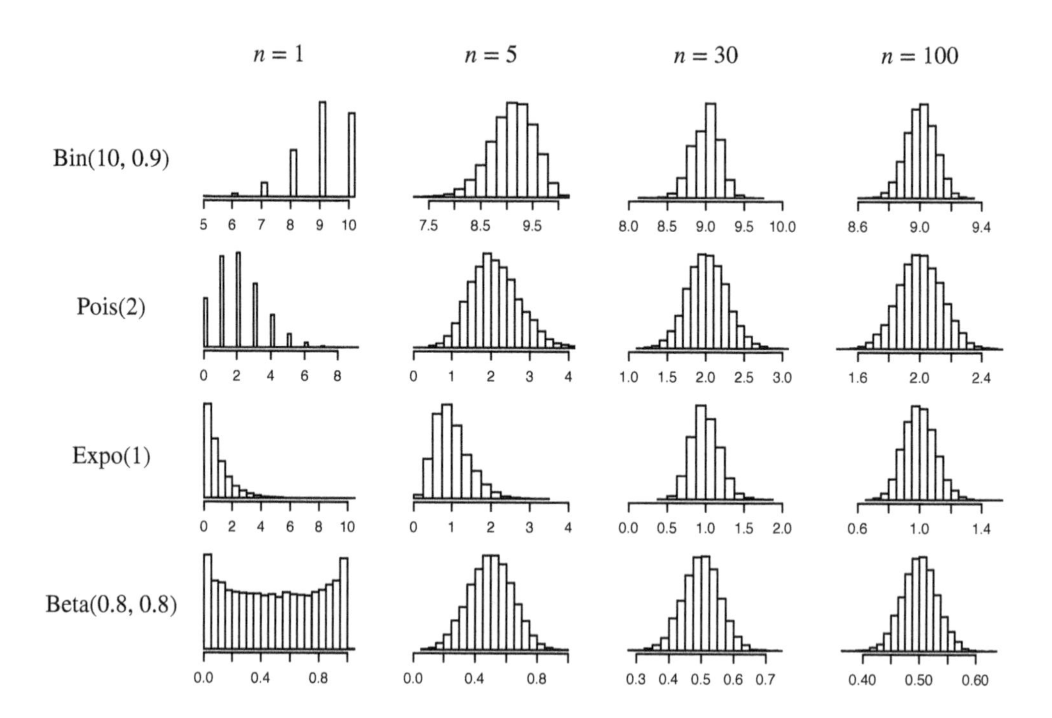

FIGURE 10.5

Central limit theorem. Histograms of the distribution of $\bar{X}_n$ for different starting distributions of the X_j (indicated by the rows) and increasing values of n (indicated by the columns). Each histogram is based on 10,000 simulated values of $\bar{X}_n$. Regardless of the starting distribution of the X_j, the distribution of $\bar{X}_n$ approaches a Normal distribution as n grows.

This is completely equivalent to the approximation for $\bar{X}_n$, but it can be useful to write it in this form because many of the named distributions we have studied can be considered as a sum of i.i.d. r.v.s. Here are three quick examples.

Example 10.3.4 (Poisson convergence to Normal). Let $Y \sim \text{Pois}(n)$. By Theorem 4.8.1, we can consider Y to be a sum of n i.i.d. Pois(1) r.v.s. Therefore, for large n,

$$Y \dot{\sim} \mathcal{N}(n, n). \quad \square$$

Example 10.3.5 (Gamma convergence to Normal). Let $Y \sim \text{Gamma}(n, \lambda)$. By Theorem 8.4.3, we can consider Y to be a sum of n i.i.d. Expo(λ) r.v.s. Therefore, for large n,

$$Y \dot{\sim} \mathcal{N}\left(\frac{n}{\lambda}, \frac{n}{\lambda^2}\right). \quad \square$$

Example 10.3.6 (Binomial convergence to Normal). Let $Y \sim \text{Bin}(n, p)$. By Theorem 3.8.8, we can consider Y to be a sum of n i.i.d. Bern(p) r.v.s. Therefore, for large n,

$$Y \dot{\sim} \mathcal{N}\left(np, np(1-p)\right).$$

This is probably the most widely used Normal approximation in statistics. To account for the discreteness of Y, we write the probability $P(Y = k)$ (which would be exactly 0 under the Normal approximation) as $P(k - 1/2 < Y < k + 1/2)$ (so that it becomes an interval of non-zero width) and apply the Normal approximation to the latter. This is known as the *continuity correction*, and it yields the following approximation for the PMF of Y:

$$P(Y = k) = P(k - 1/2 < Y < k + 1/2) \approx \Phi\left(\frac{k + 1/2 - np}{\sqrt{np(1-p)}}\right) - \Phi\left(\frac{k - 1/2 - np}{\sqrt{np(1-p)}}\right).$$

The Normal approximation to the Binomial distribution is complementary to the Poisson approximation discussed in Chapter 4. The Poisson approximation works best when p is small, while the Normal approximation works best when n is large and p is around 1/2, so that the distribution of Y is symmetric or nearly so. $\square$

We'll conclude with an example that uses *both* the LLN and the CLT.

Example 10.3.7 (Volatile stock). Each day, a very volatile stock rises 70% or drops 50% in price, with equal probabilities and with different days independent. Let Y_n be the stock price after n days, starting from an initial value of $Y_0 = 100$.

(a) Explain why $\log Y_n$ is approximately Normal for n large, and state its parameters.

(b) What happens to $E(Y_n)$ as $n \to \infty$?

(c) Use the law of large numbers to find out what happens to Y_n as $n \to \infty$.

Solution:

(a) We can write $Y_n = Y_0(0.5)^{n-U_n}(1.7)^{U_n}$ where $U_n \sim \text{Bin}(n, \frac{1}{2})$ is the number of times the stock rises in the first n days. This gives

$$\log Y_n = \log Y_0 - n \log 2 + U_n \log 3.4,$$

which is a location-scale transformation of U_n. By the CLT, U_n is approximately $\mathcal{N}(\frac{n}{2}, \frac{n}{4})$ for large n, so $\log Y_n$ is approximately Normal with mean

$$E(\log Y_n) = \log 100 - n \log 2 + (\log 3.4) \cdot E(U_n) \approx \log 100 - 0.081n$$

and variance

$$\text{Var}(\log Y_n) = (\log 3.4)^2 \cdot \text{Var}(U_n) \approx 0.374n.$$

(b) We have $E(Y_1) = (170 + 50)/2 = 110$. Similarly,

$$E(Y_{n+1}|Y_n) = \frac{1}{2}(1.7Y_n) + \frac{1}{2}(0.5Y_n) = 1.1Y_n,$$

so

$$E(Y_{n+1}) = E(E(Y_{n+1}|Y_n)) = 1.1E(Y_n).$$

Thus $E(Y_n) = 1.1^n E(Y_0) = 100 \cdot 1.1^n$, which goes to ∞ as $n \to \infty$.

(c) As in (a), let $U_n \sim \text{Bin}(n, \frac{1}{2})$ be the number of times the stock rises in the first n days. Note that even though $E(Y_n) \to \infty$, if the stock goes up 70% one day and then drops 50% the next day, then overall it has dropped 15% since $1.7 \cdot 0.5 = 0.85$. So after many days, Y_n will be very small if about half the time the stock rose 70% and about half the time the stock dropped 50%—and the law of large numbers ensures that this *will* be the case! Writing Y_n in terms of U_n/n in order to apply the LLN, we have

$$Y_n = Y_0(0.5)^{n-U_n}(1.7)^{U_n} = Y_0\left(\frac{(3.4)^{U_n/n}}{2}\right)^n.$$

Since $U_n/n \to 0.5$ with probability 1, $(3.4)^{U_n/n} \to \sqrt{3.4} < 2$ with probability 1, so $Y_n \to 0$ with probability 1.

Paradoxically, $E(Y_n) \to \infty$ but $Y_n \to 0$ with probability 1. To gain some intuition for this result, consider the extreme example where a gambler starts with \$100 and each day either quadruples their money or loses their entire fortune, with equal probabilities. Then on average the gambler's wealth doubles each day, which sounds good until one notices that eventually there will be a day when the gambler goes broke. The gambler's actual fortune goes to 0 with probability 1, whereas the expected value goes to infinity due to tiny probabilities of getting extremely large amounts of money, as in the St. Petersburg paradox. □

☣ **10.3.8** (The evil Cauchy). The central limit theorem requires that the mean and variance of the X_j be finite, and our proof of the WLLN relied on the same conditions. The Cauchy distribution introduced in Example 7.1.25 has no mean or variance, so the Cauchy distribution obeys neither the law of large numbers nor the central limit theorem. It can be shown that the sample mean of n Cauchys is still Cauchy, no matter how large n gets. So the sample mean never approaches a Normal distribution, contrary to the behavior seen in the CLT. There is also no true mean for $\bar{X}_n$ to converge to, so the LLN does not apply either.

10.4 Chi-Square and Student-t

We'll round out the chapter by introducing the last two continuous distributions in this book, both of which are closely related to the Normal distribution.

Definition 10.4.1 (Chi-Square distribution). Let $V = Z_1^2 + \cdots + Z_n^2$ where $Z_1, Z_2, \ldots, Z_n$ are i.i.d. $\mathcal{N}(0,1)$. Then V is said to have the *Chi-Square distribution with n degrees of freedom.* We write this as $V \sim \chi_n^2$.

As it turns out, the χ_n^2 distribution is a special case of the Gamma.

Theorem 10.4.2. The χ_n^2 distribution is the Gamma$(\frac{n}{2}, \frac{1}{2})$ distribution.

Proof. First, we verify that the PDF of $Z_1^2 \sim \chi_1^2$ equals the PDF of the Gamma$(\frac{1}{2}, \frac{1}{2})$: for $x > 0$,

$$F(x) = P(Z_1^2 \leq x) = P(-\sqrt{x} \leq Z_1 \leq \sqrt{x}) = \Phi(\sqrt{x}) - \Phi(-\sqrt{x}) = 2\Phi(\sqrt{x}) - 1,$$

so

$$f(x) = \frac{d}{dx}F(x) = 2\varphi(\sqrt{x})\frac{1}{2}x^{-1/2} = \frac{1}{\sqrt{2\pi x}}e^{-x/2},$$

which is indeed the Gamma$(\frac{1}{2}, \frac{1}{2})$ PDF. Then, because $V = Z_1^2 + \cdots + Z_n^2 \sim \chi_n^2$ is the sum of n independent Gamma$(\frac{1}{2}, \frac{1}{2})$ r.v.s, we have $V \sim$ Gamma$(\frac{n}{2}, \frac{1}{2})$. ■

From our knowledge of the mean and variance of a Gamma distribution, we have $E(V) = n$ and $\text{Var}(V) = 2n$. We can also obtain the mean and variance using the fact that V is the sum of i.i.d. squared Normals, along with the Normal moments derived in Chapter 6:

$$\begin{aligned} E(V) &= nE(Z_1^2) = n, \\ \text{Var}(V) &= n\text{Var}(Z_1^2) = n\left(E(Z_1^4) - (EZ_1^2)^2\right) = n(3-1) = 2n. \end{aligned}$$

To get the MGF of the Chi-Square distribution, just plug $n/2$ and $1/2$ into the more general Gamma(a, λ) MGF, which we found in Theorem 8.4.3 to be $\left(\frac{\lambda}{\lambda - t}\right)^a$ for $t < \lambda$. This gives

$$M_V(t) = \left(\frac{1}{1-2t}\right)^{n/2}, \quad t < 1/2.$$

The Chi-Square distribution is important in statistics because it is related to the distribution of the *sample variance*, which can be used to estimate the true variance of a distribution. When our random variables are i.i.d. Normals, the distribution of the sample variance after appropriate scaling is Chi-Square.

Example 10.4.3 (Distribution of sample variance). For i.i.d. $X_1, \dots, X_n \sim \mathcal{N}(\mu, \sigma^2)$, the sample variance is the r.v.

$$S_n^2 = \frac{1}{n-1}\sum_{j=1}^{n}(X_j - \bar{X}_n)^2.$$

Show that

$$\frac{(n-1)S_n^2}{\sigma^2} \sim \chi^2_{n-1}.$$

Solution:

First let's show that $\sum_{j=1}^{n}(Z_j - \bar{Z}_n)^2 \sim \chi^2_{n-1}$ for standard Normal r.v.s $Z_1, \dots, Z_n$; this is consistent with the more general result we are asked to prove and also serves as a useful stepping stone. Let's start with the following useful identity, which is a special case of the identity from the proof of Theorem 6.3.4:

$$\sum_{j=1}^{n} Z_j^2 = \sum_{j=1}^{n}(Z_j - \bar{Z}_n)^2 + n\bar{Z}_n^2.$$

Now take the MGF of both sides. By Example 7.5.9, $\sum_{j=1}^{n}(Z_j - \bar{Z}_n)^2$ and $n\bar{Z}_n^2$ are independent, so the MGF of their sum is the product of the individual MGFs. Also,

$$\sum_{j=1}^{n} Z_j^2 \sim \chi^2_n \text{ and } n\bar{Z}_n^2 \sim \chi^2_1,$$

so

$$\left(\frac{1}{1-2t}\right)^{n/2} = \left(\text{MGF of } \sum_{j=1}^{n}(Z_j - \bar{Z}_n)^2\right) \cdot \left(\frac{1}{1-2t}\right)^{1/2}.$$

This implies

$$\left(\text{MGF of } \sum_{j=1}^{n}(Z_j - \bar{Z}_n)^2\right) = \left(\frac{1}{1-2t}\right)^{(n-1)/2},$$

which is the χ^2_{n-1} MGF. Since the MGF determines the distribution, we have

$$\sum_{j=1}^{n}(Z_j - \bar{Z}_n)^2 \sim \chi^2_{n-1}.$$

For general $X_1, \dots, X_n$, use a location-scale transformation to write $X_j = \mu + \sigma Z_j$ and $\bar{X}_n = \mu + \sigma\bar{Z}_n$. When we express $\sum_{j=1}^{n}(X_j - \bar{X}_n)^2$ in terms of the Z_j, the μ cancels and the σ comes out squared:

$$\sum_{j=1}^{n}(X_j - \bar{X}_n)^2 = \sum_{j=1}^{n}(\mu + \sigma Z_j - (\mu + \sigma\bar{Z}_n))^2 = \sigma^2\sum_{j=1}^{n}(Z_j - \bar{Z}_n)^2.$$

All in all,

$$\frac{(n-1)S_n^2}{\sigma^2} = \frac{1}{\sigma^2}\sum_{j=1}^{n}(X_j - \bar{X}_n)^2 = \frac{1}{\sigma^2}\cdot\sigma^2\sum_{j=1}^{n}(Z_j - \bar{Z}_n)^2 \sim \chi^2_{n-1},$$

which is what we wanted. This also implies that $E(S_n^2) = \sigma^2$, which agrees with what we showed in Theorem 6.3.4: the sample variance is unbiased for estimating the true variance. □

The Student-t distribution is defined by representing it in terms of a standard Normal r.v. and a χ^2_n r.v.

Definition 10.4.4 (Student-t distribution). Let

$$T = \frac{Z}{\sqrt{V/n}},$$

where $Z \sim \mathcal{N}(0,1)$, $V \sim \chi^2_n$, and Z is independent of V. Then T is said to have the *Student-t distribution with n degrees of freedom*. We write this as $T \sim t_n$. Often "Student-t distribution" is abbreviated to "t distribution".

The Student-t distribution was introduced in 1908 by William Gosset, a Master Brewer at Guinness, while working on quality control for beer. He was required by the company to publish his work under a pseudonym, and he chose the name Student. The t distribution forms the basis for hypothesis testing procedures known as *t-tests*, which are extremely widely used in practice (we do not introduce the details of t-tests here since they are better left for a course on statistical inference).

The PDF of the Student-t distribution with n degrees of freedom looks similar to that of a standard Normal, except with heavier tails (much heavier if n is small, and not much heavier if n is large). The formula for the PDF is

$$f_T(t) = \frac{\Gamma((n+1)/2)}{\sqrt{n\pi}\Gamma(n/2)}(1+t^2/n)^{-(n+1)/2},$$

though we will not prove this since the derivation is messy and anyway many of the most important properties of the Student-t distribution are easier to understand by thinking about how we defined it in terms of a Normal and a χ^2_n r.v., rather than by doing tedious calculations with the PDF. Here are some of these properties.

Theorem 10.4.5 (Student-t properties). The Student-t distribution t_n has the following properties.

1. Symmetry: If $T \sim t_n$, then $-T \sim t_n$ as well.

2. Cauchy as special case: The t_1 distribution is the same as the Cauchy distribution, introduced in Example 7.1.25.

3. Convergence to Normal: As $n \to \infty$, the t_n distribution approaches the standard Normal distribution.

Proof. In the proof of each property, we appeal to Definition 10.4.4.

1. Express

$$T = \frac{Z}{\sqrt{V/n}},$$

where $Z \sim \mathcal{N}(0,1)$, $V \sim \chi^2_n$, and Z is independent of V. Then

$$-T = \frac{-Z}{\sqrt{V/n}},$$

where $-Z \sim \mathcal{N}(0,1)$, so $-T \sim t_n$.

2. Recall that the Cauchy distribution is defined as the distribution of X/Y where X and Y are i.i.d. $\mathcal{N}(0,1)$. By definition, $T \sim t_1$ can be expressed as $T = Z/\sqrt{V}$, where $\sqrt{V} = \sqrt{Z_1^2} = |Z_1|$ with Z_1 independent of Z. But by symmetry, $Z/|Z_1|$ has the same distribution as Z/Z_1, and Z/Z_1 is Cauchy. Thus the t_1 and Cauchy distributions are the same.

3. This follows from the SLLN. Consider a sequence of i.i.d. standard Normal r.v.s $Z_1, Z_2, \ldots$, and let

$$V_n = Z_1^2 + \cdots + Z_n^2.$$

By the SLLN, $V_n/n \to E(Z_1^2) = 1$ with probability 1. Now let $Z \sim \mathcal{N}(0,1)$ be independent of all the Z_j, and let

$$T_n = \frac{Z}{\sqrt{V_n/n}}$$

for all n. Then $T_n \sim t_n$ by definition, and since the denominator converges to 1, we have $T_n \to Z \sim \mathcal{N}(0,1)$. Therefore, the distribution of the T_n approaches the distribution of Z. ■

Figure 10.6 plots the Student-t PDF for various values of n, demonstrating all three properties of the above theorem: the PDFs are all symmetric around 0, the PDF for $n = 1$ looks like that of the Cauchy distribution, and as $n \to \infty$ the heavy tails become lighter, and the PDF approaches the standard Normal PDF.

10.5 Recap

Inequalities and limit theorems are two different ways to handle expectations and probabilities that we don't wish to calculate exactly. Inequalities allow us to obtain lower and/or upper bounds on the unknown value: Cauchy-Schwarz and Jensen give us bounds on expectations, while Markov, Chebyshev, and Chernoff give us bounds on tail probabilities.

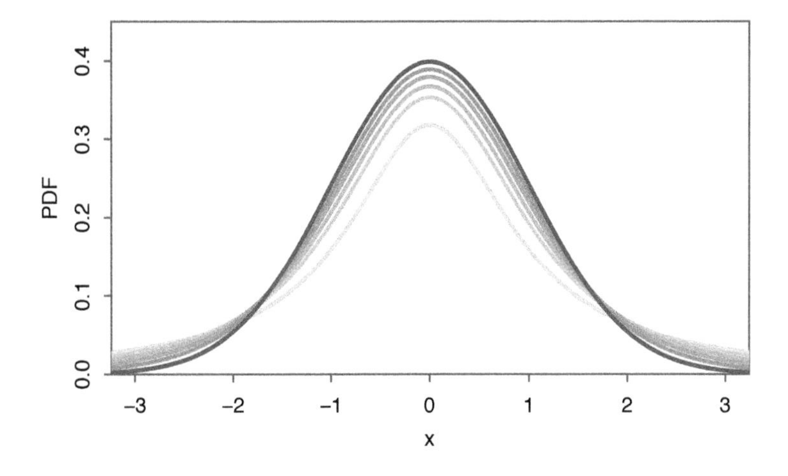

FIGURE 10.6
PDFs of Student-t distribution with (light to dark) $n = 1, 2, 3, 5, 10$ degrees of freedom, as well as the standard Normal PDF (black). As $n \to \infty$, the Student-t PDFs approach the standard Normal PDF.

Two limit theorems, the law of large numbers and the central limit theorem, describe the behavior of the sample mean $\bar{X}_n$ of i.i.d. $X_1, X_2, \dots$ with mean μ and variance σ^2. The SLLN says that as $n \to \infty$, the sample mean $\bar{X}_n$ converges to the true mean μ with probability 1. The CLT says that the distribution of $\bar{X}_n$, after standardization, converges to the standard Normal distribution:

$$\sqrt{n}\left(\frac{\bar{X}_n - \mu}{\sigma}\right) \to \mathcal{N}(0, 1).$$

This can be translated into an approximation for the distribution of $\bar{X}_n$:

$$\bar{X}_n \dot{\sim} \mathcal{N}(\mu, \sigma^2/n).$$

Equivalently, we can say that the distribution of the sum $S_n = X_1 + \cdots + X_n = n\bar{X}_n$, after standardization, converges to the standard Normal distribution:

$$\frac{S_n - n\mu}{\sigma\sqrt{n}} \to \mathcal{N}(0, 1).$$

Figure 10.7 illustrates the progression from a distribution to i.i.d. r.v.s with that distribution, from which a sample mean can be formed and studied as an r.v. in its own right. Chebyshev's inequality, the LLN, and the CLT all give important information about the behavior of the sample mean.

The Chi-Square and Student-t distributions are two important named distributions in statistics. The Chi-Square is a special case of the Gamma. The Student-t has a bell-shaped PDF with heavier tails than the Normal, and converges to the standard Normal as the degrees of freedom increase.

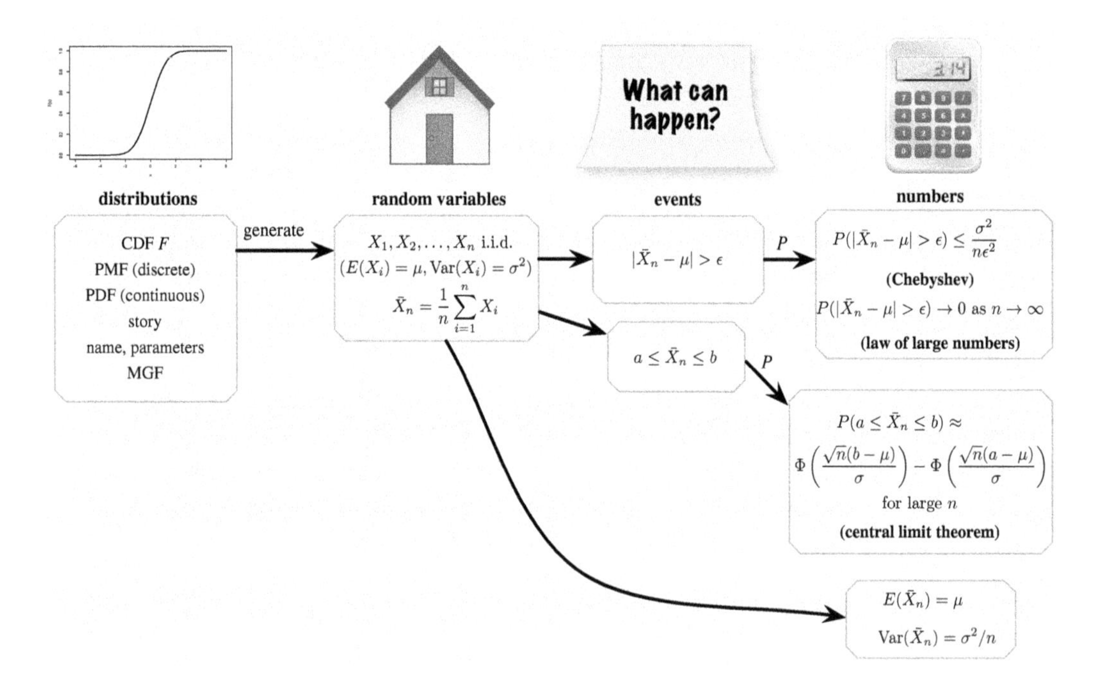

FIGURE 10.7

The sample mean $\bar{X}_n$ of i.i.d. random variables $X_1, \ldots, X_n$ is an important quantity in many problems. Chebyshev's inequality bounds the probability of the sample mean being far from the true mean. The weak law of large numbers, which follows from Chebyshev's inequality, says that for n large, the probability is very high that the sample mean will be very close to the true mean. The central limit theorem says that for n large the distribution of the sample mean will be approximately Normal.

Here, one last time, is the diagram of relationships between the named distributions, updated to include the Chi-Square distribution (as a special case of the Gamma) and Student-t distribution (with the Cauchy as a special case). We have also added arrows to show the convergence of the Poisson, Gamma, and Student-t distributions to Normality; the first two are a consequence of the central limit theorem, and the third is a consequence of the law of large numbers.

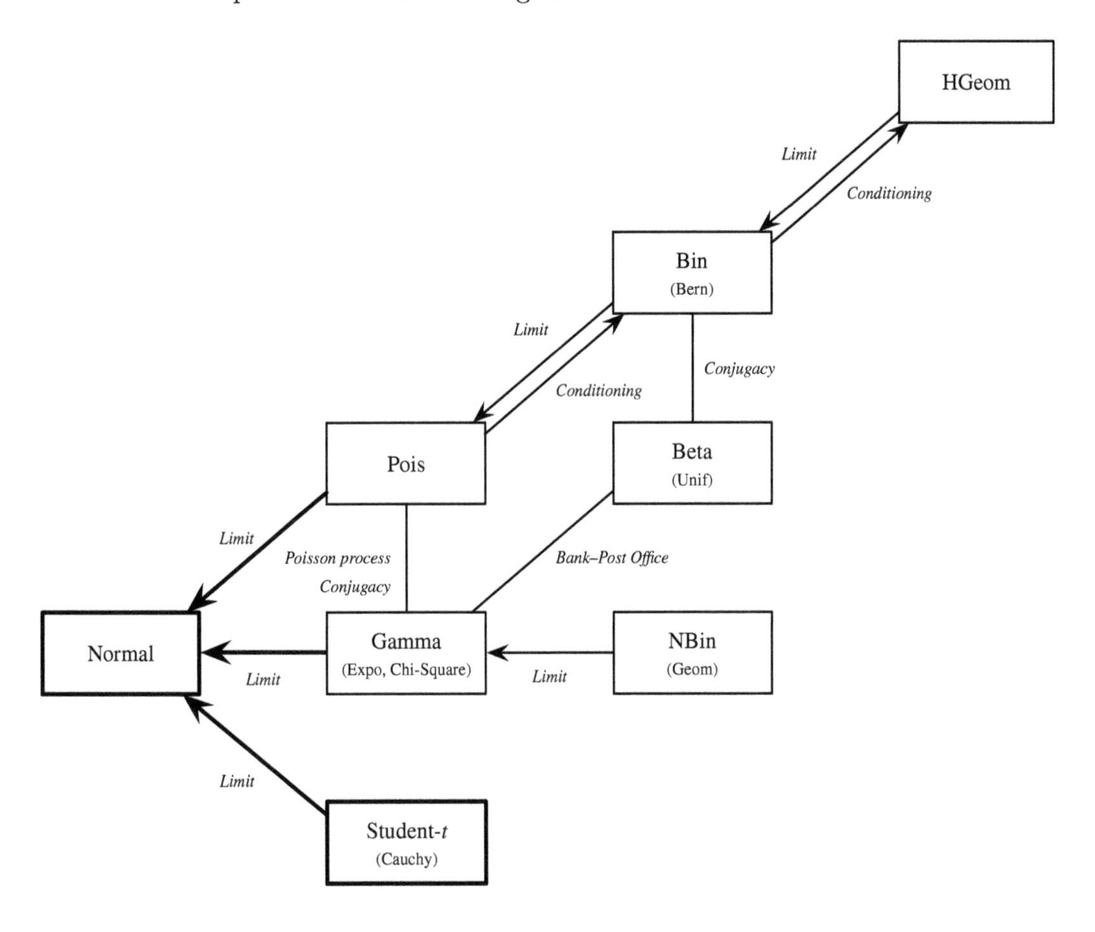

Now we see that all the named distributions are connected to one another!

10.6 R

Jensen's inequality

R makes it easy to compare the expectations of X and $g(X)$ for a given choice of g, and this allows us to verify some special cases of Jensen's inequality. For example, suppose we simulate 10^4 times from the Expo(1) distribution:

```
x <- rexp(10^4)
```

According to Jensen's inequality, $E(\log X) \leq \log EX$. The former can be approximated by `mean(log(x))` and the latter can be approximated by `log(mean(x))`, so compute both:

```
mean(log(x))
log(mean(x))
```

For the Expo(1) distribution, we find that `mean(log(x))` is approximately -0.6 (the true value is around -0.577), while `log(mean(x))` is approximately 0 (the true value is 0). This indeed suggests $E(\log X) \leq \log EX$. We could also compare `mean(x^3)` to `mean(x)^3`, or `mean(sqrt(x))` to `sqrt(mean(x))`—the possibilities are endless.

Visualization of the law of large numbers

To plot the running proportion of Heads in a sequence of independent fair coin tosses, we first generate the coin tosses themselves:

```
nsim <- 300
p <- 1/2
x <- rbinom(nsim,1,p)
```

Then we compute $\bar{X}_n$ for each value of n and store the results in `xbar`:

```
xbar <- cumsum(x)/(1:nsim)
```

The above line of code performs elementwise division of the two vectors `cumsum(x)` and `1:nsim`. Finally, we plot `xbar` against the number of coin tosses:

```
plot(1:nsim,xbar,type="l",ylim=c(0,1))
```

You should see that the values of `xbar` approach `p`, by the LLN.

Monte Carlo estimate of π

A famous example of Monte Carlo integration is the Monte Carlo estimate of π. The unit disk $\{(x, y) : x^2 + y^2 \leq 1\}$ is inscribed in the square $[-1, 1] \times [-1, 1]$, which has area 4. If we generate a large number of points that are Uniform on the square, the proportion of points falling inside the disk is approximately equal to the ratio of the disk's area to the square's area, which is $\pi/4$. Thus, to estimate π we can take the proportion of points inside the circle and multiply by 4.

In R, to generate Uniform points on the 2D square, we can independently generate the x-coordinate and the y-coordinate as Unif$(-1, 1)$ r.v.s, using the results of Example 7.1.23:

```
nsim <- 10^6
```

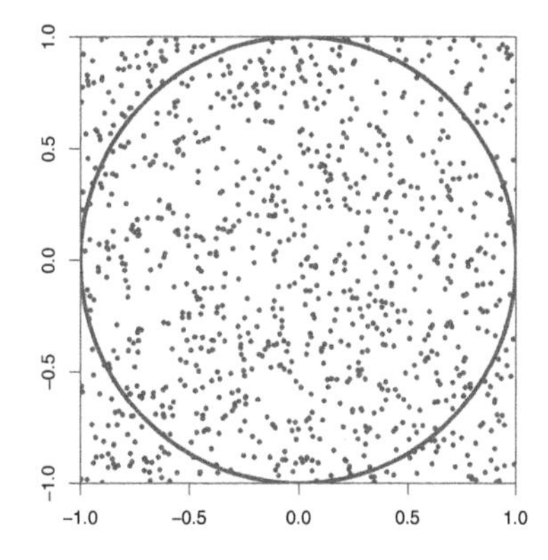

FIGURE 10.8
Monte Carlo estimate of π: Generate points that are Uniform on the 2D square $[-1,1] \times [-1,1]$, which has area 4. The proportion of points falling in the unit disk is approximately $\pi/4$.

```
x <- runif(nsim,-1,1)
y <- runif(nsim,-1,1)
```

To count the number of points in the disk, we type `sum(x^2+y^2<1)`. The vector `x^2+y^2<1` is an indicator vector whose ith element is 1 if the ith point falls inside the disk and 0 otherwise, so the sum of the elements is the number of points in the disk. To get our estimate of π, we convert the sum into a proportion and multiply by 4:

```
4*sum(x^2+y^2<1)/nsim
```

How close was your estimate to the actual value of π?

Visualizations of the central limit theorem

One way to visualize the central limit theorem for a distribution of interest is to plot the distribution of $\bar{X}_n$ for various values of n, as in Figure 10.5. To do this, we first have to generate i.i.d. $X_1, \dots, X_n$ a bunch of times from our distribution of interest. For example, suppose that our distribution of interest is Unif$(0,1)$, and we are interested in the distribution of $\bar{X}_{12}$, i.e., we set $n = 12$. In the following code, we create a matrix of i.i.d. standard Uniforms. The matrix has 12 columns, corresponding to X_1 through X_{12}. Each row of the matrix is a different realization of X_1 through X_{12}.

```
nsim <- 10^4
n <- 12
x <- matrix(runif(n*nsim), nrow=nsim, ncol=n)
```

Now, to obtain realizations of $\bar{X}_{12}$, we simply take the average of each row of the matrix `x`; we can do this with the `rowMeans` function:

```
xbar <- rowMeans(x)
```

Finally, we create a histogram:

```
hist(xbar)
```

You should see a histogram that looks approximately Normal. Since the Unif(0, 1) distribution is symmetric, the CLT kicks in quickly and the Normal approximation for $\bar{X}_n$ works well, even for $n = 12$. Changing `runif` to `rexp`, we see that for $X_j \sim$ Expo(1), the distribution of $\bar{X}_n$ remains skewed when $n = 12$, so a larger value of n is required before the Normal approximation is decent.

Another neat visualization of the CLT can be found in the `animation` package. This package has a built-in animation of a *quincunx* or *bean machine*, invented by the statistician and geneticist Francis Galton to illustrate the Normal distribution. After installing the package, try:

```
library(animation)
quincunx()
```

Can you use the central limit theorem to explain why the histogram produced by a quincunx should look approximately Normal?

Chi-Square and Student-t distributions

Although the Chi-Square is just a special case of the Gamma, it still has its own functions `dchisq`, `pchisq`, and `rchisq` in R: `dchisq(x,n)` and `pchisq(x,n)` return the values of the χ^2_n PDF and CDF at x, and `rchisq(nsim,n)` generates `nsim` i.i.d. χ^2_n r.v.s.

The Student-t distribution has functions `dt`, `pt`, and `rt`. To evaluate the PDF or CDF of the t_n distribution at x, we use `dt(x,n)` or `pt(x,n)`. To generate `nsim` i.i.d. r.v.s from the t_n distribution, we use `rt(nsim,n)`. Of course, `dt(x,1)` is the same as `dcauchy(x)`.

10.7 Exercises

Exercises marked with Ⓢ have detailed solutions at `http://stat110.net`.

Inequalities

1. Ⓢ In a national survey, a random sample of people are chosen and asked whether they support a certain policy. Assume that everyone in the population is equally likely to be surveyed at each step, and that the sampling is with replacement (sampling without replacement is typically more realistic, but with replacement will be a good approximation if the sample size is small compared to the population size). Let n be the sample size, and let $\hat{p}$ and p be the proportion of people who support the policy in the sample and in the entire population, respectively. Show that for every $c > 0$,

$$P(|\hat{p} - p| > c) \leq \frac{1}{4nc^2}.$$

2. Ⓢ For i.i.d. r.v.s $X_1, \dots, X_n$ with mean μ and variance σ^2, give a value of n (as a specific number) that will ensure that there is at least a 99% chance that the sample mean will be within 2 standard deviations of the true mean μ.

3. Ⓢ Show that for any two positive r.v.s X and Y with neither a constant multiple of the other,

$$E(X/Y)E(Y/X) > 1.$$

4. Ⓢ The famous *arithmetic mean-geometric mean* inequality says that for any positive numbers $a_1, a_2, \dots, a_n$,

$$\frac{a_1 + a_2 + \cdots + a_n}{n} \geq (a_1 a_2 \cdots a_n)^{1/n}.$$

Show that this inequality follows from Jensen's inequality, by considering $E \log(X)$ for an r.v. X whose possible values are $a_1, \dots, a_n$ (you should specify the PMF of X; if you want, you can assume that the a_j are distinct (no repetitions), but be sure to say so if you assume this).

5. Ⓢ Let X be a discrete r.v. whose distinct possible values are $x_0, x_1, \dots,$ and let $p_k = P(X = x_k)$. The entropy of X is $H(X) = \sum_{k=0}^{\infty} p_k \log_2(1/p_k)$.

(a) Find $H(X)$ for $X \sim \text{Geom}(p)$.

Hint: Use properties of logs, and interpret part of the sum as an expected value.

(b) Let X and Y be i.i.d. discrete r.v.s. Show that $P(X = Y) \geq 2^{-H(X)}$.

Hint: Consider $E(\log_2(W))$, where W is an r.v. taking value p_k with probability p_k.

6. Let X be a random variable with mean μ and variance σ^2. Show that

$$E(X - \mu)^4 \geq \sigma^4,$$

and use this to show that the kurtosis of X is at least -2.

Fill-in-the-blank inequalities

7. Ⓢ Let X and Y be i.i.d. positive r.v.s, and let $c > 0$. For each part below, fill in the appropriate equality or inequality symbol: write $=$ if the two sides are always equal, $\leq$ if the left-hand side is less than or equal to the right-hand side (but they are not necessarily equal), and similarly for $\geq$. If no relation holds in general, write ?.

(a) $E(\log(X))$ ____ $\log(E(X))$

(b) $E(X)$ ____ $\sqrt{E(X^2)}$

(c) $E(\sin^2(X)) + E(\cos^2(X))$ ____ 1

(d) $E(|X|)$ ___ $\sqrt{E(X^2)}$

(e) $P(X > c)$ ___ $\frac{E(X^3)}{c^3}$

(f) $P(X \leq Y)$ ___ $P(X \geq Y)$

(g) $E(XY)$ ___ $\sqrt{E(X^2)E(Y^2)}$

(h) $P(X + Y > 10)$ ___ $P(X > 5 \text{ or } Y > 5)$

(i) $E(\min(X, Y))$ ___ $\min(EX, EY)$

(j) $E(X/Y)$ ___ $\frac{EX}{EY}$

(k) $E(X^2(X^2 + 1))$ ___ $E(X^2(Y^2 + 1))$

(l) $E\left(\frac{X^3}{X^3+Y^3}\right)$ ___ $E(\frac{Y^3}{X^3+Y^3})$

8. Ⓢ Write the most appropriate of $\leq$, $\geq$, $=$, or ? in the blank for each part (where "?" means that no relation holds in general).

   In (c) through (f), X and Y are i.i.d. (independent identically distributed) positive random variables. Assume that the various expected values exist.

   (a) (probability that a roll of 2 fair dice totals 9) ___ (probability that a roll of 2 fair dice totals 10)

   (b) (probability that at least 65% of 20 fair coin flips land Heads) ___ (probability that at least 65% of 2000 fair coin flips land Heads)

   (c) $E(\sqrt{X})$ ___ $\sqrt{E(X)}$

   (d) $E(\sin X)$ ___ $\sin(EX)$

   (e) $P(X + Y > 4)$ ___ $P(X > 2)P(Y > 2)$

   (f) $E\left((X + Y)^2\right)$ ___ $2E(X^2) + 2(EX)^2$

9. Let X and Y be i.i.d. continuous r.v.s. Assume that the various expressions below exist. Write the most appropriate of $\leq$, $\geq$, $=$, or ? in the blank for each part (where "?" means that no relation holds in general).

   (a) $e^{-E(X)}$ ___ $E(e^{-X})$

   (b) $P(X > Y + 3)$ ___ $P(Y > X + 3)$

   (c) $P(X > Y + 3)$ ___ $P(X > Y - 3)$

   (d) $E(X^4)$ ___ $(E(XY))^2$

   (e) $\text{Var}(Y)$ ___ $E(\text{Var}(Y|X))$

   (f) $P(|X + Y| > 3)$ ___ $E|X|$

10. Ⓢ Let X and Y be positive random variables, *not necessarily independent*. Assume that the various expected values below exist. Write the most appropriate of $\leq$, $\geq$, $=$, or ? in the blank for each part (where "?" means that no relation holds in general).

   (a) $(E(XY))^2$ ___ $E(X^2)E(Y^2)$

   (b) $P(|X + Y| > 2)$ ___ $\frac{1}{10}E((X + Y)^4)$

   (c) $E(\log(X + 3))$ ___ $\log(E(X + 3))$

   (d) $E(X^2 e^X)$ ___ $E(X^2)E(e^X)$

   (e) $P(X + Y = 2)$ ___ $P(X = 1)P(Y = 1)$

   (f) $P(X + Y = 2)$ ___ $P(\{X \geq 1\} \cup \{Y \geq 1\})$

11. Ⓢ Let X and Y be positive random variables, *not necessarily independent.* Assume that the various expected values below exist. Write the most appropriate of $\leq$, $\geq$, $=$, or ? in the blank for each part (where "?" means that no relation holds in general).
(a) $E(X^3)$ ____ $\sqrt{E(X^2)E(X^4)}$

(b) $P(|X+Y|>2)$ ____ $\frac{1}{16}E((X+Y)^4)$

(c) $E(\sqrt{X+3})$ ____ $\sqrt{E(X+3)}$

(d) $E(\sin^2(X))+E(\cos^2(X))$ ____ 1

(e) $E(Y|X+3)$ ____ $E(Y|X)$

(f) $E(E(Y^2|X))$ ____ $(EY)^2$

12. Ⓢ Let X and Y be positive random variables, *not necessarily independent.* Assume that the various expressions below exist. Write the most appropriate of $\leq$, $\geq$, $=$, or ? in the blank for each part (where "?" means that no relation holds in general).
(a) $P(X+Y>2)$ ____ $\frac{EX+EY}{2}$

(b) $P(X+Y>3)$ ____ $P(X>3)$

(c) $E(\cos(X))$ ____ $\cos(EX)$

(d) $E(X^{1/3})$ ____ $(EX)^{1/3}$

(e) $E(X^Y)$ ____ $(EX)^{EY}$

(f) $E\left(E(X|Y)+E(Y|X)\right)$ ____ $EX+EY$

13. Ⓢ Let X and Y be i.i.d. positive random variables. Assume that the various expressions below exist. Write the most appropriate of $\leq$, $\geq$, $=$, or ? in the blank for each part (where "?" means that no relation holds in general).
(a) $E(e^{X+Y})$ ____ $e^{2E(X)}$

(b) $E(X^2e^X)$ ____ $\sqrt{E(X^4)E(e^{2X})}$

(c) $E(X|3X)$ ____ $E(X|2X)$

(d) $E(X^7Y)$ ____ $E(X^7E(Y|X))$

(e) $E(\frac{X}{Y}+\frac{Y}{X})$ ____ 2

(f) $P(|X-Y|>2)$ ____ $\frac{\text{Var}(X)}{2}$

14. Ⓢ Let X and Y be i.i.d. Gamma$(\frac{1}{2},\frac{1}{2})$, and let $Z\sim\mathcal{N}(0,1)$ (note that X and Z may be dependent, and Y and Z may be dependent). For (a),(b),(c), write the most appropriate of $<$, $>$, $=$, or ? in each blank; for (d),(e),(f), write the most appropriate of $\leq$, $\geq$, $=$, or ? in each blank.
(a) $P(X<Y)$ ____ 1/2

(b) $P(X=Z^2)$ ____ 1

(c) $P(Z\geq\frac{1}{X^4+Y^4+7})$ ____ 1

(d) $E(\frac{X}{X+Y})E((X+Y)^2)$ ____ $E(X^2)+(E(X))^2$

(e) $E(X^2Z^2)$ ____ $\sqrt{E(X^4)E(X^2)}$

(f) $E((X+2Y)^4)$ ____ 3^4

15. Let X, Y, Z be i.i.d. $\mathcal{N}(0,1)$ r.v.s. Write the most appropriate of $\leq$, $\geq$, $=$, or ? in each blank (where "?" means that no relation holds in general).
(a) $P(X^2 + Y^2 + Z^2 > 6)$ ____ $1/2$

(b) $P(X^2 < 1)$ ____ $2/3$

(c) $E\left(\frac{X^2}{X^2+Y^2+Z^2}\right)$ ____ $1/4$

(d) $\text{Var}(\Phi(X) + \Phi(Y) + \Phi(Z))$ ____ $1/4$

(e) $E(e^{-X})$ ____ $E(e^X)$

(f) $E(|X|e^X)$ ____ $\sqrt{E(e^{2X})}$

16. Let X, Y, Z, W be i.i.d. positive r.v.s with CDF F and $E(X) = 1$. Write the most appropriate of $\leq$, $\geq$, $=$, or ? in each blank (where "?" means that no relation holds in general).
(a) $F(3)$ ____ $2/3$

(b) $(F(3))^3$ ____ $P(X + Y + Z \leq 9)$

(c) $E\left(\frac{X^2}{X^2+Y^2+Z^2+W^2}\right)$ ____ $1/4$

(d) $E(XYZW)$ ____ $E(X^4)$

(e) $\text{Var}(E(Y|X))$ ____ $\text{Var}(Y)$

(f) $\text{Cov}(X + Y, X - Y)$ ____ 0

17. Let X, Y, Z be i.i.d. (independent, identically distributed) r.v.s with a continuous distribution. Write the most appropriate of $\leq$, $\geq$, $=$, or ? in each blank (where "?" means that no relation holds in general).
(a) $P(X < Y < Z)$ ____ $1/6$

(b) $P(X > 1)$ ____ $E(X)$

(c) $P\left(\sum_{k=0}^{2015}\left(\frac{X^2+1}{X^2+2}\right)^k > 3\right)$ ____ $P(X^2 > 1)$

(d) $E\left(\sqrt{X^2 + Y^2}\right)$ ____ $\sqrt{E(X^2) + E(Y^2)}$

(e) $\text{Var}(Y^2|Z)$ ____ $\text{Var}(X^2|X)$

(f) $\text{Var}(X - 2Y + 3Z)$ ____ $14\text{Var}(X)$

18. Let $X, Y, Z \sim \text{Bin}(n, p)$ be i.i.d. (independent and identically distributed). Write the most appropriate of $\leq$, $\geq$, $=$, or ? in each blank (where "?" means that no relation holds in general).
(a) $P(X < Y < Z)$ ____ $1/6$

(b) $P(X + Y + Z > n)$ ____ $3p$

(c) $P\left(\sum_{n=0}^{2016}\frac{(X^2+1)^n}{n!} > e^5\right)$ ____ $P(X > 2)$

(d) $E\left(e^X\right)$ ____ $(pe + 1 - p)^n$

(e) $\text{Var}(X + Y)$ ____ $n/2$

(f) $E(X \mid X + Y = n)$ ____ $n/2$

19. Let $X, Y, Z, W \sim \mathcal{N}(c, c^2)$ be i.i.d., where $c > 0$. Write the most appropriate of $\leq$, $\geq$, $=$, or ? in each blank (where "?" means that no relation holds in general).

    (a) $P(X + Y \leq Z - W)$ ____ $\Phi(-1)$

    (b) $P(X^4 - Y^8 \leq Z^4 - W^8)$ ____ $\Phi(-1)$

    (c) $E(X - c)^2$ ____ c

    (d) $E(X - c)^3$ ____ c

    (e) $E\left(\frac{1}{4}X^2 + X\right)$ ____ $\frac{1}{2}c^2 + c$

    (f) $\log\left(E(e^X)\right)$ ____ $\frac{1}{2}c^2 + c$

20. Let $X, Y \sim \text{Pois}(\lambda)$ be i.i.d., where $\lambda > 1$. Write the most appropriate of $\leq$, $\geq$, $=$, or ? in each blank (where "?" means that no relation holds in general).

    (a) $P(X \leq Y)$ ____ $1/2$

    (b) $P(X + Y \leq 1)$ ____ $3e^{-2\lambda}$

    (c) $E\left(e^{X+Y}\right)$ ____ $e^{2\lambda}$

    (d) $E(X \mid X + Y = 4)$ ____ 2

    (e) $\text{Var}(X \mid X + Y = 4)$ ____ 1

    (f) $E(X^2 - Y)$ ____ λ^2

LLN and CLT

21. Ⓢ Let $X_1, X_2, \ldots$ be i.i.d. positive random variables with mean 2. Let $Y_1, Y_2, \ldots$ be i.i.d. positive random variables with mean 3. Show that

$$\frac{X_1 + X_2 + \cdots + X_n}{Y_1 + Y_2 + \cdots + Y_n} \to \frac{2}{3}$$

with probability 1. Does it matter whether the X_i are independent of the Y_j?

22. Ⓢ Let $U_1, U_2, \ldots, U_{60}$ be i.i.d. Unif(0,1) and $X = U_1 + U_2 + \cdots + U_{60}$.

    (a) Which important distribution is the distribution of X very close to? Specify what the parameters are, and state which theorem justifies your choice.

    (b) Give a simple but accurate approximation for $P(X > 17)$. Justify briefly.

23. Ⓢ Let $V_n \sim \chi^2_n$ and $T_n \sim t_n$ for all positive integers n.

    (a) Find numbers a_n and b_n such that $a_n(V_n - b_n)$ converges in distribution to $\mathcal{N}(0, 1)$.

    (b) Show that $T_n^2/(n + T_n^2)$ has a Beta distribution (without using calculus).

24. Ⓢ Let $T_1, T_2, \ldots$ be i.i.d. Student-t r.v.s with $m \geq 3$ degrees of freedom. Find constants a_n and b_n (in terms of m and n) such that $a_n(T_1 + T_2 + \cdots + T_n - b_n)$ converges to $\mathcal{N}(0, 1)$ in distribution as $n \to \infty$.

25. Ⓢ (a) Let $Y = e^X$, with $X \sim \text{Expo}(3)$. Find the mean and variance of Y.

    (b) For $Y_1, \ldots, Y_n$ i.i.d. with the same distribution as Y from (a), what is the approximate distribution of the sample mean $\bar{Y}_n = \frac{1}{n}\sum_{j=1}^{n} Y_j$ when n is large?

26. Ⓢ (a) Explain why the Pois(n) distribution is approximately Normal if n is a large positive integer (specifying what the parameters of the Normal are).

    (b) Stirling's formula is an amazingly accurate approximation for factorials:

$$n! \approx \sqrt{2\pi n}\left(\frac{n}{e}\right)^n,$$

    where in fact the ratio of the two sides goes to 1 as $n \to \infty$. Use (a) to give a quick heuristic derivation of Stirling's formula by using a Normal approximation to the probability that a Pois(n) r.v. is n, with the continuity correction: first write $P(N = n) = P\left(n - \frac{1}{2} < N < n + \frac{1}{2}\right)$, where $N \sim \text{Pois}(n)$.

27. Ⓢ (a) Consider i.i.d. Pois(λ) r.v.s $X_1, X_2, \dots$. The MGF of X_j is $M(t) = e^{\lambda(e^t-1)}$. Find the MGF $M_n(t)$ of the sample mean $\bar{X}_n = \frac{1}{n}\sum_{j=1}^n X_j$.

    (b) Find the limit of $M_n(t)$ as $n \to \infty$. (You can do this with almost no calculation using a relevant theorem; or you can use (a) and the fact that $e^x \approx 1 + x$ if x is very small.)

28. Let $X_n \sim \text{Pois}(n)$ for all positive integers n. Use MGFs to show that the distribution of the standardized version of X_n converges to a Normal distribution as $n \to \infty$, without invoking the CLT.

29. An important concept in frequentist statistics is that of a *confidence interval* (CI). Suppose we observe data X from a distribution with parameter θ. Unlike in Bayesian statistics, θ is treated as a fixed but unknown constant; it is not given a prior distribution. A 95% confidence interval consists of a lower bound $L(X)$ and upper bound $U(X)$ such that

$$P(L(X) < \theta < U(X)) = 0.95$$

    for all possible values of θ. Note that in the above statement, $L(X)$ and $U(X)$ are random variables, as they are functions of the r.v. X, whereas θ is a constant. The definition says that the random interval $(L(X), U(X))$ has a 95% chance of containing the true value of θ.

    Imagine an army of frequentists all over the world, independently generating 95% CIs. The jth frequentist observes data X_j and makes a confidence interval for the parameter θ_j. Show that if there are n of these frequentists, then the fraction of their intervals which contain the corresponding parameter approaches 0.95 as $n \to \infty$.

    Hint: Consider the indicator r.v. $I_j = I(L(X_j) < \theta_j < U(X_j))$.

30. This problem extends Example 10.3.7 to a more general setting. Again, suppose a very volatile stock rises 70% or drops 50% in price each day, with equal probabilities and with different days independent.

    (a) Suppose a hedge fund manager always invests half of her current fortune into the stock each day. Let Y_n be her fortune after n days, starting from an initial fortune of $Y_0 = 100$. What happens to Y_n as $n \to \infty$?

    (b) More generally, suppose the hedge fund manager always invests a fraction α of her current fortune into the stock each day (in Part (a), we took $\alpha = 1/2$). With Y_0 and Y_n defined as in Part (a), find the function $g(\alpha)$ such that

$$\frac{\log Y_n}{n} \to g(\alpha)$$

    with probability 1 as $n \to \infty$, and prove that $g(\alpha)$ is maximized when $\alpha = 2/7$.

Mixed practice

31. As in Exercise 36 from Chapter 3, there are n voters in an upcoming election in a certain country, where n is a large, even number. There are two candidates, A and B. Each voter chooses randomly whom to vote for, independently and with equal probabilities.

(a) Use a Normal approximation (with continuity correction) to get an approximation for the probability of a tie, in terms of Φ.

(b) Use a first-order Taylor expansion (linear approximation) to the approximation from Part (a) to show that the probability of a tie is approximately $1/\sqrt{cn}$, where c is a constant (which you should specify).

32. Cassie enters a casino with $X_0 = 1$ dollar and repeatedly plays the following game: with probability 1/3, the amount of money she has increases by a factor of 3; with probability 2/3, the amount of money she has decreases by a factor of 3. Let X_n be the amount of money she has after playing this game n times. For example, X_1 is 3 with probability 1/3 and is 3^{-1} with probability 2/3.

(a) Compute $E(X_1)$, $E(X_2)$ and, in general, $E(X_n)$.

(b) What happens to $E(X_n)$ as $n \to \infty$?

(c) Let Y_n be the number of times out of the first n games that Cassie triples her money. What happens to Y_n/n as $n \to \infty$?

(d) What happens to X_n as $n \to \infty$?

33. A handy rule of thumb in statistics and life is as follows:

Conditioning often makes things better.

This problem explores how the above rule of thumb applies to estimating unknown parameters. Let θ be an unknown parameter that we wish to estimate based on data $X_1, X_2, \dots, X_n$ (these are r.v.s before being observed, and then after the experiment they "crystallize" into data). In this problem, θ is viewed as an unknown constant, and is not treated as an r.v. as in the Bayesian approach. Let T_1 be an estimator for θ (this means that T_1 is a function of $X_1, \dots, X_n$ which is being used to estimate θ).

A strategy for improving T_1 (in some problems) is as follows. Suppose that we have an r.v. R such that $T_2 = E(T_1|R)$ is a function of $X_1, \dots, X_n$ (in general, $E(T_1|R)$ might involve unknowns such as θ but then it couldn't be used as an estimator). Also suppose that $P(T_1 = T_2) < 1$, and that $E(T_1^2)$ is finite.

(a) Use Jensen's inequality to show that T_2 is better than T_1 in the sense that the mean squared error is less, i.e.,

$$E(T_2 - \theta)^2 < E(T_1 - \theta)^2.$$

Hint: Use Adam's law on the right-hand side.

(b) The *bias* of an estimator T for θ is defined to be $b(T) = E(T) - \theta$. An important identity in statistics, a form of the *bias-variance tradeoff*, is that mean squared error is variance plus squared bias:

$$E(T - \theta)^2 = \text{Var}(T) + (b(T))^2.$$

Use this identity and Eve's law to give an alternative proof of the result from (a).

(c) Now suppose that $X_1, X_2, \dots$ are i.i.d. with mean θ, and consider the special case $T_1 = X_1$, $R = \sum_{j=1}^{n} X_j$. Find T_2 in simplified form, and check that it has lower mean squared error than T_1 for $n \geq 2$. Also, say what happens to T_1 and T_2 as $n \to \infty$.

34. Each page of an n-page book has a Pois(λ) number of typos, where λ is unknown (but is not treated as an r.v.). Typos on different pages are independent. Thus we have i.i.d. $X_1, \dots, X_n \sim \text{Pois}(\lambda)$, where X_j is the number of typos on page j. Suppose we are interested in estimating the probability θ that a page has no typos:

$$\theta = P(X_j = 0) = e^{-\lambda}.$$

(a) Let $\bar{X}_n = \frac{1}{n}(X_1 + \dots + X_n)$. Show that the estimator $T_n = e^{-\bar{X}_n}$ is biased for θ. (That is, show $E(T_n) \neq \theta$. Estimators and bias are defined in the previous problem.)

(b) Show that as $n \to \infty$, $T_n \to \theta$ with probability 1.

(c) Show that $W = \frac{1}{n}(I(X_1 = 0) + \dots + I(X_n = 0))$ is unbiased for θ. Using the fact that $X_1|(X_1 + \dots + X_n = s) \sim \text{Bin}(s, 1/n)$, find $E(W|X_1 + \dots + X_n)$. Is the estimator $\tilde{W} = E(W|X_1 + \dots + X_n)$ also unbiased for θ?

(d) Using Eve's law or otherwise, show that $\tilde{W}$ has lower variance than W, and relate this to the previous question.

35. A binary sequence is being generated through some process (random or deterministic). You need to sequentially predict each new number, i.e., you predict whether the next number will be 0 or 1, then observe it, then predict the next number, etc. Each of your predictions can be based on the entire past history of the sequence.

(a) Suppose for this part that the binary sequence consists of i.i.d. Bern(p) r.v.s, with p known. What is your optimal strategy (for each prediction, your goal is to maximize the probability of being correct)? What is the probability that you will guess the nth value correctly with this strategy?

(b) Now suppose that the binary sequence consists of i.i.d. Bern(p) r.v.s, with p unknown. Consider the following strategy: say 1 as your first prediction; after that, say "1" if the proportion of 1's so far is at least 1/2, and say "0" otherwise. Find the limit as $n \to \infty$ of the probability of guessing the nth value correctly (in terms of p).

(c) Now suppose that you follow the strategy from (b), but that the binary sequence is generated by a nefarious entity who knows your strategy. What can the entity do to make your guesses be wrong as often as possible?

36. Ⓢ Let X and Y be independent standard Normal r.v.s and let $R^2 = X^2 + Y^2$ (where $R > 0$ is the distance from (X, Y) to the origin).

(a) The distribution of R^2 is an example of three of the important distributions we have seen. State which three of these distributions R^2 is an instance of, specifying the parameter values.

(b) Find the PDF of R.

Hint: Start with the PDF $f_W(w)$ of $W = R^2$.

(c) Find $P(X > 2Y + 3)$ in terms of the standard Normal CDF Φ.

(d) Compute $\text{Cov}(R^2, X)$. Are R^2 and X independent?

37. Ⓢ Let $Z_1, \dots, Z_n \sim \mathcal{N}(0, 1)$ be i.i.d.

(a) As a function of Z_1, create an Expo(1) r.v. X (your answer can also involve the standard Normal CDF Φ).

(b) Let $Y = e^{-R}$, where $R = \sqrt{Z_1^2 + \dots + Z_n^2}$. Write down (but do not evaluate) an integral for $E(Y)$.

(c) Let $X_1 = 3Z_1 - 2Z_2$ and $X_2 = 4Z_1 + 6Z_2$. Determine whether X_1 and X_2 are independent (be sure to mention which results you're using).

38. Ⓢ Let $X_1, X_2, \ldots$ be i.i.d. positive r.v.s. with mean μ, and let $W_n = \frac{X_1}{X_1+\cdots+X_n}$.

(a) Find $E(W_n)$.

Hint: Consider $\frac{X_1}{X_1+\cdots+X_n} + \frac{X_2}{X_1+\cdots+X_n} + \cdots + \frac{X_n}{X_1+\cdots+X_n}$.

(b) What random variable does nW_n converge to (with probability 1) as $n \to \infty$?

(c) For the case that $X_j \sim \text{Expo}(\lambda)$, find the distribution of W_n, preferably without using calculus. (If it is one of the named distributions we have studied, state its name and specify the parameters; otherwise, give the PDF.)

39. Let $X_1, X_2, \ldots$ be i.i.d. Expo(1).

(a) Let $N = \min\{n : X_n \geq 1\}$ be the index of the first X_j to exceed 1. Find the distribution of $N - 1$ (give the name and parameters), and hence find $E(N)$.

(b) Let $M = \min\{n : X_1 + X_2 + \cdots + X_n \geq 10\}$ be the number of X_j's we observe until their sum exceeds 10 for the first time. Find the distribution of $M - 1$ (give the name and parameters), and hence find $E(M)$.

Hint: Consider a Poisson process.

(c) Let $\bar{X}_n = (X_1 + \cdots + X_n)/n$. Find the exact distribution of $\bar{X}_n$ (give the name and parameters), as well as the approximate distribution of $\bar{X}_n$ for n large (give the name and parameters).

11

Markov chains

Markov chains were first introduced in 1906 by Andrey Markov (of Markov's inequality), with the goal of showing that the law of large numbers can apply to random variables that are not independent. To see where the Markov model comes from, start by considering an i.i.d. sequence of random variables $X_0, X_1, \dots, X_n, \dots$ where we think of n as time. This is the setting we worked in throughout Chapter 10, but for modeling real-world phenomena, independence can be an excessively restrictive assumption; it means that the X_n provide absolutely no information about each other. At the other extreme, allowing arbitrary interactions between the X_n makes it very difficult to compute even basic things. A Markov chain is a sequence of r.v.s that exhibits one-step dependence, in a precise sense that we shall soon define. Thus Markov chains are a happy medium between complete independence and complete dependence.

Since their invention, Markov chains have become extremely important in a huge number of fields such as biology, game theory, finance, machine learning, and statistical physics. They are also very widely used for simulations of complex distributions, via algorithms known as *Markov chain Monte Carlo* (MCMC). In this chapter we will introduce Markov chains and their properties, and in the next chapter we'll look at some examples of MCMC techniques.

11.1 Markov property and transition matrix

Markov chains "live" in both space and time: the set of possible values of the X_n is called the *state space*, and the index n represents the evolution of the process over *time*. The state space of a Markov chain can be either discrete or continuous, and time can also be either discrete or continuous (in the continuous-time setting, we would imagine a process X_t defined for all real $t \geq 0$). In this chapter we will focus exclusively on *discrete-state, discrete-time* Markov chains, with a *finite* state space. Specifically, we will assume that the X_n take values in a finite set, which we usually take to be $\{1, 2, \dots, M\}$ or $\{0, 1, \dots, M\}$.

Definition 11.1.1 (Markov chain). A sequence of random variables $X_0, X_1, X_2, \dots$ taking values in the *state space* $\{1, 2, \dots, M\}$ is called a *Markov chain* if for all $n \geq 0$,

$$P(X_{n+1} = j | X_n = i, X_{n-1} = i_{n-1}, \dots, X_0 = i_0) = P(X_{n+1} = j | X_n = i).$$

The quantity $P(X_{n+1} = j|X_n = i)$ is called the *transition probability* from state i to state j. In this book, when referring to a Markov chain we will implicitly assume that it is *time-homogeneous*, which means that the transition probability $P(X_{n+1} = j|X_n = i)$ is the same for all times n. But care is needed, since the literature is not consistent about whether to say "time-homogeneous Markov chain" or just "Markov chain".

The above condition is called the *Markov property*, and it says that given the entire past history $X_0, X_1, X_2, \ldots, X_n$, only the *most recent* term, X_n, matters for predicting X_{n+1}. If we think of time n as the present, times before n as the past, and times after n as the future, the Markov property says that given the present, the past and future are conditionally independent. The Markov property greatly simplifies computations of conditional probability: instead of having to condition on the entire past, we only need to condition on the *most recent* value.

To describe the dynamics of a Markov chain, we need to know the probabilities of moving from any state to any other state, that is, the probabilities $P(X_{n+1} = j|X_n = i)$ on the right-hand side of the Markov property. This information can be encoded in a matrix, called the *transition matrix*, whose (i, j) entry is the probability of going from state i to state j in one step of the chain.

Definition 11.1.2 (Transition matrix). Let $X_0, X_1, X_2, \ldots$ be a Markov chain with state space $\{1, 2, \ldots, M\}$, and let $q_{ij} = P(X_{n+1} = j|X_n = i)$ be the transition probability from state i to state j. The $M \times M$ matrix $Q = (q_{ij})$ is called the *transition matrix* of the chain.

Note that Q is a nonnegative matrix in which each row sums to 1. This is because, starting from any state i, the events "move to 1", "move to 2", ..., "move to M" are disjoint, and their union has probability 1 because the chain has to go somewhere.

Example 11.1.3 (Rainy-sunny Markov chain). Suppose that on any given day, the weather can either be rainy or sunny. If today is rainy, then tomorrow will be rainy with probability 1/3 and sunny with probability 2/3. If today is sunny, then tomorrow will be rainy with probability 1/2 and sunny with probability 1/2. Letting X_n be the weather on day n, $X_0, X_1, X_2, \ldots$ is a Markov chain on the state space $\{R, S\}$, where R stands for rainy and S for sunny. We know that the Markov property is satisfied because, from the description of the process, only today's weather matters for predicting tomorrow's.

The transition matrix of the chain is

$$\begin{array}{c} \\ R \\ S \end{array} \begin{array}{c} \begin{array}{cc} R & \;\; S \end{array} \\ \begin{pmatrix} 1/3 & 2/3 \\ 1/2 & 1/2 \end{pmatrix} \end{array}.$$

The first row says that starting from state R, we transition back to state R with probability 1/3 and transition to state S with probability 2/3. The second row says that starting from state S, we have a $1/2$ chance of moving to state R and a $1/2$ chance of staying in state S. We could just as well have used

$$\begin{array}{c} \\ S \\ R \end{array} \begin{array}{c} \begin{array}{cc} S & R \end{array} \\ \begin{pmatrix} 1/2 & 1/2 \\ 2/3 & 1/3 \end{pmatrix} \end{array}$$

as our transition matrix instead. In general, if there isn't an obvious ordering of the states of a Markov chain (as with the states R and S), we just need to fix an ordering of the states and use it consistently.

The transition probabilities of a Markov chain can also be represented with a diagram. Each state is represented by a circle, and the arrows indicate the possible one-step transitions; we can imagine a particle wandering around from state to state, randomly choosing which arrow to follow. Next to the arrows we write the corresponding transition probabilities.

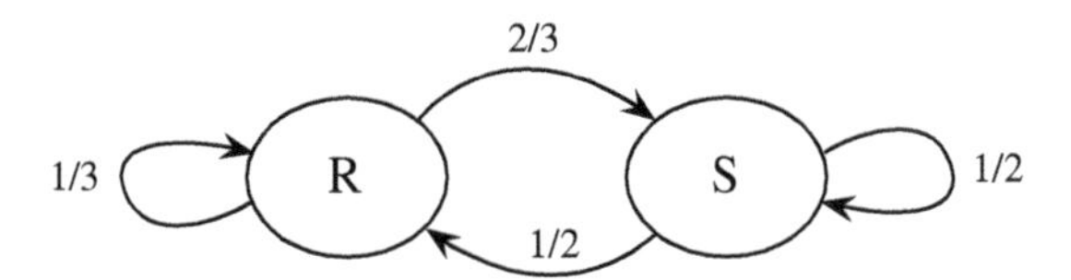

What if the weather tomorrow depended on the weather today and the weather yesterday? For example, suppose that the weather behaves as above, except that if there have been two consecutive days of rain, then tomorrow will definitely be sunny, and if there have been two consecutive days of sun, then tomorrow will definitely be rainy. Under these new weather dynamics, the X_n no longer form a Markov chain, as the Markov property is violated: conditional on today's weather, yesterday's weather can still provide useful information for predicting tomorrow's weather.

However, by enlarging the state space, we can create a new Markov chain: let $Y_n = (X_{n-1}, X_n)$ for $n \geq 1$. Then $Y_1, Y_2, \ldots$ is a Markov chain on the state space $\{(R,R),(R,S),(S,R),(S,S)\}$. You can verify that the new transition matrix is

$$\begin{array}{c} \\ (R,R) \\ (R,S) \\ (S,R) \\ (S,S) \end{array} \begin{array}{c} \begin{array}{cccc} (R,R) & (R,S) & (S,R) & (S,S) \end{array} \\ \begin{pmatrix} 0 & 1 & 0 & 0 \\ 0 & 0 & 1/2 & 1/2 \\ 1/3 & 2/3 & 0 & 0 \\ 0 & 0 & 1 & 0 \end{pmatrix} \end{array}$$

and that its corresponding graphical representation is given in the following figure.

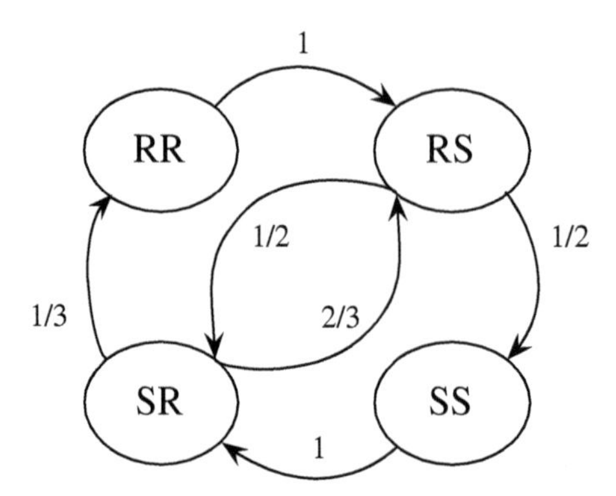

Similarly, we could handle third-order or fourth-order dependencies in the weather by further enlarging the state space to make the Markov property hold. □

Once we have the transition matrix Q of a Markov chain, we can work out the transition probabilities for longer timescales.

Definition 11.1.4 (n-step transition probability)**.** The n-step transition probability from i to j is the probability of being at j exactly n steps after being at i. We denote this by $q_{ij}^{(n)}$:

$$q_{ij}^{(n)} = P(X_n = j | X_0 = i).$$

Note that

$$q_{ij}^{(2)} = \sum_k q_{ik} q_{kj}$$

since to get from i to j in two steps, the chain must go from i to some intermediary state k, and then from k to j; these transitions are independent because of the Markov property. Since the right-hand side is the (i, j) entry of Q^2 by definition of matrix multiplication, we conclude that the matrix Q^2 gives the two-step transition probabilities. By induction, the nth power of the transition matrix gives the n-step transition probabilities:

$$q_{ij}^{(n)} \text{ is the } (i, j) \text{ entry of } Q^n.$$

Example 11.1.5 (Transition matrix of 4-state Markov chain)**.** Consider the 4-state Markov chain depicted in Figure 11.1. When no probabilities are written over the arrows, as in this case, it means all arrows originating from a given state are equally likely. For example, there are 3 arrows originating from state 1, so the transitions $1 \to 3$, $1 \to 2$, and $1 \to 1$ all have probability $1/3$. Therefore the transition matrix of the chain is

$$Q = \begin{pmatrix} 1/3 & 1/3 & 1/3 & 0 \\ 0 & 0 & 1/2 & 1/2 \\ 0 & 1 & 0 & 0 \\ 1/2 & 0 & 0 & 1/2 \end{pmatrix}.$$

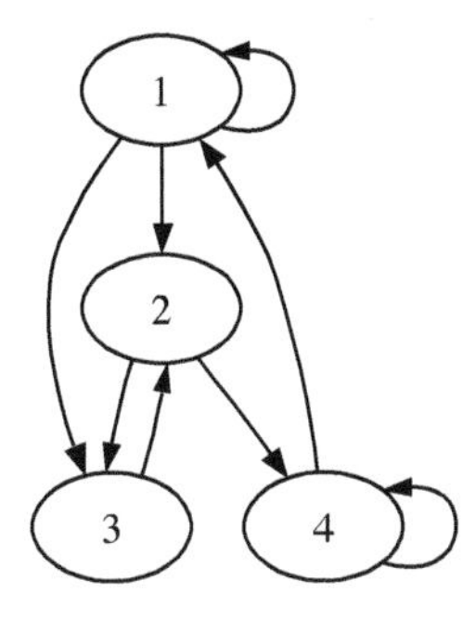

FIGURE 11.1
A 4-state Markov chain.

To compute the probability that the chain is in state 3 after 5 steps, starting at state 1, we would look at the (1,3) entry of Q^5. Here, using a computer to find Q^5,

$$Q^5 = \begin{pmatrix} 853/3888 & 509/1944 & 52/243 & 395/1296 \\ 173/864 & 85/432 & 31/108 & 91/288 \\ 37/144 & 29/72 & 1/9 & 11/48 \\ 499/2592 & 395/1296 & 71/324 & 245/864 \end{pmatrix},$$

so $q_{13}^{(5)} = 52/243$. □

Using the language of Chapter 7, the transition matrix Q encodes the *conditional distribution* of X_1 given the initial state of the chain. Specifically, the ith row of Q is the conditional PMF of X_1 given $X_0 = i$, displayed as a row vector. Similarly, the ith row of Q^n is the conditional PMF of X_n given $X_0 = i$.

To get the *marginal* distributions of $X_0, X_1, \ldots,$ we need to specify not only the transition matrix, but also the *initial conditions* of the chain. The initial state X_0 can be specified deterministically, or randomly according to some distribution. Let $(t_1, t_2, \ldots, t_M)$ be the PMF of X_0 displayed as a vector, that is, $t_i = P(X_0 = i)$. Then the marginal distribution of the chain at any time can be computed from the transition matrix, averaging over all the states using LOTP.

Proposition 11.1.6 (Marginal distribution of X_n)**.** Define $\mathbf{t} = (t_1, t_2, \ldots, t_M)$ by $t_i = P(X_0 = i)$, and view $\mathbf{t}$ as a row vector. Then the marginal distribution of X_n is given by the vector $\mathbf{t}Q^n$. That is, the jth component of $\mathbf{t}Q^n$ is $P(X_n = j)$.

Proof. By the law of total probability, conditioning on X_0, the probability that the chain is in state j after n steps is

$$P(X_n = j) = \sum_{i=1}^{M} P(X_0 = i)P(X_n = j|X_0 = i) = \sum_{i=1}^{M} t_i q_{ij}^{(n)},$$

which is the jth component of $\mathbf{t}Q^n$ by definition of matrix multiplication. ■

Example 11.1.7 (Marginal distributions of 4-state Markov chain). Again consider the 4-state Markov chain shown in Figure 11.1. Suppose that the initial conditions are $\mathbf{t} = (1/4, 1/4, 1/4, 1/4)$, meaning that the chain has equal probability of starting in each of the four states. Let X_n be the position of the chain at time n. Then the marginal distribution of X_1 is

$$\mathbf{t}Q = \begin{pmatrix} 1/4 & 1/4 & 1/4 & 1/4 \end{pmatrix} \begin{pmatrix} 1/3 & 1/3 & 1/3 & 0 \\ 0 & 0 & 1/2 & 1/2 \\ 0 & 1 & 0 & 0 \\ 1/2 & 0 & 0 & 1/2 \end{pmatrix}$$

$$= \begin{pmatrix} 5/24 & 1/3 & 5/24 & 1/4 \end{pmatrix}.$$

The marginal distribution of X_5 is

$$\mathbf{t}Q^5 = \begin{pmatrix} 1/4 & 1/4 & 1/4 & 1/4 \end{pmatrix} \begin{pmatrix} 853/3888 & 509/1944 & 52/243 & 395/1296 \\ 173/864 & 85/432 & 31/108 & 91/288 \\ 37/144 & 29/72 & 1/9 & 11/48 \\ 499/2592 & 395/1296 & 71/324 & 245/864 \end{pmatrix}$$

$$= \begin{pmatrix} 3379/15552 & 2267/7776 & 101/486 & 1469/5184 \end{pmatrix}.$$

We used a computer to perform the matrix multiplication. □

11.2 Classification of states

In this section we introduce terminology for describing the various characteristics of a Markov chain. The states of a Markov chain can be classified as *recurrent* or *transient*, depending on whether they are visited over and over again in the long run or are eventually abandoned. States can also be classified according to their *period*, which is a positive integer summarizing the amount of time that can elapse between successive visits to a state. These characteristics are important because they determine the long-run behavior of the Markov chain, which we will study in Section 11.3.

The concepts of recurrence and transience are best illustrated with a concrete example. In the Markov chain shown on the left of Figure 11.2 (previously featured in Example 11.1.5), a particle moving around between states will continue to spend time in all 4 states in the long run, since it is possible to get from any state to any other state. In contrast, consider the chain on the right of Figure 11.2, and let the particle start at state 1. For a while, the chain may linger in the triangle formed by states 1, 2, and 3, but eventually it will reach state 4, and from there it can never return to states 1, 2, or 3. It will then wander around between states 4, 5, and 6 forever. States 1, 2, and 3 are *transient* and states 4, 5, and 6 are *recurrent*.

In general, these concepts are defined as follows.

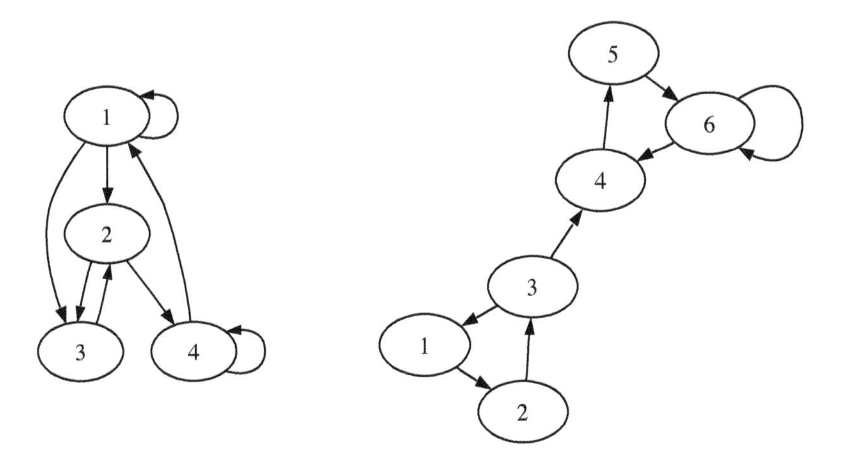

FIGURE 11.2
Left: 4-state Markov chain with all states recurrent. Right: 6-state Markov chain with states 1, 2, and 3 transient.

Definition 11.2.1 (Recurrent and transient states). State i of a Markov chain is *recurrent* if starting from i, the probability is 1 that the chain will eventually return to i. Otherwise, the state is *transient*, which means that if the chain starts from i, there is a positive probability of never returning to i.

In fact, although the definition of a transient state only requires that there be a positive probability of never returning to the state, we can say something stronger: as long as there is a positive probability of leaving i forever, the chain eventually *will* leave i forever. Moreover, we can find the distribution of the number of returns to the state.

Proposition 11.2.2 (Number of returns to transient state is Geometric). Let i be a transient state of a Markov chain. Suppose the probability of never returning to i, starting from i, is a positive number $p > 0$. Then, starting from i, the number of times that the chain returns to i before leaving forever is distributed $\text{Geom}(p)$.

The proof is by the story of the Geometric distribution: each time that the chain is at i, we have a Bernoulli trial which results in "failure" if the chain eventually returns to i and "success" if the chain leaves i forever; these trials are independent by the Markov property. The number of returns to state i is the number of failures before the first success, which matches the story of the Geometric distribution. And since a Geometric random variable always takes finite values, this proposition tells us that after some finite number of visits, the chain will leave state i forever.

If the number of states is not too large, one way to classify states as recurrent or transient is to draw a diagram of the Markov chain and use the same kind of reasoning that we used when analyzing the chains in Figure 11.2. A special case where we can immediately conclude all states are recurrent is when the chain is *irreducible*, meaning that it is possible to get from any state to any other state.

Definition 11.2.3 (Irreducible and reducible chain). A Markov chain with transition matrix Q is *irreducible* if for any two states i and j, it is possible to go from i to j in a finite number of steps (with positive probability). That is, for any states i, j there is some positive integer n such that the (i, j) entry of Q^n is positive. A Markov chain that is not irreducible is called *reducible.*

Proposition 11.2.4 (Irreducible implies all states recurrent). In an irreducible Markov chain with a finite state space, all states are recurrent.

Proof. It is clear that at least one state must be recurrent; if all states were transient, the chain would eventually leave all states forever and have nowhere to go! So assume without loss of generality that state 1 is recurrent, and consider any other state i. We know that $q_{1i}^{(n)}$ is positive for some n, by definition of irreducibility. Thus, every time the chain is at state 1, there is a positive probability that after n more steps it will be at state i.

Since the chain visits state 1 infinitely often, we know the chain *will* eventually reach state i from state 1; think of each visit to state 1 as starting a trial, where "success" is defined as reaching state i in at most n steps. From state i, the chain will return to state 1 because state 1 is recurrent, and by the same logic, it will eventually reach state i again. By induction, the chain will visit state i infinitely often. Since i was arbitrary, we conclude that all states are recurrent. ■

The converse of the proposition is false; it is possible to have a reducible Markov chain whose states are all recurrent. An example is given by the Markov chain below, which consists of two "islands" of states.

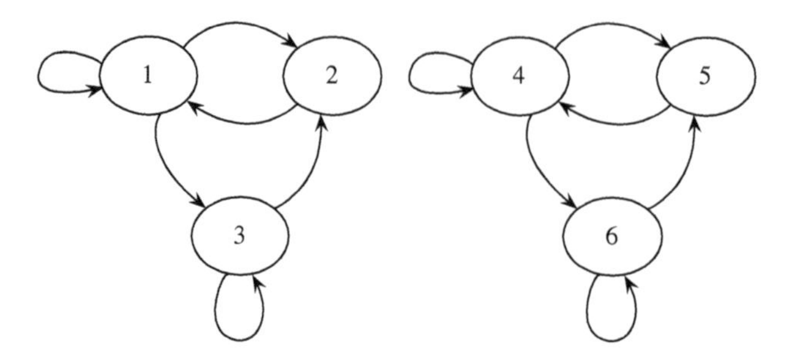

☣ **11.2.5.** Note that recurrence or transience is a property of each *state* in a Markov chain, while irreducibility or reducibility is a property of the chain as a whole.

Here are two familiar problems from earlier chapters, viewed through the lens of Markov chains. For each, we'll identify the recurrent and transient states.

Example 11.2.6 (Gambler's ruin as a Markov chain). In the gambler's ruin problem (Example 2.7.3), two gamblers, A and B, start with i and $N - i$ dollars respectively, making a sequence of bets for \$1. In each round, player A has probability p of winning and probability $q = 1 - p$ of losing. Let X_n be the wealth of gambler A at time n. Then $X_0, X_1, \ldots$ is a Markov chain on the state space $\{0, 1, \ldots, N\}$. By design, $X_0 = i$. Once the Markov chain reaches 0 or N, signifying bankruptcy for player A or player B, the chain stays in that state forever. A diagram of the chain is shown below.

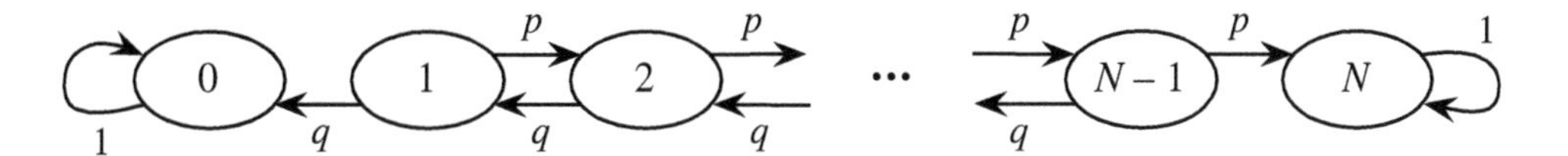

We proved in Chapter 2 that the probability that either A or B goes bankrupt is 1, so for any starting state i other than 0 or N, the Markov chain will eventually be absorbed into state 0 or N, never returning to i. Therefore, for this Markov chain, states 0 and N are recurrent, and all other states are transient. The chain is reducible because from state 0 it is only possible to go to state 0, and from state N it is only possible to go to state N. □

Example 11.2.7 (Coupon collector as a Markov chain). In the coupon collector problem (Example 4.3.12), there are C types of coupons (or toys), which we collect one by one, sampling with replacement from the C coupon types each time. Let X_n be the number of distinct coupon types in our collection after n attempts. Then $X_0, X_1, \dots$ is a Markov chain on the state space $\{0, 1, \dots, C\}$. By design, $X_0 = 0$. This chain is depicted below.

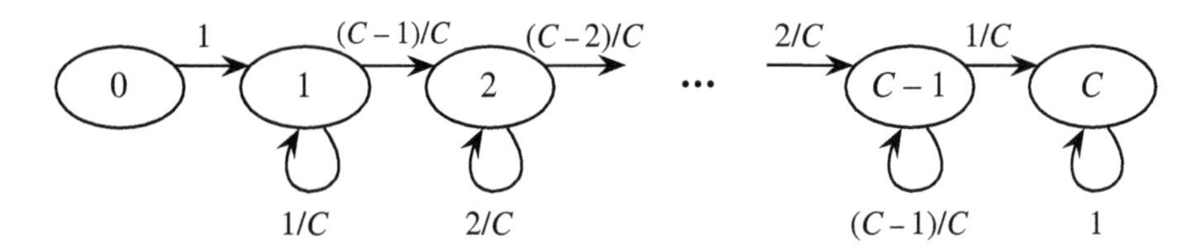

With the exception of state C, we can never return to a state after leaving it; the number of coupon types in the collection can only increase with time. Thus all states are transient except for C, which is recurrent. The chain is reducible since, e.g., it's impossible to go from state 2 back to state 1. □

Another way to classify states is according to their periods. The period of a state summarizes how much time can elapse between successive visits to the state.

Definition 11.2.8 (Period of a state, periodic and aperiodic chain). The *period* of a state i in a Markov chain is the greatest common divisor (gcd) of the possible numbers of steps it can take to return to i when starting at i. That is, the period of i is the greatest common divisor of numbers n such that the (i, i) entry of Q^n is positive. (The period of i is ∞ if it's impossible ever to return to i after starting at i.) A state is called *aperiodic* if its period equals 1, and *periodic* otherwise. The chain itself is called *aperiodic* if all its states are aperiodic, and *periodic* otherwise.

For example, let's consider again the two Markov chains from Figure 11.2, shown again in Figure 11.3. We first consider the 6-state chain on the right. Starting from state 1, it is possible to be back at state 1 after 3 steps, 6 steps, 9 steps, etc., but it

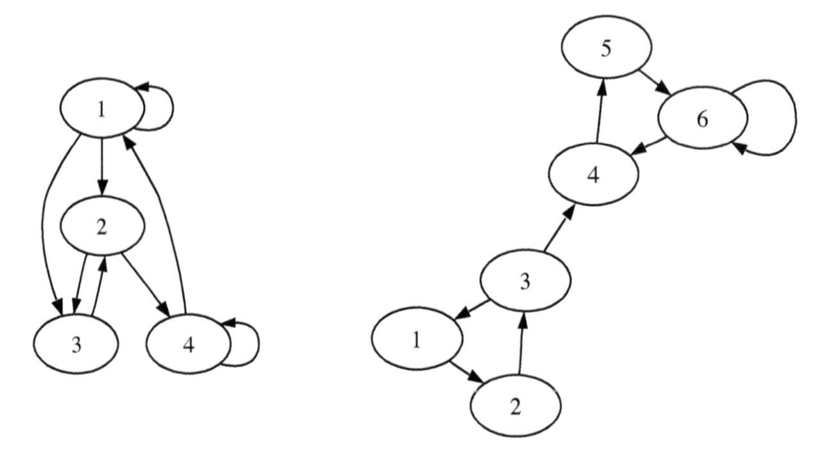

FIGURE 11.3
Left: an aperiodic Markov chain. Right: a periodic Markov chain in which states 1, 2, and 3 have period 3.

is not possible to be back at state 1 after any number of steps that is not a multiple of 3. Therefore, state 1 has period 3. Similarly, states 2 and 3 also have period 3. On the other hand, states 4, 5, and 6 have period 1, but the chain is periodic since at least one state does not have period 1. By contrast, in the chain on the left all states are aperiodic, so that chain is aperiodic.

In the gambler's ruin chain from Example 11.2.6, each state has period 2 except for 0 and N, which have period 1. In the coupon collector chain, each state has period 1 except for state 0, which has period ∞ because it's impossible to return to state 0. So neither of these chains is aperiodic.

Checking whether an irreducible chain is aperiodic is often much easier than it might seem: the next proposition shows that we only need to calculate the period of *one* state, rather than searching state by state for a state whose period is not 1.

Proposition 11.2.9 (Periods in an irreducible chain)**.** In an irreducible Markov chain, all states have the same period.

11.3 Stationary distribution

The concepts of recurrence and transience are important for understanding the long-run behavior of a Markov chain. At first, the chain may spend time in transient states. Eventually though, the chain will spend all its time in recurrent states. But what fraction of the time will it spend in each of the recurrent states? This question is answered by the *stationary distribution* of the chain, also known as the *steady-state distribution*. We will learn in this section that for irreducible and aperiodic

Markov chains, the stationary distribution describes the long-run behavior of the chain, regardless of its initial conditions. It will tell us both the long-run probability of being in any particular state, and the long-run proportion of time that the chain spends in that state.

Definition 11.3.1 (Stationary distribution). A row vector $\mathbf{s} = (s_1, \ldots, s_M)$ such that $s_i \geq 0$ and $\sum_i s_i = 1$ is a *stationary distribution* for a Markov chain with transition matrix Q if

$$\sum_i s_i q_{ij} = s_j$$

for all j. This system of linear equations can be written as one matrix equation:

$$\mathbf{s}Q = \mathbf{s}.$$

Recall that if $\mathbf{s}$ is the distribution of X_0, then $\mathbf{s}Q$ is the marginal distribution of X_1. Thus the equation $\mathbf{s}Q = \mathbf{s}$ means that if X_0 has distribution $\mathbf{s}$, then X_1 also has distribution $\mathbf{s}$. But then X_2 also has distribution $\mathbf{s}$, as does X_3, etc. That is, a Markov chain whose initial distribution is the stationary distribution $\mathbf{s}$ will stay in the stationary distribution forever.

One way to think about the stationary distribution of a Markov chain intuitively is to imagine a large number of particles, each independently bouncing from state to state according to the transition probabilities. After a while, the system of particles will approach an equilibrium where, at each time period, the number of particles leaving a state will be counterbalanced by the number of particles entering that state, and this will be true for all states. At this equilibrium, the system as a whole will appear to be *stationary*, and the proportion of particles in each state will be given by the stationary distribution. We will explore this perspective on stationary distributions more after Definition 11.4.1.

☣ **11.3.2** (Stationary distribution is marginal, not conditional). When a Markov chain is at the stationary distribution, the unconditional PMF of X_n equals $\mathbf{s}$ for all n, but the *conditional* PMF of X_n given $X_{n-1} = i$ is still encoded by the ith row of the transition matrix Q.

If a Markov chain starts at the stationary distribution, then all of the X_n are identically distributed (since they have the same marginal distribution $\mathbf{s}$), but they are not necessarily independent, since the conditional distribution of X_n given $X_{n-1} = i$ is, in general, different from the marginal distribution of X_n.

☣ **11.3.3** (Sympathetic magic). If a Markov chain starts at the stationary distribution, then the marginal distributions of the X_n are all equal. This is not the same as saying that the X_n themselves are all equal; confusing the random variables X_n with their distributions is an example of sympathetic magic.

For very small Markov chains, we may solve for the stationary distribution by hand, using the definition. The next example illustrates this for a two-state chain.

Example 11.3.4 (Stationary distribution for a two-state chain). Let

$$Q = \begin{pmatrix} 1/3 & 2/3 \\ 1/2 & 1/2 \end{pmatrix}.$$

The stationary distribution is of the form $\mathbf{s} = (s, 1-s)$, and we must solve for s in

$$\begin{pmatrix} s & 1-s \end{pmatrix} \begin{pmatrix} 1/3 & 2/3 \\ 1/2 & 1/2 \end{pmatrix} = \begin{pmatrix} s & 1-s \end{pmatrix},$$

which is equivalent to

$$\frac{1}{3}s + \frac{1}{2}(1-s) = s,$$
$$\frac{2}{3}s + \frac{1}{2}(1-s) = 1-s.$$

The only solution to these equations is $s = 3/7$, so $(3/7, 4/7)$ is the unique stationary distribution of this Markov chain.

More generally, suppose that $q_{12} = a$ and $q_{21} = b$, where $0 < a < 1$ and $0 < b < 1$. Then the transition matrix is

$$Q = \begin{pmatrix} 1-a & a \\ b & 1-b \end{pmatrix}.$$

Writing $\mathbf{s} = (s_1, s_2)$, the equation $\mathbf{s}Q = \mathbf{s}$ becomes the linear system

$$(1-a)s_1 + bs_2 = s_1,$$
$$as_1 + (1-b)s_2 = s_2$$

Both equations in this system simplify to

$$as_1 = bs_2.$$

Plugging in $s_2 = 1 - s_1$, it follows that the unique solution to this system is

$$\mathbf{s} = \left(\frac{b}{a+b}, \frac{a}{a+b}\right).$$

In short, $\mathbf{s} \propto (b, a)$, i.e., $\mathbf{s}$ is a constant times (b, a). The constant is whatever it needs to be to make the components of $\mathbf{s}$ sum to 1. As a check, note that for the Q given earlier with specific numbers, this result says that the stationary distribution should be proportional to $(b, a) = (1/2, 2/3)$, which, multiplying by 6 to clear the denominators, is equivalent to being proportional to $(3, 4)$. The stationary distribution we solved for earlier was $(3/7, 4/7)$, which is indeed proportional to $(3, 4)$. □

In linear algebra terminology, the equation $\mathbf{s}Q = \mathbf{s}$ says that $\mathbf{s}$ is a left eigenvector of Q with eigenvalue 1 (see Section A.3 of the math appendix). To get the usual kind of eigenvector (a right eigenvector), take transposes: $Q'\mathbf{s}' = \mathbf{s}'$, where the $'$ symbol denotes taking the transpose.

11.3.1 Existence and uniqueness

Does a stationary distribution always exist, and is it unique? It turns out that for a finite state space, a stationary distribution always exists. Furthermore, in irreducible Markov chains, the stationary distribution is unique.

Theorem 11.3.5 (Existence and uniqueness of stationary distribution)**.** For any irreducible Markov chain, there exists a unique stationary distribution. In this distribution, every state has positive probability.

The theorem is a consequence of a result from linear algebra called the Perron-Frobenius theorem, which is stated in Section A.3 of the math appendix.

The 4-state chain on the left of Figure 11.3 is irreducible: in terms of the picture, it is possible to go from anywhere to anywhere following the arrows; in terms of the transition matrix Q, all the entries of Q^5 are positive. Therefore, by Theorem 11.3.5, the chain has a unique stationary distribution.

On the other hand, the gambler's ruin chain from Example 11.2.6 is reducible, so the theorem does not apply. It turns out that this chain has *infinitely many* stationary distributions. In the long run, the chain will either wind up at state 0 (and stay there forever) or wind up at state N (and stay there forever). This suggests, and it is easy to check, that the degenerate distributions $\mathbf{s} = (1, 0, \dots, 0)$ and $\mathbf{t} = (0, 0, \dots, 1)$ are both stationary distributions. It follows that any weighted combination $p\mathbf{s}+(1-p)\mathbf{t}$, where $0 \leq p \leq 1$, is also a stationary distribution, since it sums to 1 and

$$(p\mathbf{s} + (1-p)\mathbf{t})Q = p\mathbf{s}Q + (1-p)\mathbf{t}Q = p\mathbf{s} + (1-p)\mathbf{t},$$

where, as usual, Q is the transition matrix.

11.3.2 Convergence

We have already informally stated that the stationary distribution describes the long-run behavior of the chain, in the sense that if we run the chain for a long time, the marginal distribution of X_n converges to the stationary distribution $\mathbf{s}$. The next theorem states that this is true as long as the chain is both irreducible and aperiodic. Then, regardless of the chain's initial conditions, the PMF of X_n will converge to the stationary distribution as $n \to \infty$. This relates the concept of stationarity to the long-run behavior of a Markov chain. The proof is omitted.

Theorem 11.3.6 (Convergence to stationary distribution)**.** Let $X_0, X_1, \dots$ be an irreducible, aperiodic Markov chain with stationary distribution $\mathbf{s}$ and transition matrix Q. Then $P(X_n = i)$ converges to s_i as $n \to \infty$. In terms of the transition matrix, Q^n converges to a matrix in which each row is $\mathbf{s}$.

Therefore, after a large number of steps, the probability that the chain is in state i is close to the stationary probability s_i, regardless of the chain's initial conditions. This makes irreducible, aperiodic chains especially nice to work with. Irreducibility means that for each (i, j) there is some power Q^m where the (i, j) entry is positive, but if we also assume aperiodicity it turns out that we can find a value of m that works for *all* i, j. More precisely, a chain is irreducible and aperiodic if and only if some power Q^m is positive in all entries.

Intuitively, the extra condition of aperiodicity is needed in order to rule out chains that just go around in circles, such as the chain in the following example, or chains where, say, some states are accessible only after an *even* number of steps while other states are accessible only after an *odd* number of steps.

Example 11.3.7 (Periodic chain). The figure below shows a periodic Markov chain where each state has period 5.

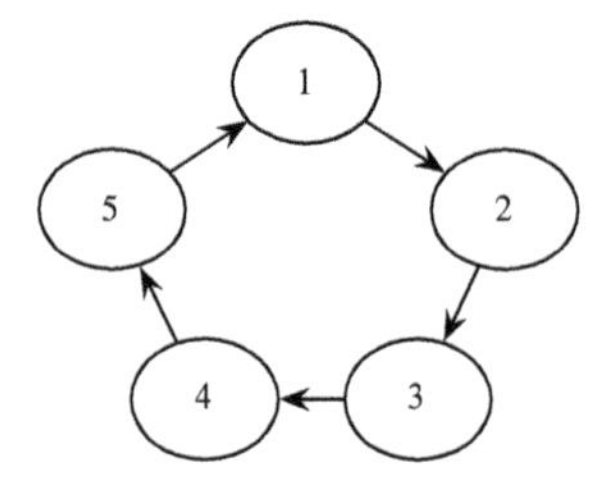

FIGURE 11.4
A periodic chain.

The transition matrix of the chain is

$$Q = \begin{pmatrix} 0 & 1 & 0 & 0 & 0 \\ 0 & 0 & 1 & 0 & 0 \\ 0 & 0 & 0 & 1 & 0 \\ 0 & 0 & 0 & 0 & 1 \\ 1 & 0 & 0 & 0 & 0 \end{pmatrix}.$$

It can be verified without much difficulty that $\mathbf{s} = (1/5, 1/5, 1/5, 1/5, 1/5)$ is a stationary distribution of this chain, and by Theorem 11.3.5, $\mathbf{s}$ is unique. However, suppose the chain starts at $X_0 = 1$. Then the PMF of X_n assigns probability 1 to the state $(n \bmod 5) + 1$ and 0 to all other states.[1] In particular, it does not converge to $\mathbf{s}$ as $n \to \infty$. Nor does Q^n converge to a matrix in which each row is $\mathbf{s}$: the chain's transitions are deterministic, so Q^n always consists of 0's and 1's. □

Lastly, the stationary distribution tells us the average time between visits to any particular state.

Theorem 11.3.8 (Expected time to return). Let $X_0, X_1, \dots$ be an irreducible Markov chain with stationary distribution $\mathbf{s}$. Let r_i be the expected time it takes the chain to return to i, given that it starts at i. Then $s_i = 1/r_i$.

Here is how the theorems apply to the two-state chain from Example 11.3.4.

[1]The notation $n \bmod 5$ denotes the remainder when n is divided by 5.

Example 11.3.9 (Long-run behavior of a two-state chain). In the long run, the chain in Example 11.3.4 will spend 3/7 of its time in state 1 and 4/7 of its time in state 2. Starting at state 1, it will take an average of 7/3 steps to return to state 1. The powers of the transition matrix converge to a matrix where each row is the stationary distribution:

$$\begin{pmatrix} 1/3 & 2/3 \\ 1/2 & 1/2 \end{pmatrix}^n \to \begin{pmatrix} 3/7 & 4/7 \\ 3/7 & 4/7 \end{pmatrix} \text{ as } n \to \infty. \qquad \square$$

11.3.3 Google PageRank

We next consider a *vastly* larger example of a stationary distribution, for a Markov chain on a state space with billions of interconnected nodes: the World Wide Web. The next example explains how the founders of Google modeled web-surfing as a Markov chain, and then used its stationary distribution to rank the relevance of webpages. For years Google described the resulting method, known as *PageRank*, as "the heart of our software".

Suppose you are interested in a certain topic, say chess, so you use a search engine to look for useful webpages with information about chess. There are millions of webpages that mention the word "chess", so a key issue a search engine needs to deal with is what order to show the search results in. It would be a disaster to have to wade through thousands of garbage pages that mention "chess" before finding informative content.

In the early days of the web, various approaches to this ranking problem were used. For example, some search engines employed people to manually decide which pages were most useful, like a museum curator. But aside from being subjective and expensive, this quickly became infeasible as the web grew. Others focused on the number of times the search term was mentioned on the site. But a page that mentions "chess" over and over again could easily be less useful than a concise reference page or a page about chess that doesn't repeatedly mention the *word*. Furthermore, this method is very open to abuse: a spam page could boost its ranking just by including a long list of words repeated over and over again.

Both of the above methods ignore the *structure* of the web: which pages link to which other pages? Taking the link structure into account led to dramatic improvements in search engines. As a first attempt, one could rank a page based on how many other pages link to it. That is, if Page A links to Page B, we consider it a *vote* for B, and we rank pages based on how many votes they have.

But this is again very open to abuse: a spam page could boost its ranking by creating thousands of other spam pages linking to it. And though it may seem democratic for each page to have equal voting power, an incoming link from a reliable page is more meaningful than a link from an uninformative page. Google PageRank, which was introduced in 1998 by Sergey Brin and the aptly named Larry Page, ranks the importance of a page not only by how many pages link to it, but also by the importance of those pages.

Consider the web as a directed network—which is what it is. Each page on the web is a node, and links between nodes represent links between pages. For example, suppose for simplicity that the web only has 4 pages, connected as shown in the figure below.

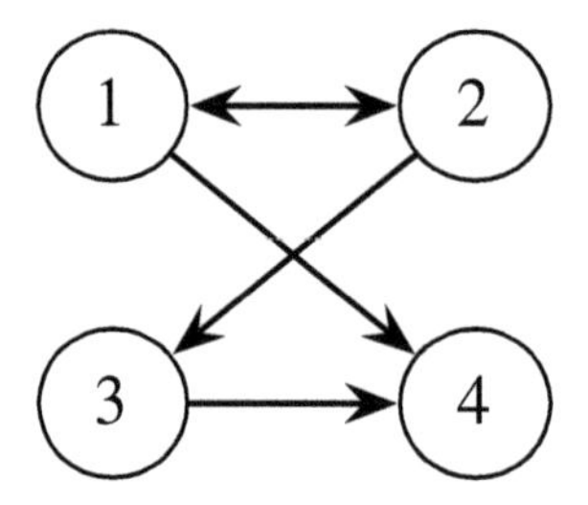

FIGURE 11.5
It's a small web after all.

Imagine someone randomly surfing the web, starting at some page and then randomly clicking links to go from one page to the next (with equal probabilities for all links on the current page). The idea of PageRank is to measure the importance of a page by the long-run fraction of time spent at that page.

Of course, some pages may have no outgoing links at all, such as page 4 above. When the web surfer encounters such a page, rather than despairing they open up a new browser window and visits a uniformly random page. Thus a page with no links is converted to a page that links to *every* page, including itself. For the example above, the resulting transition matrix is

$$Q = \begin{pmatrix} 0 & 1/2 & 0 & 1/2 \\ 1/2 & 0 & 1/2 & 0 \\ 0 & 0 & 0 & 1 \\ 1/4 & 1/4 & 1/4 & 1/4 \end{pmatrix}.$$

In general, let M be the number of pages on the web, let Q be the M by M transition matrix of the chain described above, and let $\mathbf{s}$ be the stationary distribution (assuming it exists and is unique). Think of s_j as a measure of how important page j is. Intuitively, the equation

$$s_j = \sum_i s_i q_{ij}$$

says that the score of page j should be based not only on how many other pages link to it, but on their scores. Furthermore, the "voting power" of a page gets diluted if it has a lot of outgoing links: it counts for more if page i's only link is to page j (so that $q_{ij} = 1$) than if page i has thousands of links, one of which happens to be to page j.

It is not clear that a unique stationary distribution exists for this chain, since it may not be irreducible and aperiodic. Even if it is irreducible and aperiodic, convergence to the stationary distribution could be very slow since the web is so immense. To address these issues, suppose that before each move, the web surfer flips a coin with probability α of Heads. If Heads, the web surfer clicks a random link from the current page; if Tails, the web surfer *teleports* to a uniformly random page. The resulting chain has the *Google transition matrix*

$$G = \alpha Q + (1-\alpha)\frac{J}{M},$$

where J is the M by M matrix of all 1's. Note that the row sums of G are 1 and that all entries are positive, so G is a valid transition matrix for an irreducible, aperiodic Markov chain. This means there is a unique stationary distribution $\mathbf{s}$, called *PageRank*, and the chain will converge to it! The choice of α is an important consideration; choosing α close to 1 makes sense to respect the structure of the web as much as possible, but there is a tradeoff since it turns out that smaller values of α make the chain converge much faster. As a compromise, the original recommendation of Brin and Page was $\alpha = 0.85$.

PageRank is conceptually nice, but *computing* it sounds extremely difficult, considering that $\mathbf{s}G = \mathbf{s}$ could be a system of 100 billion equations in 100 billion unknowns. Instead of thinking of this as a massive algebra problem, we can use the Markov chain interpretation: for any starting distribution $\mathbf{t}$, $\mathbf{t}G^n \to \mathbf{s}$ as $n \to \infty$. And $\mathbf{t}G$ is easier to compute than it might seem at first:

$$\mathbf{t}G = \alpha(\mathbf{t}Q) + \frac{1-\alpha}{M}(\mathbf{t}J),$$

where computing the first term isn't too hard since Q is very *sparse* (mostly 0's) and computing the second term is easy since $\mathbf{t}J$ is a vector of all 1's. Then $\mathbf{t}G$ becomes the new $\mathbf{t}$, and we can compute $\mathbf{t}G^2 = (\mathbf{t}G)G$, etc., until the sequence appears to have converged (though it is hard to *know* that it has converged). This gives an approximation to PageRank, and has an intuitive interpretation as the distribution of where the web surfer is after a large number of steps.

11.4 Reversibility

We have seen that the stationary distribution of a Markov chain is extremely useful for understanding its long-run behavior. Unfortunately, in general it may be computationally difficult to find the stationary distribution when the state space is large. This section addresses an important special case where working with eigenvalue equations for large matrices can be avoided.

Definition 11.4.1 (Reversibility). Let $Q = (q_{ij})$ be the transition matrix of a Markov chain. Suppose there is $\mathbf{s} = (s_1, \dots, s_M)$ with $s_i \geq 0, \sum_i s_i = 1$, such that

$$s_i q_{ij} = s_j q_{ji}$$

for all states i and j. This equation is called the *reversibility* or *detailed balance* condition, and we say that the chain is *reversible* with respect to $\mathbf{s}$ if it holds.

The term "reversible" comes from the fact that a reversible chain, started according to its stationary distribution, behaves in the same way regardless of whether time is run forwards or backwards. If you record a video of a reversible chain, started according to its stationary distribution, and then show the video to a friend, either in the normal way or with time reversed, your friend will not be able to determine from watching the video whether time is running forwards or backwards.

As discussed after Definition 11.3.1, we can think about the stationary distribution of a Markov chain intuitively in terms of a system consisting of a large number of particles independently bouncing around according to the transition probabilities. In the long run, the proportion of particles in any state j is the stationary probability of state j, and the flow of particles out of state j is counterbalanced by the flow of particles into state j. To see this in more detail, let n be the number of particles and $\mathbf{s}$ be a probability vector such that s_j is the current proportion of particles at state j. By definition, $\mathbf{s}$ is the stationary distribution of the chain if and only if

$$s_j = \sum_i s_i q_{ij} = s_j q_{jj} + \sum_{i:i \neq j} s_i q_{ij},$$

for all states j. This equation can be rewritten as

$$n s_j (1 - q_{jj}) = \sum_{i:i \neq j} n s_i q_{ij}.$$

The left-hand side is the approximate number of particles that will exit from state j on the next step, since there are $n s_j$ particles at state j, each of which will stay at j with probability q_{jj} and leave with probability $1 - q_{jj}$. The right-hand side is the approximate number of particles that will enter state j on the next step, since for each $i \neq j$ there are $n s_i$ particles at state i, each of which will enter state j with probability q_{ij}. So there is a balance between particles leaving state j and particles entering state j.

The reversibility condition imposes a much more stringent form of balance, in which for each *pair* of states i, j with $i \neq j$, the flow of particles from state i to state j is counterbalanced by the flow of particles from state j to state i. To see this, write the reversibility equation for states i and j as

$$n s_i q_{ij} = n s_j q_{ji}.$$

The left-hand side is the approximate number of particles that will go from state i to state j on the next step, since there are $n s_i$ particles at state i, each of which has probability q_{ij} of going to state j. Similarly, the right-hand side is the approximate number of particles that will go from state i to state j on the next step. So reversibility says that there is a balance between particles going from state i to state j, and particles going from state j to state i.

Given a transition matrix, if we can find a probability vector $\mathbf{s}$ that satisfies the reversibility condition, then $\mathbf{s}$ is automatically a stationary distribution. This should not be surprising, in light of the above discussion.

Proposition 11.4.2 (Reversible implies stationary). Suppose that $Q = (q_{ij})$ is the transition matrix of a Markov chain that is reversible with respect to a nonnegative vector $\mathbf{s} = (s_1, \ldots, s_M)$ whose components sum to 1. Then $\mathbf{s}$ is a stationary distribution of the chain.

Proof. We have

$$\sum_i s_i q_{ij} = \sum_i s_j q_{ji} = s_j \sum_i q_{ji} = s_j,$$

where the last equality is because each row sum of Q is 1. So $\mathbf{s}$ is stationary. ■

This is a powerful result because it is often easier to verify the reversibility condition than it is to solve the entire system of equations $\mathbf{s}Q = \mathbf{s}$. However, in general we may not know in advance whether it is possible to find $\mathbf{s}$ satisfying the reversibility condition, and even when it is possible, it may take a lot of effort to find an $\mathbf{s}$ that works. In the remainder of this section, we look at three types of Markov chains where it *is* possible to find an $\mathbf{s}$ that satisfies the reversibility condition. Such Markov chains are called *reversible.*

First, if Q is a symmetric matrix, then the stationary distribution is uniform over the state space: $\mathbf{s} = (1/M, 1/M, \ldots, 1/M)$. It is easy to see that if $q_{ij} = q_{ji}$, then the reversibility condition $s_i q_{ij} = s_j q_{ji}$ is satisfied when $s_i = s_j$ for all i and j.

This is a special case of a more general fact, stated below: if each column of Q sums to 1, then the stationary distribution is uniform over the state space.

Proposition 11.4.3. If each column of the transition matrix Q sums to 1, then the uniform distribution over all states, $(1/M, 1/M, \ldots, 1/M)$, is a stationary distribution. (A nonnegative matrix such that the row sums and the column sums are all equal to 1 is called a *doubly stochastic matrix.*)

Proof. Assuming each column sums to 1, the row vector $\mathbf{v} = (1, 1, \ldots, 1)$ satisfies $\mathbf{v}Q = \mathbf{v}$. It follows that $(1/M, 1/M, \ldots, 1/M)$ is stationary. ■

Second, if the Markov chain is a *random walk on an undirected network*, then there is a simple formula for the stationary distribution.

Example 11.4.4 (Random walk on an undirected network). A network is a collection of *nodes* joined by *edges*; the network is *undirected* if edges can be traversed in either direction, meaning there are no one-way streets. Suppose a wanderer randomly traverses the edges of an undirected network. From a node i, the wanderer randomly picks any of the edges at i, with equal probabilities, and then traverses the chosen edge. For example, in the network shown below, from node 3 the wanderer goes to node 1 or node 2, with probability 1/2 each.

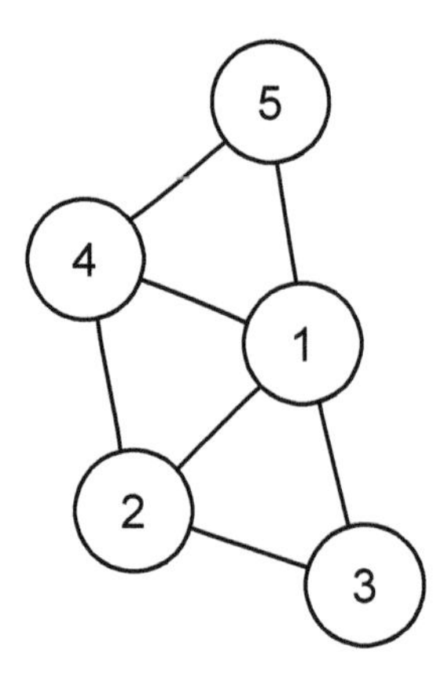

The *degree* of a node is the number of edges attached to it, and the *degree sequence* of a network with nodes $1, 2, \ldots, n$ is the vector $(d_1, \ldots, d_n)$ listing all the degrees, where d_j is the degree of node j. An edge from a node to itself is allowed (such an edge is called a *self-loop*), and counts 1 toward the degree of that node.

For example, the network above has degree sequence $\mathbf{d} = (4, 3, 2, 3, 2)$. Note that

$$d_i q_{ij} = d_j q_{ji}$$

for all i, j, since q_{ij} is $1/d_i$ if $\{i, j\}$ is an edge and 0 otherwise, for $i \neq j$. Therefore, by Proposition 11.4.2, the stationary distribution is proportional to the degree sequence. Intuitively, the nodes with the highest degrees are the most well-connected, so it makes sense that the chain spends the most time in these states in the long run. In the example above, this says that

$$\mathbf{s} = \left(\frac{4}{14}, \frac{3}{14}, \frac{2}{14}, \frac{3}{14}, \frac{2}{14} \right)$$

is the stationary distribution for the random walk.

Exercise 20 explores random walk on a *weighted* undirected network; each edge has a weight assigned to it, and the wanderer chooses where to go from i with probabilities proportional to the weights on the available edges. It turns out that this is a reversible Markov chain. More surprisingly, *every* reversible Markov chain can be represented as random walk on a weighted undirected network! □

Here is a concrete example of a random walk on an undirected network.

Example 11.4.5 (Knight on a chessboard). Consider a knight randomly moving around on a 4×4 chessboard.

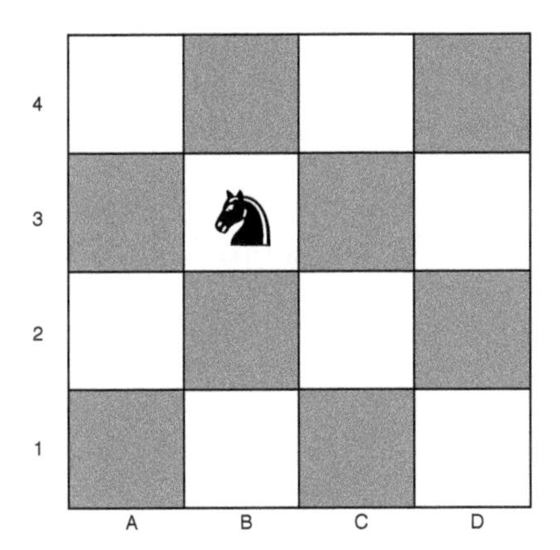

The 16 squares are labeled in a grid, e.g., the knight is currently at the square B3, and the upper left square is A4. Each move of the knight is an L-shaped jump: the knight moves two squares horizontally followed by one square vertically, or vice versa. For example, from B3 the knight can move to A1, C1, D2, or D4; from A4 it can move to B2 or C3. Note that from a light square, the knight always moves to a dark square and vice versa.

Suppose that at each step, the knight moves randomly, with each possibility equally likely. This creates a Markov chain where the states are the 16 squares. Compute the stationary distribution of the chain.

Solution:

There are only three types of squares on the board: 4 center squares, 4 corner squares (such as A4), and 8 edge squares (such as B4; exclude corner squares from being considered edge squares). We can consider the board to be an undirected network where two squares are connected by an edge if they are accessible via a single knight's move. Then a center square has degree 4, a corner square has degree 2, and an edge square has degree 3, so their stationary probabilities are $4a, 2a, 3a$ respectively for some a.

To find a, count the number of squares of each type to get $4a \cdot 4 + 2a \cdot 4 + 3a \cdot 8 = 1$, giving $a = 1/48$. Thus, each center square has stationary probability $4/48 = 1/12$, each corner square has stationary probability $2/48 = 1/24$, and each edge square has stationary probability $3/48 = 1/16$. □

Third and finally, if in each time period a Markov chain can only move one step to the left, one step to the right, or stay in place, then it is called a *birth-death chain*. All birth-death chains are reversible.

Example 11.4.6 (Birth-death chain). A *birth-death chain* on states $\{1, 2, \ldots, M\}$ is a Markov chain with transition matrix $Q = (q_{ij})$ such that $q_{ij} > 0$ if $|i - j| = 1$ and $q_{ij} = 0$ if $|i - j| \geq 2$. This says it's possible to go one step to the left and possible to go one step to the right (except at boundaries) but impossible to jump further in one step. The name stems from applications to the growth or decline of a population, where a step to the right is thought of as a birth and a step to the left is thought of as a death in the population.

For example, the chain shown below is a birth-death chain if the labeled transitions have positive probabilities, except for the loops from a state to itself, which are allowed to have 0 probability.

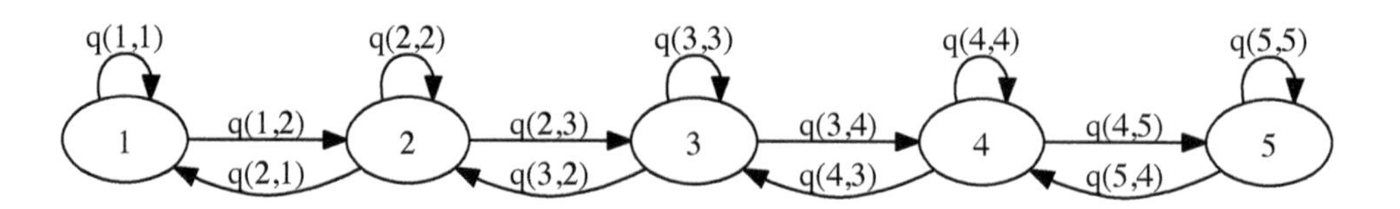

We will now show that any birth-death chain is reversible, and construct the stationary distribution. Let s_1 be a positive number, to be specified later. Since we want $s_1 q_{12} = s_2 q_{21}$, let

$$s_2 = s_1 q_{12}/q_{21}.$$

Then since we want $s_2 q_{23} = s_3 q_{32}$, let

$$s_3 = s_2 q_{23}/q_{32} = s_1 q_{12} q_{23}/(q_{32} q_{21}).$$

Continuing in this way, let

$$s_j = \frac{s_1 q_{12} q_{23} \cdots q_{j-1,j}}{q_{j,j-1} q_{j-1,j-2} \cdots q_{21}},$$

for all states j with $2 \leq j \leq M$. Choose s_1 so that the s_j sum to 1. Then the chain is reversible with respect to $\mathbf{s}$, since $q_{ij} = q_{ji} = 0$ if $|i - j| \geq 2$ and by construction $s_i q_{ij} = s_j q_{ji}$ if $|i - j| = 1$. Thus, $\mathbf{s}$ is the stationary distribution. □

The Ehrenfest chain is a birth-death chain that can be used as a simple model for the diffusion of gas molecules. The stationary distribution turns out to be a Binomial distribution.

Example 11.4.7 (Ehrenfest). There are two containers with a total of M distinguishable particles. Transitions are made by choosing a random particle and moving it from its current container into the other container. Initially, all of the particles are in the second container. Let X_n be the number of particles in the first container at time n, so $X_0 = 0$ and the transition from X_n to X_{n+1} is done as described above. This is a periodic Markov chain with state space $\{0, 1, \ldots, M\}$.

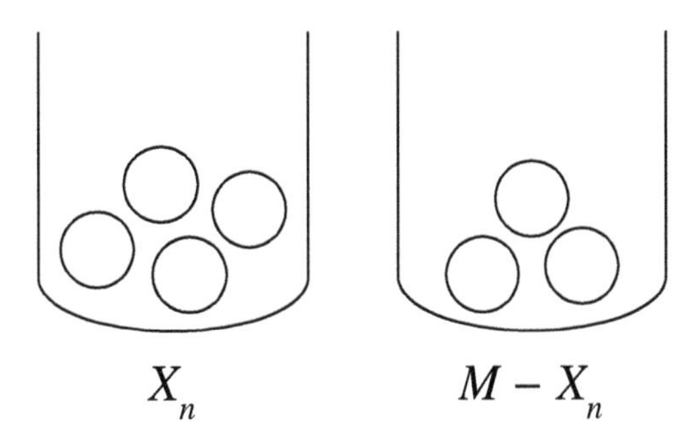

As illustrated above, the first container has X_n particles and the second has $M - X_n$ particles. We will use the reversibility condition to show that $\mathbf{s} = (s_0, s_1, \dots, s_M)$ with

$$s_i = \binom{M}{i}\left(\frac{1}{2}\right)^M$$

is the stationary distribution. Note that this is the Bin$(M, 1/2)$ PMF.

Let s_i be as claimed, and check that $s_i q_{ij} = s_j q_{ji}$. If $j = i + 1$ (with $i < M$), then

$$s_i q_{ij} = \binom{M}{i}\left(\frac{1}{2}\right)^M \frac{M-i}{M} = \frac{M!}{(M-i)!i!}\left(\frac{1}{2}\right)^M \frac{M-i}{M} = \binom{M-1}{i}\left(\frac{1}{2}\right)^M,$$

$$s_j q_{ji} = \binom{M}{j}\left(\frac{1}{2}\right)^M \frac{j}{M} = \frac{M!}{(M-j)!j!}\left(\frac{1}{2}\right)^M \frac{j}{M} = \binom{M-1}{j-1}\left(\frac{1}{2}\right)^M = s_i q_{ij}.$$

By a similar calculation, if $j = i - 1$ (with $i > 0$), then $s_i q_{ij} = s_j q_{ji}$. For all other values of i and j, $q_{ij} = q_{ji} = 0$. Therefore, $\mathbf{s}$ is stationary.

The Binomial was a natural guess for the stationary distribution because after running the Markov chain for a long time, each particle is about equally likely to be in either container. However, the PMF does *not* converge to a Binomial since the chain has period 2, with X_n guaranteed to be even when n is even, and odd when n is odd.

Happily, it turns out that another interpretation of the stationary distribution remains valid here: s_i is the long-run proportion of time that the chain spends in state i. More precisely, letting I_k be the indicator of the chain being in state i at time k, it can be shown that

$$\frac{1}{n}\sum_{k=0}^{n-1} I_k \to s_i$$

as $n \to \infty$, with probability 1. □

11.5 Recap

A Markov chain is a sequence of r.v.s $X_0, X_1, X_2, \ldots$ satisfying the Markov property, which states that given the present, the past and future are conditionally independent:

$$P(X_{n+1} = j | X_n = i, X_{n-1} = i_{n-1}, \ldots, X_0 = i_0) = P(X_{n+1} = j | X_n = i) = q_{ij}.$$

The transition matrix $Q = (q_{ij})$ gives the probabilities of moving from any state to any other state in one step. The ith row of the transition matrix is the conditional PMF of X_{n+1} given $X_n = i$. The nth power of the transition matrix gives the n-step transition probabilities. If we specify initial conditions $s_i = P(X_0 = i)$ and let $\mathbf{s} = (s_1, \ldots, s_M)$, then the marginal PMF of X_n is $\mathbf{s}Q^n$.

States of a Markov chain can be classified as recurrent or transient: recurrent if the chain will return to the state over and over, and transient if it will eventually leave forever. States can also be classified according to their periods; the period of state i is the greatest common divisor of the numbers of steps it can take to return to i, starting from i. A chain is irreducible if it is possible to get from any state to any state in a finite number of steps, and aperiodic if each state has period 1.

A stationary distribution for a finite Markov chain is a PMF $\mathbf{s}$ such that $\mathbf{s}Q = \mathbf{s}$. Under various conditions, the stationary distribution of a finite Markov chain exists and is unique, and the PMF of X_n converges to $\mathbf{s}$ as $n \to \infty$. If state i has stationary probability s_i, then the expected time for the chain to return to i, starting from i, is $r_i = 1/s_i$.

If a PMF $\mathbf{s}$ satisfies the reversibility condition $s_i q_{ij} = s_j q_{ji}$ for all i and j, it guarantees that $\mathbf{s}$ is a stationary distribution of the Markov chain with transition matrix $Q = (q_{ij})$. Markov chains for which there exists $\mathbf{s}$ satisfying the reversibility condition are called reversible. We discussed three types of reversible chains:

1. If the transition matrix is symmetric, then the stationary distribution is uniform over all states.

2. If the chain is a random walk on an undirected network, then the stationary distribution is proportional to the degree sequence, i.e.,

$$s_j = \frac{d_j}{\sum_i d_i}.$$

3. If the chain is a birth-death chain, then the stationary distribution satisfies

$$s_j = \frac{s_1 q_{12} q_{23} \cdots q_{j-1,j}}{q_{j,j-1} q_{j-1,j-2} \cdots q_{21}}$$

for $j > 1$, where s_1 is solved for at the end to make $s_1 + \cdots + s_M = 1$.

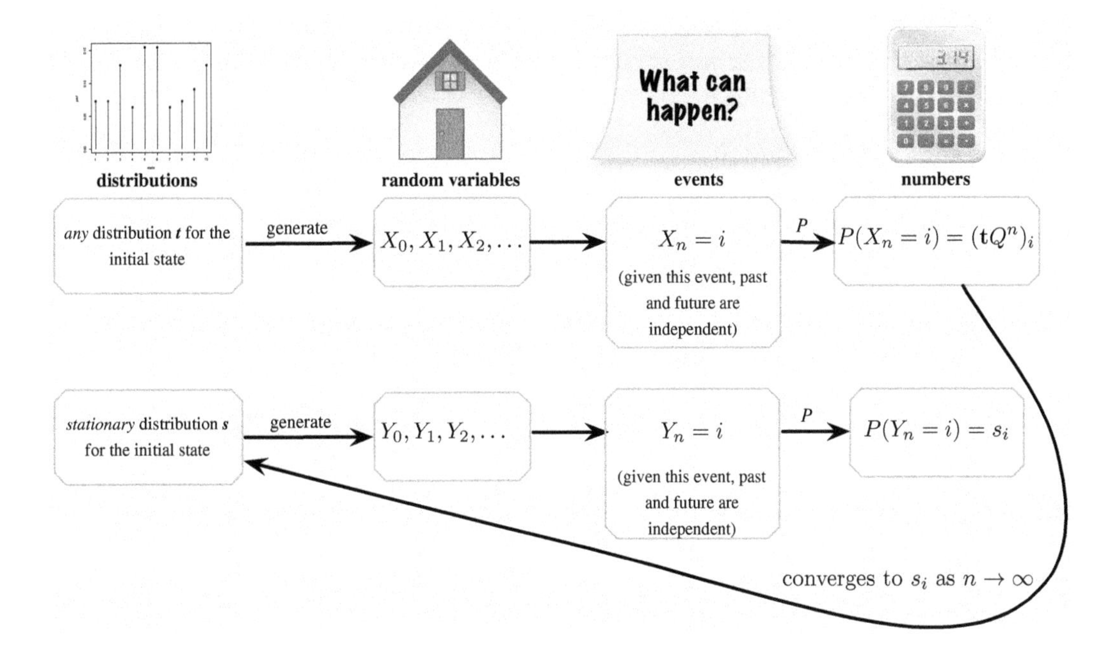

FIGURE 11.6
Given a transition matrix Q and a distribution $\mathbf{t}$ over the states, we can generate a Markov chain $X_0, X_1, \ldots$ by choosing X_0 according to $\mathbf{t}$ and then running the chain according to the transition probabilities. An important event is $X_n = i$, the event that the chain is visiting state i at time n. We can then find the PMF of X_n in terms of Q and $\mathbf{t}$, and (under conditions discussed in the chapter) the PMF will converge to the stationary distribution $\mathbf{s}$. If instead we start the chain out according to $\mathbf{s}$, then the chain will stay stationary forever.

Figure 11.6 compares two ways to run a Markov chain with transition matrix Q: choosing the initial state according to an arbitrary distribution $\mathbf{t}$ over the states, or choosing the initial state according to the stationary distribution $\mathbf{s}$. In the former case, the exact PMF after n steps can be found in terms of Q and $\mathbf{t}$, and the PMF converges to $\mathbf{s}$ (under some very general conditions discussed in this chapter). In the latter case, the chain is stationary forever.

11.6 R

Matrix calculations

Let's do some calculations for the 4-state Markov chain in Example 11.1.5, as an example of working with transition matrices in R. First, we need to specify the transition matrix Q. This is done with the `matrix` command: we type in the entries of the matrix, row by row, as a long vector, and then we tell R the number of rows and columns in the matrix (`nrow` and `ncol`), as well as the fact that we typed in the entries by row (`byrow=TRUE`):

```
Q <- matrix(c(1/3,1/3,1/3,0,
              0,0,1/2,1/2,
              0,1,0,0,
              1/2,0,0,1/2),nrow=4,ncol=4,byrow=TRUE)
```

To obtain higher order transition probabilities, we can multiply Q by itself repeatedly. The matrix multiplication command in R is `%*%` (*not* just $*$). So

```
Q2 <- Q %*% Q
Q3 <- Q2 %*% Q
Q4 <- Q2 %*% Q2
Q5 <- Q3 %*% Q2
```

produces Q^2 through Q^5. If we want to know the probability of going from state 3 to state 4 in exactly 5 steps, we can extract the $(3,4)$ entry of Q^5:

```
Q5[3,4]
```

This gives 0.229, agreeing with the value 11/48 shown in Example 11.1.5.

To compute a power Q^n without directly doing repeated matrix multiplications, we can use the command `Q %^% n` after installing and loading the `expm` package. For example, `Q %^% 42` yields Q^{42}. By exploring the behavior of Q^n as n grows, we can see Theorem 11.3.6 in action (and get a sense of how long it takes for the chain to get very close to its stationary distribution).

In particular, for n large each row of Q^n is approximately $(0.214, 0.286, 0.214, 0.286)$, so this is approximately the stationary distribution. Another way to obtain the stationary distribution numerically is to use

```
eigen(t(Q))
```

to compute the eigenvalues and eigenvectors of the transpose of Q; then the eigenvector corresponding to the eigenvalue 1 can be selected and normalized so that the components sum to 1.

Gambler's ruin

To simulate from the gambler's ruin chain from Example 11.2.6, we start by deciding the total amount of money `N`, the probability `p` of gambler A winning a given round, and the number of time periods `nsim` that we wish to simulate.

```
N <- 10
p <- 1/2
nsim <- 80
```

Next, we allocate a vector of length `nsim` called `x`, which will store the values of the Markov chain. For the initial condition, we set the first entry of `x` equal to 5; this gives both gamblers $5 to start with.

```
x <- rep(0,nsim)
x[1] <- 5
```

Now we are ready to simulate the subsequent values of the Markov chain. This is achieved with the following block of code, which we will explain step by step.

```
for (i in 2:nsim){
    if (x[i-1]==0 || x[i-1]==N){
        x[i] <- x[i-1]
    }
    else{
        x[i] <- x[i-1] + sample(c(1,-1), 1, prob=c(p,1-p))
    }
}
```

The first line and the outer set of braces constitute a *for loop*: `for (i in 2:nsim)` means that all the code inside the for loop will be executed over and over, with the value of `i` set to 2, then set to 3, then set to 4, all the way until `i` reaches the value `nsim`. Each pass through the loop represents one step of the Markov chain.

Inside the for loop, we first check to see whether the chain is already at one of the endpoints, 0 or `N`; we do this with an *if statement*. If the chain is already at 0 or `N`, then we set its new value equal to its previous value, since the chain is not allowed to escape 0 or `N`. Otherwise, if the chain is not at 0 or `N`, it is free to move left or right. We use the `sample` command to move to the right 1 unit or to the left 1 unit, with probabilities `p` and `1-p`, respectively.

To see what path was taken by the Markov chain during our simulation, we can plot the values of `x` as a function of time:

```
plot(x,type='l',ylim=c(0,N))
```

You should see a path that starts at 5 and bounces up and down before being absorbed into state 0 or state `N`.

Simulating from a finite-state Markov chain

With a few modifications, we can simulate from an arbitrary Markov chain on a finite state space. For concreteness, we will illustrate how to simulate from the 4-state Markov chain in Example 11.1.5.

As above, we can type

```
Q <- matrix(c(1/3,1/3,1/3,0,
              0,0,1/2,1/2,
```

```
          0,1,0,0,
          1/2,0,0,1/2),nrow=4,ncol=4,byrow=TRUE)
```

to specify the transition matrix Q.

Next, we choose the number of states and the number of time periods to simulate, we allocate space for the results of the simulation, and we choose initial conditions for the chain. In this example, `x[1] <- sample(1:M,1)` says the initial distribution of the chain is uniform over all states.

```
M <- nrow(Q)
nsim <- 10^4
x <- rep(0,nsim)
x[1] <- sample(1:M,1)
```

For the simulation itself, we again use `sample` to choose a number from 1 to `M`. At time `i`, the chain was previously at state `x[i-1]`, so we must use row `x[i-1]` of the transition matrix to determine the probabilities of sampling $1, 2, \ldots,$ `M`. The notation `Q[x[i-1],]` denotes row `x[i-1]` of the matrix `Q`.

```
for (i in 2:nsim){
    x[i] <- sample(M, 1, prob=Q[x[i-1],])
}
```

Since we set `nsim` to a large number, it may be reasonable to believe that the chain is close to stationarity during the latter portion of the simulation. To check this, we eliminate the first half of the simulations to give the chain time to reach stationarity:

```
x <- x[-(1:(nsim/2))]
```

We then use the `table` command to calculate the number of times the chain visited each state; dividing by `length(x)` converts the counts into proportions. The result is an approximation to the stationary distribution.

```
table(x)/length(x)
```

For comparison, the true stationary distribution of the chain is approximately $(0.214, 0.286, 0.214, 0.286)$. Is this close to what you obtained via simulation?

11.7 Exercises

Exercises marked with Ⓢ have detailed solutions at `http://stat110.net`.

Markov property

1. Ⓢ Let $X_0, X_1, X_2, \ldots$ be a Markov chain. Show that $X_0, X_2, X_4, X_6, \ldots$ is also a Markov chain, and explain why this makes sense intuitively.

2. Ⓢ Let $X_0, X_1, X_2, \ldots$ be an irreducible Markov chain with state space $\{1, 2, \ldots, M\}$, $M \geq 3$, transition matrix $Q = (q_{ij})$, and stationary distribution $\mathbf{s} = (s_1, \ldots, s_M)$. Let the initial state X_0 follow the stationary distribution, i.e., $P(X_0 = i) = s_i$.

   (a) On average, how many of $X_0, X_1, \ldots, X_9$ equal 3? (In terms of $\mathbf{s}$; simplify.)

   (b) Let $Y_n = (X_n - 1)(X_n - 2)$. For $M = 3$, find an example of Q (the transition matrix for the *original* chain $X_0, X_1, \ldots$) where $Y_0, Y_1, \ldots$ is Markov, and another example of Q where $Y_0, Y_1, \ldots$ is not Markov. In your examples, make $q_{ii} > 0$ for at least one i and make sure it is possible to get from any state to any other state eventually.

3. A Markov chain has two states, A and B, with transitions as follows:

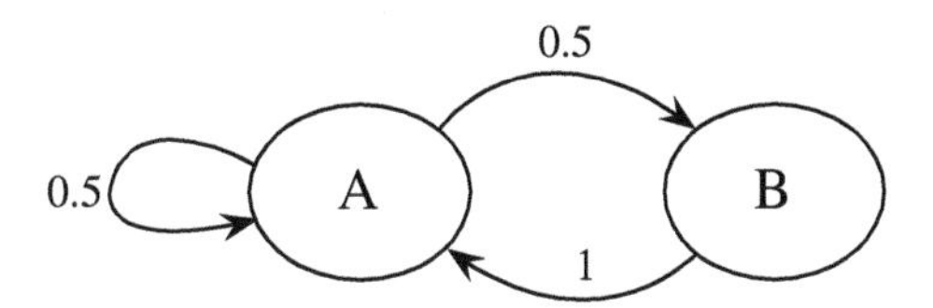

   Suppose we do not get to observe this Markov chain, which we'll call $X_0, X_1, X_2, \ldots$. Instead, whenever the chain transitions from A back to A, we observe a 0, and whenever it changes states, we observe a 1. Let the sequence of 0's and 1's be called $Y_0, Y_1, Y_2, \ldots$. For example, if the X chain starts out as

$$A, A, B, A, B, A, A, \ldots$$

   then the Y chain starts out as

$$0, 1, 1, 1, 1, 0, \ldots.$$

   (a) Show that $Y_0, Y_1, Y_2, \ldots$ is not a Markov chain.

   (b) In Example 11.1.3, we dealt with a violation of the Markov property by enlarging the state space to incorporate second-order dependence. Show that such a trick will not work for $Y_0, Y_1, Y_2, \ldots$. That is, no matter how large m is,

$$Z_n = (Y_{n-m+1}, Y_{n-m+2}, \ldots, Y_n), \quad n = m - 1, m, \ldots$$

   is still not a Markov chain.

4. There are three blocks, floating in a sea of lava. Label the blocks 1, 2, 3, from left to right. Mark the Kangaroo, a video game character, is standing on block 1. To reach safety, he must get to block 3. He can't jump directly from block 1 to block 3; his only hope is to jump from block 1 to block 2, then jump from block 2 to block 3. Each time he jumps, he has probability 1/2 of success and probability 1/2 of "dying" by falling into the lava. If he "dies", he starts again at block 1.

   Let J be the total number of jumps that Mark will make in order to get to block 3.

   (a) Find $E(J)$, using the Markov property and conditional expectation.

   (b) Explain how this problem relates to a coin tossing problem from Chapter 9.

Stationary distribution

5. Ⓢ Consider the Markov chain shown below, where $0 < p < 1$ and the labels on the arrows indicate transition probabilities.

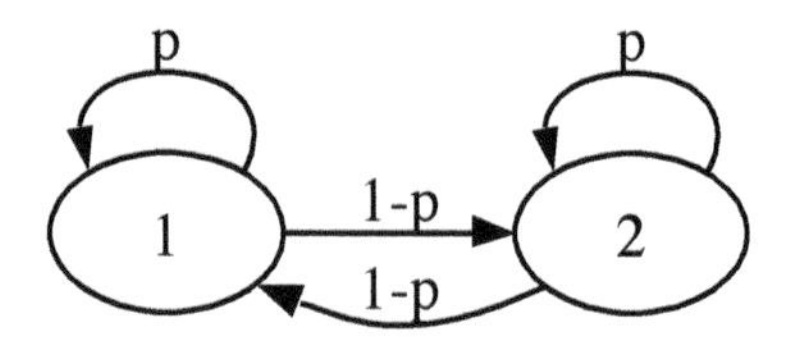

(a) Write down the transition matrix Q for this chain.

(b) Find the stationary distribution of the chain.

(c) What is the limit of Q^n as $n \to \infty$?

6. Ⓢ Consider the Markov chain shown below, where the state space is $\{1, 2, 3, 4\}$ and the labels to the right of arrows indicate transition probabilities.

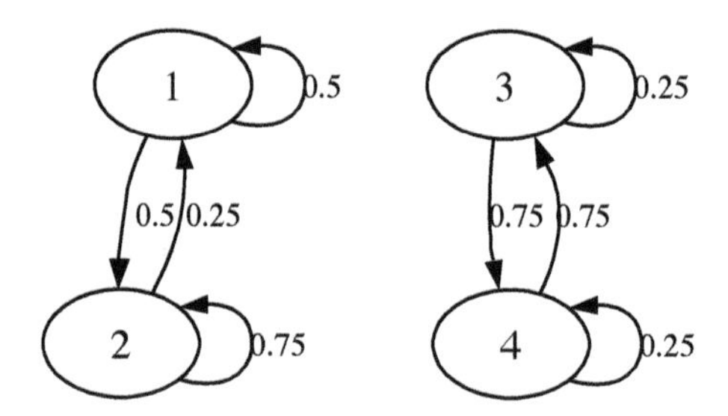

(a) Write down the transition matrix Q for this chain.

(b) Which states (if any) are recurrent? Which states (if any) are transient?

(c) Find two different stationary distributions for the chain.

7. Ⓢ Daenerys has three dragons: Drogon, Rhaegal, and Viserion. Each dragon independently explores the world in search of tasty morsels. Let X_n, Y_n, Z_n be the locations at time n of Drogon, Rhaegal, Viserion respectively, where time is assumed to be discrete and the number of possible locations is a finite number M. Their paths $X_0, X_1, X_2, \dots$; $Y_0, Y_1, Y_2, \dots$; and $Z_0, Z_1, Z_2, \dots$ are independent Markov chains with the same stationary distribution $\mathbf{s}$. Each dragon starts out at a random location generated according to the stationary distribution.

(a) Let state 0 be home (so s_0 is the stationary probability of the home state). Find the expected number of times that Drogon is at home, up to time 24, i.e., the expected number of how many of $X_0, X_1, \dots, X_{24}$ are state 0 (in terms of s_0).

(b) If we want to track all 3 dragons simultaneously, we need to consider the vector of positions, (X_n, Y_n, Z_n). There are M^3 possible values for this vector; assume that each is assigned a number from 1 to M^3, e.g., if $M = 2$ we could encode the states $(0, 0, 0), (0, 0, 1), (0, 1, 0), \dots, (1, 1, 1)$ as $1, 2, 3, \dots, 8$ respectively. Let W_n be the number between 1 and M^3 representing (X_n, Y_n, Z_n). Determine whether $W_0, W_1, \dots$ is a Markov chain.

(c) Given that all 3 dragons start at home at time 0, find the expected time it will take for all 3 to be at home again at the same time.

8. Consider the following Markov chain with $52! \approx 8 \times 10^{67}$ states. The states are the possible orderings of a standard 52-card deck. To run one step of the chain, pick 2 different cards from the deck, with all pairs equally likely, and swap the 2 cards. Find the stationary distribution of the chain.

Reversibility

9. Ⓢ A Markov chain $X_0, X_1, \dots$ with state space $\{-3, -2, -1, 0, 1, 2, 3\}$ proceeds as follows. The chain starts at $X_0 = 0$. If X_n is not an endpoint (-3 or 3), then X_{n+1} is $X_n - 1$ or $X_n + 1$, each with probability $1/2$. Otherwise, the chain gets reflected off the endpoint, i.e., from 3 it always goes to 2 and from -3 it always goes to -2. A diagram of the chain is shown below.

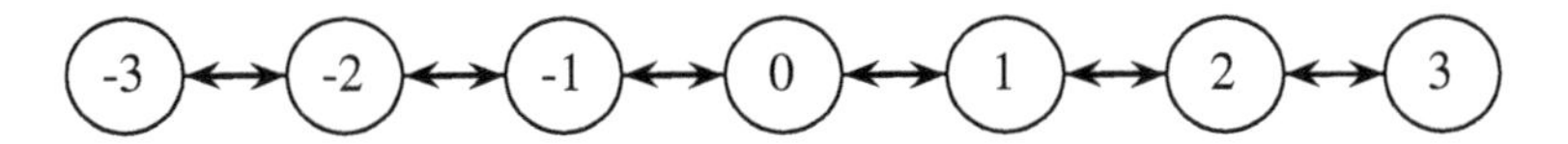

(a) Is $|X_0|, |X_1|, |X_2|, \dots$ also a Markov chain? Explain.

Hint: For both (a) and (b), think about whether the past and future are conditionally independent given the present; don't do calculations with a 7 by 7 transition matrix!

(b) Let sgn be the sign function: $\text{sgn}(x) = 1$ if $x > 0$, $\text{sgn}(x) = -1$ if $x < 0$, and $\text{sgn}(0) = 0$. Is $\text{sgn}(X_0), \text{sgn}(X_1), \text{sgn}(X_2), \dots$ a Markov chain? Explain.

(c) Find the stationary distribution of the chain $X_0, X_1, X_2, \dots$.

(d) Find a simple way to modify some of the transition probabilities q_{ij} for $i \in \{-3, 3\}$ to make the stationary distribution of the modified chain uniform over the states.

10. Ⓢ Let G be an undirected network with nodes labeled $1, 2, \dots, M$ (edges from a node to itself are not allowed), where $M \geq 2$ and random walk on this network is irreducible. Let d_j be the degree of node j for each j. Create a Markov chain on the state space $1, 2, \dots, M$, with transitions as follows. From state i, generate a proposal j by choosing a uniformly random j such that there is an edge between i and j in G; then go to j with probability $\min(d_i/d_j, 1)$, and stay at i otherwise.

(a) Find the transition probability q_{ij} from i to j for this chain, for all states i, j.

(b) Find the stationary distribution of this chain.

11. Ⓢ (a) Consider a Markov chain on the state space $\{1, 2, \dots, 7\}$ with the states arranged in a "circle" as shown below, and transitions given by moving one step clockwise or counterclockwise with equal probabilities. For example, from state 6, the chain moves to state 7 or state 5 with probability 1/2 each; from state 7, the chain moves to state 1 or state 6 with probability 1/2 each. The chain starts at state 1.

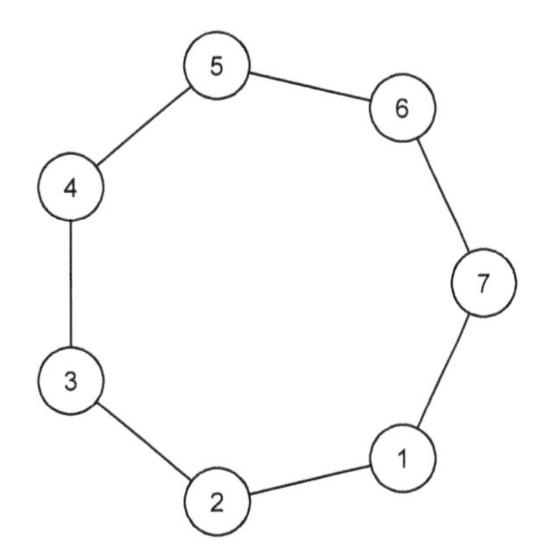

Find the stationary distribution of this chain.

(b) Consider a new chain obtained by "unfolding the circle". Now the states are arranged as shown below. From state 1 the chain always goes to state 2, and from state 7 the chain always goes to state 6. Find the new stationary distribution.

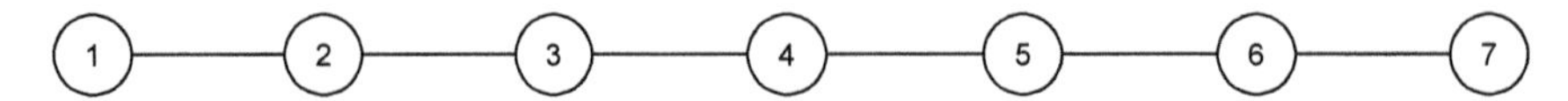

12. Ⓢ Let X_n be the price of a certain stock at the start of the nth day, and assume that $X_0, X_1, X_2, \dots$ follows a Markov chain with transition matrix Q. (Assume for simplicity that the stock price can never go below 0 or above a certain upper bound, and that it is always rounded to the nearest dollar.)

(a) A lazy investor only looks at the stock once a year, observing the values on days $0, 365, 2 \cdot 365, 3 \cdot 365, \dots$. So the investor observes $Y_0, Y_1, \dots$, where Y_n is the price after n years (which is $365n$ days; you can ignore leap years). Is $Y_0, Y_1, \dots$ also a Markov chain? Explain why or why not; if so, what is its transition matrix?

(b) The stock price is always an integer between \$0 and \$28. From each day to the next, the stock goes up or down by \$1 or \$2, all with equal probabilities (except for days when the stock is at or near a boundary, i.e., at \$0, \$1, \$27, or \$28).

If the stock is at \$0, it goes up to \$1 or \$2 on the next day (after receiving government bailout money). If the stock is at \$28, it goes down to \$27 or \$26 the next day. If the stock is at \$1, it either goes up to \$2 or \$3, or down to \$0 (with equal probabilities); similarly, if the stock is at \$27 it either goes up to \$28, or down to \$26 or \$25. Find the stationary distribution of the chain.

13. Ⓢ In chess, the king can move one square at a time in any direction (horizontally, vertically, or diagonally).

For example, in the diagram, from the current position the king can move to any of 8 possible squares. A king is wandering around on an otherwise empty 8×8 chessboard, where for each move all possibilities are equally likely. Find the stationary distribution of this chain (of course, don't list out a vector of length 64 explicitly! Classify the 64 squares into types and say what the stationary probability is for a square of each type).

14. A chess piece is wandering around on an otherwise vacant 8×8 chessboard. At each move, the piece (a king, queen, rook, bishop, or knight) chooses uniformly at random where to go, among the legal choices (according to the rules of chess, which you should look up if you are unfamiliar with them).

(a) For each of these cases, determine whether the Markov chain is irreducible, and whether it is aperiodic.

Hint for the knight: Note that a knight's move always goes from a light square to a dark square or vice versa. A *knight's tour* is a sequence of knight moves on a chessboard such that the knight visits each square exactly once. Many knight's tours exist.

(b) Suppose for this part that the piece is a rook, with initial position chosen uniformly at random. Find the distribution of where the rook is after n moves.

(c) Now suppose that the piece is a king, with initial position chosen deterministically to be the upper left corner square. Determine the expected number of moves it takes the king to return to that square, fully simplified, preferably in at most 140 characters.

(d) The stationary distribution for the random walk of the king from the previous part is not uniform over the 64 squares of the chessboard. A recipe for modifying the chain to obtain a uniform stationary distribution is as follows. Label the squares as $1, 2, \dots, 64$, and let d_i be the number of legal moves from square i. Suppose the king is currently at square i. The next move of the chain is determined as follows:

Step 1: Generate a *proposal square* j by picking uniformly at random among the legal moves from i.

Step 2: Flip a coin with probability $\min(d_i/d_j, 1)$ of Heads. If the coin lands Heads, go to j. Otherwise, stay at i.

Show that this modified chain has a stationary distribution that is uniform over the 64 squares.

15. Ⓢ Find the stationary distribution of the Markov chain shown below, *without using matrices*. The number above each arrow is the corresponding transition probability.

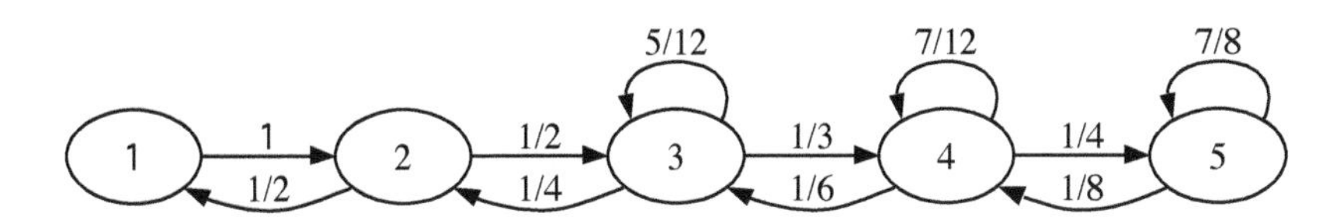

16. There are two urns with a total of $2N$ distinguishable balls. Initially, the first urn has N white balls and the second urn has N black balls. At each stage, we pick a ball at random from each urn and interchange them. Let X_n be the number of black balls in the first urn at time n. This is a Markov chain on the state space $\{0, 1, \dots, N\}$.

(a) Give the transition probabilities of the chain.

(b) Show that $(s_0, s_1, \dots, s_N)$ where

$$s_i = \frac{\binom{N}{i}\binom{N}{N-i}}{\binom{2N}{N}}$$

is the stationary distribution, by verifying the reversibility condition.

17. Find the stationary distribution of a Markov chain $X_0, X_1, X_2, \dots$ on the state space $\{0, 1, \dots, 110\}$, with transition probabilities given by

$$P(X_{n+1} = j | X_n = 0) = p, \text{ for } j = 1, 2, \dots, 110;$$

$$P(X_{n+1} = 0 | X_n = 0) = 1 - 110p;$$

$$P(X_{n+1} = j | X_n = j) = 1 - r, \text{ for } j = 1, 2, \dots, 110;$$

$$P(X_{n+1} = 0 | X_n = j) = r, \text{ for } j = 1, 2, \dots, 110,$$

where p and r are constants with $0 < p < \frac{1}{110}$ and $0 < r < 1$.

18. Determine whether the Markov chain shown below is reversible, and find the stationary distribution of the chain. The label to the left of an arrow gives the corresponding transition probability.

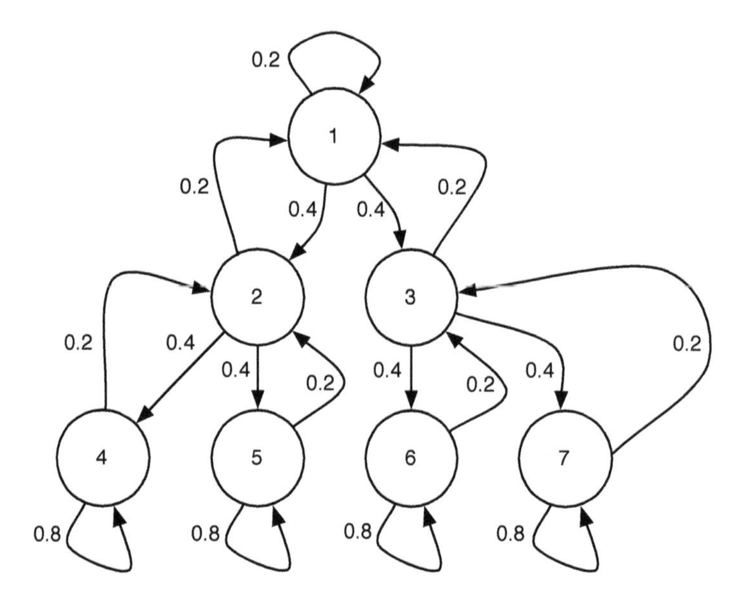

19. Nausicaa Distribution sells distribution plushies on Etsy. They have two different photos of the Evil Cauchy plushie but do not know which is more effective in getting a customer to purchase an Evil Cauchy plushie. Each visitor to their website is shown one of the two photos (call them Photo A and Photo B), and then the visitor either does buy an Evil Cauchy ("success") or does not buy one ("failure").

    Let a and b be the probabilities of success when Photo A is shown and when Photo B is shown, respectively. Even though the Evil Cauchy is irresistible, suppose that $0 < a < 1$ and $0 < b < 1$. Suppose that the following strategy is followed (note that the strategy can be followed without knowing a and b). Show the first visitor Photo A. If that visitor buys an Evil Cauchy, continue with Photo A for the next visitor; otherwise, switch to Photo B. Similarly, if the nth visitor is a "success" then show the $(n + 1)$st visitor the same photo, and otherwise switch to the other photo.

    (a) Show how to represent the resulting process as a Markov chain, drawing a diagram and giving the transition matrix. The states are A1, B1, A0, B0 (use this order for the transition matrix and stationary distribution), where, for example, being at state A1 means that the current visitor was shown Photo A and was a success.

    (b) Determine whether this chain is reversible.

    Hint: First think about which transition probabilities are zero and which are nonzero.

    (c) Show that the stationary distribution is proportional to $\left(\frac{a}{1-a}, \frac{b}{1-b}, 1, 1\right)$, and find the stationary distribution.

    (d) Show that for $a \neq b$, the stationary probability of success for each visitor is strictly better than the success probability that would be obtained by independently, randomly choosing (with equal probabilities) which photo to show to each visitor.

20. This exercise considers random walk on a *weighted* undirected network. Suppose that an undirected network is given, where each edge (i, j) has a nonnegative weight w_{ij} assigned to it (we allow $i = j$ as a possibility). We assume that $w_{ij} = w_{ji}$ since the edge from i to j is considered the same as the edge from j to i. To simplify notation, define $w_{ij} = 0$ whenever (i, j) is not an edge.

When at node i, the next step is determined by choosing an edge attached to i with probabilities proportional to the weights. For example, if the walk is at node 1 and there are 3 possible edges coming out from node 1, with weights $7, 1, 4$, then the first of these 3 edges is traversed with probability $7/12$, the second is traversed with probability $1/12$, and the third is traversed with probability $4/12$. If all the weights equal 1, then the process reduces to the kind of random walk on a network that we studied earlier.

(a) Let $v_i = \sum_j w_{ij}$ for all nodes i. Show that the stationary distribution of node i is proportional to v_i.

(b) Show that *every* reversible Markov chain can be represented as a random walk on a weighted undirected network. That is, given the transition matrix Q of a reversible Markov chain, show that we can choose the weights w_{ij} so that the random walk defined above is a Markov chain with transition matrix Q. Be sure to check that $w_{ij} = w_{ji}$.

Hint: Let $w_{ij} = s_i q_{ij}$, with $\mathbf{s}$ the stationary distribution and q_{ij} the (i, j) entry of Q.

Mixed practice

21. Ⓢ A cat and a mouse move independently back and forth between two rooms. At each time step, the cat moves from the current room to the other room with probability 0.8. Starting from room 1, the mouse moves to Room 2 with probability 0.3 (and remains otherwise). Starting from room 2, the mouse moves to room 1 with probability 0.6 (and remains otherwise).

    (a) Find the stationary distributions of the cat chain and of the mouse chain.

    (b) Note that there are 4 possible (cat, mouse) states: both in room 1, cat in room 1 and mouse in room 2, cat in room 2 and mouse in room 1, and both in room 2. Number these cases $1, 2, 3, 4$, respectively, and let Z_n be the number representing the (cat, mouse) state at time n. Is $Z_0, Z_1, Z_2, \dots$ a Markov chain?

    (c) Now suppose that the cat will eat the mouse if they are in the same room. We wish to know the expected time (number of steps taken) until the cat eats the mouse for two initial configurations: when the cat starts in room 1 and the mouse starts in room 2, and vice versa. Set up a system of two linear equations in two unknowns whose solution is the desired values.

22. (a) Alice and Bob are wandering around randomly, independently of each other, in a house with M rooms, labeled $1, 2, \dots, M$. Let d_i be the number of doors in room i (leading to other rooms, not leading outside). At each step, Alice moves to another room by choosing randomly which door to go through (with equal probabilities). Bob does the same, independently. The Markov chain they each follow is irreducible and aperiodic. Let A_n and B_n be Alice's room and Bob's room at time n, respectively, for $n = 0, 1, 2, \dots$.

    Find $\lim_{n\to\infty} P(A_n = i, B_n = j)$.

    (b) With setup as in (a), let p_{ij} be the transition probability for going from room i to room j. Let t_{ik} be the expected first time at which Alice and Bob are in the same room, if Alice starts in room i and Bob starts in room k (note that $t_{ik} = 0$ for $i = k$).

    Provide a system of linear equations, the solution of which would yield t_{ik} for all rooms i, k.

    Hint: Condition on Alice's first move and Bob's first move.

23. Let $\{X_n\}$ be a Markov chain on states $\{0, 1, 2\}$ with transition matrix

$$Q = \begin{pmatrix} 0.8 & 0.2 & 0 \\ 0 & 0.8 & 0.2 \\ 0 & 0 & 1 \end{pmatrix}.$$

The chain starts at $X_0 = 0$. Let T be the time it takes to reach state 2:

$$T = \min\{n : X_n = 2\}.$$

By drawing the Markov chain and telling a story, find $E(T)$ and $\text{Var}(T)$.

24. Consider the following Markov chain on the state space $\{1, 2, 3, 4, 5, 6\}$.

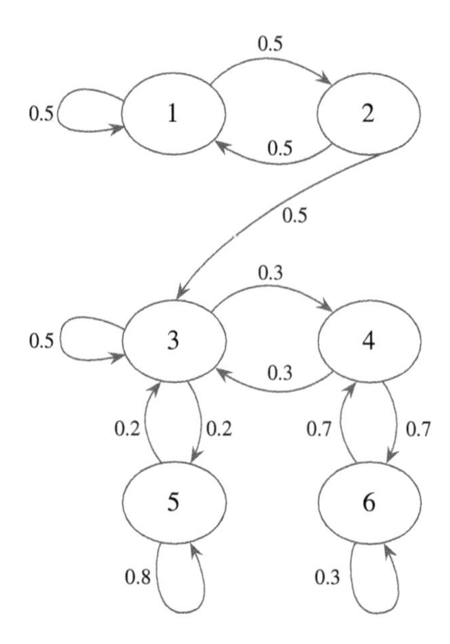

(a) Suppose the chain starts at state 1. Find the distribution of the number of times that the chain returns to state 1.

(b) In the long run, what fraction of the time does the chain spend in state 3? Explain briefly.

25. Let Q be the transition matrix of a Markov chain on the state space $\{1, 2, \ldots, M\}$, such that state M is an *absorbing state*, i.e., from state M the chain can never leave. Suppose that from any other state, it is possible to reach M (in some number of steps).

(a) Which states are recurrent, and which are transient? Explain.

(b) What is the limit of Q^n as $n \to \infty$?

(c) For $i, j \in \{1, 2, \ldots, M-1\}$, find the probability that the chain is at state j at time n, given that the chain is at state i at time 0 (your answer should be in terms of Q).

(d) For $i, j \in \{1, 2, \ldots, M-1\}$, find the expected number of times that the chain is at state j up to (and including) time n, given that the chain is at state i at time 0 (in terms of Q).

(e) Let R be the $(M-1) \times (M-1)$ matrix obtained from Q by deleting the last row and the last column of Q. Show that the (i, j) entry of $(I - R)^{-1}$ is the expected number of times that the chain is at state j before absorption, given that it starts out at state i.

Hint: We have $I + R + R^2 + \cdots = (I - R)^{-1}$, analogously to a geometric series. Also, if we partition Q as

$$Q = \left(\begin{array}{c|c} R & B \\ \hline 0 & 1 \end{array}\right)$$

where B is a $(M-1) \times 1$ matrix and 0 is the $1 \times (M-1)$ zero matrix, then

$$Q^k = \left(\begin{array}{c|c} R^k & B_k \\ \hline 0 & 1 \end{array}\right)$$

for some $(M-1) \times 1$ matrix B_k.

26. In the game called *Chutes and Ladders*, players try to be first to reach a certain destination on a board. The board is a grid of squares, numbered from 1 to the number of squares. The board has some "chutes" and some "ladders", each of which connects a pair of squares. Here we will consider the one player version of the game (this can be extended to the multi-player version without too much trouble, since with more than one player, the players simply take turns independently until one reaches the destination).

    On each turn, the player rolls a fair die, which determines how many squares forward to move on the grid, e.g., if the player is at square 5 and rolls a 3, then they advance to square 8. If the resulting square is the base of a ladder, the player gets to climb the ladder, instantly arriving at a more advanced square. If the resulting square is the top of a chute, the player instantly slides down to the bottom of the chute.

    This game can be viewed naturally as a Markov chain: given where the player currently is on the board, the past history does not matter for computing, for example, the probability of winning within the next 3 moves.

    Consider a simplified version of Chutes and Ladders, played on the 3×3 board shown below. The player starts out at square 1, and wants to get to square 9. On each move, a fair coin is flipped, and the player gets to advance 1 square if the coin lands Heads and 2 squares if the coin lands Tails. However, there are 2 ladders (shown as upward-pointing arrows) and 2 chutes (shown as downward-pointing arrows) on the board.

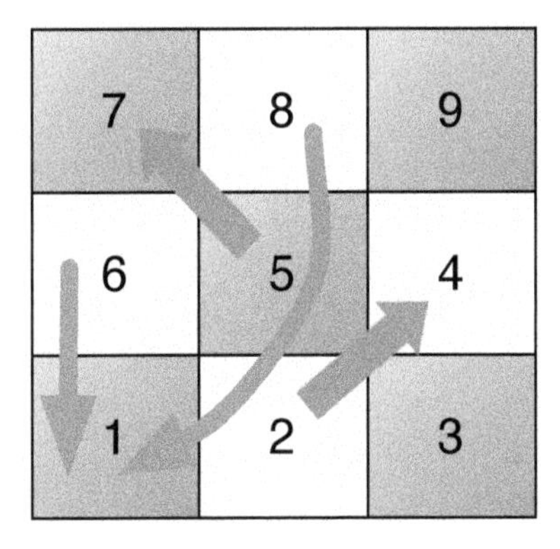

    (a) Explain why, despite the fact that there are 9 squares, we can represent the game using the 5×5 transition matrix

$$Q = \begin{pmatrix} 0 & 0.5 & 0.5 & 0 & 0 \\ 0 & 0 & 0.5 & 0.5 & 0 \\ 0.5 & 0 & 0 & 0.5 & 0 \\ 0.5 & 0 & 0 & 0 & 0.5 \\ 0 & 0 & 0 & 0 & 1 \end{pmatrix}.$$

    (b) Find the mean and variance for the number of times the player will visit square 7, *without* using matrices or any messy calculations.

    The remaining parts of this problem require matrix calculations that are best done on a computer. You can use whatever computing environment you want, but here is some information for how to do it in R. In any case, you should state what environment you used and include your code. To create the transition matrix in R, you can use the following commands:

```
a <- 0.5
Q <- matrix(c(0,0,a,a,0,a,0,0,0,0,a,a,0,0,0,0,a,a,0,0,0,0,0,a,1),nrow=5)
```

Some useful R commands for matrices are in Appendix B.2. In particular, `diag(n)` gives the $n \times n$ identity matrix, `solve(A)` gives the inverse A^{-1}, and `A %*% B` gives the product AB (note that `A*B` does *not* do ordinary matrix multiplication). Matrix powers are not built into R, but you can compute A^k using `A %^% k` after installing and loading the `expm` package.

(c) Find the median duration of the game (where duration is the number of coin flips).

Hint: Relate the CDF of the duration to powers of Q.

(d) Find the mean duration of the game (with duration defined as above).

Hint: Relate the duration to the total amount of time spent in transient states, and apply Part (e) of the previous problem.

12

Markov chain Monte Carlo

We have seen throughout this book that *simulation* is a powerful technique in probability. If you can't convince your friend that it is a good idea to switch doors in the Monty Hall problem, in one second you can simulate playing the game a few thousand times and your friend will just *see* that switching succeeds about 2/3 of the time. If you're unsure how to calculate the mean and variance of an r.v. X but you know how to generate i.i.d. draws $X_1, X_2, \ldots, X_n$ from that distribution, you can approximate the true mean and true variance using the sample mean and sample variance of the simulated draws:

$$E(X) \approx \frac{1}{n}(X_1 + \cdots + X_n) = \bar{X}_n,$$

$$\text{Var}(X) \approx \frac{1}{n-1}\sum_{j=1}^{n}(X_j - \bar{X}_n)^2.$$

The law of large numbers tells us that these approximations will be good if n is large. We can get better and better approximations by increasing n, just by running the computer for a longer time (rather than having to struggle with a possibly intractable sum or integral). As discussed in Chapter 10, this simulation approach, where we generate random values to approximate a quantity, is called a *Monte Carlo* method.

A major limitation of the Monte Carlo idea above is that we need to know how to generate $X_1, X_2, \ldots, X_n$ (hopefully efficiently, since we want to be able to make n large). For example, suppose that we want to simulate random draws from the continuous distribution with PDF f given by

$$f(x) \propto x^{3.1}(1-x)^{4.2}$$

for $0 < x < 1$ (and 0 otherwise). Staring at a *density function* does not immediately suggest how to get a *random variable* with that density. In this case we recognize the PDF of the Beta(4.1, 5.2) distribution, so assuming we had access to a Unif(0, 1) r.v., we could theoretically use universality of the Uniform. Unfortunately, finding the CDF of a Beta distribution is difficult, let alone the *inverse* of the CDF. See Example 12.1.4 for more about the Beta simulation problem.

Of course, the distributions that arise in real-world applications are often much more complicated than the Beta. For the Beta distribution, we know an expression for the normalizing constant in terms of the gamma function. But for many other distributions that are of scientific interest, the normalizing constant in the PDF or PMF is unknown, and beyond the reach of the fastest available computers and the fanciest available mathematical techniques.

This chapter introduces *Markov chain Monte Carlo* (MCMC), a powerful collection of algorithms that enable us to simulate from complicated distributions using Markov chains. The development of MCMC has revolutionized statistics and scientific computation by vastly expanding the range of possible distributions that we can simulate from, including joint distributions in high dimensions. The basic idea is to *build your own Markov chain* so that the distribution of interest is the stationary distribution of the chain.

In the previous chapter, we looked at Markov chains whose transition matrix Q was specified, and we tried to find the stationary distribution $\mathbf{s}$ of the chain. In this chapter we do the reverse: starting with a distribution $\mathbf{s}$ that we want to simulate, we will *engineer* a Markov chain whose stationary distribution is $\mathbf{s}$. If we then run this engineered Markov chain for a very long time, the distribution of the chain will approach our desired $\mathbf{s}$.

But is it possible to create a transition matrix Q with the stationary distribution we desire? Even if it is possible, is solving this problem any easier than the original problem of how to simulate random draws from that distribution? In amazing generality, MCMC shows that it *is* possible to create a Markov chain with the desired stationary distribution in an easy-to-describe way, without having to know the normalizing constant for the distribution!

MCMC is now being applied to a large number of problems in the biological, natural, and physical sciences, and many different MCMC algorithms have been developed. Here we will introduce two of the most important and most widely used MCMC algorithms: the *Metropolis-Hastings algorithm* and *Gibbs sampling.* MCMC is an enormous and growing area of statistical computing; see Brooks, Gelman, Jones, and Meng [2] for much more about its theory, methods, and applications.

12.1 Metropolis-Hastings

The Metropolis-Hastings algorithm is a general recipe that lets us start with any irreducible Markov chain on the state space of interest and then modify it into a new Markov chain that has the desired stationary distribution. This modification consists of introducing some selectiveness in the original chain: moves are *proposed* according to the original chain, but the proposal may or may not be *accepted.* For example, suppose the original chain is at a state called "Boston" and is about to transition to "San Francisco". Then for the new chain, we either accept the proposal and go to San Francisco, or we turn down the proposal and remain in Boston as the next step. With a careful choice of the probability of accepting the proposal, this simple modification guarantees that the new chain has the desired stationary distribution.

Algorithm 12.1.1 (Metropolis-Hastings). Let $\mathbf{s} = (s_1, \dots, s_M)$ be a desired stationary distribution on state space $\{1, \dots, M\}$. Assume that $s_i > 0$ for all i (if not, just delete any states i with $s_i = 0$ from the state space). Suppose that $P = (p_{ij})$ is the transition matrix for a Markov chain on state space $\{1, \dots, M\}$. Intuitively, P is a Markov chain that we know how to run but that doesn't have the desired stationary distribution.

Our goal is to modify P to construct a Markov chain $X_0, X_1, \dots$ with stationary distribution $\mathbf{s}$. We will give a Metropolis-Hastings algorithm for this. Start at any state X_0 (chosen randomly or deterministically), and suppose that the new chain is currently at X_n. To make one move of the new chain, do the following.

1. If $X_n = i$, propose a new state j using the transition probabilities in the ith row of the original transition matrix P.

2. Compute the *acceptance probability*

$$a_{ij} = \min\left(\frac{s_j p_{ji}}{s_i p_{ij}}, 1\right).$$

3. Flip a coin that lands Heads with probability a_{ij}.

4. If the coin lands Heads, accept the proposal (i.e., go to j), setting $X_{n+1} = j$. Otherwise, reject the proposal (i.e., stay at i), setting $X_{n+1} = i$.

That is, the Metropolis-Hastings chain uses the original transition probabilities p_{ij} to *propose* where to go next, then *accepts* the proposal with probability a_{ij}, staying in its current state in the event of a rejection. An especially nice aspect of this algorithm is that the normalizing constant for $\mathbf{s}$ does not need to be known, since it cancels out in s_j/s_i anyway. For example, in some problems we may want the stationary distribution to be uniform over all states (i.e., $\mathbf{s} = (1/M, 1/M, \dots, 1/M)$), but the number of states M is large and unknown, and it would be a very hard counting problem to find M. Fortunately, $s_j/s_i = 1$ regardless of M, so we can simply say $\mathbf{s} \propto (1, 1, \dots, 1)$, and we can calculate a_{ij} without having to know M.

The p_{ij} in the denominator in a_{ij} will never be 0 when the algorithm is run, since if $p_{ij} = 0$ then the original chain will never propose going from i to j. Also, if $p_{ii} > 0$ it is possible that the proposal j will equal the current state i; in that case, the chain stays at i regardless of whether the proposal is accepted. (Rejecting the proposal of staying at i but staying there anyway is like a child who just got grounded saying "yes, I will stay in my room, but not because you told me to!")

We will now show that the Metropolis-Hastings chain is reversible with stationary distribution $\mathbf{s}$.

Proof. Let Q be the transition matrix of the Metropolis-Hastings chain. We just need to check the reversibility condition $s_i q_{ij} = s_j q_{ji}$ for all i and j. This is clear for $i = j$, so assume $i \neq j$. If $q_{ij} > 0$, then $p_{ij} > 0$ (the chain can't get from i to j if it can't even *propose* going from i to j) and $p_{ji} > 0$ (otherwise the acceptance probability would be 0). Conversely, if $p_{ij} > 0$ and $p_{ji} > 0$, then $q_{ji} > 0$. So q_{ij} and q_{ji} are either both zero or both nonzero. We can assume they are both nonzero. Then

$$q_{ij} = p_{ij} a_{ij}$$

since, starting at i, the only way to get to j is first to propose doing so and then to accept the proposal. First consider the case $s_j p_{ji} \leq s_i p_{ij}$. We have

$$a_{ij} = \frac{s_j p_{ji}}{s_i p_{ij}}, \; a_{ji} = 1,$$

so

$$s_i q_{ij} = s_i p_{ij} a_{ij} = s_i p_{ij} \frac{s_j p_{ji}}{s_i p_{ij}} = s_j p_{ji} = s_j p_{ji} a_{ji} = s_j q_{ji}.$$

Symmetrically, if $s_j p_{ji} > s_i p_{ij}$, we again have $s_i q_{ij} = s_j q_{ji}$, by switching the roles of i and j in the preceding calculation. Since the reversibility condition holds, $\mathbf{s}$ is the stationary distribution of the chain with transition matrix Q. ∎

☣ **12.1.2.** The Metropolis-Hastings algorithm is an extremely general way to construct a Markov chain with a desired stationary distribution. In the above formulation, both $\mathbf{s}$ and P were very general, and nothing was stipulated about their being related (aside from being on the same state space). In practice, however, the choice of the proposal distribution is extremely important since it can make an enormous difference in how quickly the chain converges to its stationary distribution.

How to choose a good proposal distribution is a complicated topic and will not be discussed in detail here. Intuitively, a proposal distribution with a very low acceptance rate will be slow to converge (since the chain will rarely move anywhere). But a high acceptance rate may not be ideal either, since it may indicate that the chain tends to make small, timid proposals. In a large state space, such a chain will take a very long time to explore the entire space.

Here are a couple examples of how the Metropolis-Hastings algorithm can be used to simulate from distributions.

Example 12.1.3 (Zipf distribution simulation). Let $M \geq 2$ be an integer. An r.v. X has the *Zipf distribution* with parameter $a > 0$ if its PMF is

$$P(X = k) = \frac{1/k^a}{\sum_{j=1}^{M}(1/j^a)},$$

for $k = 1, 2, \ldots, M$ (and 0 otherwise). This distribution is widely used in linguistics for studying frequencies of words.

Create a Markov chain $X_0, X_1, \ldots$ whose stationary distribution is the Zipf distribution, and such that $|X_{n+1} - X_n| \leq 1$ for all n. Your answer should provide a simple, precise description of how each move of the chain is obtained, i.e., how to transition from X_n to X_{n+1} for each n.

Solution:

We can use the Metropolis-Hastings algorithm, after coming up with a proposal distribution. There are many possible proposal distributions, but one simple choice is the following random walk on $\{1, 2, \dots, M\}$. From state i with $i \neq 1, i \neq M$, move to state $i-1$ or $i+1$, with probability $1/2$ each. From state 1, stay there or move to state 2, with probability $1/2$ each. From state M, stay there or move to state $M-1$, with probability $1/2$ each. This chain is shown below.

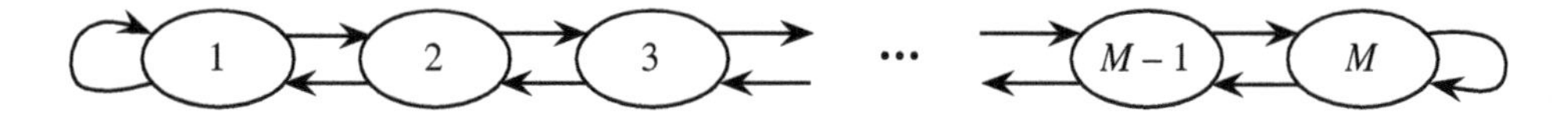

FIGURE 12.1
Proposal chain for Zipf distribution simulation.

Let P be the transition matrix of this chain. The stationary distribution for P is uniform because P is a symmetric matrix, so Proposition 11.4.3 applies. Metropolis-Hastings lets us transmogrify P into a chain whose stationary distribution is Zipf's distribution.

Let X_0 be any starting state, and generate a chain $X_0, X_1, \dots$ as follows. If the chain is currently at state i, then:

1. Generate a proposal state j according to the proposal chain P.

2. Accept the proposal with probability $\min\left(i^a/j^a, 1\right)$. If the proposal is accepted, go to j; otherwise, stay at i.

This chain is easy to implement and a move requires very little computation; note that the normalizing constant $\sum_{j=1}^{M}(1/j^a)$ is *not* needed to run the chain. □

Example 12.1.4 (Beta simulation). Let us now return to the Beta simulation problem introduced at the beginning of the chapter. Suppose that we want to generate $W \sim \text{Beta}(a, b)$, but we don't know about the `rbeta` command in R. Instead, what we have available are i.i.d. Unif$(0, 1)$ r.v.s.

(a) How can we generate W *exactly* if a and b are positive integers, using a story and universality of the Uniform?

(b) How can we generate W which is *approximately* Beta(a, b) if a and b are any positive real numbers, with the help of a Markov chain on the state space $(0, 1)$?

Solution:

(a) Applying universality of the Uniform directly for the Beta is hard, so let's first use the bank–post office story: if $X \sim \text{Gamma}(a, 1)$ and $Y \sim \text{Gamma}(b, 1)$ are independent, then $X/(X+Y) \sim \text{Beta}(a,b)$. So if we can simulate the Gamma distribution, then we can simulate the Beta distribution!

To simulate $X \sim \text{Gamma}(a,1)$, we can use $X_1 + X_2 + \cdots + X_a$ with the X_j i.i.d. Expo(1); similarly, we can simulate $Y \sim \text{Gamma}(b,1)$ as the sum of b i.i.d. Expo(1) r.v.s. Lastly, by taking the inverse of the Expo(1) CDF and applying universality of the Uniform, $-\log(1-U) \sim \text{Expo}(1)$ for $U \sim \text{Unif}(0,1)$, so we can easily construct as many Expo(1) r.v.s as we want.

(b) Let's use the Metropolis-Hastings algorithm. We have only introduced Metropolis-Hastings for finite state spaces, but the ideas are analogous for an infinite state space. A simple proposal chain we have available consists of *independent* Unif(0, 1) r.v.s. That is, the proposed state on the interval (0, 1) is always a fresh Unif(0, 1), independent of the current state. The resulting Metropolis-Hastings chain is called an *independence sampler.*

Let W_0 be any starting state, and generate a chain $W_0, W_1, \ldots$ as follows. If the chain is currently at state w, then:

1. Generate a proposal u by drawing a Unif(0, 1) r.v.

2. Accept the proposal with probability $\min\left(\frac{u^{a-1}(1-u)^{b-1}}{w^{a-1}(1-w)^{b-1}}, 1\right)$. If the proposal is accepted, go to u; otherwise, stay at w.

Again, the normalizing constant was not needed in order to run the chain. In obtaining the acceptance probability, the Beta(a,b) PDF plays the role of $\mathbf{s}$ since it's the desired stationary distribution, and the Unif(0, 1) PDF plays the role of p_{ij} (and p_{ji}) since the proposals are Unif(0, 1) r.v.s, independent of the current state.

Running the Markov chain, we have that $W_n, W_{n+1}, W_{n+2}, \ldots$ are approximately Beta(a,b) for n large. Note that these are *correlated* r.v.s., not i.i.d. draws. □

☣ **12.1.5** (MCMC produces correlated samples). A major question in running a Markov chain $X_0, X_1, \ldots$ for a Monte Carlo computation is how long to run it. In part, this is because it is usually hard to know how close the chain's distribution at time n will be to the stationary distribution. Another issue is that $X_0, X_1, \ldots$ are correlated in general. Some chains tend to get stuck in certain regions of the state space, rather than exploring the whole space. If a chain can get stuck easily, then X_n may be highly positively correlated with X_{n+1}. The *autocorrelation at lag k* is the correlation between X_n and the value k steps later, X_{n+k}, in the limit as n grows. It is desirable for the autocorrelation at lag k to approach 0 rapidly as k increases. High autocorrelation tends to mean high variances for Monte Carlo approximations.

Analysis of how long to run a chain and finding diagnostics for whether the chain has been run long enough are active research areas. Some general advice is to run your chains for a very large number of steps and to try chains from diverse starting points to see how stable the results are.

Metropolis-Hastings is often useful even for immense state spaces. It can even be useful for problems that may not sound at first like they have anything to do with simulating a distribution, such as code-breaking.

Example 12.1.6 (Code-breaking). Markov chains have recently been applied to code-breaking; this example will introduce one way in which this can be done. (For further information about such applications, see Diaconis [6] and Chen and Rosenthal [3].) A *substitution cipher* is a permutation g of the letters from a to z, where a message is enciphered by replacing each letter α by $g(\alpha)$. For example, if g is the permutation given by

```
abcdefghijklmnopqrstuvwxyz
zyxwvutsrqponmlkjihgfedcba
```

where the second row lists the values $g(a), g(b), \dots, g(z)$, then we would encipher the word "statistics" as "hgzgrhgrxh". (We could also include capital letters, spaces, and punctuation marks if desired.) The state space is all $26! \approx 4 \cdot 10^{26}$ permutations of the letters a through z. This is an extremely large space: if we had to try decoding a text using each of these permutations, and could handle one permutation per nanosecond, it would still take over 12 billion years to work through all the permutations. So a brute-force investigation that goes through each permutation one by one is infeasible; instead, we will look at *random* permutations.

(a) Consider the Markov chain that picks two different random coordinates between 1 and 26 and swaps those entries of the 2nd row, e.g., if we pick 7 and 20, then

```
abcdefghijklmnopqrstuvwxyz
zyxwvutsrqponmlkjihgfedcba
```

becomes

```
abcdefghijklmnopqrstuvwxyz
zyxwvugsrqponmlkjihtfedcba
```

Find the probability of going from a permutation g to a permutation h in one step (for all g, h), and find the stationary distribution of this chain.

(b) Suppose we have a system that assigns a positive "score" $s(g)$ to each permutation g. Intuitively, this could be a measure of how likely it would be to get the observed enciphered text, given that g was the cipher used. Use the Metropolis-Hastings algorithm to construct a Markov chain whose stationary distribution is proportional to the list of all scores $s(g)$.

Solution:

(a) The probability of going from g to h in one step is 0 unless h can be obtained from g by swapping 2 entries of the second row. Assuming that h can be obtained in this way, the probability is $\frac{1}{\binom{26}{2}}$, since there are $\binom{26}{2}$ such swaps, all equally likely.

This Markov chain is irreducible, since by performing enough swaps we can get from any permutation to any other permutation. (Imagine rearranging a deck of cards by swapping cards two at a time; it is possible to reorder the cards in any desired configuration by doing this enough times.) Note that $p(g,h) = p(h,g)$, where $p(g,h)$ is the transition probability of going from g to h. Since the transition matrix is symmetric, the stationary distribution is *uniform* over all 26! permutations of the letters a through z.

(b) For our proposal chain, we'll use the chain from (a). Starting from any state g, generate a proposal h using the chain from (a). Flip a coin with probability $\min(s(h)/s(g), 1)$ of Heads. If Heads, go to h; if Tails, stay at g.

To prove this has the desired stationary distribution, we can appeal to our general proof of Algorithm 12.1.1 or check the reversibility condition directly. For practice, we'll do the latter: we need $s(g)q(g,h) = s(h)q(h,g)$ for all g and h, where $q(g,h)$ is the transition probability from g to h in the modified chain. If $g = h$ or $q(g,h) = 0$, then the equation clearly holds, so assume $g \neq h$ and $q(g,h) \neq 0$. Let $p(g,h)$ be the transition probability from (a) (which is the probability of proposing h when at g). First consider the case that $s(g) \leq s(h)$. Then $q(g,h) = p(g,h)$ and

$$q(h,g) = p(h,g)\frac{s(g)}{s(h)} = p(g,h)\frac{s(g)}{s(h)} = q(g,h)\frac{s(g)}{s(h)},$$

so $s(g)q(g,h) = s(h)q(h,g)$. Now consider the case that $s(h) < s(g)$. By a symmetric argument (reversing the roles of g and h), we again have $s(g)q(g,h) = s(h)q(h,g)$. Thus, the stationary probability of g is proportional to its score $s(g)$.

In other words, using the Metropolis-Hastings algorithm, we started with a Markov chain that was equally likely to visit all of the ciphers in the long run and created a Markov chain whose stationary distribution sorts the ciphers according to their scores, visiting the most promising ciphers the most often in the long run. ☐

Here is another MCMC example with an immense state space and which at first sight might not seem to have much connection with simulating a distribution. This example points to the fact that MCMC can be used not only for sampling but also for *optimization*.

Example 12.1.7 (Knapsack problem). Bilbo the Burglar finds m treasures in Smaug's Lair. Bilbo is deciding which treasures to steal (or justly reclaim, depending on one's point of view); he can't take everything all at once, since the maximum weight he can carry is w pounds. Label the treasures from 1 to m, and suppose that the jth treasure is worth g_j gold pieces and weighs w_j pounds. So Bilbo must choose a vector $x = (x_1, \dots, x_m)$, where x_j is 1 if he steals the jth treasure and 0 otherwise, such that the total weight of the treasures j with $x_j = 1$ is at most w. Let C be the space of all such vectors, so C consists of all binary vectors $(x_1, \dots, x_m)$ with $\sum_{j=1}^{m} x_j w_j \leq w$.

Bilbo wishes to maximize the total worth of the treasure he takes. Finding an optimal solution is an extremely difficult problem, known as the *knapsack problem*, which has a long history in computer science. A brute force solution would be completely infeasible in general. Bilbo decides instead to explore the space C using MCMC—luckily, he has a laptop running R with him.

(a) Consider the following Markov chain. Start at $(0, 0, \dots, 0)$. One move of the chain is as follows. Suppose the current state is $x = (x_1, \dots, x_m)$. Choose a uniformly random J in $\{1, 2, \dots, m\}$, and obtain y from x by replacing x_J with $1 - x_J$ (i.e., toggle whether that treasure will be taken). If y is not in C, stay at x; if y is in C, move to y. Show that the uniform distribution over C is stationary for this chain.

(b) Show that the chain from (a) is irreducible, and that it may or may not be aperiodic (depending on $w, w_1, \dots, w_m$).

(c) The chain from (a) is a useful way to get approximately *uniform* solutions, but Bilbo is more interested in finding solutions where the value (in gold pieces) is high. In this part, the goal is to construct a Markov chain with a stationary distribution that puts much higher probability on any particular high-value solution than on any particular low-value solution. Specifically, suppose that we want to simulate from the distribution

$$s(x) \propto e^{\beta V(x)},$$

where $V(x) = \sum_{j=1}^{m} x_j g_j$ is the value of x in gold pieces and β is a positive constant. The idea behind this distribution is to give exponentially more probability to each high-value solution than to each low-value solution. Create a Markov chain whose stationary distribution is as desired.

Solution:

(a) The transition matrix is symmetric since for $x \neq y$, the transition probabilities from x to y and from y to x are either both 0 or both $1/m$. So the stationary distribution is uniform over C.

(b) We can go from any $x \in C$ to $(0, 0, \dots, 0)$ by dropping treasures one at a time. We can go from $(0, 0, \dots, 0)$ to any $y \in C$ by picking up treasures one at a time. Combining these, we can go from anywhere to anywhere, so the chain is irreducible.

To study periodicity, let's look at some simple cases. First consider the simple case where $w_1 + \dots + w_m < w$, i.e., Bilbo can carry all the treasure at the same time. Then all binary vectors of length m are allowed. So the period of $(0, 0, \dots, 0)$ is 2 since, starting at that state, Bilbo needs to pick up and then put down a treasure in order to get back to that state. In fact, if Bilbo starts at $(0, 0, \dots, 0)$, after any odd number of moves he will be carrying an odd number of treasures.

Now consider the case where $w_1 > w$, i.e., the first treasure is too heavy for Bilbo. From any $x \in C$, there is a $1/m$ chance that the chain will try to pick up the first treasure, and if that happens, the chain will stay at x. So the period of each state is 1.

(c) We can apply Metropolis-Hastings using the chain from (a) to make proposals. Start at $(0, 0, \dots, 0)$. Suppose the current state is $x = (x_1, \dots, x_m)$. Then:

1. Choose a uniformly random J in $\{1, 2, \ldots, m\}$, and obtain y from x by replacing x_J with $1 - x_J$.

2. If y is not in C, stay at x. If y is in C, flip a coin that lands Heads with probability $\min\left(1, e^{\beta(V(y)-V(x))}\right)$. If the coin lands Heads, go to y; otherwise, stay at x.

This chain will converge to the desired stationary distribution. But how should β be chosen? If β is very large, then the best solutions are given very high probability, but the chain may be very slow to converge to the stationary distribution since it can easily get stuck in *local modes*: the chain may find itself in a state which, while not globally optimal, is still better than the other states that can be reached in one step, and then the probability of rejecting proposals to go elsewhere may be very high. On the other hand, if β is close to 0, then it's easy for the chain to explore the space, but there isn't as much incentive for the chain to uncover good solutions.

An optimization technique called *simulated annealing* avoids having to choose one value of β. Instead, one specifies a sequence of β values, such that β gradually increases over time. At first, β is small and the space C can be explored broadly. As β gets larger and larger, the stationary distribution becomes more and more concentrated on the best solution or solutions. The name "simulated annealing" comes from an analogy with the annealing of metals, a process in which a metal is heated to a high temperature and then gradually cooled until it reaches a very strong, stable state; β corresponds to the reciprocal of temperature. □

As mentioned in Example 12.1.4, the Metropolis-Hastings algorithm can also be applied in a continuous state space, using PDFs instead of PMFs. This is extremely useful in Bayesian inference, where we often want to study the posterior distribution of an unknown parameter. This posterior distribution may be very complicated to work with analytically, and may have an unknown normalizing constant.

The MCMC approach is to obtain a large number of draws from a Markov chain whose stationary distribution is the posterior distribution. We can then use these draws to approximate the true posterior distribution. For example, we can estimate the posterior mean using the sample mean of these draws, and the posterior median using the sample median of the draws.

Gelman et al. [10] and McElreath [18] provide extensive introductions to Bayesian thinking and Bayesian data analysis, with wide ranges of applications and emphasis on statistical modeling and simulation. Various supplementary materials for these two books can be found at `http://www.stat.columbia.edu/~gelman/book` and `https://xcelab.net/rm/statistical-rethinking` respectively.

Example 12.1.8 (Normal-Normal conjugacy). Let $Y|\theta \sim \mathcal{N}(\theta, \sigma^2)$, where σ^2 is known but θ is unknown. Using the Bayesian framework, we treat θ as a random variable, with prior given by $\theta \sim \mathcal{N}(\mu, \tau^2)$ for some known constants μ and τ^2. That is, we have the *two-level model*

$$\theta \sim \mathcal{N}(\mu, \tau^2)$$
$$Y|\theta \sim \mathcal{N}(\theta, \sigma^2).$$

Describe how to use the Metropolis-Hastings algorithm to find the posterior mean and variance of θ after observing the value of Y.

Solution:

After observing $Y = y$, we can update our prior uncertainty for θ using Bayes' rule. Because we are interested in the posterior distribution of θ, any terms not depending on θ can be treated as part of the normalizing constant. Thus,

$$f_{\theta|Y}(\theta|y) \propto f_{Y|\theta}(y|\theta) f_\theta(\theta) \propto e^{-\frac{1}{2\sigma^2}(y-\theta)^2} e^{-\frac{1}{2\tau^2}(\theta-\mu)^2}.$$

Since we have a quadratic function of θ in the exponent, we recognize the posterior PDF of θ as a Normal PDF. The posterior distribution stays in the Normal family, which tells us that *the Normal is the conjugate prior of the Normal.* In fact, by completing the square (a rather tedious calculation which we shall omit), we can obtain an explicit formula for the posterior distribution of θ:

$$\theta|Y = y \sim \mathcal{N}\left(\frac{\frac{1}{\sigma^2}}{\frac{1}{\sigma^2} + \frac{1}{\tau^2}} y + \frac{\frac{1}{\tau^2}}{\frac{1}{\sigma^2} + \frac{1}{\tau^2}} \mu, \frac{1}{\frac{1}{\sigma^2} + \frac{1}{\tau^2}} \right).$$

Let's try to make sense of this formula.

- It says that the *posterior mean* of θ, $E(\theta|Y = y)$, is a weighted average of the prior mean μ and the observed data y. The weights are determined by how certain we are about θ before getting the data and how precisely the data are measured. If we are already very sure about θ even before getting the data, then τ^2 will be small and $1/\tau^2$ will be large, which will give a lot of weight to the prior mean μ. On the other hand, if the data are very precise, then σ^2 will be small and $1/\sigma^2$ will be large, which will give a lot of weight to the data y.

- For the posterior variance, if we define *precision* to be the reciprocal of variance, then the result simply says that the posterior precision of θ is the sum of the prior precision $1/\tau^2$ and the data precision $1/\sigma^2$.

This is all well and good, but let's suppose we didn't know how to complete the square, or that we wanted to check our calculations for specific values of y, σ^2, μ, and τ^2. We can do this by simulating from the posterior distribution of θ, using the Metropolis-Hastings algorithm to construct a Markov chain whose stationary distribution is $f_{\theta|Y}(\theta|y)$. The same method can also be applied to a wide variety of distributions that are far more complicated than the Normal to work with analytically. A Metropolis-Hastings algorithm for generating $\theta_0, \theta_1, \ldots$ is as follows.

1. If $\theta_n = x$, propose a new state x' according to some transition rule. One way to do this in a continuous state space is to generate a Normal r.v. ϵ_n with mean 0 and add it onto the current state to get the proposed state: in other words, we generate $\epsilon_n \sim \mathcal{N}(0, d^2)$ for some constant d, and then set $x' = x + \epsilon_n$. This is the analog of a transition matrix for a continuous state space. The only additional detail is deciding d; in practice, we try to choose a moderate value that is neither too large nor too small.

2. The acceptance probability is

$$a(x, x') = \min\left(\frac{s(x')p(x', x)}{s(x)p(x, x')}, 1\right),$$

where s is the desired stationary PDF (this was a PMF in the discrete case) and $p(x, x')$ is the probability *density* of proposing x' from x (this was p_{ij} in the discrete case).

In this problem, we want the stationary PDF to be $f_{\theta|Y}$, so we'll use that for s. As for $p(x, x')$, proposing x' from x is the same as having $\epsilon_n = x' - x$, so we evaluate the PDF of ϵ_n at $x' - x$ to get

$$p(x, x') = \frac{1}{\sqrt{2\pi}d} e^{-\frac{1}{2d^2}(x'-x)^2}.$$

However, since $p(x', x) = p(x, x')$, these terms cancel from the acceptance probability, leaving us with

$$a(x, x') = \min\left(\frac{f_{\theta|Y}(x'|y)}{f_{\theta|Y}(x|y)}, 1\right).$$

Once again, the normalizing constant cancels in the numerator and denominator of the acceptance probability.

3. Flip a coin that lands Heads with probability $a(x, x')$, independently of the Markov chain.

4. If the coin lands Heads, accept the proposal and set $\theta_{n+1} = x'$. Otherwise, stay in place and set $\theta_{n+1} = x$.

We ran the algorithm for 10^4 iterations with the settings $Y = 3$, $\mu = 0$, $\sigma^2 = 1$, $\tau^2 = 4$, and $d = 1$. Figure 12.2 shows a histogram of the resulting draws from the posterior distribution of θ. The posterior distribution indeed looks like a Normal curve. We can estimate the posterior mean and variance using the *sample mean* and *sample variance*. For the draws we obtained, the sample mean is 2.4 and the sample variance is 0.8. These are in close agreement with the theoretical values:

$$E(\theta|Y = 3) = \frac{\frac{1}{\sigma^2}}{\frac{1}{\sigma^2} + \frac{1}{\tau^2}} y + \frac{\frac{1}{\tau^2}}{\frac{1}{\sigma^2} + \frac{1}{\tau^2}} \mu = \frac{1}{1 + \frac{1}{4}} \cdot 3 + \frac{\frac{1}{4}}{1 + \frac{1}{4}} \cdot 0 = 2.8,$$

$$\text{Var}(\theta|Y = 3) = \frac{1}{\frac{1}{\sigma^2} + \frac{1}{\tau^2}} = \frac{1}{1 + \frac{1}{4}} = 0.8.$$

The posterior mean is closer to the observed data than to the prior mean, which makes sense because τ^2 is larger than σ^2, corresponding to a relatively high level of prior uncertainty. Using the code provided in the R section of this chapter, you can see how the posterior distribution changes for different values of y, μ, σ^2, and τ^2.

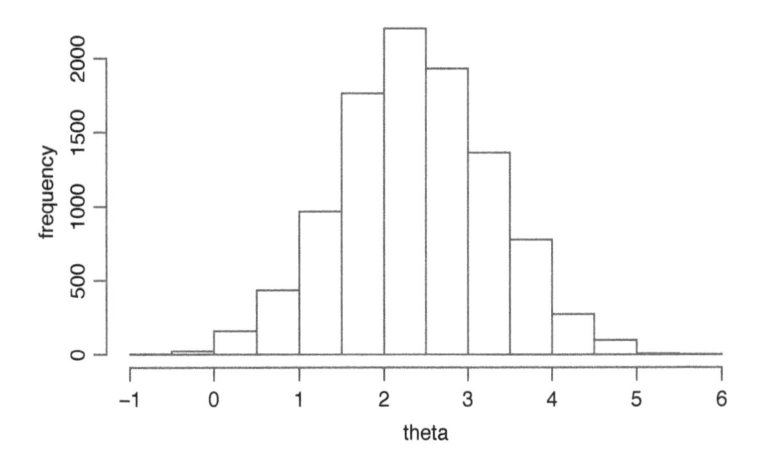

FIGURE 12.2
Histogram of 10^4 draws from the posterior distribution of θ given $Y = 3$, obtained using Metropolis-Hastings with $\mu = 0$, $\sigma^2 = 1$, and $\tau^2 = 4$. The sample mean is 2.4 and the sample variance is 0.8, in agreement with the theoretical values.

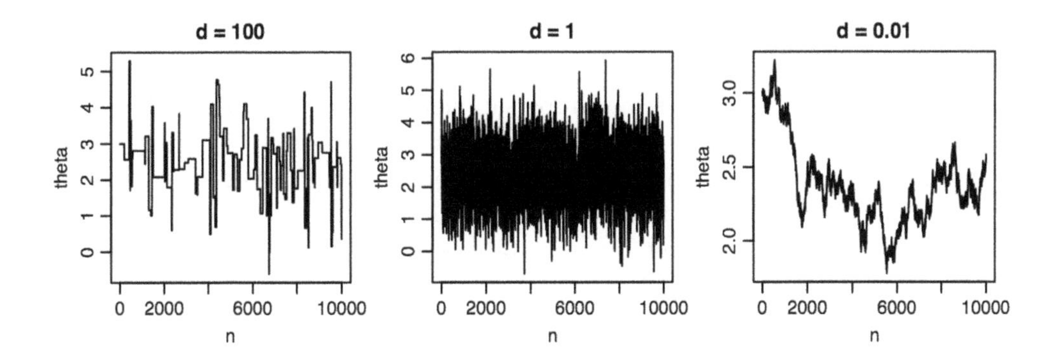

FIGURE 12.3
Trace plots of θ_n as a function of the iteration number n, for $d = 100, 1, 0.01$.

To help diagnose whether our Markov chain is adequately exploring the state space, we can make a *trace plot*, which is a plot of the samples θ_n as a function of n. Figure 12.3 shows three trace plots corresponding to three different choices for the standard deviation d of the proposals, namely $d = 100$, $d = 1$, and $d = 0.01$. The trace plot for $d = 100$ has numerous flat regions where the chain is staying in place. This indicates that d is too large, so the proposals are often rejected. On the other hand, $d = 0.01$ is too small; we can see from the trace plot that the chain takes tiny steps and is unable to venture very far from its starting point. The trace plot for $d = 1$ is just right, exhibiting neither the low acceptance rate of the $d = 100$ chain, nor the restricted mobility of the $d = 0.01$ chain.

In this example, the posterior distribution of θ is available analytically, so we used MCMC for illustrative purposes, to show that the results obtained by MCMC agree with their theoretical counterparts. But the same technique also applies in problems where the prior isn't conjugate and the posterior isn't a named distribution. □

12.2 Gibbs sampling

Gibbs sampling is an MCMC algorithm for obtaining approximate draws from a joint distribution, based on sampling from *conditional* distributions one at a time: at each stage, one variable is updated (keeping all the other variables fixed) by drawing from the conditional distribution of that variable given all the other variables. This approach is especially useful in problems where these conditional distributions are pleasant to work with.

First we will run through how the Gibbs sampler works in the bivariate case, where the desired stationary distribution is the joint PMF of discrete r.v.s X and Y. There are several forms of Gibbs samplers, depending on the order in which updates are done. We will introduce two major kinds of Gibbs sampler: *systematic scan*, in which the updates sweep through the components in a deterministic order, and *random scan*, in which a randomly chosen component is updated at each stage.

Algorithm 12.2.1 (Systematic scan Gibbs sampler). Let X and Y be discrete r.v.s with joint PMF $p_{X,Y}(x,y) = P(X = x, Y = y)$. We wish to construct a two-dimensional Markov chain (X_n, Y_n) whose stationary distribution is $p_{X,Y}$. The systematic scan Gibbs sampler proceeds by updating the X-component and the Y-component in alternation. If the current state is $(X_n, Y_n) = (x_n, y_n)$, then we update the X-component while holding the Y-component fixed, and then update the Y-component while holding the X-component fixed:

1. Draw x_{n+1} from the conditional distribution of X given $Y = y_n$, and set $X_{n+1} = x_{n+1}$.

2. Draw y_{n+1} from the conditional distribution of Y given $X = x_{n+1}$, and set $Y_{n+1} = y_{n+1}$.

Repeating steps 1 and 2 over and over, the stationary distribution of the chain $(X_0, Y_0), (X_1, Y_1), (X_2, Y_2), \dots$ is $p_{X,Y}$.

Algorithm 12.2.2 (Random scan Gibbs sampler). As above, let X and Y be discrete r.v.s with joint PMF $p_{X,Y}(x,y)$. We wish to construct a two-dimensional

Markov chain (X_n, Y_n) whose stationary distribution is $p_{X,Y}$. Each move of the random scan Gibbs sampler picks a uniformly random component and updates it, according to the conditional distribution given the other component:

1. Choose which component to update, with equal probabilities.
2. If the X-component was chosen, draw a value x_{n+1} from the conditional distribution of X given $Y = y_n$, and set $X_{n+1} = x_{n+1}, Y_{n+1} = y_n$. Similarly, if the Y-component was chosen, draw a value y_{n+1} from the conditional distribution of Y given $X = x_n$, and set $X_{n+1} = x_n, Y_{n+1} = y_{n+1}$.

Repeating steps 1 and 2 over and over, the stationary distribution of the chain $(X_0, Y_0), (X_1, Y_1), (X_2, Y_2), \dots$ is $p_{X,Y}$.

Gibbs sampling generalizes naturally to higher dimensions. If we want to sample from a d-dimensional joint distribution, the Markov chain we construct will be a sequence of d-dimensional random vectors. At each stage, we choose one component of the vector to update, and we draw from the conditional distribution of that component given the most recent values of the other components. We can either cycle through the components of the vector in a systematic order, or choose a random component to update each time.

The Gibbs sampler is less flexible than the Metropolis-Hastings algorithm in the sense that we don't get to choose a proposal distribution; this also makes it simpler in the sense that we don't *have* to choose a proposal distribution. The flavors of Gibbs and Metropolis-Hastings are rather different, in that Gibbs emphasizes conditional distributions while Metropolis-Hastings emphasizes acceptance probabilities. But the algorithms are closely connected, as we show below.

Theorem 12.2.3 (Random scan Gibbs as Metropolis-Hastings). The random scan Gibbs sampler is a special case of the Metropolis-Hastings algorithm, in which the proposal is *always* accepted. In particular, it follows that the stationary distribution of the random scan Gibbs sampler is as desired.

Proof. We will show this in two dimensions, but the proof is similar in any dimension. Let X and Y be discrete r.v.s whose joint PMF is the desired stationary distribution. Let's work out what the Metropolis-Hastings algorithm says to do, using the following proposal distribution: from (x, y), randomly update one coordinate by running one move of the random scan Gibbs sampler.

To simplify notation, write

$$P(X = x, Y = y) = p(x, y), P(Y = y|X = x) = p(y|x), P(X = x|Y = y) = p(x|y).$$

More formally, we should write $p_{Y|X}(y|x)$ instead of $p(y|x)$, to avoid issues like wondering what $p(5|3)$ means. But writing $p(y|x)$ is more compact and does not create ambiguity in this proof.

Let's compute the Metropolis-Hastings acceptance probability for going from (x, y)

to (x', y'). The states (x, y) and (x', y') must be equal in at least one component, since the proposal says to update only one component. Suppose that $x = x'$ (the case $y = y'$ can be handled symmetrically). Then the acceptance probability is

$$\frac{p(x, y')p(y|x)\frac{1}{2}}{p(x, y)p(y'|x)\frac{1}{2}} = \frac{p(x)p(y'|x)p(y|x)}{p(x)p(y|x)p(y'|x)} = 1.$$

Thus, this Metropolis-Hastings algorithm always accepts the proposal! So it's just running the random scan Gibbs sampler without modifying it. ■

Let's study some concrete examples of Gibbs samplers.

Example 12.2.4 (Graph coloring). Let G be a network (also called a *graph*): there are n *nodes*, and for each pair of distinct nodes, there either is or isn't an *edge* joining them. We have a set of k *colors*, e.g., if $k = 7$, the color set may be {red, orange, yellow, green, blue, indigo, violet}. A k*-coloring* of the network is an assignment of a color to each node, such that two nodes joined by an edge can't be the same color. For example, a 3-coloring of a network is illustrated below. Graph coloring is an important topic in computer science, with wide-ranging applications such as task scheduling and the game of Sudoku.

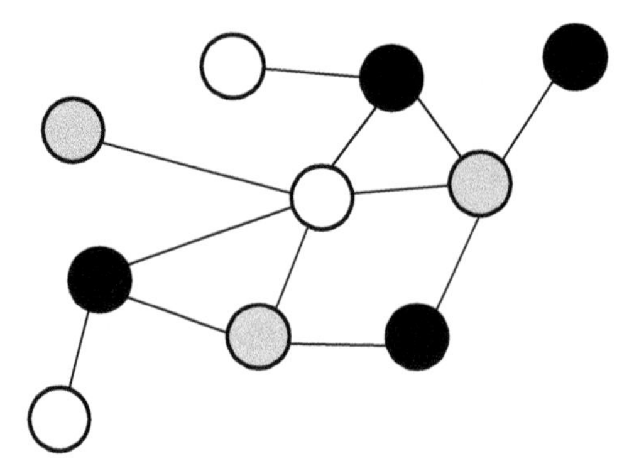

FIGURE 12.4
A 3-coloring of a network.

Suppose that it is possible to k-color G. Form a Markov chain on the space of all k-colorings of G, with transitions as follows: starting with a k-coloring of G, pick a uniformly random node, figure out what the legal colors are for that node, and then repaint that node with a uniformly random legal color (note that this random color may be the same as the current color). Show that this Markov chain is reversible, and find its stationary distribution.

Solution:

Let C be the set of all k-colorings of G, and let q_{ij} be the transition probability of going from i to j for any k-colorings i and j in C. We will show that $q_{ij} = q_{ji}$, which implies that the stationary distribution is uniform on C.

For any k-coloring i and node v, let $L(i, v)$ be the number of legal colorings for node v, keeping the colors of all other nodes the same as they are in i. If k-colorings i and j differ at more than one node, then $q_{ij} = 0 = q_{ji}$. If $i = j$, then obviously $q_{ij} = q_{ji}$. If i and j differ at exactly one node v, then $L(i, v) = L(j, v)$, so

$$q_{ij} = \frac{1}{n} \frac{1}{L(i, v)} = \frac{1}{n} \frac{1}{L(j, v)} = q_{ji}.$$

So the transition matrix is symmetric, which shows that the stationary distribution is uniform over the state space.

How is this an example of Gibbs sampling? Think of each node in the graph as a discrete r.v. that can take on k possible values. These nodes have a joint distribution, and the constraint that connected nodes cannot have the same color imposes a complicated dependence structure between nodes.

We would like to sample a random k-coloring of the entire graph; that is, we want to draw from the joint distribution of all the nodes. Since this is difficult, we instead *condition* on all but one node. If the joint distribution is to be uniform over all legal graphs, then the conditional distribution of one node given all the others is uniform over its legal colors. Thus, at each stage of the algorithm, we are drawing from the conditional distribution of one node given all the others: we're running a random scan Gibbs sampler! □

Example 12.2.5 (Darwin's finches). When Charles Darwin visited the Galápagos Islands, he kept a record of the finch species he observed on each island. Table 12.1 summarizes Darwin's data, with each row corresponding to a species and each column to an island. The presence of a 1 in entry (i, j) of the table indicates that species i was observed on island j.

	Island																	
Species	1	2	3	4	5	6	7	8	9	10	11	12	13	14	15	16	17	**Total**
1	0	0	1	1	1	1	1	1	1	1	0	1	1	1	1	1	1	14
2	1	1	1	1	1	1	1	1	1	1	0	1	0	1	1	0	0	13
3	1	1	1	1	1	1	1	1	1	1	1	1	0	1	1	0	0	14
4	0	0	1	1	1	0	0	1	0	1	0	1	1	0	1	1	1	10
5	1	1	1	0	1	1	1	1	1	1	0	1	0	1	1	0	0	12
6	0	0	0	0	0	0	0	0	0	0	1	0	1	0	0	0	0	2
7	0	0	1	1	1	1	1	1	1	0	0	1	0	1	1	0	0	10
8	0	0	0	0	0	0	0	0	0	0	0	1	0	0	0	0	0	1
9	0	0	1	1	1	1	1	1	1	1	0	1	0	0	1	0	0	10
10	0	0	1	1	1	1	1	1	1	1	0	1	0	1	1	0	0	11
11	0	0	1	1	1	0	1	1	0	1	0	0	0	0	0	0	0	6
12	0	0	1	1	0	0	0	0	0	0	0	0	0	0	0	0	0	2
13	1	1	1	1	1	1	1	1	1	1	1	1	1	1	1	1	1	17
Total	4	4	11	10	10	8	9	10	8	9	3	10	4	7	9	3	3	122

TABLE 12.1

Presence of 13 finch species (rows) on 17 islands (columns). A value of 1 in entry (i, j) indicates that species i was observed on island j. Data are from Sanderson [22].

Given these data, we might be interested in knowing whether the pattern of 0's and 1's observed in the table is anomalous in some way. For example, does there appear to be dependence between the rows and columns? Do some pairs of species frequently occur together on the same islands, more often than one would expect by chance? These patterns may shed light on the dynamics of inter-species cooperation or competition. One way to test for such patterns is by looking at a lot of random tables with the same row and column sums as the observed table, to see how the observed table compares to the random ones. This is a common technique in statistics known as a *goodness-of-fit test.*

But how do we generate random tables with the same row and column sums as Table 12.1? The number of tables satisfying these constraints is impossible to enumerate. MCMC comes to the rescue: we'll create a Markov chain on the space of all tables with these row and column sums, whose stationary distribution is uniform over all such tables.

To construct the Markov chain, we need a way to transition from one table to another without changing the row or column sums. Starting from the observed table, randomly select two rows and two columns. If the four entries at their intersection have one of the following two patterns:

$$\begin{matrix} 0 & 1 \\ 1 & 0 \end{matrix} \quad \text{or} \quad \begin{matrix} 1 & 0 \\ 0 & 1 \end{matrix}$$

then switch to the opposite pattern with probability 1/2; otherwise stay in place. For example, if we selected rows 1 and 3 and columns 1 and 17, we would switch the four entries at their intersection from $\begin{smallmatrix} 0 & 1 \\ 1 & 0 \end{smallmatrix}$ to $\begin{smallmatrix} 1 & 0 \\ 0 & 1 \end{smallmatrix}$ with probability 1/2. This is a symmetric transition rule (for all tables t and t', the transition probability from t to t' equals the transition probability from t' to t), the transitions never alter the row or column sums, and it can be shown that the Markov chain defined in this way is irreducible. Therefore the stationary distribution is uniform over all tables with the given row and column sums, as desired.

To interpret this procedure as a Gibbs sampler, consider conditioning on all entries in the table besides the four entries at the intersection of rows 1 and 3 and columns 1 and 17. If the stationary distribution is to be uniform over all tables with the given row and column sums, then the conditional distribution of these four entries must be uniform over all configurations that don't change the row and column sums, namely $\begin{smallmatrix} 0 & 1 \\ 1 & 0 \end{smallmatrix}$ and $\begin{smallmatrix} 1 & 0 \\ 0 & 1 \end{smallmatrix}$. Thus, at each stage, we are drawing from the conditional distribution of four entries given all the rest. □

As with Metropolis-Hastings, Gibbs sampling also applies to continuous distributions, replacing conditional PMFs with conditional PDFs.

Example 12.2.6 (Chicken-egg with unknown parameters). A chicken lays N eggs, where $N \sim \text{Pois}(\lambda)$. Each egg hatches with probability p, where p is unknown; we let $p \sim \text{Beta}(a, b)$. The constants λ, a, b are known.

Here's the catch: we don't get to observe N. Instead, we only observe the number of eggs that hatch, X. Describe how to use Gibbs sampling to find $E(p|X = x)$, the posterior mean of p after observing x hatched eggs.

Solution:

By the chicken-egg story, the distribution of X given p is Pois(λp). The posterior PDF of p is proportional to

$$f(p|X = x) \propto P(X = x|p)f(p) \propto e^{-\lambda p}(\lambda p)^x p^{a-1}q^{b-1},$$

where we have dropped all terms not depending on p.

This isn't a named distribution, so it might appear as though we're stuck, but we can get ourselves out of this rut by thinking conditionally. What do we wish we knew? The total number of eggs! Conditional on observing $N = n$ and knowing the true value of p, the distribution of X would be Bin(n, p). By conditioning on the total number of eggs, we recover *Beta-Binomial conjugacy* between p and X. This allows us to write down the posterior distribution right away using Story 8.3.3:

$$p|X = x, N = n \sim \text{Beta}(x + a, n - x + b).$$

The fact that conditioning on N makes matters so much nicer inspires us to use Gibbs sampling to tackle the problem. We alternate between sampling from p conditional on N and sampling from N conditional on p, as described below. Throughout, we must also condition on $X = x$, since we seek to learn about the posterior distributions of the parameters conditional on the evidence.

We make an initial guess for p and N, then iterate the following steps:

1. Conditional on $N = n$ and $X = x$, draw a new guess for p from the Beta$(x + a, n - x + b)$ distribution.

2. Conditional on p and $X = x$, the number of unhatched eggs is $Y \sim$ Pois$(\lambda(1 - p))$ by the chicken-egg story, so we can draw Y from the Pois$(\lambda(1 - p))$ distribution and set the new guess for N to be $N = x + Y$.

After many iterations, we have draws for both p and N. If we want, we can ignore the draws of N, since N was merely a device to help us sample p. But for fun, we'll plot both: Figure 12.5 shows histograms of the posterior draws of p and N when $\lambda = 10$, $a = b = 1$ (corresponding to a Unif$(0, 1)$ prior on p), and we observe $X = 7$ hatched eggs.

As for the posterior mean $E(p|X = x)$ originally asked for in the problem, we can take the sample mean of the draws of p to get a good approximation. In this case, the sample mean is 0.68. Using the code provided in the R section of this chapter, you can try changing the values of λ, a, b, and x to see how the histograms and the posterior mean are affected. The key strategy for this problem was to add the unobserved number of eggs N to the model, so that we would have pleasant conditional distributions and could use Gibbs sampling conveniently. □

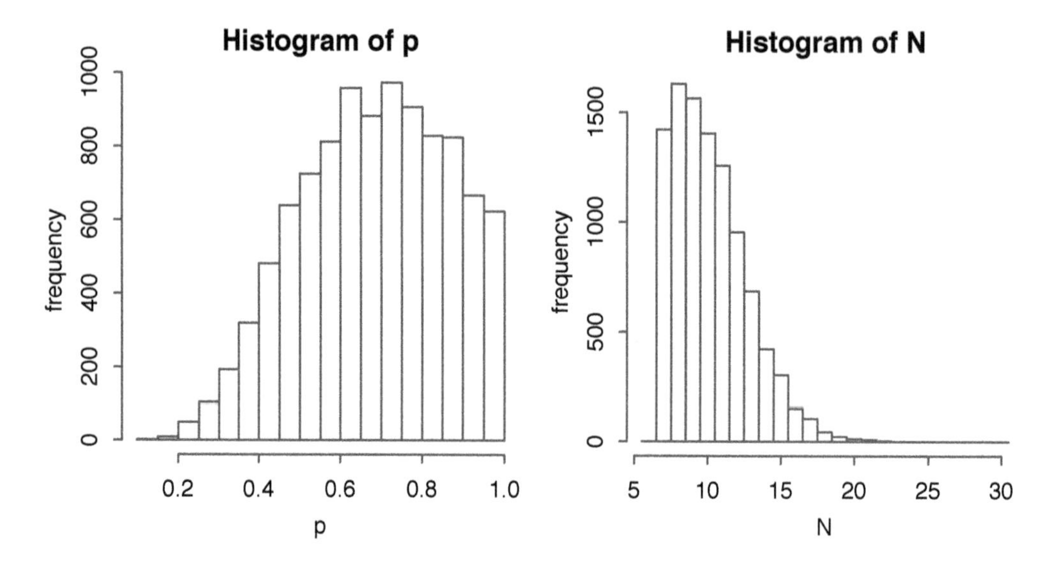

FIGURE 12.5
Histograms of 10^4 draws from the posterior distributions of p and N, where $\lambda = 10$, $a = 1$, $b = 1$, and we observe $X = 7$.

12.3 Recap

Markov chain Monte Carlo allows us to sample from complicated distributions using Markov chains. MCMC has been applied in a very wide variety of problems in recent years. The main idea behind MCMC algorithms is to construct a Markov chain whose stationary distribution is the distribution we wish to sample from. After running the Markov chain for a long time, the values that the Markov chain takes on can serve as draws from the desired distribution.

The two MCMC algorithms discussed in this chapter are Metropolis-Hastings and Gibbs sampling. The Metropolis-Hastings algorithm uses any irreducible Markov chain on the state space to generate proposals, then accepts or rejects those proposals so as to produce a modified Markov chain with the desired stationary distribution. Moreover, the resulting chain is reversible. The choice of the proposal distribution is extremely important in practice, since a bad proposal distribution may result in very slow convergence to the stationary distribution.

Gibbs sampling is a method for drawing from a d-dimensional joint distribution by updating the components of a d-dimensional Markov chain one at a time, conditional on all other components. This can be done through a *systematic scan*, which deterministically cycles through the components in a fixed order, or a *random scan*, which randomly chooses which component to update at each stage.

12.4 R

Metropolis-Hastings

Here's how to implement the Metropolis-Hastings algorithm for Example 12.1.8, the Normal-Normal model. First, we choose our observed value of Y and decide on values for the constants σ, μ, and τ:

```
y <- 3
sigma <- 1
mu <- 0
tau <- 2
```

We also need to choose the standard deviation of the proposals for step 1 of the algorithm, as explained in Example 12.1.8; for this problem, we let $d = 1$. We set the number of iterations to run, and we allocate a vector `theta` of length 10^4 which we will fill with our simulated draws:

```
d <- 1
niter <- 10^4
theta <- rep(0,niter)
```

Now for the main loop. We initialize θ to the observed value y, then run the algorithm described in Example 12.1.8:

```
theta[1] <- y
for (i in 2:niter){
    theta.p <- theta[i-1] + rnorm(1,0,d)
    r <- dnorm(y,theta.p,sigma) * dnorm(theta.p,mu,tau) /
           (dnorm(y,theta[i-1],sigma) * dnorm(theta[i-1],mu,tau))
    flip <- rbinom(1,1,min(r,1))
    theta[i] <- if(flip==1) theta.p else theta[i-1]
}
```

Let's step through each line inside the loop. The proposed value of θ is `theta.p`, which equals the previous value of θ plus a Normal random variable with mean 0 and standard deviation `d` (recall that `rnorm` takes the standard deviation and not the variance as input). The ratio `r` is

$$\frac{f_{\theta|Y}(x'|y)}{f_{\theta|Y}(x|y)} = \frac{e^{-\frac{1}{2\sigma^2}(y-x')^2}e^{-\frac{1}{2\tau^2}(x'-\mu)^2}}{e^{-\frac{1}{2\sigma^2}(y-x)^2}e^{-\frac{1}{2\tau^2}(x-\mu)^2}},$$

where `theta.p` is playing the role of x' and `theta[i-1]` is playing the role of x. The coin flip to determine whether to accept or reject the proposal is `flip`, which is a coin flip with probability `min(r,1)` of Heads (encoding Heads as 1 and Tails as 0). Finally, we set `theta[i]` equal to the proposed value if the coin flip lands Heads, and keep it at the previous value otherwise.

The vector `theta` now contains all of our simulation draws. We typically discard some of the initial draws to give the chain some time to approach the stationary distribution. The following line of code discards the first half of the draws:

```
theta <- theta[-(1:(niter/2))]
```

To see what the remaining draws look like, we can create a histogram using `hist(theta)`. We can also compute summary statistics such as `mean(theta)` and `var(theta)`, which give us the sample mean and sample variance.

Gibbs

Now let's implement Gibbs sampling for Example 12.2.6, the chicken-egg story with unknown hatching probability and invisible unhatched eggs. The first step is to decide on our observed value of X, as well as the constants λ, a, b:

```
x <- 7
lambda <- 10
a <- 1
b <- 1
```

Next we decide how many iterations to run, and we allocate space for our results, creating two vectors `p` and `N` of length 10^4 which we will fill with our simulated draws:

```
niter <- 10^4
p <- rep(0,niter)
N <- rep(0,niter)
```

Finally, we're ready to run the Gibbs sampler. We initialize p and N to the values 0.5 and $2x$, respectively, and then we run the algorithm as explained in Example 12.2.6:

```
p[1] <- 0.5
N[1] <- 2*x
for (i in 2:niter){
    p[i] <- rbeta(1,x+a,N[i-1]-x+b)
    N[i] <- x + rpois(1,lambda*(1-p[i-1]))
}
```

Again, we discard the initial draws:

```
p <- p[-(1:(niter/2))]
N <- N[-1:(niter/2))]
```

To see what the remaining draws look like, we can make histograms using `hist(p)` and `hist(N)`, which is how we created Figure 12.5. We can also compute summary statistics such as `mean(p)` or `median(p)`.

12.5 Exercises

1. Let $p(x, y)$ be the joint PMF of two discrete r.v.s X and Y. Using shorthand notation as we used with the Gibbs sampler, let $p(x)$ and $p(y)$ be the marginal PMFs of X and Y, and $p(x|y)$ and $p(y|x)$ be the conditional PMFs of X given Y and Y given X. Suppose the support of Y is the same as the support of the conditional distribution of $Y|X$.

   (a) Use the identity $p(x)p(y|x) = p(y)p(x|y)$ to find an expression for the marginal PMF $p(y)$ in terms of the conditional PMFs $p(x|y)$ and $p(y|x)$.

   Hint: Rewrite the identity as $p(x)/p(y) = p(x|y)/p(y|x)$ and take a sum.

   (b) Explain why the two conditional distributions $p(x|y)$ and $p(y|x)$ determine the joint distribution $p(x, y)$, and how this fact relates to the Gibbs sampler.

2. We have a network G with n nodes and some edges. Each node of G can either be vacant or occupied. We want to place particles on the nodes of G in such a way that the particles are not too crowded. Thus, define a feasible configuration as a placement of particles such that each node is occupied by at most one particle, and no neighbor of an occupied node is occupied.

   Construct a Markov chain whose stationary distribution is uniform over all feasible configurations. Clearly specify the transition rule of your Markov chain, and explain why its stationary distribution is uniform.

3. This problem considers an application of MCMC techniques to image analysis. Imagine a 2D image consisting of an $L \times L$ grid of black-or-white pixels. Let Y_j be the indicator of the jth pixel being white, for $j = 1, \dots, L^2$. Viewing the pixels as nodes in a network, the neighbors of a pixel are the pixels immediately above, below, to the left, and to the right (except for boundary cases).

   Let $i \sim j$ stand for "i and j are neighbors". A commonly used model for the joint PMF of $\mathbf{Y} = (Y_1, \dots, Y_{L^2})$ is

$$P(\mathbf{Y} = \mathbf{y}) \propto \exp\left(\beta \sum_{(i,j):i\sim j} I(y_i = y_j)\right).$$

   If β is positive, this says that neighboring pixels prefer to have the same color. The normalizing constant of this joint PMF is a sum over all 2^{L^2} possible configurations, so it may be very computationally difficult to obtain. This motivates the use of MCMC to simulate from the model.

   (a) Suppose that we wish to simulate random draws from the joint PMF of $\mathbf{Y}$, for a particular known value of β. Explain how we can do this using Gibbs sampling, cycling through the pixels one by one in a fixed order.

   (b) Now provide a Metropolis-Hastings algorithm for this problem, based on a proposal of picking a uniformly random site and toggling its value.

13

Poisson processes

Poisson processes serve as a simple model for events occurring in time or space: in one dimension, cars passing by a highway checkpoint; in two dimensions, flowers in a meadow; in three dimensions, stars in a region of the galaxy. Poisson processes are a primary building block for more complicated processes in time and space, which are the focus of a branch of statistics called *spatial statistics.*

Poisson processes are also useful in probability, as they tie together many of the named distributions and provide us with insightful story proofs for results that might otherwise be tedious to show. This leads to a new problem-solving strategy, which we will demonstrate: even when a problem makes no mention of Poisson processes, there are sometimes ways to *pretend* the r.v.s are coming from a Poisson process so that we can use convenient Poisson process properties.

In this chapter, we'll review the already familiar definition of the 1D Poisson process, derive and discuss three important properties of the 1D Poisson process, and then extend these properties to Poisson processes in higher dimensions.

13.1 Poisson processes in one dimension

In Section 5.6, we defined a one-dimensional Poisson process and showed that the interarrival times are i.i.d. Exponentials. In Chapter 8 we showed that the time of the jth arrival (relative to some fixed starting time) is Gamma. Let's review these results, with notation that will help us generalize to higher dimensions.

Definition 13.1.1 (1D Poisson process). A sequence of arrivals in continuous time is a *Poisson process* with rate λ if the following conditions hold:

1. The number of arrivals in an interval of length t is distributed Pois(λt).
2. The numbers of arrivals in disjoint time intervals are independent.

Usually we will assume that the timeline starts at $t = 0$, in which case we have a Poisson process on $(0, \infty)$, but we can use the same conditions to define a Poisson process on $(-\infty, \infty)$ if we want the timeline to be infinite in both directions.

Consider a Poisson process on $(0, \infty)$. Departing slightly from our notation in previous chapters, let $N(t)$ be the number of arrivals in $(0, t]$. Then the number of arrivals in $(t_1, t_2]$ is $N(t_2) - N(t_1)$, for $0 < t_1 < t_2$. Let T_j be the time of the jth arrival. Since $T_1 > t$ is the same event as $N(t) = 0$ (by the *count-time duality*, as discussed in Section 5.6),

$$P(T_1 > t) = P(N(t) = 0) = e^{-\lambda t},$$

so $T_1 \sim \text{Expo}(\lambda)$. Next let's condition on the first arrival time T_1 and look at the additional time $T_2 - T_1$ until the second arrival. Then

$$(T_2 - T_1)|T_1 \sim \text{Expo}(\lambda)$$

by the same argument as above, since we have a fresh Poisson process starting at T_1. Since the conditional distribution of $(T_2 - T_1)|T_1$ does not depend on T_1, we have that $T_2 - T_1$ is independent of T_1, so $T_2 - T_1 \sim \text{Expo}(\lambda)$ unconditionally too.

Continuing in this way, the interarrival times $T_j - T_{j-1}$ are independent, with

$$T_j - T_{j-1} \sim \text{Expo}(\lambda).$$

So a Poisson process can be described dually as a process in which arrival counts are Poissons or a process in which interarrival times are Exponentials. Also note that since T_j is the sum of j i.i.d. $\text{Expo}(\lambda)$ r.v.s,

$$T_j \sim \text{Gamma}(j, \lambda).$$

The connection with the Exponential gives us a simple way to generate n arrivals from a Poisson process.

Story 13.1.2 (Generative story for 1D Poisson process). To generate n arrivals from a Poisson process on $(0, \infty)$ with rate λ:

1. Generate n i.i.d. $\text{Expo}(\lambda)$ r.v.s $X_1, \dots, X_n$.
2. For $j = 1, \dots, n$, set $T_j = X_1 + \dots + X_j$.

Then we can take $T_1, \dots, T_n$ to be the arrival times. □

Figure 13.1 depicts three realizations of Poisson processes with rates 1, 2, and 5, respectively, plotted up to time 10. In all three cases, we can see that despite the interarrival times being i.i.d., the arrivals are *not* evenly spaced. Rather, there is a lot of variability in the interarrival times, which produces clumps of arrivals. This phenomenon is known as *Poisson clumping*. It might seem like an amazing coincidence to observe a cluster of several arrivals that are close together in time, but Poisson clumping says that having such clusters is common with Poisson processes.

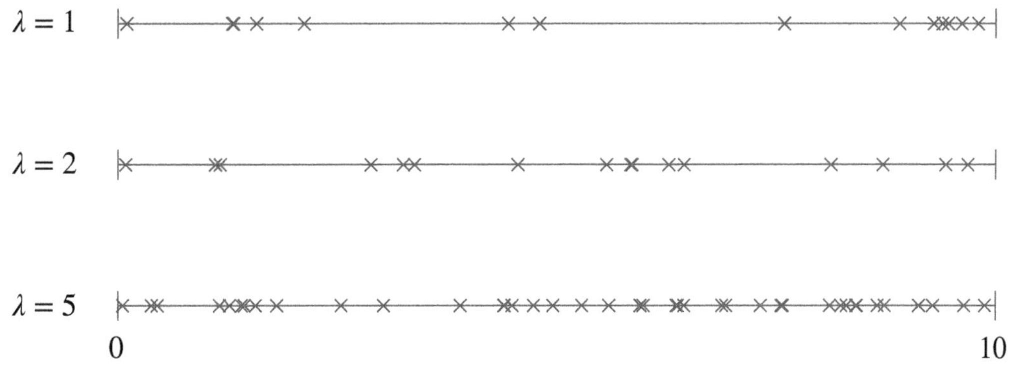

FIGURE 13.1
Simulated Poisson process in one dimension, for $\lambda = 1, 2, 5$. The arrivals are far from evenly spaced from each other, and in fact they sometimes clump together,

13.2 Conditioning, superposition, and thinning

The three most important properties to understand about the Poisson process are *conditioning*, *superposition*, and *thinning*. These correspond to properties we've already seen about the Poisson distribution, so they should already be plausible.

13.2.1 Conditioning

What happens when we take a Poisson process and condition on the total number of events in an interval? Our first result is that conditional on the total number of events in an interval, the number of events in a fixed subinterval is Binomial. This follows from Theorem 4.8.2, where we showed that we can get from the Poisson to the Binomial by *conditioning*.

Theorem 13.2.1 (Conditional counts). Let $(N(t) : t > 0)$ be a Poisson process with rate λ, and $t_1 < t_2$. The conditional distribution of $N(t_1)$ given $N(t_2) = n$ is

$$N(t_1) \mid N(t_2) = n \sim \text{Bin}\left(n, \frac{t_1}{t_2}\right).$$

Proof. Figure 13.2 illustrates the setup. The claim is that conditional on having a total of n arrivals in $(0, t_2]$, the number of arrivals in $(0, t_1]$ is Binomial, with n trials and success probability proportional to t_1.

Since $(0, t_1]$ and $(t_1, t_2]$ are disjoint, $N(t_1)$ is independent of $N(t_2) - N(t_1)$. The first is distributed $\text{Pois}(\lambda t_1)$, the second is distributed $\text{Pois}(\lambda(t_2 - t_1))$, and their sum is

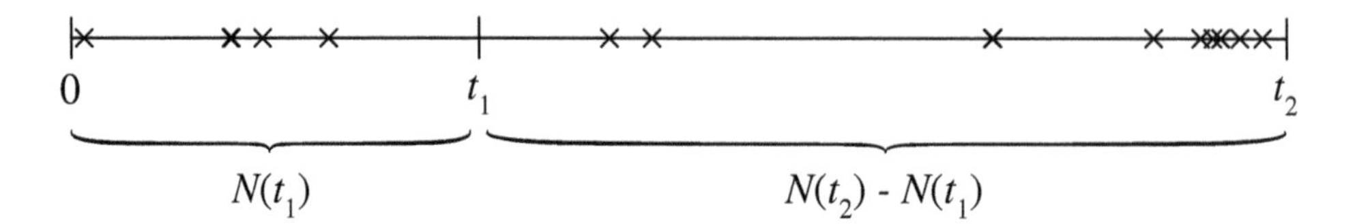

FIGURE 13.2
Conditioning. Given that n arrivals occurred in $(0, t_2]$, the conditional distribution of the number of arrivals in $(0, t_1]$ is Binomial with parameters n and t_1/t_2.

the number of arrivals in $(0, t_2]$, namely $N(t_2)$. By Theorem 4.8.2,

$$N(t_1) \mid N(t_2) = n \sim \text{Bin}\left(n, \frac{\lambda t_1}{\lambda t_1 + \lambda(t_2 - t_1)}\right),$$

which is precisely what we wanted to show. ■

Carrying the idea of conditioning on the number of arrivals further, we have the following striking result: in a Poisson process, given that $N(t) = n$, the arrival times are distributed as if we threw down n i.i.d. Unif$(0, t)$ points.

First let's look at a simple case, where there has been only one arrival.

Proposition 13.2.2. In a Poisson process of rate λ, conditional on $N(t) = 1$, the first arrival time T_1 has the Unif$(0, t)$ distribution.

Proof. Let $0 < s < t$. By a form of the count-time duality,

$$\begin{aligned} P(T_1 \leq s|N(t) = 1) &= \frac{P(T_1 \leq s, N(t) = 1)}{P(N(t) = 1)} \\ &= \frac{P(N(s) = 1, N(t) - N(s) = 0)}{P(N(t) = 1)} \\ &= \frac{(e^{-\lambda s}\lambda s)(e^{-\lambda(t-s)})}{e^{-\lambda t}\lambda t} \\ &= \frac{s}{t}. \end{aligned}$$

Thus, the conditional CDF of T_1 given $N(t) = 1$ is the Unif$(0, t)$ CDF. ■

More generally, given that $N(t) = n$, the arrival times T_j look like the order statistics of n i.i.d. Unif$(0, t)$ r.v.s. We omit the proof since in introducing order statistics in Chapter 8, we focused on their *marginal* distributions.

Theorem 13.2.3 (Conditional times). In a Poisson process of rate λ, conditional on $N(t) = n$, the joint distribution of the arrival times $T_1, \ldots, T_n$ is the same as the joint distribution of the order statistics of n i.i.d. Unif$(0, t)$ r.v.s.

From Chapter 8, we know that the order statistics of Unif(0, 1) r.v.s are Betas, so the conditional distributions of the T_j are *scaled* Betas; to get Beta distributions, we can just divide the T_j by t so that their support is $(0, 1)$:

$$t^{-1}T_j \mid N(t) = n \sim \text{Beta}(j, n - j + 1).$$

We now have another way to generate arrivals from a Poisson process, this time for a specific interval instead of a specific number of arrivals.

Story 13.2.4 (Generative story for Poisson process, take 2)**.** To generate arrivals from a Poisson process with rate λ in an interval $(0, t]$:

1. Generate the total number of events in the interval, $N(t) \sim \text{Pois}(\lambda t)$.
2. Given $N(t) = n$, generate n i.i.d. Unif$(0, t)$ r.v.s $U_1, \dots, U_n$.
3. For $j = 1, \dots, n$, set $T_j = U_{(j)}$. □

This is actually the generative story we used to create Figure 13.1, since we knew we wanted to simulate in the interval $(0, 10]$.

Example 13.2.5 (Users on a website)**.** Users visit a certain website according to a Poisson process with rate λ_1 users per minute, where an "arrival" at a certain time means that at that time someone starts browsing the site. After arriving at the site, each user browses the site for an Expo(λ_2) amount of time (and then leaves), independent of other users.

Suppose that at time 0, no one is using the site. Let N_t be the number of users who arrive in the interval $(0, t]$, and let C_t be the number of users who are *currently* browsing the site at time t.

(a) Let X be the time of arrival and Y be the time of departure for a user who arrives at a Uniform time point in $[0, t]$. Find the joint PDF of X and Y.

(b) Let p_t be the probability that a user who arrives at a Uniform time point in $(0, t]$ is still browsing the site at time t. Find p_t.

(c) Find the distribution of C_t in terms of λ_1, λ_2, and t.

(d) Little's law is a very general result, which says the following:

The long-run average number of customers in a stable system is the long-term average arrival rate multiplied by the average time a customer spends in the system.

Explain what happens to $E(C_t)$ for t large, and how this can be interpreted in terms of Little's law.

Solution:

(a) We have $X \sim \text{Unif}(0, t)$. Given $X = x$, Y is an Expo(λ_2) shifted to start at x, i.e., $(Y - x)|(X = x) \sim \text{Expo}(\lambda_2)$. So the joint PDF of X and Y is

$$f(x, y) = f_X(x) f_{Y|X}(y|x) = \frac{\lambda_2}{t} e^{-\lambda_2(y-x)}, \text{ for } 0 < x < t \text{ and } x < y.$$

(b) With notation as in (a), we want to find $p_t = P(Y > t)$. This can be done by integrating the joint PDF over all (x, y) with $y > t$:

$$\begin{aligned}
P(Y > t) &= \frac{1}{t}\int_0^t \int_t^\infty \lambda_2 e^{-\lambda_2(y-x)} dydx \\
&= \frac{1}{t}\int_0^t e^{\lambda_2 x}\left(\int_t^\infty \lambda_2 e^{-\lambda_2 y} dy\right) dx \\
&= \frac{e^{-\lambda_2 t}}{t}\int_0^t e^{\lambda_2 x} dx \\
&= \frac{1 - e^{-\lambda_2 t}}{\lambda_2 t}.
\end{aligned}$$

(c) By Theorem 13.2.3, given $N_t = n$ we have that the n arrival times in $(0, t]$ are i.i.d. and uniform in that interval. Therefore, $C_t | N_t \sim \text{Bin}(N_t, p_t)$, with p_t as above, and $N_t \sim \text{Pois}(\lambda_1 t)$. So by the chicken-egg story (Story 7.1.9),

$$C_t \sim \text{Pois}(\lambda_1 p_t t).$$

That is,

$$C_t \sim \text{Pois}\left(\frac{\lambda_1(1 - e^{-\lambda_2 t})}{\lambda_2}\right).$$

(d) As $t \to \infty$, $E(C_t) \to \lambda_1/\lambda_2$. This agrees with Little's law since it says the long-run average number of users in the system (browsing the site) is the rate at which users arrive (λ_1) times the average time a user browses in a session ($1/\lambda_2$). □

13.2.2 Superposition

The next property we will examine is *superposition*: if we take two independent Poisson processes and overlay them, we get another Poisson process. (The combined process is a Poisson process in its own right, but we can imagine that the arrivals are tagged to indicate which of the underlying processes they came from.)

Theorem 13.2.6 (Superposition). Let $(N_1(t) : t > 0)$ and $(N_2(t) : t > 0)$ be independent Poisson processes with rates λ_1 and λ_2, respectively. Then the combined process $N(t) = N_1(t) + N_2(t)$ is a Poisson process with rate $\lambda_1 + \lambda_2$.

Proof. Let's verify the two properties in the definition of Poisson process.

1. For all $t > 0$, $N_1(t) \sim \text{Pois}(\lambda_1 t)$ and $N_2(t) \sim \text{Pois}(\lambda_2 t)$, independently, so $N(t) \sim \text{Pois}((\lambda_1 + \lambda_2)t)$, by Theorem 4.8.1. The same argument applies for any interval of length t, not just intervals of the form $(0, t]$.

2. Arrivals in disjoint intervals are independent in the combined process because they are independent in the two individual processes, and the individual processes are independent of each other. ■

There is a very natural way to generate a superposition.

Story 13.2.7 (Generative story for superposition). To generate the superposition of two independent Poisson processes, $(N_1(t) : t > 0)$ with rate λ_1 and $(N_2(t) : t > 0)$ with rate λ_2:

1. Generate arrivals from the Poisson process $(N_1(t) : t > 0)$.
2. Generate arrivals from the Poisson process $(N_2(t) : t > 0)$.
3. Superpose the results of steps 1 and 2. □

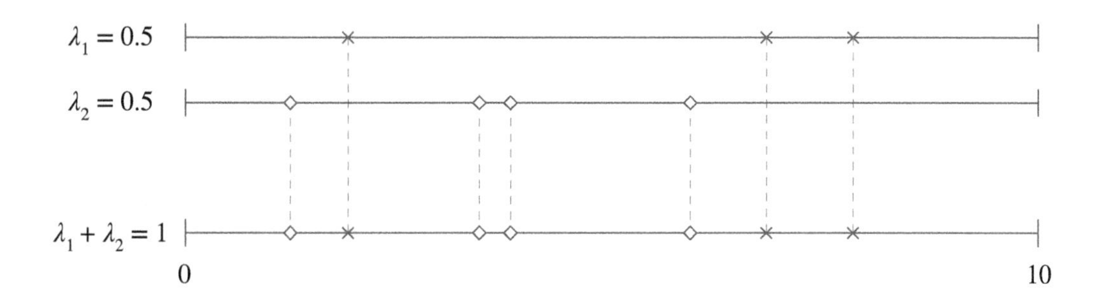

FIGURE 13.3
Superposition. The superposition of independent Poisson processes is a Poisson process, and the rates add. The top two timelines are independent Poisson processes, each with rate 0.5. The bottom timeline is the superposition of the top two Poisson processes and is itself a Poisson process with rate 1.

Figure 13.3 depicts a superposed Poisson process consisting of $\times$'s and $\diamond$'s. Let's call the $\times$'s "type-1 events" and the $\diamond$'s "type-2 events". A natural question to ask is: what is the probability of observing a type-1 event before a type-2 event?

Theorem 13.2.8 (Probability of type-1 event before type-2 event). Consider two independent Poisson processes: a Poisson process of type-1 arrivals, with rate λ_1, and a Poisson process of type-2 arrivals, with rate λ_2. In the superposition of these two processes, the probability of the first arrival being type-1 is $\lambda_1/(\lambda_1 + \lambda_2)$.

Proof. Let T be the time until the first type-1 event and let V be the time until the first type-2 event. We seek $P(T \leq V)$. We could do this with 2D LOTUS, integrating the joint PDF of T and V over the region of interest in the 2D plane. But it turns out we can avoid calculus altogether.

We have $T \sim \text{Expo}(\lambda_1)$ and $V \sim \text{Expo}(\lambda_2)$, independently. Rescaling, let

$$\tilde{T} = \lambda_1 T,\ \tilde{V} = \lambda_2 V.$$

Then $\tilde{T}, \tilde{V}$ are i.i.d. Expo(1).

Letting $U = \tilde{T}/(\tilde{T}+\tilde{V})$, we have

$$\begin{aligned}
P(T \le V) &= P\left(\frac{\tilde{T}}{\lambda_1} \le \frac{\tilde{V}}{\lambda_2}\right) \\
&= P\left(\frac{\tilde{T}}{\tilde{T}+\tilde{V}} \le \frac{\tilde{V}}{\tilde{T}+\tilde{V}} \cdot \frac{\lambda_1}{\lambda_2}\right) \\
&= P\left(U \le (1-U) \cdot \frac{\lambda_1}{\lambda_2}\right) \\
&= P\left(U \le \frac{\lambda_1}{\lambda_1+\lambda_2}\right).
\end{aligned}$$

Since $\tilde{T}, \tilde{V} \sim \text{Expo}(1)$, the bank–post office story tells us that $U \sim \text{Beta}(1,1)$. In other words, U is standard Uniform! Thus, $P(T \le V) = \lambda_1/(\lambda_1+\lambda_2)$. Note that when $\lambda_1 = \lambda_2$ this reduces to $1/2$, as it should by symmetry. ■

The above result applies to the first arrival in the combined Poisson process. After the first arrival, however, the same reasoning applies to the second arrival: by the memoryless property, the time to the next type-1 event is $\text{Expo}(\lambda_1)$ and the time to the next type-2 event is $\text{Expo}(\lambda_2)$, independent of the past. Therefore the second arrival is a type-1 arrival with probability $\lambda_1/(\lambda_1+\lambda_2)$, independent of the first arrival. Similarly, all of the arrival types can be viewed as i.i.d. coin tosses with probability $\lambda_1/(\lambda_1+\lambda_2)$ of Heads, where Heads corresponds to type-1.

This yields an alternative generative story for the superposition of two independent Poisson processes: we can first generate an $\text{Expo}(\lambda_1+\lambda_2)$ r.v. to decide when the next arrival occurs, and then independently flip a coin with probability $\lambda_1/(\lambda_1+\lambda_2)$ of Heads to decide what kind of arrival it is.

Story 13.2.9 (Generative story for superposition, take 2). To generate the superposition of two independent Poisson processes, with rates λ_1 and λ_2:

1. Generate i.i.d. $\text{Expo}(\lambda_1+\lambda_2)$ r.v.s $X_1, X_2, \dots$, and let the jth arrival be at time $T_j = X_1 + \cdots + X_j$.

2. Generate i.i.d. r.v.s $I_1, I_2, \dots \sim \text{Bern}(\lambda_1/(\lambda_1+\lambda_2))$, independent of $X_1, X_2, \dots$. Let the jth arrival be type-1 if $I_j = 1$, and type-2 otherwise.

□

This story provides us with a quick proof of a result known as the *competing risks theorem*, which seems like a surprising independence result when stated on its own but becomes very intuitive when viewed in the context of Poisson processes.

Example 13.2.10 (Competing risks). The lifetime of Fred's refrigerator is $Y_1 \sim \text{Expo}(\lambda_1)$, and the lifetime of his dishwasher is $Y_2 \sim \text{Expo}(\lambda_2)$, independent of Y_1. Show that $\min(Y_1, Y_2)$, the time of the first appliance failure, is independent of $I(Y_1 < Y_2)$, the indicator that the refrigerator failed first. This may seem surprising.

For example, if the average lifetime of his refrigerator is 15 years and the average lifetime of his dishwasher is 7 years, and then one of these appliances fails after 7 years, it would be natural to guess that it was the dishwasher that failed. But in fact, we will show that knowing the time of the first appliance failure provides no information about which appliance failed.

Solution: We will use an *embedding strategy*, using Poisson process ideas even though this problem doesn't mention Poisson processes anywhere! We will embed the r.v.s Y_1 and Y_2 into a Poisson process that we ourselves invent, in order to take advantage of the properties of Poisson processes. So let's imagine there is an entire Poisson process of refrigerator failures with rate λ_1 and a Poisson process of dishwasher failures with rate λ_2.

We can interpret Y_1 as the waiting time for the first arrival in the refrigerator process and Y_2 as the waiting time for the first arrival in the dishwasher process. This approach is valid since $(\min(Y_1, Y_2), I(Y_1 < Y_2))$ is a function of (Y_1, Y_2), so what matters is the joint distribution of (Y_1, Y_2), and the way we construct (Y_1, Y_2) does have the correct joint distribution.

Furthermore, $\min(Y_1, Y_2)$ is the waiting time for the first arrival in the superposition of the two Poisson processes, and $I(Y_1 < Y_2)$ is the indicator of this arrival being a type-1 event. But in the above generative story, the arrival times and event types in the superposed Poisson process are generated completely independently! Hence, $\min(Y_1, Y_2)$ and $I(Y_1 < Y_2)$ are independent, with $\min(Y_1, Y_2) \sim \text{Expo}(\lambda_1 + \lambda_2)$ and $I(Y_1 < Y_2) \sim \text{Bern}(\lambda_1/(\lambda_1 + \lambda_2))$. □

A direct consequence of Story 13.2.9 is that if we project a superposed Poisson process into discrete time, keeping the sequence of type-1 and type-2 arrivals but not the arrival times themselves, we are left with i.i.d. $\text{Bern}(\lambda_1/(\lambda_1+\lambda_2))$ r.v.s $I_1, I_2, \dots$, where I_j is the indicator of the jth arrival being type-1. Figure 13.4 illustrates what it means to remove the continuous-time information from the Poisson process, and Theorem 13.2.11 states the result formally.

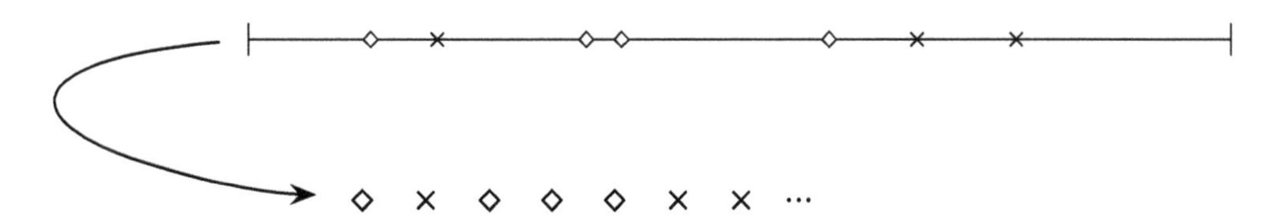

FIGURE 13.4
Projection into discrete time. Stripping out the continuous-time information from a superposed Poisson process produces a sequence of i.i.d. indicators of type-1 versus type-2 events. Here $\times$ represents type-1 events and $\diamond$ represents type-2 events.

Theorem 13.2.11 (Projection of superposition into discrete time). Consider the superposition $(N(t) : t > 0)$ of two independent Poisson processes with rates λ_1 and λ_2. For $j = 1, 2, \dots$, let I_j be the indicator of the jth event being from the Poisson process with rate λ_1. Then the I_j are i.i.d. $\text{Bern}(\lambda_1/(\lambda_1 + \lambda_2))$.

Using this result, we can prove with a story that a Gamma mixture of Poissons is Negative Binomial, which we first learned from Fred's adventures in Blotchville (Story 8.4.5). We'll consider a special case first.

Theorem 13.2.12 (Exponential mixture of Poissons is Geometric). Suppose that $X \sim \text{Expo}(\lambda)$ and $Y|X = x \sim \text{Pois}(x)$. Then $Y \sim \text{Geom}(\lambda/(\lambda+1))$.

Proof. As with the competing risks theorem, we embed X and Y into Poisson processes. Consider two independent Poisson processes, a process of failures arriving at rate 1 and another of successes arriving at rate λ. Let X be the time of the first success; then $X \sim \text{Expo}(\lambda)$. Let Y be the number of failures before the time of the first success. By the definition of a Poisson process with rate 1, $Y|X = x \sim \text{Pois}(x)$. Therefore X and Y satisfy the conditions of the theorem.

To get the marginal distribution of Y, strip out the continuous-time information! In discrete time we have i.i.d. Bernoulli trials with success probability $\lambda/(\lambda+1)$, and Y is defined as the number of failures before the first success, so by the story of the Geometric distribution, $Y \sim \text{Geom}(\lambda/(\lambda+1))$. ■

The reasoning for the general case is analogous.

Theorem 13.2.13 (Gamma mixture of Poissons is Negative Binomial). Suppose that $X \sim \text{Gamma}(r, \lambda)$ and $Y|X = x \sim \text{Pois}(x)$. Then $Y \sim \text{NBin}(r, \lambda/(\lambda+1))$.

Proof. Consider two independent Poisson processes, a process of failures arriving at rate 1 and another of successes arriving at rate λ. Let X be the time of the rth success, so $X \sim \text{Gamma}(r, \lambda)$. Let Y be the number of failures before the time of the rth success. Then $Y|X = x \sim \text{Pois}(x)$ by definition of Poisson process. We have that Y is the number of failures before the rth success in a sequence of i.i.d. Bernoulli trials with success probability $\lambda/(\lambda+1)$, so $Y \sim \text{NBin}(r, \lambda/(\lambda+1))$. ■

13.2.3 Thinning

The last property of Poisson processes we will discuss is *thinning*: if we take a Poisson process and, for each arrival, independently flip a coin to decide whether it is a type-1 event or type-2 event, we end up with two independent Poisson processes.

Theorem 13.2.14 (Thinning). Let $(N(t) : t > 0)$ be a Poisson process with rate λ, and classify each arrival as a type-1 event with probability p and a type-2 event with probability $1-p$, where these classifications are independent of each other and independent of the arrival times. Then the type-1 events form a Poisson process with rate λp, the type-2 events form a Poisson process with rate $\lambda(1-p)$, and these two processes are independent.

Proof. Let $\lambda_1 = \lambda p$ and $\lambda_2 = \lambda(1-p)$, so $\lambda = \lambda_1 + \lambda_2$ and $p = \lambda_1/(\lambda_1 + \lambda_2)$. Then we find ourselves in Story 13.2.9. But Story 13.2.7 and Story 13.2.9 are two equivalent generative stories for superposition! The probabilistic structures of the outputs of the two stories are identical, thinking of the arrivals in Story 13.2.7 from $(N_i(t) : t > 0)$ as being type-i. So we can assume instead that Story 13.2.7 was used to generate the process. Happily, in this story we know from the beginning that the type-1 process is independent of the type-2 process. ■

Thus we can superpose independent Poisson processes to get a combined Poisson process, or we can split a single Poisson process into independent Poisson processes. Figure 13.5 is an illustration of thinning. We simply flipped Figure 13.3 upside-down, which is appropriate because thinning is the flip side of superposition!

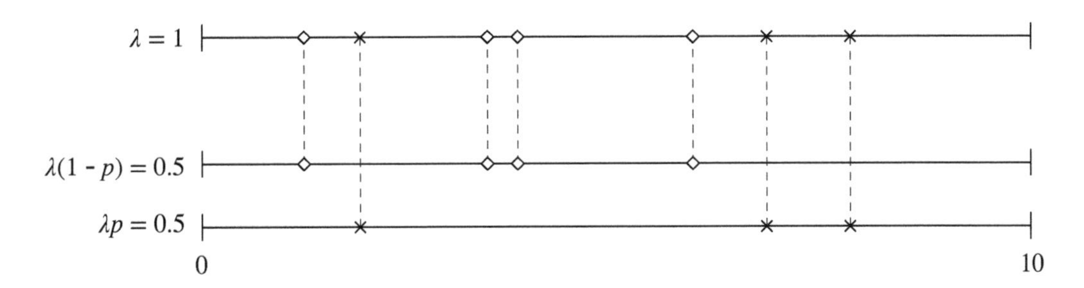

FIGURE 13.5
Thinning. Starting with a single Poisson process, if we let each arrival be type-1 with probability p and type-2 with probability $1 - p$, we obtain two independent Poisson processes with rates λp and $\lambda(1 - p)$. Here $\times$ represents type-1 events and $\diamond$ represents type-2 events, and we take $p = 0.5$.

Thinning is the Poisson process analog of the chicken-egg story. Suppose that we thin a Poisson process $(N(t) : t > 0)$ of rate λ as in the statement of Theorem 13.2.14, and let $(N_i(t) : t > 0)$ be the process of type-i arrivals. Thinking of type-1 arrivals as eggs that hatch, it follows immediately from the chicken-egg story that we have $N_1(t) \sim \text{Pois}(\lambda pt)$, $N_2(t) \sim \text{Pois}(\lambda(1-p)t)$, with $N_1(t)$ independent of $N_2(t)$.

For practice and as an example of the elegance of thinning, let's find the distribution of a random sum in two different ways: using MGFs and conditional expectation, and using thinning.

Example 13.2.15 (First Success sum of Exponentials). Let $X_1, X_2, \ldots$ be i.i.d. Expo(λ) and $N \sim \text{FS}(p)$, independent of the X_j. Find the distribution of the random sum

$$Y = \sum_{j=1}^{N} X_j.$$

Solution:

We'll solve this problem twice, first with tools from Chapter 9 and then with a Poisson process story that uses thinning. For the first method, we recognize Y as the sum of a random number of random variables, so we can find the MGF of

Y using Adam's law, conditioning on N. Recalling that the MGF of the Expo(λ) distribution is $\lambda/(\lambda - t)$ for $t < \lambda$, we have

$$\begin{aligned}
E\left(e^{tY}\right) &= E\left(E\left(e^{t\sum_{j=1}^{N} X_j}|N\right)\right) \\
&= E\left(E\left(e^{tX_1}\right)E\left(e^{tX_2}\right)\dots E\left(e^{tX_N}\right)|N\right) \\
&= E\left(E\left(e^{tX_1}\right)^N\right) \\
&= E\left(\left(\frac{\lambda}{\lambda - t}\right)^N\right).
\end{aligned}$$

Now we can use LOTUS with the FS(p) PMF,

$$P(N = k) = q^{k-1}p \text{ for } k = 1, 2, \dots$$

The LOTUS sum is

$$E\left(\left(\frac{\lambda}{\lambda - t}\right)^N\right) = \sum_{k=1}^{\infty}\left(\frac{\lambda}{\lambda - t}\right)^k q^{k-1}p,$$

which simplifies to $\lambda p/(\lambda p - t)$ for $t < \lambda p$, after doing some algebra and summing a geometric series. This is the Expo(λp) MGF, so $Y \sim$ Expo(λp).

Now let's see how Poisson processes can spare us from algebra while also providing insight into why Y is Exponential. Using the embedding strategy, since the X_j are i.i.d. Expo(λ) we are free to interpret the X_j as interarrival times in a Poisson process with rate λ. So let's imagine such a Poisson process, and let's further imagine that each of the arrivals is a *special* arrival with probability p, independently. Then we can interpret N as the number of arrivals until the first special arrival and Y as the waiting time for the first special arrival. But by the thinning property, the special arrivals form a Poisson process with rate λp. The waiting time for the first special arrival is thus distributed Expo(λp). □

Thinning works analogously with more than 2 types. In this setting, thinning is sometimes called the *coloring theorem.*

Theorem 13.2.16 (Coloring). Let $(N(t) : t > 0)$ be a Poisson process with rate λ, and C be a finite set of "colors", labeled from 1 through c. Suppose that each arrival gets randomly assigned a color from C, with color i having probability p_i. The color assignments are independent of each other and independent of the arrival times. Let $(N_i(t) : t > 0)$ be the color i process, i.e., $N_i(t)$ is the number of arrivals with color i in $(0, t]$. Then $(N_i(t) : t > 0)$ is a Poisson process with rate λp_i, for $i = 1, 2, \dots, c$, and these c monochromatic processes are independent.

Proof. We will induct on c. For $c = 1$ there is nothing to show. For $c = 2$ the result is the thinning theorem that we already proved. Now assume that the result holds for c colors, and show it when there are $c+1$ colors. Let's call color 1 *green*. Use thinning to assign each arrival to be green with probability p_1 and non-green with probability $1-p_1$. This splits the Poisson process into 2 independent processes: a green process of rate λp_1 and a non-green process of rate $\lambda(1-p_1)$. By the inductive hypothesis, we can then split the non-green process into c independent processes, one for each of colors $2, 3, \ldots, c+1$, where color j now gets probability

$$\tilde{p}_j = \frac{p_j}{1-p_1} = \frac{p_j}{p_2 + p_3 + \cdots + p_{c+1}},$$

for $j = 2, 3, \ldots, c+1$. We then have $c+1$ independent Poisson processes, one for each color, such that the process for color j has rate λp_j for $j = 1, 2, \ldots, c+1$. ■

The next example shows how coloring can help us decompose a complicated Poisson process into more manageable components.

Example 13.2.17 (Cars on a highway). Suppose cars enter a one-way highway from a common entrance, following a Poisson process with rate λ. The ith car has velocity V_i and travels at this velocity forever; no time is lost when one car overtakes another car. Assume the V_i are i.i.d. discrete r.v.s whose support is a finite set of positive values. The process starts at time 0, and we'll consider the highway entrance to be at location 0.

For fixed locations a and b on the highway with $0 < a < b$, let Z_t be the number of cars located in the interval $[a, b]$ at time t. (For instance, on an interstate highway running west to east through the midwestern United States, a could be Kansas City and b could be St. Louis; then Z_t would be the number of cars on the highway that are in the state of Missouri at time t.) Figure 13.6 illustrates the setup of the problem and the definition of Z_t.

Assume that t is large enough that $t > b/V_i$ for all possible values of V_i. Show that Z_t has a Poisson distribution with mean $\lambda(b-a)E(V_i^{-1})$.

Solution:

Since the V_i are discrete with finite support, we can enumerate the set of possible velocities $v_1, \ldots, v_m$ and their probabilities $p_1, \ldots, p_m$. After doing so, we realize that the cars entering the highway represent m types of arrivals, each corresponding to a different velocity. This suggests breaking up our overall Poisson process into simpler ones. Let's color the Poisson process m ways according to the velocity of the cars, resulting in a Poisson process with rate λp_1 for cars of velocity v_1, a Poisson process with rate λp_2 for cars of velocity v_2, and so forth.

For each of these m separate Poisson processes, we can ask: within what time interval do cars from this process have to enter the highway in order to be in $[a, b]$ at time t? This is a matter of physics, not statistics:

$$\text{distance} = \text{velocity} \cdot \text{time},$$

so a car that enters the highway at time s with velocity v will be at position $(t-s)v$

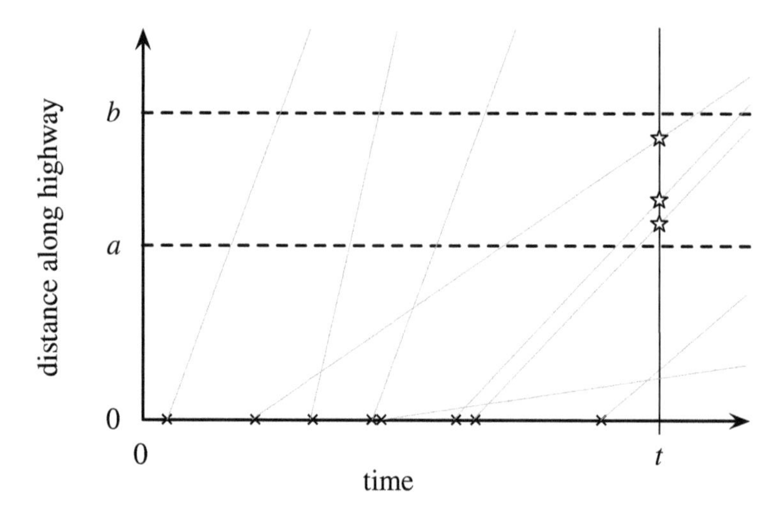

FIGURE 13.6
Cars enter a highway. Their entrance times form a Poisson process and are indicated by $\times$'s on the time axis. The ith car has velocity V_i, represented by the slope of the line emanating from the ith $\times$ symbol. We are interested in Z_t, the number of cars located in the interval $[a, b]$ at time t. Here we observe $Z_t = 3$, depicted by the three stars.

at time t. Thus, in order for the car's position to be between a and b, we require its arrival time to be between $t - b/v$ and $t - a/v$. (By our assumption that t is sufficiently large, we don't need to worry about $t - b/v$ being negative.) If the car arrives prior to time $t - b/v$, it will already have passed b by time t; if the car arrives after time $t - a/v$, it won't have reached a by time t.

We now have the answer for each of the separate Poisson processes. Within the process where cars have velocity v_j, the number of cars arriving between $t - b/v_j$ and $t - a/v_j$, which we'll call Z_{tj}, is distributed $\text{Pois}(\lambda p_j(b - a)/v_j)$: the rate of the process is λp_j, and the length of the interval $[t - b/v_j, t - a/v_j]$ is $(b - a)/v_j$.

Since the separate processes are independent, Z_{t1} through Z_{tm} are independent Poisson r.v.s. Thus,

$$Z_t = Z_{t1} + \cdots + Z_{tm} \sim \text{Pois}\left(\lambda(b - a)\sum_{j=1}^{m} \frac{p_j}{v_j}\right),$$

and $\sum_{j=1}^{m} p_j/v_j$ is the expectation of V_i^{-1} by LOTUS. This is what we wanted. $\square$

To wrap up this section, here is a table describing the correspondences between properties of the Poisson process and properties of the Poisson distribution. In the second column, $Y_1 \sim \text{Pois}(\lambda_1)$ and $Y_2 \sim \text{Pois}(\lambda_2)$ are independent.

Poisson process	**Poisson distribution**
conditioning	$Y_1 \mid Y_1 + Y_2 = n \sim \text{Bin}(n, \lambda_1/(\lambda_1 + \lambda_2))$
superposition	$Y_1 + Y_2 \sim \text{Pois}(\lambda_1 + \lambda_2)$
thinning	chicken-egg story

13.3 Poisson processes in multiple dimensions

Poisson processes in multiple dimensions are defined analogously to the 1D Poisson process: we just replace the notion of length with the notion of area or volume. For concreteness, we will now define 2D Poisson processes, after which it should also be clear by analogy how to define Poisson processes in higher dimensions.

Definition 13.3.1 (2D Poisson process). Events in the plane $\mathbb{R}^2$ are a *2D Poisson process* with intensity λ if the following conditions hold:

1. The number of events in a region A is distributed $\text{Pois}(\lambda \cdot \text{area}(A))$.
2. The numbers of events in disjoint regions are independent of each other.

As one might guess, conditioning, superposition, and thinning properties apply to 2D Poisson processes. Let $N(A)$ be the number of events in a region A, and let $B \subseteq A$. Given $N(A) = n$, the conditional distribution of $N(B)$ is Binomial:

$$N(B) \mid N(A) = n \sim \text{Bin}\left(n, \frac{\text{area}(B)}{\text{area}(A)}\right).$$

Conditional on the total number of events in the larger region A, the probability of an event falling into a subregion is proportional to the area of the subregion; thus the locations of the events are conditionally Uniform, and we can generate a 2D Poisson process in A by first generating the number of events $N(A) \sim \text{Pois}(\lambda \cdot \text{area}(A))$ and then placing the events uniformly at random in A. Figure 13.7 shows simulated 2D Poisson processes in the square $[0, 5] \times [0, 5]$ for intensities $\lambda = 1, 2, 5$.

As in the 1D case, the superposition of independent 2D Poisson processes is a 2D Poisson process, and the intensities add. We can also thin a 2D Poisson process to get independent 2D Poisson processes.

One property of 1D Poisson processes for which we haven't asserted a higher-dimensional analog is the count-time duality. The next example, the last in this chapter, will lead us to a spatial analog: a count-distance duality.

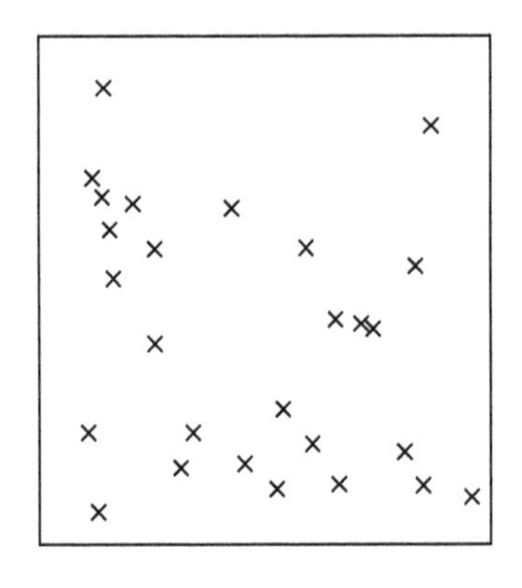
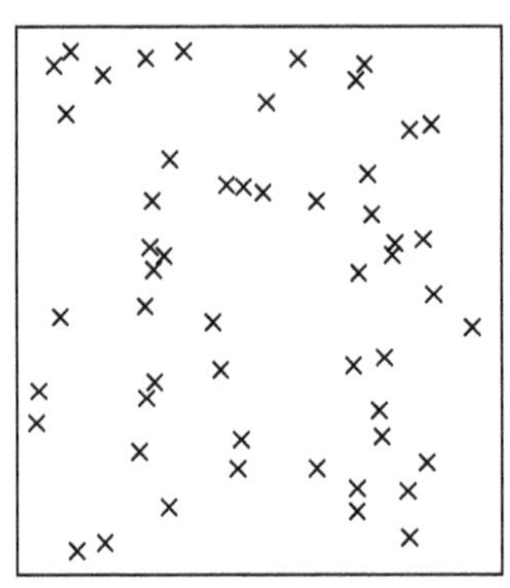
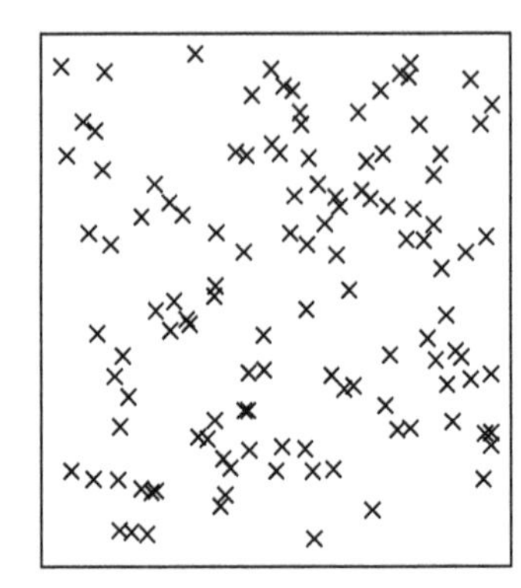

FIGURE 13.7
Simulated 2D Poisson process in the square $[0, 5] \times [0, 5]$, for $\lambda = 1, 2, 5$.

Example 13.3.2 (Nearest star). Stars in a certain universe are distributed according to a 3D Poisson process with intensity λ. If you live in this universe, what is the distribution of the distance from you to the nearest star?

Solution:

In a 3D Poisson process with intensity λ, the number of events in a region of space V is Poisson with mean $\lambda \cdot \text{volume}(V)$. Let R be the distance from you to the nearest star. The key observation is that in order for the event $R > r$ to occur, there must be no stars within a sphere of radius r around you; in fact, these two events are equivalent. For any $r > 0$, let N_r be the number of events within radius r of you, so $N_r \sim \text{Pois}(\lambda \cdot \frac{4}{3}\pi r^3)$. Then we have a *count-distance* duality:

$$R > r \text{ is the same event as } N_r = 0,$$

Therefore,

$$P(R > r) = P(N_r = 0) = e^{-\frac{4}{3}\lambda\pi r^3},$$

which gives that the CDF of R is

$$P(R \leq r) = 1 - e^{-\frac{4}{3}\lambda\pi r^3},$$

for $r > 0$ (and 0 otherwise). This is a Weibull (see Example 6.5.5). Specifically,

$$R \sim \text{Wei}\left(\frac{4\pi\lambda}{3}, 3\right).$$

□

Poisson processes have numerous extensions, some of which are explored in the exercises. We can allow λ to vary as a function of time or space instead of remaining constant; this is called an *inhomogeneous Poisson process*. We can allow λ to be a random variable; this is called a *Cox process*. Finally, we can allow the rate to increase by λ after each successive arrival; this is called a *Yule process*.

13.4 Recap

A Poisson process in one dimension is a sequence of arrivals such that the number of arrivals in any interval is Poisson (with mean proportional to the length of the interval) and disjoint intervals have independent numbers of arrivals. Some operations that we can perform with Poisson processes are conditioning, superposition, and thinning. Conditioning on the total number of arrivals in an interval allows us to view the arrivals as independent Uniforms on the interval. Superposition and thinning are complementary, and they allow us to split and merge Poisson processes when convenient. All of these properties have analogs for higher dimensions.

Poisson processes tie together many of the named distributions we have studied in this book:

- Poisson for the arrival counts,
- Exponential and Gamma for the interarrival times and arrival times,
- Binomial for the conditional counts,
- Uniform and (scaled) Beta for the conditional arrival times,
- Geometric and Negative Binomial for the discrete waiting times for special arrivals.

Poisson processes are also especially amenable to story proofs. A problem-solving strategy we used several times in this chapter is to embed r.v.s into a Poisson process in the hopes of discovering a story proof, even when the original problem appears to be unrelated to Poisson processes.

Poisson processes unite two of the important themes of this book, named distributions and stories, in a natural way. We think it is fitting to end with a topic that weaves together the story threads from throughout the book.

13.5 R

1D Poisson process

In Chapter 5, we discussed how to simulate a specified number of arrivals from a one-dimensional Poisson process by using the fact that the interarrival times are i.i.d. Exponentials. In this chapter, Story 13.2.4 tells us how to simulate a Poisson process within a specified interval $(0, L]$. We first generate the number of arrivals $N(L)$, which is distributed Pois(λL). Conditional on $N(L) = n$, the arrival times are distributed as the order statistics of n i.i.d. Unif$(0, L)$ r.v.s. The following code simulates arrivals from a Poisson process with rate 10 in the interval $(0, 5]$:

```
L <- 5
lambda <- 10
n <- rpois(1,lambda*L)
t <- sort(runif(n,0,L))
```

To visualize the Poisson process we have generated, we can plot the cumulative number of arrivals $N(t)$ as a function of t:

```
plot(t,1:n,type="s")
```

This produces a staircase plot as in Figure 13.8.

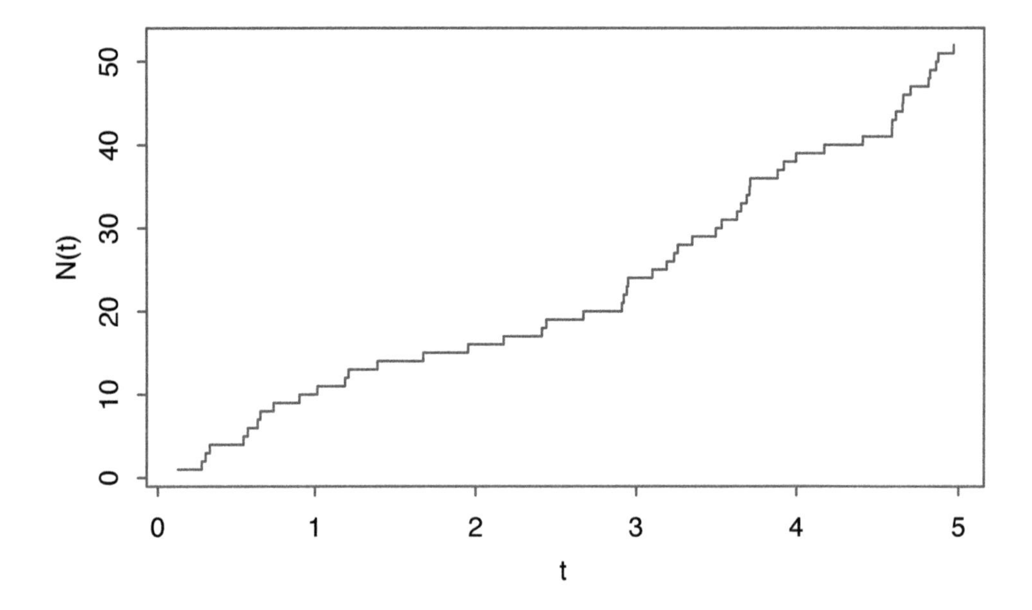

FIGURE 13.8
Number of arrivals in a Poisson process with rate 10 in the interval $(0, 5]$.

Thinning

It is straightforward to thin a Poisson process in R. The following code starts with a vector of arrival times $\mathbf{t}$ and the corresponding number of arrivals n from a Poisson process of rate λ on some interval, generated as above. For each arrival, we flip a coin with probability p of Heads; these coin tosses are stored in the vector $\mathbf{y}$. Finally, the arrivals for which the coin landed Heads are labeled as type-1; the rest are labeled as type-2. The resulting vectors of arrival times, $\mathbf{t}_1$ and $\mathbf{t}_2$, are realizations of independent Poisson processes of rates λp and $\lambda(1-p)$, by Theorem 13.2.14.

We can carry out this procedure for $p = 0.3$ as follows:

```
p <- 0.3
y <- rbinom(n,1,p)
t1 <- t[y==1]
t2 <- t[y==0]
```

2D Poisson process

Simulating a 2D Poisson process of rate λ on a square is nearly as easy as simulating a 1D Poisson process. In the square $(0, L] \times (0, L]$, the number of arrivals is distributed Pois(λL^2). Conditional on the number of arrivals, the locations of the arrivals are i.i.d. Uniform points in the square. By Example 7.1.23, the coordinates for each of these points are i.i.d. Unif$(0, L)$. So for $L = 5$, $\lambda = 10$, we can type:

```
L <- 5
lambda <- 10
n <- rpois(1,lambda*L^2)
x <- runif(n,0,L)
y <- runif(n,0,L)
```

13.6 Exercises

1. Passengers arrive at a bus stop according to a Poisson process with rate λ. The arrivals of buses are exactly t minutes apart. Show that on average, the sum of the waiting times of the riders on one of the buses is $\frac{1}{2}\lambda t^2$.

2. Earthquakes occur over time according to a Poisson process with rate λ. The jth earthquake has intensity Z_j, where the Z_j are i.i.d. with mean μ and variance σ^2. Find the mean and variance of the cumulative intensity of all the earthquakes up to time t.

3. Alice receives phone calls according to a Poisson process with rate λ. Unfortunately she has lost her cell phone charger. The battery's remaining life is a random variable T with mean μ and variance σ^2. Let $N(T)$ be the number of phone calls she receives before the battery dies; find $E(N(T))$, $\text{Var}(N(T))$, and $\text{Cov}(T, N(T))$.

4. Emails arrive in Bob's inbox according to a Poisson process with rate λ, measured in emails per hour; each email is work-related with probability p and personal with probability $1 - p$. The amount of time it takes to answer a work-related email is a random variable with mean μ_W and variance σ_W^2, the amount of time it takes to answer a personal email has mean μ_P and variance σ_P^2, and the response times for different emails are independent.

   What is the average amount of time Bob has to spend answering all the emails that arrive in a t-hour interval? What about the variance?

5. In an endless soccer match, goals are scored according to a Poisson process with rate λ. Each goal is made by team A with probability p and team B with probability $1-p$. For $j > 1$, the jth goal is a *turnaround* if it is made by a different team than the $(j-1)$st goal; for example, in the sequence AABBA..., the 3rd and 5th goals are turnarounds.

(a) In n goals, what is the expected number of turnarounds?

(b) If an A-to-B turnaround has just occurred, what is the expected time until the next B-to-A turnaround?

6. Let N_t be the number of arrivals up until time t in a Poisson process of rate λ, and let T_n be the time of the nth arrival. Consider statements of the form

$$P\left(N_t \lesseqgtr_1 n\right) = P\left(T_n \lesseqgtr_2 t\right),$$

where $\lesseqgtr_1$ and $\lesseqgtr_2$ are replaced by symbols from the list $<, \leq, \geq, >$. Which of these statements are true?

7. Claims against an insurance company follow a Poisson process with rate $\lambda > 0$. A total of N claims were received over two disjoint time periods of combined length $t = t_1 + t_2$, with t_i the length of period i.

(a) Given this information, derive the conditional probability distribution of N_1, the number of claims made in the first period, given N.

(b) The amount paid for the ith claim is X_i, with $X_1, X_2, \ldots$ i.i.d. and independent of the claims process. Let $E(X_i) = \mu$, $\text{Var}(X_i) = \sigma^2$, for $i = 1, \ldots, N$. Given N, find the mean and variance of the total claims paid in period 1. That is, find these two conditional moments of the quantity

$$W_1 = \sum_{i=1}^{N_1} X_i,$$

where (by convention) $W_1 = 0$ if $N_1 = 0$.

8. On a certain question-and-answer website, $N \sim \text{Pois}(\lambda_1)$ questions will be posted tomorrow, with λ_1 measured in questions/day. Given N, the post times are i.i.d. and uniformly distributed over the day (a day begins and ends at midnight). When a question is posted, it takes an $\text{Expo}(\lambda_2)$ amount of time (in days) for an answer to be posted, independent of what happens with other questions.

(a) Find the probability that a question posted at a uniformly random time tomorrow will not yet have been answered by the end of that day.

(b) Find the joint distribution of how many answered and unanswered questions posted tomorrow there will be at the end of that day.

9. An *inhomogeneous Poisson process* in one dimension is a Poisson process whose rate, instead of being constant, is a nonnegative function $\lambda(t)$ of time. Formally, we require that the number of arrivals in the interval $[t_1, t_2)$ be Poisson-distributed with mean $\int_{t_1}^{t_2} \lambda(t)dt$ and that disjoint intervals be independent. When $\lambda(t)$ is constant, this reduces to the definition of the ordinary or *homogeneous* Poisson process.

(a) Show that we can generate arrivals from an inhomogeneous Poisson process in the interval $[t_1, t_2)$ using the following procedure.

1. Let λ_{max} be the maximum value of $\lambda(t)$ in the interval $[t_1, t_2)$. Create a 2D rectangle $[t_1, t_2) \times [0, \lambda_{\text{max}}]$, and plot the function $\lambda(t)$ in the rectangle.
2. Generate $N \sim \text{Pois}(\lambda_{\text{max}}(t_2 - t_1))$, and place N points uniformly at random in the rectangle.
3. For each of the N points: if the point falls below the curve $\lambda(t)$, accept it as an arrival in the process, and take its horizontal coordinate to be its arrival time. If the point falls above the curve $\lambda(t)$, reject it.

Hint: Verify that the two conditions in the definition are satisfied.

(b) Suppose we have an inhomogeneous Poisson process with rate function $\lambda(t)$. Let

$N(t)$ be the number of arrivals up to time t and T_j be the time of the jth arrival. Explain why the hybrid joint PDF of $N(t)$ and $T_1, \dots, T_{N(t)}$, which constitute all the data observed up to time t, is given by

$$f(n, t_1, \dots, t_n) = \frac{e^{-\lambda_{\text{total}}} \lambda_{\text{total}}^n}{n!} \cdot n! \frac{\lambda(t_1) \dots \lambda(t_n)}{\lambda_{\text{total}}^n} = e^{-\lambda_{\text{total}}} \lambda(t_1) \dots \lambda(t_n)$$

for $0 < t_1 < t_2 < \cdots < t_n$ and nonnegative integer n, where $\lambda_{\text{total}} = \int_0^t \lambda(u)du$.

10. A *Cox process* is a generalization of a Poisson process, where the rate λ is a random variable. That is, λ is generated according to some distribution on $(0, \infty)$ and then, given that value of λ, a Poisson process with that rate is generated.

(a) Explain intuitively why disjoint intervals in a 1D Cox process are *not* independent.

(b) In a 1D Cox process where $\lambda \sim \text{Gamma}(\alpha, \beta)$, find the covariance between the number of arrivals in $[0, t)$ and the number of arrivals in $[t, t + s)$.
Hint: Condition on λ.

11. In a *Yule process* with rate λ, the rate of arrivals increases after each new arrival, so that the time of the first arrival is Expo(λ) and the time between the $(j - 1)$st and jth arrivals is Expo($j\lambda$) for $j = 2, 3, \dots$. So interarrival times are independent but not i.i.d.

(a) Show that the superposition of two independent Yule processes with the same rate λ is a Yule process, except shifted so that the interarrival times are Expo(2λ), Expo(3λ), Expo(4λ), ... rather than Expo(λ), Expo(2λ), Expo(3λ), ...

(b) Show that if we project the process from Part (a) into discrete time, the resulting sequence of type-1 and type-2 arrivals is equivalent to the following discrete-time process:

1. Start with two balls in an urn, labeled 1 and 2.
2. Draw a ball out of the urn at random, note its number, and replace it along with another ball with the same number.
3. Repeat step 2 over and over.

12. Consider the coupon collector problem: there are n toy types, and toys are collected one by one, sampling with replacement from the set of toy types each time. We solved this problem in Chapter 4, assuming that all toy types are equally likely. Now suppose that at each stage, the jth toy type is collected with probability p_j, where the p_j are not necessarily equal. Let N be the number of toys needed until we have a full set; we wish to find $E(N)$. This problem outlines an embedding method for calculating $E(N)$.

(a) Suppose that the toys arrive according to a Poisson process with rate 1, so that the interarrival times between toys are i.i.d. $X_j \sim \text{Expo}(1)$. For $j = 1, \dots, n$, let Y_j be the waiting time until the first toy of type j. What are the distributions of the Y_j? Are the Y_j independent?

(b) Explain why $T = \max(Y_1, \dots, Y_n)$, the waiting time until all toy types are collected, can also be written as $X_1 + \cdots + X_N$, where the X_j are defined as in Part (a). Use this to show that $E(T) = E(N)$.

(c) Show that $E(T)$, and hence $E(N)$, can be found by computing the integral

$$\int_0^\infty \left(1 - \prod_{j=1}^n (1 - e^{-p_j t})\right) dt.$$

Hint: Use the identity $E(T) = \int_0^\infty P(T > t)dt$, which was shown in Example 5.3.8.

A

Math

A.1 Sets

A set is a Many that allows itself to be thought of as a One.
– Georg Cantor

A *set* is a collection of objects. The objects can be anything: numbers, people, cats, courses, even other sets! The language of sets allows us to talk precisely about *events*. If S is a set, then the notation $x \in S$ indicates that x is an element or member of the set S (and $x \notin S$ indicates that x is not in S). We can think of the set as a club, with precisely defined criteria for membership. For example:

1. $\{1, 3, 5, 7, \dots\}$ is the set of all odd numbers;
2. $\mathbb{R}$ is the set of all real numbers;
3. {Worf, Jack, Tobey} is the set of Joe's cats;
4. $[3, 7]$ is the closed interval consisting of all real numbers between 3 and 7;
5. $\{HH, HT, TH, TT\}$ is the set of all possible outcomes if a coin is flipped twice (where, for example, HT means the first flip lands Heads and the second lands Tails).

To describe a set (when it's tedious or impossible to list out its elements), we can give a rule that says whether each possible object is or isn't in the set. For example, $\{(x, y) \in \mathbb{R}^2 : x^2 + y^2 \leq 1\}$ is the disk of radius 1 in the plane $\mathbb{R}^2$, centered at the origin. In this expression, the ":" is read as "such that".

A.1.1 The empty set

The smallest set, which is both subtle and important, is the *empty set*, which is the set that has no elements whatsoever. It is denoted by $\emptyset$ or by $\{\}$. Make sure not to confuse $\emptyset$ with $\{\emptyset\}$. The former has no elements, while the latter has one element. If we visualize the empty set as an empty paper bag, then we can visualize $\{\emptyset\}$ as a paper bag inside of a paper bag.

A.1.2 Subsets

If A and B are sets, then we say A is a *subset* of B (and write $A \subseteq B$) if every element of A is also an element of B. For example, the set of all integers is a subset of the set of all real numbers. It is always true that $\emptyset$ and A itself are subsets of A; these are the extreme cases for subsets. A general strategy for showing that $A \subseteq B$ is to let x be an arbitrary element of A, and then show that x must also be an element of B. A general strategy for showing that $A = B$ for two sets A and B is to show that each is a subset of the other.

A.1.3 Unions, intersections, and complements

The *union* of two sets A and B, written as $A \cup B$, is the set of all objects that are in A or B (or both). The *intersection* of A and B, written as $A \cap B$, is the set of all objects that are in both A and B. We say that A and B are *disjoint* if $A \cap B = \emptyset$. For n sets $A_1, \dots, A_n$, the union $A_1 \cup A_2 \cup \dots \cup A_n$ is the set of all objects that are in *at least one* of the A_j's, while the intersection $A_1 \cap A_2 \cap \dots \cap A_n$ is the set of all objects that are in *all* of the A_j's.

In many applications, all the sets we're working with are subsets of some set S (in probability, this may be the set of all possible outcomes of some experiment). When S is clear from the context, we define the *complement* of a set A to be the set of all objects in S that are *not* in A; this is denoted by A^c.

Unions, intersections, and complements can be visualized easily using Venn diagrams, such as the one shown below. The union is the entire shaded region, while the intersection is the football-shaped region of points that are in both A and B. The complement of A is all points in the rectangle that are outside of A.

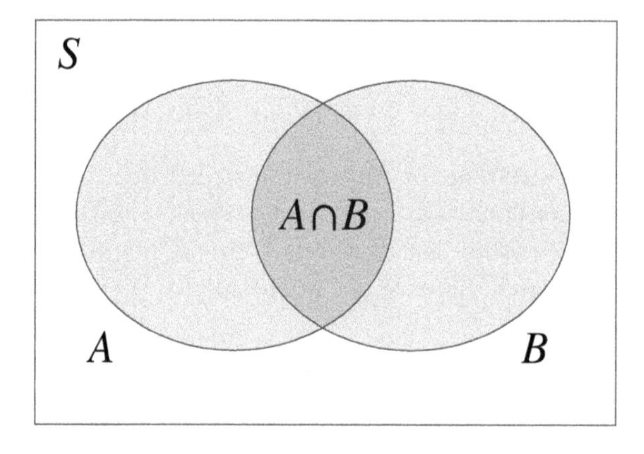

FIGURE A.1
A Venn diagram.

Note that the area of the region $A \cup B$ is the area of A plus the area of B, minus the area of $A \cap B$ (this is a basic form of the *inclusion-exclusion principle*).

De Morgan's laws give an elegant, useful duality between unions and intersections:

$$(A_1 \cup A_2 \cup \cdots \cup A_n)^c = A_1^c \cap A_2^c \cap \cdots \cap A_n^c$$

$$(A_1 \cap A_2 \cap \cdots \cap A_n)^c = A_1^c \cup A_2^c \cup \cdots \cup A_n^c$$

It is much more important to *understand* De Morgan's laws than to memorize them! The first law says that not being in at least one of the A_j is the same thing as not being in A_1, nor being in A_2, nor being in A_3, etc.

For example, let A_j be the set of people who like the jth Star Wars prequel, for $j \in \{1, 2, 3\}$. Then $(A_1 \cup A_2 \cup A_3)^c$ is the set of people for whom it is *not* the case that they like at least one of the prequels, but that is the same as $A_1^c \cap A_2^c \cap A_3^c$, the set of people who don't like *The Phantom Menace*, don't like *Attack of the Clones*, and don't like *Revenge of the Sith.*

The second law says that not being in all of the A_j is the same thing as being outside at least one of the A_j. For example, let the A_j be defined as in the previous paragraph. If it is not the case that you like all of the Star Wars prequels (making you a member of the set $(A_1 \cap A_2 \cap A_3)^c$), then there must be at least one prequel that you don't like (making you a member of the set $A_1^c \cup A_2^c \cup A_3^c$), and vice versa.

Proving the following facts about sets (not just drawing Venn diagrams, though they are very helpful for building intuition) is good practice:

1. $A \cap B$ and $A \cap B^c$ are disjoint, with $(A \cap B) \cup (A \cap B^c) = A$.
2. $A \cap B = A$ if and only if $A \subseteq B$.
3. $A \subseteq B$ if and only if $B^c \subseteq A^c$.

A.1.4 Partitions

A collection of subsets $A_1, \ldots, A_n$ of a set S is a *partition* of S if $A_1 \cup \cdots \cup A_n = S$ and $A_i \cap A_j = \emptyset$ for all $i \neq j$. In words, a partition of a set is a collection of disjoint subsets whose union is the entire set. For example, the set of even numbers $\{0, 2, 4, \ldots\}$ and the set of odd numbers $\{1, 3, 5, \ldots\}$ form a partition of the set of nonnegative integers.

A.1.5 Cardinality

A set may be finite or infinite. If A is a finite set, we write $|A|$ for the number of elements in A, which is called its *size* or *cardinality*. For example, $|\{2, 4, 6, 8, 10\}| = 5$ since there are 5 elements in this set. A very useful fact is that A and B are finite sets, then

$$|A \cup B| = |A| + |B| - |A \cap B|.$$

This is a form of the inclusion-exclusion theorem from Chapter 1. It says that to count how many elements are in the union of A and B, we can add the separate counts for each, and then adjust for the fact that we have double-counted the elements (if any) that are in both A and B.

Two sets A and B are said to have the *same size* or *same cardinality* if they can be put into one-to-one correspondence, i.e., if each element of A can be paired up with exactly one element of B, with no unpaired elements in either set. We say that A is *smaller than* B if there is *not* a one-to-one correspondence between A and B, but there *is* a one-to-one correspondence between A and some *subset* of B.

For example, suppose that we want to count the number of people in a movie theater with 100 seats. Assume that no one in the theater is standing, and no seat has more than one person in it. The obvious thing to do is to go around counting people one by one (though it's surprisingly easy to miss someone or accidentally count someone twice). But if every seat is occupied, then a much easier method is to note that there must be 100 people, since there are 100 seats and there is a one-to-one correspondence between people and seats. If some seats are empty, then there must be fewer than 100 people there.

This idea of looking at one-to-one correspondences makes sense both for finite and for infinite sets. Consider the perfect squares $1^2, 2^2, 3^2, \dots$. Galileo pointed out the paradoxical result that on the one hand it seems like there are fewer perfect squares than positive integers (since every perfect square is a positive integer, but lots of positive integers aren't perfect squares), but on the other hand it seems like these two sets have the same size since they can be put into one-to-one correspondence: pair 1^2 with 1, pair 2^2 with 2, pair 3^2 with 3, etc.

The resolution of Galileo's paradox is to realize that intuitions about finite sets don't necessarily carry over to infinite sets. By definition, the set of all perfect squares and the set of all positive integers do have the same size. Another famous example of this is *Hilbert's hotel.* For any hotel in the real world the number of rooms is finite. If every room is occupied, there is no way to accommodate more guests, other than by cramming more people into already occupied rooms.

Now consider an imaginary hotel with an infinite sequence of rooms, numbered $1, 2, 3, \dots$. Assume that all the rooms are occupied, and that a weary traveler arrives, looking for a room. Can the hotel give the traveler a room, without leaving any of the current guests without a room? Yes, one way is to have the guest in room n move to room $n+1$, for all $n = 1, 2, 3, \dots$. This frees up room 1, so the traveler can stay there.

What if *infinitely many* travelers arrive at the same time, such that their cardinality is the same as that of the positive integers (so we can label the travelers as traveler 1, traveler 2, ...)? The hotel could fit them in one by one by repeating the above procedure over and over again, but it would take forever (infinitely many moves) to accommodate everyone, and it would be bad for business to make the current guests keep moving over and over again. Can the room assignments be updated just *once* so that everyone has a room? Yes, one way is to have the guest in room n move to room to $2n$ for all $n = 1, 2, 3, \ldots$, and then have traveler n move into room $2n - 1$. In this way, the current guests occupy all the even-numbered rooms, and the new guests occupy all the odd-numbered rooms.

An infinite set is called *countably infinite* if it has the same cardinality as the set of all positive integers. A set is called *countable* if it is finite or countably infinite, and *uncountable* otherwise. The mathematician Cantor showed that not all infinite sets are the same size. In particular, the set of all real numbers is uncountable, as is any interval in the real line of positive length.

A.1.6 Cartesian product

The *Cartesian product* of two sets A and B is the set

$$A \times B = \{(a, b) : a \in A,\ b \in B\}.$$

For example, $[0, 1] \times [0, 1]$ is the square $\{(x, y) : x, y \in [0, 1]\}$, and $\mathbb{R} \times \mathbb{R} = \mathbb{R}^2$ is two-dimensional Euclidean space.

A.2 Functions

Let A and B be sets. A *function* from A to B is a deterministic rule that, given an element of A as input, provides an element of B as an output. That is, a function from A to B is a machine that takes an x in A and "maps" it to some y in B. Different x's can map to the same y, but each x only maps to one y. Here A is called the *domain* and B is called the *target*. The notation $f : A \to B$ says that f is a function mapping A into B. The *range* of f is $\{y \in B : f(x) = y \text{ for some } x \in A\}$.

It is important to distinguish between f (the function) and $f(x)$ (the value of the function when evaluated at x). That is, f is a rule, while $f(x)$ is a number for each number x. The function g given by $g(x) = e^{-x^2/2}$ is exactly the same as the function g given by $g(t) = e^{-t^2/2}$; what matters is the rule, not the name we use for the input. For this example, the rule is that x gets mapped to $e^{-x^2/2}$; this rule is also denoted by $x \mapsto e^{-x^2/2}$.

A function f from the real line to the real line is *continuous* if $f(x) \to f(a)$ as $x \to a$, for any value of a. It is called *right-continuous* if this is true when approaching from the right, i.e., $f(x) \to f(a)$ as $x \to a$ while ranging over values with $x > a$.

The set A may not be a set of numbers. In probability, it is extremely common and useful to consider functions whose domains are the set of all possible outcomes of some experiment. It may be difficult to write down a formula for the function, but the function is still valid as long as it's defined unambiguously.

A.2.1 One-to-one functions

Let f be a function from A to B. Then f is a *one-to-one* function if $f(x) \neq f(y)$ whenever $x \neq y$. That is, any two distinct inputs in A get mapped to two distinct outputs in B; for each y in B, there can be at most one x in A that maps to it.

Let f be a one-to-one function from A to B, and let C be the range of f, i.e.,

$$C = \{b \in B : f(a) = b \text{ for some } x \in A\}.$$

Then there is an *inverse function* $f^{-1} : C \to A$, defined by letting $f^{-1}(y)$ be the unique element $x \in A$ such that $f(x) = y$. A function that has an inverse is called an *invertible* function.

For example, let $f(x) = x^2$ for all real x. This is *not* a one-to-one function since, for example, $f(3) = f(-3)$. But now assume instead that the domain of f is chosen to be $[0, \infty)$, so we are defining f as a function from $[0, \infty)$ to $[0, \infty)$. Then f is one-to-one, and its inverse function is given by $f^{-1}(y) = \sqrt{y}$ for all $y \in [0, \infty)$.

The *pigeonhole principle* says that if A and B are finite sets with $|A| > |B|$, then there does not exist a one-to-one function from A to B. In other words, if k objects are placed into n boxes, where $k > n$, then there must be at least one box that contains more than one object.

For example, in the birthday problem from Chapter 1 (assuming there are 365 days in a year), with 366 or more people there is guaranteed to be at least one birthday match. There is no one-to-one function assigning a birthday to each of the people, when there are more people than there are days in a year.

A.2.2 Increasing and decreasing functions

Let $f : A \to \mathbb{R}$, where A is a set of real numbers. Then f is an *increasing* function if $x \leq y$ implies $f(x) \leq f(y)$ (for all $x, y \in A$). Note that this definition allows there to be regions where f is flat, e.g., the constant function that is equal to 42 everywhere is an increasing function, and any CDF (see Section 3.6) is an increasing function. We say that f is *strictly increasing* if $x < y$ implies $f(x) < f(y)$. For example, $f : \mathbb{R} \to \mathbb{R}$ with $f(x) = x^3$ is a strictly increasing function.

Similarly, f is a *decreasing* function if $x \leq y$ implies $f(x) \geq f(y)$, and is a *strictly decreasing* function if $x < y$ implies $f(x) > f(y)$. For example, $f : (0, \infty) \to (0, \infty)$ with $f(x) = 1/x$ is a strictly decreasing function.

A *monotone* function is a function that is either increasing or decreasing. A *strictly monotone* function is a function that is either strictly increasing or strictly decreasing. Note that any strictly monotone function is one-to-one.

A.2.3 Even and odd functions

Let f be a function from $\mathbb{R}$ to $\mathbb{R}$. We say f is an *even function* if $f(x) = f(-x)$ for all x, and we say f is an *odd function* if $-f(x) = f(-x)$ for all x. If neither of these conditions is satisfied, then f is neither even nor odd. Figure A.2 shows the graphs of two even functions and two odd functions.

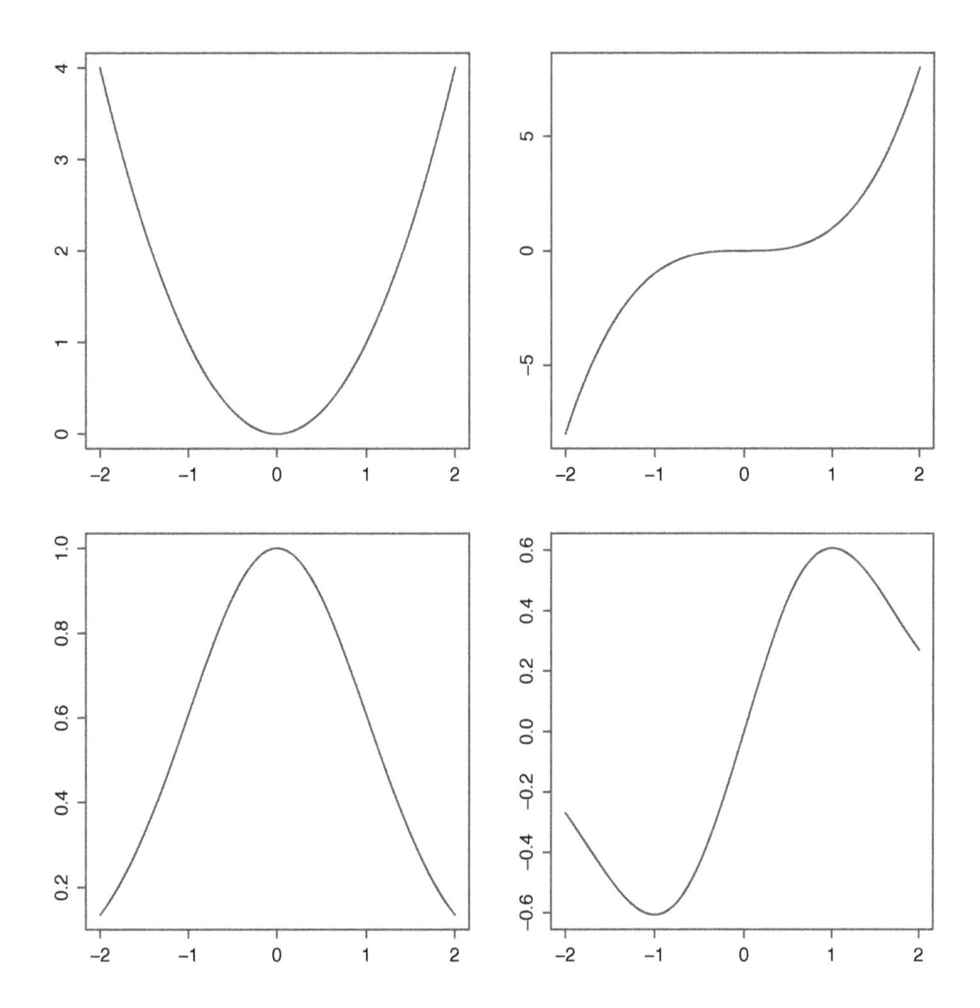

FIGURE A.2
Even and odd functions. The graphs on the left are even functions: $f(x) = x^2$ on the top and $f(x) = e^{-x^2/2}$ on the bottom. The graphs on the right are odd functions: $f(x) = x^3$ on the top and $f(x) = xe^{-x^2/2}$ on the bottom.

Even and odd functions have nice symmetry properties. The graph of an even function remains the same if you reflect it about the vertical axis, and the graph of an odd function remains the same if you rotate it 180 degrees around the origin.

Even functions have the property that for any a,

$$\int_{-a}^{a} f(x)dx = 2\int_{0}^{a} f(x)dx,$$

assuming the integral exists. This is because the area under the function from $-a$ to 0 is equal to the area under the function from 0 to a. Odd functions have the property that for any a,

$$\int_{-a}^{a} f(x)dx = 0,$$

again assuming the integral exists. This is because the area under the function from $-a$ to 0 cancels the area under the function from 0 to a.

A.2.4 Convex and concave functions

A function g whose domain is an interval I is *convex* if

$$g(px_1 + (1-p)x_2) \leq pg(x_1) + (1-p)g(x_2)$$

for all $x_1, x_2 \in I$ and $p \in (0,1)$. Geometrically, this says that if we draw a line segment connecting two points on the graph of g, then the line segment lies above the graph of g. If the derivative g' exists, then an equivalent definition is that every tangent line to the graph of g lies below the graph. If g'' exists, then an equivalent definition is that $g''(x) \geq 0$ for all $x \in I$. An example is shown in Figure A.3. A simple example is $g(x) = x^2$, whose second derivative is $g''(x) = 2 > 0$.

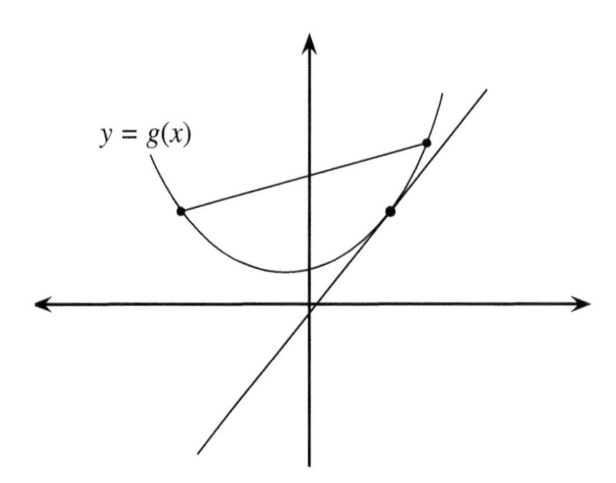

FIGURE A.3
Graph of a convex function g. We have $g''(x) \geq 0$. Any line segment connecting two points on the curve lies above the curve. Any tangent line lies below the curve.

A function g is *concave* if $-g$ is convex. If g'' exists, then g is concave if and only if $g''(x) \leq 0$ for all x in the domain. For example, $g(x) = \log(x)$ defines a concave function on $(0,\infty)$ since $g''(x) = -1/x^2 < 0$ for all $x \in (0,\infty)$.

A.2.5 Exponential and logarithmic functions

An *exponential function* is a function of the form $f(x) = a^x$ for some number $a > 0$. If $a > 1$, the function is increasing, and if $0 < a < 1$, the function is decreasing. The most common exponential function we'll work with is $f(x) = e^x$, and a very useful limit result to know is that

$$\left(1 + \frac{x}{n}\right)^n \to e^x$$

as $n \to \infty$, for any real number x. This has an interpretation in terms of a bank paying compound interest on a deposit: as compounding occurs more and more times per year, the growth rate approaches exponential growth.

Here are some properties of exponential functions:

1. $a^x a^y = a^{x+y}$.
2. $a^x b^x = (ab)^x$.
3. $(a^x)^y = a^{xy}$.

The inverse of an exponential function is a logarithmic function: for positive y, $\log_a y$ is defined to be the number x such that $a^x = y$. Throughout this book, when we write $\log y$ without explicitly specifying the base, we are referring to the *natural logarithm* (base e).

Here are some properties of logarithms:

1. $\log_a x + \log_a y = \log_a xy$.
2. $\log_a x^n = n \log_a x$.
3. $\log_a x = \frac{\log x}{\log a}$.

A.2.6 Floor function and ceiling function

The *floor function* is defined by letting $\lfloor x \rfloor$ be the greatest integer less than or equal to x. That is, it says to round down to an integer. For example, $\lfloor 3.14 \rfloor = 3$, $\lfloor -1.3 \rfloor = -2$, and $\lfloor 5 \rfloor = 5$. (Some books denote the floor function by $[x]$ but this is bad notation since it does not suggest a corresponding notation for the ceiling function, and since square brackets are also used for other purposes.)

The ceiling function is defined by letting $\lceil x \rceil$ be the smallest integer greater than or equal to x. For example, $\lceil 3.14 \rceil = 4$, $\lceil -1.3 \rceil = -1$, and $\lceil 5 \rceil = 5$.

A.2.7 Factorial function and gamma function

The *factorial function* takes a positive integer n and returns the product of the integers from 1 to n, denoted $n!$ and read "n factorial":

$$n! = n(n-1)(n-2)\cdots 1.$$

Also, we define $0! = 1$. This convention makes sense since if we think of $n!$ as the number of ways in which n people can line up, there is 1 way for $n = 0$ (this just means there is no one there, so the line is empty). It is also very helpful since, for example, we can then say that $n!/(n-1)! = n$ for all positive integers n without running into trouble when $n = 1$.

The factorial function grows extremely quickly as n grows. A famous, useful approximation for factorials is *Stirling's formula*,

$$n! \approx \sqrt{2\pi n}\left(\frac{n}{e}\right)^n.$$

The ratio of the two sides converges to 1 as $n \to \infty$. For example, direct calculation gives $52! \approx 8.066 \times 10^{67}$, while Stirling's formula says $52! \approx 8.053 \times 10^{67}$.

The *gamma function* Γ generalizes the factorial function to positive real numbers; it is defined by

$$\Gamma(a) = \int_0^\infty x^a e^{-x} \frac{dx}{x}, \quad a > 0,$$

and has the property that

$$\Gamma(n) = (n-1)!$$

for all positive integers n. Also, $\Gamma(1/2) = \sqrt{\pi}$.

An important property of the gamma function, which generalizes the fact that $n! = n \cdot (n-1)!$, is that

$$\Gamma(a+1) = a\Gamma(a)$$

for all $a > 0$. See Chapter 8 for more about the gamma function.

A.3 Matrices

Neo: What is the Matrix?
Trinity: The answer is out there, Neo, and it's looking for you, and it will find you if you want it to.
– *The Matrix* (film from 1999)

A *matrix* is a rectangular array of numbers, such as $\begin{pmatrix} 3 & 1/e \\ 2\pi & 1 \end{pmatrix}$ or $\begin{pmatrix} 1 & 1 & 0 \\ 1 & 2 & 3 \end{pmatrix}$. We say that the dimensions of a matrix are m by n (also written $m \times n$) if it has m rows and n columns (so the former example is 2 by 2, while the latter is 2 by 3). The matrix is called *square* if $m = n$. If $m = 1$, we have a *row vector*; if $n = 1$, we have a *column vector*. We write $A = (a_{ij})$ to indicate that the row i, column j entry of A is a_{ij}.

A.3.1 Matrix addition and multiplication

To *add* two matrices A and B with the same dimensions, just add the corresponding entries, e.g.,

$$\begin{pmatrix} 1 & 1 & 0 \\ 1 & 1 & 1 \end{pmatrix} + \begin{pmatrix} 1 & 0 & 0 \\ 1 & 1 & 0 \end{pmatrix} = \begin{pmatrix} 2 & 1 & 0 \\ 2 & 2 & 1 \end{pmatrix}.$$

When we *multiply* an $m \times n$ matrix A by an $n \times r$ matrix B, we obtain an $m \times r$ matrix AB. The product AB is undefined if the number of columns of A does not equal the number of rows of B.

The row i, column j entry of AB is $\sum_{k=1}^{n} a_{ik} b_{kj}$, where a_{ij} and b_{ij} are the row i, column j entries of A and B, respectively. For example, here is how to multiply a 2×3 matrix by a 3×1 vector:

$$\begin{pmatrix} 1 & 2 & 3 \\ 4 & 5 & 6 \end{pmatrix} \begin{pmatrix} 7 \\ 8 \\ 9 \end{pmatrix} = \begin{pmatrix} 1 \cdot 7 + 2 \cdot 8 + 3 \cdot 9 \\ 4 \cdot 7 + 5 \cdot 8 + 6 \cdot 9 \end{pmatrix} = \begin{pmatrix} 50 \\ 122 \end{pmatrix}.$$

Note that AB may not equal BA, even if both are defined. To multiply a matrix A by a scalar, just multiply each entry by that scalar.

The *transpose* of a matrix A is the matrix whose row i, column j entry is the row j, column i entry of A. It is denoted by A' and read as "A transpose". The rows of A are the columns of A', and the columns of A are the rows of A'. A square matrix A is *symmetric* if $A' = A$. A useful property of transposes is that if A and B are matrices such that the product AB is defined, then $(AB)' = B'A'$.

The *determinant* of a 2×2 matrix is defined by

$$\begin{vmatrix} a & b \\ c & d \end{vmatrix} = ad - bc.$$

Determinants can also be defined for $n \times n$ matrices, in a recursive manner not reviewed here.

A.3.2 Eigenvalues and eigenvectors

An *eigenvalue* of an $n \times n$ matrix A is a number λ such that

$$A\mathbf{v} = \lambda \mathbf{v}$$

for some $n \times 1$ column vector $\mathbf{v}$, where the elements of $\mathbf{v}$ are not all zero. The vector $\mathbf{v}$ is called an *eigenvector* of A, or a *right eigenvector*. (Similarly, a *left eigenvector* of A is a row vector $\mathbf{w}$ satisfying $\mathbf{w}A = \lambda\mathbf{w}$ for some λ.) This definition says that when A and $\mathbf{v}$ are multiplied, $\mathbf{v}$ just gets stretched by the constant λ.

Some matrices have no real eigenvalues, but the *Perron-Frobenius theorem* tells us that in a special case that is of particular interest to us in Chapter 11, eigenvalues exist and have nice properties. Let A be a square matrix whose entries are nonnegative and whose rows sum to 1. Further assume that for all i and j, there exists $k \geq 1$ such that the row i, column j entry of A^k is positive. Then the Perron-Frobenius theorem says that 1 is an eigenvalue of A, which is in fact the largest eigenvalue of A, and there is a corresponding eigenvector whose entries are all positive.

A.4 Difference equations

A *difference equation* describes a sequence of numbers recursively in terms of earlier terms in the sequence. For example, $a_0, a_1, \ldots$ is a *Fibonacci sequence* if

$$a_i = a_{i-1} + a_{i-2}$$

for all $i \geq 2$. There are infinitely many Fibonacci sequences, but such a sequence is uniquely determined after a_0 and a_1 are specified (these are called the *initial conditions* or *boundary conditions*). For example, $a_0 = 0, a_1 = 1$ yields the Fibonacci sequence $0, 1, 1, 2, 3, 5, 8, 13, \ldots$.

Difference equations often arise in probability, especially when applying LOTP. In this section, we will show how to solve difference equations of the form

$$p_i = p \cdot p_{i+1} + q \cdot p_{i-1},$$

where $p \neq 0$ and $q = 1 - p$. (This equation comes up in the gambler's ruin problem.) The first step is to *guess* a solution of the form $p_i = x^i$. Plugging this into the above, we have

$$x^i = p \cdot x^{i+1} + q \cdot x^{i-1},$$

which reduces to $px^2 - x + q = 0$. This is called the *characteristic equation*, and the solution to the difference equation depends on whether the characteristic equation has one or two distinct roots. If there are two distinct roots r_1 and r_2, then the solution is of the form

$$p_i = ar_1^i + br_2^i,$$

for some constants a and b. If there is only one distinct root r, then the solution is of the form

$$p_i = ar^i + bir^i.$$

In our case, the characteristic equation has roots 1 and q/p, since

$$\frac{1 \pm \sqrt{1-4p(1-p)}}{2p} = \frac{1 \pm \sqrt{(2p-1)^2}}{2p} = \frac{1 \pm |2p-1|}{2p}$$

is $(1+2p-1)/(2p) = 1$ or $(2-2p)/(2p) = q/p$. The roots are distinct if $p \neq q$ and are both equal to 1 if $p = q$. So we have

$$p_i = \begin{cases} a + b\left(\dfrac{q}{p}\right)^i, & p \neq q, \\ a + bi, & p = q. \end{cases}$$

This is called the *general solution* of the difference equation, since we have not yet specified the constants a and b. To get a *specific solution*, we need to know two points in the sequence in order to solve for a and b.

A.5 Differential equations

Differential equations are the continuous version of difference equations. A differential equation uses derivatives to describe a function or collection of functions. For example, the differential equation

$$\frac{dy}{dx} = 3y$$

describes a collection of functions that have the following property: the instantaneous rate of change of the function at any point (x, y) is equal to $3y$. This is an example of a *separable* differential equation because we can separate the x's and y's, putting them on opposite sides of the equation:

$$\frac{dy}{y} = 3dx.$$

Now we can integrate both sides, giving $\log y = 3x + c$, or equivalently,

$$y = Ce^{3x},$$

where C is any constant. This is called the *general solution* of the differential equation, and it tells us that all functions satisfying the differential equation are of the form $y = Ce^{3x}$ for some C. To get a *specific solution*, we need to specify one point on the graph, which allows us to solve for C.

A.6 Partial derivatives

If you can do ordinary derivatives, you can do partial derivatives: just hold all the other input variables constant except for the one you're differentiating with respect to. For example, let $f(x, y) = y \sin(x^2+y^3)$. Then the partial derivative with respect to x is

$$\frac{\partial f(x,y)}{\partial x} = 2xy\cos(x^2+y^3),$$

and the partial derivative with respect to y is

$$\frac{\partial f(x,y)}{\partial y} = \sin(x^2+y^3) + 3y^3\cos(x^2+y^3).$$

These are *first order* partial derivatives. If we take a partial derivative of a first order partial derivative, we get a *second order* partial derivative, such as

$$\frac{\partial^2 f(x,y)}{\partial y \partial x} = \frac{\partial}{\partial y}\left(\frac{\partial f(x,y)}{\partial x}\right) = 2x(\cos(x^2+y^3) - 3y^3\sin(x^2+y^3)).$$

We get the same result if we take the partial derivatives in the opposite order (with respect to y and then with respect to x):

$$\frac{\partial^2 f(x,y)}{\partial x \partial y} = \frac{\partial}{\partial x}\left(\frac{\partial f(x,y)}{\partial y}\right) = 2x(\cos(x^2+y^3) - 3y^3\sin(x^2+y^3)).$$

Under mild technical assumptions, it is also true for general f that

$$\frac{\partial^2 f(x,y)}{\partial x \partial y} = \frac{\partial^2 f(x,y)}{\partial y \partial x}.$$

The *Jacobian* of a transformation which maps $(x_1, \dots, x_n)$ to $(y_1, \dots, y_n)$ is the $n \times n$ matrix of all possible first order partial derivatives, given by

$$\frac{\partial \mathbf{y}}{\partial \mathbf{x}} = \begin{pmatrix} \frac{\partial y_1}{\partial x_1} & \frac{\partial y_1}{\partial x_2} & \cdots & \frac{\partial y_1}{\partial x_n} \\ \vdots & \vdots & & \vdots \\ \frac{\partial y_n}{\partial x_1} & \frac{\partial y_n}{\partial x_2} & \cdots & \frac{\partial y_n}{\partial x_n} \end{pmatrix}.$$

A.7 Multiple integrals

If you can do single integrals, you can do multiple integrals: just do more than one integral, holding variables other than the current variable of integration constant.

For example,

$$\begin{aligned}\int_0^1 \int_0^y (x-y)^2 dxdy &= \int_0^1 \int_0^y (x^2 - 2xy + y^2) dxdy \\ &= \int_0^1 \left((x^3/3 - x^2 y + xy^2) \Big|_0^y \right) dy \\ &= \int_0^1 (y^3/3 - y^3 + y^3) dy \\ &= \frac{1}{12}.\end{aligned}$$

A.7.1 Change of order of integration

We can also integrate in the other order, $dydx$ rather than $dxdy$, as long as we are careful about the limits of integration. Since we're integrating over all (x, y) with x and y between 0 and 1 such that $x \le y$, to integrate the other way we write

$$\begin{aligned}\int_0^1 \int_x^1 (x-y)^2 dydx &= \int_0^1 \int_x^1 (x^2 - 2xy + y^2) dydx \\ &= \int_0^1 \left((x^2 y - xy^2 + y^3/3) \Big|_x^1 \right) dx \\ &= \int_0^1 (x^2 - x + 1/3 - x^3 + x^3 - x^3/3) dx \\ &= \frac{1}{12}.\end{aligned}$$

A.7.2 Change of variables

In making a change of variables with multiple integrals, a Jacobian is needed. To simplify notation, we will discuss the two-dimensional version; the n-dimensional version is analogous. Suppose we make an invertible transformation from (x, y) to (u, v), where the inverse transformation is given by

$$x = g(u, v),\ y = h(u, v).$$

Let A be a region in the (x, y)-plane and B be the corresponding region in the (u, v)-plane. Then the change of variables formulas says that

$$\iint\limits_A f(x, y) dxdy = \iint\limits_B f(g(u, v), h(u, v)) \cdot \left| \left| \frac{\partial(x, y)}{\partial(u, v)} \right| \right| dudv,$$

where $\left| \left| \frac{\partial(x,y)}{\partial(u,v)} \right| \right|$ is the absolute value of the determinant of the Jacobian matrix. A few technical assumptions are needed for this result, such as that the partial derivatives that appear in the Jacobian matrix exist and are continuous, and that the determinant of the Jacobian matrix is never 0.

For example, let's find the area of a circle of radius 1. To find the area of a region, we just need to integrate 1 over that region (so the part that may be tricky is the limits of integration; the integrand is just 1). So the area is

$$\iint\limits_{x^2+y^2\leq 1} 1 dxdy = \int_{-1}^{1}\int_{-\sqrt{1-y^2}}^{\sqrt{1-y^2}} 1 dxdy = 2\int_{-1}^{1}\sqrt{1-y^2}\,dy.$$

Note that the limits for the inner variable (x) of the double integral can depend on the outer variable (y), while the outer limits are constants. The last integral can be done with a trigonometric substitution, but instead let's simplify the problem by transforming to polar coordinates: let

$$x = r\cos\theta,\ y = r\sin\theta,$$

where r is the distance from (x, y) to the origin and $\theta \in [0, 2\pi)$ is the angle. The Jacobian matrix of this transformation is

$$\frac{\partial(x,y)}{\partial(r,\theta)} = \begin{pmatrix} \cos\theta & -r\sin\theta \\ \sin\theta & r\cos\theta, \end{pmatrix}$$

so the absolute value of the determinant of the Jacobian is $r(\cos^2\theta + \sin^2\theta) = r$. That is, $dxdy$ becomes $rdrd\theta$. So the area of the circle is

$$\int_0^{2\pi}\int_0^1 rdrd\theta = \int_0^{2\pi}\frac{1}{2}d\theta = \pi.$$

For a circle of radius r, it follows immediately that the area is πr^2 since we can convert our units of measurement to the unit for which the radius is 1.

This may seem like a lot of work just to get such a familiar result, but it served as illustration and with similar methods we can get the volume of a ball in any number of dimensions! It turns out that the volume of a ball of radius 1 in $\mathbb{R}^n$ is

$$V_n = \frac{\pi^{n/2}}{\Gamma(n/2+1)},$$

where Γ is the gamma function (see Section A.2.7).

A.8 Sums

There are several kinds of sums that come up frequently in probability.

A.8.1 Binomial theorem

The *binomial theorem* states that

$$(x+y)^n = \sum_{k=0}^{n} \binom{n}{k} x^k y^{n-k},$$

for any nonnegative integer n. Here $\binom{n}{k}$ is a *binomial coefficient*, defined as the number of ways to choose k objects out of n, with order not mattering. An explicit formula for $\binom{n}{k}$ in terms of factorials for $0 \leq k \leq n$ is

$$\binom{n}{k} = \frac{n!}{(n-k)!k!}.$$

A proof of the binomial theorem is given in Example 1.4.19.

Sometimes 0^0 is said to be undefined, but there are several strong arguments for defining it to be 1, and we will take $0^0 = 1$ in this book. For one simple reason for doing so, consider the case $x = 0$, $y = 1$ in the binomial theorem. Then the left-hand side is $1^n = 1$, and we need to have $0^0 = 1$ in order to make the right-hand side also equal to 1.

A.8.2 Geometric series

A series of the form $\sum_{n=0}^{\infty} x^n$ is called a *geometric series*. For $|x| < 1$, the series converges and we have

$$\sum_{n=0}^{\infty} x^n = \frac{1}{1-x}.$$

The series diverges if $|x| \geq 1$. A series of the same form except with finitely many terms is called a *finite geometric series*. For $x \neq 1$, the sum is

$$\sum_{k=0}^{n} x^k = \frac{1-x^{n+1}}{1-x}.$$

Note that the $n+1$ in the exponent is the number of terms in the sum. If the sum starts at $k = m$ rather than $k = 0$ (where m is an integer with $0 \leq m \leq n$), then we can easily reduce to the case where the sum starts at 0. Making the substitution $j = k - m$, we have

$$\sum_{k=m}^{n} x^k = \sum_{j=0}^{n-m} x^{j+m} = x^m \sum_{j=0}^{n-m} x^j = x^m \cdot \frac{1-x^{n-m+1}}{1-x}.$$

A.8.3 Taylor series for e^x

The Taylor series for e^x is

$$\sum_{n=0}^{\infty} \frac{x^n}{n!} = e^x, \text{ for all } x.$$

A.8.4 Harmonic series and other sums with a fixed exponent

It is also useful to know that $\sum_{n=1}^{\infty} 1/n^c$ converges for $c > 1$ and diverges for $c \leq 1$. For $c = 1$, this is called the *harmonic series*. The sum of the first n terms of the harmonic series can be approximated using

$$\sum_{k=1}^{n} \frac{1}{k} \approx \log(n) + \gamma$$

for n large, where $\gamma \approx 0.577$ is the *Euler-Mascheroni constant.*

The sum of the first n positive integers is

$$\sum_{k=1}^{n} k = n(n+1)/2 = \binom{n+1}{2}.$$

It follows from this identity, for example, that the sum of the first n odd positive integers is

$$\sum_{k=1}^{n}(2k-1) = 2\sum_{k=1}^{n} k - \sum_{k=1}^{n} 1 = n(n+1) - n = n^2.$$

For squares of integers, we have

$$\sum_{k=1}^{n} k^2 = n(n+1)(2n+1)/6.$$

For cubes of integers, amazingly, the sum is the square of the sum of the first n positive integers! That is,

$$\sum_{k=1}^{n} k^3 = (n(n+1)/2)^2.$$

Exercise 22 from Chapter 1 asks for a story proof of this result.

A.9 Pattern recognition

Much of math and statistics is really about *pattern recognition*: seeing the essential structure of a problem, recognizing when one problem is essentially the same as another problem (just in a different guise), noticing symmetry, and so on. We will see many examples of this kind of thinking in this book. For example, suppose we have the series $\sum_{k=0}^{\infty} e^{tk}e^{-\lambda}\lambda^k/k!$, with λ a positive constant. The $e^{-\lambda}$ can be taken out from the sum, and then the structure of the series exactly matches up with the structure of the Taylor series for e^x. Therefore

$$\sum_{k=0}^{\infty} \frac{e^{tk}e^{-\lambda}\lambda^k}{k!} = e^{-\lambda}\sum_{k=0}^{\infty} \frac{(\lambda e^t)^k}{k!} = e^{-\lambda}e^{\lambda e^t} = e^{\lambda(e^t-1)},$$

valid for all real t.

Similarly, suppose we want the Taylor series for $1/(1-x^3)$ about $x = 0$. It would be tedious to start taking derivatives of this function. Instead, note that this function is reminiscent of the result of summing a geometric series. Therefore

$$\frac{1}{1-x^3} = \sum_{n=0}^{\infty} x^{3n},$$

for $|x^3| < 1$ (which is equivalent to $|x| < 1$). What matters is the structure, not what names we use for variables!

A.10 Common sense and checking answers

It is very easy to make mistakes in probability, so checking answers is especially important. Some useful strategies for checking answers are:

- seeing whether the answer makes sense intuitively (though as we have often seen in this book, probability has many results that seem counterintuitive at first);
- making sure your answer isn't a category error;
- making sure you're avoiding biohazards;
- trying out simple cases;
- trying out extreme cases;
- looking for alternative methods to solve the problem (including methods that may only give a bound or approximation, such as applying one of the inequalities from Chapter 10 or using R to run a simulation).

B

R

B.1 Vectors

Command	What it does
`c(1,1,0,2.7,3.1)`	creates the vector $(1, 1, 0, 2.7, 3.1)$
`1:100`	creates the vector $(1, 2, \dots, 100)$
`(1:100)^3`	creates the vector $(1^3, 2^3, \dots, 100^3)$
`rep(0,50)`	creates the vector $(0, 0, \dots, 0)$ of length 50
`seq(0,99,3)`	creates the vector $(0, 3, 6, 9, \dots, 99)$
`v[5]`	5th entry of vector v (index starts at 1)
`v[-5]`	all but the 5th entry of v
`v[c(3,1,4)]`	3rd, 1st, 4th entries of vector v
`v[v>2]`	*entries* of v that exceed 2
`which(v>2)`	*indices* of v such that entry exceeds 2
`which(v==7)`	*indices* of v such that entry equals 7
`min(v)`	minimum of v
`max(v)`	maximum of v
`which.max(v)`	indices where $\max(v)$ is achieved
`sum(v)`	sum of the entries in v
`cumsum(v)`	cumulative sums of the entries in v
`prod(v)`	product of the entries in v
`rank(v)`	ranks of the entries in v
`length(v)`	length of vector v
`sort(v)`	sorts vector v (in increasing order)
`unique(v)`	lists each element of v once, without duplicates
`tabulate(v)`	tallies how many times each element of v occurs
`table(v)`	same as `tabulate(v)`, except in table format
`c(v,w)`	concatenates vectors v and w
`union(v,w)`	union of v and w as sets
`intersect(v,w)`	intersection of v and w as sets
`v+w`	adds v and w entrywise (recycling if needed)
`v*w`	multiplies v and w entrywise (recycling if needed)

B.2 Matrices

Command	What it does
`matrix(c(1,3,5,7), nrow=2, ncol=2)`	creates the matrix $\begin{pmatrix} 1 & 5 \\ 3 & 7 \end{pmatrix}$
`dim(A)`	gives the dimensions of matrix A
`diag(A)`	extracts the diagonal of matrix A
`diag(c(1,7))`	creates the diagonal matrix $\begin{pmatrix} 1 & 0 \\ 0 & 7 \end{pmatrix}$
`rbind(u,v,w)`	binds vectors u, v, w into a matrix, as rows
`cbind(u,v,w)`	binds vectors u, v, w into a matrix, as columns
`t(A)`	transpose of matrix A
`A[2,3]`	row 2, column 3 entry of matrix A
`A[2,]`	row 2 of matrix A (as a vector)
`A[,3]`	column 3 of matrix A (as a vector)
`A[c(1,3),c(2,4)]`	submatrix of A, keeping rows $1, 3$ and columns $2, 4$
`rowSums(A)`	row sums of matrix A
`rowMeans(A)`	row averages of matrix A
`colSums(A)`	column averages of matrix A
`colMeans(A)`	column means of matrix A
`eigen(A)`	eigenvalues and eigenvectors of matrix A
`solve(A)`	A^{-1}
`solve(A,b)`	solves $A\mathbf{x} = \mathbf{b}$ for $\mathbf{x}$ (where $\mathbf{b}$ is a column vector)
`A %*% B`	matrix multiplication AB
`A %^% k`	matrix power A^k (using `expm` package)

B.3 Math

Command	What it does
`abs(x)`	$\|x\|$
`exp(x)`	e^x
`log(x)`	$\log(n)$
`log(x,b)`	$\log_b(n)$
`sqrt(x)`	$\sqrt{x}$
`floor(x)`	$\lfloor x \rfloor$
`ceiling(x)`	$\lceil x \rceil$
`factorial(n)`	$n!$
`lfactorial(n)`	$\log(n!)$ (helps prevent overflow)
`gamma(a)`	$\Gamma(a)$
`lgamma(a)`	$\log(\Gamma(a))$ (helps prevent overflow)
`choose(n,k)`	binomial coefficient $\binom{n}{k}$
`pbirthday(k)`	solves birthday problem for k people
`if (x>0) x^2 else x^3`	x^2 if $x > 0$, x^3 otherwise (piecewise)
`f <- function(x) exp(-x)`	defines the function f by $f(x) = e^{-x}$
`integrate(f, lower=0, upper=Inf)`	finds $\int_0^\infty f(x)dx$ numerically
`optimize(f,lower=0,upper=5,maximum=TRUE)`	maximizes f numerically on $[0, 5]$
`uniroot(f, lower=0, upper=5)`	searches numerically for a zero of f in $[0, 5]$

B.4 Sampling and simulation

Command	What it does
`sample(7)`	random permutation of $1, 2, \dots, 7$
`sample(52,5)`	picks 5 times from $1, 2, \dots, 52$ (don't replace)
`sample(letters,5)`	picks 5 random letters of the alphabet (don't replace)
`sample(3,5,replace=TRUE,prob=p)`	picks 5 times from $1, 2, 3$ with probabilities p (replace)
`replicate(10^4,` *experiment*`)`	simulates 10^4 runs of *experiment*

B.5 Plotting

Command	What it does
`curve(f, from=a, to=b)`	graphs the function f from a to b
`plot(x,y)`	creates scatter plot of the points (x_i, y_i)
`plot(x,y,type="l")`	creates line plot of the points (x_i, y_i)
`points(x,y)`	adds the points (x_i, y_i) to the plot
`lines(x,y)`	adds line segments through the (x_i, y_i) to the plot
`abline(a,b)`	adds the line with intercept a, slope b to the plot
`hist(x, breaks=b, col="blue")`	blue histogram of the values in x, with b bins suggested
`par(new=TRUE)`	tells R *not* to clear the palette when we make our next plot
`par(mfrow=c(1,2))`	tells R we want 2 side-by-side plots (a 1 by 2 array of plots)

B.6 Programming

Command	What it does
`x <- pi`	sets x equal to π
`x>3 && x<5`	Is $x > 3$ and $x < 5$? (TRUE/FALSE)
`x>3 || x<5`	Is $x > 3$ or $x < 5$? (TRUE/FALSE)
`if (n>3) x <- x+1`	adds 1 to x if $n > 3$
`if (n==0) x <- x+1 else x <- x+2`	adds 1 to x if $n = 0$, else adds 2
`v<-rep(0,50); for (k in 1:50) v[k]<-pbirthday(k)`	solves birthday problem up to 50 people

B.7 Summary statistics

Command	What it does
`mean(v)`	sample mean of vector v
`var(v)`	sample variance of vector v
`sd(v)`	sample standard deviation of vector v
`median(v)`	sample median of vector v
`summary(v)`	min, 1st quartile, median, mean, 3rd quartile, max of v
`quantile(v,p)`	pth sample quantile of vector v
`cov(v,w)`	sample covariance of vectors v and w
`cor(v,w)`	sample correlation of vectors v and w

B.8 Distributions

Command	What it does
`help(distributions)`	shows documentation on distributions
`dbinom(k,n,p)`	PMF $P(X = k)$ for $X \sim \text{Bin}(n, p)$
`pbinom(x,n,p)`	CDF $P(X \leq x)$ for $X \sim \text{Bin}(n, p)$
`qbinom(a,n,p)`	quantile $\min\{x : P(X \leq x) \geq a\}$ for $X \sim \text{Bin}(n, p)$
`rbinom(r,n,p)`	vector of r i.i.d. $\text{Bin}(n, p)$ r.v.s
`dgeom(k,p)`	PMF $P(X = k)$ for $X \sim \text{Geom}(p)$
`dhyper(k,w,b,n)`	PMF $P(X = k)$ for $X \sim \text{HGeom}(w, b, n)$
`dnbinom(k,r,p)`	PMF $P(X = k)$ for $X \sim \text{NBin}(r, p)$
`dpois(k,r)`	PMF $P(X = k)$ for $X \sim \text{Pois}(r)$
`dbeta(x,a,b)`	PDF $f(x)$ for $X \sim \text{Beta}(a, b)$
`dcauchy(x)`	PDF $f(x)$ for $X \sim$ Cauchy
`dchisq(x,n)`	PDF $f(x)$ for $X \sim \chi^2_n$
`dexp(x,r)`	PDF $f(x)$ for $X \sim \text{Expo}(r)$
`dgamma(x,a,r)`	PDF $f(x)$ for $X \sim \text{Gamma}(a, r)$
`dlnorm(x,m,s)`	PDF $f(x)$ for $X \sim \mathcal{LN}(m, s^2)$
`dnorm(x,m,s)`	PDF $f(x)$ for $X \sim \mathcal{N}(m, s^2)$
`dt(x,n)`	PDF $f(x)$ for $X \sim t_n$
`dunif(x,a,b)`	PDF $f(x)$ for $X \sim \text{Unif}(a, b)$
`dweibull(x,a,b)`	PDF $f(x)$ for $X \sim \text{Weibull}(b^{-a}, a)$

There are commands that are completely analogous to `pbinom`, `qbinom`, and `rbinom` for the other distributions appearing in the above table. For example, `pnorm`, `qnorm`, and `rnorm` can be used to get the CDF, the quantiles, and random generation for the Normal (note that the mean and standard deviation need to be provided as the parameters, rather than the mean and variance).

For the Multinomial, `dmultinom` can be used for calculating the joint PMF and `rmultinom` can be used for generating random vectors. For the Multivariate Normal, after installing and loading the `mvtnorm` package, `dmvnorm` can be used for calculating the joint PDF and `rmvnorm` can be used for random generation.

C

Table of distributions

Name	Param.	PMF or PDF	Mean	Variance
Bernoulli	p	$P(X=1)=p, P(X=0)=q$	p	pq
Binomial	n,p	$\binom{n}{k}p^k q^{n-k}$, $k \in \{0,1,\dots,n\}$	np	npq
FS	p	pq^{k-1}, $k \in \{1,2,\dots\}$	$1/p$	q/p^2
Geom	p	pq^k, $k \in \{0,1,2,\dots\}$	q/p	q/p^2
NBin	r,p	$\binom{r+k-1}{r-1}p^r q^k$, $k \in \{0,1,2,\dots\}$	rq/p	rq/p^2
HGeom	w,b,n	$\frac{\binom{w}{k}\binom{b}{n-k}}{\binom{w+b}{n}}$, $k \in \{0,1,\dots,n\}$	$\mu = \frac{nw}{w+b}$	$(\frac{w+b-n}{w+b-1})\mu(1-\frac{\mu}{n})$
NHGeom	w,b,r	$\frac{\binom{r+k-1}{r-1}\binom{w+b-r-k}{w-r}}{\binom{w+b}{w}}$, $k \in \{0,1,\dots,b\}$	$\frac{rb}{w+1}$	$\frac{rb(w+b+1)(w-r+1)}{(w+1)^2(w+2)}$
Poisson	λ	$\frac{e^{-\lambda}\lambda^k}{k!}$, $k \in \{0,1,2,\dots\}$	λ	λ
Uniform	$a<b$	$\frac{1}{b-a}$, $x \in (a,b)$	$\frac{a+b}{2}$	$\frac{(b-a)^2}{12}$
Normal	μ,σ^2	$\frac{1}{\sigma\sqrt{2\pi}}e^{-(x-\mu)^2/(2\sigma^2)}$	μ	σ^2
Log-Normal	μ,σ^2	$\frac{1}{x\sigma\sqrt{2\pi}}e^{-(\log x-\mu)^2/(2\sigma^2)}$, $x>0$	$\theta = e^{\mu+\sigma^2/2}$	$\theta^2(e^{\sigma^2}-1)$
Expo	λ	$\lambda e^{-\lambda x}$, $x>0$	$1/\lambda$	$1/\lambda^2$
Weibull	λ,γ	$\gamma\lambda e^{-\lambda x^\gamma}x^{\gamma-1}$, $x>0$	$\mu = \frac{\Gamma(1+1/\gamma)}{\lambda^{1/\gamma}}$	$\frac{\Gamma(1+2/\gamma)}{\lambda^{2/\gamma}} - \mu^2$
Gamma	a,λ	$\Gamma(a)^{-1}(\lambda x)^a e^{-\lambda x}x^{-1}$, $x>0$	a/λ	a/λ^2
Beta	a,b	$\frac{\Gamma(a+b)}{\Gamma(a)\Gamma(b)}x^{a-1}(1-x)^{b-1}$, $0<x<1$	$\mu = \frac{a}{a+b}$	$\frac{\mu(1-\mu)}{a+b+1}$
Chi-Square	n	$\frac{1}{2^{n/2}\Gamma(n/2)}x^{n/2-1}e^{-x/2}$, $x>0$	n	$2n$
Student-t	n	$\frac{\Gamma((n+1)/2)}{\sqrt{n\pi}\Gamma(n/2)}(1+x^2/n)^{-(n+1)/2}$	0 if $n>1$	$\frac{n}{n-2}$ if $n>2$

References

[1] Donald J. Albers and Gerald L. Alexanderson. *More Mathematical People: Contemporary Conversations.* Academic Press, 1990.

[2] Steve Brooks, Andrew Gelman, Galin Jones, and Xiao-Li Meng. *Handbook of Markov Chain Monte Carlo.* CRC Press, 2011.

[3] Jian Chen and Jeffrey S. Rosenthal. Decrypting classical cipher text using Markov chain Monte Carlo. *Statistics and Computing*, 22(2):397–413, 2012.

[4] William G. Cochran. The effectiveness of adjustment by subclassification in removing bias in observational studies. *Biometrics*, 1968.

[5] Persi Diaconis. Statistical problems in ESP research. *Science*, 201(4351):131–136, 1978.

[6] Persi Diaconis. The Markov chain Monte Carlo revolution. *Bulletin of the American Mathematical Society*, 46(2):179–205, 2009.

[7] Persi Diaconis, Susan Holmes, and Richard Montgomery. Dynamical bias in the coin toss. *SIAM Review*, 49(2):211–235, 2007.

[8] Bradley Efron and Ronald Thisted. Estimating the number of unseen species: How many words did Shakespeare know? *Biometrika*, 63(3):435, 1976.

[9] Bradley Efron and Ronald Thisted. Did Shakespeare write a newly-discovered poem? *Biometrika*, 74:445–455, 1987.

[10] Andrew Gelman, John B. Carlin, Hal S. Stern, David B. Dunson, Aki Vehtari, and Donald B. Rubin. *Bayesian Data Analysis.* CRC Press, 2013.

[11] Andrew Gelman and Deborah Nolan. You can load a die, but you can't bias a coin. *The American Statistician*, 56(4):308–311, 2002.

[12] Andrew Gelman, Boris Shor, Joseph Bafumi, and David K. Park. *Red State, Blue State, Rich State, Poor State: Why Americans Vote the Way They Do (Expanded Edition).* Princeton University Press, 2009.

[13] Gerd Gigerenzer and Ulrich Hoffrage. How to improve Bayesian reasoning without instruction: Frequency formats. *Psychological Review*, 102(4):684, 1995.

[14] Prakash Gorroochurn. *Classic Problems of Probability.* John Wiley & Sons, 2012.

[15] Richard Hamming. You and your research. *IEEE Potentials*, pages 37–40, October 1993.

[16] David P. Harrington. The randomized clinical trial. *Journal of the American Statistical Association*, 95(449):312–315, 2000.

[17] David J.C. MacKay. *Information Theory, Inference, and Learning Algorithms.* Cambridge University Press, 2003.

[18] Richard McElreath. *Statistical Rethinking: A Bayesian Course with Examples in R and Stan.* Chapman & Hall/CRC Press, 2015.

[19] Pierre Rémond de Montmort. *Essay d'Analyse sur les Jeux de Hazard.* Quilau, Paris, 1708.

[20] John Allen Paulos. *Innumeracy: Mathematical Illiteracy and Its Consequences.* Macmillan, 1988.

[21] Horst Rinne. *The Weibull Distribution: A Handbook.* CRC Press, 2008.

[22] James G. Sanderson. Testing ecological patterns. *American Scientist*, 88:332–339, 2000.

[23] Tom W. Smith, Peter Marsden, Michael Hout, and Jibum Kim. General social surveys, 1972–2012. *Sponsored by National Science Foundation. NORC, Chicago: National Opinion Research Center*, 2013.

[24] Stephen M. Stigler. Isaac Newton as a probabilist. *Statistical Science*, 21(3):400–403, 2006.

[25] Tom Stoppard. *Rosencrantz & Guildenstern Are Dead.* Samuel French, Inc., 1967.

[26] R.J. Stroeker. On the sum of consecutive cubes being a perfect square. *Compositio Mathematica*, 97:295–307, 1995.

[27] Amos Tversky and Daniel Kahneman. Causal schemas in judgments under uncertainty. In Daniel Kahneman, Paul Slovic, and Amos Tversky, editors, *Judgment under Uncertainty: Heuristics and Biases.* Cambridge University Press, 1982.

[28] Herbert S. Wilf. *generatingfunctionology.* A K Peters/CRC Press, 3rd edition, 2005.

[29] Elizabeth Wrigley-Field. Length-biased sampling. Online article at `https://www.edge.org/response-detail/27022`, 2017.

[30] Lisa Zyga. Why too much evidence can be a bad thing. Online article at `https://phys.org/news/2016-01-evidence-bad.html`, 2016.

Index

For Product Safety Concerns and Information please contact our EU
representative GPSR@taylorandfrancis.com
Taylor & Francis Verlag GmbH, Kaufingerstraße 24, 80331 München, Germany

www.ingramcontent.com/pod-product-compliance
Ingram Content Group UK Ltd.
Pitfield, Milton Keynes, MK11 3LW, UK
UKHW051941210425
457613UK00028B/264

* 9 7 8 1 1 3 8 3 6 9 9 1 7 *